Physics2000

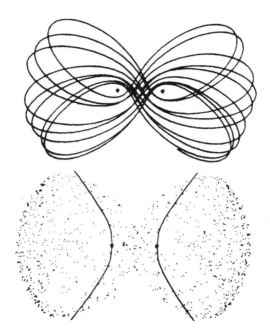

Student project by Bob Piela
explaining the hydrogen
molecule ion.

by E. R. Huggins
Department of Physics
Dartmouth College
Hanover, New Hampshire

ISBN 0-9707836-0-4 (*Physics 2000* 2–PART SET + CD)
ISBN 0-9707836-1-2 (*Physics 2000* PART 1 + CD)
ISBN 0-9707836-2-0 (*Physics 2000* PART 2 + CD)
ISBN 0-9707836-3-9 (*Physics 2000* CD)
ISBN 0-9707836-4-7 (*Calculus 2000* + CD)
ISBN 0-9707836-5-5 (*Physics 2000* Solutions)
ISBN 0-9707836-6-3 (*Physics 2000* 2–PART SET + *Calculus 2000* + CD)

Copyright © 1999 **Moose Mountain Digital Press**
Etna, New Hampshire 03750
All rights reserved

Preface

ABOUT THE COURSE

Physics2000 is a calculus based, college level introductory physics course that is designed to include twentieth century physics throughout. This is made possible by introducing Einstein's special theory of relativity in the first chapter. This way, students start off with a modern picture of how space and time behave, and are prepared to approach topics such as mass and energy from a modern point of view.

The course, which was developed during 30 plus years working with premedical students at Dartmouth College, makes very gentle assumptions about the student's mathematical background. All the calculus needed for studying Physics2000 is contained in a supplementary chapter which is the first chapter of a ***Physics Based Calculus*** text. We can cover all the necessary calculus in one reasonable length chapter because the concepts are introduced in the physics text and the calculus chapter only needs to handle the formalism. (The remaining chapters of the calculus text introduce the mathematical tools and concepts used in advanced introductory courses for physics and engineering majors. These chapters will be available at ***www.physics2000.com*** in late 2000.)

In the physics text, the concepts of velocity and acceleration are introduced through the use of strobe photographs in Chapter 3. How these definitions can be used to predict motion is discussed in Chapter 4 on calculus and Chapter 5 on the use of the computer.

Students themselves have made major contributions to the organization and content of the text. Student's enthusiasm for the use of Fourier analysis to study musical instruments led to the development of the MacScope™ program. The program makes it easy to use Fourier analysis to study such topics as the normal modes of a coupled aircart system and how the energy-time form of the uncertainty principle arises from the particle-wave nature of matter.

Most students experience difficulty when they first encounter abstract concepts like vector fields and Gauss' law. To provide a familiar model for a vector field, we begin the section on electricity and magnetism with a chapter on fluid dynamics. It is easy to visualize the velocity field of a fluid, and Gauss' law is simply the statement that the fluid is incompressible. We then show that the electric field has mathematical properties similar to those of the velocity field.

The format of the standard calculus based introductory physics text is to put a chapter on special relativity following Maxwell's equations, and then put modern physics after that, usually in an extended edition. This format suggests that the mathematics required to understand special relativity may be even more difficult than the integral-differential equations encountered in Maxwell's theory. Such fears are enhanced by the strangeness of the concepts in special relativity, and are driven home by the fact that relativity appears at the end of the course where there is no time to comprehend it. This format is a disaster.

Special relativity does involve strange ideas, but the mathematics required is only the Pythagorean theorem. By placing relativity at the beginning of the course you let the students know that the mathematics is not difficult, and that there will be plenty of time to become familiar with the strange ideas. By the time students have gone through Maxwell's equations in ***Physics2000***, they are thoroughly familiar with special relativity, and are well prepared to study the particle-wave nature of matter and the foundations of quantum mechanics. This material is not in an extended edition because there is time to cover it in a comfortably paced course.

ABOUT THE *PHYSICS2000* CD

The *Physics2000* CD contains the complete color version of the *Physics2000* text in Acrobat™ form along with a supplementary chapter covering all the calculus needed for the text. Included on the CD is the 36 minute motion picture *Time Dilation - An Experiment With Mu-Mesons*, and short movie segments of various physics demonstrations. Also a short cookbook on several basic dishes of Caribbean cooking.

The CD is available, for $10 postpaid, at the web site
www.physics2000.com
The black and white printed copy of the text, with the calculus chapter, is also available at the web site at a cost of $25. That includes the CD and shipping within the United States.

Use of the Text Material

Because we are trying to change the way physics is taught, Chapter 1 on special relativity, although copyrighted, may be used freely (except for the copyrighted photograph of Andromeda and frame of the muon film). All chapters may be printed and distributed to a class on a non profit basis.

ABOUT THE AUTHOR

E. R. Huggins has taught physics at Dartmouth College since 1961. He was an undergraduate at MIT and got his Ph.D. at Caltech. His Ph.D. thesis under Richard Feynman was on aspects of the quantum theory of gravity and the non uniqueness of energy momentum tensors. Since then most of his research has been on superfluid dynamics and the development of new teaching tools like the student-built electron gun and MacScope™. He wrote the non calculus introductory physics text ***Physics1*** in 1968 and the computer based text ***Graphical Mechanics*** in 1973. The ***Physics2000*** text, which summarizes over thirty years of experimenting with ways to teach physics, was written and class tested over the period from 1990 to 1998. All the work of producing the text was done by the author, and his wife, Anne Huggins. The text layout and design was by the author's daughter Cleo Huggins who designed eWorld™ for Apple Computer and the Sonata™ music font for Adobe Systems.

The author's eMail address is
lish.huggins@dartmouth.edu
The author welcomes any comments.

Table of Contents

PART 1

INTRODUCTION—AN OVERVIEW OF PHYSICS
- Space And Time int-2
  - The Expanding Universe int-3
- Structure of Matter int-5
  - Atoms int-5
  - Light int-7
  - Photons int-8
  - The Bohr Model int-8
- Particle-Wave Nature of Matter int-10
- Conservation of Energy int-11
- Anti-Matter int-12
- Particle Nature of Forces int-13
  - Renormalization int-14
  - Gravity int-15
- A Summary int-16
- The Nucleus int-17
- Stellar Evolution int-19
  - The Weak Interaction int-20
  - Leptons int-21
  - Nuclear Structure int-22
- A Confusing Picture int-22
- Quarks int-24
- The Electroweak Theory int-26
- The Early Universe int-27
  - The Thermal Photons int-29

CHAPTER 1 PRINCIPLE OF RELATIVITY 1-1
- The Principle of Relativity 1-2
  - A Thought Experiment 1-3
  - Statement of the Principle of Relativity ... 1-4
  - Basic Law of Physics 1-4
- Wave Motion 1-6
  - Measurement of the Speed of Waves ... 1-7
  - Michaelson-Morley Experiment 1-11
- Einstein's Principle of Relativity 1-12
  - The Special Theory of Relativity 1-13
  - Moving Clocks 1-13
  - Other Clocks 1-18
  - Real Clocks 1-20
  - Time Dilation 1-22
  - Space Travel 1-22
  - The Lorentz Contraction 1-24
  - Relativistic Calculations 1-28
  - Approximation Formulas 1-30
- A Consistent Theory 1-32
- Lack of Simultaneity 1-32
- Causality 1-36
- Appendix A 1-39
  - Class Handout 1-39

CHAPTER 2 VECTORS
- Vectors 2-2
  - Displacement Vectors 2-2
  - Arithmetic of Vectors 2-3
  - Rules for Number Arithmetic 2-4
  - Rules for Vector Arithmetic 2-4
  - Multiplication of a Vector by a Number ... 2-5
  - Magnitude of a Vector 2-6
  - Vector Equations 2-6
  - Graphical Work 2-6
- Components 2-8
  - Vector Equations in Component Form ... 2-10
- Vector Multiplication 2-11
  - The Scalar or Dot Product 2-12
  - Interpretation of the Dot Product 2-14
  - Vector Cross Product 2-15
  - Magnitude of the Cross Product 2-17
  - Component Formula for the Cross Product ... 2-17
- Right Handed Coordinate System 2-18

CHAPTER 3 DESCRIPTION OF MOTION
- Displacement Vectors 3-5
  - A Coordinate System 3-7
  - Manipulation of Vectors 3-8
  - Measuring the Length of a Vector 3-9
  - Coordinate System and Coordinate Vectors ... 3-11
- Analysis of Strobe Photographs 3-11
  - Velocity 3-11
  - Acceleration 3-13
  - Determining Acceleration
    from a Strobe Photograph 3-15
  - The Acceleration Vector 3-15
- Projectile Motion 3-16
- Uniform Circular Motion 3-17
  - Magnitude of the Acceleration for Circular Motion ... 3-18
- An Intuitive Discussion of Acceleration ... 3-20
  - Acceleration Due to Gravity 3-21
  - Projectile Motion with Air Resistance ... 3-22
- Instantaneous Velocity 3-24
  - Instantaneous Velocity from a Strobe Photograph ... 3-26

CHAPTER 4 CALCULUS IN PHYSICS

Limiting Process 4-1
    The Uncertainty Principle 4-1
Calculus Definition of Velocity 4-3
Acceleration 4-5
    Components 4-6
    Distance, Velocity and
    Acceleration versus Time Graphs 4-7
The Constant Acceleration Formulas 4-9
    Three Dimensions 4-11
Projectile Motion with Air Resistance 4-12
Differential Equations 4-14
    Solving the Differential Equation 4-14
Solving Projectile Motion Problems 4-16
    Checking Units 4-19

CHAPTER 5 COMPUTER PREDICTION OF MOTION

Step-By-Step Calculations 5-1
Computer Calculations 5-2
    Calculating and Plotting a Circle 5-2
Program for Calculation 5-4
    The DO LOOP 5-4
    The LET Statement 5-5
    Variable Names 5-6
    Multiplication 5-6
    Plotting a Point 5-6
    Comment Lines 5-7
    Plotting Window 5-7
    Practice 5-8
    Selected Printing (MOD Command) 5-10
Prediction of Motion 5-12
Time Step and Initial Conditions 5-14
An English Program for Projectile Motion 5-16
A BASIC Program for Projectile Motion 5-18
Projectile Motion with Air Resistance 5-22
    Air Resistance Program 5-24

CHAPTER 6 MASS

Definition of Mass 6-2
    Recoil Experiments 6-2
    Properties of Mass 6-3
    Standard Mass 6-3
    Addition of Mass 6-4
    A Simpler Way to Measure Mass 6-4
    Inertial and Gravitational Mass 6-5
    Mass of a Moving Object 6-5
Relativistic Mass 6-6
    Beta (β) Decay 6-6
    Electron Mass in β Decay 6-7
    Plutonium 246 6-8
    Protactinium 236 6-9
The Einstein Mass Formula 6-10
    Nature's Speed Limit 6-11
Zero Rest Mass Particles 6-11
Neutrinos 6-13
    Solar Neutrinos 6-13
    Neutrino Astronomy 6-14

CHAPTER 7 CONSERVATION OF LINEAR & ANGULAR MOMENTUM

Conservation of Linear Momentum 7-2
Collision Experiments 7-4
    Subatomic Collisions 7-7
    Example 1 Rifle and Bullet 7-7
    Example 2 7-8
Conservation of Angular Momentum 7-9
A More General Definition of Angular Momentum 7-12
Angular Momentum as a Vector 7-14
    Formation of Planets 7-17

CHAPTER 8 NEWTONIAN MECHANICS

Force 8-2
The Role of Mass 8-3
Newton's Second Law 8-4
Newton's Law of Gravity 8-5
    Big Objects 8-5
    Galileo's Observation 8-6
The Cavendish Experiment 8-7
    "Weighing" the Earth 8-8
    Inertial and Gravitational Mass 8-8
Satellite Motion 8-8
    Other Satellites 8-10
    Weight 8-11
    Earth Tides 8-12
    Planetary Units 8-14
    Table 1 Planetary Units 8-14
Computer Prediction of Satellite Orbits 8-16
    New Calculational Loop 8-17
    Unit Vectors 8-18
    Calculational Loop for Satellite Motion 8-19
    Summary 8-20
    Working Orbit Program 8-20
    Projectile Motion Program 8-21
    Orbit-1 Program 8-21
    Satellite Motion Laboratory 8-23
Kepler's Laws 8-24
    Kepler's First Law 8-26
    Kepler's Second Law 8-27
    Kepler's Third Law 8-28
Modified Gravity and General Relativity 8-29
Conservation of Angular Momentum 8-32
Conservation of Energy 8-35

CHAPTER 9 APPLICATIONS OF NEWTON'S SECOND LAW

- Addition of Forces 9-2
- Spring Forces 9-3
    - The Spring Pendulum 9-4
    - Computer Analysis of the Ball Spring Pendulum 9-8
- The Inclined Plane 9-10
- Friction 9-12
    - Inclined Plane with Friction 9-12
    - Coefficient of Friction 9-13
- String Forces 9-15
- The Atwood's Machine 9-16
- The Conical Pendulum 9-18
- Appendix The ball spring Program 9-20

CHAPTER 10 ENERGY

- Conservation of Energy 10-2
- Mass Energy 10-3
    - Ergs and Joules 10-4
- Kinetic Energy 10-5
    - Example 1 10-5
    - Slowly Moving Particles 10-6
- Gravitational Potential Energy 10-8
    - Example 2 10-10
    - Example 3 10-11
- Work 10-12
    - The Dot Product 10-13
    - Work and Potential Energy 10-14
    - Non-Constant Forces 10-14
    - Potential Energy Stored in a Spring 10-16
- Work Energy Theorem 10-18
    - Several Forces 10-19
    - Conservation of Energy 10-20
    - Conservative and Non-Conservative Forces 10-21
- Gravitational Potential Energy on a Large Scale 10-22
    - Zero of Potential Energy 10-22
    - Gravitational PotentialEnergy in a Room 10-25
- Satellite Motion and Total Energy 10-26
    - Example 4 Escape Velocity 10-28
- Black Holes 10-29
    - A Practical System of Units 10-31

CHAPTER 11 SYSTEMS OF PARTICLES

- Center of Mass 11-2
    - Center of Mass Formula 11-3
    - Dynamics of the Center of Mass 11-4
- Newton's Third Law 11-6
- Conservation of Linear Momentum 11-7
    - Momentum Version of Newton's Second Law 11-8
- Collisions 11-9
    - Impulse 11-9
    - Calibration of the Force Detector 11-10
    - The Impulse Measurement 11-11
    - Change in Momentum 11-12
    - Momentum Conservation during Collisions 11-13
    - Collisions and Energy Loss 11-14
    - Collisions that Conserve Momentum and Energy 11-16
    - Elastic Collisions 11-17
- Discovery of the Atomic Nucleus 11-19
- Neutrinos 11-20
    - Neutrino Astronomy 11-21

CHAPTER 12 ROTATIONAL MOTION

- Radian Measure 12-2
    - Angular Velocity 12-2
    - Angular Acceleration 12-3
    - Angular Analogy 12-3
    - Tangential Distance, Velocity and Acceleration ... 12-4
    - Radial Acceleration 12-5
    - Bicycle Wheel 12-5
- Angular Momentum 12-6
    - Angular Momentum of a Bicycle Wheel 12-6
    - Angular Velocity as a Vector 12-7
    - Angular Momentum as a Vector 12-7
- Angular Mass or Moment of Inertia 12-7
    - Calculating Moments of Inertia 12-8
- Vector Cross Product 12-9
    - Right Hand Rule for Cross Products 12-10
- Cross Product Definition of Angular Momentum 12-11
    - The $\vec{r} \times \vec{p}$ Definition of Angular Momentum 12-12
- Angular Analogy to Newton's Second Law 12-14
- About Torque 12-15
- Conservation of Angular Momentum 12-16
- Gyroscopes 12-18
    - Start-up 12-18
    - Precession 12-19
- Rotational Kinetic Energy 12-22
- Combined Translation and Rotation 12-24
    - Example—Objects Rolling Down an Inclined Plane 12-25
- Proof of the Kinetic Energy Theorem 12-26

CHAPTER 13 EQUILIBRIUM

Equations for equilibrium 13-2
    Example 1 Balancing Weights 13-2
Gravitational Force acting at the Center of Mass 13-4
Technique of Solving Equilibrium Problems 13-5
    Example 3 Wheel and Curb 13-5
    Example 4 Rod in a Frictionless Bowl 13-7
    Example 5 A Bridge Problem 13-9
Lifting Weights and Muscle Injuries 13-11

CHAPTER 14 OSCILLATIONS AND RESONANCE

Oscillatory Motion 14-2
The Sine Wave 14-3
    Phase of an Oscillation 14-6
Mass on a Spring; Analytic Solution 14-7
    Conservation of Energy 14-11
The Harmonic Oscillator 14-12
    The Torsion Pendulum 14-12
    The Simple Pendulum 14-15
    Small Oscillations 14-16
    Simple and Conical Pendulums 14-17
Non Linear Restoring Forces 14-19
Molecular Forces 14-20
Damped Harmonic Motion 14-21
    Critical Damping 14-23
Resonance 14-24
    Resonance Phenomena 14-26
    Transients 14-27
Appendix 14–1 Solution of the Differential Equation
for Forced Harmonic Motion 14-28
Appendix 14-2 Computer analysis
of oscillatory motion 14-30
    English Program 14-31
    The BASIC Program 14-32
    Damped Harmonic Motion 14-34

CHAPTER 15 ONE DIMENSIONAL WAVE MOTION 15-1

Wave Pulses 15-3
Speed of a Wave Pulse 15-4
Dimensional Analysis 15-6
Speed of Sound Waves 15-8
Linear and nonlinear Wave Motion 15-10
The Principle of Superposition 15-11
Sinusoidal Waves 15-12
    Wavelength, Period, and Frequency 15-13
    Angular Frequency ω 15-14
    Spacial Frequency k 15-14
    Traveling Wave Formula 15-16
    Phase and Amplitude 15-17
Standing Waves 15-18
Waves on a Guitar String 15-20
    Frequency of Guitar String Waves 15-21
    Sound Produced by a Guitar String 15-22

CHAPTER 16 FOURIER ANALYSIS, NORMAL MODES AND SOUND

Harmonic Series 16-3
Normal Modes of Oscillation 16-4
Fourier Analysis 16-6
    Analysis of a Sine Wave 16-7
    Analysis of a Square Wave 16-9
    Repeated Wave Forms 16-11
Analysis of the Coupled Air Cart System 16-12
The Human Ear 16-15
Stringed Instruments 16-18
Wind Instruments 16-20
Percussion Instruments 16-22
Sound Intensity 16-24
    Bells and Decibels 16-24
    Sound Meters 16-26
    Speaker Curves 16-27
Appendix A Fourier Analysis Lecture 16-28
    Square Wave 16-28
    Calculating Fourier Coefficients 16-28
    Amplitude and Phase 16-31
    Amplitude and Intensity 16-33
Appendix B Inside the Cochlea 16-34

CHAPTER 17 ATOMS, MOLECULES AND ATOMIC PROCESSES

Molecules 17-2
Atomic Processes 17-4
Thermal Motion 17-6
Thermal Equilibrium 17-8
Temperature 17-9
    Absolute Zero 17-9
    Temperature Scales 17-10
Molecular Forces 17-12
    Evaporation 17-14
Pressure 17-16
    Stellar Evolution 17-17
The Ideal Gas Law 17-18
    Ideal Gas Thermometer 17-20
    The Mercury Barometer
    and Pressure Measurements 17-22
Avogadro's Law 17-24
Heat Capacity 17-26
    Specific Heat 17-26
    Molar Heat Capacity 17-26
    Molar Specific Heat of Helium Gas 17-27
    Other Gases 17-27
Equipartition of Energy 17-28
    Real Molecules 17-30
Failure of Classical Physics 17-31
    Freezing Out of Degrees of Freedom 17-32
Thermal Expansion 17-33
Osmotic Pressure 17-34
Elasticity of Rubber 17-35
    A Model of Rubber 17-36

CHAPTER 18 ENTROPY

Introduction 18-2
Work Done by an Expanding Gas 18-5
Specific Heats c_v and c_p 18-6
Isothermal Expansion and PV Diagrams 18-8
    Isothermal Compression 18-9
    Isothermal Expansion of an Ideal Gas 18-9
Adiabatic Expansion 18-9
The Carnot Cycle 18-11
    Thermal Efficiency of the Carnot Cycle 18-12
    Reversible Engines 18-13
Energy Flow Diagrams 18-15
    Maximally Efficient Engines 18-15
    Reversibility 18-17
Applications of the Second Law 18-17
    Electric Cars 18-19
    The Heat Pump 18-19
    The Internal Combustion Engine 18-21
Entropy 18-22
    The Direction of Time 18-25
Appendix: Calculation of the Efficiency
of a Carnot Cycle 18-26
    Isothermal Expansion 18-26
    Adiabatic Expansion 18-26
    The Carnot Cycle 18-28

CHAPTER 19 THE ELECTRIC INTERACTION

The Four Basic Interactions 19-1
Atomic Structure 19-3
    Isotopes 19-6
The Electric Force Law 19-7
    Strength of the Electric Interaction 19-8
Electric Charge 19-8
    Positive and Negative Charge 19-10
    Addition of Charge 19-10
Conservation of Charge 19-13
    Stability of Matter 19-14
    Quantization of Electric Charge 19-14
Molecular Forces 19-15
    Hydrogen Molecule 19-16
    Molecular Forces—A More Quantitative Look 19-18
    The Bonding Region 19-19
    Electron Binding Energy 19-20
    Electron Volt as a Unit of Energy 19-21
    Electron Energy in the Hydrogen Molecule Ion .. 19-21

CHAPTER 20 NUCLEAR MATTER

Nuclear Force 20-2
    Range of the Nuclear Force 20-3
Nuclear Fission 20-3
Neutrons and the Weak Interaction 20-6
Nuclear Structure 20-7
    α (Alpha) Particles 20-8
Nuclear Binding Energies 20-9
Nuclear Fusion 20-12
Stellar Evolution 20-13
Neutron Stars 20-17
Neutron Stars
and Black Holes 20-18

PART 2

CHAPTER 23 FLUID DYNAMICS
The Current State of Fluid Dynamics 23-1
The Velocity Field 23-2
    The Vector Field 23-3
    Streamlines 23-4
    Continuity Equation 23-5
    Velocity Field of a Point Source 23-6
    Velocity Field of a Line Source 23-7
Flux 23-8
Bernoulli's Equation 23-9
Applications of Bernoulli's Equation 23-12
    Hydrostatics 23-12
    Leaky Tank 23-12
    Airplane Wing 23-13
    Sailboats 23-14
    The Venturi Meter 23-15
    The Aspirator 23-16
    Care in Applying Bernoulli's Equation ... 23-16
    Hydrodynamic Voltage 23-17
    Town Water Supply 23-18
    Viscous Effects 23-19
Vortices 23-20
    Quantized Vortices in Superfluids 23-22

CHAPTER 24 COULOMB'S LAW AND GAUSS' LAW
Coulomb's Law 24-1
    CGS Units 24-2
    MKS Units 24-2
    Checking Units in MKS Calculations 24-3
    Summary 24-3
    Example 1 Two Charges 24-3
    Example 2 Hydrogen Atom 24-4
Force Produced by a Line Charge 24-6
    Short Rod 24-9
The Electric Field 24-10
    Unit Test Charge 24-11
Electric Field lines 24-12
    Mapping the Electric Field 24-12
    Field Lines 24-13
    Continuity Equation for Electric Fields ... 24-14
    Flux 24-15
    Negative Charge 24-16
    Flux Tubes 24-17
    Conserved Field Lines 24-17
    A Mapping Convention 24-17
    Summary 24-18
    A Computer Plot 24-19

Gauss' Law 24-20
    Electric Field of a Line Charge 24-21
    Flux Calculations 24-22
    Area as a Vector 24-22
Gauss' Law for the Gravitational Field 24-23
    Gravitational Field of a Point Mass 24-23
    Gravitational Field
    of a Spherical Mass 24-24
    Gravitational Field Inside the Earth 24-24
    Solving Gauss' Law Problems 24-26
Problem Solving 24-29

CHAPTER 25 FIELD PLOTS AND ELECTRIC POTENTIAL
The Contour Map 25-1
Equipotential Lines 25-3
    Negative and Positive Potential Energy 25-4
Electric Potential of a Point Charge 25-5
Conservative Forces 25-5
Electric Voltage 25-6
    A Field Plot Model 25-10
    Computer Plots 25-12

CHAPTER 26 ELECTRIC FIELDS AND CONDUCTORS
Electric Field
Inside a Conductor 26-1
    Surface Charges 26-2
    Surface Charge Density 26-3
    Example: Field in a Hollow Metal Sphere ... 26-4
Van de Graaff generator 26-6
    Electric Discharge 26-7
    Grounding 26-8
The Electron Gun 26-8
    The Filament 26-9
    Accelerating Field 26-10
    A Field Plot 26-10
    Equipotential Plot 26-11
Electron Volt
as a Unit of Energy 26-12
    Example 26-13
    About Computer Plots 26-13
The Parallel Plate Capacitor 26-14
    Deflection Plates 26-16

CHAPTER 27 BASIC ELECTRIC CIRCUITS

Electric Current 27-2
   Positive and Negative Currents 27-3
   A Convention 27-5
Current and Voltage 27-6
   Resistors 27-6
   A Simple Circuit 27-8
   The Short Circuit 27-9
   Power 27-9
Kirchoff's Law 27-10
   Application of Kirchoff's Law 27-11
   Series Resistors 27-11
   Parallel Resistors 27-12
Capacitance and Capacitors 27-14
   Hydrodynamic Analogy 27-14
   Cylindrical Tank as a Constant Voltage Source . 27-15
   Electrical Capacitance 27-16
Energy Storage in Capacitors 27-18
   Energy Density in an Electric Field 27-19
Capacitors as Circuit Elements 27-20
The RC Circuit 27-22
   Exponential Decay 27-23
   The Time Constant RC 27-24
   Half-Lives 27-25
   Initial Slope 27-25
   The Exponential Rise 27-26
The Neon Bulb Oscillator 27-28
   The Neon Bulb 27-28
   The Neon Oscillator Circuit 27-29
   Period of Oscillation 27-30
   Experimental Setup 27-31

CHAPTER 28 MAGNETISM

Two Garden Peas 28-2
A Thought Experiment 28-4
   Charge Density on the Two Rods 28-6
   A Proposed Experiment 28-7
   Origin of Magnetic Forces 28-8
   Magnetic Forces 28-10
Magnetic Force Law 28-10
   The Magnetic Field B 28-10
   Direction of the Magnetic Field 28-11
   The Right Hand Rule for Currents 28-13
   Parallel Currents Attract 28-14
   The Magnetic Force Law 28-14
   Lorentz Force Law 28-15
   Dimensions of the
     Magnetic Field, Tesla and Gauss 28-16
   Uniform Magnetic Fields 28-16
   Helmholtz Coils 28-18

Motion of Charged Particles in Magnetic Fields 28-19
   Motion in a Uniform Magnetic Field 28-20
   Particle Accelerators 28-22
Relativistic Energy and Momenta 28-24
Bubble Chambers 28-26
   The Mass Spectrometer 28-28
   Magnetic Focusing 28-29
Space Physics 28-31
   The Magnetic Bottle 28-31
   Van Allen Radiation Belts 28-32

CHAPTER 29 AMPERE'S LAW

The Surface Integral 29-2
   Gauss' Law 29-3
The Line Integral 29-5
Ampere's Law 29-7
   Several Wires 29-10
   Field of a Straight Wire 29-11
Field of a Solenoid 29-14
   Right Hand Rule for Solenoids 29-14
   Evaluation of the Line Integral 29-15
   Calculation of $i_{enclosed}$ 29-15
   Using Ampere's law 29-15
   One More Right Hand Rule 29-16
   The Toroid 29-17

CHAPTER 30 FARADAY'S LAW

Electric Field
of Static Charges 30-2
A Magnetic Force Experiment 30-3
Air Cart Speed Detector 30-5
A Relativity Experiment 30-9
Faraday's Law 30-11
   Magnetic Flux 30-11
   One Form of Faraday's Law 30-12
   A Circular Electric Field 30-13
   Line Integral of $\vec{E}$ around a Closed Path 30-14
Using Faraday's Law 30-15
   Electric Field of an Electromagnet 30-15
   Right Hand Rule for Faraday's Law 30-15
   Electric Field of Static Charges 30-16
The Betatron 30-16
Two Kinds of Fields 30-18
   Note on our $\oint \vec{E} \cdot d\vec{\ell}$ meter 30-20
Applications of Faraday's Law 30-21
   The AC Voltage Generator 30-21
   Gaussmeter 30-23
   A Field Mapping Experiment 30-24

CHAPTER 31 INDUCTION AND MAGNETIC MOMENT

The Inductor 31-2
    Direction of the Electric Field 31-3
    Induced Voltage 31-4
    Inductance 31-5
Inductor as a Circuit Element 31-7
    The LR Circuit 31-8
The LC Circuit 31-10
    Intuitive Picture of the LC Oscillation 31-12
    The LC Circuit Experiment 31-13
Measuring the Speed of Light 31-15
Magnetic Moment 31-18
    Magnetic Force on a Current 31-18
    Torque on a Current Loop 31-20
    Magnetic Moment 31-21
    Magnetic Energy 31-22
    Summary of Magnetic Moment Equations 31-24
    Charge q in a Circular Orbit 31-24
Iron Magnets 31-26
    The Electromagnet 31-28
    The Iron Core Inductor 31-29
    Superconducting Magnets 31-30
Appendix The LC circuit and Fourier Analysis 31-31

CHAPTER 32 MAXWELL'S EQUATIONS

Gauss' Law for Magnetic Fields 32-2
Maxwell's Correction to Ampere's Law 32-4
    Example: Magnetic Field
        between the Capacitor Plates 32-6
Maxwell's Equations 32-8
Symmetry of Maxwell's Equations 32-9
Maxwell's Equations in Empty Space 32-10
    A Radiated Electromagnetic Pulse 32-10
    A Thought Experiment 32-11
    Speed of an Electromagnetic Pulse 32-14
Electromagnetic Waves 32-18
Electromagnetic Spectrum 32-20
    Components of the Electromagnetic Spectrum . 32-20
    Blackbody Radiation 32-22
    UV, X Rays, and Gamma Rays 32-22
Polarization 32-23
    Polarizers 32-24
    Magnetic Field Detector 32-26
Radiated Electric Fields 32-28
    Field of a Point Charge 32-30

CHAPTER 33 LIGHT WAVES

Superposition of Circular Wave Patterns 33-2
Huygens Principle 33-4
Two Slit Interference Pattern 33-6
    The First Maxima 33-8
Two Slit Pattern for Light 33-10
The Diffraction Grating 33-12
    More About Diffraction Gratings 33-14
    The Visible Spectrum 33-15
    Atomic Spectra 33-16
The Hydrogen Spectrum 33-17
    The Experiment on Hydrogen Spectra 33-18
    The Balmer Series 33-19
    33-19
The Doppler Effect 33-20
    Stationary Source and Moving Observer 33-21
    Doppler Effect for Light 33-22
    Doppler Effect in Astronomy 33-23
    The Red Shift and the
        Expanding Universe 33-24
A Closer Look at Interference Patterns 33-26
    Analysis of the Single Slit Pattern 33-27
Recording Diffraction Grating Patterns 33-28

CHAPTER 34 PHOTONS

Blackbody Radiation 34-2
    Planck Blackbody Radiation Law 34-4
The Photoelectric Effect 34-5
Planck's Constant h 34-8
Photon Energies 34-9
Particles and Waves 34-11
Photon Mass 34-12
    Photon Momentum 34-13
Antimatter 34-16
Interaction of Photons and Gravity 34-18
Evolution of the Universe 34-21
    Red Shift and the Expansion of the Universe 34-21
    Another View of Blackbody Radiation 34-22
Models of the universe 34-23
    Powering the Sun 34-23
    Abundance of the Elements 34-24
    The Steady State Model of the Universe 34-25
The Big Bang Model 34-26
    The Helium Abundance 34-26
    Cosmic Radiation 34-27
The Three Degree Radiation 34-27
    Thermal Equilibrium of the Universe 34-28
The Early Universe 34-29
    The Early Universe 34-29
    Excess of Matter over Antimatter 34-29
    Decoupling (700,000 years) 34-31
    Guidebooks 34-32

CHAPTER 35 BOHR THEORY OF HYDROGEN
The Classical Hydrogen Atom 35-2
    Energy Levels 35-4
The Bohr Model 35-7
    Angular Momentum in the Bohr Model 35-8
De Broglie's Hypothesis 35-10

CHAPTER 36 SCATTERING OF WAVES
Scattering of a Wave by a Small Object 36-2
Reflection of Light 36-3
X Ray Diffraction 36-4
    Diffraction by Thin Crystals 36-6
The Electron Diffraction Experiment 36-8
    The Graphite Crystal 36-8
    The Electron Diffraction Tube 36-9
    Electron Wavelength 36-9
    The Diffraction Pattern 36-10
    Analysis of the Diffraction Pattern 36-11
    Other Sets of Lines 36-12
    Student Projects 36-13
    Student project by Gwendylin Chen 36-14

CHAPTER 37 LASERS, A MODEL ATOM AND ZERO POINT ENERGY
The Laser and
Standing Light Waves 37-2
    Photon Standing Waves 37-3
    Photon Energy Levels 37-4
A Model Atom 37-4
Zero Point Energy 37-7
    Definition of Temperature 37-8
Two dimensional standing waves 37-8

CHAPTER 38 ATOMS
Solutions of Schrödinger's
Equation for Hydrogen 38-2
    The $\ell = 0$ Patterns 38-4
    The $\ell \ne 0$ Patterns 38-5
    Intensity at the Origin 38-5
    Quantized Projections of Angular Momentum 38-5
    The Angular Momentum Quantum Number 38-7
    Other notation 38-7
    An Expanded Energy Level Diagram 38-8
Multi Electron Atoms 38-9
    Pauli Exclusion Principle 38-9
    Electron Spin 38-9
The Periodic Table 38-10
    Electron Screening 38-10
    Effective Nuclear Charge 38-12
    Lithium 38-12
    Beryllium 38-13
    Boron 38-13
    Up to Neon 38-13
    Sodium to Argon 38-13
    Potassium to Krypton 38-14
    Summary 38-14
Ionic Bonding 38-15

CHAPTER 39 SPIN
The Concept of Spin 39-3
Interaction of the Magnetic Field with Spin 39-4
    Magnetic Moments and the Bohr Magneton 39-4
Insert 2 here 39-5
    Electron Spin Resonance Experiment 39-5
    Nuclear Magnetic Moments 39-6
    Sign Conventions 39-6
    Classical Picture of Magnetic Resonance 39-8
Electron Spin Resonance Experiment 39-9
Appendix Classical Picture of Magnetic Interactions 39-14

CHAPTER 40 QUANTUM MECHANICS
Two Slit Experiment 40-2
    The Two Slit Experiment from a
    Particle Point of View 40-3
    Two Slit Experiment—One Particle at a Time ... 40-3
    Born's Interpretation
    of the Particle Wave 40-6
    Photon Waves 40-6
    Reflection and Fluorescence 40-8
    A Closer Look at the Two Slit Experiment 40-9
The Uncertainty Principle 40-14
Position-Momentum Form
of the Uncertainty Principle 40-15
    Single Slit Experiment 40-16
Time-Energy Form of the Uncertainty Principle 40-19
    Probability Interpretation 40-22
    Measuring Short Times 40-22
    Short Lived Elementary Particles 40-23
The Uncertainty Principle and Energy Conservation . 40-24
Quantum Fluctuations and Empty Space 40-25
Appendix how a pulse is formed from sine waves .. 40-27

CHAPTER ON GEOMETRICAL OPTICS

Reflection from Curved Surfaces Optics-3
    The Parabolic Reflection Optics-4
Mirror Images Optics-6
    The Corner Reflector Optics-7
Motion of Light through a Medium Optics-8
    Index of Refraction Optics-9
Cerenkov Radiation Optics-10
Snell's Law Optics-11
    Derivation of Snell's Law Optics-12
Internal Reflection Optics-13
    Fiber Optics Optics-14
    Medical Imaging Optics-15
Prisms Optics-15
    Rainbows Optics-16
    The Green Flash Optics-17
    Halos and Sun Dogs Optics-18
Lenses Optics-18
    Spherical Lens Surface Optics-19
    Focal Length of a Spherical Surface Optics-20
    Aberrations Optics-21
Thin Lenses Optics-23
    The Lens Equation Optics-24
    Negative Image Distance Optics-26
    Negative Focal Length and Diverging Lenses Optics-26
    Negative Object Distance Optics-27
    Multiple Lens Systems Optics-28
    Two Lenses Together Optics-29
    Magnification Optics-30
The Human Eye Optics-31
    Nearsightedness and Farsightedness Optics-32
The Camera Optics-33
    Depth of Field Optics-34
    Eye Glasses and a Home Lab Experiment ... Optics-36
The Eyepiece Optics-37
    The Magnifier Optics-38
    Angular Magnification Optics-39
Telescopes Optics-40
    Reflecting telescopes Optics-42
    Large Reflecting Telescopes. Optics-43
    Hubble Space Telescope Optics-44
    World's Largest Optical Telescope Optics-45
    Infrared Telescopes Optics-46
    Radio Telescopes Optics-48
    The Very Long Baseline Array (VLBA) Optics-49
Microscopes Optics-50
    Scanning Tunneling Microscope Optics-51

Calculus 1   INTRODUCTION TO CALCULUS

Limiting Process .... Cal 1-3
    The Uncertainty Principle .... Cal 1-3
Calculus Definition of Velocity .... Cal 1-5
Acceleration .... Cal 1-7
    Components .... Cal 1-7
Integration .... Cal 1-8
    Prediction of Motion .... Cal 1-9
    Calculating Integrals .... Cal 1-11
    The Process of Integrating .... Cal 1-13
    Indefinite Integrals .... Cal 1-14
    Integration Formulas .... Cal 1-14
New Functions .... Cal 1-15
New Functions .... Cal 1-15
    Logarithms .... Cal 1-15
    The Exponential Function .... Cal 1-16
    Exponents to the Base 10 .... Cal 1-16
    The Exponential Function y^x .... Cal 1-16
    Euler's Number e = 2.7183. . ... Cal 1-17
Differentiation and Integration .... Cal 1-18
    A Fast Way to go Back and Forth .... Cal 1-20
    Constant Acceleration Formulas .... Cal 1-20
    Constant Acceleration Formulas
        in Three Dimensions .... Cal 1-22
More on Differentiation .... Cal 1-23
    Series Expansions .... Cal 1-23
    Derivative of the Function x^n .... Cal 1-24
    The Chain Rule .... Cal 1-25
    Remembering The Chain Rule .... Cal 1-25
    Partial Proof of the Chain Rule (optional) .... Cal 1-26
Integration Formulas .... Cal 1-27
    Derivative of the Exponential Function .... Cal 1-28
    Integral of the Exponential Function .... Cal 1-29
Derivative as the Slope of a Curve .... Cal 1-30
    Negative Slope .... Cal 1-31
The Exponential Decay .... Cal 1-32
    Muon Lifetime .... Cal 1-32
    Half Life .... Cal 1-33
    Measuring the Time Constant from a Graph ... Cal 1-34
The Sine and Cosine Functions .... Cal 1-35
    Radian Measure .... Cal 1-35
    The Sine Function .... Cal 1-36
    Amplitude of a Sine Wave .... Cal 1-37
    Derivative of the Sine Function .... Cal 1-38

Introduction
An Overview of Physics

With a brass tube and a few pieces of glass, you can construct either a microscope or a telescope. The difference is essentially where you place the lenses. With the microscope, you look down into the world of the small, with the telescope out into the world of the large.

In the twentieth century, physicists and astronomers have constructed ever larger machines to study matter on even smaller or even larger scales of distance. For the physicists, the new microscopes are the particle accelerators that provide views well inside atomic nuclei. For the astronomers, the machines are radio and optical telescopes whose large size allows them to record the faintest signals from space. Particularly effective is the Hubble telescope that sits above the obscuring curtain of the earth's atmosphere.

The new machines do not provide a direct image like the ones you see through brass microscopes or telescopes. Instead a good analogy is to the Magnetic Resonance Imaging (MRI) machines that first collect a huge amount of data, and then through the use of a computer program construct the amazing images showing cross sections through the human body. The telescopes and particle accelerators collect the vast amounts of data. Then through the use of the theories of quantum mechanics and relativity, the data is put together to construct meaningful images.

Some of the images have been surprising. One of the greatest surprises is the increasingly clear image of the universe starting out about fourteen billion years ago as an incredibly small, incredibly hot speck that has expanded to the universe we see today. By looking farther and farther out, astronomers have been looking farther and farther back in time, closer to that hot, dense beginning. Physicists, by looking at matter on a smaller and smaller scale with the even more powerful accelerators, have been studying matter that is even hotter and more dense. By the end of the twentieth century, physicists and astronomers have discovered that they are looking at the same image.

It is likely that telescopes will end up being the most powerful microscopes. There is a limit, both financial and physical, to how big and powerful an accelerator we can build. Because of this limit, we can use accelerators to study matter only up to a certain temperature and density. To study matter that is still hotter and more dense, which is the same as looking at still smaller scales of distance, the only "machine" we have available is the universe itself. We have found that the behavior of matter under the extreme conditions of the very early universe have left an imprint that we can study today with telescopes.

In the rest of this introduction we will show you some of the pictures that have resulted from looking at matter with the new machines. In the text itself we will begin to learn how these pictures were constructed.

SPACE AND TIME

The images of nature we see are images in both space and time, for we have learned from the work of Einstein that the two cannot be separated. They are connected by the speed of light, a quantity we designate by the letter c, which has the value of a billion (1,000,000,000) feet (30 cm) in a second. Einstein's remarkable discovery in 1905 was that the speed of light is an absolute speed limit. Nothing in the current universe can travel faster than the speed c.

Because the speed of light provides us with an absolute standard that can be measured accurately, we use the value of c to relate the definitions of time and distance. The meter is defined as the distance light travels in an interval of 1/299,792.458 of a second. The length of a second itself is provided by an atomic standard. It is the time interval occupied by 9,192,631,770 vibrations of a particular wavelength of light radiated by a cesium atom.

Using the speed of light for conversion, clocks often make good meter sticks, especially for measuring astronomical distances. It takes light 1.27 seconds to travel from the earth to the moon. We can thus say that the moon is 1.27 *light seconds* away. This is simpler than saying that the moon is 1,250,000,000 feet or 382,000 kilometers away. Light takes 8 minutes to reach us from the sun, thus the earth's orbit about the sun has a radius of 8 *light minutes*. Radio signals, which also travel at the speed of light, took 2 1/2 hours to reach the earth when Voyager II passed the planet Uranus (temporarily the most distant planet). Thus Uranus is 2 1/2 light hours away and our solar system has a diameter of 5 light hours (not including the cloud of comets that lie out beyond the planets.)

The closest star, Proxima Centauri, is 4.2 *light years* away. Light from this star, which started out when you entered college as a freshman, will arrive at the earth shortly after you graduate (assuming all goes well). Stars in our local area are typically 2 to 4 light years apart, except for the so called *binary stars* which are pairs of stars orbiting each other at distances as small as light days or light hours.

On a still larger scale, we find that stars form island structures called *galaxies*. We live in a fairly typical galaxy called the Milky Way. It is a flat disk of stars with a slight bulge at the center much like the Sombrero Galaxy seen edge on in Figure (1) and the neighboring spiral galaxy Andromeda seen in Figure (2). Our Milky Way is a spiral galaxy much like Andromeda, with the sun located about 2/3 of the way out in one of the spiral arms. If you look at the sky on a dark clear night you can see the band of stars that cross the sky called the Milky Way. Looking at these stars you are looking sideways through the disk of the Milky Way galaxy.

Figure 1
The Sombrero galaxy.

Figure 2
The Andromeda galaxy.

Our galaxy and the closest similar galaxy, Andromeda, are both about 100,000 light years (.1 million light years) in diameter, contain about a billion stars, and are about one million light years apart. These are more or less typical numbers for the average size, population and spacing of galaxies in the universe.

To look at the universe over still larger distances, first imagine that you are aboard a rocket leaving the earth at night. As you leave the launch pad, you see the individual lights around the launch pad and street lights in neighboring roads. Higher up you start to see the lights from the neighboring city. Still higher you see the lights from a number of cities and it becomes harder and harder to see individual street lights. A short while later all the bright spots you see are cities, and you can no longer see individual lights. At this altitude you count cities instead of light bulbs.

Similarly on our trip out to larger and larger distances in the universe, the bright spots are the galaxies for we can no longer see the individual stars inside. On distances ranging from millions up to billions of light years, we see galaxies populating the universe. On this scale they are small but not quite point like. Instruments like the Hubble telescope in space can view structure in the most distant galaxies, like those shown in Figure (3).

Figure 3
Hubble photograph of the most distant galaxies.

The Expanding Universe

In the 1920s, Edwin Hubble made the surprising discovery that, on average, the galaxies are all moving away from us. The farther away a galaxy is, the faster it is moving away. Hubble found a simple rule for this recession, a galaxy twice as far away is receding twice as fast.

At first you might think that we are at the exact center of the universe if the galaxies are all moving directly away from us. But that is not the case. Hubble's discovery indicates that the universe is expanding uniformly. You can see how a uniform expansion works by blowing up a balloon part way, and drawing a number of uniformly spaced dots on the balloon. Then pick any dot as your own dot, and watch it as you continue to blow the balloon up. You will see that the neighboring dots all move away from your dot, and you will also observe Hubble's rule that dots twice as far away move away twice as fast.

Hubble's discovery provided the first indication that there is a limit to how far away we can see things. At distances of about fourteen billion light years, the recessional speed approaches the speed of light. Recent photographs taken by the Hubble telescope show galaxies receding at speeds in excess of 95% the speed of light, galaxies close to the edge of what we call the ***visible universe***.

The implications of Hubble's rule are more dramatic if you imagine that you take a moving picture of the expanding universe and then run the movie backward in time. The rule that galaxies twice as far away are receding twice as fast become the rule that galaxies twice as far away are approaching you twice as fast. A more distant galaxy, one at twice the distance but heading toward you at twice the speed, will get to you at the same time as a closer galaxy. In fact, all the galaxies will reach you at the same instant of time.

Now run the movie forward from that instant of time, and you see all the galaxies flying apart from what looks like a single explosion. From Hubble's law you can figure that the explosion should have occurred about fourteen billion years ago.

Did such an explosion really happen, or are we simply misreading the data? Is there some other way of interpreting the expansion without invoking such a cataclysmic beginning? Various astronomers thought there was. In their ***continuous creation theory*** they developed a model of the universe that was both unchanging and expanding at the same time. That sounds like an impossible trick because as the universe expands and the galaxies move apart, the density of matter has to decrease. To keep the universe from changing, the model assumed that matter was being created throughout space at just the right rate to keep the average density of matter constant.

With this theory one is faced with the question of which is harder to accept—the picture of the universe starting in an explosion which was derisively called the ***Big Bang***, or the idea that matter is continuously being created everywhere? To provide an explicit test of the continuous creation model, it was proposed that all matter was created in the form of hydrogen atoms, and that all the elements we see around us today, the carbon, oxygen, iron, uranium, etc., were made as a result of nuclear reactions inside of stars.

To test this hypothesis, physicists studied in the laboratory those nuclear reactions which should be relevant to the synthesis of the elements. The results were quite successful. They predicted the correct or nearly correct abundance of all the elements but one. The holdout was helium. There appeared to be more helium in the universe than they could explain.

By 1960, it was recognized that, to explain the abundance of the elements as a result of nuclear reactions inside of stars, you have to start with a mixture of hydrogen and helium. Where did the helium come from? Could it have been created in a Big Bang?

As early as 1948, the Russian physicist George Gamov studied the consequences of the Big Bang model of the universe. He found that if the conditions in the early universe were just right, there should be light left over from the explosion, light that would now be a faint glow at radio wave frequencies. Gamov talked about this prediction with several experimental physicists and was told that the glow would be undetectable. Gamov's prediction was more or less ignored until 1964 when the glow was accidently detected as noise in a radio telescope. Satellites have now been used to study this glow in detail, and the results leave little doubt about the explosive nature of the birth of the universe.

What was the universe like at the beginning? In an attempt to find out, physicists have applied the laws of physics, as we have learned them here on earth, to the collapsing universe seen in the time reversed motion picture of the galaxies. One of the main features that emerges as we go back in time and the universe gets smaller and smaller, is that it also becomes hotter and hotter. The obvious question in constructing a model of the universe is how small and how hot do we allow it to get? Do we stop our model, stop our calculations, when the universe is down to the size of a galaxy? a star? a grapefruit? or a proton? Does it make any sense to apply the laws of physics to something as hot and dense as the universe condensed into something smaller than, say, the size of a grapefruit? Surprisingly, it may. One of the frontiers of physics research is to test the application of the laws of physics to this model of the hot early universe.

We will start our disruption of the early universe at a time when the universe was about a billionth of a second old and the temperature was three hundred thousand billion (3×10^{14}) degrees. While this sounds like a preposterously short time and unbelievably high temperature, it is not the shortest time or highest temperature that has been quite carefully considered. For our overview, we are arbitrarily choosing that time because of the series of pictures we can paint which show the universe evolving. These pictures all involve the behavior of matter as it has been studied in the laboratory. To go back earlier relies on theories that we are still formulating and trying to test.

To recognize what we see in this evolving picture of the universe, we first need a reasonably good picture of what the matter around us is like. With an understanding of the building blocks of matter, we can watch the pieces fit together as the universe evolves. Our discussion of these building blocks will begin with atoms which appear only late in the universe, and work down to smaller particles which play a role at earlier times. To understand what is happening, we also need a picture of how matter interacts via the basic forces in nature.

When you look through a microscope and change the magnification, what you see and how you interpret it, changes, even though you are looking at the same sample. To get a preliminary idea of what matter is made from and how it behaves, we will select a particular sample and magnify it in stages. At each stage we will provide a brief discussion to help interpret what we see. As we increase the magnification, the interpretation of what we see changes to fit and to explain the new picture. Surprisingly, when we get down to the smallest scales of distance using the greatest magnification, we see the entire universe at its infancy. We have reached the point where studying matter on the very smallest scale requires an understanding of the very largest, and vice versa.

STRUCTURE OF MATTER

We will start our trip down to small scales with a rather large, familiar example—the earth in orbit about the sun. The earth is attracted to the sun by a force called *gravity*, and its motion can be accurately forecast, using a set of rules called ***Newtonian mechanics***. The basic concepts involved in Newtonian mechanics are force, mass, velocity and acceleration, and the rules tell us how these concepts are related. (Half of the traditional introductory physics courses is devoted to learning these rules.)

Atoms

We will avoid much of the complexity we see around us by next focusing in on a single hydrogen atom. If we increase the magnification so that a garden pea looks as big as the earth, then one of the hydrogen atoms inside the pea would be about the size of a basketball. How we interpret what we see inside the atom depends upon our previous experience with physics. With a background in Newtonian mechanics, we would see a miniature solar system with the nucleus at the center and an electron in orbit. The nucleus in hydrogen consists of a single particle called the proton, and the electron is held in orbit by an electric force. At this magnification, the proton and electron are tiny points, too small to show any detail.

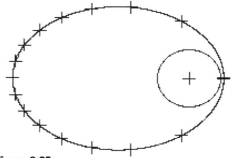

Figure 8-25a
Elliptical orbit of an earth satellite calculated using Newtonian mechanics.

There are similarities and striking differences between the gravitational force that holds our solar system together and the electric force that holds the hydrogen atom together. Both forces in these two examples are attractive, and both forces decrease as the square of the distance between the particles. That means that if you double the separation, the force is only one quarter as strong. The strength of the gravitational force depends on the *mass* of the objects, while the electric force depends upon the *charge* of the objects.

One of the major differences between electricity and gravity is that all gravitational forces are attractive, while there are both attractive and repulsive electric forces. To account for the two types of electric force, we say that there are two kinds of electric charge, which Benjamin Franklin called *positive charge* and *negative charge*. The rule is that like charges repel while opposite charges attract. Since the electron and the proton have opposite charge they attract each other. If you tried to put two electrons together, they would repel because they have like charges. You get the same repulsion between two protons. By the accident of Benjamin Franklin's choice, protons are positively charged and electrons are negatively charged.

Another difference between the electric and gravitational forces is their strengths. If you compare the electric to the gravitational force between the proton and electron in a hydrogen atom, you find that the electric force is 2270000000000000000000000000000000000000 times stronger than the gravitational force. On an atomic scale, gravity is so weak that it is essentially undetectable.

On a large scale, gravity dominates because of the cancellation of electric forces. Consider, for example, the net electric force between two complete hydrogen atoms separated by some small distance. Call them atom A and atom B. Between these two atoms there are four distinct forces, two attractive and two repulsive. The attractive forces are between the proton in atom A and the electron in atom B, and between the electron in atom A and the proton in atom B. However, the two protons repel each other and the electrons repel to give the two repulsive forces. The net result is that the attractive and repulsive forces cancel and we end up with essentially no electric force between the atoms.

Rather than counting individual forces, it is easier to add up electric charge. Since a proton and an electron have opposite charges, the total charge in a hydrogen atom adds up to zero. With no net charge on either of the two hydrogen atoms in our example, there is no net electric force between them. We say that a complete hydrogen atom is ***electrically neutral***.

While complete hydrogen atoms are neutral, they can attract each other if you bring them too close together. What happens is that the electron orbits are distorted by the presence of the neighboring atom, the electric forces no longer exactly cancel, and we are left with a small residual force called a *molecular force*. It is the molecular force that can bind the two hydrogen atoms together to form a hydrogen molecule. These molecular forces are capable of building very complex objects, like people. We are the kind of structure that results from electric forces, in much the same way that solar systems and galaxies are the kind of structures that result from gravitational forces.

Chemistry deals with reactions between about 100 different elements, and each element is made out of a different kind of atom. The basic distinction between atoms of different elements is the number of protons in the nucleus. A hydrogen nucleus has one proton, a helium nucleus 2 protons, a lithium nucleus 3 protons, on up to the largest naturally occurring nucleus, uranium with 92 protons.

Complete atoms are electrically neutral, having as many electrons orbiting outside as there are protons in the nucleus. The chemical properties of an atom are determined almost exclusively by the structure of the orbiting electrons, and their electron structure depends very much on the number of electrons. For example, helium with 2 electrons is an inert gas often breathed by deep sea divers. Lithium with 3 electrons is a reactive metal that bursts into flame when exposed to air. We go from an inert gas to a reactive metal by adding one electron.

Light

The view of the hydrogen atom as a miniature solar system, a view of the atom seen through the "lens" of Newtonian mechanics, fails to explain much of the atom's behavior. When you heat hydrogen gas, it glows with a reddish glow that consists of three distinct colors or so called *spectral lines*. The colors of the lines are bright red, swimming pool blue, and deep violet. You need more than Newtonian mechanics to understand why hydrogen emits light, let alone explain these three special colors.

In the middle of the 1800s, Michael Faraday went a long way in explaining electric and magnetic phenomena in terms of *electric* and *magnetic fields*. These fields are essentially maps of electric and magnetic forces. In 1860 James Clerk Maxwell discovered that the four equations governing the behavior of electric and magnetic fields could be combined to make up what is called a *wave equation*. Maxwell could construct his wave equation after making a small but crucial correction to one of the underlying equations.

The importance of Maxwell's wave equation was that it predicted that a particular combination of electric and magnetic fields could travel through space in a wavelike manner. Equally important was the fact that the wave equation allowed Maxwell to calculate what the speed of the wave should be, and the answer was about a billion feet per second. Since only light was known to travel that fast, Maxwell made the guess that he had discovered the theory of light, that *light consisted of a wave of electric and magnetic fields of force*.

Visible light is only a small part of what we call the *electromagnetic spectrum*. Our eyes are sensitive to light waves whose wavelength varies only over a very narrow range. Shorter wavelengths lie in the *ultraviolet* or *x ray* region, while at increasingly longer wavelengths are *infra red light*, *microwaves*, and *radio waves*. Maxwell's theory made it clear that these other wavelengths should exist, and within a few years, radio waves were discovered. The broadcast industry is now dependent on Maxwell's equations for the design of radio and television transmitters and receivers. (Maxwell's theory is what is usually taught in the second half of an introductory physics course. That gets you all the way up to 1860.)

While Maxwell's theory works well for the design of radio antennas, it does not do well in explaining the behavior of a hydrogen atom. When we apply Maxwell's theory to the miniature solar system model of hydrogen, we do predict that the orbiting electron will radiate light. But we also predict that the atom will self destruct. The unambiguous prediction is that the electron will continue to radiate light of shorter and shorter wavelength while spiraling in faster and faster toward the nucleus, until it crashes. The combination of Newton's laws and Maxwell's theory is known as *Classical Physics*. We can easily see that classical physics fails when applied even to the simplest of atoms.

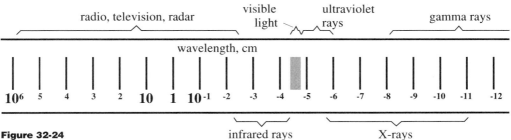

Figure 32-24
The electromagnetic spectrum.

Photons

In the late 1890's, it was discovered that a beam of light could knock electrons out of a hydrogen atom. The phenomenon became known as the *photoelectric effect*. You can use Maxwell's theory to get a rough idea of why a wave of electric and magnetic force might be able to pull electrons out of a surface, but the details all come out wrong. In 1905, in the same year that he developed his theory of relativity, Einstein explained the photoelectric effect by proposing that light consisted of a beam of particles we now call *photons*. When a metal surface is struck by a beam of photons, an electron can be knocked out of the surface if it is struck by an individual photon. A simple formula for the energy of the photons led to an accurate explanation of all the experimental results related to the photoelectric effect.

Despite its success in explaining the photoelectric effect, Einstein's photon picture of light was in conflict not only with Maxwell's theory, it conflicted with over 100 years of experiments which had conclusively demonstrated that light was a wave. This conflict was not to be resolved in any satisfactory way until the middle 1920s.

The particle nature of light helps but does not solve the problems we have encountered in understanding the behavior of the electron in hydrogen. According to Einstein's photoelectric formula, the energy of a photon is inversely proportional to its wavelength. The longer wavelength red photons have less energy than the shorter wavelength blue ones. To explain the special colors of light emitted by hydrogen, we have to be able to explain why only photons with very special energies can be emitted.

The Bohr Model

In 1913, the year after the nucleus was discovered, Neils Bohr developed a somewhat ad hoc model that worked surprisingly well in explaining hydrogen. Bohr assumed that the electron in hydrogen could travel on only certain *allowed orbits*. There was a smallest, lowest energy orbit that is occupied by an electron in cool hydrogen atoms. The fact that this was the *smallest* allowed orbit meant that the electron would not spiral in and crush into the nucleus.

Using Maxwell's theory, one views the electron as radiating light continuously as it goes around the orbit. In Bohr's picture the electron does not radiate while in one of the allowed orbits. Instead it radiates, it emits a photon, only when it jumps from one orbit to another.

To see why heated hydrogen radiates light, we need a picture of thermal energy. A gas, like a bottle of hydrogen or the air around us, consists of molecules flying around, bouncing into each other. Any moving object has extra energy due to its motion. If all the parts of the object are moving together, like a car traveling down the highway, then we call this energy of motion *kinetic energy*. If the motion is the random motion of molecules bouncing into each other, we call it *thermal energy*.

The temperature of a gas is proportional to the average thermal energy of the gas molecules. As you heat a gas, the molecules move faster, and their average thermal

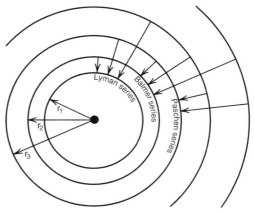

Figure 35-6
The allowed orbits of the Bohr Model.

energy and temperature rises. At the increased speed the collisions between molecules are also stronger.

Consider what happens if we heat a bottle of hydrogen gas. At room temperature, before we start heating, the electrons in all the atoms are sitting in their lowest energy orbits. Even at this temperature the atoms are colliding but the energy involved in a room temperature collision is not great enough to knock an electron into one of the higher energy orbits. As a result, room temperature hydrogen does not emit light.

When you heat the hydrogen, the collisions between atoms become stronger. Finally you reach a temperature in which enough energy is involved in a collision to knock an electron into one of the higher energy orbits. The electron then falls back down, from one allowed orbit to another until it reaches the bottom, lowest energy orbit. The energy that the electron loses in each fall, is carried out by a photon. Since there are only certain allowed orbits, there are only certain special amounts of energy that the photon can carry out.

To get a better feeling for how the model works, suppose we number the orbits, starting at orbit 1 for the lowest energy orbit, orbit 2 for the next lowest energy orbit, etc. Then it turns out that the photons in the red spectral line are radiated when the electron falls from orbit 3 to orbit 2. The red photon's energy is just equal to the energy the electron loses in falling between these orbits. The more energetic blue photons carry out the energy an electron loses in falling from orbit 4 to orbit 2, and the still more energetic violet photons correspond to a fall from orbit 5 to orbit 2. All the other jumps give rise to photons whose energy is too large or too small to be visible. Those with too much energy are ultraviolet photons, while those with too little are in the infra red part of the spectrum. The jump down to orbit 1 is the biggest jump with the result that all jumps down to the lowest energy orbit results in ultraviolet photons.

It appears rather ad hoc to propose a theory where you invent a large number of special orbits to explain what we now know as a large number of spectral lines. One criterion for a successful theory in science is that you get more out of the theory than you put in. If Bohr had to invent a new allowed orbit for each spectral line explained, the theory would be essentially worthless.

However this is not the case for the Bohr model. Bohr found a simple formula for the electron energies of all the allowed orbits. This one formula in a sense explains the many spectral lines of hydrogen. A lot more came out of Bohr's model than Bohr had to put in.

The problem with Bohr's model is that it is essentially based on Newtonian mechanics, but there is no excuse whatsoever in Newtonian mechanics for identifying any orbit as special. Bohr focused the problem by discovering that the allowed orbits had special values of a quantity called *angular momentum*.

Angular momentum is related to rotational motion, and in Newtonian mechanics angular momentum increases continuously and smoothly as you start to spin an object. Bohr could explain his allowed orbits by proposing that there was a special unique value of angular momentum—call it a *unit of angular momentum*. Bohr found, using standard Newtonian calculations, that his lowest energy orbit had one unit of angular momentum, orbit 2 had two units, orbit 3 three units, etc. Bohr could explain his entire model by the one assumption that angular momentum was *quantized*, i.e., came only in units.

Bohr's quantization of angular momentum is counter intuitive, for it leads to the picture that when we start to rotate an object, the rotation increases in a jerky fashion rather than continuously. First the object has no angular momentum, then one unit, then 2 units, and on up. The reason we do not see this jerky motion when we start to rotate something large like a bicycle wheel, is that the basic unit of angular momentum is very small. We cannot detect the individual steps in angular momentum, it seems continuous. But on the scale of an atom, the steps are big and have a profound effect.

With Bohr's theory of hydrogen and Einstein's theory of the photoelectric effect, it was clear that classical physics was in deep trouble. Einstein's photons gave a lumpiness to what should have been a smooth wave in Maxwell's theory of light and Bohr's model gave a jerkiness to what should be a smooth change in angular momentum. The bumps and jerkiness needed a new picture of the way matter behaves, a picture that was introduced in 1924 by the graduate student Louis de Broglie.

PARTICLE-WAVE NATURE OF MATTER

Noting the wave and particle nature of light, de Broglie proposed that the electron had both a wave and a particle nature. While electrons had clearly exhibited a particle behavior in various experiments, de Broglie suggested that it was the wave nature of the electron that was responsible for the special allowed orbits in Bohr's theory. De Broglie presented a simple wave picture where, in the allowed orbits, an integer number of wavelengths fit around the orbit. Orbit 1 had one wavelength, orbit 2 had two wavelengths, etc. In De Broglie's picture, electron waves in non allowed orbits would cancel themselves out. Borrowing some features of Einstein's photon theory of light waves, de Broglie could show that the angular momentum of the electron would have the special quantized values when the electron wave was in one of the special, non cancelling orbits.

With his simple wave picture, de Broglie had hit upon the fundamental idea that was missing in classical physics. The idea is that *all matter*, not just light, has a *particle wave nature*.

It took a few years to gain a satisfactory interpretation of the dual particle wave nature of matter. The current interpretation is that things like photons are in fact particles, but their motion is governed, not by Newtonian mechanics, but by the laws of wave motion. How this works in detail is the subject of our chapter on Quantum Mechanics. One fundamental requirement of our modern interpretation of the particle wave is that, for the interpretation to be meaningful, all forms of matter, without exception, must have this particle wave nature. This general requirement is summarized by a rule discovered by Werner Heisinberg, a rule known as the *uncertainty principle*. How the rule got that name is also discussed in our chapter on quantum mechanics.

In 1925, after giving a seminar describing de Broglie's model of electron waves in hydrogen, Erwin Schrödinger was chided for presenting such a "childish" model. A colleague reminded him that waves do not work that way, and suggested that since Schrödinger had nothing better to do, he should work out a real wave equation for the electron waves, and present the results in a couple of weeks.

It took Schrödinger longer than a couple of weeks, but he did succeed in constructing a wave equation for the electron. In many ways Schrödinger's wave equation for the electron is analogous to Maxwell's wave equation for light. Schrödinger's wave equation for the electron allows one to calculate the behavior of electrons in all kinds of atoms. It allows one to explain and predict an atom's electron structure and chemical properties. Schrödinger's equation has become the fundamental equation of chemistry.

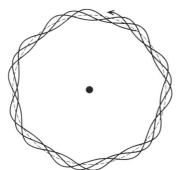

Figure 35-9
De Broglie picture of an electron wave cancelling itself out.

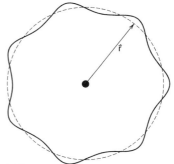

Figure 35-10
If the circumference of the orbit is an integer number of wavelengths, the electron wave will go around without any cancellation.

CONSERVATION OF ENERGY

Before we go on with our investigation of the hydrogen atom, we will take a short break to discuss the idea of conservation of energy. This idea, which originated in Newtonian mechanics, survives more or less intact in our modern particle-wave picture of matter.

Physicists pay attention to the concept of energy only because energy is conserved. If energy disappears from one place, it will show up in another. We saw this in the Bohr model of hydrogen. When the electron lost energy falling down from one allowed orbit to a lower energy orbit, the energy lost by the electron was carried out by a photon.

You can store energy in an object by doing *work* on the object. When you lift a ball off the floor, for example, the work you did lifting the ball, the energy you supplied, is stored in a form we call *gravitational potential energy*. Let go of the ball and it falls to the floor, loosing its gravitational potential energy. But just before it hits the floor, it has a lot of energy of motion, what we have called kinetic energy. All the gravitational potential energy the ball had before we dropped it has been converted to kinetic energy.

After the ball hits the floor and is finally resting there, it is hard to see where the energy has gone. One place it has gone is into thermal energy, the floor and the ball are a tiny bit warmer as a result of your dropping the ball.

Another way to store energy is to compress a spring. When you release the spring you can get the energy back. For example, compress a watch spring by winding up the watch, and the energy released as the spring unwinds will run the watch for a day. We could call the energy stored in the compressed spring *spring potential energy*. Physicists invent all sorts of names for the various forms of energy.

One of the big surprises in physics was Einstein's discovery of the equivalence of mass and energy, a relationship expressed by the famous equation $E = mc^2$. In that equation, E stands for the energy of an object, m its mass, and c is the speed of light. Since the factor c^2 is a constant, Einstein's equation is basically saying that mass is a form of energy. The c^2 is there because mass and energy were initially thought to be different quantities with different units like kilograms and joules. The c^2 simply converts mass units into energy units.

What is amazing is the amount of energy that is in the form of mass. If you could convert all the mass of a pencil eraser into electrical energy, and sell the electrical energy at the going rate of 10¢ per kilowatt hour, you would get about 10 million dollars for it. The problem is converting the mass to another, more useful, form of energy. If you can do the conversion, however, the results can be spectacular or terrible. Atomic and hydrogen bombs get their power from the conversion of a small fraction of their mass energy into thermal energy. The sun gets its energy by "burning" hydrogen nuclei to form helium nuclei. The energy comes from the fact that a helium nucleus has slightly less mass than the hydrogen nuclei out of which it was formed.

If you have a particle at rest and start it moving, the particle gains kinetic energy. In Einstein's view the particle at rest has energy due to its *rest mass*. When you start the particle moving, it gains energy, and since mass is equivalent to energy, it also gains mass. For most familiar speeds the increase in mass due to kinetic energy is very small. Even at the speeds travelled by rockets and spacecraft, the increase in mass due to kinetic energy is hardly noticeable. Only when a particle's speed gets up near the speed of light does the increase in mass become significant.

One of the first things we discussed about the behavior of matter is that nothing can travel faster than the speed of light. You might have wondered if nature had traffic cops to enforce this speed limit. It does not need one, it uses a law of nature instead. As the speed of an object approaches the speed of light, its mass increases. The closer to the speed of light, the greater increase in mass. To push a particle up to the speed of light would give it an infinite mass and therefore require an infinite amount of energy. Since that much energy is not available, no particle is going to exceed nature's speed limit.

This raises one question. What about photons? They are particles of light and therefore travel *at* the speed of light. But their energy is not infinite. It depends instead on the wavelength or color of the photon. Photons escape the rule about mass increasing with speed by starting out with no rest mass. You stop a photon and nothing is left. Photons can only exist by traveling at the speed of light.

When a particle is traveling at speeds close enough to the speed of light that its kinetic energy approaches its rest mass energy, the particle behaves differently than slowly moving particles. For example, push on a slowly moving particle and you can make the particle move faster. Push on a particle already moving at nearly the speed of light, and you merely make the particle more massive since it cannot move faster. Since the relationship between mass and energy came out of Einstein's theory of relativity, we say that particles moving near the speed of light obey ***relativistic*** mechanics while those moving slowly are ***nonrelativistic***. Light is always relativistic, and all automobiles on the earth are nonrelativistic.

ANTI-MATTER

Schrödinger's equation for electron waves is a nonrelativistic theory. It accurately describes electrons that are moving at speeds small compared to the speed of light. This is fine for most studies in chemistry, where chemical energies are much much less than rest mass energies. You can see the difference for example by comparing the energy released by a conventional chemical bomb and an atomic bomb.

Schrödinger of course knew Einstein's theory of relativity, and initially set out to derive a ***relativistic wave equation*** for the electron. This would be an equation that would correctly explain the behavior of electrons even as the speed of the electrons approached the speed of light and their kinetic energy became comparable to or even exceeded their rest mass energy.

Schrödinger did construct a relativistic wave equation. The problem was that the equation had two solutions, one representing ordinary electrons, the other an apparently impossible particle with a negative rest mass. In physics and mathematics we are often faced with equations with two or more solutions. For example, the formula for the hypotenuse c of a right triangle with sides of lengths a and b is

$$c^2 = a^2 + b^2$$

This equation has two solutions, namely $c = +\sqrt{a^2 + b^2}$ and $c = -\sqrt{a^2 + b^2}$. The negative solution does not give us much of a problem, we simply ignore it.

Schrödinger could not ignore the negative mass solutions in his relativistic wave equation for the following reason. If he started with just ordinary positive mass electrons and let them interact, the equation predicted that the negative mass solutions would be created! The peculiar solutions could not be ignored if the equation was to be believed. Only by going to his nonrelativistic equation could Schrödinger avoid the peculiar solutions.

A couple years later, Dirac tried again to develop a relativistic wave equation for the electron. At first it appeared that Dirac's equation would avoid the negative mass solutions, but with little further work, Dirac found that the negative mass solutions were still there. Rather than giving up on his new equation, Dirac found a new interpretation of these peculiar solutions. Instead of viewing them as negatively charged electrons with a negative mass, he could interpret them as positive mass particles with a positive electric charge. According to Dirac's equation, positive and negative charged solutions could be created or destroyed in pairs. The pairs could be created any time enough energy was available.

Dirac predicted the existence of this positively charged particle in 1929. It was not until 1933 that Carl Anderson at Caltech, who was studying the elementary particles that showered down from the sky (particles called *cosmic rays*), observed a positively charged particle whose mass was the same as that of the electron. Named the *positron*, this particle was immediately identified as the positive particle expected from Dirac's equation.

In our current view of matter, all particles are described by relativistic wave equations, and all relativistic wave equations have two kinds of solutions. One solution is for ordinary matter particles like electrons, protons, and neutrons. The other solution, which we now call *antimatter*, describes anti particles, the *antielectron* which is the positron, and the *antiproton* and the *antineutron*. Since all antiparticles can be created or destroyed in particle-anti particle pairs, the antiparticle has to have the opposite conserved property so that the property will remain conserved. As an example, the positron has the opposite charge as the electron so that electric charge is neither created or destroyed when electron-positron pairs appear or disappear.

While all particles have antiparticles, some particles like the photon, have no conserved properties other than energy. As a result, these particles are indistinguishable from their antiparticles.

PARTICLE NATURE OF FORCES

De Broglie got his idea for the wave nature of the electron from the particle-wave nature of light. The particle of light is the photon which can knock electrons out of a metal surface. The wave nature is the wave of electric and magnetic force that was predicted by Maxwell's theory. When you combine these two aspects of light, you are led to the conclusion that electric and magnetic forces are ultimately caused by photons. We call any force resulting from electric or magnetic forces as being due to the *electric interaction*. The photon is the particle responsible for the electric interaction.

Let us see how our picture of the hydrogen atom has evolved as we have learned more about the particles and forces involved. We started with a miniature solar system with the heavy proton at the center and an electron in orbit. The force was the electric force that in many ways resembled the gravitational force that keeps the earth in orbit around the sun. This picture failed, however, when we tried to explain the light radiated by heated hydrogen.

The next real improvement comes with Schrödinger's wave equation describing the behavior of the electron in hydrogen. Rather than there being allowed orbits as in Bohr's model, the electron in Schrödinger's picture has allowed *standing wave patterns*. The chemical properties of atoms can be deduced from these wave patterns, and Schrödinger's equation leads to accurate predictions of the wavelengths of light radiated not only by hydrogen but other atoms as well.

There are two limitations to Schrödinger's equation. One of the limitations we have seen is that it is a non relativistic equation, an equation that neglects any change in the electron's mass due to motion. While this is a very good approximation for describing the slow speed electron in hydrogen, the wavelengths of light radiated by hydrogen can be measured so accurately that tiny relativistic effects can be seen. Dirac's relativistic wave equation is required to explain these tiny relativistic corrections.

The second limitation is that neither Schrödinger's or Dirac's equations take into account the particle nature of the electric force holding hydrogen together. In the hydrogen atom, the particle nature of the electric force has only the very tiniest effect on the wavelength of the radiated light. But even these effects can be measured and the particle nature must be taken into account. The theory that takes into account both the wave nature of the electron and the particle nature of the electric force is called *quantum electrodynamics*, a theory finally developed in 1947 by Richard Feynman and Julian Schwinger. Quantum electrodynamics is the most precisely tested theory in all of science.

In our current picture of the hydrogen atom, as described by quantum electrodynamics, the force between the electron and the proton nucleus is caused by the continual exchange of photons between the two charged particles. While being exchanged, the photon can do some subtle things like create a positron electron pair which quickly annihilates. These subtle things have tiny but measurable effects on the radiated wavelengths, effects that correctly predicted by the theory.

The development of quantum electrodynamics came nearly 20 years after Dirac's equation because of certain mathematical problems the theory had to overcome. In this theory, the electron is treated as a point particle with *no size*. The accuracy of the predictions of quantum electrodynamics is our best evidence that this is the correct picture. In other words, we have no evidence that the electron has a finite size, and a very accurate theory which assumes that it does not. However, it is not easy to construct a mathematical theory in which a finite amount of mass and energy is crammed into a region of no size. For one thing you are looking at infinite densities of mass and energy.

Renormalization

The early attempts to construct the theory of quantum electrodynamics were plagued by infinities. What would happen is that you would do an initial approximate calculation and the results would be good. You would then try to improve the results by calculating what were supposed to be tiny corrections, and the corrections turned out to be infinitely large. One of the main accomplishments of Feynman and Schwinger was to develop a mathematical procedure, sort of a mathematical slight of hand, that got rid of the infinities. This mathematical procedure became known as *renormalization*.

Feynman always felt that renormalization was simply a trick to cover up our ignorance of a deeper more accurate picture of the electron. I can still hear him saying this during several seminars. It turned out however that renormalization became an important guide in developing theories of other forces. We will shortly encounter two new forces as we look down into the atomic nucleus, forces called the ***nuclear interaction*** and the ***weak interaction***. Both of these forces have a particle-wave nature like the electric interaction, and the successful theories of these forces used renormalization as a guide.

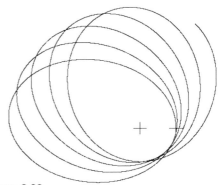

Figure 8-33
*Einstein's theory of gravity predicted that Mercury's elliptical orbit "precessed" or rotated somewhat like the rotation seen in the above orbit. Mercury's precession is **much, much smaller**.*

Gravity

The one holdout, the one force for which we do not have a successful theory, is gravity. We have come a long way since Newton's law of gravity. After Einstein developed his theory of relativity in 1905, he spent the next 12 years working on a relativistic theory of gravity. The result, known as **general relativity** is a theory of gravity that is in many ways similar to Maxwell's theory of electricity. Einstein's theory predicts, for example, that a planet in orbit about a star should emit gravitational waves in much the same way that Maxwell's theory predicts that an electron in orbit about a nucleus should emit electromagnetic radiation or light.

One of the difficulties working with Einstein's theory of gravity is that Newton's theory of gravity explains almost everything we see, and you have to look very hard in places where Newton's law is wrong and Einstein's theory is right. There is an extremely small but measurable correction to the orbit of Mercury that Newton's theory cannot explain and Einstein's theory does.

Einstein's theory also correctly predicts how much light will be deflected by the gravitational attraction of a star. You can argue that because light has energy and energy is equivalent to mass, Newton's law of gravity should also predict that starlight should be deflected by the gravitational pull of a star. But this Newtonian argument leads to half the deflection predicted by Einstein's theory, and the deflection predicted by Einstein is observed.

The gravitational radiation predicted by Einstein's theory has not been detected directly, but we have very good evidence for its existence. In 1974 Joe Taylor from the University of Massachusetts, working at the large radio telescope at Arecibo discovered a pair of neutron stars in close orbit about each other. We will have more to say about neutron stars later. The point is that the period of the orbit of these stars can be measured with extreme precision.

Einstein's theory predicts that the orbiting stars should radiate gravitational waves and spiral in toward each other. This is reminiscent of what we got by applying Maxwell's theory to the electron in hydrogen, but in the case of the pair of neutron stars the theory worked. The period of the orbit of these stars is changing in exactly the way one would expect if the stars were radiating gravitational waves.

If our wave-particle picture of the behavior of matter is correct, then the gravitational waves must have a particle nature like electromagnetic waves. Physicists call the gravitational particle the **graviton**. We think we know a lot about the graviton even though we have not yet seen one. The graviton should, like the photon, have no rest mass, travel at the speed of light, and have the same relationship between energy and wavelength.

One difference is that because the graviton has energy and therefore mass, and because gravitons interact with mass, gravitons interact with themselves. This self interaction significantly complicates the theory of gravity. In contrast photons interact with electric charge, but photons themselves do not carry charge. As a result, photons do not interact with each other which considerably simplifies the theory of the electric interaction.

An important difference between the graviton and the photon, what has prevented the graviton from being detected, is its fantastically weak interaction with matter. You saw that the gravitational force between the electron and a proton is a thousand billion billion billion times weaker than the electric force. In effect this makes the graviton a thousand billion billion billion billion times harder to detect. The only reason we know that this very weak force exists at all is that it gets stronger and stronger as we put more and more mass together, to form large objects like planets and stars.

Not only do we have problems thinking of a way to detect gravitons, we have run into a surprising amount of difficulty constructing a theory of gravitons. The theory would be known as the **quantum theory of gravity**, but we do not yet have a quantum theory of gravity. The problem is that the theory of gravitons interacting with point particles, the gravitational analogy of quantum electrodynamics, does not work. The theory is not renormalizable, you cannot get rid of the infinities. As in the case of the electric interaction the simple calculations work well, and that is why we think we know a lot about the graviton. But when you try to make what should be tiny relativistic corrections, the correction turns out to be infinite. No mathematical slight of hand has gotten rid of the infinities.

The failure to construct a consistent quantum theory of gravity interacting with point particles has suggested to some theoretical physicists that our picture of the electron and some other particles being point particles is wrong. In a new approach called *string theory*, the elementary particles are view not as point particles but instead as incredibly small one dimensional objects called *strings*. The strings vibrate, with different modes of vibration corresponding to different elementary particles.

String theory is complex. For example, the strings exist in a world of 10 dimensions, whereas we live in a world of 4 dimensions. To make string theory work, you have to explain what happened to the other six dimensions.

Another problem with string theory is that it has not led to any predictions that distinguish it from other theories. There are as yet no tests, like the deflection of starlight by the sun, to demonstrate that string theory is right and other theories are wrong.

String theory does, however, have one thing going for it. By spreading the elementary particles out from zero dimensions (points) to one dimensional objects (strings), the infinities in the theory of gravity can be avoided.

A SUMMARY

Up to this point our focus has been on the hydrogen atom. The physical magnification has not been too great, we are still picturing the atom as an object magnified to the size of a basketball with two particles, the electron and proton, that are too small to see. They may or may not have some size, but we cannot tell at this scale.

What we have done is change our perception of the atom. We started with a picture that Newton would recognize, of a small solar system with the massive proton at the center and the lighter electron held in orbit by the electric force. When we modernize the picture by including Maxwell's theory of electricity and magnetism, we run into trouble. We end up predicting that the electron will lose energy by radiating light, soon crashing into the proton. Bohr salvaged the picture by introducing his allowed orbits and quantized angular momentum, but the success of Bohr's theory only strengthened the conviction that something was fundamentally wrong with classical physics.

Louis de Broglie pointed the way to a new picture of the behavior of matter by proposing that all matter, not just light, had a particle-wave nature. Building on de Broglie's idea, Schrödinger developed a wave equation that not only describes the behavior of the electron in hydrogen, but in larger and more complex atoms as well.

While Schrödinger's non relativistic wave equation adequately explains most classical phenomena, even in the hydrogen atom, there are tiny but observable relativistic effects that Dirac could explain with his relativistic wave equation for the electron. Dirac handled the problem of all relativistic wave equations having two solutions by reinterpreting the second solution as representing antimatter.

Dirac's equation is still not the final theory for hydrogen because it does not take into account the fact that electric forces are ultimately caused by photons. The wave theory of the electron that takes the photon nature of the electric force into account is known as quantum electrodynamics. The predictions of quantum electrodynamics are in complete agreement with experiment, it is the most precisely tested theory in science.

The problems resulting from treating the electron as a point particle were handled in quantum electrodynamics by renormalization. Renormalization does not work, however, when one tries to formulate a quantum theory of gravity where the gravitational force particle—the graviton—interacts with point particles. This has led some theorists to picture the electron not as a point but as an incredibly small one dimensional object called a *string*. While string theory is renormalizable, there have been no experimental tests to show that string theory is right and the point particle picture is wrong. This is as far as we can take our picture of the hydrogen atom without taking a closer look at the nucleus.

THE NUCLEUS

To see the nucleus we have to magnify our hydrogen atom to a size much larger than a basketball. When the atom is enlarged so that it would just fill a football stadium, the nucleus, the single proton, would be about the size of a pencil eraser. The proton is clearly not a point particle like the electron. If we enlarge the atom further to get a better view of the nucleus, to the point where the proton looks as big as a grapefruit, the atom is about 10 kilometers in diameter. This grapefruit sized object weighs 1836 times as much as the electron, but it is the electron wave that occupies the 10 kilometer sphere of space surrounding the proton.

Before we look inside the proton, let us take a brief look at the nuclei of some other atoms. Once in a great while you will find a hydrogen nucleus with two particles. One is a proton and the other is the electrically neutral particles called the *neutron*. Aside from the electric charge, the proton and neutron look very similar. They are about the same size and about the same mass. The neutron is a fraction of a percent heavier than the proton, a small mass difference that will turn out to have some interesting consequences.

As we mentioned, the type of element is determined by the number of protons in the nucleus. All hydrogen atoms have one proton, all helium atoms 2 protons, etc. But for the same element there can be different numbers of neutrons in the nucleus. Atoms with the same numbers of protons but different numbers of neutrons are called different *isotopes* of the element. Another isotope of hydrogen, one that is unstable and decays in roughly 10 years, is a nucleus with one proton and two neutrons called *tritium*.

The most stable isotope of helium is helium 4, with 2 protons and 2 neutrons. Helium 3 with 2 protons and one neutron is stable but very rare. Once we get beyond hydrogen we name the different isotopes by adding a number after the name, a number representing the total number of protons and neutrons. For example the heaviest, naturally occurring atom is the isotope Uranium 238, which has 92 protons and 146 neutrons for a total of 238 nuclear particles, or *nucleons* as we sometimes refer to them.

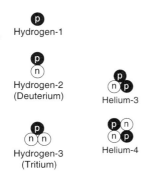

Figure 19-2
Isotopes of hydrogen and helium.

The nucleons in a nucleus pack together much like the grapes in a bunch, or like a bag of grapefruit. At our enlargement where a proton looks as big as a grapefruit, the uranium nucleus would be just over half a meter in diameter, just big enough to hold 238 grapefruit.

When you look at a uranium nucleus with its 92 positively charged protons mixed in with electrically neutral neutrons, then you have to wonder, what holds the thing together? The protons, being all positively charged, all repel each other. And because they are so close together in the nucleus, the repulsion is extremely strong. It is much stronger than the attractive force felt by the distant negative electrons. There must be another kind of force, and attractive force, that keeps the protons from flying apart.

The attractive force is not gravity. Gravity is so weak that it is virtually undetectable on an atomic scale. The attractive force that overpowers the electric repulsion is called the ***nuclear force***. The nuclear force between nucleons is attractive, and essentially blind to the difference between a proton and a neutron. To the nuclear force, a proton and a neutron look the same. The nuclear force has no effect whatsoever on an electron.

Figure 19-1
Styrofoam ball model of the uranium nucleus..

One of the important features of the nuclear force between nucleons is that it has a short range. Compared to the longer range electric force, the nuclear force is more like a contact cement. When two protons are next to each other, the attractive nuclear force is stronger than the electric repulsion. But separate the protons by more than about 4 protons diameters and the electric force is stronger.

If you make nuclei by adding nucleons to a small nucleus, the object becomes more and more stable because all the nucleons are attracting each other. But when you get to nuclei whose diameter exceeds around 4 proton diameters, protons on opposite sides of the nucleus start to repel each other. As a result nuclei larger than that become less stable as you make them bigger. The isotope Iron 56 with 26 protons and 30 neutrons, is about 4 proton diameters across and is the most stable of all nuclei. When you reach Uranium which is about 6 proton diameters across, the nucleus has become so unstable that if you jostle it by hitting it with a proton, it will break apart into two roughly equal sized more stable nuclei. Once apart, the smaller nuclei repel each other electrically and fly apart releasing electric potential energy. This process is called ***nuclear fission*** and is the source of energy in an atomic bomb.

While energy is released when you break apart the large unstable nuclei, energy is also released when you add nucleons to build up the smaller, more stable nuclei. For example, if you start with four protons (four hydrogen nuclei), turn two of the protons into neutrons (we will see how to do this shortly) and put them together to form stable helium 4 nucleus, you get a considerable release of energy. You can easily figure out how much energy is released by noting that 4 protons have a mass that is about .7 percent greater than a helium nucleus. As a result when the protons combine to form helium, about .7 percent of their mass is converted to other forms of energy. Our sun is powered by this energy release as it "burns" hydrogen to form helium. This process is called ***nuclear fusion*** and is the source of the energy of the powerful hydrogen bombs.

STELLAR EVOLUTION

Our sun is about half way through burning up the hydrogen in its hot, inner core. When the hydrogen is exhausted in another 5 billion years, the sun will initially cool and start to collapse. But the collapse will release gravitational potential energy that makes the smaller sun even hotter than it was before running out of hydrogen. The hotter core will emit so much light that the pressure of the light will expand the surface of the sun out beyond the earth's orbit, and the sun will become what is known as a red giant star. Soon, over the astronomically short time of a few million years, the star will cool off becoming a dying, dark ember about the size of the earth. It will become what is known as a ***black dwarf***.

If the sun had been more massive when the hydrogen ran out and the star started to collapse, then more gravitational potential energy would have been released. The core would have become hotter, hot enough to ignite the helium to form the heavier nucleus carbon. Higher temperatures are required to burn helium because the helium nuclei, with two protons, repel each other with four times the electric repulsion than hydrogen nuclei. As a result more thermal energy is required to slam the helium nuclei close enough for the attractive nuclear force to take over.

Once the helium is burned up, the star again starts to cool and contract, releasing more gravitational potential energy until it becomes hot enough to burn the carbon to form oxygen nuclei. This cycle keeps repeating, forming one element after another until we get to Iron 56. When you have an iron core and the star starts to collapse and gets hotter, the iron does not burn. You do not get a release of energy by making nuclei larger than iron. As a result the collapse continues resulting in a huge implosion.

Once the center collapses, a strong shock wave races out through the outer layers of the star, tearing the star apart. This is called a ***supernova explosion***. It is in these supernova explosions with their extremely high temperatures that nuclei larger than iron are formed. All the elements inside of you that are down the periodic table from iron were created in a supernova. ***Part of you has already been through a supernova explosion***.

What is left behind of the core of the star depends on how massive the star was to begin with. If what remains of the core is 1.4 times as massive as our sun, then the gravitational force will be strong enough to cram the electrons into the nuclei, turning all the protons into neutrons, and leaving behind a ball of neutrons about 20 kilometers in diameter. This is called a ***neutron star***. A neutron star is essentially a gigantic nucleus held together by gravity instead of the nuclear force.

If you think that squeezing the mass of a star into a ball 20 kilometers in diameter is hard to picture (at this density all the people on the earth would fit into the volume of a raindrop), then consider what happens if the remaining core is about six times as massive as the sun. With such mass, the gravitational force is so strong that the neutrons are crushed and the star becomes smaller and smaller.

The matter in a neutron star is about as rigid as matter can get. The more rigid a substance is, the faster sound waves travel through the substance. For example, sound travels considerably faster through steel than air. The matter in a neutron star is so rigid, or shall we say so incompressible, that the speed of sound approaches the speed of light.

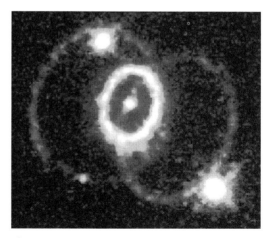

Figure 4
1987 supernova as seen by the Hubble telescope.

When gravity has crushed the neutrons in a neutron star, it has overcome the strongest resistance any known force can possibly resist. But, as the collapse continues, gravity keeps getting stronger. According to our current picture of the behavior of matter, a rather unclear picture in this case, the collapse continues until the star becomes a point with no size. Well before it reaches that end, gravity has become so strong that light can no longer escape, with the result that these objects are known as **black holes**.

We have a fuzzy picture of what lies at the center of a black hole because we do not have a quantum theory of gravity. Einstein's classical theory of gravity predicts that the star collapses to a point, but before that happens we should reach a state where the quantum effects of gravity are important. Perhaps string theory will give us a clue as to what is happening. We will not learn by looking because light cannot get out.

The formation of neutron stars and black holes emphasizes an important feature of gravity. On an atomic scale, gravity is the weakest of the forces we have discussed so far. The gravitational force between an electron and a proton is a thousand billion billion billion billion (10^{39}) times weaker than the electric force. Yet because gravity is long range like the electric force, and has no cancellation, it ends up dominating all other forces, even crushing matter as we know it, out of existence.

Figure 5
Hubble telescope's first view of a lone neutron star in visible light. This star is no greater than 16.8 miles (28 kilometers) across.

The Weak Interaction

In addition to gravity, the electric interaction and the nuclear force, there is one more basic force or interaction in nature given the rather bland name the *weak interaction*. While considerably weaker than electric or nuclear forces, it is far far stronger than gravity on a nuclear scale.

A distinctive feature of the weak interaction is its very short range. A range so short that only with the construction of the large accelerators since 1970 has one been able to see the weak interaction behave more like the other forces. Until then, the weak interaction was known only by reactions it could cause, like allowing a proton to turn into a neutron or vice versa.

Because of the weak interaction, an individual neutron is not stable. Within an average time of about 10 minutes it decays into a proton and an electron. Sometimes neutrons within an unstable nucleus also decay into a proton and electron. This kind of nuclear decay was observed toward the end of the nineteenth century when knowledge of elementary particles was very limited, and the electrons that came out in these nuclear decays were identified as some kind of a ray called a *beta ray*. (There were alpha rays which turned out to be helium nuclei, beta rays which were electrons, and gamma rays which were photons.) Because the electrons emitted during a neutron decay were called beta rays, the process is still known as the ***beta decay*** process.

The electron is emitted when a neutron decays in order to conerve electric charge. When the neutral neutron decays into a positive proton, a negatively charged particle must also be emitted so that the total charge does not change. The lightest particle available to carry out the negative charge is the electron.

Early studies of the beta decay process indicated that while electric charge was conserved, energy was not. For example, the rest mass of a neutron is nearly 0.14 percent greater than the rest mass of a proton. This mass difference is about four times larger than the rest mass of the electron, thus there is more than enough mass energy available to create the electron when the neutron decays. If energy is conserved, you would expect that the energy left over after the electron is created would appear as kinetic energy of the electron.

Careful studies of the beta decay process showed that sometimes the electron carried out the expected amount of energy and sometimes it did not. These studies were carried out in the 1920s, when not too much was known about nuclear reactions. There was a serious debate about whether energy was actually conserved on the small scale of the nucleus.

In 1929, Wolfgang Pauli proposed that energy was conserved, and that the apparently missing energy was carried out by an elusive particle that had not yet been seen. This elusive particle, which became known as the *neutrino* or "little neutral one", had to have some rather peculiar properties. Aside from being electrically neutral, it had to have essentially no rest mass because in some reactions the electron was seen to carry out all the energy, leaving none to create a neutrino rest mass.

The most bizarre property f the neutrino was its undetectability. It had to pass through matter leaving no trace. It was hard to believe such a particle could exist, yet on the other hand, it was hard to believe energy was not conserved. The neutrino was finally detected thirty years later and we are now quite confident that energy is conserved on the nuclear scale.

The neutrino is elusive because it interacts with matter only through the weak interaction (and gravity). Photons interact via the strong electric interaction and are quickly stopped when they encounter the electric charges in matter. Neutrinos can pass through light years of lead before there is a good chance that they will be stopped. Only in the collapsing core of an exploding star or in the very early universe is matter dense enough to significantly absorb neutrinos. Because neutrinos have no rest mass, they, like photons, travel at the speed of light.

Leptons

We now know that neutrinos are emitted in the beta decay process because of another conservation law, the *conservation of leptons*. The leptons are a family of light particles that include the electron and the neutrino. When an electron is created, an anti neutrino is also created so that the number of leptons does not change.

Actually there are three distinct conservation laws for leptons. The lepton family consists of six particles, the electron, two more particles with rest mass and three different kinds of neutrino. The other massive particles are the *muon* which is 207 times as massive as the electron, and the recently discovered *tau* particle which is 3490 times heavier. The three kinds of neutrino are the *electron type neutrino*, the *muon type neutrino* and the *tau type neutrino*. The names come from the fact that each type of particle is separately conserved. For example when a neutron decays into a proton and an electron is created, it is an *anti electron type neutrino* that is created at the same time to conserve electron type particles.

In the other common beta decay process, where a proton turns into a neutron, a positron is created to conserve electric charge. Since the positron is the anti particle of the electron, its opposite, the electron type neutrino, must be created to conserve leptons.

Nuclear Structure

The light nuclei, like helium, carbon, oxygen, generally have about equal numbers of protons and neutrons. As the nuclei become larger we find a growing excess of neutrons over protons. For example when we get up to Uranium 238, the excess has grown to 146 neutrons to 92 protons.

The most stable isotope of a given element is the one with the lowest possible energy. Because the weak interaction allows protons to change into neutrons and vice versa, the number of protons and neutrons in a nucleus can shift until the lowest energy combination is reached.

Two forms of energy that play an important role in their proess are the extra mass energy of the neutrons, and the electric potential energy of the protons. It takes a lot f to shove two protons together against their electric repulsion. The work you do in shoving them together is stored as electric potential energy which will be released if you let go and the particles fly apart. This energy will not be released, however, if the protons are latched together by the nuclear force. But in that case the electric potential energy can be released by turning one of the protons into a neutron. This will happen if enough electric potential energy is available not only to create the extra neutron rest mass energy, but also the positron required to conserve electric charge.

The reason that the large nuclei have an excess of neutrons over protons is that electric potential energy increases faster with increasing number of protons than neutron mass energy does with increasing numbers of neutrons. The amount of extra neutron rest mass energy is more or less proportional to the number of neutrons. But the increase in electric potential energy as you add a proton depends on the number of protons already in the nucleus. The more protons already there, the stronger the electric repulsion when you try to add another proton, and the greater the potential energy stored. As a result of this increasing energy cost of adding more protons, the large nuclei find their lowest energy balance having an excess of neutrons.

A CONFUSING PICTURE

By 1932, the basic picture of matter looked about as simple as it can possibly get. The elementary particles were the proton, neutron, and electron. Protons and neutrons were held together in the nucleus by the nuclear force, electrons were bound to nuclei by the electric force to form atoms, a residual of the electric force held atoms together to form molecules, crystals and living matter, and gravity held large chunks of matter together for form planets, stars and galaxies. The rules governing the behavior of all this was quantum mechanics on a small scale, which became Newtonian mechanics on the larger scale of our familiar world. There were a few things still to be straightened out, such as the question as to whether energy was conserved in beta decays, and in fact why beta decays occurred at all, but it looked as if these loose ends should be soon tied up.

The opposite happened. By 1960, there were well over 100 so called elementary particles, all of them unstable except for the familiar electron, proton and neutron. Some lived long enough to travel kilometers down through the earth's atmosphere, others long enough to be observed in particle detectors. Still others had such short lifetimes that, even moving at nearly the speed of light, they could travel only a few proton diameters before decaying. With few exceptions, these particles were unexpected and their behavior difficult to explain. Where they were expected, they were incorrectly identified.

One place to begin the story of the progression of unexpected particles is with a prediction made in 1933 by Heidi Yukawa. Yukawa proposed a new theory of the nuclear force. Noting that the electric force was ultimately caused by a particle, Yukawa proposed that the nuclear force holding the protons and neutrons together in the nucleus was also caused by a particle, a particle that became known as the ***nuclear force meson***. The zero rest mass photon gives rise to the long range electric force. Yukawa developed a wave equation for the nuclear force meson in which the range of the force depends on the rest mass of the meson. The bigger the rest mass of the meson, the shorter the range. (Later in the text, we will use the uncertainty principle to explain this relationship between the range of a force and the rest mass of the particle causing it.)

From the fact that iron is the most stable nucleus, Yukawa could estimate that the range of the nuclear force is about equal to the diameter of an iron nucleus, about four proton diameters. From this, he predicted that the nuclear force meson should have a rest mass bout 300 times the rest mass of the electron (about 1/6 the rest mass of a proton).

Shortly after Yukawa's prediction, the muon was discovered in the rain of particles that continually strike the earth called cosmic rays. The rest mass of the muon was found to be about 200 times that of the electron, not too far off the predicted mass of Yukawa's particle. For a while the muon was hailed as Yukawa's nuclear force meson. But further studies showed that muons could travel considerable distances through solid matter. If the muon were the nuclear force meson, it should interact strongly with nuclei and be stopped rapidly. Thus the muon was seen as not being Yukawa's particle. Then there was the question of what role the muon played. Why did nature need it?

In 1947 another particle called the π meson was discovered. (There were actually three π mesons, one with a positive charge, the π^+, one neutral, the π°, and one with a negative charge, the π^-.) The π mesons interacted strongly with nuclei, and had the mass close to that predicted by Yukawa, 274 electron masses. The π mesons were then hailed as Yukawa's nuclear force meson.

However, at almost the same time, another particle called the K meson, 3.5 times heavier than the π meson, was discovered. It also interacted strongly with nuclei and clearly played a role in the nuclear force. The nuclear force was becoming more complex than Yukawa had expected.

Experiments designed to study the π and K mesons revealed other particles more massive than protons and neutrons that eventually decayed into protons and neutrons. It became clear that the proton and neutron were just the lightest members of a family of proton like particles. The number of particles in the proton family was approaching 100 by 1960. During this time it was also found that the π and K mesons were just the lightest members of another family of particles whose number exceeded 100 by 1960. It was rather mind boggling to think of the nuclear force as being caused by over 100 different kinds of mesons, while the electric force had only one particle, the photon.

One of the helpful ways of viewing matter at that time was to identify each of the particle decays with one of the four basic forces. The very fastest decays were assumed to be caused by the strong nuclear force. Decays that were about 100 times slower were identified with the slightly weaker electric force. Decays that took as long as a billionth of a second, a relatively long lifetime, were found to be caused by the weak interaction. The general scheme was the weaker the force, the longer it took to cause a particle decay.

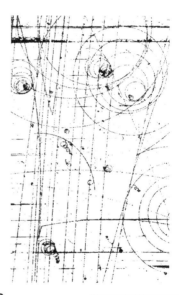

Figure 6
First bubble chamber photograph of the Ω^- particle. The $\Omega^-, \Xi^0, \Lambda^0$ and p^+ are all members of the proton family, the K's and π's are mesons, the γ's are photons and the e^- and e^+ are electrons and positrons. Here we see two examples of the creation of an electron-positron pair by a photon.

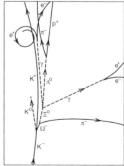

QUARKS

The mess seen in 1960 was cleaned up, brought into focus, primarily by the work of Murray Gell-Mann. In 1961 Gell-Mann and Yval Neuman found a scheme that allowed one to see symmetric patterns in the masses and charges of the various particles. In 1964 Gell-Mann and George Zweig discovered what they thought was the reason for the symmetries. The symmetries would be the natural result if the proton and meson families of particles were made up of smaller particles which Gell-Mann called *quarks*.

Initially Gell-Mann proposed that there were three different kinds of quark, but the number has since grown to six. The lightest pair of the quarks, the so called *up quark* and *down quark* are found in protons and neutrons. If the names "up quark" and "down quark" seem a bit peculiar, they are not nearly as confusing as the names *strange quark*, *charm quark*, *bottom quark* and *top quark* given the other four members of the quark family. It is too bad that the Greek letters had been used up naming other particles.

In the quark model, all members of the proton family consist of three quarks. The proton and neutron, are made from the *up* and *down* quarks. The proton consists of two *up* and one *down* quark, while the neutron is made from one *up* and two *down* quarks. The weak interaction, which as we saw can change protons into neutrons, does so by changing one of the proton's *up* quarks into a *down* quark.

The π meson type of particles, which were thought to be Yukawa's nuclear force particles, turned out instead to be quark-antiquark pairs. The profusion of what were thought to be elementary particles in 1960 resulted from the fact that there are many ways to combine three quarks to produce members of the proton family or a quark and an antiquark to create a meson. The fast elementary particle reactions were the result of the rearrangement of the quarks within the particle, while the slow reactions resulted when the weak interaction changed one kind of quark into another.

A peculiar feature of the quark model is that quarks have a *fractional charge*. In all studies of all elementary particles, charge was observed to come in units of the amount of charge on the electron. The electron had (–1) units, and the neutron (0) units. All of the more than 100 "elementary" particles had either +1, 0, or –1 units of change. Yet in the quark model, quarks had a charge of either (+2/3) units like the *up* quark or (-1/3) units like the *down* quark. (The anti particles have the opposite charge, -2/3 and +1/3 units respectively.) You can see that a proton with two *up* and one *down* quark has a total charge of (+2/3 +2/3 -1/3) = (+1) units, and the neutron with two *down* and one *up* quark has a total charge (-1/3 -1/3 +2/3) = (0) units.

The fact that no one had ever detected an individual quark, or ever seen a particle with a fractional charge, made the quark model hard to accept at first. When Gell-Mann initially proposed the model in 1963, he presented it as a mathematical construct to explain the symmetries he had earlier observed.

The quark model gained acceptance in the early 1970s when electrons at the Stanford high energy accelerator were used to probe the structure of the proton. This machine had enough energy, could look in sufficient detail to detect the three quarks inside. The quarks were real.

In 1995, the last and heaviest of the six quarks, the *top* quark, was finally detected at the Fermi Lab Accelerator. The *top* quark was difficult to detect because it is 185 times as massive as a proton. A very high energy accelerator was needed to create and observe this massive particle.

With the quark model, our view of matter has become relatively simple again: there are two families of particles called quarks and leptons. Each family contains six particles. It is not a coincidence that there are the same number of particles in each family. In the current theory of matter called the **standard model**, each pair of leptons is intimately connected to a pair of quarks. The electron type leptons are associated with the *down* and *up* quarks, the muon and muon type neutrino with the *strange* and *charm* quarks, and the tau and tau type neutrino with the *bottom* and *top* quarks.

Are there more than six quarks and six leptons? Are there still heavier lepton neutrino pairs associated with still heavier quarks? That the answer is no, that six is the limit, first came not from accelerator experiments, but from studies of the early universe. Here we have a question concerning the behavior of matter on the very smallest of scales of distance, at the level of quarks inside proton like particles, and we find the answer by looking at matter on the very largest of scales, the entire universe. The existence of more than six leptons and quarks would have altered the relative abundance of hydrogen, deuterium, and helium remaining after the big bang. It would have led to an abundance that is not consistent with what we see now. Later experiments with particle accelerators confirmed the results we first learned from the early universe.

Our picture of the four basic interactions has also become clearer since the early 1930s. The biggest change is in our view of the nuclear force. The basic nuclear force is now seen to be the force between quarks that holds them together to form protons, neutrons and other particles. What we used to call the nuclear force, that short range force binding protons and neutrons together in a nucleus, is now seen as a residual effect of the force between quarks. The old nuclear force is analogous to the residual electric force that binds complete atoms together to form molecules.

As the electric interaction is caused by a particle, the photon, the nuclear force is also caused by particles, eight different ones called **gluons**. The nuclear force is much more complex than the electric force because gluons not only interact with quarks, they also interact with themselves. This gives rise to a very strange force between quarks. Other forces get weaker as you separate the interacting particles. The nuclear force between quarks gets stronger! As a result quarks are confined to live inside particles like protons, neutrons and mesons. This is why we have still never seen an individual quark or an isolated particle with a fractional charge.

Figure 28-28
Fermi Lab accelerator magnets.

Figure 28-29
Fermi Lab accelerator where the top quark was first observed.

THE ELECTROWEAK THEORY

Another major advance in our understanding of the nature of the basic interactions came in 1964 when Steven Weinberg, Abdus Salam and Sheldon Glashow discovered a basic connection between the electric and weak interactions. Einstein had spent the latter part of his life trying without success to unify, find a common basis for, the electric and the gravitational force. It came somewhat as a surprise that the electric and weak interactions, which appear so different, had common origins. Their theory of the two forces is known as the *electroweak theory.*

In the electroweak theory, if we heat matter to a temperature higher than 1000 billion degrees, we will find that the electric and weak interaction are a single force. If we then let the matter cool, this single electroweak force splits into the two separate forces, the electric interaction and the weak interaction. This splitting of the forces is viewed as a so called *phase transition*, a transition in the state of matter like the one we see when water turns to ice at a temperature of 0°C. The temperature of the phase transition for the electroweak force sounds impossibly hot, but it is attainable if we build a big enough accelerator. The cancelled superconducting supercollider was supposed to allow us to study the behavior of matter at these temperatures.

One of the major predictions of the electroweak theory was that after the electric and weak interactions had separated, electric forces should be caused by zero rest mass photons and the weak interaction should be caused by three rather massive particles given the names W^+, W^- and Z^0 mesons. These mesons were found, at their predicted mass, in a series of experiments performed at CERN in the late 1970s.

We have discussed Yukawa's meson theory of forces, a theory in which the range of the force is related to the rest mass of the particle responsible for the force. As it turns out, Yukawa's theory does not work for nuclear forces for which it was designed. The gluons have zero rest mass but because of their interaction, gives rise to a force unlike any other. What Yukawa's theory does describe fairly well is the weak interaction. The very short range of the weak interaction is a consequence of the large masses of the weak interaction mesons W^+, W^- and Z^0. (The W mesons are 10 times as massive as a proton, the Z^0 is 11 times as massive.)

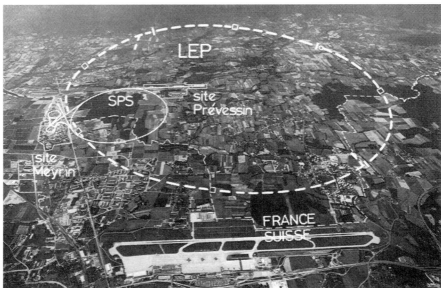

Figure 28-30
Paths for the large particle accelerators at CERN. The Geneva airport is in the foreground.

THE EARLY UNIVERSE

In the reverse motion picture of the expanding universe, the universe becomes smaller and smaller and hotter as we approach the big bang that created it. How small and how hot are questions we are still studying. But it now seems that with reasonable confidence we can apply the laws of physics to a universe that is about one nanosecond old and at a temperature of three hundred thousand billion degrees. This is the temperature of the electroweak transition where the weak and electric interactions become separate distinct forces. We have some confidence in our knowledge of the behavior of matter at this temperature because this temperature is being approached in the largest of the particle accelerators.

3×10^{14} degrees

At three hundred thousand billion degrees the only structures that survive the energetic thermal collisions are the elementary particles themselves. At this time the universe consists of a soup of quarks and anti quarks, leptons and anti leptons, gravitons and gluons. Photons and the weak interaction mesons W^+, W^- and Z° are just emerging from the particle that gave rise to the electroweak force. The situation may not actually be that simple. When we get to that temperature we may find some of the exotic elementary particles suggested by some recent attempts at a quantum theory of gravity.

10^{13} degrees

When the universe reaches the ripe old age of a millionth of a second, the time it takes light to travel 1000 feet, the temperature has dropped to 10 thousand billion degrees. At these temperatures the gluons are able to hold the quarks together to form protons, neutrons, mesons, and their anti particles. It is still much too hot, however, for protons and neutrons to stick together to form nuclei.

When we look closely at the soup of particles at 10 thousand billion degrees, there is activity in the form of the annihilation and creation of particle-antiparticle pairs. Proton-antiproton pairs, for example, are rapidly annihilating, turning into photons and mesons. But just as rapidly photons and mesons are creating proton-antiproton pairs.

In the next 10 millionths of a second the universe expands and cools to a point where the photons and mesons no longer have enough energy to recreate the rapidly annihilating proton and neutron pairs. Soon the protons and neutrons and their antiparticles will have essentially disappeared from the universe.

Matter particles survive

The protons and neutrons will have almost disappeared but not quite. For some reason, not yet completely understood, there was a tiny excess of protons over antiprotons and neutrons over antineutrons. The estimate is that there were 100,000,000,001 matter particles for every 100,000,000,000 antimatter particles. It was the tiny excess of matter over antimatter that survived the proton and neutron annihilation.

3×10^{10} degrees

After this annihilation, nothing much happens until the universe approaches the age of a tenth of a second and the temperature has dropped to 30 billion degrees. During this time the particles we see are photons, neutrinos and antineutrinos and electrons and positrons. These particles exist in roughly equal numbers. The electron-positron pairs are rapidly annihilating to produce photons, but the photons are equally rapidly creating electron positron pairs.

38% neutrons

There are still the relatively few protons and neutrons that survived the earlier annihilation. The weak interaction allows the protons to turn into neutrons and vice versa, with the result there are roughly equal numbers of protons and neutrons. The numbers are not quite equal, however, because at those temperatures there is a slightly greater chance for the heavier neutron to decay into a lighter proton than vice versa. It is estimated that the ratio of neutrons to protons has dropped to 38% by the time the universe is .11 seconds old. The temperature is still too high for protons and neutrons to combine to form nuclei.

Neutrinos escape at one second

As we noted, neutrinos are special particles in that the only way they interact with matter is through the weak interaction. Neutrinos pass right through the earth with only the slightest chance of being stopped. But the early universe is so dense that the neutrinos interact readily with all the other particles.

When the universe reaches an age of about one second, the expansion has reduced the density of matter to the point that neutrinos can pass undisturbed through matter. We can think of the neutrinos as decoupling from matter and going on their own independent way. From a time of one second on, the only thing that will happen to the neutrinos is that they will continue to cool as the universe expands. At an age of 1 second, the neutrinos were at a temperature of 10 billion degrees. By today they have cooled to only a few degrees above absolute zero. This is our prediction, but these cool neutrinos are too elusive to have been directly observed.

24% neutrons

Some other interesting things are also beginning to happen at the time of 1 second. The photons have cooled to a point that they just barely have enough energy to create electron-positron pairs to replace those that are rapidly annihilating. The result is that the electrons and positrons are beginning to disappear. At these temperatures it is also more favorable for neutrons to turn into protons rather than vice versa, with the result that the ratio of neutrons to protons has dropped to 24%.

3×10^9 degrees (13.8 seconds)

When the temperature of the universe has dropped to 3 billion degrees, at the time of 13.8 seconds, the energy of the photons has dropped below the threshold of being able to create electron-positron pairs and the electrons and the positrons begin to vanish from the universe. There was the same tiny excess of electrons over positrons as there had been of protons over antiprotons. Only the excess of electrons will survive.

Positrons annihilated

After about three minutes the positrons are gone and from then on the universe consists of photons, neutrinos, anti neutrinos and the few matter particles. The neutrinos are not interacting with anything, and the matter particles are outnumbered by photons in a ration of 100,000,000,000 to one. The photons essentially dominate the universe.

Deuterium bottleneck

At the time of 13.8 seconds the temperature was 3 billion degrees, cool enough for helium nuclei to survive. But helium nuclei cannot be made without first making deuterium, and deuterium is not stable at that temperature. Thus while there are still neutrons around, protons and neutrons still cannot form nuclei because of this deuterium bottleneck.

Helium created

When the universe reaches an age of three minutes and 2 seconds, and the ratio of neutrons to protons has dropped to 13%, finally deuterium is stable. These surviving neutrons are quickly swallowed up to form deuterium which in turn combine to form the very stable helium nuclei. Since there are equal numbers of protons and neutrons in a helium nucleus, the 13% of neutrons combined with an equal number of protons to give 26% by weight of helium nuclei and 74% protons or hydrogen nuclei.

By the time the helium nuclei form, the universe has become too cool to burn the helium to form heavier elements. The creation of the heavier elements has to wait until stars begin to form one third of a million years later.

The formation of elements inside of stars was the basis of the continuous creation theory. As we mentioned, one could explain the abundance of all the elements except helium as being a by product of the evolution of stars. To explain the helium abundance it was necessary to abandon his continuous creation theory and accept that there might have been a big bang after all.

The Thermal Photons

After the electron positron pairs had vanished, what is left in the universe are the photons, neutrinos, anti neutrinos, and the few matter particles consisting of protons, helium nuclei and a trace of deuterium and lithium. There are enough electrons to balance the charge on the hydrogen and helium nuclei, but the photons are energetic enough to break up any atoms that might try to form. The neutrinos have stopped interacting with anything and the matter particles are outnumbered by photons in a ratio of 100 billion to one. At this time the photons dominate the universe.

One way to understand why the universe cools as it expands is to picture the expansion of the universe as stretching the wavelength of the photons. Since the energy of a photon is related to its wavelength (the longer the wavelength the lower the energy), this stretching of wavelengths lowers the photon energies. Because the photons dominate the young universe, when the photons lose energy and cool down, so does everything else that the photons are interacting with.

.7 million years

Until the universe reaches the age of nearly a million years, the photons are knocking the matter particles around, preventing them from forming whole atoms or gravitational structures like stars. But at the age of .7 million years the temperature has dropped to 3000 degrees, and something very special happens at that point. The matter particles are mostly hydrogen, and if you cool hydrogen below 3000 degrees it becomes transparent. The transition in going from above 3000 degrees to below, is like going from inside the surface of the sun to outside. We go from an opaque, glowing universe to a transparent one.

Transparent universe

When the universe becomes transparent, the photons no longer have any effect on the matter particles and the matter can begin to form atoms, stars, and galaxies. Everything we see today, except for the primordial hydrogen and helium, was formed after the universe became transparent.

Think about what it means that the universe became transparent at an age of .7 million years. In our telescopes, as we look at more and more distant galaxies, the light from these galaxies must have taken more and more time to reach us. As we look farther out we are looking farther back in time. With the Hubble telescope we are now looking at galaxies formed when the universe was less than a billion years old, less than 10% of its current age.

Imagine that you could build a telescope even more powerful than the Hubble, one that was able to see as far out, as far back to when the universe was .7 million years old. If you could look that far out what would you see? You would be staring into a wall of heated opaque hydrogen. You would see this wall in every direction you looked. If you tried to see through the wall, you would be trying to look at the universe at earlier, hotter times. It would be as futile as trying to look inside the sun with a telescope.

Although this wall at .7 million light years consists of essentially the same heated hydrogen as the surface of the sun, looking at it would not be the same as looking at the sun. The light from this wall has been traveling toward us for the last 14 billion years, during which time the expansion of the universe has stretched the wavelength and cooled the photons to a temperature of less than 3 degrees, to a temperature of 2.74 degrees above absolute zero to be precise.

Photons at a temperature of 2.74 degrees can be observed, not by optical telescopes but by radio antennas instead. In 1964 the engineers Arno Penzias and Robert Wilson were working with the radio antenna that was communicating with the Telstar satellite. The satellite was a large aluminized balloon that was supposed to reflect radio signals back to earth. The radio antenna had to be very sensitive to pick up the weak reflected signals.

In checking out the antenna, Penzias and Wilson were troubled by a faint noise that they could not eliminate. Further study showed that the noise was characteristic of a thermal bath of photons whose temperature was around 3 degrees. After hearing a seminar on the theory of the big bang and on the possibility that there might be some light remaining from the explosion, Penzias and Wilson immediately realized that the noise in their antenna was that light. Their antenna in effect was looking at light from the time the universe became transparent. At that time, only a few astronomers and physicists were taking the big bang hypothesis seriously. The idea of the universe beginning in an explosion seemed too preposterous. After Penzias and Wilson saw the light left over from the hotter universe, no other view has been acceptable.

The fact that the universe became transparent at an age of .7 million years, means that the photons, now called the *cosmic background radiation*, travelled undisturbed by matter. By studying these photons carefully, which we are now doing in various rocket and satellite experiments, we are in a sense, taking an accurate photograph of the universe when the universe was .7 million years old.

Figure 34-11
Penzias and Wilson, and the Holmdel radio telescope.

This photograph shows an extremely uniform universe. The smoothness shows us that stars and galaxies had not yet begun to form. In fact the universe was so smooth that it is difficult to explain how galaxies did form in the time between when the universe went transparent and when we see galaxies in the most distant Hubble telescope photographs. The COBE (Cosmic Background Explorer) satellite was able to detect tiny fluctuations in the temperatures of the background radiation, indicating that there was perhaps just enough structure in the early hot universe to give us the stars, galaxies and clustering of galaxies we see today.

One of the questions you may have had reading our discussion of the early universe, is how do we know that the photons, and earlier the particle-anti particle pairs outnumbered the matter particles by a ratio of 100 billion to one? How do we estimate the tiny excess of matter over anti matter that left behind all the matter we see today? The answer is that the thermal photons we see today outnumber protons and neutrons by a factor of 100 billion to one and that ratio should not have changed since the universe was a few minutes old.

We also mentioned that it would be futile to try to look under the surface of the sun using a telescope. That is true if we try to use a photon telescope. However we can, in effect, see to the very core of the sun using neutrinos. In the burning of hydrogen to form helium, for each helium nucleus created, two protons are converted to neutrons via the weak interaction. In the process two neutrinos are emitted. As a result the core of the sun is a bright source of neutrinos which we can detect and study here on earth.

While it would be futile to use photons to see farther back to when the universe was about .7 million years old, we should be able to see through that barrier using neutrinos. The universe became transparent to neutrinos at the end of the first second. If we could detect these neutrinos, we would have a snapshot of the universe as it looked when it was one second old. Thus far, we have not found a way to detect these cosmic background neutrinos.

Chapter 1
Principle of Relativity

The subject of this book is the behavior of matter—the particles that make up matter, the interactions between particles, and the structures that these interactions create. There is a wondrous variety of activity, as patterns and structures form and dissipate, and all of this activity takes place in an arena we call space and time. The subject of this chapter is that arena—space and time itself.

Initially, one might think that a chapter on space and time would either be extraordinarily dull, or too esoteric to be of any use. From the **it's too dull** point of view, distance is measured by meter sticks, and there are relationships like the Pythagorean theorem and various geometric and trigonometric rules already familiar to the reader. Time appears to be less challenging—it is measured by clocks and seems to march inexorably forward.

On the **too esoteric** side are the theories like Einstein's General Theory of Relativity which treats gravity as a distortion of space and time, the Feynman-Wheeler picture of antimatter as being matter traveling backward in time, and recent "super symmetry" theories which assume a ten dimensional space. All of these theories are interesting, and we will briefly discuss them. We will do that later in the text after we have built up enough of a background to understand why these theories were put forth.

What can we say in an introductory chapter about space and time that is interesting, or useful, or necessary for a physics text? Why not follow the traditional approach and begin with the development of Newton's theory of mechanics. You do not need a very sophisticated picture of space and time to understand Newtonian mechanics, and this theory explains an enormous range of phenomena, more than you can learn in one or several years. There are three main reasons why we will not start off with the Newtonian picture. The first is that the simple Newtonian view of space and time is approximate, and the approximation fails badly in many examples we will discuss in this text. By starting with a more accurate picture of space and time, we can view these examples as successful predictions rather than failures of the Newtonian theory.

The second reason is that the more accurate picture of space and time is based on the simplest, yet perhaps most general law in all of physics—the principle of relativity. The principle of relativity not only underlies all basic theories of physics, it was essential in the discovery of many of these theories. Of all possible ways matter could behave, only a very, very few are consistent with the principle of relativity, and by concentrating on these few we have been able to make enormous strides in understanding how matter interacts. By beginning the text with the principle of relativity, the reader starts off with one of the best examples of a fundamental physical law.

Our third reason for starting with the principle of relativity and the nature of space and time, is that it is fun. The math required is simple – only the Pythagorean theorem. Yet results like clocks running slow, lengths contracting, the existence of an ultimate speed, and *questions of causality, are stimulating topics. Many of these results are counter intuitive. Your effort will not be in struggling with mathematical formulas, but in visualizing yourself in new and strange situations. This visualization starts off slowly, but you will get used to it and become quite good at it. By the end of the course the principle of relativity, and the consequences known as Einstein's special theory of relativity will be second nature to you.*

THE PRINCIPLE OF RELATIVITY

In this age of jet travel, the principle of relativity is not a strange concept. It says that ***you cannot feel motion in a straight line at constant speed***. Recall a smooth flight where the jet you were in was traveling at perhaps 500 miles per hour. A moving picture is being shown and all the window shades are closed. As you watch the movie are you aware of the motion of the jet? Do you feel the jet hurtling through the air at 500 miles per hour? Does everything inside the jet crash to the rear of the plane because of this immense speed?

No—the only exciting thing going on is the movie. The smooth motion of the jet causes no excitement whatsoever. If you spill a diet Coke, it lands in your lap just as it would if the plane were sitting on the ground. The problem with walking around the plane is the food and drink cart blocking the aisles, not the motion of the plane. Because the window shades are closed, you cannot even be sure that the plane is moving. If you open your window shade and look out, and if it is daytime and clear, you can look down and see the land move by. Flying over the Midwestern United States you will see all those square 40 acre plots of land move by, and this tells you that you are moving. If someone suggested to you that maybe the farms were moving and you were at rest, you would know that was ridiculous, the plane has the jet engines, not the farms.

Despite the dull experience in a jumbo jet, we often are able to sense motion. There is no problem in feeling motion when we start, stop, or go around a sharp curve. But starting, stopping, and going around a curve are not examples of motion at constant speed in a straight line, the kind of motion we are talking about. ***Changes in speed or in direction of motion are called accelerations***, and we can feel accelerations. (Note: In physics a decrease in speed is referred to as a negative acceleration.)

Even without accelerations, even when we are moving at constant speed in a straight line, we can have a strong sense of motion. Driving down a freeway at 60 miles per hour in a low-slung, open sports car can be a notable, if not scary, experience.

This sense of motion can be misleading. The first wide screen moving pictures took the camera along on a roller coaster ride. Most people in the audience found watching this ride to be almost as nerve wracking as actually riding a roller coaster. Some even became sick. Yet the audience was just sitting at rest in the movie theater.

Exercise 1

Throughout this text we will insert various exercises where we want you to stop and think about or work with the material. At this point we want you to stop reading and think about various times you have experienced motion. Then eliminate all those that involved accelerations, where you speeded up, slowed down, or went around a curve. What do you have left, and how real were the sensations?

One of my favorite examples occurred while I was at a bus station in Boston. A number of busses were lined up side by side waiting for their scheduled departure times. I recall that after a fairly long wait, I observed that we were moving past the bus next to us. I was glad that we were finally leaving. A few seconds later I looked out the window again; the bus next to us had left and we were still sitting in the station. I had mistaken that bus's motion for our own!

A Thought Experiment

Not only can you feel accelerated motion, you can easily see relative motion. I had no problem seeing the bus next to us move relative to us. My only difficulty was in telling whether they were moving or we were moving.

An example of where it is more obvious who is moving is the example of the jet flying over the Midwestern plains. In the daytime the passengers can see the farms go by; it is easy to detect the relative motion of the plane and the farms. And it is quite obvious that it is the plane moving and the farms are at rest. Or is it?

To deal with this question we will go through what is called a **thought experiment** where we solve a problem by imagining a sometimes contrived situation, and then figure out what the consequences would be if we were actually in that situation. Galileo is well known for his use of thought experiments to explain the concepts of the new mechanics he was discovering.

For our thought experiment, imagine that we are going to take the Concorde supersonic jet from Boston, Massachusetts to San Francisco, California. The jet has been given special permission to fly across the country at supersonic speeds so that the trip, which is scheduled to leave at noon, takes only three hours.

When we arrive in San Francisco we reset our watches to Pacific Standard Time to make up for the 3 hour difference between Boston and San Francisco. We reset our watches to noon. When we left, it was noon and the sun was overhead. When we arrive it is still noon and the sun is still overhead. One might say that the jet flew fast enough to follow the sun, the 3 hour trip just balancing the 3 hours time difference.

But there is another view of the trip shown in Figure (1). When we took off at noon, the earth, the airplane and sun were lined up as shown in Figure (1a). Three hours later the earth, airplane and sun are still lined up as shown in Figure (1b). The only difference between (1a) and (1b) is that the earth has been rotating for three hours so that San Francisco, rather than Boston is now under the plane. The view in Figure (1) is what an astronaut approaching the earth in a spacecraft might see.

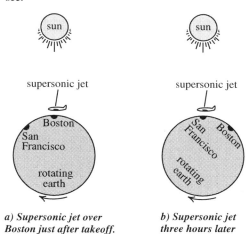

a) *Supersonic jet over Boston just after takeoff.*

b) *Supersonic jet three hours later over San Francisco.*

Figure 1
One view of a three hour trip from Boston to San Francisco. It is possible, even logical, to think of the jet as hovering at rest while the earth turns underneath.

For someone inside the jet, looking down at the Midwestern farms going by, who is really moving? Are the farms *really* at rest and the plane moving? Or is the plane at rest and the farms going by? Figure (1) suggests that the latter point of view may be more accurate, at least from the perspective of one who sees the bigger picture including the earth, airplane, and sun.

But, you might ask, what about the jet engines and all the fuel that is being expended to move the jet at 1000 miles/hour? Doesn't that prove that it is the jet that is moving? Not necessarily. When the earth rotates, it drags the atmosphere around with it creating a 1000 mi/hr wind that the plane has to fly through in order to stand still. Without the jet engines and fuel, the plane would be dragged back with the land and never reach San Francisco.

This thought experiment has one purpose. To loosen what may have been a firmly held conviction that when you are in a plane or car, you are moving and the land that you see go by must necessarily be at rest. Perhaps, under some circumstances it is more logical to think of yourself at rest and the ground as moving. *Or, perhaps it does not make any difference*. The principle of relativity allows us to take this last point of view.

Statement of the Principle of Relativity

Earlier we defined uniform motion as motion at constant speed in a straight line. And we mentioned that the principle of relativity said that you could not feel this uniform motion. Since it is not exactly clear what is meant by "feeling" uniform motion, a more precise statement of the principle of relativity is needed, a statement that can be tested by experiment. The following is the definition we will use in this text.

> *Imagine that you are in a capsule and you may have any equipment you wish inside the capsule. The principle of relativity states that <u>there is no experiment you can perform that will allow you to tell whether or not the capsule is moving with uniform motion</u>—motion in a straight line at constant speed.*

In the above definition the capsule can use anything you want as an example—a jet plane, a car, or a room in a building. Generally, think of it as a sealed capsule like the jet plane where the moving picture is being shown and all of the window shades are shut. Of course you can look outside, and you may see things going by. But, as shown in Figure (1), seeing things outside go by does not prove that you and the capsule are moving. That cannot be used as evidence of your own uniform motion.

Think about what kind of experiments you might perform in the sealed capsule to detect your uniform motion. One experiment is to drop a coin on the floor. If you are at rest, the coin falls straight down. But if you are in a jet travelling 500 miles per hours and the flight is smooth, and you drop a coin, the coin still falls straight down. Dropping a coin does not distinguish between being at rest or moving at 500 miles per hour; this is one experiment that does not violate the principle of relativity.

There are many other experiments you can perform. You could use gyroscopes, electronic circuits, nuclear reactions, gravitational wave detectors, anything you want. The principle of relativity states that none of these will allow you to detect your uniform motion.

Exercise 2

Think about what you might put inside the capsule and what experiments you might perform to detect the motion of the capsule. Discuss your ideas with others and see if you can come up with some way of violating the principle of relativity.

Basic Law of Physics

We mentioned that one of the incentives for beginning the text with the principle of relativity is that it is an excellent example of a basic law of physics. It is simple and easy to state—there is no experiment that you can perform that allows you to detect your own uniform motion. Yet it is general—there is no experiment that can be done at any time, at any place, using anything, that can detect your uniform motion. And most important, it is completely subject to experimental test on an all-or-nothing basis. Just one verifiable experiment detecting one's own uniform motion, and the principle of relativity is no longer a basic law. It may become a useful approximation, but not a basic law.

Once a fundamental law like the principle of relativity is discovered or accepted, it has a profound effect on the way we think about things. In this case, if there is no way that we can detect our own uniform motion, then we might as well ignore our motion and always assume that we are at rest. Nature is usually easier to explain if we take the point of view that we are at rest and that other people and things are moving by. It is the principle of relativity that allows us to take this self-centered point of view.

It is a shock, a lot of excitement is generated, when what was accepted as a basic law of physics is discovered not to be exactly true. The discovery usually occurs in some obscure corner of science where no one thought to look before. And it will probably have little effect on most practical applications. But the failure of a basic law changes the way we think.

Suppose, for example, that it was discovered that the principle of relativity did not apply to the decay of an esoteric elementary particle created only in the gigantic particle accelerating machines physicists have recently built. This violation of the principle of relativity would have no practical effect on our daily lives, but it would have a profound psychological effect. We would then know that our uniform motion could be detected, and therefore on a fundamental basis we could no longer take the point of view that we are at rest and others are moving. There would be legitimate debates as to who was moving and who was at rest. We would search for a formulation of the laws of physics that made it intuitively clear who was moving and who was at rest.

This is almost what happened in 1860. In that year, James Clerk Maxwell summarized the laws of electricity and magnetism in four short equations. He then solved these equations to predict the existence of a wave of electric and magnetic force that should travel at a speed of approximately 3×10^8 meters per second. The predicted speed, which we will call c, could be determined from simple measurements of the behavior of an electric circuit.

Before Maxwell, no one had considered the possibility that electric and magnetic forces could combine in a wavelike structure that could travel through space. The first question Maxwell had to answer was what this wave was. Did it really exist? Or was it some spurious solution of his equations?

The clue was that the speed c of this wave was so fast that only light had a comparable speed. And more remarkably the known speed of light, and the speed c of his wave were very close—to within experimental error they were equal. As a consequence Maxwell proposed that he had discovered the theory of light, and that this wave of electric and magnetic force was light itself.

Maxwell's theory explained properties of light such as polarization, and made predictions like the existence of radio waves. Many predictions were soon verified, and within a few years there was little doubt that Maxwell had discovered the theory of light.

One problem with Maxwell's theory is that by measurements of the speed of light, it appears that one should be able to detect one's own uniform motion. In the next section we shall see why. This had two immediate consequences. One was a change in the view of nature to make it easy to see who was moving and who was not. The second was a series of experiments to see if the earth were moving or not.

In the resulting view of nature, all of space was filled with an invisible substance called *ether*. Light was pictured as a wave in the ether medium just as ocean waves are waves in the medium of water. The experiments, initiated by Michaelson and Morley, were designed to detect the motion of the earth by measuring how fast the earth was moving through the ether medium.

The problem with the ether theory was that all experiments designed to detect ether, or to detect motion through it, seemed to fail. The more clever the experiment, the more subtle the apparent reason for the failure. We will not engage in any further discussion of the ether theory, because ether still has never been detected. But we will take a serious look in the next section at how the measurement of the speed of a pulse of light should allow us to detect our own uniform motion. And then in the rest of the chapter we will discuss how a young physicist, working in a patent office in 1905, handled the problem.

Figure 2
Rain drops creating circular waves on the surface of a puddle. (Courtesy Bill Jack Rodgers, Los Alamos Scientific Laboratory.)

Figure 3
This ocean wave traveled hundreds of miles from Hurricane Bertha to the Maine coast (July 31, 1990).

WAVE MOTION

We do not need to know the details of Maxwell's theory to appreciate how one should be able to use the theory to violate the principle of relativity. All we need is an understanding of some of the basic properties of wave motion.

The most familiar examples of wave motion are the waves on the surface of water. We have seen the waves that spread out in circles when a stone is dropped in a pond, or rain hits a puddle in a sidewalk as shown in Figure (2). And most of us have seen the ocean waves destroying themselves as they crash into the beach. The larger ocean waves often originate at a storm far out to sea, and have traveled hundreds or even a thousand miles to reach you (see Figure (3)). The very largest ocean waves, created by earthquakes or exploding volcanoes have been known to travel almost around the earth.

Although you cannot see them, sound waves are a more familiar form of wave motion. Sound moving through air, waves moving over water, and light, all have certain common features and ways of behaving which we classify as *wave motion*. In later chapters we will study the subject of wave motion in considerable detail. For now we will limit our discussion to a few of the features we need to understand the impact of Maxwell's theory.

Two examples of wave motion that are easy to study are a wave pulse traveling down a rope as indicated in Figure (4) or down a stretched Slinky® (the toy coil that 'climbs' down stairs) as shown in Figure (5). The advantage of using a stretched Slinky is that the waves travel so slowly that you can study them as they move. It turns out that the speed of a wave pulse depends upon the medium along which, or through which, it is traveling. For example, the speed of a wave pulse along a rope or Slinky is given by the formula

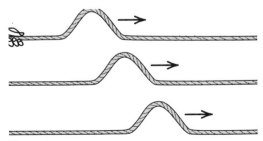

Figure 4
Wave pulse traveling along a rope.

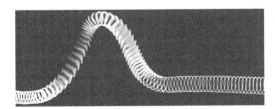

Figure 5
Wave pulse traveling along a Slinky.

$$\begin{matrix}\text{Speed of} \\ \text{wave pulse}\end{matrix} = \sqrt{\frac{\tau}{\mu}} \qquad (1)$$

where τ is the tension in the rope or Slinky, and μ the mass per unit length. Do not worry about precise definitions of tension or mass, the important point is that there is a formula for the speed of the wave pulse, a formula that depends only on the properties of the medium along which the pulse is moving.

The speed does not depend upon the shape of the pulse or how the pulse was created. For example, the Slinky pulse travels much more slowly than the pulse on the rope because the suspended Slinky has very little tension τ. We can slow the Slinky wave down even more by hanging crumpled pieces of lead on each end of the coils of the Slinky to increase its mass per unit length μ.

Another kind of wave we can create in the Slinky is the so called *compressional wave* shown in Figure (6). Here the end of the Slinky was pulled back and released, giving a moving pulse of compressed coils. The formula for the speed of the compression wave is still given by Equation (1), if we interpret τ as the stiffness (*Youngs modulus*) of the suspended Slinky.

If we use a loudspeaker to produce a compressional pulse in air, we get a sound wave that travels out from the loudspeaker at the speed of sound. The formula for the speed of a sound wave is

$$\begin{matrix}\text{Speed of} \\ \text{sound}\end{matrix} = \sqrt{\frac{B}{\rho}} \qquad (2)$$

where B is the *bulk modulus* which can be thought of as the rigidity of the material, and the mass per unit length μ is replaced by the mass per unit volume ρ.

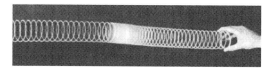

Figure 6
To create a compressional wave on a suspended Slinky, pull the end back a bit and let go.

A substance like air, which is relatively compressible, has a small rigidity B, while substances like steel and granite are very rigid and have large values of B. As a result sound travels much faster in steel and granite than in air. For air at room temperature and one atmosphere of pressure, the speed of sound is 343 meters or 1125 feet per second. Sound travels about 20 times faster in steel and granite. Again the important point is that the speed of a wave depends on the properties of the medium through which it is moving, and not on the shape of the wave or the way it was produced.

Measurement of the Speed of Waves

If you want to know how fast your car is traveling you look at the speedometer. Some unknown machinery in the car makes the needle of the speedometer point at the correct speed. Since the wave pulses we are discussing do not have speedometers, we have to carry out a series of measurements in order to determine their speed. In this section we wish to discuss precisely how the measurements can be made using meter sticks and clocks so that there will be no ambiguity, no doubt about precisely what we mean when we talk about the speed of a wave pulse. We will use the Slinky wave pulse as our example, because the wave travels slowly enough to actually carry out these measurements in a classroom demonstration.

The first experiment, shown in Figure (7), involves two students and the instructor. One student stands at the end of the stretched Slinky and releases a wave pulse like that shown in Figure (6). The instructor holds a meter stick up beside the Slinky as shown. The other student has a stopwatch and measures the length of time it takes the pulse to travel from the front to the back of the stick. (She presses the button once when the pulse reaches the front of the meter stick, presses it again when the pulse gets to the back, and reads the elapsed time T.) The speed of the pulse is then defined to be

$$\begin{matrix}\text{Speed of} \\ \text{Slinky pulse}\end{matrix} = \frac{1 \text{ meter}}{T \text{ seconds}} \qquad (3)$$

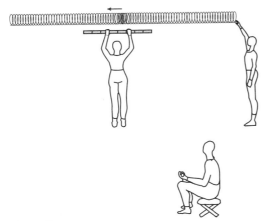

Figure 7
Experiment to measure the speed of a wave pulse on a suspended Slinky. Here the instructor holds the meter stick at rest.

Later in the course, when we have discussed ways of measuring tension τ and mass per unit length μ, we can compare the experimental result we get from Equation (3) with the theoretically predicted result of Equation (1). With a little practice using the stopwatch, it is not difficult to get reasonable agreement between theory and experiment.

In our second experiment, shown in Figure (8a) everything is the same except that the instructor has been replaced by a student, let us say it is Bill, holding the meter stick and running toward the student who releases the wave pulse. Again the second student measures the length of time it takes the pulse to travel from the front to the back of the meter stick. Let us call this the time T_1. This time T_1 is less than T because Bill and the meter stick are moving toward the pulse.

To Bill, the pulse passes his one meter long stick in a time T_1, therefore the speed of the pulse past him is

$$v_1 = \frac{1 \text{ meter}}{T_1 \text{ seconds}} = \frac{\text{speed of pulse}}{\text{relative to Bill}} \quad (4a)$$

Bill should also have carried the stopwatch so that v_1 would truly represent his measurement of the speed of the pulse. But it is too awkward to hold the meter stick, and run and observe when the pulse is passing the ends of the stick.

The speed v_1 measured by Bill is not the same as the speed v measured by the instructor in Figure (7). v_1 is greater than v because Bill is moving toward the wave pulse. This is not surprising: if you are on a freeway and everyone is traveling at a speed v = 55 miles per hour, the oncoming traffic in the opposite lane is traveling past you at a speed of 110 miles per hour because you are moving toward them.

In Figure (8b) we again have the same situation as in Figure (7) except that Bill is now replaced by Joan who is running away from the student who releases the pulse. Joan is moving in the same direction as the pulse and it takes a longer time T_2 for the pulse to pass her. (Assume that Joan is not running faster than the pulse.) The speed of the pulse relative to Joan is

$$v_2 = \frac{1 \text{ meter}}{T_2 \text{ seconds}} = \frac{\text{speed of pulse}}{\text{relative to Joan}} \quad (4b)$$

Joan's speed v_2 will be considerably less than the speed v observed by the instructor.

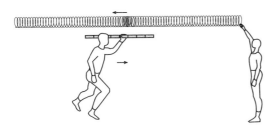

Figure 8a
Bill runs toward the source of the pulse while measuring its speed.

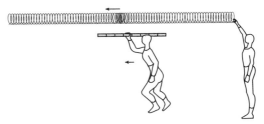

Figure 8b
Joan runs away from the source of the pulse while measuring its speed.

In these three experiments, the instructor is special (wouldn't you know it). Only the instructor measures the speed v predicted by theory, only for the instructor is the speed given by $v = \sqrt{\tau/\mu}$. Both the students Bill and Joan observe different speeds, one larger and one smaller than the theoretical value.

What is special about the instructor? In this case the instructor gets the predicted answer because she is at rest relative to the Slinky. If we hadn't seen the experiment, but just looked at the answers, we could tell that the instructor was at rest because her result agreed with the predicted speed of a Slinky wave. Bill got too high a value because he was moving toward the pulse; Joan, too low a value because she was moving in the same direction.

The above set of experiments is not strikingly profound. In a sense, we have developed a new and rather cumbersome way to tell who is not moving relative to the Slinky. But the same procedures can be applied to a series of experiments that gives more interesting results. In the new series of experiments, we will use a pulse of light rather than a wave pulse on a Slinky.

Since the equipment is not likely to be available among the standard set of demonstration apparatus, and since it will be difficult to run at speeds comparable to the speed of light we will do this as a thought experiment. We will imagine that we can measure the time it takes a light pulse to go from the front to the back of a meter stick. We will imagine the kind of results we expect to get, and then see what the consequences would be if we actually got those results.

The apparatus for our new thought experiment is shown in Figure (9). We have a laser which can produce a very short pulse of light – only a few millimeters long. The meter stick now has photo detectors and clocks mounted on each end, so that we can accurately record the times at which the pulse of light passed each end. These clocks were synchronized, so the time difference is the length of time T it takes the pulse of light to pass the meter stick.

Before the experiment, the instructor gives a short lecture to the class. She points out that according to Maxwell's theory of light, a light wave should travel at a speed c given by the formula

$$c = \frac{1}{\sqrt{\mu_0 \varepsilon_0}} \qquad (5)$$

where μ_0 and ε_0 are constants in the theory of electricity. She says that later on in the year, the students will perform an experiment in which they measure the value of the product $\mu_0\varepsilon_0$. This experiment involves measuring the size of coils of wire and plates of aluminum, and timing the oscillation of an electric current sloshing back and forth between the plates and the coil. The important point is that these measurements do not involve light. It is analogous to the Slinky where the predicted speed $\sqrt{\tau/\mu}$ of a Slinky wave involved measurements of the stiffness τ and mass per unit length μ, and had nothing to do with observations of a Slinky wave pulse.

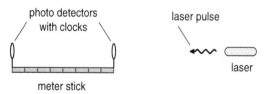

Figure 9
Apparatus for the thought experiment. Now we wish to measure the speed of a laser wave pulse, rather than the speed of a Slinky wave pulse. The photo detectors are used to measure the length of time the laser pulse takes to pass by the meter stick.

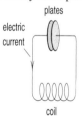

Figure 10a
Plates and coil for measuring the experimental value of $\mu_0\varepsilon_0$.

Figure 10b
The plates and coil we use in the laboratory.

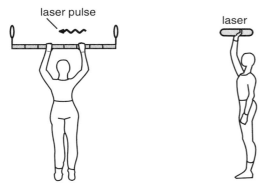

Figure 11
Experiment to measure the speed of a light wave pulse from a laser. Here the instructor holds the meter stick at rest.

Although she is giving out the answer to the lab experiment, she points out that the value of c from these measurements is

$$c = \frac{1}{\sqrt{\mu_0 \varepsilon_0}} = 3 \times 10^8 \text{ meters/second} \quad (6)$$

which is a well-known but uncomfortably large and hard to remember number. However, she points out, 3×10^8 meters is almost exactly one billion (10^9) feet. If you measure time, not in seconds, but in billionths of a second, or *nanoseconds*, where

$$1 \text{ nanosecond} \equiv 10^{-9} \text{ seconds} \quad (7)$$

then since light travels only one foot in a nanosecond, the speed of light is simply

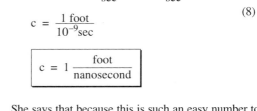

$$c = 3 \times 10^8 \frac{\text{meter}}{\text{sec}} = 10^9 \frac{\text{feet}}{\text{sec}}$$

$$c = \frac{1 \text{ foot}}{10^{-9} \text{sec}} \quad (8)$$

$$\boxed{c = 1 \frac{\text{foot}}{\text{nanosecond}}}$$

She says that because this is such an easy number to remember, she will use it throughout the rest of the course.

The lecture on Maxwell's theory being over, the instructor starts in on the thought experiment. In the first run she stands still, holding the meter stick, and the student with the laser emits a pulse of light as shown in Figure (11). The pulse passes the 3.28 foot length of the meter stick in an elapsed time of 3.28 nanoseconds, for a measured speed

$$v(\text{light pulse}) = \frac{3.28 \text{ feet}}{3.28 \text{ nanoseconds}}$$

$$= 1 \frac{\text{foot}}{\text{nanosecond}} \quad (9)$$

The teacher notes, with a bit of complacency, that she got the predicted speed of 1 foot/nanosecond. Again, the instructor is special.

Then the instructor invites Bill to hold the meter stick and run toward the laser as shown in Figure (12a). Since this is a thought experiment, she asks Bill to run at nearly the speed of light, so that the time should be cut in half and Bill should see light pass him at nearly a speed of 2c.

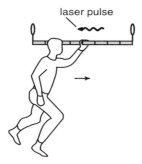

Figure 12a
Bill runs toward the source of the pulse while measuring its speed.

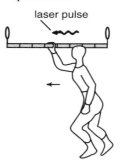

Figure 12b
Joan runs away from the source of the pulse while measuring its speed.

Then she invites Joan to hold the meter stick and run at about half the speed of light in the other direction as shown in Figure (12b). One would expect that the light would take twice as long to pass Joan as it did the instructor and that Joan should obtain a value of about c/2 for the speed of light.

Suppose it turned out this way. Suppose that the instructor got the predicted answer 1 foot/nanosecond, while Bill who is running toward the pulse got a higher value and Joan, running with the pulse got a lower value. Just as in our Slinky pulse experiment we could say that the instructor was at rest while both Bill and Joan were moving.

But, moving relative to what? In the Slinky experiment, the instructor was at rest relative to the Slinky – the medium through which a Slinky wave moves. Light pulses travel through empty space. Light comes to us from stars 10 billion light years away, almost across the entire universe. The medium through which light moves is empty space.

If the experiment came out the way we described, the instructor would have determined that she was at rest relative to empty space, while Bill and Joan would have determined that they were moving. ***They would have violated the principle of relativity***, which says that you cannot detect your own motion relative to empty space.

The alert student might argue that the pulses of light come out of the laser like bullets from a gun at a definite muzzle velocity, and that all the instructor, Bill and Joan are doing is measuring their speed relative to the laser. Experiments have carefully demonstrated that the speed of a pulse of light depends in no way on the motion of the emitter just as the Slinky pulse depended in no way on how the student started the pulse. Maxwell's theory predicts that light is a wave, and many experiments have verified the wave nature of light, including the fact that its speed does not depend on how it was emitted.

From the logical simplicity of the above thought experiment, from the ease with which we should be able to violate the principle of relativity (if we could accurately measure the speed of a pulse of light passing us), it is not surprising that after Maxwell developed this theory of light, physicists did not take the principle of relativity seriously, at least for the next 45 years.

Michaelson-Morley Experiment

The period from 1860 to 1905 saw a number of attempts to detect one's own or the earth's motion through space by measuring the speed of pulses of light. Actually it was easier and far more accurate to compare the speeds of light traveling in different directions. If you were moving forward through space (like Bill in our thought experiment), you should see light coming from in front of you traveling faster than light from behind or even from the side.

Michaelson and Morley used a device called a Michaelson interferometer which compared the speeds of pulses of light traveling at right angles to each other. A detailed analysis of their device is not hard, just a bit lengthy. But the result was that the device should be able to detect small differences in speeds, small enough differences so that the motion of the earth through space should be observable -- even the motion caused by the earth orbiting the sun.

At this point we can summarize volumes of the history of science by pointing out that no experiment using the Michaelson interferometer, or any device based on measuring or comparing the speed of light pulses, ever succeeded in detecting the motion of the earth.

Exercise 3

Units of time we will often use in this course are the millisecond, the microsecond, and the nanosecond, where

$$1 \text{ millisecond} = 10^{-3} \text{ seconds (one thousandth)}$$
$$1 \text{ microsecond} = 10^{-6} \text{ seconds (one millionth)}$$
$$1 \text{ nanosecond} = 10^{-9} \text{ seconds (one billionth)}$$

How many feet does light travel in
  a) one millisecond (1ms)?
  b) one microsecond (1µs)?
  c) one nanosecond (1ns)?

EINSTEIN'S PRINCIPLE OF RELATIVITY

In 1905 Albert Einstein provided a new perspective on the problems we have been discussing. He was apparently unaware of the Michaelson-Morley experiments. Instead, Einstein was familiar with Maxwell's equations for electricity and magnetism, and noted that these equations had a far simpler form if you took the point of view that you are at rest. He suggested that these equations took this simple form, not just for some privileged observer, but for everybody. If the principle of relativity were correct after all, then everyone, no matter how they were moving, could take the point of view that they were at rest and use the simple form of Maxwell's equations.

How did Einstein deal with measurements of the speed of light? We have seen that if someone, like Bill in our thought experiment, detects a pulse of light coming at them at a speed faster than c = 1 foot/nanosecond, then that person could conclude that they themselves were moving in the direction from which the light was coming. They would have thereby violated the principle of relativity.

Einstein's solution to that problem was simple. He noted that any measurement of the speed of a pulse of light that gave an answer different from c = 1 foot/nanosecond could be used to violate the principle of relativity. Thus if the principle of relativity were correct, *all measurements of the speed of light must give the answer c*.

Let us put this in terms of our thought experiment. Suppose the instructor observed that the light pulse passed the 3.24 foot long meter stick in precisely 3.24 nanoseconds. And suppose that Bill, moving at nearly the speed of light toward the laser, also observed that the light took 3.24 nanoseconds to pass by his meter stick. And suppose that Joan, moving away from the laser at half the speed of light, also observed that the pulse of light took 3.24 nanoseconds to pass by her meter stick. If the instructor, Bill and Joan all got precisely the same answer for the speed of light, then none of their results could be used to prove that one was at rest and the others moving. Since their answer of 3.24 feet in 3.24 nanoseconds or 1 foot/nanosecond is in agreement with the predicted value $c = 1/\sqrt{\mu_0 \varepsilon_0}$ from Maxwell's theory, they could all safely assume that they were at rest. At the very least, their measurements of the speed of the light pulse could not be used to detect their own motion.

As we said, the idea is simple. You always get the answer c whenever you measure the speed of a light pulse moving past you. But the idea is horrendous. Einstein went against more than 200 years of physics and centuries of observation with this suggestion.

Suppose, for example, we heard about a freeway where all cars traveled at precisely 55 miles per hour – **no exceptions**. Hearing about this freeway, our three people in the thought experiment decide to test the rule. The instructor sets up measuring equipment in the median strip and observes that the rule is correct. Cars in the north bound lane travel north at 55 miles per hour, and cars in the south bound lane go south at 55 miles per hour.

For his part of the experiment, Bill gets into one of the north bound cars. Since Bill knows about the principle of relativity he takes the point of view that he is at rest. If the 55 miles per hour speed is truly a fundamental law, then he, who is at rest, should see the south bound cars pass at 55 miles per hour.

Likewise, Joan, who is in a south bound car, can take the point of view that she is at rest. She knows that if the 55 miles per hour speed limit is a fundamental law, then north bound cars must pass her at precisely 55 miles per hour. If the instructor, Bill and Joan all observe that every car on the freeway always passes them at the same speed of 55 miles per hour, then none of them can use this observation to detect their own motion.

Freeways do not work that way. Bill will see south bound cars passing him at 110 miles per hour. And Joan will see north bound cars passing at 110 miles per hour. From these observations Bill and Joan will conclude that in fact they are moving – at least relative to the freeway.

Measurements of the speed of a pulse of light differ, however, in two significant ways from measurements of the speed of a car on a freeway. First of all, light moves through empty space, not relative to anything.

Secondly, light moves at enormous speeds, speeds that lie completely outside the realm of common experience. Perhaps, just perhaps, the rules we have learned so well from common experience, do not apply to this realm. The great discoveries in physics often came when we look in some new realm on the very large scale, or the very small scale, or in this case on the scale of very large, unfamiliar speeds.

The Special Theory of Relativity

Einstein developed his special theory of relativity from two assumptions:

1) *The principle of relativity is correct.*

2) *Maxwell's theory of light is correct.*

As we have seen, the only way Maxwell's theory of light can be correct and not violate the principle of relativity, is that every observer who measures the speed of light, must get the predicted answer $c = 1/\sqrt{\mu_0 \varepsilon_0} = 1$ foot/nanosecond. Temporarily we will use this as the statement of Einstein's second postulate:

2a) *Everyone, no matter how he or she is moving, must observe that light passes them at precisely the speed c.*

Postulates (1) and (2a) salvage both the principle of relativity and Maxwell's theory, but what else do they predict? We have seen that measurements of the speed of a pulse of light do not behave in the same way as measurements of the speed of cars on a freeway. Something peculiar seems to be happening at speeds near the speed of light. What are these peculiar things? How do we find out?

To determine the consequences of his two postulates, Einstein borrowed a technique from Galileo and used a series of thought experiments. Einstein did this so clearly, explained the consequences so well in his 1905 paper, that we will follow essentially the same line of reasoning. The main difference is that Einstein made a number of strange predictions that in 1905 were hard to believe. But these predictions were not only verified, they became the cornerstone of much of 20th century physics. We will be able to cite numerous tests of all the predictions.

Moving Clocks

Our first thought experiment for Einstein's special relativity will deal with the behavior of clocks. We saw that the measurement of the speed of a pulse of light required a timing device, and perhaps the peculiar results can be explained by the peculiar behavior of the timing device.

Also the peculiar behavior seems to happen at high speeds near the speed of light, not down at freeway speeds. Thus the question we would like to ask is what happens to a clock that is moving at a high speed, near the speed of light?

That is a tough question. There are many kinds of clocks, ranging from hour glasses dripping sand, to the popular digital quartz watches, to the atomic clocks used by the National Bureau of Standards. The oldest clock, from which we derive our unit of time, is the motion of the earth on its axis each 24 hours. We have both the problem of deciding which kind of clock we wish to consider moving at high speeds, and then figure out how that clock behaves.

The secret of working with thought experiments is to keep everything as simple as possible and do not try to do too much at once. If we want to understand what happens to a moving clock, we should start with the simplest clock we can find. If we cannot understand that one, we will imagine an even simpler one.

A clock that is fairly easy to understand is the old grandfather's clock shown in Figure (13), where the timing device is the swinging pendulum. There are also wheels, gears, and hands, but these merely count swings of the pendulum. The pendulum itself is what is important. If you shorten the pendulum it swings faster and the hands go around faster.

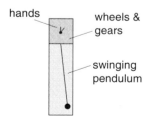

Figure 13
Grandfather's clock.

We could ask what we would see if we observed a grandfather's clock moving past us at a high speed, near the speed of light. The answer is likely to be "I don't know". The grandfather's clock, with its swinging pendulum mechanism, is still too complicated.

A simpler timing device was considered by Einstein, namely a bouncing pulse of light. Suppose, we took the grandfather's clock of Figure (13), and replaced the pendulum by two mirrors and a pulse of light as shown in Figure (14). Space the mirrors 1 foot apart so that the pulse of light will take precisely one nanosecond to bounce either up or down. Leave the rest of the machinery of the grandfather's clock more or less intact. In other words have the wheels and gears now count bounces of the pulse of light rather than swings of the pendulum. And recalibrate the face of the clock so that for each bounce, the hand advances one nanosecond. (The marvelous thing about thought experiments is that you can get away with this. You do not have to worry about technical feasibility, only logical consistency.)

The advantage of replacing the pendulum with a bouncing light pulse is that, so far, the only thing whose behavior we understand when moving at nearly the speed of light is light itself. We know that light always moves at the speed c in all circumstances, to any observer. If we use a bouncing light pulse as a timing device, and can figure out how the pulse behaves, then we can figure out how the clock behaves.

For our thought experiment it is convenient to construct two identical light pulse clocks as shown in Figure (15). We wish to take great care that they are identical, or at least that they run at precisely the same rate. Once they are finished, we adjust them so that the pulses bounce up and down together for weeks on end.

Now we get to the really hypothetical part of our thought experiment. We give one of the clocks to an astronaut, and we keep the other for reference. The astronaut is instructed to carefully pack his clock, accelerate up to nearly the speed of light, unpack his clock, and go by us at a constant speed so that we can compare our reference clock to his moving clock.

Before we describe what we see, let us take a look at a brief summary of the astronaut's log book of the trip. The astronaut writes, "I carefully packed the light pulse clock because I did not want it damaged during the accelerations. My ship can maintain an acceleration of 5gs, and even then it took about a month to get up to our final speed of just over half the speed of light."

"Once the accelerations were over and I was coasting, I took the light pulse clock out of its packing and set it up beside the window, so that the class could see the clock as it went by. Before the trip I was worried that I might have some trouble getting the light pulse into the clock, but it was no problem at all. I couldn't even tell that I was moving! The light pulse went in and the clock started ticking just the way it did back in the lab, before we started the trip."

"It was not long after I started coasting, that the class went by. After that, I packed everything up again, decelerated, and returned to earth."

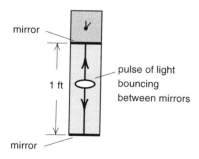

Figure 14
Light pulse clock. We can construct a clock by having a pulse of light bounce between two mirrors. If the mirrors are one foot apart, then the time between bounces will be one nanosecond. The face of the clock displays the number of bounces.

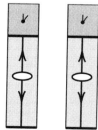

Figure 15
Two identical light pulse clocks.

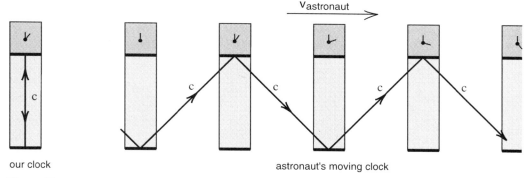

Figure 16
In order to stay in the astronaut's moving clock, the light pulse must follow a longer, saw-tooth, path.

What we saw as the astronaut went by is illustrated in the sketch of Figure (16). On the left is our reference clock, on the right the astronaut's clock moving by. You will recall that the astronaut had no difficulty getting the light pulse to bounce, and as a result we saw his clock go by with the pulse bouncing inside.

For his pulse to stay in his clock, his pulse had to travel along the saw-tooth path shown in Figure (16). The saw-tooth path is longer than the up and down path taken by the pulse in our reference clock. His pulse had to travel farther than our pulse to tick off one nanosecond.

Here is what is peculiar. If Einstein's postulate is right, if the speed of a pulse of light is always c under any circumstances, then our pulse bouncing up and down, and the astronaut's pulse traveling along the saw tooth path are both traveling at the same speed c. Since the astronaut's pulse travels farther, the astronaut's clock must take longer to tick off a nanosecond. ***The astronaut's clock must be running slower!***

Because there are no budget constraints in a thought experiment, we are able to get a better understanding of how the astronaut's clock was behaving by having the astronaut repeat the trip, this time going faster, about .95 c. What we saw is shown in Figure (17). The astronaut's clock is moving so fast that the saw tooth path is stretched way out. The astronaut's pulse takes a long time to climb from the bottom to the top mirror in his clock, his nanoseconds take a long time, and his clock runs very slowly.

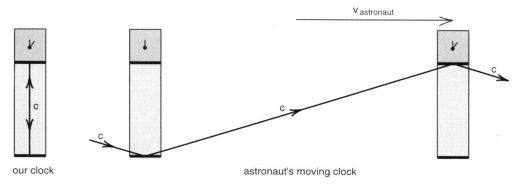

Figure 17
When the astronaut goes faster, his light pulse has to go farther in order to register a bounce. Since the speed of light does not change, it takes longer for one bounce to register, and the astronaut's moving clock runs slower.

It does not take too much imagination to see that if the astronaut came by at the speed of light c, the light pulse, also traveling at a speed c, would have to go straight ahead just to stay in the clock. It would never be able to get from the bottom to the top mirror, and his clock would never tick off a nanosecond. His clock would stop!

Exercise 4

Discuss what the astronaut should have seen when the class of students went by. In particular, draw the astronaut's version of Figure (16) and describe the situation from the astronaut's point of view.

It is not particularly difficult to calculate the amount by which the astronaut's clock runs slow. All that is required is the Pythagorean theorem. In Figure (18), on the left, we show the path of the light pulse in our reference clock, and on the right the path in the astronaut's moving clock. Let T be the length of time it takes our pulse to go from the bottom to the top mirror, and T' the longer time light takes to travel along the diagonal line from his bottom mirror to his top mirror. We can think of T as the length of one of our nanoseconds, and T' as the length of one of the astronaut's longer nanoseconds.

The distance an object, moving at a speed v, travels in a time T, is vT. (If you go 30 miles per hour for 3 hours, you travel 90 miles.) Thus, in Figure (18), the distance our light pulse travels in going from the bottom to the top mirror is cT as shown. The astronaut's light pulse, which takes a time T' to travel the diagonal path, must have gone a distance cT' as shown.

During the time T', while the astronaut's light pulse is going along the diagonal path, the astronaut's clock, which is traveling at a speed v, moves forward a distance vT' as shown. This gives us a right triangle whose base is vT', whose hypotenuse is cT', and whose height, determined from our clock, is cT. According to the Pythagorean theorem, these sides are related by

$$(cT')^2 = (vT')^2 + (cT)^2 \quad (10)$$

Carrying out the squares, and collecting the terms with T' on one side, we get

$$c^2T'^2 - v^2T'^2 = (c^2 - v^2)T'^2 = c^2T^2$$

$$T'^2 = \frac{c^2T^2}{c^2 - v^2} = \frac{c^2T^2 \times (1/c^2)}{(c^2 - v^2) \times (1/c^2)} = \frac{T^2}{1 - v^2/c^2}$$

Taking the square root of both sides gives

$$\boxed{T' = \frac{T}{\sqrt{1 - v^2/c^2}}} \quad (11)$$

Equation (11) gives a precise relationship between the length of our nanosecond T and the astronaut's longer nanosecond T'. We see that the astronaut's basic time unit T' is longer than our basic time unit T by a factor $1/\sqrt{1 - v^2/c^2}$.

The factor $1/\sqrt{1 - v^2/c^2}$ appears in a number of calculations involving Einstein's special theory of relativity. As a result, it is essential to develop an intuitive feeling for this number. Let us consider several examples to begin to build this intuition. If v = 0, then

$$T' = \frac{T}{\sqrt{1 - v^2/c^2}} = \frac{T}{\sqrt{1 - 0}}$$

$$T' = \frac{T}{1} = T \quad (v = 0) \quad (12)$$

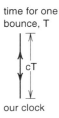
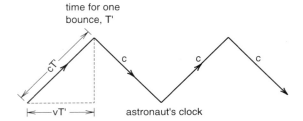

Figure 18

In our clock, the light pulse travels a distance cT in one bounce. In the astronaut's clock, the pulse travels a distance cT' while the clock moves forward a distance vT' during one bounce.

and we see that a clock at rest keeps the same time as ours. If the astronaut goes by at one tenth the speed of light, v = .1 c, and we get

$$T' = \frac{T}{\sqrt{1 - (.1c)^2/c^2}} = \frac{T}{\sqrt{1 - .01}}$$

$$T' = \frac{T}{\sqrt{.99}} = 1.005T \quad (v = c/10) \quad (13)$$

In this case the astronaut's seconds lengthen only by a factor 1.005 which represents only a .5% increase. If the astronaut's speed is increased to half the speed of light, we get

$$T' = \frac{T}{\sqrt{1 - (.5c)^2/c^2}} = \frac{T}{\sqrt{1 - .25}}$$

$$T' = \frac{T}{\sqrt{.75}} = 1.15T \quad (v = c/2) \quad (14)$$

Now we are getting a 15% increase in the length of the astronaut's seconds.

When we work with atomic or subatomic particles, it is not difficult to accelerate these particles to speeds close to the speed of light. Shortly we will consider a particle called a muon, that is traveling at a speed v = .994 c. For this particle we have

$$T' = \frac{T}{\sqrt{1 - (.994c)^2/c^2}} = \frac{T}{\sqrt{1 - .988}}$$

$$T' = \frac{T}{\sqrt{.012}} = 9T \quad (v = .994c) \quad (15)$$

Here we are beginning to see some large effects. If the astronaut were traveling this fast, his seconds would be 9 times longer than ours, his clock would be running only 1/9th as fast.

If we go all the way to v = c, Equation (11) gives

$$T' = \frac{T}{\sqrt{1 - c^2/c^2}} = \frac{T}{\sqrt{1 - 1}}$$

$$T' = \frac{T}{\sqrt{0}} = \infty \quad (v = c) \quad (16)$$

In this case, the astronaut's seconds would be infinitely long and the astronaut's clock would stop. This agrees with our earlier observation that if the astronaut went by at the speed of light, the light pulse in his clock would have to go straight ahead just to stay in the clock. It would not have time to move up or down, and therefore not be able to tick off any seconds.

So far we have been able to use a pocket calculator to evaluate $1/\sqrt{1 - v^2/c^2}$. But if the astronaut were flying in a commercial jet plane at a speed of 500 miles per hour, you have problems because $1/\sqrt{1 - v^2/c^2}$ is so close to 1 that the calculator cannot tell the difference. In a little while we will show you how to do such calculations, but for now we will just state the answer.

$$\frac{1}{\sqrt{1 - v^2/c^2}} = 1 + 2.7 \times 10^{-13} \quad \substack{\text{for a speed} \\ \text{of 500 mi/hr}} \quad (17)$$

To put this result in perspective suppose the astronaut flew on the jet for what we thought was a time T = 1 hour or 3600 seconds. The astronaut's light pulse clock would show a longer time T' given by

$$T' = \frac{T}{\sqrt{1 - v^2/c^2}}$$

$$= (1 + 2.7 \times 10^{-13}) \times (3600 \text{ seconds})$$

$$= 1 \text{ hour} + .97 \times 10^{-9} \text{ seconds}$$

Since $.97 \times 10^{-9}$ seconds is close to a nanosecond, we can write

$$T' \approx 1 \text{ hour} + 1 \text{ nanosecond} \quad (18)$$

The astronaut's clock takes 1 hour plus 1 nanosecond to move its hand forward 1 hour. We would say that his light pulse clock is losing a nanosecond per hour.

Students have a tendency to memorize formulas, and Equation (11), $T' = T/\sqrt{1 - v^2/c^2}$ looks like a good candidate. ***But don't!*** If you memorize this formula, you will mix up T' and T, forgetting which seconds belong to whom. ***There is a much easier way to always get the right answer.***

For any speed v less than or equal to c (which is all we will need to consider) the quantity $\sqrt{1 - v^2/c^2}$ is always a number less than or equal to 1, and $1/\sqrt{1 - v^2/c^2}$ is always greater than or equal to 1. For the examples we have considered so far, we have

Table 1

v	$\sqrt{1 - v^2/c^2}$	$1/\sqrt{1 - v^2/c^2}$
0	1	1
500 mi/hr	$1 - 2.7 \times 10^{-13}$	$1 + 2.7 \times 10^{-13}$
c/10	.995	1.005
c/2	.87	1.15
.994c	1/9	9
c	0	∞

You also know intuitively that for the moving light pulse clock, the light pulse travels a longer path, and therefore the moving clock's seconds are longer.

If you remember that $\sqrt{1 - v^2/c^2}$ appears somewhere in the formula, all you have to do is ask yourself what to do with a number less than one to make the answer bigger; clearly, you have to divide by it.

As an example of this way of reasoning, note that if a moving clock's seconds are longer, then the rate of the clock is slower. The number of ticks per unit time is less. If we want to talk about the rate of a moving clock, do we multiply or divide by $\sqrt{1 - v^2/c^2}$? To get a reduced rate, we multiply by $\sqrt{1 - v^2/c^2}$ since that number is always less than one. Thus we can say that the rate of the moving clock is reduced by a factor $\sqrt{1 - v^2/c^2}$.

The factor $\sqrt{1 - v^2/c^2}$ will appear numerous times throughout the text. But in every case you should have an intuitive idea of whether the quantity under consideration should increase or decrease. If it increases, divide by the $\sqrt{1 - v^2/c^2}$, and if it decreases, multiply by $\sqrt{1 - v^2/c^2}$. This approach gives the right answer, reduces memorization, and eliminates obscure notation like T' and T.

Other Clocks

So far we have an interesting but limited result. We have predicted that if someone carrying a light pulse clock moves by us at a speed v, we will see that their light pulse clock runs slow by a factor $\sqrt{1 - v^2/c^2}$. Up until now we have said nothing about any other kind of clock, and we have the problem that no one has actually constructed a light pulse clock. But we can easily generalize our result with another thought experiment fairly similar to the one we just did.

For the new thought experiment let us rejoin the discussion between the astronaut and the class of students. We begin just after the students have told the astronaut what they saw. "I was afraid of that," the astronaut replies. "I never did trust that light pulse clock. I am not at all surprised that it ran slow. But now my digital watch, it's really good. It is based on a quartz crystal and keeps really good time. It wouldn't run slow like the light pulse clock."

"I'll bet it would," Bill interrupts. "How much?," the astronaut responds indignantly. "The cost of one more trip," Bill answers.

In the new trip, the astronaut is to place his digital watch right next to the light pulse clock so that the astronaut and the class can see both the digital watch and the light pulse clock at the same time. The idea is to compare the rates of the two timing devices.

"Look what would happen," Bill continues, "if your digital watch did not slow down. When you come by, your digital watch would be keeping "God's time" as you call it, while your light pulse clock would be running slower."

"The important part of this experiment is that because the faces of the two clocks are together, if we see them running at different rates, you will too. *You would notice that here on earth, when you are at rest, the two clocks ran at the same rate. But when you were moving at high speed, they would run at different rates. You could use this difference in rates to detect your own motion, and therefore violate the principle of relativity.*"

The astronaut thought about this for a bit, and then responded, "I'll grant that you are partly right. On my previous trips, after the accelerations ceased and I started coasting toward the class, I did not feel any motion. I had no trouble unpacking the equipment and setting it up. The light pulse went in just as it had back in the lab, and I was sure that the light pulse clock was working just fine. I certainly would have noticed any difference in the rates of the two clocks."

"Are you insinuating," the astronaut continued, "that the reason I did not detect my light pulse running slow was because my digital watch was also running slow?"

"Almost," replied Bill, "but you have other timing devices in your capsule. You shave once a day because you do not like the feel of a beard. This is a cyclic process that could be used as the basis of a new kind of clock. If your shaving cycle clock did not slow down just like the light pulse clock, you could time your shaving cycle with the light pulse clock and detect your motion. You would notice that you had to shave more times per light pulse month when you were moving than when you were at rest. This would violate the principle of relativity."

"*Wow*," the astronaut exclaimed, "*if the principle of relativity is correct, and the light pulse clock runs slow, then every process, all timing devices in my ship have to run slow in precisely the same way so that I cannot detect the motion of the ship.*"

The astronaut's observation highlights the power and generality of the principle of relativity. It turns a limited theory about the behavior of one special kind of clock into a general theory about the behavior of all possible clocks. If the light pulse clock in the astronaut's capsule is running slow by a factor $\sqrt{1 - v^2/c^2}$, then all clocks must run slow by exactly the same factor so that the astronaut cannot detect his motion.

Real Clocks

Our theory still has a severe limitation. We have to assume that the light pulse clock runs slow. But no one has yet built a light pulse clock. Thus our theory is still based on thought experiments and conjectures about the behavior of light. If we had just one real clock that ran slow by a factor $\sqrt{1 - v^2/c^2}$, then the principle of relativity would guarantee that all other clocks ran slow in precisely the same way. Then we would not need any conjectures about the behavior of light. The principle of relativity would do it all!

In 1905 when Einstein proposed the special theory of relativity, he did not have any examples of moving clocks that were observed to run slow. He had to rely on his intuition and the two postulates. It was not until the early 1930s, in studies of the behavior of an elementary particle called the ***muon***, that experimental evidence was obtained showing that a real moving clock actually ran slow.

A muon at rest has a half life of 2.2 microseconds or 2,200 nanoseconds. That means that if we start with 1000 muons, 2.2 microseconds later about half will have decayed and only about 500 will be left. Wait another 2.2 microseconds and half of the remaining muons will decay and we will have only about 250 left, etc. If we wait 5 half lives, just over 10 microseconds, only one out of 32 of the original particles remain ($1/2 \times 1/2 \times 1/2 \times 1/2 \times 1/2 = 1/32$).

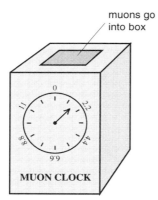

Figure 19
In our muon clock, every time half of the muons inside decay, we replace them and move the hand on the face forward by 2.2 microseconds.

Muons are created when cosmic rays from outer space strike the upper atmosphere. Few cosmic rays make it down to the lower atmosphere, so that most muons are created in the upper atmosphere, several miles up. The interesting results, observed in the 1930s was that there were almost as many high energy muons striking the surface of the earth as there were several miles up. This indicated that most of the high energy muons seemed to be surviving the several mile trip down through the earth's atmosphere.

Suppose we have a muon traveling at almost the speed of light, almost 1 foot per nanosecond. To go a mile, 5280 feet, would take 5,280 nanoseconds or about 5 microseconds. Therefore a 2 mile trip takes at least 10 microseconds, which is 5 half lives. One would expect that in this 2 mile trip, only one out of every 32 muons that started the trip would survive. Yet the evidence was that most of the high energy muons, those traveling close to the speed of light, survived. How did they do this?

We can get an idea of why the muons survive when we realize that the muon half life can be used as a timing device for a clock. Imagine that we have a box with a dial on the front as shown in Figure (19). We set the hand to 0 and put 1000 muons in the box. We wait until half the muons decay, whereupon we advance the hand 2.2 microseconds, replace the decayed muons so that we again have 1000 muons, and then wait until half have decayed again. If we keep repeating this process the hand will advance one muon half life in each cycle. Here we have a clock based on the muon half life rather than the swings of a pendulum or the vibrations of a quartz crystal.

The fact that most high energy muons raining down through the atmosphere survive the trip means that their half life is in excess of 10 microseconds, much longer than the 2.2 microsecond half life of a muon at rest. A clock based on these moving muons would run much slower than a muon clock at rest. Thus the experimental observation that the muons survive the trip down through the atmosphere gives us our first example of a real clock that runs slow when moving.

In the early 1960s, a motion picture was made that carefully studied the decay of muons in the trip down from the top of Mount Washington in New Hampshire to sea level (the sea level measurements were made in Cambridge, Massachusetts), a trip of about 6000 feet. Muons traveling at a speed of v = .994c were studied and from the number surviving the trip, it was determined that the muon half life was lengthened to about 20 microseconds, a factor of 9 times longer than the 2.2 microsecond lifetime of muons at rest. Since $1/\sqrt{1 - v^2/c^2} = 9$ for v = .994c, a result we got back in Equation (16), we see that the moving picture provides an explicit example of a moving clock that runs slow by a factor $\sqrt{1 - v^2/c^2}$.

At the present time there are two ways to observe the slowing down of real clocks. One is to use elementary particles like the muon, whose lifetimes are lengthened significantly when the particle moves at nearly the speed of light. The second way is to use modern atomic clocks which are so accurate that one can detect the tiny slowing down that occurs when the clock rides on a commercial jet. We calculated that a clock traveling 500 miles per hour should lose one nanosecond every hour. This loss was detected to an accuracy of 1% when physicists at the University of Maryland in the early 1980s flew an atomic clock for 15 hours over Chesapeake Bay.

In more recent times atomic clocks have become so accurate that the slowing down of the clock has become a nuisance. When these clocks are moved from one location to another, they have to be corrected for the time that was lost due to their motion. For these clocks, even a one nanosecond error is too much.

Thus today the slowing down of moving clocks is no longer a hypothesis but a common observational fact. The slowing down by $\sqrt{1 - v^2/c^2}$ has been seen both for clocks moving at the slow speeds of a commercial jet and the high speeds travelled by elementary particles. We now have real clocks that run slow by a factor $\sqrt{1 - v^2/c^2}$ and no longer need to hypothesize about the behavior of light pulses. *All of our conjectures in this chapter hinge on the principle of relativity alone.*

Movie
To play the movie, click the cursor in the photo to the left. Use up or down arrows on the keyboard to raise or lower volume. Left and right arrows step one frame forward or back and esc *stops it. The movie is 36 minutes long. The Movie* Time Dilation: An Experiment with Mu-Mesons *is presented with the permission of Education Development Center Inc., Newton, Massachusetts.*

Figure 19a -- Muon Lifetime Movie
The lifetimes of 568 muons, traveling at a speed of .994c, were plotted as vertical lines. If the muon's clocks did not run slow, these lines would show how far the muons could travel before decaying. One can see that very few of the muons would survive the trip from the top of Mt. Washington to sea level. Yet the majority do survive.

Time Dilation

If all moving clocks run slow, does *time itself* run slow for the moving observer? That raises the question of how we define time. If time is nothing more than what we measure by clocks, and all clocks run slow, we might as well say that time runs slow. And we can give this effect a name like time dilation, the word dilation referring to the stretching out of seconds in a moving clock.

But time is such a personal concept, it plays such a basic role in our lives, that it seems almost demeaning that time should be nothing more than what we measure by clocks. We have all had the experience that time runs slow when we are bored, and fast when we are busy. Time is associated with all aspects of our life, including death. Can such an important concept be abstracted to be nothing more than the results of a series of measurements?

Let us take the following point of view. Let physicists' time be that which is measured by clocks. Physicists' time is what runs slow for an object moving by. If your sense of time does not agree with physicists' time, think of that as a challenge. Try to devise some experiment to show that your sense of time is measurably different from physicists' time. If it is, you might be able to devise an experiment that violates the principle of relativity.

Space Travel

In human terms, time dilation should have its greatest effect on space travelers who need to travel long distances and therefore must go at high speeds. To get an idea of the distances involved in space travel, we note that light takes 1.25 seconds to travel from the earth to the moon (the moon is 1.25 billion feet away), and 8 minutes to travel from the sun to the earth. We can say that the moon is 1.25 light seconds away and the sun is 8 light minutes distant.

Currently Neptune is the most distant planet (Pluto will be the most distant again in a few years). When Voyager II passed Neptune, the television signals from Voyager, which travel at the speed of light, took 2.5 hours to reach us. Thus our solar system has a radius of 2.5 light hours. It takes 4 years for light to reach us from the nearest star from our sun; stars are typically one to a few light years apart.

If you look up at the sky at night and can see the Milky Way, you will see part of our galaxy, a spiral structure of stars that looks much like our neighboring galaxy Andromeda shown in Figure (20). Galaxies are about 100,000 light years across, and typically spaced about a

Figure 20
The Andromeda galaxy, about a million light years away, and about 1/10 million light years in diameter.

million light years apart. As we will see there are even larger structures in space; there are interesting things to study on an even grander scale.

Could anyone who is reading this text survive a trip to explore our neighboring galaxy Andromeda, or just survive a trip to some neighboring star, say, only 200 light years away?

Before Einstein's theory, one would guess that the best way to get to a distant star would be to go so fast that the trip would not take very long. But now we have a problem. In Einstein's theory, the speed of light is a special speed. If we had the astronaut carry our light pulse clock at a speed greater than the speed of light, the light pulse could not remain in the clock. The astronaut would also notice that he could not keep the light pulse in the clock, and could use that fact to detect his own uniform motion. In other words, the principle of relativity implies that we or the astronaut cannot travel faster than the speed of light.

That the speed of light is a limiting speed is common knowledge to physicists working with elementary particles. Small particle accelerators about a meter in diameter can accelerate electrons up to speeds approaching $v = .9999c$. The two mile-long accelerator at Stanford University, which holds the speed record for accelerating elementary particles here on earth, can only get electrons up to a speed $v = .999999999c$. The speed of light is Nature's speed limit, how this speed limit is enforced is discussed in Chapter 6.

Does Einstein's theory preclude the possibility that we could visit a distant world in our lifetime; are we confined to our local neighborhood of stars by Nature's speed limit? The behavior of the muons raining down through the atmosphere suggests that we are not confined. The muons, you will recall, live only 2.2 microseconds (on the average) when at rest. Yet the muons go much farther than the 2200 feet that light could travel in a muon half life. They survive the trip down through the atmosphere because their clocks are running slow.

If humans could accompany muons on a trip at a speed $v = .994c$, the human clocks should also run slow, their lifetimes should also expand by the same factor of 9. If the human clocks did not run slow and the muon clocks did, the difference in rates could be used to detect uniform motion in violation of the principle of relativity.

The survival of the muons suggest that we should be able to travel to a distant star in our own lifetime. Suppose, for example, we wish to travel to the star Zeta (we made up that name) which is 200 light years away. If we traveled at the speed $v = .994c$, our clocks should run slow by a factor $1/\sqrt{1 - v^2/c^2} = 1/9$, and the trip should only take us $200 \times 1/9 = 22.4$ years. We would be only 22.4 years older when we get there. A healthy, young crew should be able to survive that.

The Lorentz Contraction

A careful study of this proposed trip to star Zeta uncovers a consequence of Einstein's theory that we have not discussed so far. To see what this effect is, to see that it is just as real as the slowing down of moving clocks, we will treat this proposed trip as a new thought experiment which will be analyzed from several points of view.

In this thought experiment, the instructor and the class, who participated in the previous thought experiments, decide to travel to Zeta at a speed of v = .994c. They have a space ship constructed which on the inside looks just like their classroom, so that classroom discussions can be continued during the trip.

On the earth, a permanent government subagency of NASA is established to record transmissions from the space capsule and maintain an earth bound log of the trip. Since the capsule, traveling at less than the speed of light, will take over 200 years to get to Zeta, and since the transmissions upon arrival will take 200 years to get back, the NASA agency has to remain in operation for over 400 years to complete its assignment. NASA's summary of the trip, written in the year 2406, reads as follows: "The spacecraft took off in the year 2001 and spent four years accelerating up to a speed of v = .994c. During this acceleration everything was packed away, but when they got up to the desired speed, the rocket engines were shut off and they started the long coast to the Zeta. This coast started with a close fly-by of the earth in late January of the year 2005. The NASA mission control officer who recorded the fly-by noted that his great, great, great, grandchildren would be alive when the spacecraft reached its destination."

The mission control officer then wrote down the following calculations that were later verified in detail. "The spacecraft is traveling at a speed v = .994c, so that it will take 1/.994 times longer than it takes a pulse of light to reach the star. Since the star is 200 light years away, the spacecraft should take 200/.994 = 201.2 years to get there.

But the passengers inside are also moving at a speed v = .994c, their clocks and biological processes run slow by a factor $\sqrt{1 - v^2/c^2} = 1/9$, and the amount of time they will age is

$$\text{amount of time space travelers age} = 201.2 \text{ years} \times \frac{1}{9} = 22.4 \text{ years}$$

Even the oldest member of the crew, the instructor, will be able to survive."

The 2406 entry continued; "During the intervening years we maintained communication with the capsule and everything seemed to go well. There were some complaints about our interpretation of what was happening but that did not matter, everything worked out just as we had predicted. The spacecraft flew past Zeta in March of the year 2206, and we received the communications of the arrival this past March. The instructor said she planned to retire after they decelerated and the spacecraft landed on a planet orbiting Zeta. She was not quite sure what her class of middle aged students would do."

NASA's predictions may have come true, but from the point of view of the class in the capsule, not everything worked out the way NASA said it did.

As NASA mentioned, a few years were spent accelerating the space capsule to the speed v = .994c. The orbit was chosen so that just after the engines were shut off and the coast to Zeta began, the spacecraft would pass close to the earth for one final good-by.

There was quite a change from the acceleration phase to the coasting phase. During the acceleration everything had to be securely fastened, and there was the constant vibration of the engines. But when the engines were shut off, you couldn't feel motion any more; everything floated as in the TV pictures of the early astronauts orbiting the earth.

When the coasting started, the instructor and class settled down to the business of monitoring the trip. The first step was to test the principle of relativity. Was there any experiment that they could do inside the capsule that could detect the motion of the capsule? Various experiments were tried, but none demonstrated that the capsule itself was moving. As a result the students voted to take the point of view that they and the capsule were at rest, and the things outside were moving by.

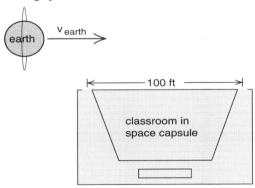

Figure 21
To measure the speed of the earth as it passes by, the class measures the time it takes a small satellite to pass by the windows in the back of the classroom. The windows are 100 feet apart.

Very shortly after the engines were shut off, the earth went by. This was expected, and the students were ready to measure the speed of the earth as it passed. There were two windows 100 feet apart on the back wall of the classroom, as shown in Figure (21). When the earth came by, there was an orbiting spacecraft, essentially at rest relative to the earth, that passed close to the windows. The students measured the time it took the front edge of this orbiting craft to travel the 100 feet between the windows. They got 100.6 nanoseconds and therefore concluded that the orbiting craft and the earth itself were moving by at a speed

$$v_{earth} = \frac{100 \text{ feet}}{100.6 \text{ nanoseconds}}$$

$$= .994 \frac{\text{feet}}{\text{nanosecond}} = .994 \, c$$

So far so good. That was supposed to be the relative speed of the earth.

In the first communications with earth, NASA mission control said that the space capsule passed by the earth at noon, January 17, 2005. Since all the accurate clocks had been dismantled to protect them from the acceleration, and only put back together when the coasting started, the class was not positive about what time it was. They were willing to accept NASA's statement that the fly-by occurred on January 17, 2005. From then on, however, the class had their own clocks in order—light pulse clocks, digital clocks and an atomic clock. From then on they would keep their own time.

For the next 22 years the trip went smoothly. There were numerous activities, video movies, etc., to keep the class occupied. Occasionally, about once every other month, a star went by. As each star passed, its speed v was measured and the class always got the answer v = .994c. This confirmed that the earth and the neighboring stars were all moving together like bright dots on a huge moving wall.

The big day was June 13, 2027, the 45th birthday of Jill who was eighteen when the trip was planned. This was the day, 22.4 years after the earth fly-by, that Zeta went by. The students made one more speed measurement and determined that Zeta went by at a speed v = .994c. An arrival message was sent to NASA, one day was allowed for summary discussions of the trip, and then the deceleration was begun.

After a toast to Jill for her birthday, Bill began the conversation. "Over the past few years, the NASA communications and even our original plans for the trip have been bothering me. The star charts say that Zeta is 200 light years from the earth, but that cannot be true."

"Look at the problem this way." Bill continues. "The earth went by us at noon on January 17, 2005, just 22.4 years ago. When the earth went by, we observed that it took 100.6 nanoseconds to pass by our 100 foot wide classroom. Thus the earth went by at a speed v = .994 feet/nanosecond, or .994c. Where is the earth now, 22.4 years later? How far could the earth have gotten, traveling at a speed .994c for 22.4 years? My answer is

$$\frac{\text{distance of earth}}{\text{from spaceship}} = .994 \frac{\text{light year}}{\text{year}} \times 22.4 \text{ years}$$

$$= 22 \text{ light years}$$

"You're right!" Joan interrupted, "Even if the earth had gone by at the speed of light, it would have gone only 22.4 light years in the 22.4 years since fly-by. The star chart must be wrong."

The instructor, who had just entered the room, said, "I object to that remark. As a graduate student I sat in on part of a course in astronomy and they described how the distance to Zeta was measured." The instructor drew a sketch, Figure (22), and continued. "Here is the earth in its orbit about the sun, and two observations, six months apart, are made of Zeta. You see that the two positions of the earth and the star form a triangle. Telescopes can accurately measure the two angles I labeled θ_1 and θ_2, and the distance across the earth's orbit is accurately known to be 16 light minutes. If you know two angles and one side of a triangle, then you can calculate the other sides from simple geometry. One reason for choosing a trip to Zeta is that we had accurate measurements of the distance to that star. We knew that it was 200 light years away, and we knew that traveling at a speed .994c, we could survive the trip in our lifetime."

Bill responded, "I think you entered the room too late and missed my argument. Let me summarize it. *Point 1*: the earth went by a little over 22 years ago. *Point 2*: we actually measured that the earth was traveling by us at almost the speed of light. *Point 3*: even light cannot go farther than 22 light years in 22 years. The earth can be no farther than about 22 light years away. *Point 4*: Zeta passed by us today, thus **the distance from the earth to Zeta is about 22 light years, not 200 light years!**"

"But what about NASA's calculations and all their plans," the instructor said, interrupting a bit nervously.

"We do not care what NASA thinks," responded Bill. "We have had no acceleration since the earth went by. Thus the principle of relativity guarantees that we can take the point of view that we are at rest and that it is the earth and NASA that are moving. From our point of view, the earth is 22 light years away. What NASA thinks is their business."

Joan interrupts, "Let us not argue on this last day. Let's figure out what is happening. There is something more important here than just how far away the earth is."

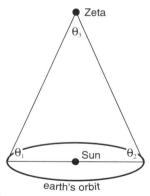

Figure 22
Instructor's sketch showing how the distance from the earth to the star Zeta was measured. (For a star 200 light years away, θ_3 is 4.5 millionths of a degree.)

"Remember in the old lectures on time dilation where the astronaut carried a light pulse clock. We used the peculiar behavior of that clock and the principle of relativity to deduce that time ran slow for a moving observer."

"Now for us, NASA is the moving observer. More than that, the earth, sun, and the stars, including Zeta, have all passed us going in the same direction and the same speed v = .994c. We can think of them as all in the same huge space ship. Or we can think of the earth and the stars as painted dots on a very long rod. A very long rod moving past us at a speed v = .994c. See my sketch (Figure 23)."

"To NASA, and the people on earth, this huge rod, with the sun at one end and Zeta at the other, is 200 light years long. Our instructor showed us how earth people measured the length of the rod. But as Bill has pointed out, to us this huge rod is only 22 light years long. That moving rod is only 1/9th as long as the earth people think it is."

"But," Bill interrupts, the factor of 1/9 is exactly the factor $\sqrt{1 - v^2/c^2}$ by which the earth people thought our clocks were running slow. Everyone sees something peculiar. The earth people see our clocks running slow by a factor $\sqrt{1 - v^2/c^2}$, and we see this hypothetical rod stretched from the sun to Zeta contracted by a factor $\sqrt{1 - v^2/c^2}$."

"But I still worry about the peculiar rod of Joan's," Bill continues, "what about real rods, meter sticks, and so forth? Will they also contract?"

At this point Joan sees the answer to that. "Remember, Bill, when we first discussed moving clocks, we had only the very peculiar light pulse clock that ran slow. But then we could argue that all clocks, no matter how they are constructed, had to run slow in exactly the same way, or we could violate the principle of relativity."

"We have just seen that my 'peculiar' rod, as you call it, contracts by a factor $\sqrt{1 - v^2/c^2}$. We should be able to show with some thought experiments that all rods, no matter what they are made of, must contract in exactly the same way as my peculiar one or we could violate the principle of relativity."

"That's easy," replies Bill. Just imagine that we string high tensile carbon filament meter sticks between the sun and Zeta. I estimate (after a short calculation) that it should take only 6×10^{17} of them. As we go on our trip, it doesn't make any difference whether the meter sticks are there or not, everything between the earth and Zeta passes by in 22 years. We still see 6×10^{17} meter sticks. But each one must have shortened by a factor $\sqrt{1 - v^2/c^2}$ so that all of them fit in the shortened distance of 22 light years. It does not make any difference what the sticks are made of."

Jim, who had not said much up until now, said, "OK, from your arguments I see that the length of the meter sticks, the length in the direction of motion must contract by a factor $\sqrt{1 - v^2/c^2}$, but what about the width? Do the meter sticks get skinnier too?"

The class decided that Jim's question was an excellent one, and that a new thought experiment was needed to decide.

Let's try this," suggested Joan. "Imagine that we have a space ship 10 feet in diameter and we build a brick wall with a circular hole in it 10 feet in diameter (Figure (24)). Let us assume that widths, as well as lengths, contract. To test the hypothesis, we hire an astronaut to

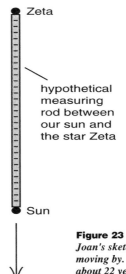

Figure 23
Joan's sketch of the Sun and Zeta moving by. This "object" passed by in about 22 years, moving at nearly the speed of light. Thus the "object" was about 22 light years long.

fly the 10 foot diameter capsule through the 10 foot hole at nearly the speed of light, say at v = .994c. If widths contract like lengths, the capsule should contract to 10/9 of a foot; it should be just over 13 inches in diameter when it gets to the 10 foot hole. It should have no trouble getting through."

"But look at the situation from the astronaut's point of view. He is sitting there at rest, and a brick wall is approaching him at a speed v = .994c. He has been told that there is a 10 foot hole in the wall, but he has also been told that the width of things contracts by a factor $\sqrt{1 - v^2/c^2}$. That means that the diameter of the hole should contract from 10 feet to 13 inches. He is sitting there in a 10 foot diameter capsule, a brick wall with a 13 inch hole is approaching him, and he is supposed to fit through. No way! He bails out and looks for another job."

"That's a good way to do thought experiments, Joan," replied the instructor. "Assume that what you want to test is correct, and then see if you can come up with an inconsistency. In this case, by assuming that widths contract, you predicted that the astronaut should easily make it through the hole in the wall. But the astronaut faced disaster. The crash, from the astronaut's point of view would have been an unfortunate violation of the principle of relativity, which he could use as evidence of his own uniform motion."

"To sum it up," the instructor added, "we now have time dilation where moving clocks run slow by a factor $\sqrt{1 - v^2/c^2}$, and we see that moving lengths contract by the same factor. Only lengths in the direction of motion contract, widths are unchanged."

Figure 24
Do diameters contract?

Leaving our thought experiment, it is interesting to note that the discovery of the contraction of moving lengths occurred before Einstein put forth the special theory of relativity. In the 1890s, physicist George Fitzgerald assumed that the length of one of the arms in Michaelson's interferometer, the arm along the direction of motion, contracted by a factor $\sqrt{1 - v^2/c^2}$. This was just the factor needed to keep the interferometer from detecting the earth's motion in the Michaelson-Morley experiments. It was a short while later that H.A. Lorentz showed that if the atoms in the arm of the Michaelson interferometer were held together by electric forces, then such a contraction would follow from Maxwell's theory of electricity. The big step, however, was Einstein's assumption that the principle of relativity is correct. Then, if one object happens to contract when moving, all objects must contract in exactly the same way so that the contraction could not be used to detect one's own motion. This contraction is called the Lorentz-Fitzgerald contraction, or Lorentz contraction, for short.

Relativistic Calculations

Although we have not quite finished with our discussion of Einstein's special theory of relativity, we have covered two of the important consequences, time dilation and the Lorentz contraction, which will play important roles throughout the text. At this point we will take a short break to discuss easy ways to handle calculations involving these relativistic effects. Then we will take another look at Einstein's theory to see if there are any more new effects to be discovered.

After our discussion of time dilation, we pointed out the importance of the quantity $\sqrt{1 - v^2/c^2}$ which is a number always less than 1. If we wanted to know how much longer a moving observer's time interval was, we divided by $\sqrt{1 - v^2/c^2}$ to get a bigger number. If we wanted to know how much less was the frequency of a moving clock, we multiplied by $\sqrt{1 - v^2/c^2}$ to get a smaller number.

With the Lorentz contraction we have another effect that depends upon $\sqrt{1 - v^2/c^2}$. If we see an object go by us, the object will contract in length. To predict its contracted length, we multiply the uncontracted length by $\sqrt{1 - v^2/c^2}$ to get a smaller number. If, on the other hand, an object moving by us had a contracted length l, and we stop the object, the contraction is undone and the length increases. We get the bigger uncontracted length by dividing by $\sqrt{1 - v^2/c^2}$.

As we mentioned earlier, first determine intuitively whether the number gets bigger or smaller, then either multiply by or divide by the $\sqrt{1 - v^2/c^2}$ as appropriate. This always works for time dilation, the Lorentz contraction, and, as we shall see later, relativistic mass.

We will now work some examples involving the Lorentz contraction to become familiar with how to handle this effect.

Example 1   Muons and Mt Washington

In the Mt. Washington experiment, muons travel 6000 feet from the top of Mt. Washington to sea level at a speed $v = .994c$. Most of the muons survive despite the fact that the trip should take about 6 microseconds (6000 nanoseconds), and the muon half life is $\tau = 2.2$ microseconds for muons at rest.

We say that the muons survive the trip because their internal timing device runs slow and their half life expands by a factor $1/\sqrt{1 - v^2/c^2} = 9$. The half life of the moving muons should be

$$\text{half life of moving muons} = \frac{\tau}{\sqrt{1 - v^2/c^2}}$$

$$= 2.2 \text{ microseconds} \times 9$$

$$= 19.8 \text{ microseconds}$$

This is plenty of time for the muons to make the trip.

From the muon's point of view, they are sitting at rest and it is Mt. Washington that is going by at a speed $v = .994c$. The muon's clocks aren't running slow, instead the height of Mt. Washington is contracted.

To calculate the contracted length of the mountain, start with the 6000 foot uncontracted length, multiply by $\sqrt{1 - v^2/c^2} = 1/9$ to get

$$\text{contracted height of Mt. Washington} = 6000 \text{ feet} \times \frac{1}{9}$$

$$= 667 \text{ feet}$$

Traveling by at nearly the speed of light, the 667 foot high Mt. Washington should take about 667 nanoseconds or .667 microseconds to go by. Since this is considerably less than the 2.2 microsecond half life of the muons, most of them should survive until sea level comes by.

Example 2   Slow Speeds

Joan walks by us slowly, carrying a meter stick pointing in the direction of her motion. If her speed is $v = 1$ foot/second, what is the contracted length of her meter stick as we see it?

This is an easy problem to set up. Since her meter stick is contracted, we multiply 1 meter times the $\sqrt{1 - v^2/c^2}$ with $v = 1$ foot/second. The problem comes in evaluating the numbers. Noting that 1 nanosecond $= 10^{-9}$ seconds, we can use the conversion factor 10^{-9} seconds/nanosecond to write

$$v = 1 \frac{\text{ft}}{\text{sec}} \times 10^{-9} \frac{\text{sec}}{\text{nanosecond}}$$

$$= 10^{-9} \frac{\text{ft}}{\text{nanosecond}} = 10^{-9} c$$

Thus we have

$$\frac{v}{c} = 10^{-9} \, , \; \frac{v^2}{c^2} = 10^{-18}$$

and for Joan's slow walk we have

$$\sqrt{1 - v^2/c^2} = \sqrt{1 - 10^{-18}} \tag{19}$$

If we try to use a calculator to evaluate the square root in Equation (19), we get the answer 1. For the calculator, the number 10^{-18} is so small compared to 1, that it is ignored. It is as if the calculator is telling us that when Joan's meter stick is moving by at only 1 foot/second, there is no noticeable contraction.

But there is some contraction, and we may want to know the contraction no matter how small it is. Since calculators cannot handle numbers like $1 - 10^{-18}$, we need some other way to deal with such expressions. For this, there is a convenient set of approximation formulas which we will now derive.

Approximation Formulas

The approximation formulas deal with numbers close to 1, numbers that can be written in the form $(1 + a)$ or $(1 - a)$ where a is a number much less than 1. For example the square root in Equation (19) can be written as

$$\sqrt{1 - 10^{-18}} = \sqrt{1 - \alpha}$$

where $\alpha = 10^{-18}$ is truly a number much less than 1.

The idea behind the approximation formulas is that if a is much less than 1, a^2 is very much less than 1 and can be neglected. To see how this works, let us calculate $(1 + a)^2$ and see how we can neglect a^2 terms even when a is as large as .01. An exact calculation is

$$(1 + \alpha)^2 = 1 + 2\alpha + \alpha^2$$

which for $\alpha = .01$, $\alpha^2 = .0001$ is

$$(1 + \alpha)^2 = 1 + .02 + .0001$$
$$= 1.0201$$

If we want to know how much $(1 + \alpha)^2$ differs from 1, but do not need too much precision, we could round off 1.0201 to 1.02 to get

$$(1 + \alpha)^2 \approx 1.02$$

(The symbol $\approx$ means "approximately equal to"). But in replacing 1.0201 by 1.02, we are simply dropping the α^2 term in Equation (19). We can write

$$(1 + \alpha)^2 \approx 1 + 2\alpha = 1 + .02 = 1.02 \quad (20)$$

Equation (20) is our first example of an approximation formula.

In Equation (20) the smaller a is the better the approximation. If a = .0001 we have

$$(1.0001)^2 = 1.00020001 \quad \text{(exact)}$$

Equation (20) gives

$$(1 + .0001)^2 \approx 1 + .0002 = 1.0002$$

and we see that the neglected α^2 terms become less and less important.

Some useful approximation formulas are the following

$$(1 + \alpha)^2 \approx 1 + 2\alpha \quad (20)$$

$$(1 - \alpha)^2 \approx 1 - 2\alpha \quad (21)$$

$$\frac{1}{1 + \alpha} \approx 1 - \alpha \quad (22)$$

$$\frac{1}{1 - \alpha} \approx 1 + \alpha \quad (23)$$

$$\sqrt{1 - \alpha} \approx 1 - \frac{\alpha}{2} \quad (24)$$

$$\frac{1}{\sqrt{1 - \alpha}} \approx 1 + \frac{\alpha}{2} \quad (25)$$

We have already derived Equation (20). Equation (21) follows from (20) if we replace α by $-\alpha$.

Equation (22) can be derived as follows. Multiply the quantity $1 - a$ by $(1+a)/(1+a)$ which is 1 to get

$$1 - \alpha = (1 - \alpha) \times \frac{(1 + \alpha)}{(1 + \alpha)} = \frac{(1 - \alpha^2)}{1 + \alpha} \approx \frac{1}{1 + \alpha}$$

In the last step we dropped the α^2 terms.

To derive the approximate formula for a square root, start with

$$\left(1 - \frac{\alpha}{2}\right) \times \left(1 - \frac{\alpha}{2}\right) = 1 - 2\frac{\alpha}{2} + \frac{\alpha^2}{4} \approx 1 - \alpha \quad (26)$$

taking the square root of Equation (26) gives

$$1 - \frac{\alpha}{2} \approx \sqrt{1 - \alpha}$$

which is the desired result. Again we only neglected α^2 terms.

To derive Equation (25), first use Equation (24) to get

$$\frac{1}{\sqrt{1 - \alpha}} \approx \frac{1}{1 - \frac{\alpha}{2}}$$

Then use Equation (23), with a replaced by a/2 to get

$$\frac{1}{1-\frac{\alpha}{2}} \approx 1 + \frac{\alpha}{2}$$

which is the desired result.

For those who are interested, the approximation formulas we have written are the first term of the so called *binomial expansion*:

$$(1+\alpha)^n = 1 + n\alpha + \frac{n(n-1)}{2}\alpha^2 + \cdots \quad (27)$$

where the coefficients of α, α^2, etc. are known as the *binomial coefficients*. If you need more accurate approximations, you can use Equation (27) and keep terms in α^2, α^3, etc. For all the work in this text, the first term is adequate.

Exercise 5
Show that Equations (20) through (25) are all examples of the first order binomial expansion

$$(1+\alpha)^n \approx 1 + n\alpha \quad (27a)$$

We are now ready to apply our approximation formulas to evaluate $\sqrt{1-10^{-18}}$ that appeared in Equation (17). Since $\alpha = 10^{-18}$ is very small compared to 1, we have

$$\sqrt{1-10^{-18}} = \sqrt{1-\alpha} \approx 1 - \frac{\alpha}{2} = 1 - \frac{10^{-18}}{2}$$

Thus the length of Joan's meter stick is

$$\text{length of Joan's contracted meter stick} = 1 \text{ meter} \times \sqrt{1-v^2/c^2}$$

$$= 1 \text{ meter}\left(1 - \frac{10^{-18}}{2}\right)$$

$$= 1 \text{ meter} - 5 \times 10^{-19} \text{ meters}$$

Exercise 6
We saw that time dilation in a commercial jet was not a big effect either—clocks losing only one nanosecond per hour in a jet traveling at 500 miles per hour. This was not an unnoticed effect, however, because modern atomic clocks can detect this loss.

In our derivation of the one nanosecond loss, we stated in Equation (17) that

$$\frac{1}{\sqrt{1-v^2/c^2}} \approx 1 + 2.7 \times 10^{-13} \quad \text{for a speed of 500 miles/hour} \quad (17)$$

Starting with

$$v = 500 \frac{\text{miles}}{\text{hour}} \times 5280 \frac{\text{feet}}{\text{mile}} \times \frac{1}{3600 \text{ sec/hour}}$$

use the approximation formulas to derive the result stated in Equation (17).

Exercise 7
Here is an exercise where you do not need the approximation formulas, but which should get you thinking about the Lorentz contraction. Suppose you observe that the Mars-17 spacecraft, traveling by you at a speed of $v = .995c$, passes you in 20 nanoseconds. Back on earth, the Mars-17 spacecraft is stored horizontally in a hanger that is the same length as the spacecraft. How long is the hanger?

A CONSISTENT THEORY

As we gain experience with Einstein's special theory of relativity, we begin to see a consistent pattern emerge. We are beginning to see that there is general agreement on what happens, even if different observers have different opinions as to *how* it happens. A good example is the Mt. Washington experiment observing muons traveling from the top of Mt. Washington to sea level. Everyone agrees that the muons made it. The muons are actually seen down at sea level. How they made it is where we get the differing points of view. We say that they made it because their clocks ran slow. They say they made it because the mountain was short. Time dilation is used from one point of view, the Lorentz contraction from another.

Do we have a complete, consistent theory now? In any new situation will we always agree on the predicted outcome of an experiment, even if the explanations of the outcome differ? Or are there some new effects, in addition to time dilation and the Lorentz contraction, that we will have to take into account?

The answer is that there is one more effect, called the **lack of simultaneity** which is a consequence of Einstein's theory. When we take into account this lack of simultaneity as well as time dilation and the Lorentz contraction, we get a completely consistent theory. Everyone will agree on the predicted outcome of any experiment involving uniform motion. No other new effects are needed to explain inconsistencies.

The lack of simultaneity turns out to be the biggest effect of special relativity, it involves two factors of $\sqrt{1 - v^2/c^2}$. But in this case the formulas are not as important as becoming familiar with some of the striking consequences. We will find ourselves dealing with problems such as whether we can get answers to questions that have not yet been asked, or whether gravity can crush matter out of existence. Strangely enough, these problems are related.

LACK OF SIMULTANEITY

One of the foundations of our intuitive sense of time is the concept of simultaneity. "Where were you when the murder was committed," the prosecutor asks. "At the time of the murder," the defendant replies, "I was eating dinner across town at Harvey's Restaurant." If the defendant can prove that the murder and eating dinner at Harvey's were simultaneous events, the jury will set him free. Everyone knows what simultaneous events are, or do they?

One of the most unsettling consequences of Einstein's theory is that the simultaneity of two events depends upon the point of view of the observer. Two events that from our point of view occurred simultaneously, may not be simultaneous to an observer moving by. Worse yet, two events that occurred one after the other to us, may have occurred in the reverse order to a moving observer.

To see what happens to the concept of simultaneous events, we will return to our thought experiment involving the instructor and the class. The action takes place on the earth before the trip to the star Zeta, and Joan has just brought in a paperback book on relativity.

"I couldn't understand that book either," the instructor says to Joan, "he starts with Einstein's analogy of trains and lightening bolts, but then switches to wind and sound waves, which completely confused me. There are many popular attempts to explain Einstein's theory, but most do not do very well when it comes to the lack of simultaneity."

"One of the problems with these popular accounts," the instructor continues, "is that we have to imagine too much. In today's lecture I will try to avoid that. In class we are going to carry out a real experiment involving two simultaneous events. We are going to discuss that experiment until everyone in class is completely clear about what happened. No imagining yet, just observe what actually occurred. When there are no questions left, then we will look at our real experiment from the point of view of someone moving by. At that point the main features of Einstein's theory are easy to see."

"The apparatus for our experiment is set up here on the lecture bench (Figure 25). On the left side of the bench I have a red flash bulb and on the right side a green flash bulb. These flash bulbs are attached to batteries and photocells so that when a light beam strikes their base, they go off."

"In the center of the desk is a laser and in front of it a beam splitter that uses half silvered mirrors. When I turn the laser on, the laser beam comes out, strikes the beam splitter, and divides into two beams. One beam travels to the left and sets off the red flash bulb, while the other beam goes to the right and triggers the green flash bulb. I will call the beams emerging from the beam splitter 'trigger beams' or 'trigger pulses'."

"Let us analyze the experiment before we carry it out," the instructor continues. "We will use the Einstein postulate that the speed of light is c to all observers. Thus the left trigger pulse travels at a speed c and so does the right one as I showed on the sketch. Since the beam splitter is in the center of the desk, the trigger pulses which start out together, travel the same distance at the same speeds to reach the flash bulbs. As a result the flash bulbs must go off simultaneously."

"The flashing of the flashbulbs are an example of what I mean by **simultaneous events**," the instructor adds with emphasis. "I know that they will be simultaneous events because of the way I set up the experiment".

"OK, let's do the experiment."

While the instructor is adjusting the apparatus, one of the flashbulbs goes off accidentally which amuses the class, but finally the apparatus is ready, the laser beam turned on, and both bulbs fire.

"Well, were they simultaneous flashes?" the instructor asks the class.

"I guess so," Bill responds, a bit hesitantly.

"How do you know," the instructor asks.

"Because you set it up that way," answers Bill.

Turning and pointing a finger at Joan who is sitting on the right side of the room nearer the green flash bulb (as in Figure 26), the instructor says, "Joan, for you which flash was first?" Joan thought for a second and replied, "The green bulb is closer, I should have seen the green light first."

"But which occurred first?" the instructor interrupts.

"What are you trying to get at?" Joan asks.

Figure 25
Lecture demonstration experiment in which two flashbulbs are fired simultaneously by trigger signals from a laser. The laser and beam splitter are at the center of the lecture bench, so that the laser light travels equal distances to reach the red and green bulbs. A photocell, battery and relay are mounted in each flashbulb base.

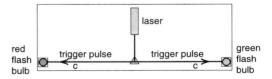

Figure 26
Although Joan sees the light from the green flash first, she knows that the two flashes were simultaneous because of the way the experiment was set up.

Joan

"Let me put it this way," the instructor responds. "Around 1000 BC, the city of Troy fell to the invading Greek army. About the same time, a star at the center of the Crab Nebula exploded in what is known as a supernova explosion. Since the star is 2000 light years away, the light from the supernova explosion took 2000 years to get here. The light arrived on July 4, 1057, about the time of the Battle of Hastings. Now which are simultaneous events? The supernova explosion and the Battle of Hastings, or the supernova explosion and the fall of Troy"

"I get the point," replied Joan. "Just because they saw the light from the supernova explosion at the time of the Battle of Hastings, does not mean that the supernova explosion and that battle occurred at the same time. We have to calculate back and figure out that the supernova explosion occurred about the time the Greeks were attacking Troy, 2000 years before the light reached us."

"As I sit here looking at your experiment," Joan continues, "I see the light from the green flash before the light from the red flash, but I am closer to the green bulb than the red bulb. If I measure how much sooner the green light arrives, then measure the distances to the two bulbs, and do some calculations, I'll probably find that the two flashes occurred at the same time."

"It is much easier than that." the instructor exclaimed, "Don't worry about when the light reaches you, just look at the way I set up the experiment – two trigger pulses, starting at the same time, traveling the same distance at the same speed. The flashes must have occurred simultaneously. I chose this experiment because it is so easy to analyze when you look at the trigger pulses."

"Any other questions?" the instructor asks. But by this time the class is ready to go on. "Now let us look at the experiment from the point of view of a Martian moving to the right a high speed v (Figure 27a). The Martian sees the lecture bench, laser, beam splitter and two flash bulbs all moving to the left as shown (Figure 27b). The lecture bench appears shortened by the Lorentz contraction, but the beam splitter is still in the middle of the bench. What is important is that the trigger pulses, being light, both travel outward from the beam splitter *at a speed c* .

As the bench passes by, the Martian sees that the green flash bulb quickly runs into the trigger pulse like this (▣—c→ v←▯). But on the other side there is a race between the trigger pulse and the red flash bulb, (v←▯   ←c—▣), and the race continues for a long time after the green bulb has fired. For the Martian, the green bulb actually fired first, and the two flashes were not simultaneous."

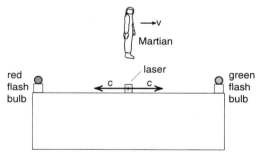

Figure 27a
In our thought experiment, a Martian astronaut passes by our lecture bench at a high speed v.

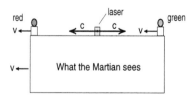

Figure 27b
The Martian astronaut sees the green flashbulb running into its trigger signal and firing quickly. The red flashbulb is running away from it's trigger signal, and therefore will not fire for a long time. Clearly the green flash occurs first.

"How much later can the red flash occur?" asks Bill.

The instructor replied, "The faster the bench goes by, the closer the race, and the longer it takes the trigger pulse to catch the red flash bulb. It isn't too hard to calculate the time difference. In the notes I handed out before class, I calculated that if the Martian sees our 12 foot long lecture bench go by at a speed

$$v = .9999999999999999999999999999992c \quad (28)$$

then the Martian will determine that the red flash occurred *one complete earth year* after the green flash. Not only are the two flashes not simultaneous, there is no fundamental limit as to how far apart in time that the two flashes can occur."

The reader will find the instructor's class notes in Appendix A of this chapter.

At this point Joan asks a question. "Suppose an astronaut from the planet Venus passed our experiment traveling the other way. Wouldn't she see the red flash first?"

"Let's draw a sketch," the instructor replies. The result is in Figure (28b). "The Venusian astronaut sees the lecture bench moving to the left. Now the red flash bulb runs into the trigger signal, and the race is with the green flash bulb. If the Venusian were going by at the same speed as the Martian (Equation 30) then the green flash would occur one year after the red one."

"With Einstein's theory, not only does the simultaneity of two events depend upon the observer's point of view, *even the order of the two events*—which one occurred first—depends upon how the observer is moving!"

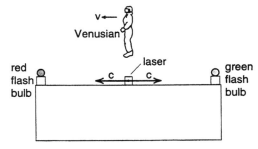

Figure 28a
Now a Venusian astronaut passes by our lecture bench at a high speed v in the other direction.

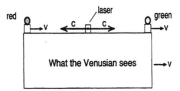

Figure 28b
The Venusian astronaut sees the red flashbulb running into its trigger signal and firing quickly. The green flashbulb is running away from its trigger signal, and therefore will not fire for a long time. Clearly the red flash occurs first.

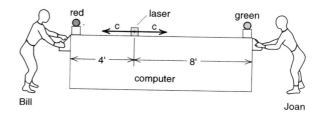

Figure 29
To test the speed of the computer, Bill thinks of a question, and types it in, when he sees the red flash. Joan checks to see if the answer arrives at the same time as the green flash.

CAUSALITY

"You can reverse the order of two events that are years apart!" Bill exclaimed. "Couldn't something weird happen in that time?"

"What about cause and effect," asked Joan. "If you can reverse the order of events, can't you reverse cause and effect? Can't the effect come before the cause?"

"In physics," the instructor responds, "there is a principle called *causality* which says that you cannot reverse cause and effect. Causality is not equivalent to the principle of relativity, but it is closely related, as we can see from the following thought experiment."

"Suppose," she said, "we read an ad for a brand new IBM computer that is really fast. The machine is so fast that when you type a question in at one end, the answer is printed out at the other end, 4 nanoseconds later. We look at the ad, see that the machine is 12 feet long, and order one to replace our lecture bench. After the machine is installed, we decide to test the accuracy of the ad. Do we really get answers in 4 nanoseconds? To find out, we set up the laser, beam splitter and flash bulbs on the computer instead of the lecture bench. The main difference in the setup is that the laser and beam splitter have been moved from the center, over closer to the end where we type in questions. We have set it up so that the trigger pulse travels 4 feet to the red bulb and 8 feet to the green bulb as shown in the sketch (Figure 29). Since the trigger pulse takes 4 nanoseconds to get to the red bulb, and 8 nanoseconds to reach the green bulb, the red flash will go off 4 nanoseconds before the green one. We will use these 4 nanoseconds to time the speed of the computer."

"Bill," the instructor says, motioning to him, "you come over here, and when you see the red flash ***think*** of a question. Then type it into the machine. Do not think of the question until after you see the red flash, but then think of it and type it in quickly. We will assume that you can do that in much less than a nanosecond. You can always do that kind of thing in a thought experiment."

"OK, Joan," the instructor says, motioning to Joan, "you come over here and look for the answer to Bill's question. If the ad is correct, if the machine is so fast that the answer comes out in 4 nanoseconds, then the answer should arrive when the green flash goes off."

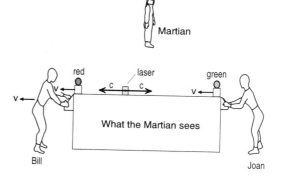

Figure 30
To a Martian passing by, our computer is moving to the left at a speed near the speed of light. The race between the red bulb and its trigger signal takes so long that the green bulb fires first. As a result, Joan sees the answer to a question that Bill has not yet thought of. (This is what could happen if information travels faster than the speed of light.)

The instructor positions Bill and Joan and the equipment as shown in Figure (29), turns on the laser and fires the flash bulbs.

"Did you type in the question," the instructor asks Bill, "when the red flash occurred?"

"Of course," responds Bill, humoring the instructor.

"And did the answer arrive at the same time as the green flash," the instructor asks Joan.

"Sure," replies Joan, "why not?"

"Suppose it did," replied the instructor. "Suppose the ad is right, and the answer is printed when the green flash goes off. Let us now look at this situation from the point of view of a Martian who is traveling to the right at a very high speed. The situation to the Martian looks like this (Figure 30). Although the red bulb is closer to the beam splitter, it is racing away from the trigger pulse (v ← 🔴 ← c 📧). If the computer is going by fast enough, the race between the red bulb and its trigger pulse will take much longer than the head-on collision between the green bulb and its trigger pulse. The green flash will occur before the red flash."

"And I," interrupts Joan, "will see the answer to a question that Bill has not even thought of yet!"

"I thought you would be in real trouble if you could reverse the order of events," Joan added.

"It is not really so bad," the instructor continued. "If the ad is right, if the 12 foot long computer can produce answers that travel across the machine in 4 nanoseconds, we are in deep trouble. In that case we could see answers to questions that have not yet been asked. That machine can be used to violate the principle of causality. But there was something peculiar about that machine. When the answer went through the machine, information went through the machine at three times the speed of light. Light takes 12 nanoseconds to cross the machine, while the answer went through in 4 nanoseconds."

"Suppose," asks Bill, "that the answer did not travel faster than light. Suppose it took 12 nanoseconds instead of 4 nanoseconds for the answer to come out." The instructor replied, "To measure a 12 nanosecond delay with our flash bulb apparatus, we would have to set the beam splitter right up next to the red bulb like this (Figure 31) in order for the trigger signal to reach the green bulb 12 nanoseconds later. But with this setup, the red bulb flashes as soon as the laser is turned on. No one, no matter how they are moving by, sees a race between the red bulb and the trigger signal. Everybody agrees that the green flash occurs after the red flash."

"You mean," interrupts Joan, "that *you cannot violate causality if information does not travel faster than the speed of light?*"

"That's right," the instructor replies, "that's one of the important and basic consequences of Einstein's theory."

"That's interesting," adds Bill. "It would violate the principle of relativity if we observed the astronaut's capsule, or probably any other object, traveling faster than the speed of light. The speed of light is beginning to play an important role."

Figure 31
If the answer to Bill's question takes 12 nanoseconds to travel through the 12 foot long computer, then this is the setup required to check the timing. The red bulb fires instantaneously, and everyone agrees that the red flash occurs first, and the answer appears later.

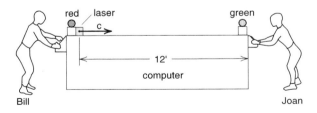

"That's pretty far out," replied Joan. "I didn't know that physics could say anything about how information – *ideas* – moved."

Jim, who had been sitting in the back of the classroom and not saying much, raised his hand. "At the beginning of the course when we were talking about sound pulses, you said that the more rigid the material, the faster the speed of sound in the material. You used Slinky pulses in your demonstrations because a Slinky is so compressible that a Slinky pulse travels slowly. You can't compress air as easily as a Slinky, and sound pulses travel faster in air. Since steel is very rigid, sound goes very fast in steel."

"During these discussions about the speed of light, I have been wondering. Is there any kind of material that is so rigid that sound waves travel at the speed of light?"

"What made you ask that?" the instructor asked.

"I've been reading a book about the life and death of stars," Jim replied. I just finished the chapter on neutron stars, and they said that the nuclear matter in a neutron star was very incompressible. It had to be to resist the strong gravitational forces. I was wondering, how fast is the speed of sound in this nuclear matter?"

"Up close to the speed of light," the instructor replied.

"If the nuclear matter were even more rigid, more incompressible, would the speed of sound exceed the speed of light?" Jim asked.

"It can't," the instructor replied.

"Then," Jim asked, "doesn't that put a limit on how incompressible, how rigid matter can be?"

"That looks like one of the consequences of Einstein's theory," the instructor replies.

"Then that explains what they were trying to say in the next chapter on black holes. They said that if you got too much matter concentrated in a small region, the gravitational force would become so great that it would crush the matter out of existence."

"I didn't believe it, because I thought that the matter would be squeezed down into a new form that is a lot more incompressible than nuclear matter, and the collapse of the star would stop. But now I am beginning to see that there may not be anything much more rigid than nuclear matter. Maybe black holes exist after all."

"Will you tell us about neutron stars and black holes?" Joan asks eagerly.

"Later in the course," the instructor responds.

APPENDIX A

Class Handout

To predict how long it takes for the trigger pulse to catch the red bulb in Figure (27b), let l be the uncontracted half length of the lecture bench (6 feet for our discussion). To the Martian, that half of the lecture bench has contracted to a length $l\sqrt{1 - v^2/c^2}$.

In the race, the red bulb traveling at a speed v, starts out a distance $l\sqrt{1 - v^2/c^2}$ ahead of the trigger pulse, which is traveling at a speed c. Let us assume that the race lasts a time t and that the trigger pulse catches the red bulb a distance x from where the trigger pulse started. Then we have

$$x = ct \qquad (30)$$

In the same time t, the green bulb only travels a distance $x - l\sqrt{1 - v^2/c^2}$, but this must equal vt;

$$vt = x - l\sqrt{1 - v^2/c^2} \qquad (31)$$

Using Equation (30) in (31) gives

$$vt = ct - l\sqrt{1 - v^2/c^2}$$

$$t(c - v) = l\sqrt{1 - v^2/c^2}$$

Solving for t gives

$$t = \frac{l\sqrt{1 - v^2/c^2}}{c - v} = \frac{l\sqrt{(1 + v/c)(1 - v/c)}}{c(1 - v/c)}$$

$$= \frac{l\sqrt{1 + v/c}}{c\sqrt{1 - v/c}}$$

If v is very close to c, then $(1 + v/c) \approx 2$ and we get

$$t \approx \frac{l\sqrt{2}}{c}\left(\frac{1}{\sqrt{1 - v/c}}\right)$$

If we plug in the numbers $t = 3 \times 10^7$ seconds (one earth year), $l = 6$ feet, $c = 10^9$ feet/second, we get

$$\sqrt{1 - v/c} = \frac{l\sqrt{2}}{ct} = \frac{6 \text{ ft}\sqrt{2}}{10^9 \text{ ft/sec} \times 3 \times 10^7 \text{ sec}}$$

$$= 2.8 \times 10^{-16}$$

Squaring this gives

$$(1 - v/c) = 8 \times 10^{-32}$$

Thus if

$$v = (1 - 8 \times 10^{-32})c$$

$$= .99999999999999999999999999999992c$$

then *the race will last a whole year*. On the other side, the trigger signal runs into the green flash bulb in far *less than a nanosecond* because the lecture bench is highly Lorentz contracted.

Chapter 2
Vectors

In the first chapter on Einstein's special theory of relativity, we saw how much we could learn from the simple concept of uniform motion. Everything in the special theory can be derived from (1) the idea that you cannot detect your own uniform motion, and (2) the existence of a real clock that runs slow by a factor $\sqrt{1-v^2/c^2}$.

We are now about to study more complicated kinds of motion where either the speed, the direction of motion, or both, are changing. Our work with non-uniform motion will be based to a large extent on a concept discovered by Galileo about 300 years before Einstein developed the special theory of relativity. It is interesting that after studying complex forms of motion for over 300 years, we still had so much to learn about simple uniform motion. But the history of science is like that. Major discoveries often occur when we see the simple underlying features after a long struggle with complex situations. If our goal is to present scientific ideas in the orderly progression from the simple to the complex, we must expect that the historical order of their discovery will not necessarily follow the same route.

Galileo was studying the motion of projectiles, trying to predict where cannon balls would land. He devised a set of experiments involving cannon balls rolling along slightly inclined planes. These experiments effectively slowed down the action, allowing Galileo to see the way the speed of a falling object changed as the object fell. To explain his results Galileo invented the concept of **acceleration** and pointed out that the simple feature of projectile motion is that projectiles move with constant or uniform acceleration. We can think of this as one step up in complexity from the uniform motion discussed in the previous chapter.

To study motion today, we have many tools that were not available to Galileo. In the laboratory we can slow down the action, or stop it, using strobe photographs or television cameras. To describe and analyze motion we have a number of mathematical tools, particularly the concept of vectors and the subject of calculus. And to predict motion, to predict not only where cannon balls land but also the trajectory of a spacecraft on a mission to photograph the solar system, we now have digital computers.

As we enter the study of more complex forms of motion, you will notice a shift in the way ideas are presented. Throughout the text, our goal is to construct a modern view of nature starting as much as possible from the basic underlying ideas. In our study of special relativity, the underlying idea, the principle of relativity, is more accurately expressed in terms of your experience flying in a jet than it is by any formal set of equations. As a result we were able to extract the content of the theory in a series of discussions that drew upon your experience.

In most other topics in physics, common experience is either not very helpful or downright misleading. If you have driven a car, you know where the accelerator pedal is located and have some idea about what *acceleration is. But unless you have already learned it in a physics course, your view of acceleration will bear little relationship to the concept of acceleration developed by Galileo and now used by physicists. It is perhaps unfortunate that we use the word acceleration in physics, for we often have to spend more time dismantling the students' previous notions of acceleration than we do building the concept as used in physics. And sometimes we fail.*

The physical ideas that we will study are often simply expressed in terms of mathematical concepts like a vector, a derivative, or an integral. This does not mean that we will drop physical intuition and rely on mathematics. Instead we will use them both to our best advantage. In some examples, the physical situation is obvious, and can be used to provide insight into the related mathematics. The best way, for example, to obtain a solid grip on calculus is to see it applied to physics problems. On the other hand, the concept of a vector, whose mathematical properties are easily developed, is an extremely powerful tool for explaining many phenomena in physics.

VECTORS

In this chapter we will study the vector as a mathematical object. The idea is to have the concept of vectors in our bag of mathematical tools ready for use in our study of more complex motion, ready to be applied to the ideas of velocity, acceleration and later, force and momentum.

In a sense, we will develop a *new math* for vectors. We will begin with a definition of ***displacement vectors***, and will then explain how two vectors are added. From this, we will develop a set of rules for the arithmetic of vectors. In some ways, the rules are the same as those for numbers, but in other ways they are different. We will see that most of the rules of arithmetic apply to vectors and that learning the vector convention is relatively simple.

Displacement Vectors

A displacement vector is a mathematical way of expressing the separation or displacement between two objects. To see what is involved in describing the separation between objects, consider a map such as the one in Figure (1), which shows the position of the two cities, New York and Boston. If we are driving on well-marked roads, it is sufficient, when planning a trip, to know that these two cities are separated by a distance of 190 miles. However, the pilot of a small plane flying from New York to Boston in a fog must know in what direction to fly; he must also know that Boston is located at an angle of 54 degrees east of north from New York.

Figure 1
Displacement vectors. Boston and Corning, N. Y., have equal displacements from New York and Pittsburgh, respectively. These displacements are located at different parts of the map, but they are the same displacement.

The statement that Boston is located a distance of 190 miles and at an angle of 54 degrees east of north from New York provides sufficient information to allow a pilot leaving New York to reach Boston in the thickest fog. The separation or displacement between the two cities is completely described by giving both the distance and the direction.

Looking again at Figure (1), we see that Corning, N.Y., is located 190 miles, at an angle of 54 degrees east of north, from Pittsburgh. The very same instructions, travel 190 miles at an angle of 54 degrees, will take a pilot from either Pittsburgh to Corning or New York to Boston. If we say that these instructions define what we mean by the word *displacement,* then we see that Corning has the same displacement from Pittsburgh as Boston does from New York. (For our discussion we will ignore the effects of the curvature of the earth.) The displacement itself is completely described when we give both the distance and direction, and does not depend upon the point of origin.

The displacement we have been discussing can be represented graphically by an arrow pointing in the direction of the displacement (54 degrees east of north), and whose length *represents* the distance (190 mi). An arrow that represents a displacement is called a displacement vector, or simply a *vector*. One thing you should note is that a vector that defines a distance and a direction does not depend on its point of origin. In Figure (1) we have drawn two arrows; but they both represent the same displacement, and thus are *the same vector.*

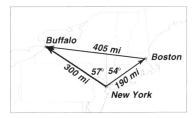

Figure 2
Addition of vectors. The vector sum of the displacement from New York to Boston plus the displacement from Boston to Buffalo is the displacement from New York to Buffalo.

Arithmetic of Vectors

Suppose that a pilot flies from New York to Boston and then to Buffalo. To his original displacement from New York to Boston he adds a displacement from Boston to Buffalo. What is the sum of these two displacements? After these displacements he will be 300 miles from New York at an angle 57 degrees west of north, as shown in Figure (2). This is the ***net displacement*** from New York, which is what we mean by the sum of the first two displacements.

If the pilot flies to five different cities, he is adding together five displacements, which we can represent by the vectors $\vec{a}$, $\vec{b}$, $\vec{c}$, $\vec{d}$, and $\vec{e}$ shown in Figure (3). (An arrow placed over a symbol is used to indicate that the symbol represents a vector.) Since the pilot's net displacement from his point of origin, represented by the bold vector, is simply the sum of his previous five displacements, we will say that the bold vector is the sum of the other five vectors. We will write this sum as ($\vec{a} + \vec{b} + \vec{c} + \vec{d} + \vec{e}$), but remember that the addition of vectors is defined graphically as illustrated in Figure (3).

If the numbers 405 and 190 are added, the answer is 595. But, as seen in Figure (2), if you add the vector representing the 405-mile displacement from Boston to Buffalo to the vector representing the 190-mile displacement from New York to Boston, the result is a vector representing a 300-mile displacement. Clearly, there is a difference between adding numbers and vectors. The plus sign between two numbers has a different meaning from that of the plus sign between two vectors.

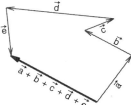

Figure 3
The sum of five displacements $\vec{a}$, $\vec{b}$, $\vec{c}$, $\vec{d}$, and $\vec{e}$ equals the vector $\vec{a}+\vec{b}+\vec{c}+\vec{d}+\vec{e}$.

Although vectors differ from numbers, some similarities between the two can be noted, particularly with regard to the rules of arithmetic. First, we will review the rules of arithmetic for numbers, and then see which of these rules also apply to vectors.

Rules for Number Arithmetic

1. Commutative law. In adding two numbers, a and b, the order of addition makes no difference.

$$a + b = b + a$$

2. Associative law. In adding three or more numbers, a, b, and c, we have

$$(a + b) + c = a + (b + c)$$

That is, if we first add a to b, and then add c, we get the same result as if we had added a to the sum $(b + c)$.

3. The negative of a number is defined by

$$a + (-a) = 0$$

where $(-a)$ is the negative of a.

4. Subtraction is defined as the addition of the negative number.

$$a - b = a + (-b)$$

These rules are so obvious when applied to numbers that it is hard to realize that they are rules. Let us apply the foregoing rules to vectors, using the method of addition of displacements.

Rules for Vector Arithmetic

1. The commutative law implies that

$$\vec{a} + \vec{b} = \vec{b} + \vec{a}$$

Figure (4) verifies this rule graphically. The reader should be able to see that $\vec{a} + \vec{b}$ and $\vec{b} + \vec{a}$ are the same vectors.

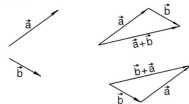

Figure 4

2. The associative law applied to vectors would imply

$$(\vec{a} + \vec{b}) + \vec{c} = \vec{a} + (\vec{b} + \vec{c})$$

From Figure (5) you should convince yourself that this law works.

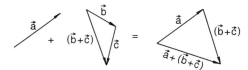

Figure 5

3. The negative of a vector is defined by

$$\vec{a} + \vec{-a} = 0$$

The only way to get a *zero displacement* is to return to the point of origin. Thus, the negative of a vector is a vector of the same length but pointing in the opposite direction (Figure 6).

Figure 6

4. The subtraction of vectors is now easy. If we want $\vec{a} - \vec{b}$, we just find $\vec{a} + \vec{-b}$. That is

$$\vec{a} - \vec{b} = \vec{a} + \vec{-b}$$

To subtract, we just add the negative vector as shown in Figure (7).

Figure 7

Multiplication of a Vector by a Number

Suppose we multiply a vector $\vec{a}$ by the number 5. What do we mean by the result $5\vec{a}$? Let us again try to follow the rules of arithmetic to answer this question. In arithmetic we were taught that

$$5a = a+a+a+a+a$$

Let us try the same rule for vectors.

$$5\vec{a} = \vec{a} + \vec{a} + \vec{a} + \vec{a} + \vec{a}$$

With this definition we see that $5\vec{a}$ is a vector in the same direction as $\vec{a}$ but five times as long (see Figure 8).

Figure 8

We may also multiply a vector by a negative number (see Figure 9); the minus sign just turns the vector around. For example,

$$-3\vec{a} = 3(-\vec{a}) = -\vec{a} + -\vec{a} + -\vec{a}$$

Figure 9

When we multiply a vector by a positive number, we merely change the length of the vector; multiplication by a negative number changes the length *and* reverses the direction.

Magnitude of a Vector

Often we will want to discuss only the length or magnitude of a vector, regardless of the direction in which it is pointing. For example, if we represent the displacement of Boston from New York by the vector $\vec{s}$, then the magnitude of $\vec{s}$ (the length of this displacement) is 190 miles. We use a vertical bar on each side of the vector to represent the magnitude; thus, we write $|\vec{s}| = 190$ mi (see Figure 10).

Figure 10

Vector Equations

Just as we can solve algebraic equations involving numbers, we can do the same for vectors. Suppose, for example, we would like to find the vector $\vec{x}$ in the vector equation

$$2\vec{a} + 3\vec{b} + 2\vec{x} = \vec{c}$$

Solving this equation the same way we would any other, we get

$$\vec{x} = 1/2\vec{c} - \vec{a} - 3/2\vec{b}$$

Graphically, we find $(1/2)\vec{c}$, $-\vec{a}$, and $(-3/2)\vec{b}$; we then vectorially add these quantities together to get the vector $\vec{x}$.

Graphical Work

In the early sections of this text, we shall do a fair amount of graphical work with vectors. As we can see from the previous examples, the main problem in graphical work is to move a vector accurately from one part of the page to another. This is easily done with a plastic triangle and ruler as described in the following example.

Example 1

The vector $\vec{s}$ starts from point a and we would like to redraw it starting from point b, as shown in Figure (11).

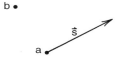

Solution: We want to draw a line through b that is parallel to $\vec{s}$. This can be done with a straightedge and triangle as shown in Figure (12).

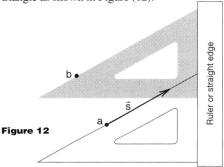

Figure 12

Place the straight edge and triangle so that one side of the triangle lies along the straight edge and the other along the vector $\vec{s}$. Then slide the triangle along the straight edge until the side of the triangle that was originally along $\vec{s}$ now passes through b. Draw this line through b. If nothing has slipped, the line will be parallel to $\vec{s}$ as shown in Figure (13).

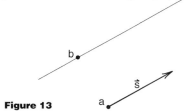

Figure 13

We now have the direction of $\vec{s}$ starting from b. Thus, we have only to put in the length. This is most easily done by marking the length of $\vec{s}$ on the edge of a piece of paper and reproducing this length, starting from b as shown in Figure (14).

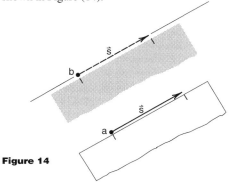

Figure 14

By being careful, using a sharp pencil, and practicing, you should have no difficulty in performing accurate and rapid graphical work. The practice can be gained by doing Problems 1 through 5. (Note that it is essential to distinguish a vector from a number. Therefore, when you are solving problems or working on a laboratory experiment, *it is recommended that you always place an arrow over the symbol representing a vector.*)

Exercise 1 Commutative law

The vectors $\vec{a}$, $\vec{b}$, and $\vec{c}$ of Figure 15 are shown enlarged on the tear out page 2-19. Using that page for your work, find

(a) $\vec{a} + \vec{b} + \vec{c}$ (in black);

(b) $\vec{b} + \vec{c} + \vec{a}$ (in red);

(c) $\vec{c} + \vec{a} + \vec{b}$ (in blue).

(Label all your work.)

Does the commutative law work?

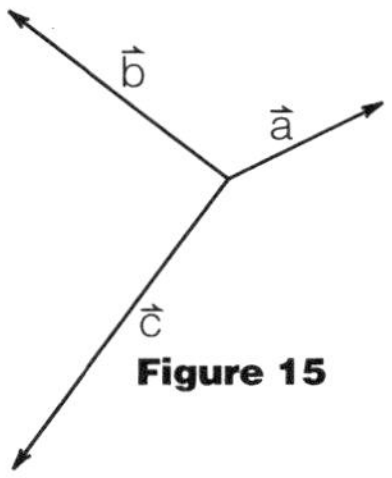

Figure 15

Exercise 2 Associative law

Use the tear out page 2-20 for the vectors of Figure (16), find

(a) $\vec{a} + \vec{b}$ (in black);

(b) $(\vec{a} + \vec{b}) + \vec{c}$ (in red);

(c) $(\vec{b} + \vec{c})$ (in black);

(d) $\vec{a} + (\vec{b} + \vec{c})$ (in blue).

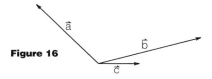

Figure 16

Exercise 3 Subtraction

Use the tear out page 2-21 for the three vectors $\vec{a}$, $\vec{b}$, and $\vec{c}$ shown in Figure (17), find the following vectors graphically, labeling your results.

(a) $\vec{a} + \vec{b}$

(b) $\vec{a} - \vec{b}$

(c) $\vec{b} - \vec{a}$

(d) $(\vec{a} - \vec{b}) + (\vec{b} - \vec{a})$

(e) $\vec{b} + \vec{c} - \vec{a}$

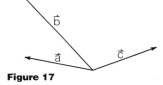

Figure 17

Exercise 4 Equations

Suppose that a physical law is given by the vector equation

$$\vec{P}_i = \vec{P}_f$$

Suppose that $\vec{P}_f$ is the sum of two vectors; that is,

$$\vec{P}_f = \vec{P}_{f1} + \vec{P}_{f2}$$

Given the two vectors $\vec{P}_i$ and $\vec{P}_{f1}$ (Figure 18), find $\vec{P}_{f2}$. (These vectors are found on the tear out page 2-22.)

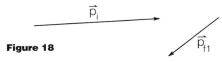

Figure 18

Exercise 5

Assume that the vectors $\vec{P}_f$, $\vec{P}_{f1}$, and $\vec{P}_{f2}$ are related by the *vector* law:

$$\vec{P}_f = \vec{P}_{f1} + \vec{P}_{f2}$$

In addition, the *magnitudes* of the vectors are related by

$$\left|P_f\right|^2 = \left|P_{f1}\right|^2 + \left|P_{f2}\right|^2$$

If you are given $\vec{P}_f$ and only the *direction* of $\vec{P}_{f1}$ (Figure 19), find $\vec{P}_{f1}$ and $\vec{P}_{f2}$ graphically. (These vectors are found on the tear out page 2-22.)

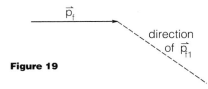

Figure 19

COMPONENTS

Another way to work with vectors, one that is especially convenient for solving numerical problems, is through the use of a coordinate system and components. To illustrate this method, suppose we were giving instructions to a pilot on how to fly from New York to Boston. One way, which we have mentioned, would be to tell the pilot both the direction and the distance she must fly, as "fly at an angle of 54 degrees east of north for a distance of 190 miles." But we could also tell her "fly 132 miles due east and then fly 112 miles due north." This second routing, which describes the displacement in terms of its easterly and northerly components, as illustrated in Figure (20), is less direct, but will also lead the pilot to Boston.

We can use the same alternate technique to describe a vector drawn on a piece of paper. In Figure (20), we drew two lines to indicate easterly and northerly directions. We have drawn the same lines in Figure (21), but now we will say that these lines represent the *x and y directions*. The lines themselves are called the *x and y axes*, respectively, and form what is called a *coordinate system*.

Just as the displacement from New York to Boston had both an easterly and northerly component, the vector $\vec{a}$ in Figure (21) has both an *x and a y component*. In fact, the vector $\vec{a}$ is just the sum of its component vectors $\vec{a}_x$ and $\vec{a}_y$:

$$\vec{a} = \vec{a}_x + \vec{a}_y \tag{1}$$

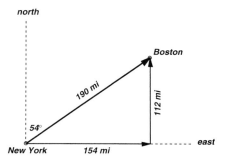

Figure 20
Two ways to reach Boston from New York.

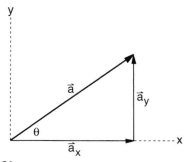

Figure 21
Component vectors. The sum of the component vectors $\vec{a}_x$ and $\vec{a}_y$ is equal to the vector $\vec{a}$.

Trigonometry can be used to find the length or magnitude of the component vectors; we get

$$a_x \equiv |\vec{a_x}| = |\vec{a}| \cos \theta \quad (2)$$

$$a_y \equiv |\vec{a_y}| = |\vec{a}| \sin \theta \quad (3)$$

Often we will represent the magnitude of a component vector by not using the arrow, as was done in the foregoing equations. (The equal sign with three bars, $a_x \equiv |\vec{a_x}|$, simply means that a_x *is defined to be the same symbol as* $|\vec{a_x}|$.) It is common terminology to call the magnitude of a component vector simply the *component*; for example, $a_x \equiv |\vec{a_x}|$ is called *the x component of the vector* $\vec{a}$.

Addition of Vectors by Adding Components

An important use of components is as a means for handling vectors numerically rather than graphically. We will show how this works by using an example of the addition of vectors by adding components.

Consider the three vectors shown in Figure (22). Since each vector is the vector sum of its individual components vectors, we have

$$\vec{a} = \vec{a_x} + \vec{a_y} \qquad \vec{b} = \vec{b_x} + \vec{b_y} \qquad \vec{c} = \vec{c_x} + \vec{c_y}$$

By adding all three vectors $\vec{a}$, $\vec{b}$, and $\vec{c}$ together, we get

$$\vec{a} + \vec{b} + \vec{c} = (\vec{a_x} + \vec{a_y}) + (\vec{b_x} + \vec{b_y}) + (\vec{c_x} + \vec{c_y})$$

The right-hand side of this equation may be rearranged to give

$$\vec{a} + \vec{b} + \vec{c} = (\vec{a_x} + \vec{b_x} + \vec{c_x}) + (\vec{a_y} + \vec{b_y} + \vec{c_y}) \quad (4)$$

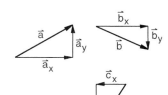

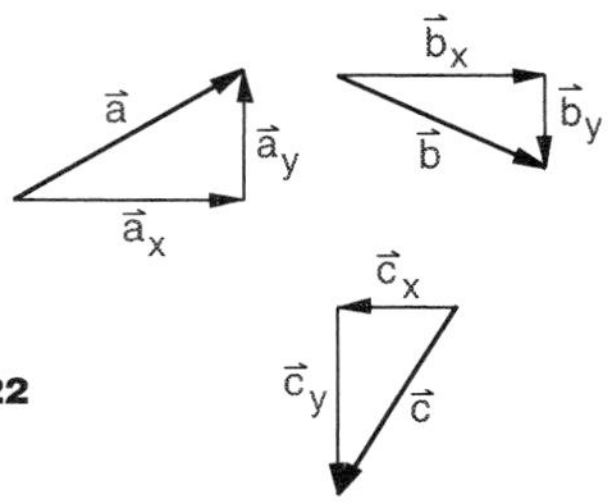

Figure 22

Equation 4 gives us a new way to add vectors, as illustrated in Figure (23). Previously we would have added the vectors $\vec{a}$, $\vec{b}$, and $\vec{c}$ directly, as shown in Figure (24). The new rule shows how we can first add the x components $(\vec{a_x} + \vec{b_x} + \vec{c_x})$ as shown in Figure (23a), then separately add the y components $(\vec{a_y} + \vec{b_y} + \vec{c_y})$ as shown in Figure (23b), and then add these vector sums vectorially, as shown in Figure (23c), to get the vector $(\vec{a} + \vec{b} + \vec{c})$.

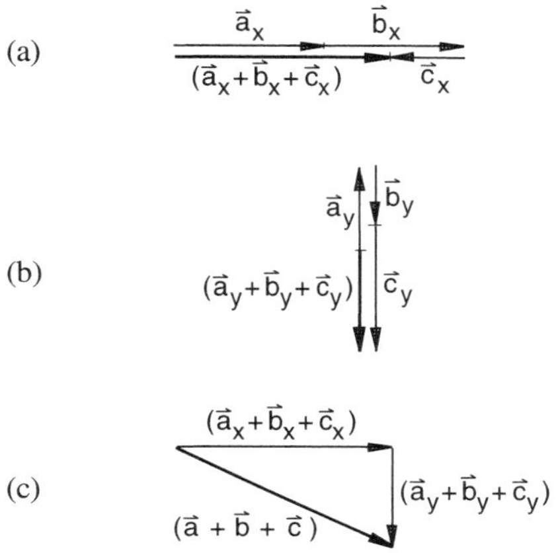

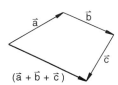

Figure 23

Figure 24

The advantage of using components is that we can numerically add or subtract the lengths of vectors that point in the same direction. Thus, to add 500 vectors, we would compute the lengths of all the x components and add (or subtract) these together. We would then add the lengths of the y components, and finally, we would vectorially add the resulting x and y components. Since the x and y components are at right angles, we may find the total length and final direction by using the Pythagorean theorem and trigonometry, as shown in Figure (25).

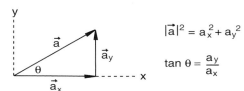

$$|\vec{a}|^2 = a_x^2 + a_y^2$$

$$\tan \theta = \frac{a_y}{a_x}$$

Figure 25

It is not necessary to always choose the x components horizontally and the y components vertically. We may choose a coordinate system (x', y') tilted at an angle, as shown in Figure (26). To use the language of the mathematician, $a_{x'}$ is the component of (or projection of) $\vec{a}$ in the direction x'. We see that the vector sum of all the component vectors still adds up to the vector itself.

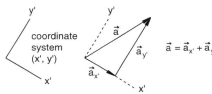

$$\vec{a} = \vec{a}_{x'} + \vec{a}_{y'}$$

Figure 26

Exercise 6
Imagine you are given the vectors $\vec{a}$, $\vec{b}$, and $\vec{c}$ and the two sets of coordinate axes (x_1, y_1) and (x_2, y_2) shown in Figure (27). Using the vectors found on the tear out page 2-23

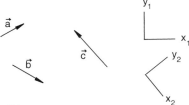

Figure 27

a) Find $(\vec{a} + \vec{b} + \vec{c})$ by direct addition of vectors.

b) Choose x_1 and y_1 as your coordinate axes. Find (in red) the x_1 and y_1 components of $\vec{a}$, $\vec{b}$, $\vec{c}$. Then
  (i) Find $(\vec{a}_{x1} + \vec{b}_{x1} + \vec{c}_{x1})$
  (ii) Find $(\vec{a}_{y1} + \vec{b}_{y1} + \vec{c}_{y1})$
  (iii) Find $(\vec{a}_{x1} + \vec{b}_{x1} + \vec{c}_{x1}) + (\vec{a}_{y1} + \vec{b}_{y1} + \vec{c}_{y1})$.

How does this compare with $(\vec{a} + \vec{b} + \vec{c})$?

c) Repeat part (b) for the coordinate axis (x_2, y_2).

Vector Equations in Component Form
Often we will run into a situation where we have a vector equation of the form

$$\vec{c} = \vec{a} + \vec{b}$$

but you have to solve the equation using components. This is easy to do, because to go from a vector equation to component equations, just rewrite the equation three (or two) times, once for each component. The above equation becomes

$$c_x = a_x + b_x$$
$$c_y = a_y + b_y$$
$$c_z = a_z + b_z$$

VECTOR MULTIPLICATION

We have seen how the rules work for vector addition, subtraction, and the multiplication of a vector by a number. Does it make any sense to multiply two vectors together? In considering the multiplication of the two vectors, the first question to answer is: what is the result? What kind of a thing do we get if we multiply a vector pointing east by a vector pointing north? Do we get a vector pointing in some third direction? Do we get a number that does not point? Or do we get some quantity more complex than a vector? And perhaps a more important question – why would one want to multiply two vectors together?

We will see in the study of physics that there are various reasons why we will want to multiply vectors, and we can get various answers. One kind of multiplication produces a number; this is called *scalar* multiplication or the ***dot product***. We will see examples of scalar multiplication shortly. A few chapters later we will encounter the ***vector cross product*** where the result of the multiplication of two vectors is itself a vector, one that points in a direction perpendicular to the two vectors being multiplied together. Finally there is a form of multiplication that leads to a quantity more complex than a vector, an object called a ***tensor*** or a ***matrix***. A tensor is an object that maintains the directional nature of both vectors involved in the product. Tensors are useful in the formal mathematical description of the basic laws of physics, but are not needed and will not be used in this text.

The names *scalar*, *vector*, and *tensor* describe a hierarchy of mathematical quantities. Scalars are numbers like, 1, 3, and -7, that have a magnitude but do not point anywhere. Vectors have both a magnitude and a direction. Tensors have the basic properties of both vectors used to construct them. In fact there are higher rank tensors that have the properties of 3, 4, or more vectors. People working with Einstein's generalized gravitational theory have to work all the time with tensors.

One of the remarkable discoveries of the twentieth century is that there is a close relationship between the mathematical properties of scalars, vectors, and tensors, and the physical properties of the various elementary particles. Later on we will discuss particles such as the π meson now used in cancer research, the photon which is the particle of light (a beam of light is a beam of photons), and the graviton, the particle hypothesized to be responsible for the gravitational force. It turns out that the physical properties of the π meson resemble the mathematical properties of a scalar, the properties of the photon are described by a vector (we will see this later in the text), and it requires a tensor to describe the graviton (that is why people working with gravitational theories have to work with tensors).

One of the surprises of physics and mathematics is that there are particles like the electron, proton and neutron, the basic constituents of atoms, that are not described by scalars, vectors, or tensors. To describe these particles, a new kind of a mathematical object had to be invented—an object called the ***spinor***. The spinor describing the electron has properties half way between a scalar and a vector. No one knew about the existence of spinors until the discovery was forced by the need to explain the behavior of electrons. In this text we will not go into the mathematics of spinors, but we will encounter some of the unusual properties that spinors have when we study the behavior of electrons in atoms. In a very real sense the spinor nature of electrons is responsible for the periodic table of elements and the entire field of chemistry.

In this text we can discuss a great many physical concepts using only scalars or vectors, and the two kinds of vector products that give a scalar or vector as a result. We will first discuss the scalar or dot product which is some ways is already a familiar concept, and then the vector or cross product which plays a significant role later in the text.

The Scalar or Dot Product

In a scalar product, we start with two vectors, multiply them together, and get a number as a result. What kind of a mathematical process does that involve? The Pythagorean theorem provides part of the answer.

Suppose that we have a vector $\vec{a}$ whose x and y components are a_x and a_y as shown in Figure (28). Then the magnitude or length $|\vec{a}|$ of the vector is given by the Pythagorean theorem as

$$|\vec{a}|^2 = (a_x)^2 + (a_y)^2 \qquad (4)$$

Figure 28

In some sense $|\vec{a}|^2$ is the product of the vector $\vec{a}$ with itself, and the answer is a number that is equal to the square of the length of the vector $\vec{a}$.

Now suppose that we use a different coordinate system x', y' shown in Figure (29) but have the same vector $\vec{a}$. In this new coordinate system the length of the vector $\vec{a}$ is given by the formula

$$|\vec{a}|^2 = \left(a'_x\right)^2 + \left(a'_y\right)^2 \qquad (5)$$

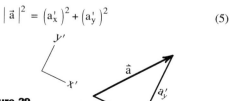

Figure 29

The components a'_x and a'_y are different from a_x and a_y, but we know that the length of $|\vec{a}|$ has not changed, thus $|\vec{a}|^2$ must be the same in Equations (4) and (5). We have found a quantity $|\vec{a}|^2$ which has the same value in all coordinate systems even though the pieces a_x^2 and a_y^2 change from one coordinate system to another. This is the key property of what we will call the scalar product.

To formalize this concept, we will define the *scalar product* of the vector $\vec{a}$ with itself as being the square of the length of $|\vec{a}|$. We will denote the scalar product by using the ***dot*** symbol to denote scalar multiplication:

$$\begin{array}{c}\text{Scalar product}\\ \text{of } \vec{a} \text{ with itself}\end{array} \equiv \vec{a} \cdot \vec{a} \equiv |\vec{a}|^2 \qquad (6)$$

From Equations (4) and (6) we have in the (x, y) coordinate system

$$\vec{a} \cdot \vec{a} = a_x^2 + a_y^2 \qquad (7)$$

In the (x', y') coordinate system we get

$$\vec{a} \cdot \vec{a} = \left(a'_x\right)^2 + \left(a'_y\right)^2 \qquad (8)$$

The fact that the length of the vector $\vec{a}$ is the same in both coordinate systems means that this scalar or "dot" product of $\vec{a}$ with itself has the same value even though the components or pieces a_x^2, a_y^2 or $\left(a'_x\right)^2$, $\left(a'_y\right)^2$ are different. In a more formal language, we can say that the scalar product $\vec{a} \cdot \vec{a}$ is unchanged by, or ***invariant*** under changes in the coordinate system. Basically we can say that there is physical meaning to the quantity $\vec{a} \cdot \vec{a}$ (i.e. the length of the vector) that does not depend upon the coordinate system used to measure the vector.

Exercise 7

Find the dot product $\vec{a} \cdot \vec{a}$ for a vector with components a_x, a_y, a_z in three dimensional space. How does the Pythagorean theorem enter in this case?

The example of calculating $\vec{a} \cdot \vec{a}$ above gives us a clue to guessing a more general definition of dot or scalar products when we have to deal with the product of two different vectors $\vec{a}$ and $\vec{b}$. As a guess let us try as a definition

$$\vec{a} \cdot \vec{b} \equiv a_x b_x + a_y b_y \tag{9}$$

or in three dimensions

$$\vec{a} \cdot \vec{b} \equiv a_x b_x + a_y b_y + a_z b_z \tag{10}$$

This definition of a dot product does not represent the length of either $\vec{a}$ or $\vec{b}$ but perhaps $\vec{a} \cdot \vec{b}$ has the special property that its value is independent of the choice of coordinate system, just as $\vec{a} \cdot \vec{a}$ had the same value in any coordinate system. To find out we need to calculate the quantity $\left(a'_x b'_x + a'_y b'_y + a'_z b'_z\right)$ in another coordinate system and see if we get the same answer. We will do a simple case to show that this is true, and leave the more general case to the reader.

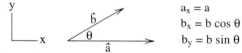

Figure 30

$a_x = a$
$b_x = b \cos \theta$
$b_y = b \sin \theta$

Suppose we have two vectors $\vec{a}$ and $\vec{b}$ separated by an angle q as shown in Figure (30). Let the lengths $|\vec{a}|$ and $|\vec{b}|$ be denoted by a and b respectively. Choosing a coordinate system (x, y) where the x axis lines up with $\vec{a}$, we have

$a_x = a, \ a_y = 0$
$b_x = b \cos \theta, \ b_y = b \sin \theta$

and the dot product, Equation 9, gives

$$\left(\vec{a} \cdot \vec{b}\right) = a_x b_x + a_y b_y$$
$$= ab \cos \theta + 0$$

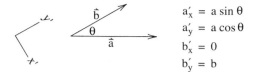

$a'_x = a \sin \theta$
$a'_y = a \cos \theta$
$b'_x = 0$
$b'_y = b$

Figure 31

Next choose a coordinate system (x', y') rotated from (x, y) by an angle $(90 - \theta)$ as shown in Figure (31). Here b lies along the y' axis and the dot product is given by

$$\vec{a} \cdot \vec{b} = a'_x b'_x + a'_y b'_y$$
$$= ab \cos \theta + 0$$

Again we get the result

$$\vec{a} \cdot \vec{b} = ab \cos \theta \tag{11}$$

Equation (11) holds no matter what coordinate system we use, as you can see by working the following exercise.

Exercise 8

Choose a coordinate system (x'', y'') where the x axis is an angle ϕ below the horizontal as shown in Figure (32). First calculate the components $a''_x, a''_y, b''_x, b''_y$ and then show that you still get

$$\vec{a} \cdot \vec{b} \equiv a''_x b''_x + a''_y b''_y = ab \cos \theta$$

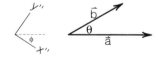

Figure 32

To do this problem, you need the following relationships.

$\sin(\theta + \phi) = \sin(\theta)\cos(\phi) + \cos(\theta)\sin(\phi)$
$\cos(\theta + \phi) = \cos(\theta)\cos(\phi) - \sin(\theta)\sin(\phi)\ )$
$\sin^2(\phi) + \cos^2(\phi) = 1 \quad$ for any angle ϕ

(This problem is much messier than the example we did.)

Interpretation of the Dot Product

When $\vec{a}$ and $\vec{b}$ are the same vector, then we had $\vec{a} \cdot \vec{a} = |\vec{a}|^2$ which is just the square of the length of the vector. If $\vec{a}$ and $\vec{b}$ are different vectors but parallel to each other, then $\theta = 0°$, $\cos \theta = 1$, and we get

$$\vec{a} \cdot \vec{b} = ab$$

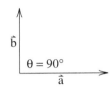

In other words the dot product of parallel vectors is just the product of the lengths of the vectors.

Another extreme is when the vectors are perpendicular to each other. In this case $\theta = 90°$, $\cos \theta = 0$ and $\vec{a} \cdot \vec{b} = 0$. The dot product of perpendicular vectors is zero. In a sense the dot product of two vectors measures the parallelism of the vectors. If the two vectors are parallel, the dot product is equal to the full product ab. If they are perpendicular, we get nothing. If they are at some intermediate angle, we get a number between ab and zero.

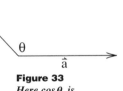

Increasing θ more, we see that if the vectors are separated by an angle between 90° and 180° as in Figure (33), then the $\cos \theta$ and the dot product are negative. A negative dot product indicates an anti-parallelism. The extreme case is $\theta = 180°$ where $\vec{a} \cdot \vec{b} = -ab$.

Figure 33
Here $\cos \theta$ is negative.

Physical Use of the Dot Product

We have seen that the dot product $\vec{a} \cdot \vec{b}$ is given by the simple formula $\vec{a} \cdot \vec{b} = ab \cos \theta$ and it has the special property that

$$\vec{a} \cdot \vec{b} \equiv a_x b_x + a_y b_y + a_z b_z$$

has the same value in any coordinate system even though the components a_x, b_x etc., are different in different coordinate systems. The fact that $\vec{a} \cdot \vec{b}$ is the same number in different coordinate systems means that it is truly a number with no dependence on direction. That is what we mean by a *scalar quantity*. This is a special property because $\vec{a} \cdot \vec{b}$ is made up of the vectors $\vec{a}$ and $\vec{b}$ that *do* depend upon direction and whose values *do* change when we go to different coordinate systems.

In physics there are quantities like displacements $\vec{x}$, velocities $\vec{v}$, forces $\vec{F}$ that all behave like vectors. All point somewhere and have components that depend upon our choice of direction. Yet we will deal with other quantities like energy which does not point anywhere. Energy has a magnitude but no direction. Yet our formulas for energy involve the vectors $\vec{x}$, $\vec{v}$, and $\vec{F}$. How can we construct numbers or scalars from vectors? The answer is - take dot or scalar products of the vectors. This is the mathematical reason why most of our formulas for energy will involve dot products.

Vector Cross Product

The other kind of vector product we will use in this course is the *vector cross product* where we *multiply two vectors $\vec{a}$ and $\vec{b}$ together to get a third vector $\vec{c}$*. The notation is

$$\vec{a} \times \vec{b} = \vec{c} \qquad (12)$$

where the name *cross product* comes from the cross we place between the vectors we are multiplying together. When you first encounter the cross product, it does not seem particularly intuitive. But we use it so much in later chapters that you will get quite used to it. Perhaps the best procedure is to skim over this material now, and refer back to it later when we start using it in various physics applications.

To define the cross product $\vec{a} \times \vec{b} = \vec{c}$, we have to define not only the magnitude but also the direction of the resulting vector $\vec{c}$. Starting with two vectors $\vec{a}$ and $\vec{b}$ pointing

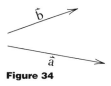

Figure 34

in different directions as in Figure 34, what unique direction is there for $\vec{c}$ to point? Should $\vec{c}$ point half way between $\vec{a}$ and $\vec{b}$, or should it be closer to $\vec{a}$ because $\vec{a}$ is longer than $\vec{b}$? No, there is nothing particularly unique or obvious about any of the directions in the plane defined by $\vec{a}$ and $\vec{b}$. The only truly unique direction is perpendicular to this plane. We will say that $\vec{c}$ points in this unique direction as shown in Figure 35.

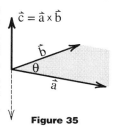

Figure 35

The direction perpendicular to the plane of $\vec{a}$ and $\vec{b}$ is not quite unique. The vector $\vec{c}$ could point either up or down as indicated by the solid or dotted vector in Figure 35. To select between these two choices, we use what is called the *right hand rule* which can be stated as follows: ***Point the fingers of your right hand in the direction of the first of the two vectors in the cross product $\vec{a} \times \vec{b}$ (in this case the vector $\vec{a}$). Then curl your fingers until they point in the direction of the second vector (in this case $\vec{b}$), as shown in Figure 36. If you orient your right hand so that this curling is physically possible, then your thumb will point in the direction of the cross product vector $\vec{c}$.***

Exercise 9

What direction would the vector $\vec{c}$ point if you used your left hand rather than your right hand in the above rule?

We said that the vector cross product was not a particularly intuitive concept when you first encounter it. In the above exercise, you see that if by accident you use your left hand rather than your right hand, $\vec{c} = \vec{a} \times \vec{b}$ will point the other way. One can reasonably wonder how a cross product could appear in any law of physics, for why would nature prefer right hand rules over left handed rules. It seems unbelievable that any basic concept should involve anything as arbitrary as the right hand rule.

There are two answers to this problem. One is that in most cases, nature has no preference for right handedness over left handedness. In these cases it turns out that any law of physics that involves right hand rules turns out to involve an even number of them so that any physical prediction does not depend upon whether you used a right hand rule or a left hand rule, as long as you use the same rule throughout. Since there are more right handed people than left handed people, the right hand rule has been chosen as the standard convention.

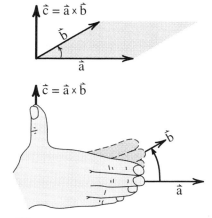

Figure 36
Right hand rule for the vector cross product.

Exercise 10

There is left and right handedness in the direction of the threads on a screw or bolt. In Figure (37a) we show a screw with a right handed thread. By this, we mean that if we turn the screw in the direction that we can curl the fingers of our right hand, the screw will move through wood in the direction that the thumb of our right hand points.

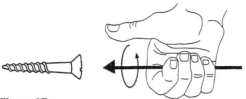

Figure 37a
Right handed thread.

In Figure (37b), we have a left hand thread. If we turn the screw in the direction we can curl the fingers of our left hand, the screw will move through the direction pointed by our left thumb.

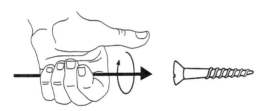

Figure 37b
Left handed thread

For this exercise find some screws and bolts, and determine whether the threads are right handed or left handed. Manufacturers use one kind of thread predominately over the other. Which is the predominant thread? Can you locate examples of the other kind of thread? (The best place to look for the other kind of thread is in the mechanism of some water faucets. Can you find a water faucet where one side uses a right hand thread and the other a left hand thread? If you find one, determine which is the right and which the left hand thread.)

Until 1956 it was believed that the basic laws of physics did not distinguish between left and right handedness. The fact that there are more right handed than left handed people, or that the DNA used by living organisms had a right handed spiral structure (like a right handed thread) was simply an historical accident. But then in 1956 it was discovered that the elementary particle called the *neutrino* was fundamentally left handed. Neutrinos spin like a top. If a neutrino is passing by you and you point the thumb of your left hand in the direction the neutrino is moving, the fingers of your left hand curl in the direction that the neutrino is spinning. Or we may say that the neutrino turns in the direction of a left handed thread, as shown in Figure 38.

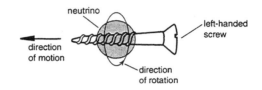

Figure 38
The neutrino is inherently a left handed object. When one passes by you, it spins in the direction that the threads on a left handed screw turn.

Another particle, called the anti-neutrino, is right handed. If you point the thumb of your right hand in the direction of motion of an anti-neutrino, the fingers of your right hand can curl in the direction that the anti-neutrino rotates. T.D. Lee and N.C. Yang received the 1957 Nobel prize in physics for their discovery that some basic phenomena of physics can be used to distinguish between left and right handedness.

The idea of right or left handedness in the laws of physics will appear in several of our later discussions of the basic laws of physics. The point for now is that having a quantity like the vector cross product that uses the right hand convention may be a useful tool to distinguish between left and right handedness.

Exercise 11

Go back to Figure 34 where we show the vectors $\vec{a}$ and $\vec{b}$, and draw the vector $\vec{c}\,' = \vec{b} \times \vec{a}$. Use the right hand rule as we stated it to determine the direction of $\vec{c}\,'$. From your result, decide what happens when you reverse the order in which you write the vectors in a cross product. Which of the arithmetic rules does this violate?

Magnitude of the Cross Product

Now that we have the right hand rule to determine the direction of $\vec{c} = \vec{a} \times \vec{b}$, we now need to specify the magnitude of $\vec{c}$.

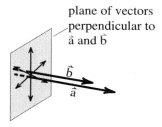

Figure 39

A clue as to a consistent definition of the magnitude of $\vec{c}$ is the fact that when $\vec{a}$ and $\vec{b}$ are parallel, they do not define a plane. In this special case there is an entire plane perpendicular to both $\vec{a}$ and $\vec{b}$, as shown in Figure 39. Thus there is an infinite number of directions that $\vec{c}$ could point and still be perpendicular to both $\vec{a}$ and $\vec{b}$. We can avoid this mathematical ambiguity only if $\vec{c}$ has zero magnitude when $\vec{a}$ and $\vec{b}$ are parallel. We do not care where $\vec{c}$ points if it has no length.

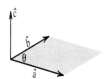

Figure 40

The simplest formula for the magnitude $|\vec{c}| = |\vec{a} \times \vec{b}|$, that is related to the product of $\vec{a}$ and $\vec{b}$, yet has zero length when $\vec{a}$ and $\vec{b}$ are parallel is

$$|\vec{c}| = |\vec{a} \times \vec{b}| = ab\sin\theta \qquad (13)$$

where $a \equiv |\vec{a}|$ and $b \equiv |\vec{b}|$ are the lengths of $\vec{a}$ and $\vec{b}$ respectively, and θ is the angle between them. Equation 13 is the definition we will use for the magnitude of the vector cross product.

In Equation 13, we see that not only is the cross product zero when the vectors are parallel, but is a maximum when the vectors are perpendicular. In the sense that the dot product $\vec{a} \cdot \vec{b}$ was a measure of the parallelism of the vectors $\vec{a}$ and $\vec{b}$, the cross product is a measure of their perpendicularity. If $\vec{a}$ and $\vec{b}$ are perpendicular, then the length of c is just the product ab. As the vectors become parallel the length of c reduces to zero.

Component Formula for the Cross Product

Sometimes one needs the formula for the components of $\vec{c} = \vec{a} \times \vec{b}$ expressed in terms of the components of $\vec{a}$ and $\vec{b}$. The result is a mess, and is remembered only by those who frequently use cross products. The answer is

$$c_x = a_y b_z - a_z b_y$$
$$c_y = a_z b_x - a_x b_z$$
$$c_z = a_x b_y - a_y b_x \qquad (14)$$

These formulas are not so bad if you are doing a computer calculation and you are letting the computer evaluate the individual components.

Exercise 12

Assume that $\vec{a}$ points in the x direction and $\vec{b}$ is in the xy plane as shown in Figure 41. By the right hand rule, $\vec{c}$ will point along the z axis as shown. Use Equation 14 to calculate the magnitude of c_z and compare your result with Equation 13.

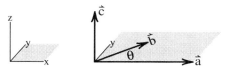

Figure 41

RIGHT HANDED COORDINATE SYSTEM

Notice in Figure 41, we have drawn an (x, y, z) coordinate system where z rises up from the xy plane. We could have drawn z down and still have three perpendicular directions. Why did we select the upward direction for z?

The answer is that the coordinate system shown in Figure 41 is a **right hand coordinate system**, defined as follows. Point the fingers of your right hand in the direction of the first coordinate axis (x). Then curl your fingers toward the second coordinate axis (y). If you have oriented your right hand so that you can curl your fingers this way, then your thumb points in the direction of the third coordinate axis (z).

The importance of using a right handed coordinate system is that Equation 14 for the cross product expressed as components works only for a right handed coordinate system. If by accident you used a left handed coordinate system, the signs in the equation would be reversed.

Exercise 13

Decide which of the (x, y, z) coordinate systems are right handed and which are left handed.

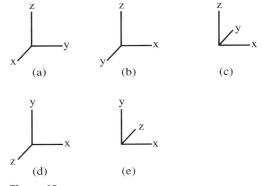

Figure 42

Figure 15
Vectors for Exercise 1, page 7. Find
(a) $\vec{a} + \vec{b} + \vec{c}$ (in black);
(b) $\vec{b} + \vec{c} + \vec{a}$ (in red);
(c) $\vec{c} + \vec{a} + \vec{b}$ (in blue).

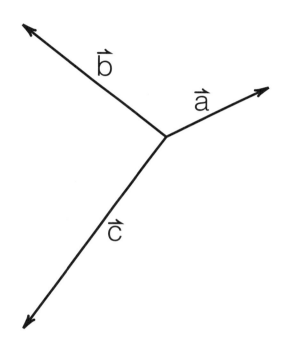

Figure 16
Vectors for Exercise 2, page 7. Find
(a) $\vec{a} + \vec{b}$ (in black);
(b) $(\vec{a} + \vec{b}) + \vec{c}$ (in red);
(c) $(\vec{b} + \vec{c})$ (in black);
(d) $\vec{a} + (\vec{b} + \vec{c})$ (in blue).

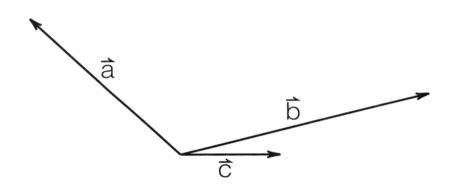

Figure 17
Vectors for Exercise 3, page 7. Find

(a) $\vec{a} + \vec{b}$

(b) $\vec{a} - \vec{b}$

(c) $\vec{b} - \vec{a}$

(d) $(\vec{a} - \vec{b}) + (\vec{b} - \vec{a})$

(e) $\vec{b} + \vec{c} - \vec{a}$

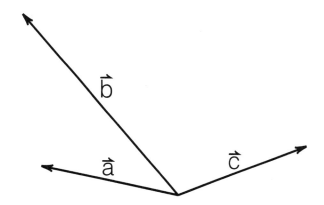

Figure 18
Vectors for Exercise 4, page 7.

$$\vec{P_i} = \vec{P_f}$$

Suppose that $\vec{P_f}$ is the sum of two vectors; that is,

$$\vec{P_f} = \vec{P_{f1}} + \vec{P_{f2}}$$

Given the two vectors $\vec{P_i}$ and $\vec{P_{f1}}$ (Figure 18), find $\vec{P_{f2}}$.

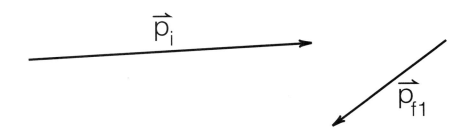

Figure 19
Vectors for Exercise 5, page 8.

$$\vec{P_f} = \vec{P_{f1}} + \vec{P_{f2}}$$

In addition, the *magnitudes* of the vectors are related by

$$|\vec{P_f}|^2 = |\vec{P_{f1}}|^2 + |\vec{P_{f2}}|^2$$

If you are given $\vec{P_f}$ and only the *direction* of $\vec{P_{f1}}$, find $\vec{P_{f1}}$ and $\vec{P_{f2}}$ graphically.

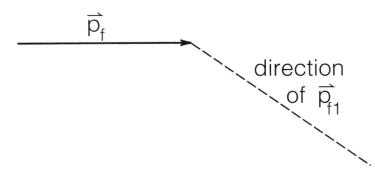

Figure 27
Vectors for Exercise 6, page 10.

(a) Find $\vec{a} + \vec{b} + \vec{c}$ by direct addition of vectors.

(b) Choose x_1 and y_1 as your coordinate axes. Find (in red) the x_1 and y_1 components of $\vec{a}, \vec{b}, \vec{c}$. Then

  (i) Find $\vec{a}_{x1} + \vec{b}_{x1} + \vec{c}_{x1}$

  (ii) Find $\vec{a}_{y1} + \vec{b}_{y1} + \vec{c}_{y1}$

  (iii) Find $(\vec{a}_{x1} + \vec{b}_{x1} + \vec{c}_{x1}) + (\vec{a}_{y1} + \vec{b}_{y1} + \vec{c}_{y1})$.

  How does this compare with $(\vec{a} + \vec{b} + \vec{c})$?

(c) Repeat part B for the coordinate axis (x_2, y_2). (you can use the back side of this page.)

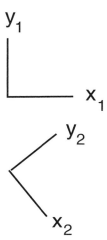

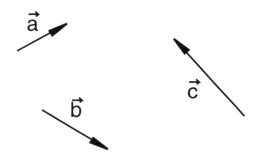

Figure 27
Vectors for Exercise 6, page 10, repeated.

(a) Find $\vec{a} + \vec{b} + \vec{c}$ by direct addition of vectors.

(b) Choose x_1 and y_1 as your coordinate axes. Find (in red) the x_1 and y_1 components of $\vec{a}, \vec{b}, \vec{c}$. Then

   (i) Find $\vec{a}_{x1} + \vec{b}_{x1} + \vec{c}_{x1}$

   (ii) Find $\vec{a}_{y1} + \vec{b}_{y1} + \vec{c}_{y1}$

   (iii) Find $(\vec{a}_{x1} + \vec{b}_{x1} + \vec{c}_{x1}) + (\vec{a}_{y1} + \vec{b}_{y1} + \vec{c}_{y1})$.

   How does this compare with $(\vec{a} + \vec{b} + \vec{c})$?

(c) Repeat part B for the coordinate axis (x_2, y_2).

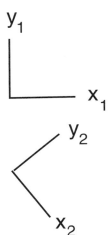

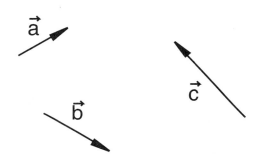

Figure 1
Marcel Duchamp, *Nude Decending a Staircase*
Philadelphia Museum of Art: Louise and Walter
Arensberg Collection

Chapter 3
Description of Motion

On the facing page is a reproduction (Figure 1) of Marcel Duchamp's painting, Nude Descending a Staircase, which was first displayed in New York at The International Exhibition of Modern Art, generally known as the Armory Show, in 1913. The objective of the painting, to convey a sense of motion, is achieved by repeating the stylized human form five times as it descends the steps. At the risk of obscuring the artistic qualities of the painting, we may imagine this work as a series of five flash photographs taken in sequence as the model walked downstairs.

In the next few chapters, a similar technique will be used to describe motion. We now have devices available, such as the stroboscope (called the strobe), that produce short bursts of light at regular intervals; with the strobe, we can photograph the successive positions of an object, such as a ball moving on the end of a string (see Figure 2). Although we do not have the artist's freedom of expression to convey the concept of motion by using a strobe photograph, we do obtain a more accurate measure of the motion.

Figure 2
Strobe photograph showing the motion of a ball on the end of a string.

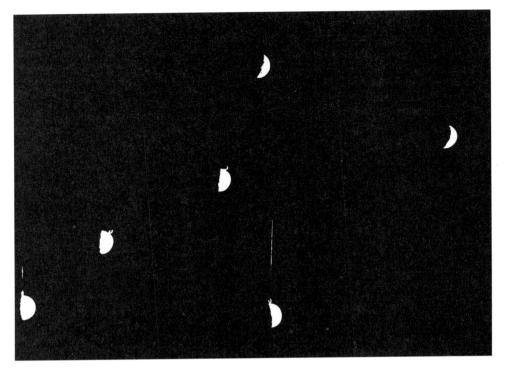

Figure 3
Strobe photograph of a moving object. In this photograph, the time between flashes is so long that the motion is difficult to understand.

The photograph in Figure (2) was taken with the strobe flashing five times per second while the ball was moving slowly. As a result, we see a smooth curve and have a fairly complete idea of the ball's entire motion. When we run the strobe at a rate of five flashes per second but move the ball more rapidly in a complicated pattern, the result is as shown in Figure (3). From this picture it is difficult to guess the ball's path; thus Figure (3) provides us with a poor representation of the motion of the ball. But if we turn the strobe up from 5 to 15 flashes per second (as in Figure 4), the rapid and complicated motion of the ball is easily understood.

The motion of any object can be described by locating its position at successive intervals of time. A strobe photograph is particularly useful because it shows the position at equal time intervals through-out the picture; that is, in Figure (2) at intervals of 1/5 sec and in Figure (4) at intervals of 1/15 sec. For this text, we will use a special symbol, Δt, to represent the time interval between flashes of the strobe. The t stands for time, while the Δ (Greek letter delta) indicates that these are short time intervals between flashes. Thus, $\Delta t = 1/5$ sec in Figures (2) and (3), and $\Delta t = 1/15$ sec in Figure (4).

For objects that are moving slowly along fairly smooth paths, we can use fairly long time intervals Δt between strobe flashes and their motion will be adequately described. As the motion becomes faster and more complicated, we turn the strobe up to a higher flashing rate to follow the object, as in Figure (4). To study complicated motion in more detail, we locate the position of the object after shorter and shorter time intervals Δt.

DISPLACEMENT VECTORS

When we represent the motion of an object by a strobe photograph, we are in fact representing this motion by a series of displacements, the successive displacements of the object in equal intervals of time. Mathematically, we can describe these displacements by a series of displacement vectors, as shown in Figure (5). This illustration is a reproduction of Figure (2) with the successive displacement vectors drawn from the center of the images.

Figure 4
Strobe photograph of a similar motion. In this photograph, the time between flashes was reduced and the motion is more easily understood.

Figure 5
Displacement vectors. The displacement between flash number 1 and flash number 2 is represented by the displacement vector $\vec{s}_1$ and so on. The entire path taken by the ball is represented by the series of eight displacement vectors.

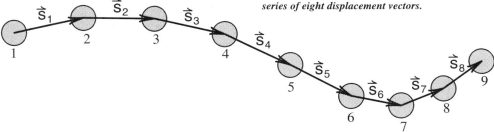

3-6 Description of Motion

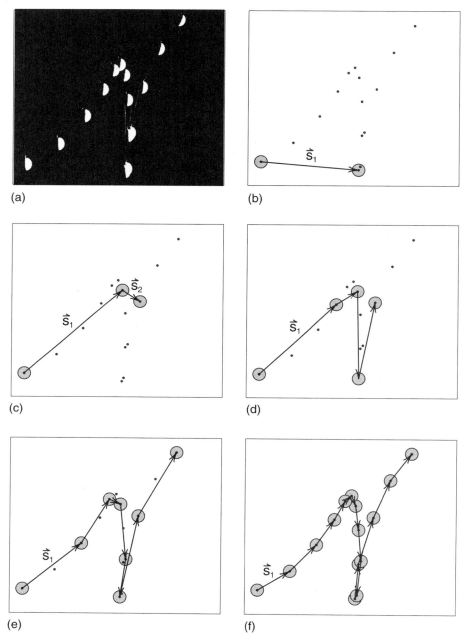

Figure 6
Representation of the path of a ball for various Δt. As the shorter and shorter Δt is used, the path of the ball is more accurately represented, as in figures (b) through (d).

In a sense we are approximating the path of the ball by a series of straight lines along the path. This is reasonably accurate provided that Δt is short enough, as shown in Figure (6).

In Figure (6), (a) is the strobe photograph shown in Figure (4), taken at a strobe interval of Δt = 1/15 sec; (b) shows how this photograph would have looked if we had set the strobe for Δt = 10/15 sec, or 2/3 sec. Only one out of ten exposures would have been produced. If we had represented the path of the ball by the vector $\vec{s}_1$ it would have been a gross misrepresentation. In (c), which would be the strobe picture at Δt = 6/15 sec, we see that the ball is no longer moving in a straight line, but still $\vec{s}_1$ and $\vec{s}_2$ provide a poor representation of the true motion. Cutting Δt in half to get (d), Δt = 3/15 sec, we would discover that there is a kink in the path of the ball. While taking the picture, we would have had to be careful in noticing the sequence of positions in order to draw the correct displacement vectors.

Reducing Δt to 2/15 sec (e), would give us a more detailed picture of the kink. This is not too different from (d); moreover, we begin to suspect that the seven displacement vectors in (e) represent the path fairly accurately. When we reduce Δt to 1/15 sec (f), we get more pictures of the same kink and the curve becomes smoother. It now appears that in most places the 14 displacement vectors form a fairly accurate picture of the true path. We notice, however, that the very bottom of the kink is cut off abruptly; here, shorter time intervals are needed to get an accurate picture of the motion.

A Coordinate System

In the strobe photographs discussed so far, we have a precise idea of the time scale, 1/5 second between flashes in Figure (2), 1/15 second in Figure (4), but no idea about the distance scale. As a result we know the direction of the succeeding displacement vectors, but do not know their magnitude.

One way to introduce a distance scale is to photograph the motion in front of a grid as shown in Figure (7). With this setup we obtain photographs like that shown in Figure (8), where we see the strobe motion of a steel ball projectile superimposed on the grid. The grid is illuminated by room lights which are dimmed to balance the exposure of the grid and the strobe flashes.

Figure 7
Experimental setup for taking strobe photographs. A Polaroid camera is used record the motion of a ball moving in front of a grid. The grid, made of stretched fish line, is mounted in front of a black painted wall.

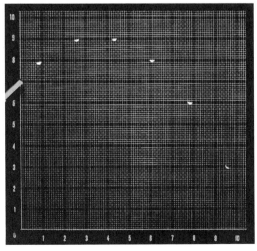

Figure 8
Strobe photograph of a steel ball projectile. The strobe flashes were 1/10 second apart.

Using techniques like that illustrated in Figure (9) to locate the centers of the images, we can transfer the information from the strobe photograph to graph paper and obtain the results shown in Figure (10). Figure (10) is the end result of a fair amount of tedious lab work, and the starting point for our analysis. For those who do not have strobe facilities, or the time to extract the information from a strobe photograph, we will include in the text a number of examples already transferred to graph paper in the form of Figure (10).

Using a television camera attached to an Apple II computer, we can, in under 2 minutes, obtain results that look like Figure (10). We will include a few of these *computer strobe photographs* in our examples of motion. However the computer strobe is not yet commercially available because we plan to use a computer with more modern graphics capabilities. It is likely that within a few years, one will be able to easily and quickly obtain results like those in Figure (10). The grid, which has now become the graph paper in Figure (10), serves as our coordinate system for locating the images.

Manipulation of Vectors

Figure (10) represents the kind of experimental data upon which we will base our description of motion. We have, up to now, described the motion of the projectile in terms of a series of displacement vectors labeled $\vec{s}_{-1}, \vec{s}_0, \cdots \vec{s}_3$ as shown. To go further, to introduce concepts like velocity and acceleration, we need to perform certain routine operations on these displacement vectors, like adding and subtracting them. A number of vector operations were discussed in Chapter 2, let us briefly review here those that we need for the analysis of strobe photographs. We will also introduce the concept of a **coordinate vector** which will be useful in much of our work.

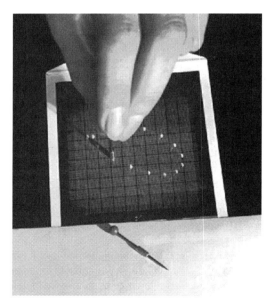

Figure 9
Using a pin and cylinder to locate the center of the ball. Move the cylinder until it just covers the image of the ball and then gently press down on the pin. The pin prick will give an accurate location for the center.

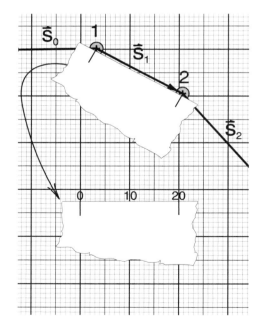

Figure 11
Measuring the length of the vector $\vec{S}_1$.

Measuring the Length of a Vector

One of the first pieces of information we need from a strobe photograph is the magnitude or length of the displacement vectors we have drawn. Figure (11) illustrates the practical way to obtain the lengths of the individual vectors from a graph like Figure (10). Take a piece of scrap paper and mark off the length of the vector as shown in the upper part of the figure. Then rotate the paper until it is parallel to the grid lines, and note the distance between the marks.

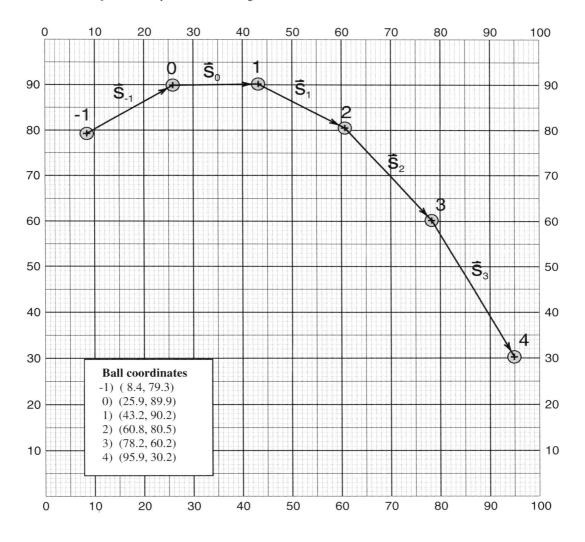

Figure 10
Strobe photograph transferred to graph paper. Using the pin and cylinder of Figure (9), we located the coordinates of the center of each image in Figure (8), and then reconstructed the strobe photograph as shown. We can now perform our analysis on the large graph paper rather than the small photograph.

In Figure (11), we see that the marks are 20 small grid spacings apart. In Figure (10), we see that each grid spacing represents a distance of 1 centimeter. Thus in Figure (11), the vector $\vec{s}_1$ has a magnitude of 20 centimeters. We can write this formally as

$$|\vec{s}_1| = 20 \text{ cm}$$

This technique may seem rather simple, but it works well and you will use it often.

Graphical Addition and Subtraction

Since we are working with experimental data in graphical form, we need to use graphical techniques to add and subtract vectors. These techniques, originally introduced in Chapter 2, are reviewed here in Figures (12) and (13). Figure (12a) and (12b) show the addition of two vectors by placing them head to tail. Think of the vectors $\vec{A}$ and $\vec{B}$ as separate trips; the sum $(\vec{A} + \vec{B})$ is our net displacement as we take the trips $\vec{A}$ and $\vec{B}$ in succession. To subtract $\vec{B}$ from $\vec{A}$, we simply add $(-\vec{B})$ to $\vec{A}$ as shown in Figure (12c).

To perform vector addition and subtraction, we need to move the vectors from one place to another. This is easily done with a triangle and a straight edge as indicated in Figure (13). The triangle and straight edge allows you to draw a parallel line; then mark a piece of paper as in Figure (11), to make the new vector have the same length as the old one.

For those who are mathematically inclined, this simple graphical work with vectors may seem elementary, especially compared to the exercises encountered in an introductory calculus course. But, as we shall see, this graphical work emphasizes the basic concepts. We will have many opportunities later to extract sophisticated formulas from these basic graphical operations.

For these exercises, you may use the practice graph on page 3-28, and the tear out sheet on page 3-29.

Exercise 1

Find the magnitudes of the vectors $\vec{s}_0$, $\vec{s}_1$, $\vec{s}_2$, and $\vec{s}_3$ in Figure (10).

Exercise 2

Explain why the vector $\vec{s}_{04}$, given by

$$\vec{s}_{04} = \vec{s}_0 + \vec{s}_1 + \vec{s}_2 + \vec{s}_3$$

has a magnitude of 91.3 cm which is quite a bit less than the sum of the lengths $|s_0| + |s_1| + |s_2| + |s_3|$.

Exercise 3

Use graphical methods to find the vector $(\vec{s}_3 - \vec{s}_2)$. (The result should point vertically downward and have a length of about 10 cm.)

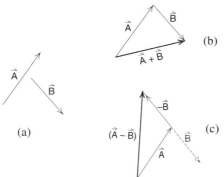

Figure 12
Addition and subtraction of vectors.

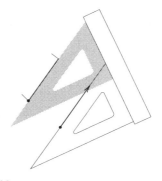

Figure 13
Moving vectors around. (This was discussed in Figure 2-12.)

Coordinate System and Coordinate Vectors

A coordinate system allows us to convert graphical work into a numerical calculation that can, for example, be carried out on a computer. Figure (14) illustrates two convenient ways of describing the location of a point. One is to give the x and y coordinates of the point (x,y), and the other is to use a *coordinate vector* $\vec{R}$ which we define as a vector that is drawn *from the origin of the coordinate system* to the point of interest.

Figure (15) illustrates the way an arbitrary vector $\vec{S}$ can be expressed in terms of coordinate vectors. From the diagram we see that $\vec{R}_2$ is the vector sum of $\vec{R}_1 + \vec{S}$, thus we can solve for the vector $\vec{S}$ to get the result

$$\vec{S} = \vec{R}_2 - \vec{R}_1.$$

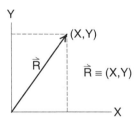

Figure 14
The coordinate vector $\vec{R}$, which starts at the origin, locates the point (x,y).

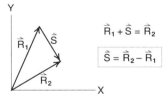

Figure 15
Expressing the vector $\vec{S}$ in terms of coordinate vectors.

ANALYSIS OF STROBE PHOTOGRAPHS

In our analysis of the strobe photograph of projectile motion, Figure (10), we are representing the path of the ball by a series of displacement vectors $\vec{S}_0 ... \vec{S}_3$ (We will think of the photograph as starting at point (0). The point labeled (-1) will be used later in our calculation of the instantaneous velocity at point (0). In a sense, we "know" that the ball actually went along a smooth continuous curve, and we could have represented the curve more accurately by reducing Δt as we did in Figure (6). But with many images to mark the trajectory, each displacement vector $\vec{S}_i$ becomes too short for accurate graphical work. In taking a strobe photograph, one must reach a compromise where the displacement vectors $\vec{S}_i$ are long enough to work with, but short enough to give a reasonable picture of the motion.

Velocity

The series of displacement vectors in Figure (10) show not only the trajectory of the projectile, but because the images are located at equal time intervals, we also have an idea of the speed of the projectile along its path. A long displacement vector indicates a higher speed than a short one. For each of the displacement vectors we can calculate what one would call the average speed of the projectile during that interval.

The idea of an average speed for a trip should be fairly familiar. If, for example, you went on a trip for a total distance of 90 miles, and you took 2 hours, you divide 90 miles by 2 hours to get an average speed of 45 miles per hour. For more detailed information about your speed, you break the trip up into small segments. For example, if you wanted to know how fast you were moving down the interstate highway, you measure how long it takes to pass two consecutive mile markers. If it took one minute, then your average speed during this short time interval is one mile divided by 1/60 hour which is 60 miles per hour. If you broke the whole trip down into 1 minute intervals, measured how far you went during each interval, and calculated your average speed for each interval, you would have a fairly complete record of your speed during your trip. It is this kind of record that we get from a strobe photograph of the motion of an object.

In physics, we use a concept that contains more information than simply the speed of the object. We want to know not only how many miles per hour or centimeters per second an object is moving, but also what direction the object is moving. This information is all contained in the concept of a *velocity vector*.

To construct a velocity vector for the projectile shown in Figure (10), when, for example, the ball is at position 1, we take the displacement vector $\vec{S}_1$, divide it by the strobe time interval Δt, to get what we will call the velocity vector $\vec{v}_1$:

$$\vec{v}_1 \equiv \frac{\vec{S}_1}{\Delta t} \qquad (1)$$

In Equation (1), what we have done is multiply the vector $\vec{S}_1$ by the number $(1/\Delta t)$ to get $\vec{v}_1$. From our earlier discussion of vectors we know that multiplying a vector by a number gives us a vector that points in the same direction, but has a new length. Thus $\vec{v}_1$ is a vector that points in the same direction as $\vec{S}_1$, but it now has a length given by

$$|\vec{v}_1| = \frac{|\vec{S}_1|}{\Delta t} = \frac{20 \text{ cm}}{.1 \text{ sec}} = 200 \frac{\text{cm}}{\text{sec}}$$

where we used $|\vec{S}_1| = 20$ cm from Figure (11) and we knew that $\Delta t = .1$ sec for this strobe photograph.

Not only have we changed the length of $\vec{S}_1$ by multiplying by $(1/\Delta t)$, we have also changed the dimensions from that of a distance (cm) to that of a speed (cm/sec). Thus the velocity vector $\vec{v}_1$ contains two important pieces of information. *It points in the direction of the motion of the ball, and has a length or magnitude equal to the speed of the ball.*

(Physics texts get rather picky over the use of the words speed and velocity. The word **speed** is reserved for the magnitude of the velocity, like 200 cm/sec. The word **velocity** is reserved for the velocity vector as defined above; the **velocity vector**

describes both the direction of motion and the speed. We will also use this convention throughout the text.)

In our discussion of strobe photographs, we noted that if we used too long a time interval Δt, we got a poor description of the motion as in Figures (6b) and (6c). As we used shorter time intervals as in Figures (6d, e, and f), we got a better and better picture of the path.

We have the same problem in dealing with the velocity of an object. If we use a very long Δt, we get a crude, average, description of the object's velocity. As we use a shorter and shorter Δt, our description of the velocity, Equation (1), becomes more and more precise.

Since, in this chapter, we will be working with experimental data obtained from strobe photographs, there is a practical limit on how short a time interval Δt we can use and have vectors big enough to work with. We will see that, for the kinds of motion that we encounter in the introductory physics lab, a reasonably short Δt like .1 sec gives reasonably accurate results.

If you make more precise measurements of the position of an object you generally find that as you use shorter and shorter Δt to measure velocity, you reach a point where the velocity vector no longer changes. What happens is that you reach a point where, if you cut Δt in half, the particle goes in the same direction but only half as far. Thus both the displacement $\vec{S}_1$ and the time interval Δt are both cut in half, and the ratio $\vec{v}_1 = \vec{S}_1/\Delta t$ is unchanged. This limiting process, where we see that the velocity vector changes less and less as Δt is reduced, is demonstrated graphically in our discussion of instantaneous velocity at the end of the chapter.

Exercise (4)

What is the magnitude of the velocity vector $\vec{v}_3$, for the ball in Figure (10). Give your answer in cm/sec.

Equation (1) is well suited for graphical work but for numerical calculations it is convenient to express $\vec{S}_i$ in terms of the coordinate vectors $\vec{R}_i$. This is done in Figure (16), where we see that the vector sum $\vec{R}_i + \vec{S}_i = \vec{R}_{i+1}$ thus $\vec{S}_i = \vec{R}_{i+1} - \vec{R}_i$ and Equation (1) becomes

$$\vec{v}_i = \frac{(\vec{R}_{i+1} - \vec{R}_i)}{\Delta t} \qquad (2)$$

If we call $(\vec{R}_{i+1} - \vec{R}_i)$ the "change in the position $\vec{R}$ during the time Δt", and denote this change by $\Delta\vec{R}$, Equation (2) becomes

$$\vec{v}_i = \frac{(\vec{R}_{i+1} - \vec{R}_i)}{\Delta t} = \frac{\Delta\vec{R}}{\Delta t} \qquad (2a)$$

which is perhaps a more familiar notation for those who have already studied calculus. In a calculus course, one would define the velocity $\vec{v}_i$ by taking the limit as $\Delta t \to 0$ (i.e., by turning the strobe flashing rate "all the way up"). In our experimental work with strobe photographs, we reduce Δt only to the point where we have a reasonable representation of the path; using too short a time interval makes the experimental analysis impossible.

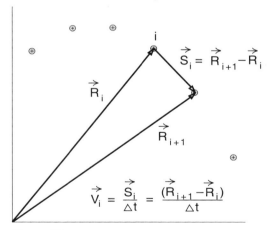

Figure 16
Expressing the velocity vector $\vec{v}_i$, in terms of the coordinate vectors $\vec{R}_i$ and $\vec{R}_{i+1}$.

Acceleration

In Chapter 1 on Einstein's special theory of relativity, we limited our discussion to uniform motion, motion in a straight line at constant speed. If we took a strobe photograph of an object undergoing uniform motion, we would get a result like that shown in Figure (17). All the velocity vectors would point in the same direction and have the same length. We will, from now on, call this *motion with constant velocity*, meaning that the velocity vector is constant, unchanging.

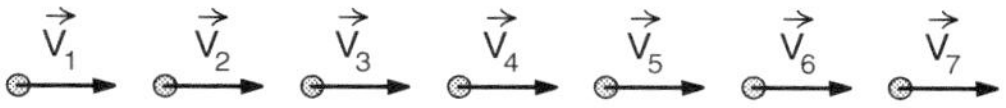

Figure 17
Motion with constant velocity.

From the principle of relativity we learned that there is something very special about motion with constant velocity—we cannot feel it. Recall that one statement of the principle of relativity was that there is no experiment that you can perform to detect your own uniform motion relative to empty space. You cannot tell, for example, whether the room you are sitting in is at rest or hurdling through space at a speed of 100,000 miles per hour.

Although we cannot feel or detect our own uniform motion, we can easily detect non uniform motion. We know what happens if we slam on the brakes and come to a sudden stop—everything in the car falls forward. A strobe photograph of a car using the brakes might look like that shown in Figure (18a). Each successive velocity vector gets shorter and shorter until the car comes to rest.

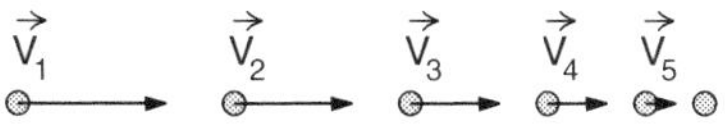

Figure 18a
Put on the brakes, and your velocity changes.

Another way the velocity of a car can change is by going around a corner as illustrated in Figure (18b). In that figure the speed does not change, each velocity vector has the same length, but the directions are changing. It is also easy to detect this kind of change in velocity—all the packages in the back seat of your car slide to one side of the seat.

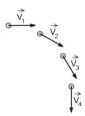

Figure 18b
When you drive around a corner, your speed may not change, but your velocity vector changes in direction.

The point we want to get at is, *what do we feel when our velocity changes?* Consider two examples. In the first, we are moving at constant velocity, due east at 60 miles per hour. A strobe photograph showing our initial and final velocity vectors $\vec{v}_i$ and $\vec{v}_f$ would look like that in Figure (19a). If we define the *change in velocity* $\Delta\vec{v}$ by the equation

$$\Delta\vec{v} \equiv \vec{v}_f - \vec{v}_i$$

then from Figure (19b) we see that $\Delta\vec{v} = 0$ for uniform motion.

For the second example, suppose we are traveling due south at 60 miles per hour, and a while later are moving due east at 60 miles per hour, as indicated in Figure (20a). Now we have a non zero change in velocity $\Delta\vec{v}$ as indicated in Figure (20b).

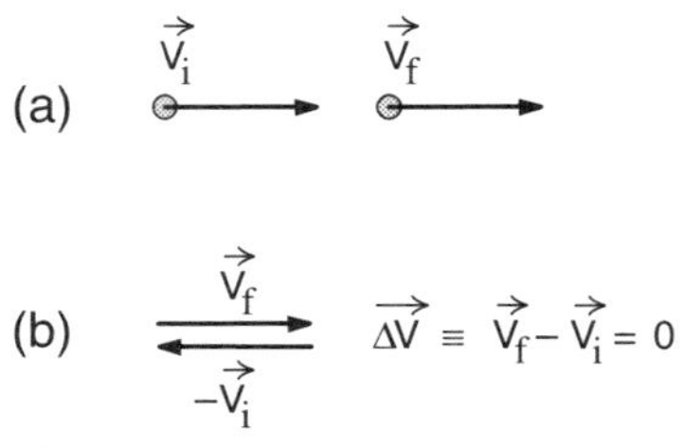

In our two examples, we find that if we have uniform motion which we cannot feel, the change in velocity $\Delta\vec{v}$ is zero. If we have non uniform motion, $\Delta\vec{v}$ is not zero and we can feel that. *Is it $\Delta\vec{v}$, the change in velocity, that we feel?*

Almost, but not quite. Let us look at our second example, Figure (20), more carefully. There are two distinct ways that our velocity can change from pointing south to pointing east. In one case there could have been a gradual curve in the road. It may have taken several minutes to go around the curve and we would be hardly aware of the turn.

In the other extreme, we may have been driving south, bounced off a stalled truck, and within a fraction of a second finding ourselves traveling due east. In both cases our change in velocity $\Delta\vec{v} = \vec{v}_f - \vec{v}_i$ is the same, as shown in Figure (20b). But the effect on us is terribly different. The difference in the two cases is that the change in velocity $\Delta\vec{v}$ occurred much more rapidly when we struck the truck than when we went around the curve. What we feel is not $\Delta\vec{v}$ alone, but how fast $\Delta\vec{v}$ happens.

If we take the change in velocity $\Delta\vec{v}$ and divide it by the time Δt over which the change takes place, then the smaller Δt, the more rapidly the change takes place, the bigger the result. This ratio $\Delta\vec{v}/\Delta t$ which more closely represents what we feel than $\Delta\vec{v}$ alone, is given the special name *acceleration*.

(a) $\vec{v}_i \longrightarrow \quad \vec{v}_f \longrightarrow$

(b) $\begin{matrix} \vec{v}_f \\ -\vec{v}_i \end{matrix} \quad \Delta\vec{v} \equiv \vec{v}_f - \vec{v}_i = 0$

Figure 19
We see that $\Delta\vec{v} = 0$ for motion with constant velocity.

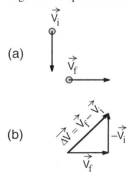

Figure 20
$\Delta\vec{v} \neq 0$ *when we change our direction of motion.*

The physicists' use of the word acceleration for the quantity $\vec{\Delta v}/\Delta t$ presents a problem for students. The difficulty is that we have grown up using the word *acceleration*, and already have some intuitive feeling for what that word means. Unfortunately this intuition usually does not match what physicists mean by acceleration. Perhaps physicists should have used a different name for $\vec{\Delta v}/\Delta t$, but this did not happen. The problem for the student is therefore not only to develop a new intuition for the quantity $\vec{\Delta v}/\Delta t$, but also to discard previous intuitive ideas of what acceleration might be. This can be uncomfortable.

The purpose of the remainder of this chapter is to develop a new intuition for the physics definition of acceleration. To do this we will consider three examples of motion; projectile motion, uniform circular motion, and projectile motion with air resistance. In each of these cases, which can be carefully studied in the introductory lab or simulated, we will use strobe photographs to determine how the acceleration vector $\vec{\Delta v}/\Delta t$ behaves. In each case we will see that there is a simple relationship between the behavior of the acceleration vector and the forces pulling or pushing on the object. This relationship between force and acceleration, which is the cornerstone of mechanics, will be discussed in a later chapter. Here our goal is to develop a clear picture of acceleration itself.

Determining Acceleration from a Strobe Photograph

We will use strobe photographs to provide an explicit experimental definition of acceleration. In the next chapter we will see how the strobe definitions go over to the calculus definition that you may have already studied. We prefer to start with the strobe definition, not only because it provides a more intuitive approach to the concept, but also because of its experimental origin. With an experimental definition we avoid some conceptual problems inherent in calculus. It turns out, surprisingly, that some of the concepts involved in the calculus definition of acceleration are inconsistent with physics. We can more clearly understand these inconsistencies when we use an experimental definition of acceleration as the foundation for our discussion.

The Acceleration Vector

The quantity $\vec{\Delta v}/\Delta t$, which we call acceleration, is usually denoted by the vector $\vec{a}$

$$\vec{a} = \frac{\vec{\Delta v}}{\Delta t} \quad (3)$$

where $\vec{\Delta v}$ is the change in the velocity vector during the time Δt. To see how to apply Equation (3) to a strobe photograph, suppose that Figure (21) represents a photograph of a particle moving with some kind of non uniform velocity. Labeling the image positions 1, 2, 3, etc. and the corresponding velocity vectors $\vec{v}_1, \vec{v}_2, \vec{v}_3 \cdots$, let us consider what the particle's acceleration was during the time it went from position 2 to position 3. At position 2 the particle's velocity was $\vec{v}_2$. When it got to position 3 its velocity was $\vec{v}_3$. The time it took for the velocity to change from $\vec{v}_2$ to $\vec{v}_3$, a change $\vec{\Delta v} = \vec{v}_3 - \vec{v}_2$, was the strobe time Δt. Thus according to Equation (3), the particle's acceleration during the interval 2 to 3, which we will call $\vec{a}_3$, is given by

$$\vec{a}_3 \equiv \frac{\vec{\Delta v}_{23}}{\Delta t} = \frac{\vec{v}_3 - \vec{v}_2}{\Delta t} \quad (4a)$$

(One could object to using the label $\vec{a}_3$ for the acceleration during the interval 2 to 3. But a closer inspection shows that $\vec{a}_3$ is an accurate name. Actually the velocity $\vec{v}_3$ is the average velocity in the interval 3 to 4, and $\vec{v}_2$ is the average velocity in the interval 2 to 3. Thus $\vec{\Delta v} = \vec{v}_3 - \vec{v}_2$ is a change in velocity centered on position 3. As a result Equation (4a) gives surprisingly accurate results when working with experimental strobe photographs. In any case such errors become vanishingly small when we use sufficiently short Δt's.)

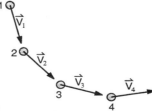

Figure 21
Determining $\vec{a}$ for non uniform motion.

If we have a strobe photograph with many images, then by extending Equation (4a), the acceleration at position i is

$$\boxed{\vec{a}_i \equiv \frac{\Delta \vec{v}_i}{\Delta t} = \frac{\vec{v}_i - \vec{v}_{i-1}}{\Delta t}} \quad \begin{array}{l}\text{strobe definition}\\ \text{of acceleration}\end{array} \quad (4)$$

We will call Equation (4) our **strobe definition of acceleration**. Implicit in this definition is that we use a short enough Δt so that all the kinks in the motion are visible, but a long enough Δt so that we have vectors long enough to work with.

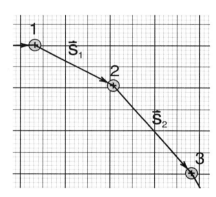

Figure 10a
A section of the projectile motion photograph, Figure (10), showing the displacement vectors $\vec{S}_1$ and $\vec{S}_2$.

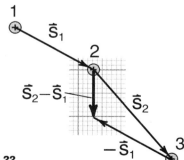

Figure 22
The vector $\vec{S}_2 - \vec{S}_1$ points straight down and has a length of about 10 cm.

PROJECTILE MOTION

As our first example in the use of our strobe definition of acceleration, let us calculate the acceleration of the ball at position 2 in our strobe photograph, Figure (10), of projectile motion.

The first problem we face is that Equation (4) expresses the acceleration vector $\vec{a}_2$ in terms of the velocity vectors $\vec{v}_1$ and $\vec{v}_2$, while the strobe photograph shows only the displacement vectors $\vec{S}_1$ and $\vec{S}_2$, as seen in Figure (10a), a segment of Figure (10) reproduced here.

The easiest way to handle this problem is to use the formulas

$$\vec{v}_1 = \frac{\vec{S}_1}{\Delta t}; \qquad \vec{v}_2 = \frac{\vec{S}_2}{\Delta t}$$

in Equation (4a) to express $\vec{a}_2$ directly in terms of the known vectors $\vec{S}_1$ and $\vec{S}_2$. The result is

$$\vec{a}_2 = \frac{\vec{v}_2 - \vec{v}_1}{\Delta t} = \frac{(\vec{S}_1/\Delta t) - (\vec{S}_2/\Delta t)}{\Delta t}$$

$$\boxed{\vec{a}_2 = \frac{\vec{S}_2 - \vec{S}_1}{\Delta t^2}} \quad \begin{array}{l}\text{experimental}\\ \text{measurement}\\ \text{of acceleration}\end{array} \quad (5)$$

Equation (5) tells us that we can calculate the acceleration vector $\vec{a}_2$ by first constructing the vector $\vec{S}_2 - \vec{S}_1$, and then dividing by Δt^2. That means that $\vec{a}_2$ **points in the direction of the vector $\vec{S}_2 - \vec{S}_1$, and has a length equal to the length $|\vec{S}_2 - \vec{S}_1|$ (in cm) divided by Δt^2**. As a result the magnitude of the acceleration vector has the dimensions of cm/sec^2.

Let us apply Equation (5) to our projectile motion photograph, Figure (10), to see how all this works. The first step is to use *vector subtraction* to construct the *vector* $\vec{S}_2 - \vec{S}_1$. This is done in Figure (22). First we draw the vectors $\vec{S}_1$ and $\vec{S}_2$, and then construct the vector $-\vec{S}_1$ as shown. (The vector $-\vec{S}_1$ is the same as $\vec{S}_1$ except that it points in the opposite direction.) Then we add the vectors $\vec{S}_2$ and $-\vec{S}_1$ to get the vector $\vec{S}_2 - \vec{S}_1$ by the usual technique of vector addition as shown

$$\vec{S}_2 - \vec{S}_1 = \vec{S}_2 + (-\vec{S}_1) \quad (6)$$

Note that even if $\vec{S}_2$ and $\vec{S}_1$ had the same length, the difference $\vec{S}_2 - \vec{S}_1$ would not necessarily be zero because this is vector subtraction, ***NOT NUMERICAL SUBTRACTION.***

Once we have constructed the vector $\vec{S}_2 - \vec{S}_1$, we know the direction of the acceleration vector $\vec{a}_2$ because it points in the ***same direction*** as $\vec{S}_2 - \vec{S}_1$. In Figure (22), we see that $\vec{S}_2 - \vec{S}_1$ points straight down, thus $\vec{a}_2$ points straight down also.

Now that we have the direction of $\vec{a}_2$, all that is left is to calculate its magnitude or length. This magnitude is given by the formula

$$|\vec{a}_2| = \frac{|\vec{S}_2 - \vec{S}_1|}{\Delta t^2} = \frac{\text{length of vector } \vec{S}_2 - \vec{S}_1}{\Delta t^2} \quad (7)$$

To get the length of the $\vec{S}_2 - \vec{S}_1$, we can use the technique shown in Figure (11). Mark off the length of the vector $\vec{S}_2 - \vec{S}_1$ on a piece of scrap paper, and then use the grid to see how many centimeters apart the marks are. In this case, where $\vec{S}_2 - \vec{S}_1$ points straight down, we immediately see that $\vec{S}_2 - \vec{S}_1$ is about 10 cm long. Thus the magnitude of $\vec{a}_2$ is given by

$$|\vec{a}_2| = \frac{|\vec{S}_2 - \vec{S}_1|}{\Delta t^2} = \frac{10 \text{ cm}}{(.1 \text{ sec})^2} = 1000 \frac{\text{cm}}{\text{sec}^2} \quad (8)$$

where we knew $\Delta t = .1$ sec for the strobe photograph in Figure (10).

Our conclusion is that, at position 2 in the projectile motion photograph, the ball had an acceleration $\vec{a}_2$ that pointed straight down, and had a magnitude of about 1000 cm/sec².

For this exercise, you may use the tear out sheet on page 3-30.

Exercise 5

(**Do this now before reading on**.) Find the acceleration vectors $\vec{a}_0, \vec{a}_1$, and $\vec{a}_3$ for the projectile motion in Figure (10). From your results, what can you say about the acceleration of a projectile?

UNIFORM CIRCULAR MOTION

To give the reader some time to think about the above exercise on projectile motion, we will change the topic for a while and analyze what is called *uniform circular motion*. In uniform circular motion, the particle travels like a speck of dust sitting on a revolving turntable.

The explicit example we would like to consider is a golf ball with a string attached, being swung in a circle over the instructor's head, as indicated in Figure (23a). We could photograph this motion, but it is very easy to simulate a strobe photograph of uniform circular motion by drawing a circle with a compass, and marking off equal intervals as shown in Figure (23b). In that figure we have also sketched in the displacement vectors as we did in our analysis of the projectile motion photograph.

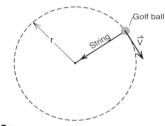

Figure 23a
Swinging a golf ball around at constant speed in a circle.

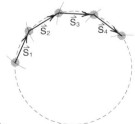

Figure 23b
Simulating a strobe photograph of a golf ball swinging at constant speed in a circle. We marked off equal distances using a compass.

Figure (23b) shows the kind of errors we have to deal with in using a strobe to study motion. Clearly the golf ball travels along the smooth circular path rather than the straight line segments marked by the vectors. As we use shorter and shorter Δt our approximation of the path gets better and better, but soon the vectors get too short for accurate graphical work. Choosing images spaced as in Figure (23b) gives vectors a reasonable length, and a reasonable approximation of the circular path.

(It will turn out that when we use our strobe definition of acceleration, most errors caused by using a finite Δt cancel, and we get a very accurate answer. Thus we do not have to worry much about how far apart we draw the images.)

Now that we have the displacement vectors we can construct the acceleration vectors $\vec{a}_1, \vec{a}_2, \cdots$ using Equation (5). The construction for $\vec{a}_2$ is shown in Figure (24). To the vector $\vec{S}_2$ we add the vector $-\vec{S}_1$ to get the vector $\vec{S}_2 - \vec{S}_1$ as shown.

The first thing we note is that the ***vector $\vec{S}_2 - \vec{S}_1$ points toward the center of the circle!*** Thus the acceleration vector $\vec{a}_2$ given by

$$\vec{a}_2 = \frac{\vec{S}_2 - \vec{S}_1}{\Delta t^2} \quad (9)$$

also points toward the center of the circle.

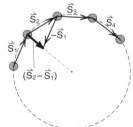

Figure 24
We find that the vector $\vec{S}_2 - \vec{S}_1$, and therefore the acceleration, points toward the center of the circle.

Exercise (6)
(Do this now.) Find the direction of at least 4 more acceleration vectors around the circle. In each case show that $\vec{a}_i$ points toward the center of the circle.

We said earlier that the physicists' definition of acceleration, which becomes $\vec{a}_i = (\vec{S}_i - \vec{S}_{i-1})/\Delta t^2$, does not necessarily agree with your own intuitive idea of acceleration. We have just discovered that, using the physicists' definition, a particle moving at constant speed along a circular path ***accelerates toward the center of the circle***. Unless you had a previous physics course, you would be unlikely to guess this result. It may seem counter intuitive. But, as we said, we are using these examples to develop an intuition for the physics definition of acceleration. Whether you like it or not, according to the physics definition, a particle moving at constant speed around a circle, is accelerating toward the center. In a little while, the reason for this will become clear.

Magnitude of the Acceleration for Circular Motion

Although perhaps not intuitive, we have gotten a fairly simple result for the direction of the acceleration vector for uniform circular motion. The center is the only unique point for a circle, and that is where the acceleration vector points. The next thing we need to know is how long the acceleration vectors are; what is the magnitude of this center pointing acceleration. From the strobe definition, the magnitude $|\vec{a}_2|$ is $\vec{a}_2 = |\vec{S}_2 - \vec{S}_1|/\Delta t^2$, a rather awkward result that appears to depend upon the size of Δt that we choose. However with a bit of geometrical construction we can re-express this result in terms of the particle's speed v and the circle's radius r. The derivation is messy, but the result is simple. This is one case, where, when we finish the derivation, we recommend that the student memorize the answer rather than try to remember the derivation. Uniform circular motion appears in a number of important physics problems, thus the formula for the magnitude of the acceleration is important to know.

In Figure (25a) we have constructed two triangles, which are shown separately in Figures (25b) and (25c). As seen in Figure (25b), the big triangle which goes from the center of the circle to positions (1) and (2) has two equal sides of length r, the radius of the circle, and one side whose length is equal to the particle's speed v times the strobe time Δt.

The second triangle, shown in Figure (25c), has sides of length $|\vec{S}_2|$ and $|-\vec{S}_1|$, but both of these are of length $v\Delta t$ as shown. The third side is of length $|\vec{S}_2 - \vec{S}_1|$, the length we need for our calculation of the magnitude of the acceleration vector.

The trick of this calculation is to note that the angles labeled θ in Figures (29b, c) are the same angle, so that these two triangles are similar isosceles triangles. The proof that these angles are equal is given in Figure (26) and its caption.

With similar triangles we can use the fact that the ratios of corresponding sides are equal. Equating the ratio of the short side to the long side of the triangle of Figure (25b), to the ratio of the short side to the long side of the triangle in Figure (25c), we get

$$\frac{v\Delta t}{r} = \frac{|\vec{S}_2 - \vec{S}_1|}{v\Delta t} \qquad (10)$$

Multiplying Equation (10) through by v and dividing both sides by Δt gives

$$\frac{v^2}{r} = \frac{|\vec{S}_2 - \vec{S}_1|}{\Delta t^2} = |\vec{a}_1| \qquad (11)$$

we got $|\vec{S}_2 - \vec{S}_1|/\Delta t^2$ on the right side, but this is just the magnitude of the acceleration vector $\vec{a}_1$. Since the same derivation applies to any position around the circle, we get the simple and general result that, for a particle moving with uniform circular motion, the particle's acceleration $\vec{a}$ points toward the center of the circle, and has a magnitude

$$\boxed{|\vec{a}| = \frac{v^2}{r}} \qquad \text{acceleration of a particle in uniform circular motion} \qquad (12)$$

where v is the speed of the particle and r is the radius of the circle. As we said, this simple result should be memorized.

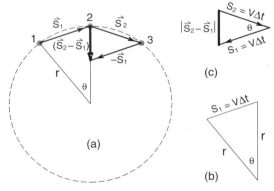

Figure 25
Derivation of the formula for the magnitude of the acceleration of a particle with uniform circular motion.

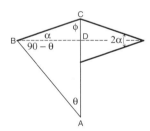

Figure 26
That the two angles labeled θ in Figure (25) are the same, may be seen in the following geometrical construction. Since the sum of the angles in any triangle is 180°, we get $\alpha + \varphi + 90° = 180°$ (from triangle BCD). Because BAC is an isosceles triangle, $(90° - \theta) + \alpha = \varphi$. Eliminating φ we get $2\alpha = \theta$, which is the result we expected.

AN INTUITIVE DISCUSSION OF ACCELERATION

We have now studied two examples of non uniform motion, the projectile motion seen in the strobe photograph of Figure (10), and the circular motion of a golf ball on the end of a string, a motion we illustrated in Figure (23). In each case we calculated the acceleration vector of the particle at different points along the trajectory. Let us now review our results to see if we can gain some understanding of why the acceleration vector behaves the way it does.

If you worked Exercise (5) correctly, you discovered that all the acceleration vectors are the same, at least to within experimental accuracy. As shown in Figure (27), as the steel ball moves along its trajectory, its acceleration vector points downward toward the earth, and has a constant magnitude of about 1000 cm/sec².

As shown in Figure (28), the golf ball being swung at constant speed around in a circle on the end of a string, accelerates toward the center of the circle, in the direction of the string pulling on the ball. The magnitude of the acceleration has the constant value.

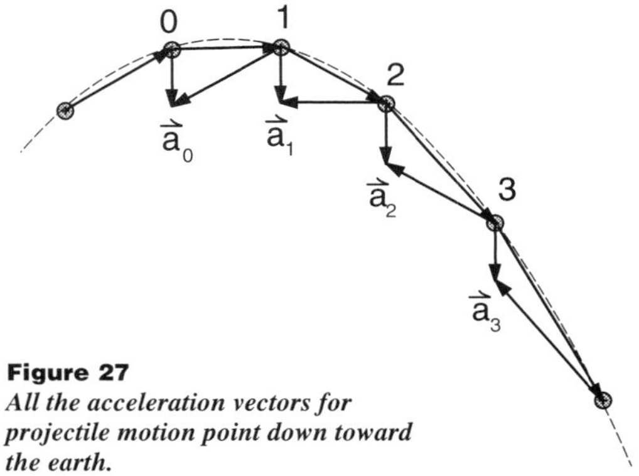

Figure 27
All the acceleration vectors for projectile motion point down toward the earth.

We said that the string was pulling on the ball. To see that this is true, try swinging a ball on the end of a string (or a shoe on the end of a shoelace) in a circle. To keep the ball (or shoe) moving in a circle, you have to pull in on the string. In turn, the string pulls in on the ball (or shoe). If you no longer pull in on the string, i.e., let go, the ball or shoe flies away and no longer undergoes circular motion. The string pulling on the ball is necessary in order to have circular motion.

What is the common feature of projectile and circular motion? In both cases the object accelerates in the direction of the force acting on the object. When you throw a steel ball in the air, the ball does not escape earth's gravity. As the ball moves through the air, gravity is constantly pulling down on the ball. The result of this gravitational pull or force is to accelerate the ball in the direction of the gravitational force. That is why the projectile motion acceleration vectors point down toward the earth.

When we throw a ball a few feet up in the air, it does not get very far away from the surface of the earth. In other words we expect the gravitational pull to be equally strong throughout the trajectory. If the ball's acceleration is related to the gravitational pull, then we expect the acceleration to also be constant throughout the trajectory. Thus it is not surprising that all the vectors have the same length in Figure (27).

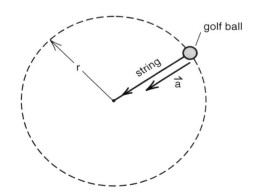

Figure 28
The golf ball accelerates in the direction of the string that is pulling on it.

In the case of circular motion, the string has to pull in on the golf ball to keep the ball moving in a circle. As a result of this pull of the string toward the center of the circle, the ball accelerates toward the center of the circle. Again the acceleration is in the direction of the force on the object.

This relationship between force and acceleration, which we are just beginning to see in these two examples, forms the cornerstone of what is called *classical* or *Newtonian mechanics*. We have more details to work out, but we have just glimpsed the basic idea of much of the first half of this course. To give historical credit for these ideas, it was Galileo who first saw the importance of the concept of acceleration that we have been discussing, and Isaac Newton who pinned down the relationship between force and acceleration.

Acceleration Due to Gravity

Two more topics, both related to projectile motion, will finish our discussion in this section. The first is the fact that, if we can neglect air resistance, all projectiles near the surface of the earth have the same downward acceleration $\vec{a}$. If a steel ball and a feather are dropped in a vacuum, they fall together with the same acceleration. This acceleration, which is caused by gravity, is called the *acceleration due to gravity* and is denoted by the symbol $\vec{g}$. The vector $\vec{g}$ points down toward the earth, and, at the surface of the earth, has a magnitude.

$$|\vec{g}| \equiv g = 980 \text{ cm/sec}^2 \quad \begin{array}{l}\textit{acceleration due to}\\ \textit{gravity at the sur-}\\ \textit{face of the earth}\end{array} \quad (13)$$

This is quite consistent with our experimental result of about 1000 cm/sec^2 that we got from the analysis of the strobe photograph in Figure (10).

If we go up away from the earth, the acceleration due to gravity decreases. At an altitude of 1,600 miles, the acceleration is down to half its value, about 500 cm/sec^2. On other planets g has different values. For example, on the moon, g is only about 1/6 as strong as it is here on the surface of the earth, i.e.

$$g_{moon} = 167 \text{ cm/sec}^2 \quad (14)$$

From the relationship we have seen between force and acceleration we can understand why a projectile that goes only a few feet above the surface of the earth should have a constant acceleration. The gravitational force does not change much in those few feet, and therefore we would not expect the acceleration caused by gravity to change much either.

On the other hand there is no obvious reason, at this point, why in the absence of air, a steel ball and a feather should have the same acceleration. Galileo believed that all projectiles, in the absence of air resistance, have the same acceleration. But it was not until Newton discovered both the laws of mechanics (the relationship between acceleration and force) and the law of gravity, that it became a physical prediction that all projectiles have the same gravitational acceleration.

In the early part of the 20th century, Einstein went a step farther than Newton, and used the fact that all objects have the same gravitational acceleration to develop a geometrical interpretation of the theory of gravity. The gravitational force was reinterpreted as a curvature of space, with the natural consequence that a curvature of space affects all objects in the same way. This theory of gravity, known as *Einstein's general theory of relativity*, was a result of Einstein's effort to make the theory of gravity consistent with the principle of relativity.

It is interesting how the simplest ideas, the principle of relativity, and the observation that the gravitational acceleration is the same for all objects, are the cornerstones of one of the most sophisticated theories in physics, in this case Einstein's general theory of relativity. Even today, over three quarters of a century since Einstein developed the theory, we still do not understand what many of the predictions or consequences of Einstein's theory will be. It is exciting, for these predictions may help us understand the behavior of the universe from its very beginning.

Exercise (7)

The first earth satellite, Sputnik 1, traveled in a low, nearly perfect, circular orbit around the earth as illustrated in Figure (29).

(a) What was the direction of Sputnik 1's acceleration vector as it went around the earth?

(b) What was the direction of the force of gravity on Sputnik 1 as the satellite went around the earth?

(c) How is this problem related to the problem of the motion of the golf ball on the end of a string? Give an answer that your roommate, who has not had a physics course, would understand.

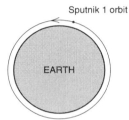

Figure 29
Sputnik 1's circular orbit.

Projectile Motion with Air Resistance

Back to a more mundane subject, we wish to end this discussion of acceleration with the example of projectile motion with air resistance. Most introductory physics texts avoid this topic because they cannot deal with it effectively. Using calculus, one can handle only the simplest, most idealized examples, and even then the analysis is beyond the scope of most texts. But using strobe photographs it is easy to analyze projectile motion with air resistance, and we learn quite a bit from the results.

What turned out to be difficult, was to find an example where air resistance affected the motion of a projectile enough to produce a noticeable effect. We found that a golf ball and a ping pong ball have almost the same acceleration when thrown in the air, despite the considerable difference in weight or mass. Only when we used the rough surfaced Styrofoam balls used for Christmas tree ornaments did we finally get enough air resistance to give a significant effect.

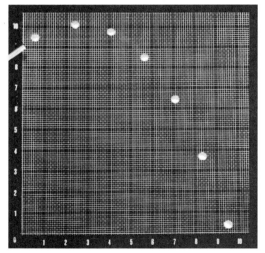

Figure 30a
Motion of a Styrofoam ball. This is the lightest ball we could find.

A strobe photograph of the projectile motion of the Styrofoam ball is seen in Figure (30a), and an analysis showing the resulting acceleration vectors in Figure (30b). In Figure (30b) we have also drawn the acceleration vectors $\vec{g}$ that the ball would have had if there had not been any air resistance. We see that the effect of air resistance is to bend back and shorten the acceleration vectors.

Figure (31) is a detailed analysis of the Styrofoam's acceleration at point (3). (We used an enlargement of the strobe photograph to improve the accuracy of our work, such detailed analysis is difficult using small Polaroid photographs.) In Figure (31) $\vec{v}_3$ is the velocity of the ball, $\vec{g}$ is the acceleration due to gravity, and $\vec{a}_3$ the ball's actual acceleration. The vector $\vec{a}_{air}$, which represents the change in $\vec{a}$ caused by air resistance is given by the vector equation

$$\vec{a}_3 = \vec{g} + \vec{a}_{air} \qquad (15)$$

The important feature of Figure (31) is that $\vec{a}_{air}$ is oppositely directed to the ball's velocity $\vec{v}_3$. To understand why, imagine that you are the stick figure riding on the ball in Figure (31). You will feel a wind in your face, a wind directed oppositely to $\vec{v}_3$. This wind will push on the ball in the direction opposite to $\vec{v}_3$, i.e., in the direction of $\vec{a}_{air}$. Thus we conclude that the acceleration $\vec{a}_{air}$ is created by the force of the wind on the ball.

What we learn from this example is that *if we have two forces simultaneously acting on an object, each force independently produces an acceleration, and the net acceleration is the vector sum of the independent accelerations*. In this case the independent accelerations are caused by gravity and the wind. The net acceleration $\vec{a}_3$ of the ball is given by the vector Equation (15), $\vec{a}_3 = \vec{g} + \vec{a}_{air}$. As we will see in later chapters, this vector addition of accelerations plays a fundamental role in mechanics.

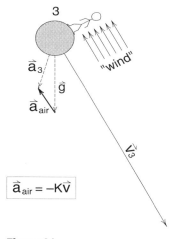

Figure 31
When we do a detailed comparison of $\vec{a}$ and $\vec{g}$ at point 3, we see that the air resistance produces an acceleration $\vec{a}_{air}$ that points in the direction of the wind felt by the ball.

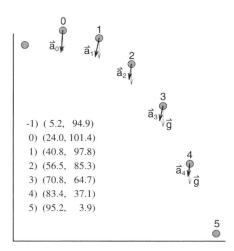

-1) (5.2, 94.9)
0) (24.0, 101.4)
1) (40.8, 97.8)
2) (56.5, 85.3)
3) (70.8, 64.7)
4) (83.4, 37.1)
5) (95.2, 3.9)

Figure 30b
Acceleration of the Styrofoam ball.

INSTANTANEOUS VELOCITY

In calculus, instantaneous velocity is defined by starting with the equation $\vec{v}_i = (\vec{R}_{i+1} - \vec{R}_i)/\Delta t$ and then taking the limiting value of $\vec{v}_i$ as we use shorter and shorter time steps Δt. This corresponds in a strobe photograph to using a higher and higher flashing rate which would give increasingly short displacement vectors $\vec{S}_i$. In the end result one pictures the instantaneous velocity being defined at each point along the continuous trajectory of the object.

The effect of using shorter and shorter Δt is illustrated in Figure (32). In each of these sketches the dotted line represents the smooth continuous trajectory of the ball. In Figure (32a) where $\Delta t = 0.4$ sec and there are only two images the only possible definition of $\vec{v}_0$ is the displacement between these images, divided by Δt as shown. Clearly Δt is too large here for an accurate representation of the ball's motion.

A better description of motion is obtained in Figure (32b) where $\Delta t = 0.1$ sec as in the original photograph. We used this value of in our analysis of the projectile motion, Figure (10). Reducing Δt by another factor 1/4 gives the results shown in Figure (32c). At this point the images provide a detailed picture of the path and $\vec{v}_0 = \vec{S}_0/\Delta t$ is now tangent to the path at (0). A further decrease in Δt would produce a negligible change in $\vec{v}_0$.

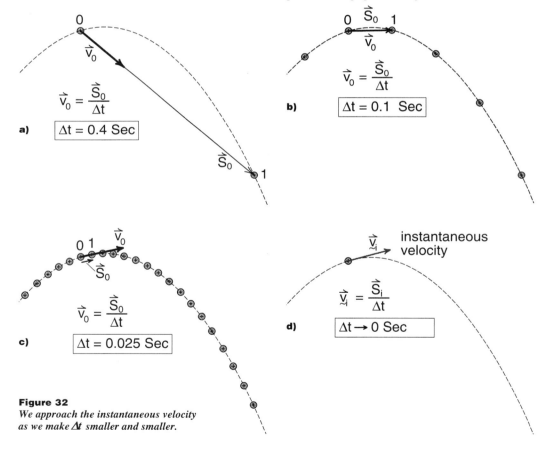

Figure 32
We approach the instantaneous velocity as we make Δt smaller and smaller.

The instantaneous velocity at point (0) is the final value of $\vec{v}_0$, the value illustrated in Figure (32d) which no longer changes as Δt is reduced. This is an abstract concept in that we are assuming such a final value exists. We are assuming that we always reach a point where using a stroboscope with a still higher flashing rate produces no observable change in the value of $\vec{v}_0$. This assumption, which has worked quite well in the analysis of large objects such as ping pong balls and planets, has proven to be false when investigated on an atomic scale. According to the quantum theory which replaces classical mechanics on an atomic scale when one uses a sufficiently short Δt in an attempt to measure velocity, the measurement destroys the experiment rather than giving a better value of $\vec{v}_0$.

Instantaneous Velocity from a Strobe Photograph

In the case of projectile motion (i.e., motion with constant acceleration) there is a simple yet precise method for determining an object's instantaneous velocity $\vec{\underline{v}}_i$ from a strobe photograph. (*Vectors representing instantaneous velocity will be underlined in order to distinguish them from the vectors representing the strobe definition of velocity.*) This method, which also gives quite good approximate values for other kinds of motion, will be used in our computer calculations for determining the initial velocity of the object.

To see what the method is, consider Figure (33) where we have drawn the vector obtained from Figure (32d). We have also drawn a line from the center of image (–1) to the center of image (+1) and notice that $\vec{\underline{v}}_i \Delta t$ is parallel to and precisely half as long as this line. Thus we can construct $\vec{\underline{v}}_i \Delta t$ by connecting the preceding and following images and taking half of that line.

The vector constructed by the above rule is actually the average of the preceding velocity vector $\vec{v}_{-1}$ and the following vector $\vec{v}_0$.

$$\vec{\underline{v}}_i = \frac{\vec{v}_{-1} + \vec{v}_0}{2} \qquad (16)$$

as illustrated in Figure (33). (Note that the vector sum $(\vec{v}_{-1} + \vec{v}_0)\Delta t$ is the same as the line $2\vec{\underline{v}}_i \Delta t$ which connects the preceding and following image.) This is a reasonable estimate of the ball's instantaneous velocity because $\vec{v}_{-1}$ is the average velocity during the time Δt before the ball got to (0), and $\vec{v}_0$ the average velocity during the interval after leaving (0). The ball's velocity at (0) should have a value intermediate between $\vec{v}_{-1}$ and $\vec{v}_0$, which is what Equation (16) says.

The constant acceleration formula

$$\vec{S} = \vec{\underline{v}}_i t + \frac{1}{2}\vec{a}t^2 \qquad (17)$$

which may be familiar from a high school physics course, provides a direct application of the concept of instantaneous velocity. (Remember that this is not a general formula; it applies only to motion with constant acceleration where the vector $\vec{a}$ changes neither in magnitude or direction.) As illustrated in Figure (34) the total displacement of the projectile

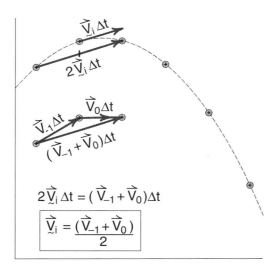

Figure 33

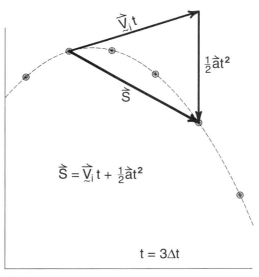

Figure 34
Illustration of the constant acceleration formula as a vector equation.

during a time t (here t = 3Δt) is the vector sum of $\vec{v}_i t$ and $1/2\vec{a}t^2$. To draw this figure, we used

$$\vec{v}_i t = \vec{v}_i(3\Delta t) = 3(\vec{v}_i \Delta t)$$

and obtained $\vec{v}_i \Delta t$ from our method of determining instantaneous velocity. We also used

$$\tfrac{1}{2}\vec{a}t^2 = \tfrac{1}{2}\vec{a}(3\Delta t)^2 = \tfrac{9}{2}\vec{a}\Delta t^2$$

where we obtained $\vec{a}\Delta t^2$ from the relation

$$\vec{a} = \frac{\vec{S}_2 - \vec{S}_1}{\Delta t^2}$$

$$\vec{a}\Delta t^2 = (\vec{S}_2 - \vec{S}_1)$$

as illustrated in Figure (35).

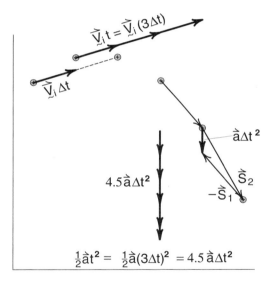

Figure 35
How to construct the vectors $\vec{v}_i t$ and $1/2\,\vec{a}t^2$ from a strobe photograph.

For these exercises, use the tear out sheet on pages 3-31,32.

Exercise 8

Use Equation 17 to predict the displacements of the ball

(a) Starting at position (0) for a total time t = 4Δt.

(b) Starting at position (1) for a total time t = 3Δt.

Do the work graphically as we did in Figures 33-35.

Exercise 9

The other constant acceleration formula is

$$\vec{v}_f = \vec{v}_i + \vec{a}t$$

where $\vec{v}_i$ is the initial velocity, and $\vec{v}_f$ the object's velocity a time t later. Apply this equation to Figure 10 to predict the ball's instantaneous velocity $\vec{v}_f$ at point (3) for a ball starting at point (0). Check your prediction by graphically determining the instantaneous velocity at point (3).

Show your results on graph paper.

Exercise 10

Show that the constant acceleration formulas would correctly predict projectile motion even if time ran backward. (For example, assume that the ball went backward as shown in Figure (36), and repeat Exercise 8b, going from position 3 to position 0.)

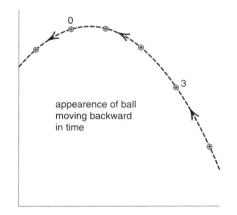

Figure 36
Run the motion of the ball backward in time, and it looks like it was launched from the lower right.

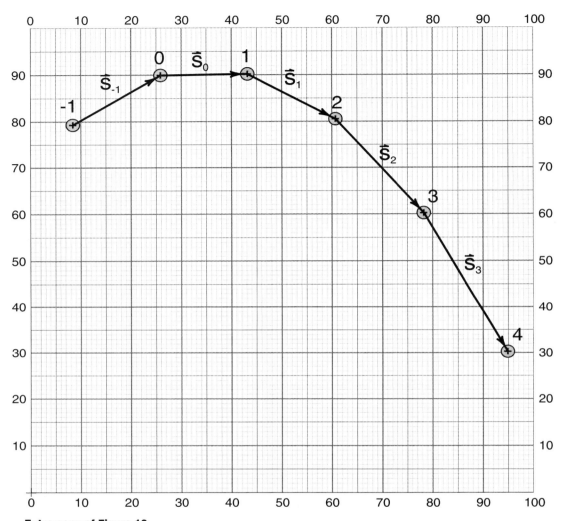

Extra copy of Figure 10
Use this graph for practice with vectors.

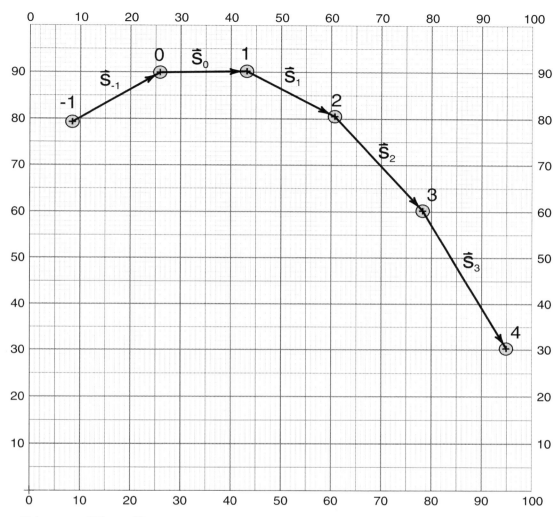

Extra copy of Figure 10
Use this graph for homework that you pass in. This page may be torn out.

3-30   Description of Motion   Tear out page

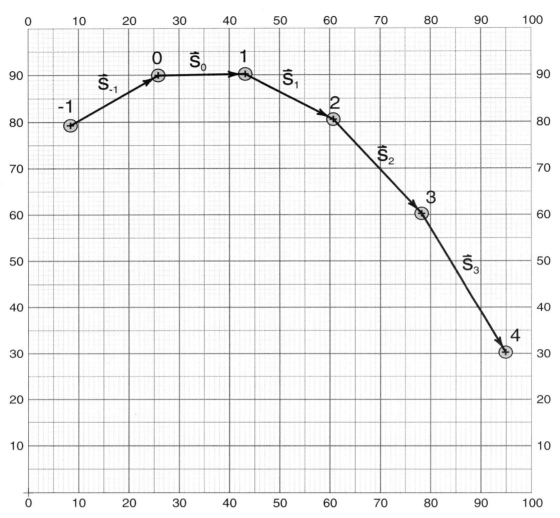

Extra copy of Figure 10
Use this graph for homework that you pass in. This page may be torn out.

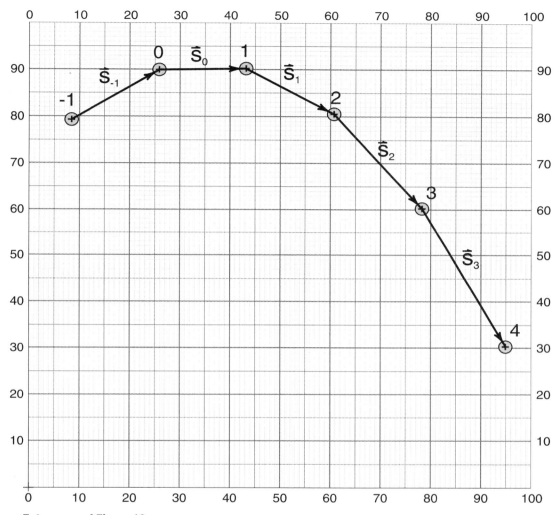

Extra copy of Figure 10
Use this graph for homework that you pass in. This page may be torn out.

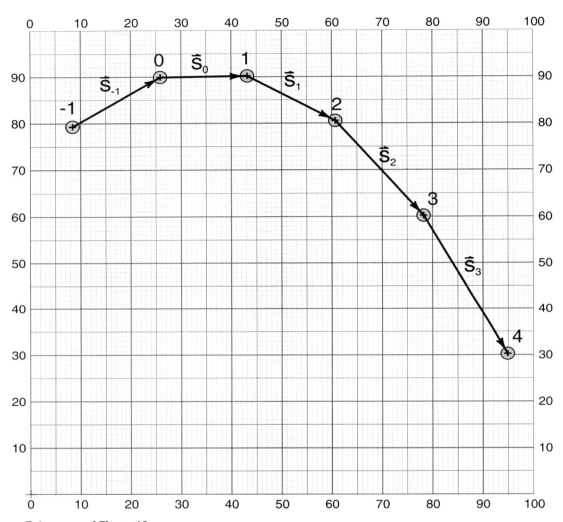

Extra copy of Figure 10
Use this graph for homework that you pass in. This page may be torn out.

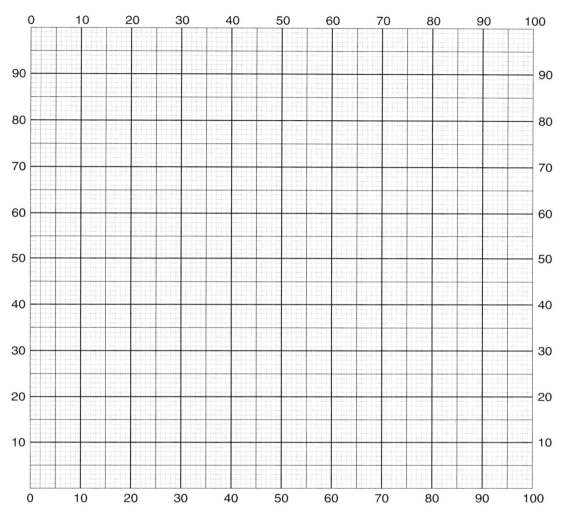

Spare graph paper
Use this graph paper if you want practice with vectors. This is not a tear out page.

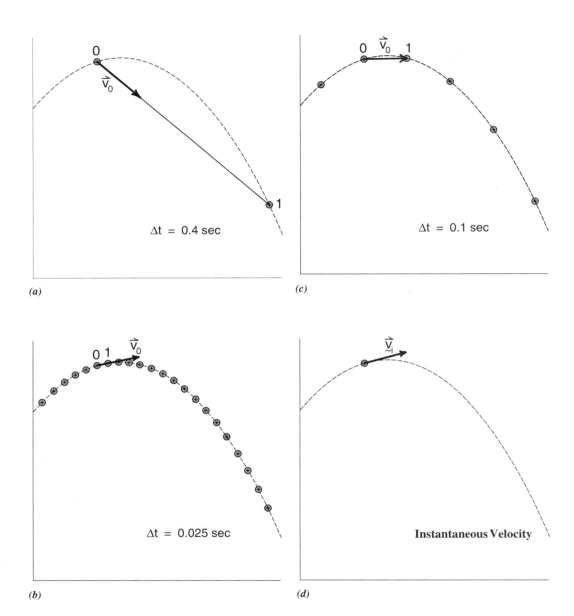

Figure 1
Transition to instantaneous velocity.

Chapter 4
Calculus in Physics

This chapter, which discusses the use of calculus in physics, is for those who have had a calculus course which they remember fairly well. For those whose calculus is weak or poorly remembered, or for those who have not studied calculus, you should replace this chapter with Chapter 1 of *Calculus 2000*. That chapter can be found at the end of Physics 2000 Part 2.

In the previous chapter we used strobe photographs to define velocity and acceleration vectors. The basic approach was to turn up the strobe flashing rate as we did in going from Figure (3-3) to (3-4) until all the kinks are clearly visible and the successive displacement vectors give a reasonable description of the motion. We did not turn the flashing rate too high, for the practical reason that the displacement vectors became too short for accurate work. Calculus corresponds to conceptually turning the strobe all the way up.

LIMITING PROCESS

In our discussion of instantaneous velocity we conceptually turned the strobe all the way up as illustrated in Figures (2-32a) through (2-32d), redrawn here in Figure (1). In these figures, we initially see a fairly large change in $\vec{v}_0$ as the strobe rate is increased and Δt reduced. But the change becomes smaller and it looks as if we are approaching some final value of $\vec{v}_0$ that does not depend on the size of Δt, provided Δt is small enough. It looks as if we have come close to the final value in Figure (1c).

The progression seen in Figure (1) is called a ***limiting process***. The idea is that there really is some true value of $\vec{v}_0$ which we have called the instantaneous velocity, and that we approach this true value for sufficiently small values of Δt. This is a calculus concept, and in the language of calculus, we are ***taking the limit as Δt goes to zero***.

The Uncertainty Principle

For over 200 years, from the invention of calculus by Newton and Leibnitz until 1924, the limiting process and the resulting concept of instantaneous velocity was one of the cornerstones of physics. Then in 1924 Werner Heisenberg discovered what he called the ***uncertainty principle*** which places a limit on the accuracy of experimental measurements.

Heisenberg discovered something very new and unexpected. He found that the act of making an experimental measurement unavoidably affects the results of an experiment. This had not been known previously because the effect on large objects like golf balls is undetectable. But on an atomic scale where we study small systems like electrons moving inside an atom, the effect is not only observable, it can dominate our study of the system.

One particular consequence of the uncertainly principle is that the more accurately we measure the position of an object, the more we disturb the motion of the object. This has an immediate impact on the concept of instantaneous velocity. If we turn the strobe all the way up, reduce Δt to zero, we are in effect trying to measure the position of the object with infinite precision. The consequence would be an infinitely big disturbance of the motion of the object we are studying. If we actually could turn the strobe all the way up, we would destroy the object we were trying to study.

It turns out that the uncertainty principle can have a significant impact on a larger scale of distance than the atomic scale. Suppose, for example, that we constructed a chamber 1 cm on a side, and wished to study the projectile motion of an electron inside. Using Galileo's idea that objects of different mass fall at the same rate, we would expect that the motion of the electron projectile should be the same as more massive objects. If we took a strobe photograph of the electron's motion, we would expect get results like those shown in Figure (2). This figure represents projectile motion with an acceleration $g = 980 \, cm/sec^2$ and $\Delta t = .01 sec$, as the reader can easily check.

When we study the uncertainty principle in Chapter 30, we will see that a measurement that is accurate enough to show that Position (2) is below Position (1), could disturb the electron enough to reverse its direction of motion. The next position measurement could find the electron over where we drew Position (3), or back where we drew Position (0), or anywhere in the region in between. As a result we could not even determine what direction the electron is moving. This uncertainty would not be the result of a sloppy experiment, it is the best we can do with the most accurate and delicate measurements possible.

The uncertainty principle has had a significant impact on the way physicists think about motion. Because we now know that the measuring process affects the results of the measurement, we see that it is essential to provide experimental definitions to any physical quantity we wish to study. A conceptual definition, like turning the strobe all the way up to define instantaneous velocity, can lead to fundamental inconsistencies.

Even an experimental definition like our strobe definition of velocity can lead to inconsistent results when applied to something like the electron in Figure (2). But these inconsistencies are real. Their existence is telling us that the very concept of velocity is beginning to lose meaning for these small objects.

On the other hand the idea of the limiting process and instantaneous velocity is very convenient when applied to larger objects where the effects of the uncertainty principle are not detectable. In this case we can apply all the mathematical tools of calculus developed over the past 250 years. The status of instantaneous velocity has changed from a basic concept to a useful mathematical tool. Those problems for which this mathematical tool works are called problems in *classical physics*; and those problems for which the uncertainty principle is important, are in the realm of what we call *quantum physics*.

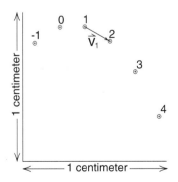

Figure 2
Hypothetical electron projectile motion experiment.

CALCULUS DEFINITION OF VELOCITY

With the above perspective on the physical limitations on the limiting process, we can now return to the main topic of this chapter—the use of calculus in defining and working with velocity and acceleration.

In discussing the limiting process in calculus, one traditionally uses a special set of symbols which we can understand if we adopt the notation shown in Figure (3). In that figure we have drawn the coordinate vectors $\vec{R}_i$ and $\vec{R}_{i+1}$ for the i th and (i + 1) th positions of the object. We are now using the symbol $\overrightarrow{\Delta R}_i$ to represent the displacement of the ball during the i to i+1 interval. The vector equation for $\overrightarrow{\Delta R}_i$ is

$$\overrightarrow{\Delta R}_i = \vec{R}_{i+1} - \vec{R}_i \qquad (1)$$

In words, Equation (1) tells us that $\overrightarrow{\Delta R}_i$ is the change, during the time Δt, of the position vector $\vec{R}$ describing the location of the ball.

The velocity vector $\vec{v}_i$ is now given by

$$\vec{v}_i = \frac{\overrightarrow{\Delta R}_i}{\Delta t} \qquad (2)$$

This is just our old strobe definition $\vec{v}_i = \vec{S}_i/\Delta t$, but using a notation which emphasizes that the displacement $\vec{S}_i = \overrightarrow{\Delta R}_i$ is the *change in position* that occurs during the time Δt. The Greek letter Δ (delta) is used both to represent the idea that the quantity $\overrightarrow{\Delta R}_i$ or Δt is small, and to emphasize that both of these quantities change as we change the strobe rate.

The limiting process in Figure (1) can be written in the form

$$\underline{\vec{v}}_i \equiv \underset{\Delta t \to 0}{\text{Limit}} \frac{\overrightarrow{\Delta R}_i}{\Delta t} \qquad (3)$$

where the word "Limit" with $\Delta t \to 0$ underneath, is to be read as "limit as Δt goes to zero". For example we would read Equation (3) as "*the instantaneous velocity $\underline{\vec{v}}_i$ at position i is the limit, as Δt goes to zero, of the ratio $\overrightarrow{\Delta R}_i/\Delta t$.*"

For two reasons, Equation (3) is not quite yet in standard calculus notation. One is that in calculus, only the limiting value, in this case, the instantaneous velocity, is considered to be important. Our strobe definition $\vec{v}_i = \overrightarrow{\Delta R}_i/\Delta t$ is only a step in the limiting process. Therefore when we see the vector $\vec{v}_i$, we should assume that it is the limiting value, and no special symbol like the underline is used. For this reason we will drop the underline and write

$$\vec{v}_i = \underset{\Delta t \to 0}{\text{Limit}} \frac{\overrightarrow{\Delta R}_i}{\Delta t} \qquad (3a)$$

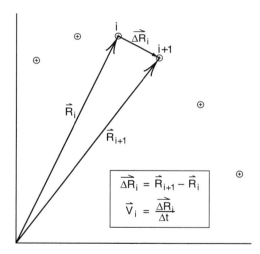

Figure 3
Definitions of $\Delta \vec{R}_i$ and $\vec{V}_i$.

The second change deals with the fact that when Δt goes to zero we need an infinite number to time steps to get through our strobe photograph, and thus it is not possible to locate a position by counting time steps. Instead we measure the time t that has elapsed since the beginning of the photograph, and use that time to tell us where we are, as illustrated in Figure (4). Thus instead of using $\vec{v}_i$ to represent the velocity at position i, we write $\vec{v}(t)$ to represent the velocity at time t. Equation (3) now becomes

$$\vec{v}(t) = \underset{\Delta t \to 0}{\text{Limit}} \frac{\overrightarrow{\Delta R}(t)}{\Delta t} \qquad (3b)$$

where we also replaced $\overrightarrow{\Delta R}_i$ by its value $\overrightarrow{\Delta R}(t)$ at time t.

Although Equation (3b) is in more or less standard calculus notation, the notation is clumsy. It is a pain to keep writing the word Limit with a $\Delta t \to 0$ underneath. To streamline the notation, we replace the Greek letter Δ with the English letter d as follows

$$\boxed{\underset{\Delta t \to 0}{\text{Limit}} \frac{\overrightarrow{\Delta R}(t)}{\Delta t} \equiv \frac{d\vec{R}(t)}{dt}} \qquad (4)$$

(The symbol $\equiv$ means *defined equal to*.) To a mathematician, the symbol $d\vec{R}(t)/dt$ is just shorthand notation for the limiting process we have been describing. But to a physicist, there is a different, more practical meaning. Think of dt as a short Δt, short enough so that the limiting process has essentially occurred, but not too short to see what is going on. In Figure (1), a value of dt less than .025 seconds is probably good enough.

If dt is small but finite, then we know exactly what the $d\vec{R}(t)$ is. It is the small but finite displacement vector at the time t. It is our old strobe definition of velocity, with the added condition that dt is such a short time interval that the limiting process has occurred. From this point of view, which we will use throughout this text, dt is a real time interval, and $d\vec{R}(t)$ a real vector which we can work with in a normal way. The only thing special about these quantities is that when we see the letter d instead of Δ, we must remember that a limiting process is involved. In this notation, the calculus definition of velocity is

$$\boxed{\vec{v}(t) = \frac{d\vec{R}(t)}{dt}} \qquad (5)$$

where $\vec{R}(t)$ and $\vec{v}(t)$ are the particle's coordinate vector and velocity vector respectively as shown in Figure (5). Remember that this is just fancy shorthand notation for the limiting process we have been describing.

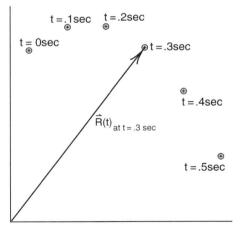

Figure 4
Rather than counting individual images, we can locate a position by measuring the elapsed time t. In this figure, we have drawn the displacement vector $\vec{R}(t)$ at time t = .3 sec.

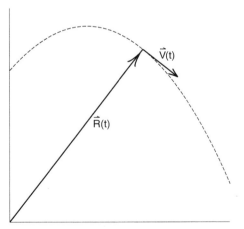

Figure 5
Instantaneous position and velocity at time t.

ACCELERATION

In the analysis of strobe photographs, we defined both a velocity vector $\vec{v}$ and an acceleration vector $\vec{a}$. The definition of $\vec{a}$, shown in Figure (2-12) reproduced here in Figure (6) was

$$\vec{a}_i \equiv \frac{\vec{v}_{i+1} - \vec{v}_i}{\Delta t} \qquad (6)$$

In our graphical work we replaced $\vec{v}_i$ by $\vec{S}_i/\Delta t$ so that we could work directly with the displacement vectors $\vec{S}_i$ and experimentally determine the behavior of the acceleration vector for several kinds of motion.

Let us now change this graphical definition of acceleration over to a calculus definition, using the ideas just applied to the velocity vector. First, assume that the ball reached position i at time t as shown in Figure (6). Then we can write

$$\vec{v}_i = \vec{v}(t)$$

$$\vec{v}_{i+1} = \vec{v}(t+\Delta t)$$

to change the time dependence from a count of strobe flashes to the continuous variable t. Next, define the vector $\overline{\Delta \vec{v}}(t)$ by

$$\overline{\Delta \vec{v}}(t) \equiv \vec{v}(t+\Delta t) - \vec{v}(t) \quad \left(= \vec{v}_{i+1} - \vec{v}_i \right) \qquad (7)$$

We see that $\overline{\Delta \vec{v}}(t)$ is the change in the velocity vector as the time advances from t to t+Δt.

The strobe definition of $\vec{a}_i$ can now be written

$$\vec{a}(t) \begin{pmatrix} strobe \\ definition \end{pmatrix} = \frac{\vec{v}(t+\Delta t) - \vec{v}(t)}{\Delta t} \equiv \frac{\overline{\Delta \vec{v}}(t)}{\Delta t} \qquad (8)$$

Now go through the limiting process, turning the strobe up, reducing Δt until the value of $\vec{a}(t)$ settles down to its limiting value. We have

$$\vec{a}(t) \begin{pmatrix} calculus \\ definition \end{pmatrix} = \underset{\Delta t \to 0}{\text{Limit}} \left(\frac{\vec{v}(t+\Delta t) - \vec{v}(t)}{\Delta t} \right)$$

$$= \underset{\Delta t \to 0}{\text{Limit}} \left(\frac{\overline{\Delta \vec{v}}(t)}{\Delta t} \right) \qquad (9)$$

Finally use the shorthand notation d/dt for the limiting process:

$$\boxed{\vec{a}(t) = \frac{d\vec{v}(t)}{dt}} \qquad (10)$$

Equation (10) does not make sense unless you remember that it is notation for all the ideas expressed above. Again, physicists think of dt as a short but finite time interval, and $d\vec{v}(t)$ as the small but finite change in the velocity vector during the time interval dt. It's our strobe definition of acceleration with the added requirement that Δt is short enough that the limiting process has already occurred.

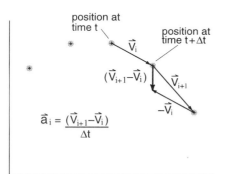

Figure 6
Experimental definition of the acceleration vector.

Components

Even if you have studied calculus, you may not recall encountering formulas for the derivatives of vectors, like $d\vec{R}(t)/dt$ and $d\vec{v}(t)/dt$ which appear in Equations (5) and (10). To bring these equations into a more familiar form where you can apply standard calculus formulas, we will break the vector Equations (5) and (10) down into component equations.

In the chapter on vectors, we saw that any vector equation like

$$\vec{A} = \vec{B} + \vec{C} \tag{11}$$

is equivalent to the three component equations

$$\begin{align} A_x &= B_x + C_x \\ A_y &= B_y + C_y \\ A_z &= B_z + C_z \end{align} \tag{12}$$

The advantage of the component equations was that they are simply numerical equations and no graphical work or trigonometry is required.

The limiting process in calculus does not affect the decomposition of a vector into components, thus Equation (5) for $\vec{v}(t)$ and Equation (10) for $\vec{a}(t)$ become

$$\vec{v}(t) = d\vec{R}(t)/dt \tag{5}$$
$$v_x(t) = dR_x(t)/dt \tag{5a}$$
$$v_y(t) = dR_y(t)/dt \tag{5b}$$
$$v_z(t) = dR_z(t)/dt \tag{5c}$$

and

$$\vec{a}(t) = d\vec{v}(t)/dt \tag{10}$$
$$a_x(t) = dv_x(t)/dt \tag{10a}$$
$$a_y(t) = dv_y(t)/dt \tag{10b}$$
$$a_z(t) = dv_z(t)/dt \tag{10c}$$

Often we use the letter x for the x coordinate of the vector $\vec{R}$ and we use y for R_y and z for R_z. With this notation, Equation (5) assumes the shorter and perhaps more familiar form

$$v_x(t) = dx(t)/dt \tag{5a'}$$
$$v_y(t) = dy(t)/dt \tag{5b'}$$
$$v_z(t) = dz(t)/dt \tag{5c'}$$

At this point the notation has become deceptively short. You now have to remember that x(t) stands for the x coordinate of the particle at a time t.

We have finally boiled the notation down to the point where it would be familiar from any calculus course. If we restrict our attention to one dimensional motion along the x axis. Then all we have to concern ourselves with are the x component equations

$$\boxed{\begin{aligned} v_x(t) &= \frac{dx(t)}{dt} \\ a_x(t) &= \frac{dv_x(t)}{dt} \end{aligned}} \tag{10a}$$

Distance, Velocity and Acceleration versus Time Graphs

One of the ways to build an intuition for Equations (5a) and (10a) is through the use of graphs of position, velocity and acceleration versus time. Suppose, for example, we had a particle moving at constant speed in the x direction, the uniform motion that the principle of relativity tells us that we cannot detect. Graphs of distance x(t), velocity v(t) and acceleration a(t) for this motion are shown in Figure (7).

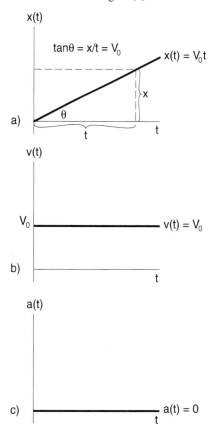

Figure 7
Motion with constant velocity.

If the particle is moving at constant speed
$$v_x(t) = v_0 \tag{11}$$
then the graph of velocity versus time is a straight horizontal line of height v_0 as shown in Figure (7b).

If you travel away from home at constant speed, then your distance from home is proportional to the time you have traveled. If you start at $t = 0$, then at time t your distance from home is
$$x(t) = v_0 t \tag{12}$$
This is graphed as the straight line as shown in Figure (7a). The slope of this line, the tangent of the angle θ is x/t, which from Equation (12) is v_0.

When a particle moves at constant velocity, there is no change in the succeeding velocity vectors, thus the acceleration a(t) is zero for all time
$$a(t) = 0 \tag{13}$$
as shown in Figure (7c).

In summary, we have seen that for this example of uniform motion in the x direction
$$x(t) = v_0 t \tag{12}$$
$$v(t) = v_0 \tag{11}$$
$$a(t) = 0 \tag{13}$$
Now let us see if these results agree with our calculus definitions (5a) and (10a). From Equation (5a) we get
$$v(t) = \frac{dx(t)}{dt} = \frac{d}{dt}(v_0 t) \tag{14}$$
The v_0 being constant comes outside and we have
$$v(t) = v_0 \frac{dt}{dt} = v_0 \tag{15}$$
where we used dt/dt = 1. Our calculus result agrees with Equation (12).

From Equation (10), we get
$$a(t) = \frac{dv(t)}{dt} = \frac{d}{dt}(v_0) = 0 \tag{16}$$
because the derivative of a constant is zero.

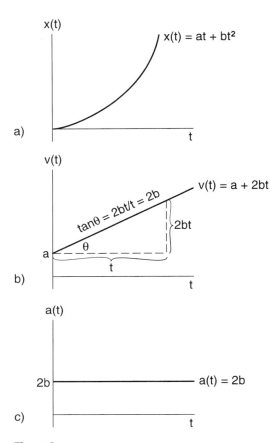

Figure 8
Motion with constant acceleration.

What we should begin to see from this example, is that if we have the formula for x(t) then it is easy to use calculus to figure out the particle's velocity and acceleration. Let us consider one more example. Suppose x(t) is given by the formula

$$x(t) = at + bt^2 \tag{17}$$

where a and b are constants. Then the calculus formulas (5a) and (10a) give

$$v(t) = \frac{dx(t)}{dt} = a + b\frac{d}{dt}(t^2)$$

$$= a + 2bt \tag{18}$$

where we used $d(t^2)/dt = 2t$. Equation (10a) gives

$$a(t) = \frac{dv(t)}{dt} = 2b \tag{19}$$

The results in Equations (17), (18) and (19) are graphed in Figures (8a, b and c) The position vs time a straight line with a slope 2b, and the acceleration is a constant 2b. Figure (8) therefore represents an example of motion with constant acceleration.

THE CONSTANT ACCELERATION FORMULAS

Unfortunately life is not as simple as one might think from the preceding example. If you have the formula for x(t), then you can calculate v(t) and a(t) very easily by differentiation. But usually you have to go the other way. From the physics you figure out what the acceleration is, then you have to work back to get v(t) and finally x(t). At best, this reverse process involves integration which is typically quite a bit harder than differentiation.

Let us work out an example where we know the acceleration and have to integrate to get the velocity and position. We will take the easiest non trivial case where the acceleration is constant. The result will be the constant acceleration formulas.

If we know a(t), the first step is to *solve* equation (10a) by turning it into an integral equation as follows

$$a(t) = \frac{dv(t)}{dt} \qquad (10a)$$

First multiply both sides by dt. (Remember that physicists keep dt very small but finite, so that we can move it around.) We get

$$dv(t) = a(t)dt \qquad (20)$$

Now integrate both sides of Equation (20) from time t = 0 up to time t = T. (This is called a definite integral.) We get

$$\int_0^T dv(t) = \int_0^T a(t)dt \qquad (21)$$

The integral on the left is simply v(t) evaluated between 0 and T.

$$\int_0^T dv(t) = v(t)\Big|_0^T = v(t) - v(0) \qquad (22)$$

On the right side of Equation (21), we set $a(t) = a_0$ (for constant acceleration) to get

$$\int_0^T a(t)dt = \int_0^T a_0 dt = a_0 \int_0^T dt$$

$$= a_0 t \Big|_0^T = a_0 T - a_0 \times 0 = a_0 T \qquad (23)$$

Using Equations (22) and (23) in (21) we get

$$v(T) - v(0) = a_0 T \qquad (24)$$

The next step is to recognize that Equation (24) applies to any time T, so that we can replace T by t to get

$$v(t) = v(0) + a_0 t \qquad (25)$$

To emphasize that v(0), the particle's speed at time t = 0, is not a variable, we will use the notation $v(0) \equiv v_0$ and Equation (25) becomes

$$\boxed{v(t) = v_0 + a_0 t} \qquad (26)$$

(If the steps we have used to derive Equation (26) were familiar and comfortable, then your calculus background is in good shape and you should not have much of a problem with calculus in reading this text. If, on the other hand what we did was strange, if the notation was unfamiliar and the steps unpredictable, a review of calculus is indicated. What we have done in the derivation of Equation (26) is use the concept of a definite integral. We will use definite integrals throughout the course and now is the time to learn how to use them. You should also be sure that you can do simple differentiations like $d/dt(at^2) = 2at$.)

To get the other constant acceleration formula, start with Equation (5a)

$$v(t) = \frac{dx(t)}{dt} \quad (5a)$$

and multiply through by dt to get

$$dx(t) = v(t)dt \quad (27)$$

Again integrate both sides from $t = 0$ to $t = T$ to get

$$\int_0^T dx(t) = \int_0^T v(t)dt \quad (28)$$

We can immediately do the integral on the left hand side

$$\int_0^T dx(t) = x(t)\Big|_0^T = x(T) - x(0) \quad (29)$$

At this point we cannot do the integral on the right side of Equation (28) until we know explicitly how v(t) depends on the variable t. If, however, the acceleration is constant, we can use Equation (26) for v(t) to get

$$\int_0^T v(t)dt = \int_0^T (v_0 + a_0 t)dt$$

$$= \int_0^T v_0 dt + \int_0^T a_0 t\, dt \quad (30)$$

$$= v_0 \int_0^T dt + a_0 \int_0^T t\, dt$$

Knowing that

$$\int_0^T t\, dt = \frac{t^2}{2} \quad (31)$$

we get

$$\int_0^T v(t)dt = v_0 t\Big|_0^T + a_0 \frac{t^2}{2}\Big|_0^T$$

$$= v_0 T + a_0 \frac{T^2}{2} \quad (32)$$

Using Equations (29) and (32) in (28) gives

$$x(T) - x_0 = v_0 T + \frac{1}{2} a_0 T^2 \quad (33)$$

Since Equation (33) applies for any arbitrary time T, we can replace T by t to get

$$\boxed{x(t) = x_0 + v_0 t + \frac{1}{2} a_0 t^2} \quad (34)$$

where we have written x_0 for $x(0)$, the position of the particle at time $t = 0$.

Three Dimensions

Equations (26) and (34) are the constant acceleration formulas for motion in one dimension, along the x axis. (We can, of course, choose the x axis to point any way we want.)

If we want to describe motion in three dimensions with constant acceleration, we repeat the steps leading to Equations (26) and (34), but starting with (5b) and (10b) for motion along the y axis, and (5c) and (10c) for motion along the z axis. The steps are essentially identical, and we end up with the six equations

$$x(t) = x_0 + v_x(0)t + \frac{1}{2}a_x t^2 \tag{35a}$$

$$y(t) = y_0 + v_y(0)t + \frac{1}{2}a_y t^2 \tag{35b}$$

$$z(t) = z_0 + v_z(0)t + \frac{1}{2}a_z t^2 \tag{35c}$$

$$v_x(t) = v_x(0)t + a_x t \tag{36a}$$

$$v_y(t) = v_y(0)t + a_y t \tag{36b}$$

$$v_z(t) = v_z(0)t + a_z t \tag{36c}$$

where we have temporarily gone back to the notation $x(0)$ for x_0, $v_x(0)$ for v_{x0}, etc., and a_x, a_y, and a_z are the x, y, z components of the assumed **constant** acceleration.

In Chapter 3 we introduced a notation that allowed us to conveniently express a vector $\vec{S}$ in terms of its components S_x, S_y and S_z, by writing the components, separated by commas, inside a parenthesis as follows

$$\vec{S} \equiv (S_x, S_y, S_z) \tag{37}$$

Using this notation, we define the following vectors by their components

$$\vec{R}(t) \equiv \big(x(t), y(t), z(t)\big) \quad \text{coordinate vector} \tag{38}$$

$$\vec{v}(t) \equiv \big(v_x(t), v_y(t), v_z(t)\big) \quad \text{velocity vector} \tag{39}$$

$$\vec{a} \equiv (a_x, a_y, a_z) \quad \text{constant acceleration} \tag{40}$$

With this vector notation, the six constant acceleration formulas (35a, b, c) and (36 a, b, c) reduce to the two vector equations

$$\vec{x}(t) = \vec{x}(0) + \vec{v}(0)t + \frac{1}{2}\vec{a}t^2 \tag{35}$$

$$\vec{v}(t) = \vec{v}_0(0) + \vec{a}t \tag{36}$$

or using the notation $\vec{R}_0 = \vec{R}(0)$, $\vec{v}_0 = \vec{v}(0)$, we have

$$\boxed{\vec{R}(t) = \vec{R}_0 + \vec{v}_0 t + \frac{1}{2}\vec{a}t^2} \tag{35'}$$

$$\boxed{\vec{v}(t) = \vec{v}_0 + \vec{a}t} \tag{36'}$$

These are the set of vector equations that we tested in our studies in Chapter 3 of instantaneous velocity with constant acceleration.

We have gone through all the details of the derivation of Equations (35) and (36), because they represent one of the major successes of the use of calculus in the prediction of motion. Whenever a particle's acceleration $\vec{a}$ is constant, and we know $\vec{a}$, $\vec{R}_0$, and $\vec{v}_0$, we can use these equations to predict the particle's position $\vec{R}(t)$ and velocity $\vec{v}(t)$ at any time t in the future.

PROJECTILE MOTION WITH AIR RESISTANCE

In our experimental study of projectile motion, we saw that when we used a styrofoam projectile, air resistance affected the acceleration of the projectile. From the point of view that we are riding on the ball, we would feel a wind in our face, blowing in a direction $-\vec{v}$, opposite to the velocity $\vec{v}$ of the projectile. The effect of this wind was to blow the acceleration vector back as shown in Figure (3-28), reproduced here as Figure (9).

We saw that the experimental vector $\vec{a}_3$ was the acceleration $\vec{g}$ we would have in the absence of air resistance, plus a correction $\vec{a}_{air}$ which pointed in the direction of the wind, in the $-\vec{v}$ direction as shown.

The magnitude $|\vec{a}_{air}|$ cannot accurately be determined from the strobe photograph. About all we can tell is that $|\vec{a}_{air}|$ is zero if the ball is at rest, and increases as the speed $|\vec{v}|$ of the ball increases. The simplest guess is that $|\vec{a}_{air}|$ is proportional to $|\vec{v}|$ and we have the formula

$$\vec{a}_{air} = -K\vec{v} \quad \text{\textit{simple guess}} \quad (41)$$

Our strobe photograph does not eliminate the possibility that $|\vec{a}_{air}|$ is more complicated, something like

$$|\vec{a}_{air}| = K_2|\vec{v}|^2 \quad (42)$$

or perhaps some combination like

$$|\vec{a}_{air}| = K_1|\vec{v}| + K_2|\vec{v}|^2 \quad (43)$$

It turns out that the motion of a sphere through a liquid (in our case a Styrofoam ball through air) has been studied extensively by both physicists and engineers. For slow speeds the motion is like Equation (41) but as the speed increases it looks more like Equation (43) and soon becomes even more complicated. The only simple fact is that $|\vec{a}_{air}|$ always points in the direction $-\vec{v}$, in the direction of the wind in our face (until vortex shedding occurs).

As an exercise to test the ability of calculus to predict motion, let us assume that our simple guess $\vec{a}_{air} = -K\vec{v}$ is good enough. We would then like to solve the calculus Equations (5) and (10) for the case where the acceleration is not constant, but is given by the formula

$$\vec{a} = \vec{g} - K\vec{v} \quad (44)$$

where $\vec{g}$ is the constant acceleration due to gravity, $\vec{v}$ is the instantaneous velocity of the particle, and K is what we will call the air resistance constant. Equation (44) is pictured in Figure (9).

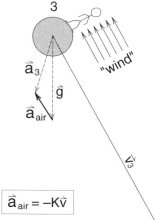

Figure 9
The acceleration produced by air resistance.

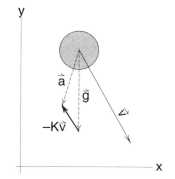

Figure 10

Our first step is to introduce a coordinate system as shown in Figure (10), and break the motion up into x and y components. Since the acceleration $\vec{g}$ due to gravity points down, we have $g_x = 0$ and the vector Equation (44) can be written as the two component equations

$$a_x = -Kv_x \qquad (g_x = 0) \qquad (44a)$$

$$a_y = g - Kv_y \qquad (g = -980 \text{ cm/sec}^2) \qquad (44b)$$

The calculus Equations (10a, b) that we have to solve become

$$a_x = \frac{dv_x}{dt} = -Kv_x \qquad (45)$$

$$a_y = \frac{dv_y}{dt} = g - Kv_y \qquad (46)$$

Let us focus on the simpler of the two equations, Equation (45) for the horizontal velocity of the projectile. We want to solve the equation

$$\boxed{\frac{dv_x(t)}{dt} + Kv_x(t) = 0} \qquad (45')$$

Suppose we try to solve Equation (45) using the same steps we used to predict v_x for constant acceleration (Equations 20 through 26). Multiplying through by dt gives

$$dv_x(t) = -Kv_x(t)dt \qquad (47)$$

Integrating from $t = 0$ to $t = T$ gives

$$\int_0^T dv_x(t) = -\int_0^T Kv_x(t)dt \qquad (48)$$

We can do the integral on the left, and remove the K from the integral on the right giving

$$v_x(t)\Big|_0^T = v_x(T) - v_x(0) = -K\int_0^T v_x(t)dt \qquad (49)$$

Now we are in trouble, because we have to integrate $v_x(t)$ in order to find $v_x(t)$. We can't do the integral until we know the answer, and we have to do the integral to get the answer. It boils down to the fact that the techniques we used to solve the calculus equations for constant acceleration do not work now. As soon as the acceleration is not constant, we have a much more difficult problem.

DIFFERENTIAL EQUATIONS

Equation (45) is an example of what is called a ***differential equation***. (An equation with derivatives in it.) Only in very special cases, as in our example of constant acceleration, can these equations be solved in a straightforward manner by integration. In slightly more complicated cases, these equations can be solved by certain standard tricks that one learns in an advanced calculus course on differential equations. We will use one of these tricks to solve Equation (45).

In general, however, differential equations cannot be solved without numerical methods that are now handled by digital computers. If, for example we assumed that the air resistance was proportional to v^2 as in Equation (42), then Equation (45) for the x component of velocity would be replaced by

$$\frac{dv_x(t)}{dt} + K_2 v_x(t)^2 = 0 \qquad (45a)$$

Equation (45a) is what is called a ***non linear*** differential equation, the word non linear coming from the appearance of the square of the unknown variable $v_x(t)$.

At the current time, there is no general way to solve non linear differential equations except by computer. Non linear differential equations have marvelously complicated features like *chaotic behavior* that have been discussed extensively in the popular press in the last few years. It is currently a hot research topic.

The point of this discussion is that when we use calculus to predict motion, a very slight increase in the complexity of the problem can lead to enormous increases in the difficulty in solving the problem. When the projectile's acceleration was constant, we could easily solve the calculus equations to get the constant acceleration formulas. If the air resistance has the simple form $\vec{a}_{air} = -K\vec{v}$, then we have to solve a differential equation, but we can still get an answer, a formula that predicts the motion of the particle. If we go up one step in complexity, if $|\vec{a}_{air}|$ is proportional to the square of the speed, then we have a non linear differential equation that we cannot solve without numerical or approximation techniques.

Calculus gives marvelous results when we can solve the problem. We get formulas describing the motion at all future times. But we are extremely limited in the kind of problems that can be solved. Simple physical modifications of a problem can turn an easy problem into an unsolvable one.

Before inventing calculus, Isaac Newton invented a simple step-by-step method that we will discuss in the next chapter. Newton's step-by-step method has the great advantage that slight complications in the physical setup lead to only slightly more work in obtaining a solution. We will see that it is almost no harder to predict projectile motion with air resistance, even with v^2 terms, than it is to predict projectile motion without air resistance. The step-by-step method will allow us to handle problems in this course, realistic problems, that do not have a calculus solution.

There are two disadvantages to the step-by-step method, however. One, is that you get a numerical answer, like an explicit orbit, rather than a general result. In contrast, the constant acceleration formulas describe all possible trajectories for motion with constant acceleration.

The second problem is that in the step-by-step method, a simple calculation is repeated many times, perhaps thousands or millions of times to obtain an accurate answer. Before digital computers, lifetimes were spent doing this kind of calculation by hand to predict the motion of the moon. But modern digital computers have changed all that. In minutes, the digital computer running your word processor can do what used to be months of work.

Solving the Differential Equation

We have essentially finished what we wanted to say about applying calculus to the problem of projectile motion with air resistance. The gist is that adding air resistance turns a simple problem into a hard one. Even for the simplest form of air resistance, $\vec{a}_{air} = -K\vec{v}$, we end up with the differential equation

$$\frac{dv_x(t)}{dt} + K v_x(t) = 0 \qquad (45)$$

which cannot be solved directly with integration.

Later in the course we will encounter several other differential equations, one having the same form as Equation (45). When we meet these equations, we will show you how to solve them.

At this time, we do not really need the solution to Equation (45). This equation does not represent a basic physics problem because our formula for air resistance is an approximation of limited validity. We include a solution for those who are interested, who want to see the problem completed now. Those for whom calculus is new or rusty may wish to skip to the next chapter.

The reason that differential equations are hard to solve is that the solutions are curves or functions rather than numbers. For example, the solution to the ordinary Equation $x^2 = 4$ is the pair of numbers $x = +2$ and $x = -2$. But Equation (45) has the decaying exponential curve shown in Figure (11) for a solution. What this curve tells us is that the v_x or the horizontal motion, dies out in time and the projectile will eventually have only y motion. After enough time the ball will be falling straight down.

One of the standard techniques for solving differential equations is to guess the answer and then plug your guess into the equation to see if you are right. When you take a course in solving differential equations, you learn how to make educated guesses. If you had been through such a course, you would guess that Equation (45) should have an exponentially decaying solution, and try a solution of the form

$$v_x(t) = v_{x0} e^{-\alpha t} \qquad \text{(guess)} \qquad (50)$$

where α and v_{x0} are constants whose values we wish to find.

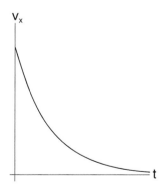

Figure 11
Air resistance causes the horizontal component of the velocity to decay exponentially.

Differentiating Equation (50) gives

$$\frac{dv_x(t)}{dt} = -v_{x0} \alpha e^{-\alpha t} \qquad (51)$$

where we used the fact that

$$\frac{de^{-\alpha t}}{dt} = -\alpha e^{-\alpha t} \qquad (52)$$

Substituting Equations (50) and (51) into Equation (45) gives

$$-v_{x0} \alpha e^{-\alpha t} + K v_{x0} e^{-\alpha t} = 0 \qquad (53)$$

First note that the exponential function $e^{-\alpha t}$ cancelled out. This indicates that we have guessed the correct function.

Next note that v_{x0} cancels. This means that any value of v_{x0} in Equation (50) is a possible solution. The particular value we want will be determined by the experimental situation.

What we have left is

$$\alpha = K \qquad (54)$$

Thus the differential Equation (45) has the exponentially decaying solution

$$\boxed{v_x(t) = v_{x0} e^{-Kt}} \qquad (55)$$

where the decay rate is the air resistance constant K.

For those of you who have actually had a course in solving differential equations, see if you can solve for the vertical motion of the projectile. The differential equation you have to solve is

$$\frac{dv_y(t)}{dt} = g - K v_y(t) \qquad (46)$$

The answer for long times turns out to be simple – the projectile ends up coasting at a constant terminal velocity. See if you can get that result.

The answer is $v_y = (g/K)(1 - e^{-Kt})$ if $v_y = 0$ at $t = 0$.

Appendix A

SOLVING PROJECTILE MOTION PROBLEMS

In high school physics texts and most college level introductory physics texts, there is considerable emphasis on solving projectile motion problems. A good reason for this is that these problems provide practice in problem solving techniques such as drawing clear sketches, developing an orderly approach, and checking units. Not such a good reason is that, in texts that rely solely on algebra and calculus, the only thing they can solve in the early stages are projectile motion or circular motion problems.

The disadvantage of over emphasizing projectile motion problems is that students begin to use the projectile motion formulas as a general way of predicting motion, using the formulas in circumstances where they do not apply. The important point to remember is that the formulas $\vec{v} = \vec{v}_i + \vec{a}t$ and $\vec{x} = \vec{v}_i t + 1/2\vec{a}t^2$ are very limited in scope. They apply only when the acceleration $\vec{a}$ is constant, a not very likely circumstance in the real world. The acceleration $\vec{a}$ is not constant for circular motion, projectile motion with air resistance, satellite motion, the motion of electrons in a magnetic field, and most interesting physics problems.

From the point of view that solving projectile motion problems is basically for practice in problem solving techniques, we will show you an orderly way of handling these problems. The approach which we will illustrate using several examples should allow you with practice to handle any constant acceleration problems test makers throw at you. In these examples, we are demonstrating not only how the problem is solved but also how you should go about doing it.

(Note -- in this appendix, all velocity vectors are instantaneous velocities, thus we will not bother underlining them.)

Example A1
A boy throws a ball straight up into the air and catches it (at the same height from which he threw it) 2 sec later. How high did the ball go?

Solution: To solve all projectile problems, we use the equations

$$\vec{S} = \vec{v}_i t + \frac{1}{2}\vec{a}t^2$$

$$\vec{v}_f = \vec{v}_i + \vec{a}t$$

However these are vector equations. Using a coordinate system in which the y axis is in the vertical direction and the x axis is in the horizontal direction, we get the following equations.

Vertical motion:

$$S_y = v_{iy}t + \frac{1}{2}a_y t^2$$

$$v_{fy} = v_{iy} + a_y t$$

Horizontal motion:

$$S_x = v_{ix}t + \frac{1}{2}a_x t^2$$

$$v_{fx} = v_{ix} + a_x t$$

Now projectiles near the surface of the earth accelerate downward at a rate of nearly 980 cm/sec². This value varies slightly at different points on the surface of the earth, but is always quite close to 980 cm/sec². This acceleration due to gravity is usually designated g; since it is directed downward in the *minus* y direction, we have

$$a_y = -g = -980 \text{ cm/sec}^2 \, (-32 \text{ ft/sec}^2)$$

$$a_x = 0$$

As a result, we get the equations

Vertical motion:

(a) $S_y = v_{iy}t - \frac{1}{2}gt^2$ \hfill (A1a)

(b) $v_{fy} = v_{iy} - gt$ \hfill (A1b)

Horizontal motion:

(c) $S_x = v_{ix}t$ \hfill (A1c)

(d) $v_{fx} = v_{ix}$ \hfill (A1d)

Horizontal motion and vertical motion are entirely independent of each other. We see, for example, from Equation (A1d), that the horizontal speed of a projectile does not change; but this has already been obvious from the strobe photographs.

Now let us apply Equation (A1) to the situation where the boy throws the ball straight up and catches it 2 sec later. Since there is no horizontal motion, we only need equations (A1a, b).

One good technique for solving projectile problems is to work up to and back from the top of the trajectory. The reason is that at the top of the trajectory, Equations (A1a, b) are very easily applied.

In our problem, the ball spent half its time going up and half its time falling; thus, the fall took 1 sec. The distance that it fell is

$$S_y = v_{iy}t - \frac{1}{2}gt^2$$

where t is 1 sec, and since we are starting at the top of the trajectory. We get

$$S_y = -\frac{1}{2}gt^2 = -\frac{1}{2} \times 32 \text{ ft/sec}^2 \times 1 \text{ sec}^2$$

$$S_y = -16 \text{ ft}$$

The minus sign indicated the ball fell 16 ft below the top of the trajectory.

Example A2

A ball is thrown directly upward at a speed of 48 ft/sec. How high does it go?

Solution: First, find the time it takes to reach to top of its trajectory. We have

$$v_{iy} = 48 \text{ ft/sec}$$

$$v_{fy} = 0 \quad \textit{at the top of the trajectory}$$

From Equation (4-A1b) we have

$$v_{fy} = v_{iy} - gt$$

or

$$t = \frac{v_{iy}}{g} = \frac{48 \text{ ft/sec}}{32 \text{ ft/sec}^2} = 1.5 \text{ sec}$$

Now we can use Equation A1a to calculate how high the ball goes.

$$S_y = v_{iy}t - \frac{1}{2}gt^2$$

We have $v_{iy} = 48$ ft/sec, t = 1.5 sec to reach the top; thus S_y, the distance to the top, is

$$S_y = 48 \frac{\text{ft}}{\text{sec}} \times 1.5 \text{ sec} - \frac{1}{2} \times 32 \frac{\text{ft}}{\text{sec}^2} \times (1.5 \text{ sec})^2$$

$$S_y = 72 \text{ ft} - 36 \text{ ft} = 36 \text{ ft}$$

Example A3

An outfielder throws a ball at a speed of 96 ft/sec at an angle of 30° above the horizontal. How far away from the outfielder does the ball strike the ground?

Solution: When solving problems, the first step is to draw a neat diagram of the situation, as in Figure (A1). The first calculation is to find the x and y components of $\vec{v}_i$. From our diagram we see that

$$v_{ix} = v_i \cos\theta$$
$$= 96 \text{ ft/sec} \times 0.864$$
$$= 83 \text{ ft/sec}$$

$$v_{iy} = v_i \sin\theta$$
$$= 96 \text{ ft/sec} \times 0.50$$
$$= 48 \text{ ft/sec}$$

where $\cos 30° = 0.864$ and $\sin 30° = 0.50$.

Now we are in a position to separate the problem into two parts – vertical motion and horizontal motion. These may be treated as two independent problems.

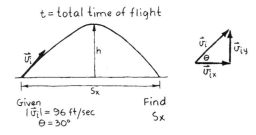

Figure A1
Sketch of the problem. On the sketch, label the symbols used, show what is given, and state what you are to find. It is generally better to work the problem in terms of letters, substituting numbers only at the end, or at convenient breaks in the problem.

Vertical motion. A ball is thrown straight up at a speed $v_{iy} = 48$ ft/sec ; how long a time t does it take to come back to the ground?

Horizontal motion. A ball travels horizontally at a speed $v_{ix} = 83$ ft/sec . If it travels for a time t (result of vertical motion problem) how far does it travel?

We see that the vertical motion problem is exactly the one we solved in Example A-2, $v_{iy} = 48$ ft/sec in both cases. Thus, using the same solution, we find that the ball takes 1.5 sec to go up and another 1.5 sec to come down, for a total time of

$$t = 3 \text{ sec}$$

Now solve the horizontal motion from

$$S_x = v_{ix} t$$

We get

$$S_x = 83 \text{ ft/sec} \times 3 \text{ sec} = 249 \text{ ft}$$

which is the answer.

Checking Units

It is easy to make a mistake when working a problem. One of the best ways to avoid mistakes is to write out the dimensions of each number used in the calculation; if the answer has the wrong dimension, you will know there is a mistake somewhere. For example, in the preliminary edition of this text the following formula accidentally appeared.

$$\vec{S} = \vec{v}_i + \frac{1}{2}\vec{a}t^2$$

Putting in the dimensions, we find

$$\vec{S}\, ft = \vec{v}_i \frac{ft}{sec} + \frac{1}{2}\vec{a}\frac{ft}{sec^2} \times (t\, sec)^2$$

or

$$(\vec{S})\, ft = (\vec{v}_i)\frac{ft}{sec} + \left(\frac{1}{2}\vec{a}t^2\right) ft$$

Clearly the $(\vec{v}_i)$ ft/sec has the wrong dimensions, since we cannot add ft/sec to ft. Thus, through a check of the dimensions we would immediately spot an error in this formula, even if we had no idea what the formula is about. To correct this formula, the $\vec{v}_i$ must be multiplied by t sec so that the result is $(\vec{v}_i)$ft/sec×t sec equals $(\vec{v}_i t)$ ft.

As another instance, in the solution of Example (A3) we had

$$t = \frac{v_{iy}}{g}$$

At this point you might begin to worry that you have made a mistake; your doubts will be dispelled, however, once dimensions are inserted

$$t\, sec = \frac{v_{iy}\, ft/sec}{g\, ft/sec^2} = \frac{v_{iy}}{g}\, sec$$

Exercise A1
A 22-caliber rifle with a muzzle velocity of 600 ft/sec is fired straight up. How high does the bullet go? How long before it hits the ground?

Exercise A2
(The rifle of Exercise A1 is fired at an angle of 45°. How far does the bullet travel? (Give answer in ft and in mi.)

Exercise A3
A right fielder is 200 ft from home plate. Just at the time he throws the ball into home plate, a runner leaves third base and takes 3.5 sec to reach home plate. If the maximum height reached by the ball is 64 ft, did the runner make it to home plate in time? (Problem from J. Orear, *Fundamental Physics*, Wiley, New York, 1961.)

Exercise A4
A steel ball is bouncing up and down on a steel plate with a period of oscillation of 1 sec. How high does it bounce? (Problem from J. Orear, *Fundamental Physics*, Wiley, New York, 1961.)

Exercise A5
A small rocket motor is capable of providing an acceleration of 0.01 g to a space capsule. If the capsule starts from a far-out space station and the rocket motor runs continuously, how far away is the capsule at the end of 1 year? What is the capsule's speed relative to the space station at the end of the year?

Exercise A6
A car traveling at 60 mi/hr strikes a tree. Inside the car the driver travels 1 ft from the time the car struck the tree until he is at rest. What is the deceleration of the driver if his deceleration is constant? Give the answer in ft/sec² and in g's.

Exercise A7

During volcanic eruptions, chunks of solid rock can be blasted out of the volcano. These projectiles are called volcanic blocks. Figure (2) shows a cross-section of Mt. Fuji, in Japan.

At what initial speed v_0 would a block have to be ejected, at 45°, in order to fall at the foot of the volcano as shown.

What is the time of flight?

(Problem from Halliday and Resnick.)

Hint Use the vector equation

$$\vec{S} = \vec{v}_i t + \frac{1}{2} \vec{a} t^2$$

which is illustrated in Figure 3-34 reproduced to the right. In this problem $\vec{S}$ is the total displacement of the rock, from the time it left the volcano until it hit the ground. Separate the vector equation into x and y components.

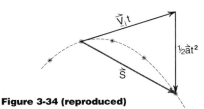

Figure 3-34 (reproduced)

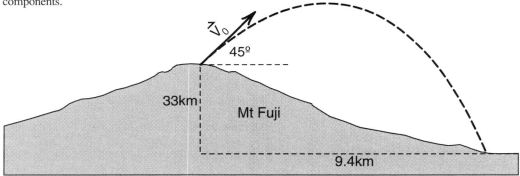

Figure A2
The farthest out blocks are the ones ejected at the greatest speed v_0 at an angle of 45°. By noting that the most distant blocks are 9.4 km away, you can thus determine the maximum speed at which the blocks were ejected.

Chapter 5
Computer Prediction of Motion

STEP-BY-STEP CALCULATIONS

In the last chapter we saw that for the special case of constant acceleration, calculus allowed us to obtain a rather remarkable set of formulas that predicted the object's motion for all future times (as long as the acceleration remained constant). We ran into trouble, however, when the situation got a bit more complicated. Add a little air resistance and the analysis using calculus became considerably more difficult. Only for the very simplest form of air resistance are we able to use calculus at all.

On the other hand, adding a little air resistance had only a little effect on the actual projectile motion. Without air resistance the projectile's acceleration vectors pointed straight down and were all the same length, as seen in Figure (3-27). Include some air resistance using the Styrofoam projectile, and the acceleration vectors tilted slightly as if blown back by the wind one would feel riding along with the ball, as seen in Figure (3-31). Since projectile motion with air resistance is almost the same as that without, one would like a method of predicting motion that is almost the same for the two cases, a method that becomes only a little harder if the physical problem becomes only a little more complex.

The clue for developing such a method is to note that in our analysis of strobe photographs, we have been breaking the motion into short time intervals of length Δt. During each of these time intervals, not much happens. In particular, the Styrofoam projectile's acceleration vector did not change much. Only over the span of several intervals was there a significant change in the acceleration vector. This suggests that we could predict the motion by assuming that the Styrofoam ball's acceleration vector was essentially constant during each time interval, and at the end of each time step correct the acceleration vector in order to predict the motion for the next time step. In this way, by a series of short calculations, we can predict the motion over a long time period. This is a rough outline of the step-by-step method of predicting motion that was originally developed by Isaac Newton and that we will discuss in this chapter.

The problem with the step-by-step prediction of motion is that it quickly gets boring. You are continually repeating the same calculation with only a small change in the acceleration vector. Worse yet, to get very accurate results you should take very many, very small, time steps. Each calculation is almost identical to the previous one, and the process becomes tedious. If these calculations are done by hand, one needs an enormous incentive in order to obtain meaningful results.

COMPUTER CALCULATIONS

Because of the tedium involved, step-by-step calculations were used only in desperate circumstances until the invention of the digital computer in the middle of the twentieth century. The digital computer is most effective and easiest to use when we have a repetitive calculation involving many, very similar steps. It is the ideal device for handling the step-by-step calculations described above. With a digital computer we can use very small time steps to get very accurate results, doing thousands or millions of steps to predict far into the future. We can cover the same range of prediction as the calculus-derived formulas, but not encounter significant difficulties when there is a slight change in the problem, such as the addition of air resistance.

To illustrate how to use the computer to handle a repetitive problem, we will begin with the calculation and plotting of the points on a circle. We will then go back to our graphical analysis of strobe photographs and see how that analysis can be turned into a series of steps for a computer prediction of motion.

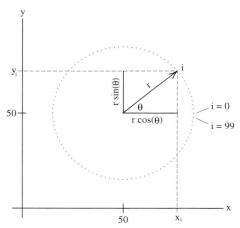

Figure 1
Points on a circle.

Calculating and Plotting a Circle

Figure (1) shows 100 points on the circumference of a circle of radius r. To make this example somewhat similar to the analysis of strobe photographs, we will choose a circle of radius r = 35 cm, centered at x = 50, y = 50, so that the entire circle will fit in the region x = 0 to 100, y = 0 to 100, as shown. The i th point around the circle has x and y coordinates given by

$$x_i = r \cos \theta_i$$
$$y_i = r \sin \theta_i \qquad (1)$$

where θ_i, the angle to the i th point, is given by

$$\theta_i = \frac{360}{100} * i \text{ degrees} = \frac{2\pi}{100} * i \text{ radians}$$

(We know that it is easier to draw a circle using a compass than it is to calculate and plot all these individual points. But if we want something more complicated than a circle, like an ellipse or Lissajous figure, we cannot use a compass. Then we have to calculate and plot individual points as we are doing.)

If we wrote out the individual steps required to calculate and plot these 100 points, the result might look like the following:

i = 0

$$\theta_0 = (2\pi/100)*0 = 0 \text{ radians}$$

$$x_0 = 50 + r \cos(\theta_0) = 50 + 35 \cos(0)$$
$$= 50 + 35*1 = 85$$

$$y_0 = 50 + r \sin(\theta_0) = 50 + 35 \sin(0)$$
$$= 50 + 35*0 = 50$$

Plot a point at (x = 85, y = 50)

i = 1

$\theta_1 = (2\pi/100)*1 = .0628$ radians

$x_1 = 50 + r\cos(\theta_1) = 50 + 35\cos(.0628)$

$= 50 + 35*.9980 = 84.93$

$y_1 = 50 + r\sin(\theta_1) = 50 + 35\sin(.0628)$

$= 50 + 35*.0628 = 52.20$

Plot a point at (x = 84.93, y = 52.20)

...

i = 50

$\theta_{50} = (2\pi/100)*50 = \pi$ radians

$x_{50} = 50 + r\cos(\theta_{50}) = 50 + 35\cos(\pi)$

$= 50 + 35*(-1) = 15$

$y_{50} = 50 + r\sin(\theta_{50}) = 50 + 35\sin(\pi)$

$= 50 + 35*0 = 50$

Plot a point at (x = 15, y = 50)

...

In the above, not only will it be tedious doing the calculations, it is even tedious writing down the steps. That is why we only showed three of the required 100 steps.

The first improvement is to find a more efficient way of writing down the steps for calculating and plotting these points. Instead of spelling out all of the details of each step, we would like to write out a short set of instructions, which, if followed carefully, will give us all the steps indicated above. Such instructions might look as follows:

1) **Let r = 35**
2) **Start with i = 0**
3) **Let $\theta_i = (2\pi/100)*i$**
4) **Let $x_i = 50 + r\cos\theta_i$**
5) **Let $y_i = 50 + r\sin\theta_i$**
6) **Plot a point at x_i, y_i**
7) **Increase i by 1**
8) **If i is less than 100, then go back to step 3 and continue in sequence**
9) **If you got here, i = 100 and you are done**

Figure 2
A program for calculating the points around a circle.

Exercise 1

Follow through the instructions in Figure (2) and see that you are actually creating the individual steps shown earlier.

PROGRAM FOR CALCULATION

The set of instructions shown in Figure (2) could be called a plan or *program* for doing the calculation. A similar set of instructions typed into a computer is called a *computer program*. Our instructions in Figure (2) would not be of much use to a person who spoke only German. But if we translated the instructions into German, then the German speaking person could follow them. Similarly, this particular set of instructions is not of much use to a computer, but if we translate them into a language the computer "understands", the computer can follow the instructions.

The computer language we will use in this course is called BASIC, a language developed at Dartmouth College for use in instruction. The philosophy in the design of BASIC is that it be as much as possible like an ordinary spoken language so that students can concentrate on their calculations rather than worry about details of operating the computer. Like human languages, the computer language has evolved over time, becoming easier to use and clearer in meaning. The version of BASIC we will use is called **True BASIC**, a modern version of BASIC written by the original developers of the language.

The way we will begin teaching you the language BASIC is to translate the set of instructions in Figure (2) into BASIC. We will do this in several steps, introducing a few new ideas at a time, just as you learn a few rules of grammar at a time when you are learning a foreign language. We will know that we have arrived at the actual language BASIC when the computer can successfully run the program. It is not unlike testing your knowledge of a foreign language by going out in the street and seeing if the people in that country understand you.

The DO LOOP

In a sense, the set of instruction in Figure (2) is already in the form of sloppy BASIC, or you might say pidgin BASIC. We only have to clean up a few grammatical rules and it will work well. The first problem we will address is the statement in instruction #8.

8) If i is less than 100, then go back to step 3 and continue in sequence

There are two problems with this instruction. One is that it is long and wordy. Computer languages are usually designed with shorter, crisper instructions. The second problem is that the instruction relies on numbering instructions, as when we say "go back to step 3". There is no problem with numbering instructions in very short programs, but clarity suffers in long programs. The name "step 3" is not a particularly descriptive name; it does not tell us why we should go back there and not somewhere else. It is much better to state that we have a cyclic calculation, and that we should go back to the beginning of this particular cycle.

The grammatical construction we will use, one of the variations of the so-called "DO LOOP", has the following structure. We mark the beginning of the cyclic process with the word "DO", and end it with the command "LOOP UNTIL...". Applied to our instructions in Figure (2), the DO LOOP would look as follows:

LET r = 35
LET i = 0
DO
    LET $\theta_i = (2\pi/100)*i$
    LET $x_i = 50 + r\cos\theta_i$
    LET $y_i = 50 + r\sin\theta_i$
    Plot a point at x_i, y_i
    Increase i by 1
LOOP UNTIL i = 100
All done

Figure 3
Introducing the DO LOOP.

In the instructions in Figure (3), we begin by establishing that r = 35 and that i will start with the value 0. Then we mark the beginning of the cyclic calculation with the command **DO**, and end it with the command "**LOOP UNTIL i = 100**". The idea is that we keep repeating all the stuff between the "DO" line and the "LOOP..." line until our value of i has been incriminated up to the value i = 100. When i reaches 100, then the loop command is ignored and we have finished both the loop and the calculation.

The LET Statement

Another major grammatical rule is needed before Figure (3) becomes a BASIC program that can be read by the computer. That involves a deeper understanding of the **LET** statement that appears in many of the instructions.

One example of a LET statement is the following

$$\text{LET } i = i + 1 \tag{2}$$

At first sight, statement (2) looks a bit peculiar. If we think of it as an equation, then we would cancel the i's and be left with

$$\text{LET } = 1$$

which is clearly nonsense. Thus the LET statement is not really an equation, and we have to find out what it is.

The LET statement combines the computer's ability to do calculations and to store numbers in memory. To understand the memory, think of the mail boxes at the post office. Above each box there is a name like "*Jones*", and Jones' mail goes inside the box. In the computer, each memory cell has a name like "i", and a number goes inside the cell. Unlike a mail box, which can hold several letters, a computer memory cell can store only one number at a time.

The rule for carrying out a LET statement like

$$\text{LET } i = i + 1$$

is to first evaluate the right hand side and store the results in the memory cell mentioned on the left side. In this example the computer evaluates i + 1 by first looking in cell "i" to see what number is stored there. It then adds 1 to that stored value to get the value (i + 1). To finish the command, it looks for a cell labeled "i", removes the number stored there and replaces it with the value just calculated. The net result of all this is that the numerical value stored in cell i is increased by 1.

There is a good mnemonic that helps you remember how a LET statement works. In the command LET i = i + 1, the computer takes the old value of i, adds one to get the new value, and stores that in cell i. If we write the LET statement as

$$\text{LET } i_{new} = i_{old} + 1$$

then it is clear what the computer is doing, and we are not tempted to cancel the i's. In this text we will often use the subscripts "old" and "new" to remind us what the computer is to do. When we actually type in the commands, we will omit the subscripts "old" and "new", because the computer does that automatically when performing a LET command.

With this understanding of the **LET** statement, our program for calculating the points on a circle becomes

LET r = 35
LET i = 0
DO
    LET $\theta_i = (2\pi/100)*i$
    LET $x_i = 50 + r\cos\theta_i$
    LET $y_i = 50 + r\sin\theta_i$
    Plot a point at x_i, y_i
    LET $i_{new} = i_{old} + 1$
LOOP UNTIL i = 100
    All done

Figure 4
Handling the LET statement.

In Figure (4), we begin our repetitive **DO LOOP** by calculating a new value of the angle θ. This new value is stored in the memory cell labeled θ, and later used to calculate new values of $x = r\cos\theta$ and $y = r\sin\theta$. Since we are using the updated values of θ, we can drop the subscripts i on the variables θ_i, x_i, y_i. After we plot the point at the new coordinate (x, y), we calculate the next value of i with the command LET $i = i + 1$, and then go back for the next calculation.

To get a working BASIC program, there are a few other small changes that are easily seen if we compare our program in Figure (4) with the working BASIC program in Figure (5). Let us look at each of the changes.

```
! --------- Plotting window
!      (x axis = 1.5 times y axis)
   SET WINDOW -40,140,-10,110

! --------- Initial conditions
   LET r = 35
   LET i = 0

! --------- Calculational loop
   DO
      LET Theta = (2*Pi/100)*i
      LET x = 50 + r*COS(Theta)
      LET y = 50 + r*SIN(Theta)
      PLOT x,y;
      LET i = i+1
   LOOP UNTIL i = 100

END
```

Figure 5
Listing of the BASIC program.

Variable Names
Our command

$$\text{LET } \theta = (2*\pi/100)*i$$

has been rewritten in the form

$$\text{LET Theta} = (2*Pi/100)*i$$

Unfortunately, only a few special symbols are available in the font chosen by True BASIC. When we want a symbol like θ and it is not available, we can spell it out as we have done.

We have spelled out the name "Pi" for π, because BASIC understands that the letters "Pi" stand for the numerical value of π. ("Pi" is what is called a reserved word in True BASIC.)

Multiplication
We are used to writing an expression like

$$r\cos(\theta)$$

and assuming that the variable r multiplies the function cos(q). In BASIC you must always use an "*" for multiplication, thus the correct way to write $r\cos(\theta)$ is

$$r*\cos(\theta)$$

Similarly we had to write 2*Pi rather than 2Pi in the line defining Theta.

Plotting a Point
Our command

   Plot a point at (x, y)

becomes in BASIC

   PLOT x,y

It is not as descriptive as our command, but it works the same way.

Comment Lines

In a number of places in the BASIC program we have added lines that begin with an exclamation point "!". These are called "comment lines" and are included to make the program more readable. A comment line has no effect on the operation of the program. The computer ignores anything on a line following an exclamation point. Thus the two lines

LET i = i + 1

LET i = i + 1   ! Increment i

are completely equivalent. (If you write a command that does something peculiar, you can explain it by adding a comment as we did above.)

Plotting Window

The only really new thing in the BASIC program of Figure (5) is the **SET WINDOW** command. We are going to plot a number of points whose x and y values all fall within the range between 0 and 100. We have to tell the computer what kind of scale to use when plotting these points.

In the command

SET WINDOW -40, 140, -10, 110

the computer adjusts the plotting scales so that the computer screen starts at -40 and goes to +140 along the horizontal axis, and ranges from -10 to +110 along the vertical axis, as shown in Figure (6).

This setting gives us plenty of room to plot anything in the range 0 to 100 as shown by the dotted square in Figure (6).

When we are plotting a circle, we would like to have it look like a circle and not get stretched out into an ellipse. In other words we would like a horizontal line 10 units long to have the same length as a vertical line 10 units long. True BASIC for the Macintosh computer could have easily have done this because Macintosh pixels are square, so that equal horizontal and vertical distances should simply contain equal numbers of pixels. (A pixel is the smallest dot that can be drawn on the screen. A standard Macintosh pixel is 1/72 of an inch on a side, a dimension consistent with typography standards.)

However True BASIC also works with IBM computers where there is no standard pixel size or shape. To handle this lack of standardization, True BASIC left it up to the user to guess what choice in the SET WINDOW command will give equal x and y dimensions. This is an unfortunate compromise.

If you are using a Macintosh MacPlus, Classic or SE, one of the computers with the 9" screens, set the horizontal dimension 1.5 times bigger than the vertical one, use the full screen as an output window, and the dimensions will match (circles will be circular and squares square.) If you have any other screen or computer, you will have to keep adjusting the SET WINDOW command until you get the desired results. (Leave the y axis range from –10 to +110, and adjust the x axis range. For the 15" screen of the iMac, we got a round circle plot for x values from –33 to +133.)

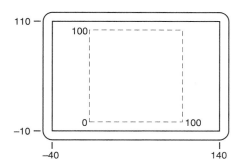

Figure 6
Using the SET WINDOW command.

Practice

The best way to learn how to handle BASIC programs is to start with a working program like the one in Figure (5), and make small modifications and see what happens. Below are a series of exercises designed to give you this practice, while at the same time introducing some techniques that will be useful in the analysis of strobe photographs. When you finish these exercises, you will be ready to use BASIC as a tool for predicting the motion of projectiles, both without and with air resistance, which is the subject of the remainder of the chapter.

Exercise 2 A Running Program

Get a copy of True BASIC (preferably version 2.0 or later), launch it, and type in the program shown in Figure (5). Type it in just as we have printed it, with the same indentations at the beginning of the lines, and the same comments. Then run the program. You should get an output window that has the circle of dots shown in Figure (7).

If something has gone wrong, and you do not get this output, first check that you have typed exactly what we printed in Figure 5. If that doesn't work, get help from a friend, advisor, computer center, whatever. Sometimes the hardest part of programming is turning on the equipment and getting things started properly.

Once you get your circle of dots, save a copy of the program.

Exercise 3 Plotting a Circular Line

It's pretty hard to see the dots in Figure (7). The output can be made more visible if lines are drawn connecting the dots to give us a circular line. In BASIC it is very easy to connect the dots you are plotting. You simply add a semicolon after the PLOT command. I.e., change the command

    PLOT x, y   ! Plots dots

to the command

    PLOT x, y;   ! Plots lines

The result is shown in Figure (8).

Modify your program by changing the PLOT command as shown, and see that your output looks like Figure (8).

(Optional—There is a short gap in the circle on the right hand side. Can you modify your program to eliminate this gap?)

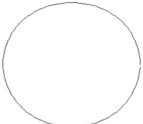

Figure 8
The circle of lines plotted by adding a semicolon to the end of the PLOT command.

Figure 7
The circle of dots plotted by the program shown in Figure (5).

Exercise 4 Labels and Axes

Although we have succeeded in drawing a circle, the output is fairly bare. It is impossible to tell, for example, that we have a circle of radius 35, centered at x = y = 50. We can get this information into the output by drawing axis and labeling them. This can be done by adding the following lines near the beginning of the program, just after the SET WINDOW command

```
! --------- Draw & label axes
   BOX LINES 0,100,0,100
   PLOT TEXT, AT -3,0 : "0"
   PLOT TEXT, AT -13,96: "y=100"
   PLOT TEXT, AT 101,0 : "x=100"
```

The results of adding these lines are shown in Figure (9). The BOX LINES command drew a box around the region of interest, and the three PLOT TEXT lines gave us the labels seen in the output.

Add the 5 lines shown above to your program and see that you get the results shown in Figure (9). Save a copy of that version of the program using a new name. Then find out how the BOX LINES and PLOT TEXT commands work by making some changes and seeing what happens.

Exercise 5a Numerical Output

Sometimes it is more useful to see the numerical results of a calculation than a plot. This can easily be done by replacing the PLOT command by a PRINT command.

To do this, go back to your original circle plotting program (the one shown in Figure (5) which we asked you to save), and change the line

   PLOT x, y

to the two lines

   PRINT "x = "; x,   "y = ";y

   !PLOT x, y

What we have done is added the PRINT line, and then put an exclamation point at the beginning of the PLOT line so that the computer would ignore the PLOT command. (We left the PLOT line in so that we could use it later.) If we ran the program we get a whole bunch of printing, part of which is shown in Figure (10). Do this and see that you get the same results.

```
x =   79.5515      y =   68.7539
x =   78.3156      y =   70.5725
x =   76.968       y =   72.3098
x =   75.5139      y =   73.9591
x =   73.9591      y =   75.5139
x =   72.3098      y =   76.968
x =   70.5725      y =   78.3156
x =   68.7539      y =   79.5515
x =   66.8614      y =   80.6707
x =   64.9023      y =   81.6689
x =   62.8844      y =   82.5422
x =   60.8156      y =   83.287
x =   58.7041      y =   83.9004
x =   56.5583      y =   84.3801
x =   54.3867      y =   84.724
x =   52.1977      y =   84.9309
x =   50.          y =   85
x =   47.8023      y =   84.9309
```

Figure 10
If we print the coordinates of every point, we get too much output.

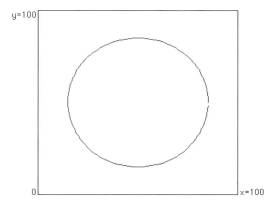

Figure 9
A box, drawn by the BOX LINES command makes a good set of axes. You can then plot text where you want it.

Selected Printing (MOD Command)

The problem with the output in Figure (10) is that we print out the coordinates of every point, and we may not want that much information. It may be more convenient, for example, if we print the coordinates for every tenth point. To do this, we use the following trick. We replace the PRINT command

PRINT "x = ";x, "y = ";y

by the command

IF MOD(i,10) THEN PRINT "x = ";x, "y = ";y

To understand what we did, remember that each time we go around the loop, the variable i is incriminated by 1. The first time i = 0, then it equals 1, then 2, etc.

The function MOD( ), stands for the mathematical term "modulus". If we count modulus 3, for example, we count: 0, 1, 2, and then go back to zero when we hit 3. Comparing regular counting with counting MOD 3, we get:

regular counting: 0 1 2 3 4 5 6 7 8 9

counting MOD 3:   0 1 2 0 1 2 0 1 2 0

Counting MOD 10, we go:
0, 1, 2, 3, 4, 5, 6, 7, 8, 9, 0, 1, 2, 3, ... etc. Every time we get up to a power of ten, we go back to zero.

The command MOD(i, 10) means evaluate the number i counting modulus 10. Thus when i gets to 10, MOD(i, 10) goes back to zero. When i gets to 20, MOD(i, 10) goes back to zero again. Thus as i increases, MOD(i, 10) goes back to zero every time i hits a power of 10.

In the command

IF MOD(i,10) THEN PRINT "x = ";x, "y = ";y

no printing occurs until i increases to a power of ten. Then we do get a print. The result is that with this command the coordinates of every tenth point are printed, and there is no printing for the other points, as we see in Figure (11).

Exercise 5b

Take your program from Exercise (5a), modify the print command with the MOD statement, and see that you get the results shown in Figure (11). Then figure out how to print every 5th point or every 20th point. See if it works.

```
x =  85        y =  50
x =  78.3156   y =  70.5725
x =  60.8156   y =  83.287
x =  39.1844   y =  83.287
x =  21.6844   y =  70.5725
x =  15        y =  50.
x =  21.6844   y =  29.4275
x =  39.1844   y =  16.713
x =  60.8156   y =  16.713
x =  78.3156   y =  29.4275
```

Figure 11
The coordinates of every tenth point is printed when we use the MOD command.

Exercise 6 Plotting Crosses

Our last exercise will be to have the computer plot both a circle, and a set of crosses located at every tenth point along the circle as shown in Figure (12). This is about as fancy a plot as we will need in the course, so that you are almost through practicing the needed fundamentals.

To plot the crosses seen in Figure (12) we added what is called a "subroutine" shown at the bottom of Figure (13).

To get the program shown in Figure (13), go back to the program of Exercise (3) (we asked you to save it), and add the command

IF MOD(i, 10) = 0 THEN CALL CROSS

where "CROSS" is the name of the subroutine at the bottom of Figure (13). You can see that the IF MOD(i,10) = 0 part of the command has the subroutine called at every tenth point.

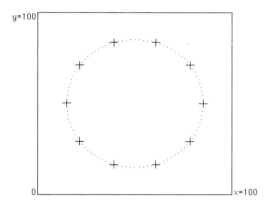

Figure 12
Here we use the MOD command and a subroutine to plot a cross at every tenth dot.

Next add in the subroutine lines as shown in Figure (13) (your program should look just like Figure 13) and see if you get the results shown in Figure (12). When you have a running program, figure out how to make the crosses bigger or smaller. How can you plot twice as many crosses?

```
! --------- Plotting window
!           (x axis = 1.5 times y axis)
  SET WINDOW -40,140,-10,110

! --------- Draw & label axes
  BOX LINES 0,100,0,100
  PLOT TEXT, AT -3,0 : "0"
  PLOT TEXT, AT -13,96: "x=100"
  PLOT TEXT, AT 101,0 : "y=100"

! --------- Initial conditions
  LET r = 35
  LET i = 0

! --------- Calculational loop
  DO
     LET Theta = (2*PI/100)*i
     LET x = 50 + r*COS(Theta)
     LET y = 50 + r*SIN(Theta)
     PLOT x,y
     IF MOD(i,5) = 0 THEN CALL CROSS
     LET i = i+1
  LOOP UNTIL i = 100

! --------- Subroutine "CROSS" draws a cross at x,y.
  SUB CROSS
      PLOT LINES: X-4,Y; X+4,Y
      PLOT LINES: X,Y-4; X,Y+4
  END SUB

END
```

Figure 13
The complete BASIC program for drawing the picture shown in Figure (12).

PREDICTION OF MOTION

Now that we have the techniques to handle a repetitive calculation we can return to the problem of using the step-by-step method to predict the motion of a projectile. The idea is that we will convert our graphical analysis of strobe photographs, discussed in Chapter 3, into a pair of equations that predict the motion of the projectile one step at a time. We will then see how these equations can be applied repeatedly to predict motion over a long period of time.

Figure (14a) is essentially our old Figure (3-16) where we used a strobe photograph to define the velocity of the projectile in terms of the projectile's coordinate vectors $\vec{R}_i$ and $\vec{R}_{i+1}$. The result was

$$\vec{v}_i = \frac{\vec{S}_i}{\Delta t} = \frac{\vec{R}_{i+1} - \vec{R}_i}{\Delta t} \quad (4)$$

If we multiply Equation (4) through by Δt and rearrange terms, we get

$$\vec{R}_{i+1} = \vec{R}_i + \vec{v}_i \Delta t \quad (5)$$

which is the vector equation pictured in Figure (14a).

Equation (5) can be interpreted as an equation that predicts the projectile's new position $\vec{R}_{i+1}$ in terms of the old position $\vec{R}_i$, the old velocity vector $\vec{v}_i$, and the time step Δt. To emphasize this predictive nature of Equation (5), let us rename $\vec{R}_{i+1}$ the new vector $\vec{R}_{new}$, and the old vectors $\vec{R}_i$ and $\vec{v}_i$, as $\vec{R}_{old}$ and $\vec{v}_{old}$. With this renaming, the equation becomes

$$\vec{R}_{new} = \vec{R}_{old} + \vec{v}_{old} * \Delta t \quad (6)$$

which is illustrated in Figure (14b).

Equation (6) predicts the new position of the ball using the old position and velocity vectors. To use Equation (6) over again to predict the next new position of the ball, we need updated values for $\vec{R}$ and $\vec{v}$. We already have $\vec{R}_{i+1}$ or $\vec{R}_{new}$ for the updated coordinate vector; what we still need is an updated velocity vector $\vec{v}_{i+1}$ or $\vec{v}_{new}$.

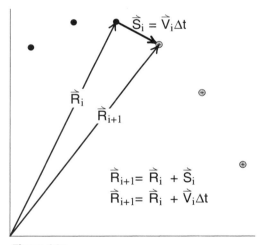

Figure 14a
To predict the next position $\vec{R}_{i+1}$ of the ball, we add the ball's displacement $\vec{S}_i = \vec{v}_i \Delta t$ to the present position $\vec{R}_i$.

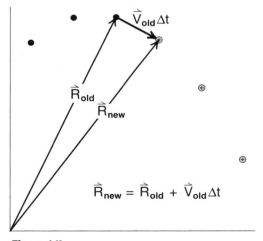

Figure 14b
So that we do not have to number every point in our calculation, we label the current position "old", and the next position "new".

To obtain the updated velocity, we use Figure (3-17), drawn again as Figure (15a), where the acceleration vector $\vec{a}_i$ was defined by the equation

$$\vec{a}_i = \frac{\vec{v}_{i+1} - \vec{v}_i}{\Delta t} \tag{7}$$

Multiplying through by Δt and rearranging terms, Equation (7) becomes

$$\vec{v}_{i+1} = \vec{v}_i + \vec{a}_i \Delta t \tag{8}$$

which expresses the new velocity vector in terms of the old velocity $\vec{v}_i$ and the old acceleration $\vec{a}_i$, as illustrated in Figure (15a). Changing the subscripts from $i+1$ and i to "new" and "old" as before, we get

$$\vec{v}_{new} = \vec{v}_{old} + \vec{a}_{old} * \Delta t \tag{9}$$

as our basic equation for the projectile's new velocity.

We have now completed one step in our prediction of the motion of the projectile. We start with the old position and velocity vectors $\vec{R}_{old}$ and $\vec{v}_{old}$, and used Equations (6) and (9) to get the new vectors $\vec{R}_{new}$ and $\vec{v}_{new}$. To predict the next step in the motion, we change the names of $\vec{R}_{new}$, $\vec{v}_{new}$ to $\vec{R}_{old}$ and $\vec{v}_{old}$ and repeat Equations (6) and (9). As long as we know the acceleration vector $\vec{a}_i$ at each step, we can predict the motion as far into the future as we want.

There are two important criteria for using this step-by-step method of predicting motion described above. One is that we must have an efficient method to handle the repetitive calculations involved. That is where the computer comes in. The other is that we must know the acceleration at each step. In the case of projectile motion, where $\vec{a}$ is constant, there is no problem. We can also handle projectile motion with air resistance if we can use formulas like

$$\vec{a}_{air} = -K\vec{v}$$
$$\vec{a} = \vec{g} + \vec{a}_{air}$$

shown in Figure (3-31). To handle more general problems, we need a new method for determining the acceleration vector. That new method was devised by Isaac Newton and will be discussed in the chapter on Newtonian Mechanics. In this chapter we will focus on projectile motion with or without air resistance so that we know the acceleration vectors throughout the motion.

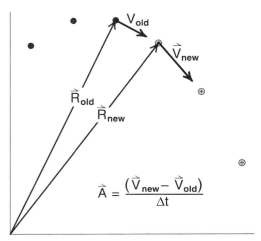

Figure 15a
Once we get to the "new" position, we will need the new velocity vector $\vec{v}_{new}$ in order to predict the next new position.

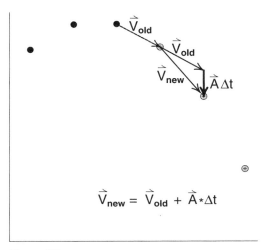

Figure 15b
The value of $\vec{v}_{new}$ is obtained from the definition of acceleration $\vec{A} = (\vec{v}_{new} - \vec{v}_{old})/\Delta t$.

TIME STEP AND INITIAL CONDITIONS

Equation (6) and (9) are the basic components of our step-by-step process, but there are several details to be worked out before we have a practical program for predicting motion. Two of the important ones are the choice of a time step Δt, and the initial conditions that get the calculations started.

In our strobe photographs we generally used a time step $\Delta t = .1$ second so that we could do effective graphical work. If we turn the strobe up and use a shorter time step, then the images are so close together, the arrows representing individual displacement vectors are so short, that we cannot accurately add or subtract them. Yet if we turn the strobe down and use a longer Δt, our analysis becomes too coarse to be accurate. The choice $\Delta t = .1$ sec is a good compromise.

When we are doing numerical calculations, however, we are not limited by graphical techniques and can get more accurate results by using shorter time steps. We will see that for the analysis of our strobe photographs, time steps in the range of .01 second to .001 second work well. Much shorter time steps, like a millionth of a second, greatly increase the computing time required while not giving more accurate results. If we use ridiculously short time steps like a nanosecond, the computer must do so many calculations that the round-off error in the computer calculations begins to accumulate and the answers get worse, not better. Just as with graphical work there is an optimal time step. (Later we will have some exercises where you try various time steps to see which give the best results.)

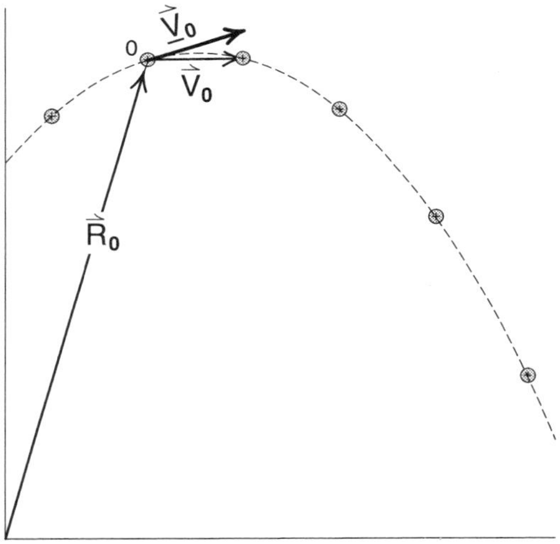

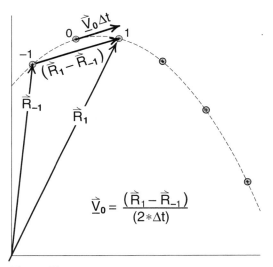

Figure 16
By using a very short time Step dt in our computer calculation, we will closely follow the continuous path shown by the dotted lines. Thus we should use the instantaneous velocity vector $\vec{v}_0$, rather than the strobe velocity $\vec{v}_0$ as our initial velocity.

Figure 17
The displacement $\vec{v}_0 \Delta t$ is just half the displacement $(\vec{R}_1 - \vec{R}_{-1})$. This is an exact result for projectile motion, and quite accurate for most strobe photographs.

When we use a short time step of .01 seconds or less for analyzing our projectile motion photographs, we are close to what we have called the instantaneous velocity illustrated in Figure (3-32). But, as shown in Figure (16), the instantaneous velocity $\underline{\vec{v}}_0$ and the strobe velocity $\vec{v}_0$ are quite different if the strobe velocity was obtained from a strobe photograph using $\Delta t = .1$ second. To use the computer to predict the motion we see in our strobe photographs, we need the initial position $\vec{R}_0$ and the initial velocity $\vec{v}_0$ as the start for our step-by-step calculation. If we are going to use a very short time step in our computer calculation, then our first velocity vector should be the instantaneous velocity $\underline{\vec{v}}_0$, not the strobe velocity .

This does not present a serious problem, because back in Chapter 3, Figure (3-33) reproduced here as Figure (17), we showed a simple method for obtaining the ball's instantaneous velocity from a strobe photograph. We saw that the instantaneous velocity $\underline{\vec{v}}_0$ was the average of the previous and following strobe velocities $\vec{v}_{-1}$ and $\vec{v}_1$:

$$\underline{\vec{v}}_0 = \frac{\vec{v}_{-1} + \vec{v}_1}{2} \qquad (10)$$

where $\vec{v}_{-1} = \vec{S}_{-1}/\Delta t$ and $\vec{v}_0 = \vec{S}_0/\Delta t$.

However, the sum of the two displacement vectors $(\vec{S}_{-1} + \vec{S}_0)$ is just the difference between the coordinate vectors $\vec{R}_1$ and $\vec{R}_{-1}$ as shown in Figure (17). Thus the instantaneous velocity of the ball at Position (0) in Figure (17) is given by the equation

$$\underline{\vec{v}}_0 = \frac{\vec{R}_1 - \vec{R}_{-1}}{\Delta t} \qquad (11)$$

If we use Equation (11) as the formula for the initial velocity in our step-by-step calculation, we are starting with the instantaneous velocity at Position (0) and can use very short time intervals in the following steps.

To avoid confusing the longer strobe time step and the shorter computer time step that we will be using in the same calculation, we will give them two different names as follows. We will use Δt for the longer strobe time step, which is needed for calculating the initial instantaneous velocity, and the name **dt** for the short computer time step.

Δt = time between strobe flashes

dt = computer time step   (12)

This choice of names is more or less consistent with calculus, where Δt is a small but finite time interval and dt is infinitesimal.

AN ENGLISH PROGRAM FOR PROJECTILE MOTION

We are now ready to write out a program for predicting the motion of a projectile. The first version will be what we call an "English" program -- one that we can easily read and understand. Once we have checked that the program does what we want it to do, we will see what modifications are necessary to translate the program into BASIC.

English Program

```
! --------- Initial conditions
    LET Δt = .1
    LET R_old = R_0
    LET V_old = (R_1 - R_-1)/(2*Δt)
    LET T_old = 0

! --------- Computer Time Step
    LET dt = .01

! --------- Calculational loop
    DO
        LET R_new = R_old + V_old*dt
        LET A = g
        LET V_new = V_old + A*dt
        LET T_new = T_old + dt
        PLOT R
    LOOP UNTIL T > 1
```

Figure 18

The first version of the English projectile motion program is shown in Figure (18). This program is designed to predict the motion of the steel ball projectile shown in Figure (3-8) and used for the drawings seen in Figures (15) and (16).

In the program we begin with a statement of the initial conditions – the starting point for the analysis of the motion. In this photograph, the strobe time step is $\Delta t = .1$ seconds, and we are beginning the calculations at the position labeled $\vec{R}_0$ in Figure (16). The instantaneous velocity at that point is given by the formula $\vec{v}_0 = (\vec{R}_1 - \vec{R}_{-1})/2\Delta t$ as shown in Figure (17). These results appear in the program in the lines

$$\text{LET } \Delta t = .1$$
$$\text{LET } \vec{R}_{old} = \vec{R}_0$$
$$\text{LET } \vec{V}_{old} = (\vec{R}_1 - \vec{R}_{-1})/(2*\Delta t)$$

Our new thing we are going to do in this program is keep track of the time by including the variable T in our calculations. We begin by setting T = 0 in the initial conditions, and then increment the clock by a computer time step dt every time we go around the calculation loop. This way T will keep track of the elapsed time throughout the calculations. The clock is initialized by the command

$$\text{LET } T_{old} = 0$$

The computer time step dt plays a significant role in the program because we will want to adjust dt so that each calculational step is short enough to give accurate results, but not so short to waste large amounts of computer time. We will start with the value dt = .01 seconds, as shown by the command

$$\text{LET } dt = .01$$

Later we will try different time steps to see if the results change or are stable.

The important part of the program is the calculational loop which is repeated again and again to give us the step-by-step calculations. The calculations begin with the command

$$\text{LET } \vec{R}_{new} = \vec{R}_{old} + \vec{V}_{old}*dt$$

which is the calculation pictured in Figure (14b). Here we are using the short computer time step dt so that $\vec{R}_{new}$ will be the position of the ball dt seconds after it was at $\vec{R}_{old}$.

The next line

$$\text{LET } \vec{A} = \vec{g}$$

simply tells us that for this projectile motion the ball's acceleration has the constant value $\vec{g}$. (Later, when we predict projectile motion with air resistance, we change this line to include the acceleration produced by the air resistance.)

To calculate the new velocity vector, we use the command

$$\text{LET } \vec{V}_{new} = \vec{V}_{old} + \vec{A}*dt$$

which is pictured in Figure (15b). Again we are using the short computer time step dt rather than the longer strobe rate Δt.

The last two lines inside the calculational loop are

$$\text{LET } T_{new} = T_{old} + dt$$

$$\text{PLOT } \vec{R}$$

The first of these increments the clock so that T will keep track of the elapsed time. Then we plot a point at the position $\vec{R}$ so that we can get a graph of the motion of the ball.

The calculational loop itself is bounded by the DO and LOOP UNTIL commands:

DO

...

...

...

LET $T_{new} = T_{old} + dt$

...

LOOP UNTIL T > 1

Remember that with a DO – UNTIL loop there is a test to see if the condition, here T > 1, is met. If T has not reached 1, we go back to the beginning of the loop and repeat the calculations. Because of the command LET $T_{new} = T_{old} + dt$, T increments by dt each time around. At some point T will get up to one, the condition will be met, and we leave the loop. At that point the program is finished. (We chose the condition T > 1 to stop the calculation because the projectile spends less than one second in the strobe photograph. Later we may use some other criterion to stop the calculation.)

A BASIC PROGRAM FOR PROJECTILE MOTION

The program in Figure (18) is quite close to a BASIC program. We have the LET statements and the Do – LOOP commands that appeared in our working BASIC program back in Figure (5). The only problem is that BASIC unfortunately does not understand vector equations. In order to translate Figure (18) into a workable BASIC program, we have to convert all the vector equations into numerical equations.

To do this conversion, we write the vector equation out as three component equations as shown below.

$$\vec{A} = \vec{B} + \vec{C} \tag{13}$$

becomes

$$Ax = Bx + Cx \tag{14a}$$

$$Ay = By + Cy \tag{14b}$$

$$Az = Bz + Cz \tag{14c}$$

We saw this decomposition of a vector equation into numerical or scalar equations in Chapter 2 on vectors and Chapter 4 on calculus. (It should have been in Chapter 2 but was accidently left out. It will be put in.) If the motion is in two dimensions, say in the x–y plane, then we only need the x and y component Equations (14a) and (14b).

Let us apply this rule to translate the vector LET statement

$$\text{LET } \vec{R}_{new} = \vec{R}_{old} + \vec{V}_{old}*dt \tag{15}$$

into two numerical LET statements. If we use the notation

$$\vec{R} = (Rx, Ry) \quad ; \quad \vec{V} = (Vx, Vy)$$

we get, dropping the subscripts "new" and "old",

$$Rx = Rx + Vx*dt \tag{16a}$$

$$Ry = Ry + Vy*dt \tag{16b}$$

We can drop the subscripts "new" and "old" because in carrying out the LET statement the computer must use the old values of Rx and Vx to evaluate the sum Rx + Vx*dt, and this result which is the new value of Rx is stored in the memory cell labeled "Rx".

To translate the initial conditions, we used the experimental values of the ball's coordinates given in Figure (3-10), the steel ball projectile motion strobe photograph we have been using for all of our drawings. These coordinates are reproduced below in Figure (19).

Ball coordinates
-1) (8.3, 79.3)
0) (25.9, 89.9)
1) (43.2, 90.2)
2) (60.8, 80.5)
3) (78.2, 60.2)
4) (95.9, 30.2)

Figure 19
Experimental coordinates of the steel ball projectile, from Figure (3-10).

Using the fact that $\vec{R}_0 = (25.9, 89.9)$, we can write the equation

$$\text{LET } \vec{R}_{old} = \vec{R}_0$$

as the two equations

$$\text{LET } Rx = 25.9$$

$$\text{LET } Ry = 89.9$$

In a similar way we use the experimental values for $\vec{R}_1$ and $\vec{R}_{-1}$ to evaluate the initial value of $\vec{V}_{old}$.

In Figure (20) we have converted the vector LET statements into scalar ones to obtain a workable BASIC program. We have also included the vector statements to the right so that you can see that the English and BASIC programs are essentially the same. We also added the SET WINDOW command so that the output could be plotted.

In Figure (21), we show the output from the Basic program of Figure (20). It looks about as bad as Figure (7), the output from our first circle plotting program. In the following exercises we will add axes, plot points closer together, and plot crosses every tenth of a second. In addition, we will get numerical output that can be compared directly with the experimental values shown in Figure(19).

BASIC Program

```
! --------- Plotting window
!           (x axis = 1.5 times y axis)
  SET WINDOW -40,140,-10,110


! --------- Initial conditions
  LET DeltaT = .1
  LET Rx = 25.9
  LET Ry = 89.9
  LET Vx = (43.2 - 8.4)/(2*.1)
  LET Vy = (90.2 - 79.3)/(2*.1)
  LET T = 0

! --------- Computer Time Step
  LET dt = .01

! --------- Calculational loop
  DO
     LET Rx = Rx + Vx*dt
     LET Ry = Ry + Vy*dt
     LET Ax = 0
     LET Ay = -980
     LET Vx = Vx + Ax*dt
     LET Vy = Vy + Ay*dt
     LET T = T + dt
     PLOT Rx,Ry
  LOOP UNTIL T > 1

  END
```

English Program

! --------- Initial conditions
    LET $\Delta t = .1$
    LET $\vec{R}_{old} = \vec{R}_0$
    LET $\vec{V}_{old} = (\vec{R}_1 - \vec{R}_{-1})/(2*\Delta t)$
    LET $T_{old} = 0$

! --------- Computer Time Step
    LET $dt = .01$

! --------- Calculational loop
    DO
        LET $\vec{R}_{new} = \vec{R}_{old} + \vec{V}_{old}*dt$
        LET $\vec{A} = \vec{g}$
        LET $\vec{V}_{new} = \vec{V}_{old} + \vec{A}*dt$
        LET $T_{new} = T_{old} + dt$
        PLOT $\vec{R}$
    LOOP UNTIL $T > 1$

END

Figure 20
Projectile Motion program in both BASIC and English.

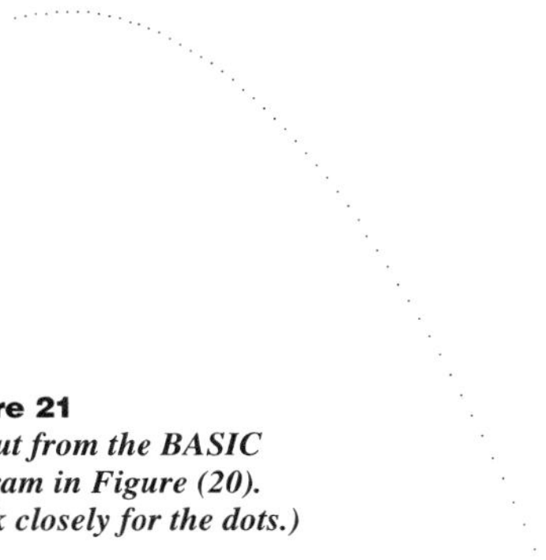

Figure 21
Output from the BASIC program in Figure (20). (Look closely for the dots.)

Exercise 7

Start BASIC, type the BASIC projectile motion program shown in Figure 20, and run it. Keep fixing it up until it gives output that looks like that shown in Figure 21.

Exercise 8 Changing the Time Step

Reduce the time step to dt = .001 seconds. The plot should become essentially a continuous line.

Exercise 9 Numerical Output

Change the plot command to a print statement to see numerical output. You can do this by turning the PLOT command into a comment, and adding a PRINT command as shown below.

!PLOT Rx,Ry

PRINT "Rx = ";Rx, "Ry = ";Ry

Just as in Exercise 5, you will get too much output when you run the program. If you have done Exercise 8, the coordinates of the ball will be printed every thousandth of a second. Yet from the strobe photograph, you have data for tenth second intervals. The next two exercises are designed to reduce the output.

Exercise 10 Attempt to reduce output

Replace the PRINT command of exercise 9 by the command

IF MOD(T,.1) = 0 THEN PRINT "Rx = ";Rx, "Ry = ";Ry

The idea is to pull the same trick we used in reducing the output in Exercise 5, going from Figure 10 to Figure 11. In The above MOD statement, we would hope that we would get output every time T gets up to a multiple of 0.1.

Try the modification of the PRINT command using MOD(T,.1) as shown above. When you do you will not get any output. The MOD(T,.1) command does not work, because the MOD function generally works only with integers. We will fix the problem in the next exercise.

Exercise 11 Reducing Numerical Output

Because the MOD function works reliably only with integers, we will introduce a counter variable i like we had in our circle plotting program.

First we must initialize i. We can do that at the same time we initialize dt as shown.

! --------- Computer Time Step and Counter

LET dt = .01

LET i = 0

Then we will increment i by 1 each time we go around the calculational loop, using the now familiar command LET i = i+1. If we are using a time step dt = .001 then we have to go around the calculational loop 100 times to reach a time interval of .1 seconds. To do this, our print command should start with IF MOD(i,100) = 0... Thus, inside the calculational loop, the Print command of Exercise 9 should be replaced by

LET i = i+1

IF MOD(i,100) = 0 THEN PRINT "RX = ";RX, "RY = ";RY

Make the changes shown above, run your program, and see that you get the output shown below in Figure 22. Compare these results with the experimental values shown in Figure 19.

```
Rx =   43.35      Ry =    90.499
Rx =   60.8       Ry =    81.298
Rx =   78.25      Ry =    62.297
Rx =   95.7       Ry =    33.496
Rx =  113.15      Ry =    -5.105
Rx =  130.6       Ry =   -53.506
Rx =  148.05      Ry =  -111.707
Rx =  165.5       Ry =  -179.708
Rx =  182.95      Ry =  -257.509
Rx =  200.4       Ry =  -345.11
```

Figure 22
Numerical output from the projectile motion program, printed at time intervals of .1 seconds. These predicted results should be compared with the experimental results seen in Figure 3-10.

Exercise 12 Plotting Crosses

Now we have the MOD statement to reduce the printing output, we can use the same trick to plot crosses in the output at .1 second intervals. All we have to do is restore the PLOT command, change the MOD statement to

IF MOD(i,100) = 0 THEN CALL CROSS

and add a cross plotting subroutine which should now look like

! --------- Subroutine "CROSS" draws a cross at Rx,Ry.

    SUB CROSS

        PLOT LINES: Rx-2,Ry; Rx+2,Ry

        PLOT LINES: Rx,Ry-2; Rx,Ry+2

    END SUB

The only change from the CROSS subroutine in the circle plotting program is that the cross is now centered at coordinates (Rx,Ry) rather than (x,y) as before.

The complete cross plotting is shown in Figure (23), and the results are plotted in Figure (24). Modify your projectile motion program to match Figure (23), and see that you get the same results. (How did we stop the plotting outside the square box?)

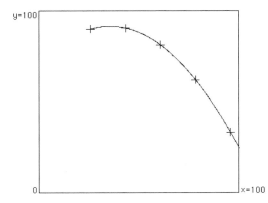

Figure 24
Output from our BASIC projectile motion program of Figure 23.

Projectile Motion Program

```
! --------- Plotting window
!           (x axis = 1.5 times y axis)
  SET WINDOW -40,140,-10,110

! --------- Draw & label axes
  BOX LINES 0,100,0,100
  PLOT TEXT, AT -3,0 : "0"
  PLOT TEXT, AT -13,96: "y=100"
  PLOT TEXT, AT 101,0 : "x=100"

! --------- Initial conditions
  LET DeltaT = .1
  LET Rx = 25.9
  LET Ry = 89.9
  LET Vx = (43.2 - 8.3)/(2*.1)
  LET Vy = (90.2 - 79.3)/(2*.1)
  LET T = 0
  CALL CROSS

! --------- Computer Time Step
  LET dt = .001
  LET i = 0

! --------- Calculational loop
  DO
     LET Rx = Rx + Vx*dt
     LET Ry = Ry + Vy*dt
     LET Ax = 0
     LET Ay = -980
     LET Vx = Vx + Ax*dt
     LET Vy = Vy + Ay*dt
     LET T = T + dt
     LET i = i+1
     IF MOD(i,100) = 0 THEN CALL CROSS
     PLOT Rx,Ry
  LOOP UNTIL RX > 100

! --------- Subroutine "CROSS" draws
           ! a cross at Rx,Ry.
  SUB CROSS
      PLOT LINES: Rx-2,Ry;   Rx+2,Ry
      PLOT LINES: Rx,Ry-2;   Rx,Ry+2
  END SUB
END
```

Figure 23
Projectile motion program that plots crosses every tenth of a second.

PROJECTILE MOTION WITH AIR RESISTANCE

Projectile motion is an example of a very special kind of motion where the acceleration vector is constant – does not change in either magnitude or direction. In this special case we can easily use calculus to predict motion far into the future.

But let the acceleration vector change even by a small amount, as in the case of projectile motion with air resistance, and a calculus solution becomes difficult or impossible to obtain. This illustrates the important role the acceleration vector plays in the prediction of motion, but overemphasizes the importance of motion with constant acceleration.

With a computer solution, very little additional effort is required to include the effects of air resistance. We will be able to adjust the acceleration for different amounts or kinds of air resistance. The point is to develop an intuition for the role played by the acceleration vector. We will see that *if we know a particle's acceleration, have a formula for it, and know how the particle started moving, we can predict where the particle will be at any time in the future.*

Once we have gained experience with this kind of prediction, we can then focus our attention on the core problem in mechanics, namely finding a general method for determining the acceleration vector. As we mentioned, the general method was discovered by Newton and will be discussed shortly in the chapter on Newtonian Mechanics.

In our study of the effects of air resistance, we will use as our main example the styrofoam ball projectile shown in Figures (3-30a, b) and reproduced here as Figures (25a, b). To obtain the coordinates listed in Figure (25b), each image was enlarged and studied separately. As a result, these coordinates should be accurate to within half a millimeter (except for possible errors due to parallax in taking the photograph).

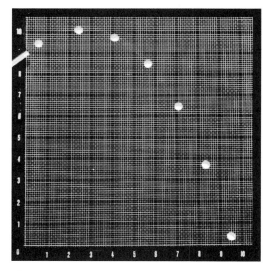

Figure 25a
The styrofoam projectile of Figure (3-30a). We have printed a negative of the photograph to show the grid lines more distinctly.

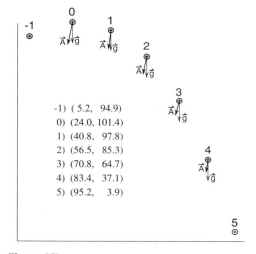

Figure 25b
To obtain as accurate a value as we could for each ball coordinate, each image was enlarged and studied separately.

Figure (26), a reproduction of Figure (3-31), is a detailed analysis of the ball's acceleration at Position (3). As shown in Figure (26) we can write the formula for the ball's acceleration vector $\vec{A}$ in the form

$$\vec{A} = \vec{g} + \vec{A}_{air} \qquad (17)$$

where one possible formula for $\vec{A}_{air}$ is

$$\vec{A}_{air} = -K\vec{V} \qquad (18)$$

$\vec{V}$ being the instantaneous velocity of the ball.

In Equation (17), $\vec{A}_{air}$ is defined as the change from the normal acceleration $\vec{g}$ the projectile would have without air resistance. As we see, $\vec{A}_{air}$ points opposite to $\vec{V}$, which is the direction of the wind we would feel if we were riding on the ball. Figure (26) suggests the physical interpretation that this wind is in effect blowing the acceleration vector back. It suggests that acceleration vectors can be pushed or pulled around, which is the underlying idea of Newtonian mechanics. In Figure (26) the earth is pulling down on the ball which gives rise to the component $\vec{g}$ of the ball's acceleration, and the wind is pushing back to give rise to the component $\vec{A}_{air}$.

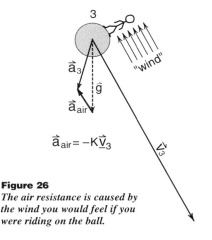

Figure 26
The air resistance is caused by the wind you would feel if you were riding on the ball.

The simplest formula we can write which has $\vec{A}_{air}$ pointing in the $-\vec{V}$ direction is Equation (18), $\vec{A}_{air} = -K\vec{V}$, where K is a constant that we have to find from the experiment. If some choice of the constant K allows us to accurately predict all the experimental points in Figure (25), then we will have verified that Equation (18) is a reasonably accurate description of the effects of air resistance.

It may happen, however, that one choice of K will lead to an accurate prediction of one position of the ball, while another choice leads to an accurate prediction of another point, but no value of K gives an accurate prediction of all the points. If this happens, equation (18) may be inadequate, and we may need a more complex formula.

The next level of complexity is that K itself depends on the speed of the ball. Then $\vec{A}_{air}$ would have a magnitude related to V^2, V^3, or something worse. In this case the air resistance is "nonlinear" and exact calculus solutions are not possible. But, as we see in Exercise 15, we can still try out different computer solutions.

In reality, when a sphere moves through a fluid like air or water, the resistance of the fluid can become very complex. At high enough speeds, the sphere can start shedding vortices, the fluid can become turbulent, and the acceleration produced by the fluid may no longer be directed opposite to the instantaneous velocity of the sphere. In Exercise 13 we take a close look at $\vec{A}_{air}$ for all interior positions for the projectile motion shown in Figure (25). We find that to within experimental accuracy, for our styrofoam projectile $\vec{A}_{air}$ does point in the $-\vec{V}$ direction. Thus a formula like Equation (18) is a good starting point. We can also tell from the experimental data whether K is constant and what a good average value for K should be.

Air Resistance Program

Figure (20) was our BASIC program for projectile motion. We would now like to modify that program so that we can predict the motion of the Styrofoam ball shown in Figure (25). To do this, we must change the command

$$\text{LET } \vec{A} = \vec{g}$$

to the new command

$$\text{LET } \vec{A} = \vec{g} - K\vec{V} \qquad (19)$$

and try different values for K until we get the best agreement between prediction and experiment.

A complete program with this modification is shown in Figure (27). In this program we see that Equation (19) has been translated into the two component equations

LET Ax = 0   − K∗Vx

LET Ay = −980 − K∗Vy

In addition, we are printing numerical output at .1 sec intervals so that we can accurately compare the predicted results with the experimental ones. In the line

$$\text{LET } K = \ldots$$

which appears in the Initial Conditions, we are to plug in various values of K until we get the best agreement that we can between theory and experiment.

Finding K does not have to be complete guesswork. In Exercise 13 we ask you to do a graphical analysis of the Styrofoam ball's acceleration at several positions using the enlargements provided. From these results you should choose some best average value for K and use that as your initial guess for K in your computer program. Then fine tune K until you get the best agreement you can. We ask you to do this in Exercise 14.

Once you have a working program that predicts the motion of the Styrofoam ball in Figure (25), you can easily do simulations of different strengths of air resistance. What if you had a steel ball being projected through a viscous liquid like honey? The viscous liquid might have the same effect as air, except that the resistance constant K should be much larger. With the computer, you can simply use larger and larger values of K to see the effects of increasing the air or fluid viscosity. We ask you to do this in Exercise 15. This is a very worthwhile exercise, for as the fluid viscosity increases, as you increase K, you get an entirely new kind of motion. There is a change in the qualitative character of the motion which you can observe by rerunning the program with different values of K.

```
! --------- Plotting window
!           (x axis = 1.5 times y axis)
    SET WINDOW -40,140,-10,110

! --------- Initial conditions
    LET DeltaT = .1
    LET Rx = ...
    LET Ry = ...                    Use initial values
    LET Ux = ...                    from Figure (25).
    LET Uy = ...
    LET T = 0
    LET K = ...                     Try different
                                    values of K
! --------- Computer Time Step
    LET dt = .001
    LET i = 0

! --------- Calculational loop
    DO
        LET Rx = Rx + Ux*dt
        LET Ry = Ry + Uy*dt
        LET Ax = 0      - K*Ux
        LET Ay = -980   - K*Uy      New formula for A
        LET Ux = Ux + Ax*dt
        LET Uy = Uy + Ay*dt
        LET T = T + dt
        LET i = i+1
        IF MOD(i,100) = 0 THEN PRINT "Rx = ";Rx, "Ry = ";Ry
    LOOP UNTIL T > 1

    END
```

Figure 27
BASIC program for projectile motion with air resistance. It is left to the reader to insert appropriate initial conditions, and choose values of the air resistance constant K.

In Exercise 16, we show you one way to modify the air resistance formulas to include nonlinear effects, i.e., to allow $\vec{A}_{air}$ to depend on V^2 as well as V. What we do is first use the Pythagorean theorem to calculate the magnitude V of the ball's speed and then use that in a more general formula for $\vec{A}_{air}$. The English lines for this are

LET $V = \sqrt{V_x^2 + V_y^2}$

LET $\vec{A} = \vec{g} - K(1 + K2*V)\vec{V}$ (20)

where we now try to find values of K and K2 that improve the agreement between prediction and experiment. The translation of these lines into BASIC is shown in Exercise 16.

Exercise 13 Graphical Analysis

Figures (28 a,b,c,d,e) are accurate enlargements of sections of Figure (25b). In each case we show three positions of the Styrofoam projectile so that you can determine the ball's instantaneous velocity $\vec{V}$ at the center position. Using the section of grid you can determine the magnitude of both $\vec{V}\Delta t$ and $\vec{A}_{air}\Delta t^2$. From that, and the fact that $\Delta t = .1$ sec, you can then determine the size of the air resistance constant K using the equation $\vec{A}_{air} = -K\vec{V}$.

Do this for each of the diagrams, positions 0 through 4 and then find a reasonable average value of K. How constant is K? Do you have any explanation for changes in K?

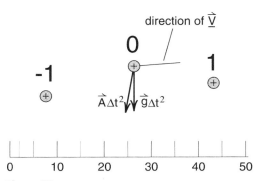

Figure 28a
Blowup of position 0 in Figure 25b.

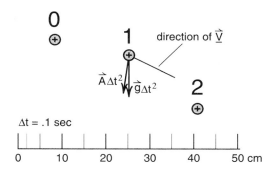

Figure 28b
Blowup of position 1 in Figure 25b.

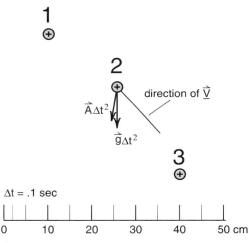

Figure 28c
Blowup of position 2 in Figure 25b.

Exercise 14 Computer Prediction

Starting with the Basic program shown in Figure (27) use the experimental values shown in Figure (25b), reproduced below, to determine the initial conditions for the motion of the ball. Then use your best value of K from Exercise 13 as your initial value of K in the program. By trial and error, find what you consider the best value of K to bring the predicted coordinates into reasonable agreement with experiment.

$$
\begin{aligned}
-1) &\ (\ 5.35,\ \ 94.84) \\
0) &\ (24.03, 101.29) \\
1) &\ (40.90,\ \ 97.68) \\
2) &\ (56.52,\ \ 85.15) \\
3) &\ (70.77,\ \ 64.56) \\
4) &\ (83.48,\ \ 36.98) \\
5) &\ (95.18,\ \ \ 3.86)
\end{aligned}
$$

Exercise 15 Viscous Fluid

After you get your program of Exercise 14 working, allow the program to print out numerical values for up to T = 15 seconds. After about 10 seconds, the nature of the motion is very different than it was at the beginning. Explain the difference. (You may be able to see the difference better by printing Vx and Vy rather than Rx and Ry.)

You will see the same phenomenon much faster if you greatly increase the air resistance constant K. Redo your program to plot the output, drawing crosses every .1 seconds. Then rerun the program for ever increasing values of K. Explain what you see.

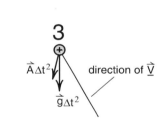

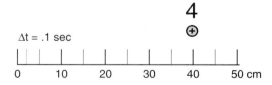

Figure 28d
Blowup of position 3 in Figure 25b.

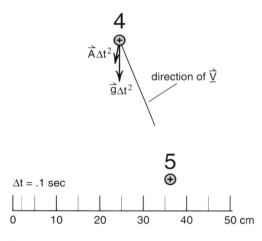

Figure 28e
Blowup of position 4 in Figure 25b.

Exercise 16 Nonlinear Air Resistance (optional)

In Exercise 14, you probably found that you were not able to precisely predict all the ball positions using one value of K. In this exercise, you allow K to depend on the ball's speed v in order to try to get a more accurate prediction. One possibility is to use the following formulas for $\vec{A}_{air}$, which we mentioned earlier:

$$\text{LET } V = \sqrt{V_x^2 + V_y^2}$$

$$\text{LET } \vec{A} = \vec{g} - K(1 + K2*V)\vec{V} \qquad (20)$$

With Equations (20), you can now adjust both K and K2 to get a better prediction. These equations are translated into BASIC as follows.

$$\text{LET } V = \text{SQR}(Vx*Vx + Vy*Vy)$$

$$\text{LET } Ax = 0 \quad - K*(1 + K2*V)*Vx$$

$$\text{LET } Ay = -980 - K*(1 + K2*V)*Vy$$

Make these modifications in the program of Exercise 14, and see if you can detect evidence for some V^2 dependence in the air resistance.

Exercise 17 Fan Added

In Figure (30), on the next page, we show the results of placing a rack of small fans to the right of the styrofoam ball's trajectory in order to increase the effect of air resistance. Now, someone riding with the ball should feel not only the wind due to the motion of the ball, but also the wind of the fans, as shown in Figure (29).

Our old air resistance formula

$$\text{LET } \vec{A} = \vec{g} + K(-\vec{V}\text{ball})$$

should probably be replaced by a command like

$$\text{LET } \vec{A} = \vec{g} + K(-\vec{V}\text{ball} + \vec{V}\text{fan})$$

Translated into BASIC, this would become

$$\text{LET } Ax = 0 \quad + K*(-Vx - V\text{fan})$$

$$\text{LET } Ay = -980 + K*(-Vy + 0 \quad) \qquad (21)$$

where $\vec{V}\text{ball} = (Vx, Vy)$ is the current velocity of the ball, and $\vec{V}\text{fan} = (-V\text{fan}, 0)$ is the wind caused by the fan. We assume that this wind is aimed in the –x direction and has a magnitude Vfan. We now have two unknown parameters K and Vfan which we can adjust to match the experimental results shown in Figure (30).

Do this, starting with the value of K that you got from the analysis of the styrofoam projectile in Figure 25b (Exercise 13 or 14). Does your resulting value for Vfan seem reasonable? Can you detect any systematic error in your analysis? For example, should Vfan be stronger near the fans, and get weaker as you move left?

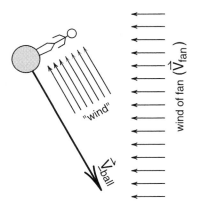

Figure 29
Additional wind created by fan.

COMPOSITE STROBE PHOTOGRAPH

IMAGE CENTERS

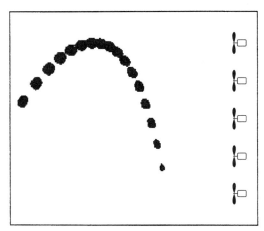
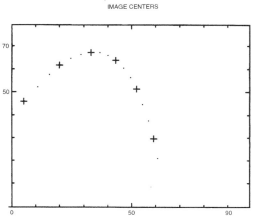

Figure 30
Styrofoam projectile with a bank of fans. In order to get more air resistance, we added a bank of small fans as shown. This Strobe "photograph" was taken with the Apple II Strobe system.

Figure 31
In this diagram, the Apple II computer has calculated and plotted the centers of each of the images seen in the composite strobe photograph on the left.

```
           DELTA T = 3/30 SECOND  (TIME BETWEEN CROSSES)

CROSS#1    X = 5.3    Y = 46.2      CROSS#2   X = 20.3   Y = 62.1
CROSS#3    X = 32.9   Y = 67.7      CROSS#4   X = 43.4   Y = 64.3
CROSS#5    X = 52.2   Y = 51.6      CROSS#6   X = 59.1   Y = 30.3

           INITIAL IMAGES

           IMAGE#1    X = 5.3       Y = 46.2
           IMAGE#2    X = 10.8      Y = 52.8
           IMAGE#3    X = 15.8      Y = 58.1
           IMAGE#4    X = 20.3      Y = 62.1
           IMAGE#5    X = 24.8      Y = 65.1
           IMAGE#6    X = 28.8      Y = 66.9
           IMAGE#7    X = 32.9      Y = 67.7
```

Figure 32
The Apple II also prints out the Coordinates of each image. The time Δt between crosses is 1/10 sec. Between the dots there is a 1/30 sec time interval. The coordinates of the initial 7 dots are printed to help determine the initial instantaneous velocity of the ball.

Chapter 6
Mass

By now we have learned how to use either calculus or the computer to predict the motion of an object whose acceleration is known. But in most problems we do not know the acceleration, at least initially. Instead we may know the forces acting on the object, or something about the object's energy, and use this information to predict motion. This approach, which is the heart of the subject of mechanics, involves mass, a concept which we introduce in this chapter.

In the metric system, mass is measured in grams or kilograms, quantities that should be quite familiar to the reader. It may be surprising that we devote an entire chapter to something that is measured daily by grocery store clerks in every country in the world. But the concept of mass plays a key role in the subject of mechanics. Here we focus on developing an experimental definition of mass, a definition that we can use without modification throughout our discussion of physics.

After introducing the experimental definition, we will go through several experiments to determine how mass, as we defined it, behaves. In low speed experiments, the kind we can do using air tracks in demonstration lectures, the results are straightforward and are what one expects. But when we consider what would happen if similar experiments were carried out with one of the objects moving at speeds near the speed of light, we predict a very different behavior for mass. This new behavior is summarized by the Einstein mass formula, a strikingly simple result that one might guess, but which we cannot quite derive from the definition of mass, and the principle of relativity alone. What is needed in addition is the law of conservation of linear momentum which we will discuss in the next chapter.

One of the striking features of Einstein's special theory of relativity is the fact that nothing, not even information, can travel faster than the speed of light. We can think of nature as having a speed limit c. In our world, speed limits are hard to enforce. We will see that the Einstein mass formula provides nature with an automatic way of enforcing its speed limit.

Einstein's mass formula appears to predict that no particle can quite reach the speed of light. We end the chapter with a discussion of how to handle particles, like photons and possibly neutrinos, that do travel at precisely the speed of light.

DEFINITION OF MASS

In everyday conversation the words *mass* and **weight** are used interchangeably. Physicists use the words mass and weight for two different concepts. Briefly, we can say that the weight of an object is the force that the object exerts against the ground, and we can measure weight with a device such as a bathroom scale. The weight of an object can change in different circumstances. For example, an astronaut who weighs 180 pounds while standing on the ground, floats freely in an orbiting space capsule. If he stood on a bathroom scale in an orbiting space craft, the reading would be zero, and we would say he is weightless. On the other hand the mass of the astronaut is the same whether he is in orbit or standing on the ground. An astronaut in orbit does not become massless. Mass is not what you measure when you stand on the bathroom scales.

What then is mass? One definition, found in the dictionary, describes mass as the property of a body that is a measure of the amount of material it contains. Another definition, which is closer to the one we will use, says that the more massive an object, the harder it is to budge.

Both of these definitions are too vague to tell us how to actually measure mass. In this section we will describe an experimental definition of mass, one that provides an explicit prescription for measuring mass. Then, using this prescription, we will perform several experiments to see how mass behaves.

Recoil Experiments

As a crude experiment suppose that the two skaters shown in Figure (6-1), a father and a child, stand in front of each other at rest and then push each other apart. The father hardly moves, while the child goes flying off. The father is more massive, harder to budge. No matter how hard or gently the skaters push apart, the big one always recoils more slowly than the smaller one. We will use this observation to define mass.

In a similar but more controlled experiment, we replace the skaters by two carts on what is called an *air track*. An air track consists of a long square metal tube with a series of small holes drilled on two sides as shown in Figure (6-2). A vacuum cleaner run backwards blows air into the tube, and the air escapes out through the small holes. The air carts have V-shaped bottoms which ride on a thin film of air, allowing the carts to move almost without friction along the track.

To represent the two skaters pushing apart on nearly frictionless ice, we set up two carts with a spring between them as shown in Figure (6-3a). A thread is tied between the carts to keep the spring compressed. When we burn the thread, the carts fly apart as shown

Figure 1
Two skaters, a father and a son, standing at rest on frictionless ice, push away from each other. The smaller, less massive child recoils faster than the more massive father.

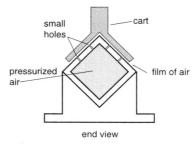

Figure 2
End view of an air track. Pressurized air from the back side of a vacuum cleaner is fed into a square hollow metal tube, and flows out through a series of small holes. A cart, riding on a film of air, can move essentially without friction along the track.

in Figure (6-3b). If the two carts are made of similar material, but one is bigger than the other, the big one will recoil at lesser speed than the small one. We say that the big cart, the one that comes out more slowly, has more mass than the small one.

Because we can precisely measure the speeds v_A and v_B of the recoiling air carts, we can use the experiment pictured in Figures (6-3a,b) to define the mass of the carts. Let us call m_A and m_B the masses of carts A and B respectively. The simplest formula relating the masses of the carts to the recoil speeds, a formula that has the more massive cart recoiling at less speed is

$$\boxed{\frac{m_A}{m_B} = \frac{v_B}{v_A}} \quad \textit{recoil definition of mass} \quad (1)$$

In words, Equation 1 says that the ratio of the masses is inversely proportional to the recoil speeds. I.e., if m_A is the small mass, the v_B is the small speed.

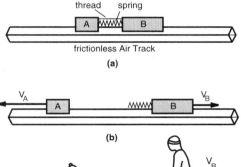

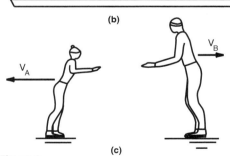

Figure 3
Recoil experiment. To simulate the two skaters pushing apart, we place two carts on an air track with a compressed spring between them. The carts are held together by a string. When the string is burned, the carts fly apart as did the skaters. The more massive cart recoils at a smaller speed $(v_B < v_A)$.

Properties of Mass

Since we now have an explicit prescription for measuring mass, we should carry out some experiments to see if this definition makes sense. Our first test is to see if the mass ratio m_A / m_B changes if we use different strength springs in the recoil experiment. If the ratio of recoil speeds v_B / v_A, and therefore the mass ratio, depends upon what kind of spring we use, then our definition of mass may not be particularly useful.

In the appendix to this chapter, we describe apparatus that allows us to measure the recoil speeds of the carts with fair precision. To within an experimental accuracy of 5% to 10% we find that the ratio v_B / v_A of the recoil speeds does not depend upon how hard the spring pushes the carts apart. When we use a stronger spring, both carts come out faster, in such a way that the speed ratio is unchanged. Thus to the accuracy of this experiment we conclude that the mass ratio does not depend upon the strength of the spring used.

Standard Mass

So far we have talked about the ratio of the masses of the two carts. What can we say about the individual masses m_A or m_B alone? There is a simple way to discuss the masses individually. What we do is select one of the masses, for instance m_B, as the **standard mass**, and measure all other masses in terms of m_B. To express m_A in terms of the standard mass m_B, we multiply both sides of Equation (1) through by m_B to get

$$m_A = m_B \left(\frac{v_B}{v_A} \right) \quad \begin{array}{l} \textit{formula for } m_A \\ \textit{in terms of the} \\ \textit{standard mass } m_B \end{array} \quad (2)$$

For a standard mass, the world accepts that the platinum cylinder kept by the International Bureau of Weights and Measures near Paris, France, is precisely one kilogram. If we reshaped this cylinder into an air cart and used it for our standard mass, then we would have the following explicit formula for the mass of cart A recoiled from the standard mass.

$$m_A = (1 \text{ kilogram}) \times \left(\frac{v_{std}}{v_A}\right) \quad \begin{array}{l}\text{using the}\\ \text{one kilogram}\\ \text{cylinder for our}\\ \text{standard mass}\end{array} \quad (3)$$

where v_{std} is the recoil speed of the standard mass. Once we have determined the mass of one of our own carts, using the standard mass and Equation (3), we can then use that cart as our standard and return the platinum cylinder to the French.

Of course the French will not let just anybody use their standard kilogram mass. What they did was to make accurate copies of the standard mass, and these copies are kept in individual countries, one of them by the National Institute of Standards and Technology in Washington, DC which then makes copies for others in the United States to use.

Addition of Mass

Consider another experiment that can be performed using air carts. Suppose we have our standard cart of mass m_B, and two other carts which we will call C and D. Let us first recoil carts C and D from our standard mass m_B, and determine that C and D have masses m_C and m_D given by

$$m_C = m_B\left(\frac{v_B}{v_C}\right) \quad ; \quad m_D = m_B\left(\frac{v_B}{v_D}\right)$$

Now what happens if, as shown in Figure (6-4), we tie carts C and D together and recoil them from cart B. How is the mass m_{C+D} of the combination of the two carts related to the individual masses m_C and m_D? If we perform the experiment shown in Figure (6-4), we find that

$$m_{C+D} = m_C + m_D \quad \text{mass adds} \quad (4)$$

The experimental result, shown in Equation (4), is that mass adds. The mass of the two carts recoiled together is the sum of the masses of the individual carts. This is the reason we can associate the concept of mass with the quantity of matter. If, for example, we have two identical carts, then together the two carts have twice as much matter and twice as much mass.

Exercise 1

In physics labs, one often finds a set of brass cylinders of various sizes, each cylinder with a number stamped on it, representing its mass in grams. The set usually includes a 50-gm, 100-gm, 200-gm, 500-gm, and 1000-gm cylinder. Suppose that you were given a rod of brass and a hacksaw; describe in detail how you would construct a set of these standard masses. At your disposal you have a frictionless air track, two carts of unknown mass that ride on the track, the standard 1000-gm mass from France (which can be placed on one of the carts), and various things like springs, thread, and matches.

A Simpler Way to Measure Mass

The preceding problem illustrates two things. One is that with an air track, carts, and a standard mass, we can use our recoil definition to measure the mass of an object. The second is that the procedure is clumsy and rather involved. What we need is a simpler way to measure mass.

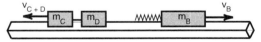

Figure 4
Addition of mass. If we tie two carts C and D together and recoil the pair from our standard mass m_A, and use the formula

$$m_{C+D} = m_B\left(\frac{v_B}{v_{C+D}}\right)$$

for the combined mass m_{C+D}, we find from experiment that $m_{C+D} = m_C + m_D$. In other words the mass of the pair of carts is the sum of the masses of the individual carts, or we can say that mass adds.

The simpler way involves the use of a balance, which is a device with a rod on a pivot and two pans suspended from the rod, as shown in Figure (6-5). If the balance is properly adjusted, we find from experiment that if equal masses are placed in each pan, the rod remains balanced and level. This means that if we place an unknown mass in one pan, and add brass cylinders of known mass to the other pan until the rod becomes balanced, the object and the group of cylinders have the same mass. To determine the mass of the object, all we have to do is add up the masses of the individual cylinders.

Inertial and Gravitational Mass

The pan balance of Figure (6-5) is actually comparing the downward gravitational force on the contents of the two pans. If the gravitational forces are equal, then the rod remains balanced. What we are noting is that there are equal gravitational forces on equal masses. This is an experimental result, not an obvious conclusion. For example, we could construct two air carts, one from wood and one from platinum. Keep adjusting the size of the carts until their recoil speeds are equal, i.e., until they have equal recoil masses. Then put these carts on the pan balance of Figure (6-5). Although the wood cart has a much bigger volume than the platinum one, we will find that the two carts still balance. The gravitational force on the two carts will be the same despite their large difference in size.

In 1922, the Swedish physicist Etvös did some very careful experiments, checking whether two objects, which had the same mass from a recoil type of experiment would experience the same gravitational force as measured by a pan balance type of experiment. He demonstrated that we would get the same result to one part in a billion. In 1960, R. H. Dicke improved Etvös' experiments to an accuracy of 1 part in 10^{11}.

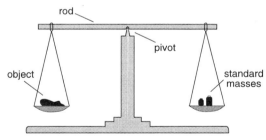

Figure 5
Schematic drawing of a pan balance. If the balance is correctly adjusted and if equal masses are placed in the pans, the rod will remain level. This allows us to determine an unknown mass simply by comparing it to a known one.

It is common terminology to call what we measure in a recoil experiment the ***inertial mass*** of the object, and what we measure using a pan balance the ***gravitational mass***. The experiments of Etvös and Dicke demonstrate that inertial mass and gravitational mass are equivalent to each other to one part in 10^{11}. Is this a coincidence, or is there some fundamental reason why these two definitions of mass turn out to be equivalent? Einstein addressed this question in his formulation of a relativistic theory of gravity known as Einstein's ***General Theory of Relativity***. We will have more to say about that later.

Mass of a Moving Object

One reason we chose the recoil experiment of Figure (3) as our experimental definition of mass is that it allows us to study the mass of moving objects, something that is not possible with a pan balance.

From the air track experiments we have discussed so far, we have found two results. One is that the ratio of the recoil speeds, and therefore the ratio of the masses of the two objects, does not depend upon the strength of the spring or the individual speeds v_A and v_B. If we use a stronger spring so that m_A emerges twice as fast, m_B also emerges twice as fast so that the ratio m_A/m_E is unchanged.

In addition, we found that mass adds. If carts C and D have masses m_C and m_D when recoiled individually from cart B, then they have a combined mass $m_{C,D} = m_C + m_D$ when they are tied together and both recoiled from cart B.

RELATIVISTIC MASS

In our air track experiments, we found that the ratio of the recoil speeds did not depend upon the strength of the spring we used. However, when the recoil speeds approach the speed of light, this simple result can no longer apply. Because of nature's speed limit c, the ratio of the recoil speeds must in general change with speed.

To see why the recoil speed ratio must change, imagine an experiment involving the recoil of two objects of very different size, for example a bullet being fired from a gun as shown in Figure (6). Suppose, in an initial experiment not much gunpowder is used and the bullet comes out at a speed of 100 meters per second and the gun recoils at a speed of 10 cm/sec = .1 m/sec. For this case the speed ratio is 1000 to 1 and we say that the gun is 1000 times as massive as the bullet.

In a second experiment we use more gun powder and the bullet emerges 10 times faster, at a speed of 1000 meters per second. If the ratio of 1000 to 1 is maintained, then we predict that the gun should recoil at a speed of 1 meter per second. If we did the experiment, the prediction would be true.

But, as a thought experiment, imagine we used such powerful gun powder that the gun recoiled at 1% the speed of light. If the speed ratio remained at 1000 to 1, we would predict that the bullet would emerge at a speed 10 times the speed of light, an impossible result. The bullet cannot travel faster than the speed of light, the speed ratio cannot be greater than 100 to 1, and thus the ratio of the masses of the two objects must have changed.

In the next section we will discuss experiments in which, instead of a bullet being fired by a gun, an electron is ejected by an atomic nucleus. The electron is such a small particle that it is often ejected at speeds approaching the speed of light. The nuclei we will consider are so much more massive that they recoil at low speeds familiar to us, speeds like that of a jet plane or earth satellite. At these low speeds the mass of an object does not change noticeably with speed. Thus in these electron recoil experiments, the mass of the nuclei is not changing due to its motion. Any change in the ratio of recoil speeds is due to a change in the mass of the electron as the speed of the electron approaches the speed of light.

We will see that as we push harder and harder on the electron, trying to make it go faster than the speed of light, *the mass of the electron increases instead*. It is precisely this increase in mass that prevents the electron emerging at a speed greater than the speed of light and this is how nature enforces the speed limit c.

Beta (β) Decay

The electron recoils we just mentioned occur in a process called β (beta) decay. In a β decay, a radioactive or unstable nucleus transforms into the nucleus of another element by ejecting an electron at high speeds as illustrated in Figure (7). In the process the nucleus itself recoils as shown.

Figure 6
To discuss higher speed recoils, consider a bullet being fired from a gun. We are all aware that the bullet emerges at a high speed, but the gun itself also recoils. (The recoil of the gun becomes obvious the first time you fire a shotgun.) In this setup, the gunpowder is analogous to the spring, and the gun and bullet are analogous to the two carts.

Figure 7
Radioactive decay of a nucleus by β decay. In this process the unstable nucleus ejects an electron, often at speeds v_e near the speed of light.

The name *β decay* is historical in origin. When Ernest Rutherford (who later discovered the atomic nucleus) was studying radioactivity in the late 1890s, he noticed that radioactive materials emitted three different kinds of radiation or rays, which he arbitrarily called α (alpha) rays, β (beta) rays and γ (gamma) rays, after the first three letters of the Greek alphabet. Further investigation over the years revealed that α rays were beams of helium nuclei, which are also known as α particles. The β rays turned out to be beams of electrons, and for this reason a nuclear decay in which an electron is emitted is known as a *β decay*. The γ rays turned out to be particles of light which we now call photons. (The particle nature of light will be discussed in a later section of this chapter.)

In the 1920s, studies of the β decay process raised serious questions about some fundamental laws of physics. It appeared that in the β decay, energy was sometimes lost. (We will discuss energy and the basic law of conservation of energy in Chapter 9.) In the early 1930s, Wolfgang Pauli proposed that in β decay, two particles were emitted—an electron and an undetectable one which later became known as the neutrino. (We will discuss neutrinos at the end of this chapter.) Pauli's hypothesis was that the missing energy was carried out by the unobservable neutrino. Thirty years later the neutrino was finally detected and Pauli's hypothesis verified.

Some of the time the neutrino created in a β decay carries essentially no energy and has no effect on the behavior of the electron and the nucleus. When this is the case, we have the genuine 2-particle recoil experiment illustrated in Figure (7). This is a recoil experiment in which one of the particles emerges at speeds near the speed of light.

Electron Mass in β Decay

Applying our definition of mass to the β decay process of Figure (7) we have

$$\frac{m_e}{m_n} = \frac{v_n}{v_e} \qquad \overset{v_e}{\longleftarrow} \; m_e \quad (m_n) \; \longrightarrow v_n \quad (5)$$

where m_e and v_e are the mass and recoil speed of the electron and m_n and v_n of the nucleus. We are assuming that the nucleus was originally at rest before the β decay.

To develop a feeling for the speeds and masses involved in the β decay process, we will analyze two examples of the β decay of a radioactive nucleus. In the first example, which we introduce as an exercise to give you some practice calculating with Equation (5), we can assume that the electron's mass is unchanged and still predict a reasonable speed for the ejected electron. In the second example, the assumption that the electron's mass is unchanged leads to nonsense.

Plutonium 246

We will begin with the decay of a radioactive nucleus called ***Plutonium 246***. This is not a very important nucleus. We have selected it because of the way in which it β decays.

The number 246 appearing in the name tells us the number of protons and neutrons in the nucleus. Protons and neutrons have approximately the same mass m_p which has the value

$$m_p = 1.67 \times 10^{-27} \text{ kg} \quad \textit{mass of proton} \tag{6}$$

The Plutonium 246 nucleus has a mass 246 times as great, thus

$$\begin{aligned} m_{Plutonium\,246} &= 246 \times m_p \\ &= 4.10 \times 10^{-25} \text{ kg} \end{aligned} \tag{7}$$

An electron at rest or moving at slow speeds has a mass $(m_e)_0$ given by

$$(m_e)_0 = 9.11 \times 10^{-31} \text{ kg} \tag{8}$$

This is called the ***rest mass*** of an electron. We have added the subscript zero to remind us that this is the mass of a slowly moving electron, one traveling at speeds much less than the speed of light.

Exercise 2 β Decay of Plutonium 246

A Plutonium 246 nucleus has an average lifetime of just over 11 days, upon which it decays by emitting an electron. If the nucleus is initially at rest, and the decay is one in which the neutrino plays no role, then the nucleus will recoil at the speed

$$v_n = 572 \, \frac{\text{meters}}{\text{second}} \quad \begin{array}{l} \textit{recoil speed of} \\ \textit{Plutonium 246} \\ \textit{in a β decay} \end{array} \tag{9}$$

This recoil speed is not observed directly, but enough is known about the Plutonium 246 β decay that this number can be accurately calculated. Note that a speed of 572 meters/second is a bit over 1000 miles per hour, the speed of a supersonic jet.

Your exercise is to predict the recoil speed v_e of the electron assuming that the mass of the electron m_e is the same as the mass $(m_e)_0$ of an electron at rest.

Your answer should be

$$v_e = .86 \, c \tag{10}$$

where

$$c = 3 \times 10^8 \, \frac{\text{meters}}{\text{second}} \tag{11}$$

is the speed of light.

The above exercise, which you should have done by now, shows that we do not get into serious trouble if we assume that the mass of the electron did not change due to the electron's motion. The predicted recoil speed $v_e = .86c$ is a bit too close to the speed of light for comfort, but the calculation does not exhibit any obvious problems. This is not true for the following example.

Protactinium 236

An even more obscure nucleus is Protactinium 236 which has a lifetime of about 12 minutes before it β decays. The Protactinium β decay is, however, much more violent than the Plutonium 246 decay we just discussed. If the Protactinium 236 nucleus is initially at rest, and the neutrino plays no significant role in the decay, then the recoil velocity of the nucleus is

$$v_n = 5170 \frac{\text{meters}}{\text{second}} \quad \begin{array}{l}\textit{recoil speed of}\\ \textit{Protactinium 236}\\ \textit{nucleus}\end{array} \quad (12)$$

This is nine times faster than the recoil speed of the Plutonium 246 nucleus.

Exercise 3 Protactinium 236 β decay.

Calculate the recoil speed of the electron assuming that the mass of the recoiling electron is the same as the mass of an electron at rest. What is wrong with the answer?

You do not have to work Exercise 3 in detail to see that we get a into trouble if we assume that the mass of the recoiling electron is the same as the mass of an electron at rest. We made this assumption in Exercise 2, and predicted that the electron in the Plutonium 246 β decay emerged at a speed of .86 c. Now a nucleus of about the same mass recoils 9 times faster. If the electron mass is unchanged, it must also recoil 9 times faster, or over seven times the speed of light. This simply does not happen.

Exercise 4 Increase in Electron Mass.

Reconsider the Protactinium 236 decay, but this time assume that the electron emerges at essentially the speed of light ($v_e = c$). (This is not a bad approximation, it actually emerges at a speed $v = .99c$). Use the definition of mass, Equation 5, to calculate the mass of the recoiling electron. Your answer should be

$$m_e = 6.8 \times 10^{-30} \text{kg} = 7.47 \times (m_e)_0 \quad (13)$$

In Exercise 4, you found that by assuming the electron could not travel faster than the speed of light, the electron mass had increased by a factor of 7.47. The emerging electron is over 7 times as massive as an electron at rest! Instead of emerging at 7 times the speed of light, the electron comes out with 7 times as much mass.

Exercise 5 A Thought Experiment.

To illustrate that there is almost no limit to how much the mass of an object can increase, imagine that we perform an experiment where the earth ejects an electron and the earth recoils at a speed of 10 cm/sec. (A β decay of the earth.) Calculate the mass of the emitted electron. By what factor has the electron's mass increased?

THE EINSTEIN MASS FORMULA

A combination of the recoil definition of mass with the observation that nothing can travel faster than the speed of light, leads to the conclusion that the mass of an object must increase as the speed of the object approaches the speed of light. Determining the formula for how mass increases is a more difficult job. It turns out that we do not have enough information at this point in our discussion to derive the mass formula. What we have to add is a new basic law of physics called the *law of conservation of linear momentum*.

We will discuss the conservation of linear momentum in the next chapter, and in the appendix to that chapter, derive the formula for the increase in mass with velocity. We put the derivation in an appendix because it is somewhat involved. But the answer is very simple, almost what you might guess.

In our discussion of moving clocks in Chapter 1, we saw that the length T' of the astronaut's second increased according to the formula

$$T' = \frac{T}{\sqrt{1-v^2/c^2}} \qquad (1\text{-}11)$$

where T was the length of one of our seconds. For slowly moving astronauts where $v \ll c$, we have $T' \approx T$ and the length of the astronaut's seconds is nearly the same as ours. But as the astronaut approaches the speed of light, the number $\sqrt{1-v^2/c^2}$ becomes smaller and smaller, and the astronaut's seconds become longer and longer. If the astronaut goes at the speed of light, $1/\sqrt{1-v^2/c^2}$ becomes infinitely large, the astronaut's seconds become infinitely long, and time stops for the astronaut.

Essentially the same formula applies to the mass of a moving object. If an object has a mass m_o when at rest or moving slowly as in air cart experiments (we call m_o the *rest mass* of the object), then when the object is moving at a speed v, its mass m is given by the formula

$$\boxed{m = \frac{m_o}{\sqrt{1-v^2/c^2}}} \qquad \textit{Einstein mass formula} \qquad (14)$$

a result first deduced by Einstein.

Equation (14) has just the properties we want. When the particle is moving slowly as in our air cart recoil experiments, $v \ll c$, $\sqrt{1-v^2/c^2} \approx 1$ and the mass of the object does not change with speed. But as the speed of the object approaches the speed of light, the $\sqrt{1-v^2/c^2}$ approaches zero, and $m = m_o/\sqrt{1-v^2/c^2}$ increases without bounds. If we could accelerate an object up to the speed of light, it would acquire an infinite mass.

Exercise 6
At what speed does the mass of an object double (i.e., at what speed does $m = 2\,m_0$?) (Answer: $v = .866\,c$.)

Exercise 7
Electrons emerging from the Stanford Linear Accelerator have a mass 200,000 times greater than their rest mass. What is the speed of these electrons? (The answer is $v = .9999999999875\,c$. Use the approximation formulas discussed in Chapter 1 to work this problem.)

Exercise 8
A car is traveling at a speed of $v = 68$ miles per hour. (68 miles/hr = 100 ft/second = 10^{-7} ft/nanosecond = $10^{-7}\,c$.) By what factor has its mass increased due to its motion. (Answer: $m/m_o = 1.000000000000005$.)

Nature's Speed Limit

When the police try to enforce a 65 mile/hr speed limit, they have a hard job. They have to send out patrol cars to observe the traffic, and chase after speeders. Even with the most careful surveillance, many drivers get away with speeding.

Nature is more clever in enforcing its speed limit c. By having the mass of an object increase as the speed of the object approaches c, it becomes harder and harder to change the speed of the object. If you accelerated an object up to the speed of light, its mass would become infinite, and it would be impossible to increase the particle's speed.

Historically it was noted that massive objects were hard to get moving, but when you got them moving, they were hard to stop. This tendency of a massive object to keep moving at constant velocity was given the name *inertia*. That is why our recoil definition of mass, which directly measures how hard it is to get an object moving, measures what is called *inertial mass*. Nature enforces its speed limit c by increasing a particle's inertia to infinity at c, making it impossible to accelerate the particle to higher speeds. Because of this scheme, no one speeds and no police are necessary.

ZERO REST MASS PARTICLES

If you think about it for a while, you may worry that nature's enforcement of its speed limit c is too effective. With the formula $m = m_o/\sqrt{1-v^2/c^2}$, we expect that nothing can reach the speed of light, because it would have an infinite mass, which is impossible.

What is light? It travels at the speed of light. If light consists of a beam of particles, and these particles travel at the speed c, then the formula $m = m_o/\sqrt{1-v^2/c^2}$ suggests that these particles have an infinite mass, which is impossible.

Then perhaps light does not consist of particles, and is therefore exempt from Einstein's formula. Back in Newton's time there was considerable debate over the nature of light. Isaac Newton supported the idea that light consisted of beams of particles. Red light was made up of red particles, green light of green particles, blue light of blue particles, etc. Christian Huygens, a well known Dutch physicist of the time, proposed that light was made up of waves, and that the different colors of light were simply waves with different wavelengths. Huygens developed the theory of wave motion in order to support his point of view. We will discuss Huygen's theory later in the text.

In 1801, about 100 years after the time of Newton and Huygens, Thomas Young performed an experiment that settled the debate they started. With his so called *two slit experiment* Young conclusively demonstrated that light was a wave phenomena.

Another century later in 1905, the same year that he published the special theory of relativity, Einstein also published a paper that conclusively demonstrated that light consisted of beams of particles, particles that we now call ***photons***. (Einstein received the Nobel Prize in 1921 for his paper on the nature of light. At that time his special theory of relativity was still too controversial to be awarded the prize.)

Thus by 1905 it was known that light was both a particle and a wave. How this could happen, how to picture something as both a particle and a wave was not understood until the development of quantum mechanics in the period 1923 through 1925.

Despite the fact that light has a wave nature, it is still made up of beams of particles called photons, and these particles travel at precisely the speed c. If we apply Einstein's mass formula to photons, we get for the photon mass m_{photon}

$$m_{photon} = \left.\frac{m_0}{\sqrt{1-v^2/c^2}}\right|_{v=c} = \frac{m_0}{\sqrt{1-1}} = \frac{m_0}{0} \quad (15)$$

where m_0 is the rest mass of the photon.

At first sight it looks like we are in deep trouble with Equation (15). Division by zero usually leads to a disaster called infinity. There is one exception to this disaster. If the rest mass m_0 of the photon is zero, then we get

$$m_{photon} = \frac{m_0}{0} = \frac{0}{0} \quad (16)$$

The number 0/0 is not a disaster, it is simply undefined. It can be 1 or 2.7, or 6×10^{-23}. It can be any number you want. (How many nothings fit into nothing? As many as you want.) In other words, if the rest mass m_0 of a photon is zero, the Einstein mass formula says nothing about the photon's mass m_{photon}. Photons do have mass, but the Einstein mass formula does not tell us what it is. (Einstein presented a new formula for the photon's mass in his 1905 paper. He found that the photon's mass was proportional to the frequency of the light wave.)

We will study Einstein's theory of photons in detail later in the text. All we need to know now is that light consists of particles called photons, these particles travel ***at*** the speed of light, and these particles have no rest mass. If you stop photons, which you do all the time when light strikes your skin, no particles are left. There is no residue of stopped photons on your skin. All that is left is the heat energy brought in by the light.

A photon is an amazing particle in that it exists only when moving at the speed of light. There is no lapse of time for photons; they cannot become old. (They cannot spontaneously decay like muons, because their half life would be infinite.) There are two different worlds for particles. Particles with rest mass cannot get up to the speed of light, while particles without rest mass travel only at the speed of light.

NEUTRINOS

Another particle that may have no rest mass is the neutrino. According to current theory there should be three different kinds of neutrinos, but for now we will not distinguish among them.

In our discussion of the β decay process, we mentioned that when a radioactive nucleus decays by emitting an electron, a neutrino is also emitted. Most of the time the energy given up by the nucleus is shared between the electron and the neutrino, thus the electrons carried out only part of the energy. The very existence of the neutrino was predicted from the fact that some energy appeared to be missing in β decay reactions and it was Pauli who suggested that this energy was carried out by an undetected particle.

Neutrinos are difficult to detect. They can pass through immense amounts of matter without being stopped or deflected. In comparison photons are readily absorbed by matter. As any scuba diver knows, even in the clearest ocean, a good fraction of the sunlight is absorbed by the time you get down to a depth of 50 or more feet. At that depth most of the red light has been absorbed and objects have a grayish blue cast. In muddy water photons are absorbed much more rapidly, and opaque objects like your skin stop photons in the distance of a few atomic diameters.

On the other hand, neutrinos can pass through the earth with almost no chance of being stopped. As a writer discussing the 1987 supernova explosion phrased it, the neutrinos from the supernova explosion swept through the earth, the earth being far more transparent to the neutrinos than a thin sheet of the clearest glass to light.

Neutrinos are now detected by what one might call a brute force technique. Aim enough neutrinos at a big enough detector and a few will be stopped and observed. The first time neutrinos were detected was in an experiment by Clyde Cowan and Fred Reines, performed in 1956, almost 30 years after Pauli had proposed the existence of the particle. Noting that nuclear reactors are a prodigious source of neutrinos, Cowan and Reines succeeded in detecting neutrinos by building a detector the size of a railroad tank car and placing it next to the reactor at Savannah River, Georgia.

The largest neutrino detectors now in use were originally built to detect the spontaneous decay of the proton (a process that has not yet been observed). They consist of a swimming pool sized tank of water surrounded by arrays of photocells, all located in deep mines to shield them from cosmic rays. If a proton decays, either spontaneously or because it was struck by a neutrino, a tiny flash of light is emitted in the subsequent particle reaction. The flash of light is then detected by one of the photocells.

Solar Neutrinos

Aside from nuclear reactors, another powerful source of neutrinos is the sun. Beta decay processes and neutrino emission are intimately associated with the nuclear reactions that power the sun. As a result neutrinos emerge from the small hot core at the center of the sun where the nuclear reactions are taking place. The sun produces so many neutrinos that we can detect them here on earth.

There is a good reason to look for these solar neutrinos. The neutrinos created in the core of the sun pass directly through the outer layers of the sun and reach us eight minutes after they were created in a nuclear reaction. In contrast, light from the hot bright core of the sun takes the order of 14,000 years to diffuse its way out to the surface of the sun. If for some reason the nuclear reactions in the sun slowed down and the core cooled, it would be about 14,000 years before the surface of the sun cooled. But the decrease in neutrinos could be detected here on earth within 8 minutes. Looking at the solar neutrinos provides a way of looking at the future of the sun 14,000 years from now.

Solar neutrinos have been studied and counted since the 1960s. Computer models of the nuclear reactions taking place in the sun make explicit predictions about how many neutrinos should be emitted. The neutrino detectors observe only about 1/3 to 1/2 that number. There have been a number of experiments using various kinds of detectors, and all the experiments show this deficiency.

If the deficiency is really an indication that the nuclear reactions in the sun's core have slowed, then we can expect a cooling of the sun within 14,000 years, a cooling that might have a significant impact on the earth's climate. On the other hand there may be some part of the nuclear reactions in the sun that we do not fully understand, with the result that the computer predictions are in error. We are not sure yet which is correct; the solar neutrino deficiency is one of the current areas of active research.

Neutrino Astronomy

An event on the night of February 23, 1987 changed the role of neutrinos in modern science. On that night neutrinos were detected from the supernova explosion in the Magellanic cloud, a small neighboring galaxy. This was the first time neutrinos were detected from an astronomical source other than our sun. The information we obtained from this observation represented what one could call the birth of neutrino astronomy.

A supernova is an exploding star, an event so powerful that, for a short period of time of about 10 seconds, the star radiates more power than all the rest of the visible universe. And this energy is radiated in the form of neutrinos.

The supernova explosion occurs when the core of a large star runs out of nuclear fuel and collapses. (This only happens to stars several times larger than our sun.) The gravitational energy released in the collapse is what provides the energy for the explosion. We know that sometimes a neutron star is formed at the center of the collapsed core, and computer simulations predict

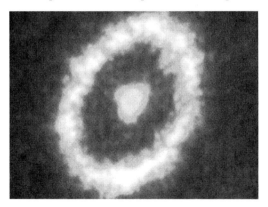

Figure 8
1987 Supernova at age 3 1/2 *years, photographed by the Hubble telescope. The ring is gas blown off by the explosion.*

that much of the energy released in the collapse is carried out in a burst of neutrinos. The core material is so dense that even the neutrinos have some difficulty getting out. They take about 10 seconds to diffuse out of the core, and as a result the neutrino pulse is about 10 seconds long.

The collapsing core also creates a shock wave that spreads out through the outer layers of the star, reaching the surface in about three hours. When the shock wave reaches the surface, the star suddenly brightens and we can see from the light that the star has exploded.

The details about the core collapse, the neutrino burst and the shock wave are all from computer models of supernova explosions, models developed over the past 25 years. Whenever you model a physical process, you like to test your model with the real process. Computer models of supernova explosions are difficult to test because there are so few supernova explosions. The last explosion in our galaxy, close enough to study in detail, occurred in 1604, shortly before the invention of the telescope.

The supernova explosion on February 23, 1987 was not only close enough to be studied, several fortunate coincidences provided much detailed information. The first coincidence was the fact that theoretical physicists had predicted in the 1960s that the proton might spontaneously decay (with a half life of about 10^{32} years.) To detect this weak spontaneous decay, several large detectors were constructed. As we mentioned, these large detectors were also capable of detecting neutrinos. On February 23, at 7:36 AM universal time, the detectors in the Kamokande lead mine in Japan, the Morton Thekol salt mine near Cleveland, Ohio and at Baksam in the Soviet Union all detected a 10 second wide pulse of neutrinos. Since the Magellanic cloud and the supernova are visible only from the southern hemisphere and all the neutrino detectors are in the northern hemisphere, all the detected neutrinos had to pass through the earth. The 10 second width of the pulse verified earlier computer models about the diffusion of neutrinos out of the collapsing core.

The exact time of the arrival of the light from the supernova explosion is harder to pin down, but some fortunate coincidences occurred there too. The supernova was first observed by a graduate student Ian Sheldon working at the Las Campanas Observatory in Chile. Ian was photographing the large Magellanic cloud on the night of February 23, 1987, and noted that a plate that he had exposed that night had a bright stellar object that was not on the plate exposed the night before. The object was so bright it should be visible to the naked eye. Ian went outside, looked up, and there it was.

Once the supernova had been spotted, there was an immediate search for more precise evidence of when the explosion had occurred. A study of the records of the neutrino detectors turned up the ten second neutrino pulse at 7:36 AM on February 23. Three hours after that Robert McNaught, an observer in Siding Spring, Australia, had exposed two plates of the large Magellanic clouds. When the plates were developed later, the supernova was visible. One hour before McNaught exposed his plates, Albert Jones, an amateur astronomer in New Zealand happened to be observing at the precise spot where the supernova occurred and saw nothing unusual. Thus the light from the supernova explosion arrived at some time between two and three hours after the neutrino pulse.

The fact that the photons from the supernova explosion arrived two to three hours later than the neutrinos, is not only a good test of the computer models of the supernova explosion, it also provides an excellent check on the rest mass of the neutrino. The 1987 supernova occurred 160,000 light years away from the earth. After the explosion, neutrinos and photons raced toward the earth. The neutrinos had a 3 hour head start, and after traveling for 160,000 years, the neutrinos were still 2 hours, and perhaps 3 hours ahead. That is as close a race as you can expect to find. From this we can conclude that neutrinos travel at, or very, very close to the speed of light. And therefore their mass must be precisely zero, or very close to it.

Chapter 7
Conservation of Linear and Angular Momentum

The truly basic laws of physics, like the principle of relativity, not only have broad applications, but are often easy to describe. The principle of relativity says that there is a quantity, namely your own uniform motion, that you cannot detect. The hard part is working out the implications of the simple idea.

*In this chapter we discuss two more basic laws of physics, laws that apply with no known exceptions to objects as large as galaxies and as small as subatomic particles. These are the **laws of the conservation of linear momentum** and the **law of the conservation of angular momentum**. These are the first of several so-called **conservation** laws that we will encounter in our study of physics. A conservation law states that there is some quantity which does not change in a given set of experiments.*

We will introduce our first example of a conservation law by going back to the results of the aircart recoil experiments that we used in the last chapter to define mass. In analyzing these results, we will see that there is a quantity, which we will call linear momentum, which does not change when the spring is released and the carts recoil. We will then look at a wider class of experiments, in which objects not only recoil, but collide at different angles. Again we will see that linear momentum does not change.

We will also see that linear momentum is conserved not just for familiar objects like billiard balls, but also for objects as small as protons colliding in a hydrogen bubble chamber.

In the appendix to this chapter we will show how the recoil definition of mass, when combined with the law of conservation of linear momentum and the principle of relativity, leads to Einstein's relativistic mass formula $m = m_0 /\sqrt{1 - v^2/c^2}$.

The second conservation law deals with angular momentum. The concept of angular momentum is a bit more subtle than that of linear momentum. As a result, in this chapter we will focus on developing an intuitive feeling for the concept. A more formal mathematical treatment will be put off until a later chapter where the formalism is needed.

The law of conservation of angular momentum has many applications that range from the astronomical scale to the subatomic scale of distance. The very existence of planets in the solar system is a consequence of the conservation of angular momentum. The law also allows us to understand the behavior of atomic nuclei in a magnetic field, a behavior that is involved in the creation of the marvelous images seen in magnetic resonance imaging apparatus. On the very smallest scale of distance, angular momentum turns out to be one of the basic intrinsic properties of all elementary particles.

CONSERVATION OF LINEAR MOMENTUM

In our discussion of the recoil definition of mass in the last chapter, we looked at a number of experiments that were performed to determine how mass behaves. One of the crucial observations was that, at least with carts on an air track, the ratio of the recoil speeds did not change as we changed the strength of the spring pushing the carts apart. If the big cart came out moving half as fast as the small one when we used a weak spring, then it still came out half as fast when we used a strong spring. With the strong spring both speeds were greater, but the ratio was still the same.

We used this unchanging ratio in our definition of mass. If, as shown in Figure (6-3) reproduced here, cart A recoils at a speed v_A and cart B at a speed v_B, then the mass ratio m_A/m_B was defined by Equation (6-1) as

$$\frac{m_A}{m_B} = \frac{v_B}{v_A} \qquad (6\text{-}1)$$

What we will do now is manipulate Equation (6-1) until we end up with a quantity that does not change when the spring is released.

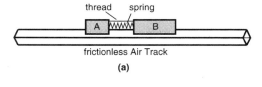

(a)

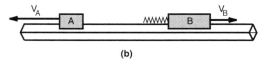

(b)

Figure 6-3
In Chapter 6 we defined the ratio of the mass of the carts m_A/m_B to be equal to the inverse ratio of the recoil speeds v_B/v_A.

Multiplying Equation (6-1) through by m_B and v_A gives

$$m_A v_A = m_B v_B \tag{1}$$

Next, note that v_A and v_B are the magnitudes of the recoil velocity vectors $\vec{v}_A$ and $\vec{v}_B$. Since $\vec{v}_A$ and $\vec{v}_B$ point in opposite directions, we can write Equation (1) as a vector equation in the form

$$m_A \vec{v}_A = -m_B \vec{v}_B \tag{2}$$

where the minus sign handles the fact that $\vec{v}_A$ and $\vec{v}_B$ are oppositely directed.

Now move the $-m_B \vec{v}_B$ to the left hand side to give the result

$$\boxed{m_A \vec{v}_A + m_B \vec{v}_B = 0} \quad \textit{after recoil} \tag{3}$$

where $\vec{v}_A$ and $\vec{v}_B$ are the cart's velocity vectors after the recoil.

We will now introduce a new interpretation. Let us define the **linear momentum** of a particle as the product of the particle's mass times its velocity. Using the letter p to denote linear momentum, we have

$$\boxed{\vec{p} \equiv m\vec{v}} \quad \textit{definition of linear momentum} \tag{4}$$

Note that linear momentum $\vec{p}$ is a vector because it is the product of a number, the mass m, times a vector, the velocity $\vec{v}$.

Looking back at Equation (3), we see that $m_A \vec{v}_A$ is the linear momentum of cart A after the recoil, and $m_B \vec{v}_B$ is the linear momentum of cart B after the recoil. Equation (3) tells us that the sum of these two linear momenta is zero.

Before the string was cut and the carts released, both carts were sitting at rest. Before the release, we have

$$\vec{v}_A = 0, \quad \vec{v}_B = 0 \quad \textit{before release} \tag{5}$$

Thus the sum of the linear momenta before the release was

$$\boxed{m_A \vec{v}_A + m_B \vec{v}_B = 0} \quad \textit{before recoil} \tag{6}$$

Comparing Equations (3) and (6) we see that the sum of the linear momenta of the two carts was not changed by the recoil. This sum was zero before the carts were released, and it is zero afterward.

We will call the sum of the linear momentum of the two carts the total linear momentum $\vec{p}_{tot}$ of the system of two carts.

$$\vec{p}_{tot} \equiv \vec{p}_A + \vec{p}_B = m_A \vec{v}_A + m_B \vec{v}_B$$
definition of total linear momentum (7)

Then we can restate our observation that Equation (3) and (6) look the same by saying that the total linear momentum $\vec{p}_{tot}$ of the system was unchanged by the recoil. Another way of phrasing it is to say that in the recoil experiment, the total linear momentum of the carts is conserved.

At this point, we do not have a new law of physics, instead, we have merely reformulated our definition of mass. But the result turns out to be far more general than we have seen so far. The general law may be stated as follows. *If we have a system of particles, and there is no net external force acting on them, then the total linear momentum of the system of particles is conserved.*

So far, we have not said much about forces and how to recognize them. Thus we will, for now, limit our discussion of the conservation of linear momentum to examples where it is fairly clear that there is no net external force or influence. In our recoil experiment, gravity is pulling down on the carts, the air is pushing up, and the two effects cancel. The air track was explicitly designed so that there would be no net force on the cart.

In our recoil experiment, our system consists of the two carts and the spring. When the thread is burned, the spring exerts a force on both carts, but the spring is part of the system. The spring forces are internal, not external forces, and therefore cannot change the linear momentum of the system. If the linear momentum was zero before the thread was burned, it must still be zero afterward.

COLLISION EXPERIMENTS

A more common example of where external forces can be ignored and linear momentum is conserved is during the collision of two objects like billiard balls. While two objects are colliding, the forces between the objects, the internal forces, are usually much greater than any outside external forces. As a result, just before, during, and just after the collision, external forces can be neglected and linear momentum is conserved.

In an experiment, that is easily carried out in the introductory physics lab, two steel balls are suspended by strings from the ceiling as shown in Figure (1). One of the two balls is pulled back and released. It strikes the ball at rest and the two balls bounce off as seen in the strobe photograph of the motion, Figure (2). The strobe photograph is analyzed in Figure (3) and the resulting momentum vectors are plotted in Figure (4).

What we see in the strobe photograph is ball 2 at rest and ball 1 coming in, attaining a velocity $\vec{v}_{1i}$ just before the collision. After the collision, balls 1 and 2 bounce off in different directions, with velocity vectors $\vec{v}_{1f}$ and $\vec{v}_{2f}$ respectively.

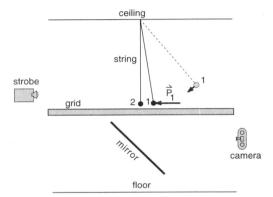

Figure 1
Experimental setup to study the conservation of linear momentum during the collision of two balls.

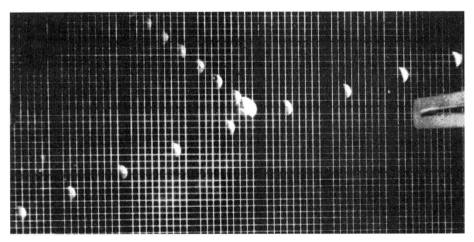

Figure 2
Strobe photograph taken using the setup of Figure (1). The data for the experiment are $m_1 = 70.3$ gm (the ball initially released), $m_2 = 240$ gm (ball initially at rest), $\Delta t = 1/10$ sec (period between flashes). Spacing between grid lines = 1 cm. (Photograph from a student lab notebook.)

In Figure (4) we have plotted the momentum $\vec{p}_{1i}$ of ball 1 just before the collision

$$\vec{p}_{1i} = m\vec{v}_{1i} = \begin{matrix}\text{momentum of}\\ \text{ball 1 before}\\ \text{the collision}\end{matrix} \quad (8)$$

and also plotted the momenta $\vec{p}_{1f}$ and $\vec{p}_{2f}$ of balls 1 and 2 after the collision. The vector sum of these two momenta is the total momentum $\vec{p}_f$ being carried out by the two balls

$$\vec{p}_f \equiv \vec{p}_{1f} + \vec{p}_{2f} = \begin{matrix}\text{momentum being}\\ \text{carried out of}\\ \text{the collision}\end{matrix} \quad (9)$$

From Figure (4) we see that $\vec{p}_f$ is equal to $\vec{p}_{1i}$ the momentum brought into the collision by ball 1. Since the same amount of momentum came out of the collision as was carried in, the total linear momentum did not change. The total linear momentum of the system of the two balls was conserved during the collision.

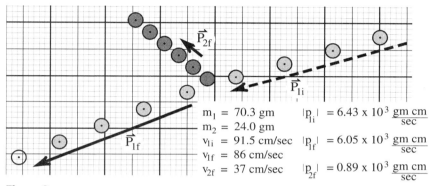

$m_1 = 70.3$ gm
$m_2 = 24.0$ gm
$v_{1i} = 91.5$ cm/sec
$v_{1f} = 86$ cm/sec
$v_{2f} = 37$ cm/sec

$|p_{1i}| = 6.43 \times 10^3 \frac{\text{gm cm}}{\text{sec}}$
$|p_{1f}| = 6.05 \times 10^3 \frac{\text{gm cm}}{\text{sec}}$
$|p_{2f}| = 0.89 \times 10^3 \frac{\text{gm cm}}{\text{sec}}$

Figure 3
Analysis of Figure (2). Ball 1 enters with a momentum $\vec{p}_{1i}$ and collides with Ball 2 which is initially at rest. After the collision, Balls 1 and 2 emerge with momenta $\vec{p}_{1f}$ and $\vec{p}_{2f}$ respectively. (Each large square on this graph paper represents a distance of 10 cm.)

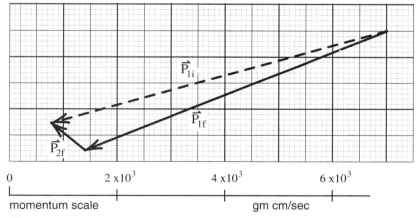

Figure 4
Here we see that the momentum $\vec{p}_{1i}$ brought in by Ball 1 is equal to the momentum $\vec{p}_{1f} + \vec{p}_{2f}$ carried out after the collision.

Exercise 1

Figures (5 a and b) show the collision between two balls of equal mass $m_1 = m_2 = 73$ grams. Again $\Delta t = 1/10$ sec. Using one of these figures, construct a graph similar to Figure (4), and compare the momentum brought in by Ball 1 with the momentum carried out by the two balls after the collision.

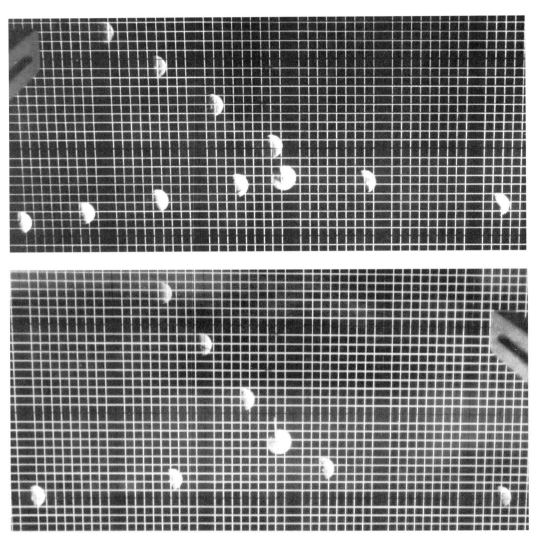

Figures 5 a, b
Strobe photographs of the collision of two equal mass balls.

Subatomic Collisions

In the study of subatomic particles, you cannot photograph or image the particles themselves, the best you can do is study the tracks left behind in a particle detector. A common particle detector, developed by Don Glaser in the early 1950s, is the bubble chamber. We will discuss the bubble chamber in more detail in later chapters. However, the basic idea is that the bubble chamber is filled with liquid hydrogen, and that a charged particle moving through the liquid hydrogen leaves a track that can be made visible as a string of bubbles.

Bubble chambers are used primarily to study the collisions between subatomic particles. In Figure (6) we have a bubble chamber photograph in which an incoming proton from a particle accelerator moves through the liquid hydrogen until it strikes a hydrogen nucleus, namely another proton. The two protons emerge from the collision, coming out at right angles as shown.

After we discuss the law of conservation of energy, we will show that if two identical particles collide, one of them being initially at rest, then if both energy and linear momentum are conserved during the collision the particles must emerge at right angles. Thus we can use the right angle between the emerging proton tracks in Figure (6) as experimental evidence that linear momentum is conserved even among the interactions of subatomic particles.

The following examples and exercises are chosen to show some of the more practical applications of the conservation of linear momentum.

Example 1 *Rifle and Bullet*

A 2-kilogram rifle fires a 10-gram bullet at a speed of 400 meter/sec. What is the recoil velocity of the gun? In this case, the rifle and bullet are initially at rest and have zero total linear momentum. Just after the bullet leaves the gun, before any external forces have had time to act on the system, the total momentum of the system (gun plus bullet) is still zero. We get

$$\vec{p}_{gun} + \vec{p}_{bullet} = 0$$

$$m_g \vec{v}_g = -m_b \vec{v}_b$$

where the minus sign indicates that $\vec{v}_g$ is in the opposite direction to the motion of the bullet. Solving for the magnitude v_g of the recoil velocity, we get

$$v_g = \frac{m_b}{m_g} v_b = \frac{10 \text{ gm}}{2000 \text{ gm}} \times 400 \frac{\text{meters}}{\text{sec}}$$

$$v_g = 2 \frac{\text{meters}}{\text{sec}}$$

Thus, we see that the initial recoil velocity of the gun is 2 meters/sec.

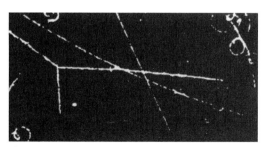
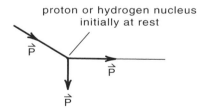

Figure 6
Collision between two protons. When a charged elementary particle passes through the liquid hydrogen in a bubble chamber, it leaves a trail of bubbles that can be photographed. Here we see a proton coming into the picture from the upper left, and striking the nucleus of one of the hydrogen atoms in the liquid hydrogen. The hydrogen nucleus is itself a proton, and after the collision the two protons emerge as shown in the sketch.

In this example we applied the law of conservation of linear momentum over such a short time that outside forces did not have time to act on the system. The conservation of linear momentum applies over longer times, but we must enlarge our concept of the system, as seen in Example (2),

Example 2

A 78 kilogram hunter standing on nearly frictionless ice fires the gun of the preceding example. What is the recoil velocity of the hunter?

Our system now consists of the bullet, gun and hunter. Initially the total linear momentum of the system is zero. After the bullet is fired, and after the gun is firmly lodged against the shoulder of the hunter, the gun and hunter together recoil at a velocity v_h. Applying the law of conservation of linear momentum, we have (remembering that v_g now equals v_h)

$$\vec{P}_{hunter} + \vec{P}_{gun} + \vec{P}_{bullet} = 0$$

$$m_h \vec{v}_h + m_g \vec{v}_h = -m_b \vec{v}_b$$

$$v_h = \frac{m_b}{m_h + m_g} v_b = \frac{10 \text{ gm}}{(78 + 2) \text{ kg}} \times 400 \frac{\text{meters}}{\text{sec}}$$

$$v_h = \frac{10 \text{ gm}}{80,000 \text{ gm}} \times 400 \frac{\text{meters}}{\text{sec}} = \frac{1}{20} \frac{\text{meter}}{\text{sec}}$$

$$v_h = 5 \frac{\text{cm}}{\text{sec}}$$

Exercise 2

a) Starting from the preceding two examples, further enlarge the system. Assume that the hunter is standing firmly on the earth when the gun is fired. Taking the point of view that the earth is initially at rest, calculate the recoil velocity of the gun, hunter, and earth. ($m_{earth} = 6 \times 10^{27}$ gm)

b) After the bullet strikes the ground, what is the velocity of the earth, assuming it was at rest before the gun was fired?

Exercise 3 Frictionless Ice

Suppose you are sitting in the middle of a completely frictionless surface, such as an idealized pond of ice. Propose a method of getting out of such a predicament. (Problem from J. Orear, Fundamental Physics, Wiley, New York, 1961.)

Exercise 4 Bullet and Block

A 10-gram bullet traveling 300 meters/sec strikes and lodges in a 3-kilogram block of wood initially at rest on a pond of ice. What is the final velocity of the block and bullet after the collision?

Exercise 5 Two Skaters Throwing Ball

Two skaters, each of mass 60 kilogram, are standing a slight distance apart on nearly frictionless ice. Initially at rest, they throw a 1-kilogram ball back and forth between them; each time the ball travels at a speed of 10 meters/sec over the ice.

(a) What is the recoil velocity of the first skater immediately after he throws the ball for the first time?

(b) After the second skater catches the ball for the first time, what is his recoil velocity?

(c) After the second skater has thrown the ball back for the first time, what is his recoil velocity?

(d) After the ball has made 10 complete round trips and the first skater is holding the ball, what is the velocity of each skater? What is the total momentum of the system of the two skaters and the ball?

Exercise 6 Rocket

An 11-ton rocket consists of 10 tons of fuel. If the fuel is discharged as exhaust gasses that travel at an average speed of 1 mile/sec (relative to the earth), how fast will the rocket be traveling when the fuel is used up? Neglect gravity and air resistance. (Hint: Consider the total momentum of all the exhaust gas.)

CONSERVATION OF ANGULAR MOMENTUM

Anyone who has watched figure skating in the winter Olympics has seen an example of the conservation of angular momentum. When a figure skater like the one shown in Figure (7) starts her spin, her arms are outstretched and she is turning slowly. As she brings her arms in, she turns faster and faster until the maneuver is completed. She starts the spin with a certain amount of angular momentum, and that amount does not change, is conserved, throughout the spin. To understand the concept of angular momentum, we have to see why the skater had the same angular momentum when rotating slowly with her arms outstretched and rotating rapidly with her arms pulled close to her body.

Even those of us who are not skilled figure skaters can repeat the skaters experience of a spin using a rotating platform and two iron dumbbells. When done as a classroom demonstration, this is sometimes known as the "three dumbbell experiment". The instructor stands on the rotating platform and holds the dumbbells out as shown in Figure (8a). A student helps in the demonstration by starting the instructor rotating slowly. The instructor then pulls in his arms and rotates even faster than a figure skater because of the mass in the dumbbells (Figure 8b). However unless the instructor is skilled at this demonstration, he is likely to make a far less graceful exit from the spin than do the Olympic figure skaters.

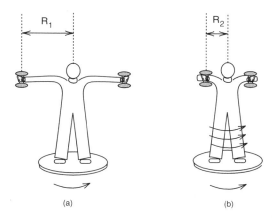

(a)   (b)

Figure 8
The "three dumbbell" experiment. The instructor, standing on a platform that is free to rotate, holds two dumbbells out at arm length as shown in (a). With a slight push from a student, the instructor starts to rotate slowly. The instructor then pulls his arms in, and the rotation increases significantly (b).

Figure 7
Figure skater doing a spin. As the skater pulls her arms in, she turns faster and faster. This is an example of the conservation of angular momentum.

In a more controlled and idealized experiment, we can set a ball swinging in a circle at the end of a string as shown in Figure (9). Let the other end of the string pass down through the small end of a plastic funnel mounted on a board as shown in Figure (9a). If we pull down on the string to reduce the radius r of the circle around which the ball is traveling, the speed v of the ball increases. An analysis of this motion shows a simple result—the product of the radius of the circle times the speed of the ball remains constant. If the ball is initially moving in a circle of radius r_1 and a speed v_1 as shown in Figure (9b), and we reduce the radius to a length r_2 as shown in Figure (9c), the new speed v_2 is given by the equation

$$v_1 r_1 = v_2 r_2 \tag{10}$$

Since r_2 is smaller than r_1, v_2 must be bigger than v_1 to keep the product constant.

The angular momentum of a ball of mass m traveling at a speed v in a circle of radius r is defined to be the product mvr. Using the letter ℓ to represent angular momentum, we have

$$\boxed{\ell \equiv mvr} \quad \begin{array}{l}\text{angular momentum of a}\\ \text{mass m traveling at a speed}\\ \text{v in a circle of radius r}\end{array} \tag{11}$$

In Figure (9a), the ball has an angular momentum

$$\ell_1 = mv_1 r_1 \tag{12}$$

after the string is shortened, the angular momentum ℓ_2 is

$$\ell_2 = mv_2 r_2 \tag{13}$$

Since the mass m of the ball did not change, and because $v_1 r_1 = v_2 r_2$, we see that

$$\ell_1 = \ell_2 \tag{14}$$

and in this example angular momentum did not change. The ball's angular momentum was conserved while we pulled on the string and the ball sped up. It was also conserved when the figure skater pulled in her arms during the spin, and the instructor pulled in on the dumbbells. The only difference in the three examples is that we have a more complex formula for angular momentum for the figure skater and instructor.

Exercise 7
What are the dimensions of angular momentum when mass is measured in grams, length in centimeters, and time in seconds?

Exercise 8
Figure (9d) is a strobe photograph from a student lab notebook. The string was suddenly pulled down, shortening the radius of the ball's circular orbit. Using this photograph, show that the angular momentum of the ball did not change.

Exercise 9
Neglecting the mass of the spokes, what is the angular momentum of a bicycle wheel of mass m and radius r, spinning with a period T? (T is the length of time the wheel takes to go around once.)

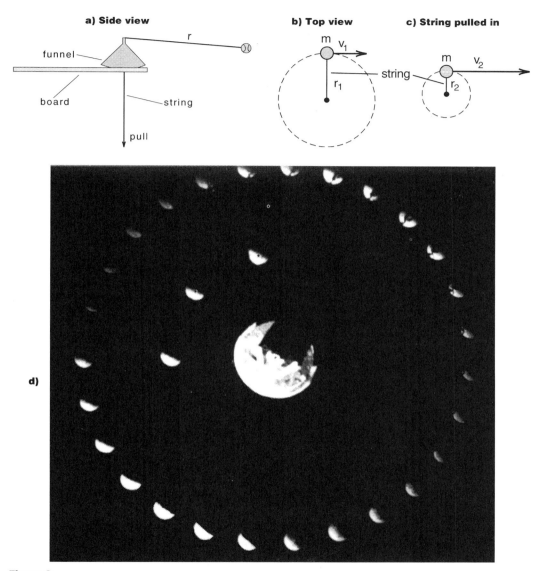

Figure 9
A more controlled demonstration of the conservation of angular momentum. One end of a string is tied to a ball of mass m, and the other is fed down through a plastic funnel mounted on the end of a board shown in (a). The ball is then swung in a circle of radius r_1, and speed v_1 as shown in (b). Then pull down on the free end of the string, to reduce the radius of the circle to r_2. It takes a fairly strong tug, but the speed of the ball increases to v_2 as shown in (c). From the experimental results shown in (d), you can check that $r_1 v_1 = r_2 v_2$. Since the angular momentum of the ball is proportional to rv, angular momentum was conserved in this experiment. (Photo from lab of G. Sheldon.)

A MORE GENERAL DEFINITION OF ANGULAR MOMENTUM

The concept of angular momentum applies to more general situations than mass traveling in a circle. For a more general definition of angular momentum, consider the situation shown in Figure (10). In Figure (10a) a ball with linear momentum $\vec{p} = m\vec{v}$ is traveling along a path that will take it a distance $r_\perp$ from some point labeled O. To make the situation more realistic, imagine that there is a light rod pivoted at point O with a hook at the other end. The length of the rod is $r_\perp$, so that the hook will just catch the ball as the ball passes by (Figure 10b).

Once the ball has been hooked, it will travel in a circle as shown in Figure (10c). If the rod is perpendicular to the path of the ball when the ball is hooked (as shown in Figure 10a) then there will be no disruption in the speed of the ball and the ball will move around the circle at the same speed v.

Once the ball is traveling in a circle we know that its angular momentum about the pivot O is given by Equation (11) as $\ell = mvr_\perp$. In our generalization of the definition of angular momentum, the ball has this same amount of angular momentum before it was hooked as it did after.

The more general definition is as follows. Consider the path of the ball shown in Figure (10a) and let us use the name *lever arm* for the distance of closest approach from the path to the axis at point O. This lever arm is the perpendicular distance to the path, the distance we have labeled $r_\perp$. Then as our new definition of angular momentum, we say that the magnitude ℓ of the ball's angular momentum is equal to the product of the magnitude of the linear momentum $p = mv$ times the length of the lever arm $r_\perp$

$$\boxed{\ell = pr_\perp} \quad (15)$$

Applying Equation (15) to the situation shown in Figure (10a), before the ball is hooked, we see that as the ball heads toward the hook, its initial angular momentum ℓ_i is

$$\ell_i = p_i r_\perp = mv\, r_\perp \quad (16)$$

Since this is the same as the angular momentum after the ball is hooked and traveling in a circle, the angular momentum is unchanged, is conserved, during the process of being captured by the hook.

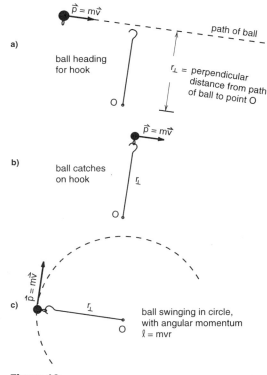

Figure 10
We say that the ball initially has an angular momentum $mvr_\perp$, in (a), that remains unchanged when the ball is caught by the hook and travels in c circle, in (c).

When you have a conserved quantity like angular momentum, it takes on a reality that goes beyond the formulas that define it. With our generalized definition of angular momentum, we find that angular momentum can be passed from one object to another. For example suppose a student is standing motionless on the rotating platform as in Figure (11a), and the instructor tosses a softball off to the side of the student as shown. The softball has a lever arm $r_\perp$ and therefore an angular momentum $mvr_\perp$ about the axis of the rotating platform. If the student reaches out and grabs the ball, she acquires the angular momentum of the ball and starts rotating as shown in Figure (11b). If she brings the ball closer in toward her body, she will rotate faster because the ball has a shorter lever arm.

Exercise 10

If you are standing directly over the axis of the platform and a ball is thrown directly toward you, as shown in Figure (12), do you start to rotate after catching the ball? Try the experiment yourself and see if the prediction is correct.

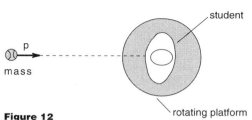

Figure 12
How much angular momentum does the student catch in this case?

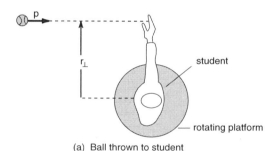

(a) Ball thrown to student

(b) Student, gaining angular momentum $|p|r_\perp = mvr_\perp$ from ball, starts to rotate

Figure 11
Student catching angular momentum from a ball thrown off to the side.

ANGULAR MOMENTUM AS A VECTOR

Our definition of angular momentum is clearly not yet complete. Even when we are standing on a rotating platform so that we can freely rotate only about the axis of the platform, there are still two different directions we can rotate—clockwise and counter clockwise. The definition of angular momentum must somehow account for these two directions of rotation.

The study of rotations can be complex, particularly if you allow rotations in three dimensions, about any of the three coordinate axes x, y or z. The rotating platform used in Figure (8) and (11) greatly simplifies the situation by restricting our motion to rotation about one axis, the axis of the platform which is conventionally called the z axis as shown in Figure (13).

One way to distinguish positive and negative rotation is by using a right-hand rule. The rule is to point the thumb of your right hand along the positive z axis, and then say that the direction of positive rotation is the direction that the fingers of your right hand curl, as seen in Figure (13). Looking down (with the z axis pointing up toward us), we find that positive rotation is counter clockwise and negative rotation is clockwise.

Exercise 11
What would be the direction of positive rotation if we used a left hand convention. Draw a sketch and explain.

The next generalization of our definition of angular momentum is best illustrated by the use of the rotating bicycle wheel mounted on a handle as shown in Figure (14). To make an effective demonstration, the tire of the bicycle wheel has been replaced by wire wrapped along the rim of the wheel to give the wheel added mass. If we spin the wheel all the mass m on the rim is moving at the same speed v and has the same lever arm r about the axis of the wheel. Thus the angular momentum of the rotating wheel is $\ell = \text{mv r}$.

The next step will at first seem arbitrary, and perhaps downright silly. What we are going to do now is to define the angular momentum of the wheel as a vector, of length $\ell = \text{mv r}$, *pointing along the axis of the wheel*. Which way it points is defined by a right-hand rule. As shown in Figure (14), curl the fingers of your right hand in the direction that the wheel is rotating and the angular momentum vector $\vec{\ell}$ points along the axis in the direction of your thumb.

This method of turning angular momentum into a vector seems doubly arbitrary. First of all, the angular momentum vector $\vec{\ell}$ points perpendicular

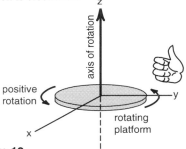

Figure 13
Definitions of the z axis and of positive and negative rotation. For convenience we will call the axis about which our platform rotates the "z axis". To distinguish the two kinds of rotation, we will use the right hand rule. Curl the finger of your right hand in the direction the platform is rotating. If your thumb points up, we say the rotation is positive. If your thumb points down, the rotation is negative.

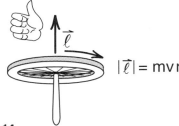

Figure 14
A further generalization of the concept of angular momentum is to say that it is a vector $\vec{\ell}$. The magnitude $|\vec{\ell}|$ is just our old definition $\text{mv} r_\perp$. We now say that the angular momentum vector $\vec{\ell}$ points in the direction of the axis of rotation, in the direction given by the right hand rule. (Curl the fingers of your right hand in the direction the wheel is rotating, and your thumb points in the direction of $\vec{\ell}$.)

to the plane in which the motion of the wheel is occurring, and then one arbitrarily selects a right-handed instead of a left-hand convention to decide which way along the axis the vector $\vec{\ell}$ should point. It is hard to believe that such arbitrary choices could have any relationship to physical reality.

But it works, as we can easily demonstrate using the bicycle wheel and the rotating platform. In the first demonstration, have a student stand at rest on the rotating platform and let the instructor spin the bicycle wheel and orient it so that the bicycle wheel's angular momentum vector $\vec{\ell}$ is pointing up as shown in Figure (15a). The instructor then hands the bicycle wheel, and its angular momentum, to the student as shown.

On the rotating platform motion is restricted to rotation about the z axis (axis of the platform), thus the only component of angular momentum that is of interest is the z component. In Figure (15a) the bicycle has a positive z component of angular momentum $\ell_z = mvr$, and the student has none. Thus the total angular momentum of the wheel and student is

$$(\ell_z)_{total} = (\ell_z)_{bicycle\,wheel} + (\ell_z)_{student}$$
$$= +mvr + 0 \quad (15)$$
$$= +mvr$$

The special thing that happens when you stand on a freely rotating platform is that you cannot change your own z component of angular momentum. If no one off the platform passes or tosses in some angular momentum, your z component of angular momentum is conserved and there is no way you can change it.

Have the student turn the bicycle wheel upside down as shown in Figure (15b). Now $\vec{\ell}$ of the bicycle wheel is pointing down so that the bicycle wheel now has a negative z component of angular momentum

$$(\ell_z)_{bicycle\,wheel} = -mvr \quad (16)$$

Since the total ℓ_z of the student and bicycle wheel must be conserved, the student must gain a positive z component of angular momentum

$$(\ell_z)_{student} = +2mvr \quad (17)$$

so that the sum remains + mvr.

What happens when the student turns the bicycle wheel over is that she starts rotating counter clockwise; she gains a positive z component of angular momentum. If at any time she turns the wheel back up, she will stop rotating.

Figure 15a
A student, at rest on the rotatable platform, is handed a rotating bicycle wheel whose angular momentum vector $\vec{\ell}$ is up as shown.

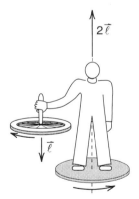

Figure 15b
The student turns the bicycle wheel over. If angular momentum is conserved, the student must gain an angular momentum $2\vec{\ell}$ directed up as shown.

Figure 15 Movie
The student turns the bicycle wheel over.

Figure 16a
The student at rest is handed a bicycle wheel pointed sideways. In this orientation the wheel has no z component of angular momentum. Once the student has the wheel, the z component of the angular momentum of the student plus wheel is conserved.

Figure 16b
What happens to the student if she turns the wheel up?

Figure 16c
What happens to the student if she turns the wheel down?

If the instructor hands the student the rotating wheel oriented horizontally as shown in Figure (16a), and the student is initially at rest, then neither the student or the bicycle wheel have a z component of angular momentum. The total z component is zero and must remain that way no matter what the student does. In the following exercise, you are to predict what will happen if she turns the wheel up or turns it down.

Exercise 12
Explain what will happen if the student orients the bicycle wheel up as shown in Figure (16b). What happens when she turns it down as shown in Figure (16c).

Exercise 13
The student at rest on the rotating platform is handed a bicycle wheel at rest. She spins the bicycle wheel and orients it so that the bicycle wheel's angular momentum vector points up. Explain carefully what happens to the student.

If you watch, or better yet try these angular momentum demonstrations yourself, you begin to believe that there is really a quantity called angular momentum that you can pass around and manipulate. Now our emphasis is on gaining an intuitive feeling for the concept, later we will come back to the topic with more mathematical machinery. But it is important to already have an intuitive grasp of the concept or the mathematical machinery will not make sense.

Figure 17
Detail of Hubble photograph of the Eagle nebula. Each nub is a star surrounded by its own gas cloud. (See page 18 for a more complete photograph of the nebula.)

Formation of Planets

Applications of the law of conservation of angular momentum are not confined to classroom demonstrations. We will end this chapter with a discussion of two astronomical applications. One deals with the formation of planets, and the other their motion about the sun.

Stars are formed in the large clouds of gas that stretch throughout the galaxy. An example of such a gas cloud is the Eagle Nebula shown in Figure (17). A particularly active area of star formation is in the nebula in the constellation of Orion shown in Figure (18). The Orion constellation rises after sunset in early winter, and the nebula is located in the middle of the sword dangling from the three bright stars of Orion's belt. Using binoculars, one can see the nebula as a bright patch of gas. The gas is illuminated by the newly formed stars inside.

A star forms when a lump of gas in the cloud begins to collapse due to the gravitational attraction between the gas particles. Judging from the spacing between stars in the neighborhood of the sun, the sun was formed from the collapse of a region of gas about two light years in radius. As the cloud collapses the gravitational attraction between the particles becomes stronger and stronger. The particles rush toward each other at a faster and faster rate, and finally collapse into a hot ball. The more mass in the collapse, the more gravitational energy released, and the hotter the resulting ball. If the ball is of the order of one tenth the mass of the sun or larger, the ball will be hot enough to ignite nuclear reactions and a star is born.

The lumpiness in the gas clouds is probably related to a large scale turbulence in the flow of the gas in the cloud. Such lumpines is usually associated with rotational motion (vorticies), thus as a lump of gas breaks away from the rest of the cloud, it is likely to have some rotation. While the rotational velocity may be small for a two light year diameter lump, as the lump contracts, the speed increasas due to the conservation of angular momentum. If there is enough angular momentum in the gas cloud, a point is reached where the rotation inhibits further collapse of the cloud.

The only way the cloud can continue to collapse is to leave some mass and most of the angular momentum outside. Computer models indicate that a rotating disk of gas forms outside the newly born star, a disk that contains most of the original angular momentum. After a while the gas in the disk condenses into planets that orbit the star. As a result of being formed from a rotating disk of gas, we expect most planets to lie in a plane, and go around the star in the same direction, which the sun and planets do. If the collapsing gas had no angular momentum, the disk would not form and there would be no planets.

Figure 18
The nebula in the constellation of Orion. This is a particularly active area for the formation of new stars.

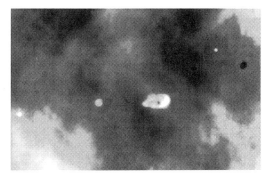

Figure 18a
Hubble photograph of forming stars in the Orion nebula. They are still surrounded by clouds of gas.

Exercise 14

Most of the angular momentum of the solar system is taken up by the distant massive planets Jupiter, Saturn and Uranus. If Jupiter were originally formed from a ring of dust 2 light-years in radius, what must have been the initial rotational speed of these particles? (The distance from the sun to Jupiter is 43 light-minutes or 2.6×10^3 light-seconds. Jupiter travels at an orbital speed of 1.3×10^6 cm/sec for a period of nearly 12 years. 1 year $\approx \pi \times 10^7$ sec)

Figure 17b
Hubble photograph of the Eagle nebula. The nubs at the top of the tallest column are young stars with their own gas clouds. Extremely bright light from a star in the background is pushing away all gas that is not gravitationally attached to a star.

Chapter 8
Newtonian Mechanics

In Chapters 4 and 5 we saw how to use calculus and the computer in order to predict the motion of a projectile. We saw that if we knew the initial position and velocity of an object, and had a formula for its acceleration vector, then we could predict its position far into the future.

To go beyond a discussion of projectile motion, to develop a general scheme for predicting motion, two new concepts are needed. One is mass, discussed in chapter 6, and the other is force, to be introduced now. We will see that once we know the forces acting on an object, we can obtain a formula for the object's acceleration and then use the techniques of Chapters 4 and 5 to predict motion. This scheme was developed in the late 1600s by Isaac Newton and is known as Newtonian Mechanics.

FORCE

The concept of a force—a push or a pull—is not as strange or unfamiliar as the acceleration vector we have been discussing. When you push on an object you are exerting a force on that object. The harder you push, the stronger the force. And the direction you push is the direction of the force. From this we see that force is a quantity that has a magnitude and a direction. As a result, it is reasonable to assume that a force is described mathematically by a vector, which we will usually designate by the letter $\vec{F}$.

It is often easy to see when forces are acting on an object. What is more subtle is the relationship between force and the resulting acceleration it produces. If I push on a big tree, nothing happens. I can push as hard as I want and the tree does not move. (No bulldozers allowed.) But if I push on a chair, the chair may move. The chair moves if I push sideways but not if I push straight down.

The ancient Greeks, in particular, Aristotle, thought that there was a direct relationship between force and velocity. He thought that the harder you pushed on an object, the faster it went. There is some truth in this if you are talking about pushing a stone along the ground or pulling a boat through water. But these examples, which were familiar problems in ancient time, turn out to be complex situations, involving friction and viscous forces.

Only when Galileo focused on a problem without much friction – projectile motion – did the important role of the acceleration vector become apparent. Later, Newton compared the motion of a projectile (the apple that supposedly fell on his head) with the motion of the planets and the moon, giving him more examples of motion without friction. These examples led Newton to the discovery that force is directly related to acceleration, not velocity.

In our discussion of projectile motion, and projectile motion with air resistance, we have begun to see the relation between force and acceleration. While a projectile is in flight, and we can neglect air resistance, the projectile's acceleration is straight down, in the direction of the earth as shown in Figure (1). As we stand on the earth, we are being pulled down by gravity. While the projectile is in flight, it is also being pulled down by gravity. It is a reasonable guess that the projectile's downward acceleration vector $\vec{g}$ is caused by the gravitational force of the earth.

When we considered the motion of a particle at constant speed in a circle as shown in Figure (2), we saw that the particle's acceleration vector pointed toward the center of the circle. A simple physical example of this circular motion was demonstrated when we tied a golf ball to a string and swing it over our head.

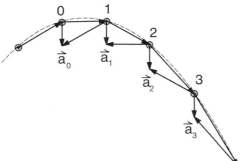

Figure 1
The earth's gravitational force produces a uniform downward gravitational acceleration. (Figure 3-27)

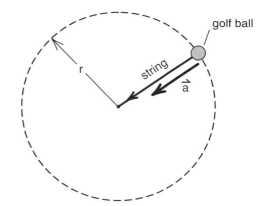

Figure 2
The acceleration of the ball is in the same direction as the force exerted by the string. (Figure 3-28)

While swinging the golf ball, it was the string pulling on the ball that kept the ball moving in a circle. (Let go of the string and the ball goes flying off.) The string is capable of pulling only along the length of the string, which in this case is toward the center of the circle. Thus the force exerted by the string is in the direction of the golf ball's acceleration vector. This makes our second example in which the particle's acceleration vector points in the same direction as the force exerted on it.

The example of projectile motion with air resistance, shown in Figure (3), presented a more complex situation. In our study of the motion of a Styrofoam projectile, we had two forces acting on the ball. There was the downward force of gravity, and also the force exerted by the wind we would feel if we were riding along with the ball. We saw that gravity and the wind each produced an acceleration vector, and that the ball's actual acceleration was the vector sum of the two individual accelerations. This is an important clue as to how we should handle situations where more than one force is acting on an object.

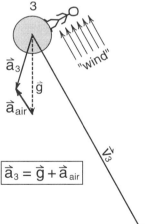

Figure 3
Gravity and the wind each produce an acceleration, $\vec{g}$ and $\vec{a}_{air}$ respectively. The net acceleration of the ball is the vector sum of the two accelerations.

THE ROLE OF MASS

Our three examples, projectile motion, motion in a circle, and projectile motion with air resistance, all demonstrate that a force produces an acceleration in the direction of the force. The next question is – how much acceleration? Clearly not all forces have the same effect. If I shove a child's toy wagon, the wagon might accelerate rapidly and go flying off. The same shove applied to a Buick automobile will not do very much.

There is clearly a difference between a toy wagon and a Buick. The Buick has much more mass than the wagon, and is much less responsive to my shove.

In our recoil definition of mass discussed in Chapter 6 and illustrated in Figure (4), we defined the ratio of two masses as the inverse ratio of their recoil speeds

$$\frac{m_1}{m_2} = \frac{v_2}{v_1}$$

The intuitive idea is that the more massive the object, the slower it recoils. The more mass, the less responsive it is to the shove that pushed the carts apart.

Think about the spring that pushes the cart apart in our recoil experiment. Once we burn the thread holding the carts together, the spring pushes out on both carts, causing them to accelerate outward. If the spring is pushing equally hard on both carts (later we will see that it must), then we see that the resulting acceleration and final velocities are inversely proportional to the mass of the cart. If m_1 is twice as massive as m_2, it gets only half as much acceleration from the same spring force. Our recoil definition and experiments on mass suggests that the effectiveness of a force in producing an acceleration is inversely proportional to the object's mass. For a given force, if you double the mass, you get only half the acceleration. That is the simplest relationship between force and mass that is consistent with our general experience, and it turns out to be the correct one.

Figure 4
Definition of mass. When two carts recoil from rest, the more massive cart recoils more slowly.

NEWTON'S SECOND LAW

We have seen that a force $\vec{F}$ acting on a mass m, produces an acceleration $\vec{a}$ that 1) is in the direction of $\vec{F}$, and 2) has a magnitude inversely proportional to m. The simplest equation consistent with these observations is

$$\vec{a} = \frac{\vec{F}}{m} \qquad (1)$$

Equation (1) turns out to be the correct relationship, and is known as ***Newton's Second Law of Mechanics***. (The ***First Law*** is a statement of the special case that, if there are no forces, there is no acceleration. That was not obvious in the late 1600s, and was therefore stated as a separate law.) A more familiar form of Newton's second law, seen in all introductory physics texts is

$$\boxed{\vec{F} = m\vec{a}} \qquad (1a)$$

If there is any equation that is essentially an icon for the introductory physics course, Equation (1a) is it.

At this point Equation (1) or (1a) serves more as a definition of force than a basic scientific result. We can, for example, see from Equation (1a) that force has the dimensions of mass times acceleration. In the MKS system of units this turns out to be kg(m/sec^2), a collection of units called the ***newton***. Thus we can say that we push on an object with a force of so many newtons. In the CGS system, the dimensions of force are gm(cm/sec^2), a set of units called a ***dyne***. A dyne turns out to be a very small unit of force, of the order of the force exerted by a fly doing push-ups. The newton is a much more convenient unit. The real confusion is in the English system of units where force is measured in *pounds*, and the unit of mass is a *slug*. We will carefully avoid doing Newton's law calculations in English units so that the student does not have to worry about pounds and slugs.

At a more fundamental level, we can use Equation (1) to ***detect the existence of a force by the acceleration it produces***. In projectile motion, how do we know that there is a gravitational force $\vec{F}_g$ acting on the projectile? Because of the gravitational acceleration. The acceleration $\vec{a}$ due to gravity is equal to $\vec{g}$ (9.8 m/sec^2 directed downward), thus we can say that the gravitational force $\vec{F}_g$ that produces this acceleration is

$$\boxed{\vec{F}_g = m\vec{g}} \qquad \begin{array}{l}\textit{gravitational force} \\ \textit{on a mass m}\end{array} \qquad (2)$$

where m is the mass of the projectile.

Figure 5
The gravitational force between small masses is proportional to the product of the masses, and inversely proportional to the square of the separation between them.

NEWTON'S LAW OF GRAVITY

Newton went beyond using the second law to define force; he also discovered a basic law for the gravitational force between objects. With Newton's law of gravity combined with Newton's second law, we can make detailed predictions about how projectiles, satellites, planets, and solar systems behave. This combination, where one has an explicit formula for gravitational forces, and the second law to predict what accelerations these forces produce, was one of the most revolutionary scientific discoveries ever made.

Newton's so-called ***universal law of gravitation*** can most simply be stated as follows. If we have two small masses of mass m_1 and m_2, separated by a distance r as shown in Figure (5), then the force between them is proportional to the product m_1m_2 of their masses, and inversely proportional to the square of the distance r between them. This can be written as an equation of the form

$$\left| \vec{F}_g \right| = G \frac{m_1 m_2}{r^2} \qquad \begin{array}{l} \textit{Newton's law} \\ \textit{of gravity} \end{array} \qquad (3)$$

where the proportionality constant G is a number that must be determined by experiment.

Equation (3) itself is not the whole story, we must make several more points. First, and very important, is the fact that gravitational forces are always attractive; m_1 is pulled directly toward m_2, and m_2 directly toward m_1. Second, the strength of these forces are equal, even if m_2 is much bigger than m_1, the force of m_2 on m_1 is the same in strength as the force of m_1 on m_2. That is why we used the same symbol $\vec{F}_g$ for the two attractive forces in Figure (5).

Newton's law of gravity is called the ***universal law of gravitation*** because Equation (3) is supposed to apply to all masses anywhere in the universe, with the same numerical constant G everywhere. G is called the ***universal gravitational constant***, and has the numerical value, in the MKS system of units

$$G = 6.67 \times 10^{-11} \frac{m^3}{kg\ sec^2} \qquad \begin{array}{l} \textit{universal} \\ \textit{gravitational} \\ \textit{constant} \end{array} \qquad (4)$$

We will discuss shortly how this number was first measured.

Exercise 1
Combine Newton's second law $\vec{F} = m\vec{a}$ with the law of gravity $\left| \vec{F}_g \right| = Gm_1m_2/r^2$ and show that the dimensions for G in Equation (4) are correct.

Big Objects

In our statement of Newton's law of gravity, we were careful to say that Equation (3) applied to two small objects. To be more explicit, we mean that the two objects m_1 and m_2 should be small in dimensions compared to the separation r between them. We can think of Equation (3) as applying to two *point particles* or *point masses*.

What happens if one or both of the objects are large compared to their separation? Suppose, for example, that you would like to calculate the gravitational force between you and the earth as you stand on the surface of the earth. The correct way to do this is to realize that you are attracted, gravitationally, to every rock, tree, every single piece of matter in the entire earth as indicated in Figure (6). Each of these pieces of matter is pulling on you, and together they produce a net gravitational force $\vec{F}_g$ which is the force $m\vec{g}$ that we saw in our discussion of projectile motion.

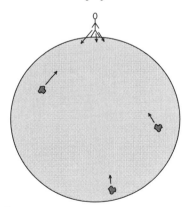

Figure 6
You are attracted to every piece of matter in the earth.

It appears difficult to add up all the individual forces exerted by every chunk of matter in the entire earth, to get the net force $\vec{F}_g$. Newton also thought that this was difficult, and according to some historical accounts, invented calculus to solve the problem. Even with calculus, it is a fairly complicated problem to add up all of these forces, but the result turns out to be very simple. *For any uniformly spherical object, you get the correct answer in Newton's law of gravity if you think of all the mass as being concentrated at a point at the center of the sphere.* (This result is an accidental consequence of the fact that gravity is a $1/r^2$ force, i.e., that it is inversely proportional to the square of the distance. We will have much more to say about this accident in later chapters.)

Since the earth is nearly a uniformly spherical object, you can calculate the gravitational force between you and the earth by treating the earth as a point mass located at its center, 4000 miles below you, as indicated in Figure (7).

Galileo's Observation

As we mentioned earlier, Galileo observed that, in the absence of air resistance, all projectiles should have the same acceleration no matter what their mass. This leads to the striking result that, in a vacuum, a steel ball and a feather fall at the same rate. Now we can see that this is a consequence of Newton's second law combined with Newton's law of gravity.

Using the results of Figure (7), i.e., calculating $\vec{F}_g$ by replacing the earth by a point mass m_e located a distance r_e below us, we get

$$F_g = \frac{Gmm_e}{r_e^2} \tag{5}$$

for the strength of the gravitational force on a particle of mass m at the surface of the earth. Combining this with Newton's second law

$$\vec{F}_g = m\vec{g} \quad \text{or} \quad F_g = mg \tag{6}$$

we get

$$mg = \frac{Gmm_e}{r^2} \tag{7}$$

The important result is that the particle's mass m cancels out of Equation (7), and we are left with the formula

$$\boxed{g = \frac{Gm_e}{r_e^2}} \tag{8}$$

for the acceleration due to gravity. We note that g depends on the earth mass m_e, the earth radius r_e, and the universal constant G, but *not on the particle's mass m*. Thus objects of different mass should have the same acceleration.

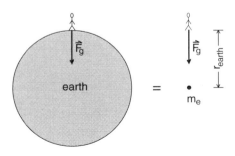

Figure 7
The gravitational force of the entire earth acting on you is the same as the force of a point particle with a mass equal to the earth mass, located at the earth's center, one earth radius below you.

Galileo Was Right!

By WILLIAM HINES

HOUSTON — Galileo was right and Apollo 15 Astronaut David R. Scott can prove it.

The 17-th century Italian mathematician believed contrary to then prevailing scientific opinion that gravity exerts an equal influence on all things, regardless of their size, shape or weight.

He even went so far as to say that if it weren't for the resistance offered by air, a cannonball and a feather would fall at the same speed.

It wasn't possible for Galileo to demonstrate the truth of that assertion, although legend says he performed a compromise experiment by dropping a large iron ball and a small one off the leaning tower of Pisa and that they hit the ground at the same instant.

Just before getting back into the Lunar Landing Craft Falcon fcr takeoff from the Moon, Scott demonstrated that Galileo was right.

Holding a metal geological hammer in his right hand and a feather ("A falcon feather," he explained) in his right, Scott faced the Apollo television camera and released the two objects.

Falling slowly in the eak gravity of the airless Moon, the hammer and the feather reached the surface at precisely the same instant.

"How about that?" the delighted Scott remarked. "Mr. Galileo was correct in his findings."

THE CAVENDISH EXPERIMENT

A key feature of Newton's law of gravitation is that all objects attract each other via gravity. Yet in practice, the only gravitational force we ever notice is the force of attraction to the earth. What about the gravitational force between two students sitting beside each other, or between your two fists when you hold them close to each other? The reason that you do not notice these forces is that the gravitational force is incredibly weak, weak compared to other forces that hold you, trees, and rocks together. Gravity is so weak that you would never notice it except for the fact that you are on top of a huge hunk of matter called the earth. The earth mass is so great that, even with the weakness of gravity, the resulting force between you and the earth is big enough to hold you down to the surface.

The gravitational force between two reasonably sized objects is not so small that it cannot be detected, it just requires a very careful experiment that was first performed by Henry Cavendish in 1798. In the Cavendish experiment, two small lead balls are mounted on the end of a light rod. This rod is then suspended on a fine glass fiber as shown in Figure (8a).

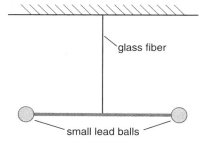

a) Side view of the small balls.

As seen in the top view in Figure (8b), two large lead balls are placed near the small ones in such a way that the gravitational force between each pair of large and small balls will cause the rod to rotate in one direction. Once the rod has settled down, the large lead balls are moved to the position shown in Figure (8c). Now the gravitational force causes the rod to rotate the other way. By measuring the angle that the rod rotates, and by measuring what force is required to rotate the rod by this angle, one can experimentally determine the strength of the gravitational force $\vec{F}_g$ between the balls. Then by using Newton's law of gravity

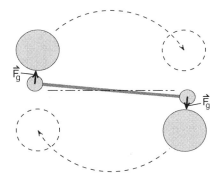

b) Top view showing two large lead balls.

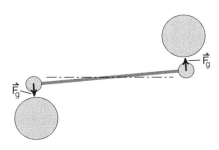

c) Top view with large balls rotated to new position.

Figure 8
The Cavendish experiment. By moving the large lead balls, the small lead balls are first pulled one way, then the other. By measuring the angle the stick holding the small balls is rotated, one can determine the gravitational force $\vec{F}_g$.

$$F_g = G\frac{m_1 m_2}{r^2}$$

Figure 9

applied to Figure (9), one can solve for G in terms of the known quantities F_g, m_1, m_2 and r^2. This was the way that Newton's universal constant G, given in Equation (4) was first measured.

"Weighing" the Earth

Once you know G, you can go back to the formula (8) for the acceleration g due to gravity, and solve for the earth mass m_e to get

$$m_e = \frac{gr_e^2}{G} = \frac{9.8 \text{ m/sec}^2 \times (6.37 \times 10^6 \text{m})^2}{6.67 \times 10^{-11} \text{m}^3/\text{kg sec}^2}$$

$$= 6.0 \times 10^{24} \text{kg} \qquad (9)$$

As a result, Cavendish was able to use his value for G to determine the mass of the earth. This was the first determination of the earth's mass, and as a result the Cavendish experiment became known as the experiment that "weighed the earth".

Exercise 2

The density of water is 1 gram/cm³. The average density of the earth's outer crust is about 3 times as great. Use Cavendish's result for the mass of the earth to decide if the entire earth is like the crust. (Hint —the volume of a sphere of radius r is $4/3\pi r^3$). Relate your result to what you have read about the interior of the earth.

Inertial and Gravitational Mass

The fact that, in the absence of air resistance, all projectiles have the same acceleration— the fact that the m's canceled in Equation (7), has a deeper consequence than mere coincidence. In Newton's second law, the m in the formula $\vec{F} = m\vec{a}$ is the mass defined by the recoil definition of mass discussed in Chapter 6. Called *inertial mass*, it is the concept of mass that we get from the law of conservation of linear momentum.

In Newton's law of gravity, the projectile's mass m in the formula $F_g = Gmm_e/r_e^2$ is what we should call the *gravitational mass* for it is defined by the gravitational interaction. It is the experimental observation that the m's cancel, the observation that all projectiles have the same acceleration due to gravity, that tells us that the inertial mass is the same as gravitational mass. This equivalence of inertial and gravitational mass has been tested with extreme precision to one part in a billion by Etvös in 1922 and to even greater accuracy by R. H. Dicke in the 1960s.

SATELLITE MOTION

The key idea that led Newton to his universal law of gravitation was that the moon, while traveling in its orbit about the earth, was subject to the same kind of force as an apple falling from a tree. We have seen that a projectile in flight, such as an apple, accelerates down toward the center of the earth. The moon, in its nearly circular orbit around the earth, also accelerates toward the center of the earth, as illustrated in Figure (10). Newton proposed that the accelerations of the falling apple and of the orbiting moon were both caused by the gravitational pull of the earth.

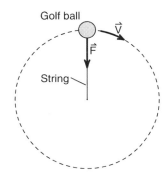

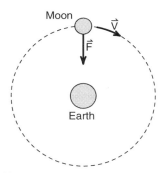

Figure 10
When we swing a golf ball in a circle, the ball accelerates toward the center of the circle, in the direction it is pulled by the string. Similarly, the moon, in its circular orbit about the earth, accelerates toward the center of the earth, in the direction it is pulled by the earth's gravity.

The moon, being farther away from the center of the earth should be expected to feel a weaker gravitational force and therefore have a weaker acceleration. From direct calculation Newton could determine how much weaker the moon's acceleration was, and thus determine how the gravitational acceleration and force decreases with distance.

To repeat Newton's calculation, we know that the apple on the surface of the earth has an acceleration $g_{apple} = 9.8 \text{ m/sec}^2$. To determine the magnitude of the moon's orbital acceleration toward the earth, $g_{moon\,orbit}$, we can use the formula derived in Chapter 3 for uniform circular motion, namely

$$|\vec{a}| = g_{moon\,orbit} = \frac{v^2}{r} \qquad \begin{array}{c}\textit{uniform}\\ \textit{circular}\\ \textit{motion}\end{array} \qquad (3\text{-}12)$$

To calculate the speed v of the moon, we note that the moon takes 27.32 days or 2.36×10^6 seconds for one complete orbit. The radius of the moon orbit is 3.82×10^8 meters, so that

$$\begin{aligned}v_{moon} &= \frac{\text{orbital circumference}}{\text{time for one orbit}} = \frac{2\pi r}{t_{orbit}} \\ &= \frac{2\pi \times (3.82 \times 10^8 \text{ meters})}{2.36 \times 10^6 \text{ sec}} \\ &= 1.02 \times 10^3 \,\frac{m}{sec} \end{aligned} \qquad (10)$$

or very close to 1 kilometer per second. Substituting this value of v into the formula v^2/r, gives

$$\begin{aligned}g_{moon\,orbit} &= \frac{\left(1.02 \times 10^3 \text{ m/sec}\right)^2}{3.82 \times 10^8 \text{ m}} \\ &= 2.70 \times 10^{-3} \,\frac{m}{sec^2}\end{aligned} \qquad (11)$$

The ratio of the moon's orbital acceleration to the apple's acceleration

$$\begin{aligned}\frac{g_{moon\,orbit}}{g_{apple}} &= \frac{2.70 \times 10^{-3} \text{ m/sec}^2}{9.8 \text{ m/sec}^2} \\ &= 2.71 \times 10^{-4}\end{aligned} \qquad (12)$$

I.e., the moon's acceleration is 27 thousand times weaker than the apple's.

To understand the meaning of this result, let us look at the square of the ratio of the distances from the apple to the center of the earth, and the moon to the center of the earth. We have

$$\left(\frac{r_{apple\,to\,center\,of\,earth}}{r_{moon\,orbit}}\right)^2 = \left(\frac{6.37 \times 10^6 \text{ m}}{3.82 \times 10^8 \text{ m}}\right)^2$$

$$= 2.78 \times 10^{-4} \qquad (13)$$

which, to the accuracy of our work, is the same as the ratio of accelerations.

Equating the results in Equations (12) and (13), we get

$$\frac{g_{moon\,orbit}}{g_{apple}} = \frac{r_e^2}{r_{moon\,orbit}^2}$$

$$g_{moon\,orbit} = \frac{g_{apple} \times r_e^2}{r_{moon\,orbit}^2} \propto \frac{1}{r_{moon\,orbit}^2} \qquad (14)$$

Where $g_{apple} \times r_e^2$ can be thought of as a constant.

From such calculations Newton saw that the gravitational acceleration of the moon, and thus the gravitational force, decreased as the square of the distance from the moon to the center of the earth. This was how Newton deduced that gravity was a $1/r^2$ force law.

Exercise 3

How far above the surface of the earth do you have to be so that, in free fall, your acceleration is half that of objects near the surface of the earth?

Other Satellites

To explain to the world the similarity of projectile and satellite motion, that both the apple and the moon were simply falling toward the center of the earth, Newton drew the sketch shown in Figure (11). In the sketch, Newton shows a projectile being fired horizontally from the top of a mountain, and shows what would happen if there were no air resistance. If the horizontal velocity were not too great, the projectile would go a short distance along the typical parabolic path we have studied in the strobe labs. As the projectile is fired faster it would travel farther before hitting the ground. Finally we reach a point where the projectile keeps falling toward the earth, but the earth keeps falling away and the projectile goes all the way around the earth without hitting it.

Another perspective of the same idea is illustrated in Figures (12) and (13). Figure (12) is a strobe photograph showing two steel balls launched simultaneously, one being dropped straight down and the other being fired horizontally. The photograph clearly demonstrates that the downward motion of the two projectiles is the same. By using the constant acceleration formulas with $g = 32$ ft/sec^2, we can easily calculate that at the end of one second both projectiles will have fallen 16 ft, and at the end of two seconds a distance of 64 ft.

In Figure (13), we have sketched the curved surface of the earth. Due to this curvature, the surface of the earth will be 16 ft below a horizontal line out at a distance of 4.9 miles, and 64 ft below at a distance of 9.8 miles. This effect can be seen from a small boat as you leave shore. When you are 10 miles off shore, you cannot see lighthouses under 64 ft tall, unless you climb your own mast. (For landlubbers sunning on the beach, sailboats with 64 ft high masts disappear from sight at a distance of 10 miles.)

Comparing Figures (12) and (13), we see that in the absence of air resistance, if a projectile were fired horizontally at a speed of 4.9 miles per second, during the first second it would fall 16 ft, but the earth would have also fallen 16 ft, and the projectile would be no closer to the surface. By the end of the 2nd second the projectile would have fallen 64 ft, but still not have come any closer to the surface of the earth. Such a projectile would keep traveling around the earth, never hitting the surface. It would fall all the way around, becoming an earth satellite.

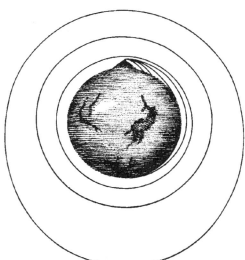

Figure 11
Newton's sketch showing that the difference between projectile and satellite motion is that satellites travel farther. Both are accelerating toward the center of the earth.

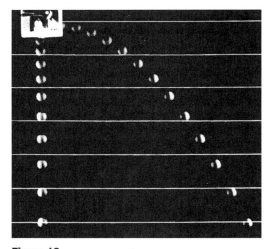

Figure 12
Two projectiles, released simultaneously. The horizontal motion has no effect on the vertical motion: they both fall at the same rate.

Exercise 4

An earth satellite in a low orbit, for instance 100 miles up, is so close to the surface of the earth (100 miles is so small compared to the earth's radius of 4000 miles) that the satellite's acceleration is essentially the same as the acceleration of projectiles here on earth. Use this result to predict the period T of the satellite's orbit. (Hint – the satellite travels one earth circumference $2\pi r_e$ in one period T. This allows you to calculate the satellite's speed v. You then use the formula v^2/r for the magnitude of the satellite's acceleration.)

Weight

The popular press often talks about the astronauts in spacecraft orbiting the earth as being *weightless*. This is verified by watching them on television floating around inside the space capsule. You might jump to the conclusion that because the astronauts are floating around in the capsule, they do not feel the effects of gravity. This is true in the same sense that when you jump off a high diving board, you do not feel the effects of gravity—until you hit the water. While you are falling, you are *weightless* just like the astronauts.

The only significant difference between your fall from the high diving board, and the astronaut's weightless experience in the space capsule, is that the astronaut's experience lasts longer. As the space capsule orbits the earth, the capsule and the astronauts inside are in continuous free fall. They have not escaped the earth's gravity, it is gravity that keeps them in orbit, accelerating toward the center of the earth. But because they are in free fall, they do not feel the acceleration, and are considered to be *weightless*.

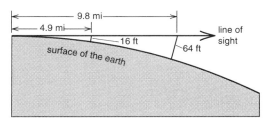

Figure 13
The curvature of the earth causes the horizon to fall away 64 feet at a distance of 9.8 miles.

If the astronaut in an orbiting space capsule is weightless, but still subject to the gravitational force of the earth, we cannot directly associate the word *weight* with the effects of gravity. In order to come up with a definition of the word weight that has some scientific value, and is reasonably consistent with the use of the word in the popular press, we can define the weight of an object as the magnitude of the force the object exerts on the bathroom scales. Here on earth, if you have an object of mass m and you set it on the bathroom scales, it will exert a downward gravitational force of magnitude

$$F_g = mg$$

Thus we say that the object has a weight W given by

$$W = mg \tag{15}$$

For example, a 60 kg boy standing on the scales exerts a gravitational force

$$W(60 \text{ kg boy}) = 60 \text{ kg} \times 9.8 \frac{m}{\sec^2}$$
$$= 588 \text{ newtons}$$

We see that weight has the dimensions of a force, which in the MKS system is newtons. If the same boy stood on the same scales in an orbiting spacecraft, both the boy and the scales would be in free fall toward the center of the earth, the boy would exert no force on the scales, and he would therefore be weightless.

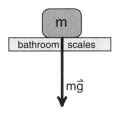

Figure 14
We will define the weight of an object as the force it exerts on the bathroom scales.

Although we try to make the definition of the word *weight* consistent with the popular use of the word, we do not actually succeed. In almost any country except the United States, when you buy a steak, the butcher will weigh it in grams. The grocer will tell you that a banana weighs 200 grams. You are not likely find a grocer who tells you the weight of an object in newtons. It is a universal convention to tell you the mass in grams or kilograms, and say that that is the weight. About the only place will you will find the word *weight* to mean a force, as measured in newtons, is in a physics course.

(In the English system of units, a pound is a force, so that it is correct to say that our 60 kg mass boy weighs 132 lbs. That, of course, leaves us with the question of what mass is in the English units. From the formula F = mg, we see that m = F/g, or an object that weighs 32 lbs has a mass 32 lbs/32ft/sec^2 = 1. As we mentioned earlier, this unit mass in the English units is called a *slug*. This is the last time we will mention slugs in this text.)

Earth Tides

An aspect of Newton's law of gravity that we have not said much about is the fact that gravity is a mutual attraction. As we mentioned, two objects of mass m_1 and m_2 separated by a distance r, attract *each other* with a gravitational force of magnitude $|\vec{F}_g| = Gm_1m_2/r^2$. The point we want to emphasize now is that the force on *each* particle has the same strength F_g.

Let us apply this idea to you, here on the surface of the earth. Explicitly, let us assume that you have just jumped off a high diving board as illustrated in Figure (15), and have not yet hit the water. While you are falling, the earth's gravity exerts a downward force $\vec{F}_g$ which produces your downward acceleration $\vec{g}$.

According to Newton's law of gravity, you are exerting an equal and opposite gravitational force $\vec{F}_g$ on the earth. Why does nobody talk about this upward force you are exerting on the earth? The answer, shown in the following exercise, is that even though you are pulling up on the earth just as hard as the earth is pulling down on you, the earth is so much more massive that your pull has no detectable effect.

Exercise 5

Assume that the person in Figure (15) has a mass of 60 kilograms. The gravitational force he exerts on the earth causes an upward acceleration of the earth a_{earth}. Show that $a_{earth} = 10^{-22}$ m/sec^2.

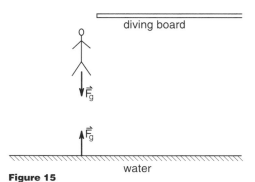

Figure 15
As you fall toward the water, the earth is pulling down on you, and you are pulling up on the earth. The two forces are of equal strength.

More significant than the force of the diver on the earth is the force of the moon on the earth. It is well known that the ocean tides are caused by the moon's gravity acting on the earth. On the night of a full moon, high tide is around midnight when the moon is directly overhead. The time of high tide changes by about an hour a day in order to stay under the moon.

The high tide under the moon is easily explained by the idea that the moon's gravity sucks the ocean water up into a bulge under the moon. As the earth rotates and we pass under the bulge, we see a high tide. This explains the high tide at midnight on a full moon.

The problem is that there are 2 high tides a day about 12 hours apart. The only way to understand two high tides is to realize that there are two bulges of ocean water, one under the moon and one on the opposite side of the earth, as shown in Figure (16). In one 24 hour period we pass under both bulges.

Why is there a bulge on the backside? Why isn't the water all sucked up into one big bulge underneath the moon?

The answer is that the moon's gravity not only pulls on the earth's water, but on the earth itself. The force of gravity that the moon exerts on the earth is just the same strength as the force the earth exerts on the moon. Since the earth is more massive, the effect on the earth is not as great, but it is noticeable. The reason for the second bulge of water on the far side of the earth is that the center of the earth is closer to the moon than the water on the back side, and therefore accelerates more rapidly toward the moon than the water on the back side. The water on the back side gets left behind to form a bulge.

The result, the fact that there are two high tides a day, the fact that there is a second bulge on the back side, is direct experimental evidence that the earth is accelerating toward the moon. It is direct evidence that the moon's gravity is pulling on the earth, just as the earth's gravity is pulling on the moon.

As a consequence of the earth's acceleration, the moon is not traveling in a circular orbit centered precisely on the center of the earth. Instead both the earth and the moon are traveling in circles about an axis point located on a line joining the earth's and moon's centers. This axis point is located much closer to the center of the earth than that of the moon, in fact it is located inside the earth about 3/4 of the way toward the earth's surface as shown in Figure (17).

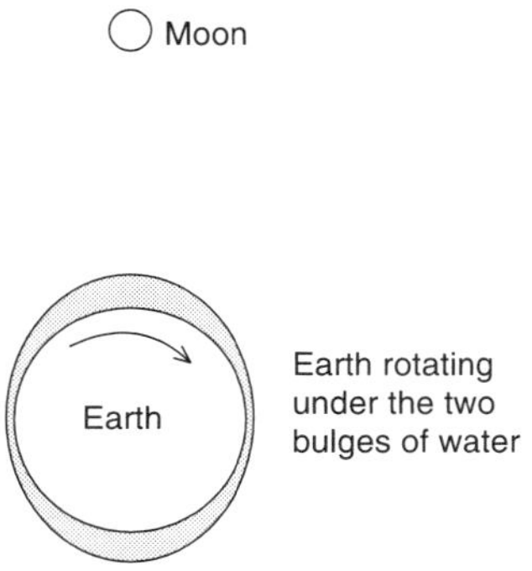

Figure 16
The two ocean bulges cause two high tides per day.

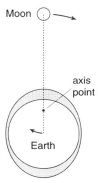

Figure 17
Both the earth and the moon travel in circular orbits about an axis point located about 1/4 of the way down below the earth's surface.

Planetary Units

In introductory physics texts, it has become almost an article of religion that all calculations shall be done using MKS units. This has some advantages – we do not have to talk about pounds and slugs, but practicing physicists seldom follow this rule. Physicists studying the behavior of elementary particles, for example, routinely use a system of units that simplify their calculations, units in which the speed of light and other fundamental constants have the numerical value 1. Using these special units they can quickly solve simple problems and gain an intuitive feeling for which quantities are important and which quantities are not.

In our work with projectiles in the lab the CGS system of units was excellent. The projectiles typically went distances from 10 to 100 cm, in times of the order of 1 second, and had masses of the order of 100 gm. There were no large exponents involved.

Now that we are studying the motion of earth satellites, we are faced with large exponents in quantities like the earth mass and the gravitational constant G which are 5.98×10^{24} kg and 6.67×10^{-11} m^3/kg sec^2 respectively. The calculations we have done so far using these numbers have required a calculator, and we have had to work hard to gain insight from the results.

Table 1 Planetary Units

Constant	Symbol	Planetary units	MKS units
Gravitational Constant	G	20	$6.67 \times 10^{-11} \frac{m^3}{kg\ sec^2}$
Acceleration due to gravity at the earth's surface	g_e	20	9.8 m/sec^2
Earth mass	m_e	1	5.98×10^{24} kg
Moon mass	m_{moon}	.0123	7.36×10^{22} kg
Sun mass	m_{sun}	3.3×10^5	1.99×10^{30} kg
Metric ton	ton	1.67×10^{-22}	1000 kg
Earth radius	r_e	1	6.37×10^6 m
Moon radius	r_{moon}	.2725	1.74×10^6 m
Sun radius	r_{sun}	109	6.96×10^8 m
Earth orbit radius	$r_{earth\ orbit}$	23400	1.50×10^{11} m
Moon orbit radius	$r_{moon\ orbit}$	60	3.82×10^8 m
Hour	hr	1 hr	3600 sec
Moon period	lunar month (siderial)	656 hrs	2.36×10^6 sec (= 27.32 days)
Year	yr	8.78×10^3 hrs	3.16×10^7 sec

We now wish to introduce a new set of units, which we will call *planetary units*, that makes satellite calculations much simpler and more intuitive. One way to design a new set of units is to first decide what will be our unit mass, our unit length, and our unit time, and then work out all the conversion factors so that we can convert a problem into our new units. For working earth satellite problems, we have found that it is convenient to take the earth mass as the unit mass, the earth radius as the unit length, and the hour as the unit time.

$$m_{earth} = 1 \quad earth\,mass$$
$$R_{earth} = 1 \quad earth\,radius$$
$$hour = 1$$

With these choices, speed, for example, is measured in (earth radii)/ hr, etc.

This system of units has a number of advantages. We can set m_e and r_e equal to 1 in the gravitational force formulas, greatly simplifying the results. We know immediately that a satellite has crashed if its orbital radius becomes less than 1. Typical satellite periods are a few hours and typical satellite speeds are from 1 to 10 earth radii per hour. What may be a bit surprising is that both the acceleration due to gravity at the surface of the earth, g, and Newton's universal gravitational constant G, have the same numerical value of 20.

Table 1 shows the conversion from MKS to planetary units of common quantities encountered in the study of satellites moving in the vicinity of the earth and the moon.

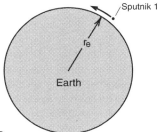

Figure 18
A satellite in a low earth orbit.

Exercise 6
We will have you convert Newton's universal gravitational constant G into planetary units. Start with

$$G = 6.67 \times 10^{-11} \frac{meters^3}{kg\,sec^2}$$

Then multiply or divide by the conversion factors

$$3600 \frac{sec}{hr}$$

$$5.98 \times 10^{24} \frac{kg}{earth\,mass}$$

$$6.37 \times 10^6 \frac{meters}{earth\,radii}$$

until all the dimensions in the formula for G are converted to planetary units. (I.e., convert from seconds to hours, kg to earth mass, and meters to earth radii.) If you do the conversion correctly, you should get the result

$$G = 20 \frac{(earth\,radii)^3}{(earth\,mass)\,hr^2}$$

Exercise 7
Explain why g and G have the same numerical value in planetary units.

As an advertisement for how easy it is to use planetary units in satellite calculation, let us repeat Exercise (4) using these units. In that exercise we wished to calculate the period of Sputnik 1, a satellite traveling in a low earth orbit. We were to assume that Sputnik's orbital radius was essentially the earth's radius r_e as shown in Figure (18), and that Sputnik's acceleration toward the center of the earth was essentially the same as the projectiles we studies in the introductory lab, i.e., $g_e = 9.8\,m/sec^2$.

Using the formula
$$a = \frac{v^2}{r}$$
we get
$$g_e = \frac{v_{Sputnik}^2}{r_e} = \frac{v_{Sputnik}^2}{1}$$
Therefore
$$v_{Sputnik} = \sqrt{g_e} = \sqrt{20} \; \frac{\text{earth radii}}{\text{hr}}$$

Now the satellite travels a total distance $2\pi r_e$ to go one orbit, therefore the time it takes is

$$\text{Sputnik period} = \frac{2\pi r_e}{v_{Sputnik}} = \frac{2\pi}{\sqrt{20}} = 1.4 \text{ hrs}$$

Compare the algebra that we just did with what you had to go through to get an answer in Exercise (4). (You should have gotten the same answer, 1.4 hrs, or 84 minutes, or 5,040 seconds. This is in good agreement with the observed time for low orbit satellites.) If you have watched satellite launches on television, you may recall waiting about an hour and a half before the satellite returned.

Exercise 8

A satellite is placed in a circular orbit whose radius is $2r_e$ (it is one earth radii above the surface of the earth.)

(a) What is the acceleration due to gravity at this altitude?

(b) What is the period of this satellite's orbit?

(c) What is the shortest possible period any earth satellite can have? Explain your answer.

Exercise 9

Communication satellites are usually placed in circular orbits over the equator, at an altitude so that they take precisely 24 hours to orbit the earth. In this way they hover over the same point on the earth and can be in continuous communication with the same transmitters and receivers. This orbit is called the *Clarke orbit*, named after the science fiction writer Arthur Clark who first emphasized the importance of such an orbit. Calculate the radius of the Clark orbit.

COMPUTER PREDICTION OF SATELLITE ORBITS

In this chapter we have discussed two special kinds of motion that a projectile or satellite can have. One is the parabolic trajectory of a projectile thrown across the room – motion that is easily described by calculus and the constant acceleration formulas. The other is the orbital motion of the moon and man-made satellites that are in circular orbits. These orbits can be analyzed using the fact that their acceleration is directed toward the center of the circle and has a magnitude v^2/r.

These two examples are deceptively simple. Newton's diagram, Figure (11), shows that there is a continuous range of orbital shapes starting from simple projectile motion out to circular orbital motion and beyond. For all these orbital shapes, we know the projectile's acceleration is the gravitational acceleration toward the center of the earth. But to go from a knowledge of the acceleration to predicting the shape of the orbit is not necessarily an easy task.

There are no simple formulas like the constant acceleration formulas that allow us to predict where the satellite will be at any time in the future. Using advanced calculus techniques one can show that the orbits should have the shape of conic sections, one example being the elliptical orbits discovered by Kepler. But if we go to more complicated problems like trying to predict the motion of the Apollo 8 spacecraft from the earth to the moon and back, then a calculus approach is completely inadequate.

On the other hand these problems are easily handled using the step-by-step method of predicting motion, the method, discussed in Chapter 5, that we implement using the computer. With a slight modification of our old projectile motion program, we can predict what will happen to an earth satellite no matter how it is launched and what orbit it has. Adding a few more lines to the program allows us to send the satellite to the moon and back.

Once we are familiar with a basic satellite motion program, we can easily add new features. We can, for example, change the exponent in the gravitational force law from $1/r^2$ to $1/r^{2.1}$ to see what happens if the gravitational force law is modified. Similar modifications were in fact predicted by Einstein's general theory of relativity, thus we will be able to observe the kind of effects that were used to verify Einstein's theory.

New Calculational Loop

In Chapter 5, we set up the machinery to do computer calculations. This involved learning the LET statement, constructing loops, plotting crosses, etc. Although this may have been a bit painful (but perhaps not as painful as learning calculus), we do not have to do much of that again. We can use essentially the same machinery to predict satellite orbits. The only significant change is in the calculational loop where we predict the particle's new position and velocity.

In the projectile motion program, the English version of the calculational loop was, from Figure (5-18)

! --------- Calculational Loop
   DO
      LET $\vec{R}_{new} = \vec{R}_{old} + \vec{V}_{old} * dt$
      LET $\vec{A} = \vec{g}$
      LET $\vec{V}_{new} = \vec{V}_{old} + \vec{A} * dt$
      LET $T_{new} = T_{old} + dt$
      PLOT $\vec{R}$
   LOOP UNTIL T > 1

Figure 19

This loop expresses the method of predicting motion that we developed from the analysis of strobe photographs. The idea behind the command

$$\text{LET } \vec{R}_{new} = \vec{R}_{old} + \vec{V}_{old} * dt$$

is illustrated in Figure (5-15a) reproduced here. The new position of the particle is obtained from the old position by adding the vector $\vec{V}_{old} * dt$ to the old coordinate vector $\vec{R}_{old}$.

Once we get to the new position of the particle, we need the new velocity vector in order to calculate the next new position. The new velocity vector is obtained from the command

$$\text{LET } \vec{V}_{new} = \vec{V}_{old} + \vec{A} * dt$$

as illustrated in Figure (5-15b). The DO–LOOP part of the program tells us to keep repeating this step-by-step process until we get as much of the trajectory as we want (in this case until one second has elapsed).

The calculational loop of Figure (19) works for projectile motion because we always know the projectile's acceleration $\vec{A}$ which is given by the line

$$\boxed{\text{LET } \vec{A} = \vec{g}} \quad \text{projectile motion} \quad (16)$$

This is the line that characterizes projectile motion, the line that tells the computer that the projectile has a constant acceleration $\vec{g}$.

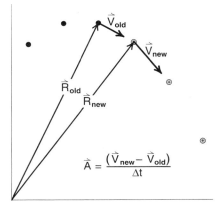

Figure 5-15a
Predicting the next new position.

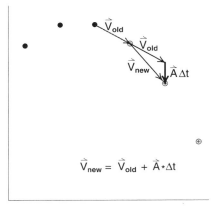

Figure 5-15b
Predicting the next new velocity.

The only fundamental change we need to make in going from projectile motion to satellite motion is to change our command for the particle's acceleration $\vec{A}$. Instead of assuming that the particle's acceleration is constant, we use Newton's law of gravity $|\vec{F}_g| = Gm_1m_2/r^2$ to calculate the force acting on the satellite, and then Newton's second law $\vec{A} = \vec{F}_g/m$ to obtain the resulting acceleration.

There are of course some other details. We have to find a way to express the vector nature of the gravitational force – i.e., to tell the computer which way the gravitational force is pointing, and we are going to change our plotting scale since we are no longer working in front of a 100 cm by 100 cm grid. But essentially we are replacing the command

$$\text{LET } \vec{A} = \vec{g}$$

by the new lines

$$\text{LET } \vec{F}_g = GM_em/R^2 \quad \text{\textit{with instructions for a direction}}$$

$$\text{LET } \vec{A} = \vec{F}_g/m$$

and then using the same old program.

Unit Vectors

We have no problem describing the direction of the gravitational force on the satellite—the force is directed toward the center of the earth. But how do we tell the computer that? What mathematical technique can we use to express the direction of $\vec{F}_g$?

The technique that we will use throughout the course is the use of the ***unit vector***. *A unit vector is a dimensionless vector of length 1 that points in the direction of interest*. If we want a vector of length 5 newtons that points in the same direction, then we multiply our unit vector by the number 5 newtons to get the desired result. (Recall that multiplying a vector by a number, for example n, gives a vector n times as long, pointing in the same direction.)

There is an easy way to construct unit vectors. If we can find some vector that points in the desired direction, we divide that vector by its own length, and we end up with a vector of length 1, the required unit vector.

In our satellite motion problem, the gravitational force $\vec{F}_g$ points toward the center of the earth. Thus to define the direction of $\vec{F}_g$, we need a unit vector that points toward the center of the earth. In Figure (20a) we show the coordinate vector $\vec{R}$ which defines the position of the satellite in a coordinate system whose origin is at the center of the earth. In Figure (20b) we see the vector $-\vec{R}$, which points from the satellite to the center of the earth, the same direction as the gravitational force. Therefore we would like to turn $-\vec{R}$ into a unit vector, which we do by dividing by the length of R, namely the distance from the center of the earth to the satellite.

Since we will often use unit vectors in this text, we will designate them by a special symbol. Instead of an arrow over the letter, we will use what is called a *caret* by typographers, or more familiarly a *hat* by physicists. Thus our unit vector in the $-\vec{R}$ direction will be denoted by $-\hat{R}$ and is given by the formula

$$\boxed{-\hat{R} = \frac{-\vec{R}}{R}} \quad \text{\textit{unit vector in the }} -\vec{R} \text{ \textit{direction}} \quad (16)$$

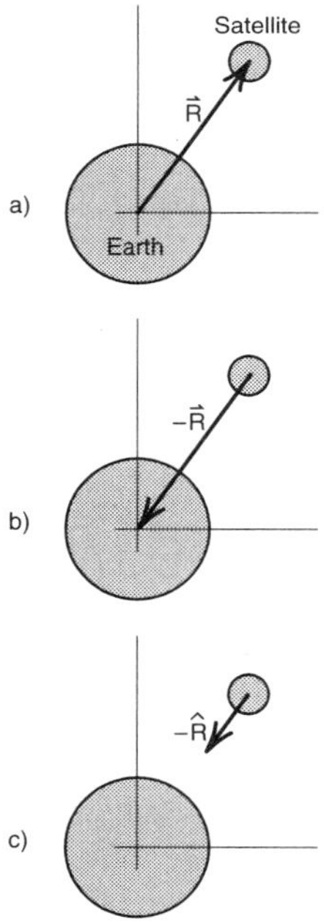

Figure 20
The unit vector $-\hat{R}$

In Equation (16), the length R is given by the Pythagorean theorem

$$R = \sqrt{R_x^2 + R_y^2} \tag{16a}$$

R_x and R_y being the x and y coordinates of the satellite.

With the unit vector $-\hat{R}$, we can now write an explicit formula for the gravitational force vector $\vec{F}_g$. We multiply the unit vector $-\hat{R}$ by the magnitude GMm/R^2 of the gravitational force to get

$$\boxed{\vec{F}_g = \frac{GM_em}{R^2}(-\hat{R})} \tag{17}$$

Calculational Loop for Satellite Motion

We are now ready to go in an orderly way from the calculational loop for projectile motion to a calculational loop for satellite motion. We can focus our attention on the following three lines of the projectile motion calculation loop (Figure 21) because the other lines remain unchanged.

$$\text{LET } \vec{R}_{new} = \vec{R}_{old} + \vec{V}_{old} * dt$$
$$\text{LET } \vec{A} = \vec{g}$$
$$\text{LET } \vec{V}_{new} = \vec{V}_{old} + \vec{A} * dt$$

Figure 21

The first step is to replace LET $\vec{A} = \vec{g}$ by Newton's law of gravity and Newton's second law as shown in Figure (22).

$$\text{LET } \vec{R}_{new} = \vec{R}_{old} + \vec{V}_{old} * dt$$
$$\text{LET } \vec{F}_g = (-\hat{R})GM_em/R^2$$
$$\text{LET } \vec{A} = \vec{F}_g/m$$
$$\text{LET } \vec{V}_{new} = \vec{V}_{old} + \vec{A} * dt$$

Figure 22

Because BASIC is limited to working with numerical commands rather than vectors (an unfortunate limitation), the next step is to make sure that we can translate each of these vector commands into the separate x and y components. We will do this separately for each of the 4 lines.

The command

$$\text{LET } \vec{R}_{new} = \vec{R}_{old} + \vec{V}_{old} * dt$$

for $\vec{R}_{new}$ becomes

$$\text{LET } Rx = Rx + Vx * dt \tag{18a}$$
$$\text{LET } Ry = Ry + Vy * dt \tag{18b}$$

where we drop the subscripts "new" and "old" because the computer automatically takes the old values on the right side of the LET statement, calculates a new value, and stores the new value in the memory cell named on the left side of the LET statement. (See our discussion of the LET statement on page 5-5).

In Equations (18a) and (18b) we obtain numerical values for the new coordinates Rx and Ry of the satellite. However, we will also need to know the distance R from the satellite to the center of the earth (in order to construct the unit vector $-\hat{R}$). The value of R is easily determined by adding the command

$$\text{LET } R = SQR(Rx*Rx + Ry*Ry) \tag{18c}$$

where SQR is BASIC's way of saying square root.

The translation of the command for $\vec{F}_g$ only requires the translation of the unit vector $\hat{R}$ into x and y coordinates. Remembering that $\hat{R} = \vec{R}/R$, we get

$$\hat{R}_x = Rx/R ; \quad \hat{R}_y = Ry/R \tag{19}$$

thus the translation of the LET statement for $\vec{F}_g$ can be written as

$$\text{LET } Fg = G * Me * M / (R*R)$$
$$\text{LET } Fgx = (-Rx / R) * Fg$$
$$\text{LET } Fgy = (-Ry / R) * Fg$$

The computer can handle these lines because it already knows the new values of Rx, Ry and R from Equations (18a, b, and c).

The translation of LET statements for $\vec{A}$ and $\vec{V}_{new}$ are straightforward. We get

LET Ax = Fgx / M
LET Ay = Fgy / M

LET Vx = Vx + Ax * dt
LET Vy = Vy + Ay * dt
LET V   = SQR(Vx*Vx + Vy*Vy)

We included a calculation of the magnitude V of the satellite's speed for future use. We may, for example, want to construct a unit vector in the $-\vec{V}$ direction to represent the direction of air resistance on a reentering satellite. We have found it convenient to routinely calculate the magnitude of any vector whose x and y coordinates we have just calculated.

Summary

To summarize our translation, we started with the vector commands

$$\text{LET } \vec{R}_{new} = \vec{R}_{old} + \vec{V}_{old} * dt$$
$$\text{LET } \vec{F}_g = (-\hat{R})GM_e m/R^2$$
$$\text{LET } \vec{A} = \vec{F}_g/m$$
$$\text{LET } \vec{V}_{new} = \vec{V}_{old} + \vec{A} * dt$$

and ended up with the BASIC commands

LET Rx = Rx + Vx * dt
LET Ry = Ry + Vy * dt
LET R   = SQR (Rx*Rx + Ry*Ry)

LET Fg = G * Me * M / (R*R)
LET Fgx = (–Rx / R) * Fg
LET Fgy = (–Ry / R) * Fg

LET Ax = Fgx / M
LET Ay = Fgy / M

LET Vx = Vx + Ax * dt
LET Vy = Vy + Ay * dt
LET V   = SQR(Vx*Vx + Vy*Vy)

Working Orbit Program

We are now ready to convert a working projectile motion program, Figure (5-23) reproduced here, into a working orbital motion program. In addition to converting the calculational loop as we have just discussed, we need to change some constants and plotting ranges, but the general structure of the program will be unchanged.

Plotting Window

We will initially consider satellite motion that stays reasonably close to the earth, within several earth radii. Using planetary units, and placing the earth at the center of the plot, we can get a reasonable range of orbits if we let Rx vary for example from - 9 to +9 earth radii. If we have a standard 9" Macintosh screen, the x dimension should be 1.5 times the y dimension, thus Ry should go only from -6 to +6. The following command sets up this plotting window

SET WINDOW   -9, 9, -6, 6

To show where the earth is located, we can use the following lines to plot a cross at the center of the earth

LET Rx = 0
LET Ry = 0
CALL CROSS

Constants and Initial Conditions

In going from the projectile motion to the satellite motion program, we have to change the constants and initial conditions. Using planetary units, our constants G, Me, and m are

LET G = 20
LET Me = 1
LET m = .001

(Our choice of the satellite mass m does not matter because it cancels out of the calculation.)

For initial conditions, we will start the satellite .1 earth radii above the surface of the earth on the + x axis;

LET Rx = 1.1
LET Ry = 0
LET R = SQR(Rx*Rx + Ry*Ry)
CALL CROSS

Projectile Motion Program

```
! --------- Plotting window
!          (x axis = 1.5 times y axis)
   SET WINDOW -40,140,-10,110

! --------- Draw & label axes
   BOX LINES 0,100,0,100
   PLOT TEXT, AT -3,0 : "0"
   PLOT TEXT, AT -13,96: "y=100"
   PLOT TEXT, AT 101,0 : "x=100"


! --------- Initial conditions
   LET DeltaT = .1
   LET Rx = 25.9
   LET Ry = 89.9
   LET Vx = (43.2 - 8.3)/(2*.1)
   LET Vy = (90.2 - 79.3)/(2*.1)
   LET T = 0
   CALL CROSS

! --------- Computer Time Step
   LET dt = .001
   LET i = 0

! --------- Calculational loop
   DO
      LET Rx = Rx + Vx*dt
      LET Ry = Ry + Vy*dt
      LET Ax = 0
      LET Ay = -980
      LET Vx = Vx + Ax*dt
      LET Vy = Vy + Ay*dt
      LET T = T + dt
      LET i = i+1
      IF MOD(i,100) = 0 THEN CALL CROSS
      PLOT Rx,Ry
   LOOP UNTIL RX > 100

! --------- Subroutine "CROSS" draws
          ! a cross at Rx,Ry.
   SUB CROSS
      PLOT LINES: Rx-2,Ry;  Rx+2,Ry
      PLOT LINES: Rx,Ry-2;  Rx,Ry+2
   END SUB
END
```

Figure 23
Projectile motion program that plots crosses every tenth of a second.

Orbit-1 Program

```
!--------- Plotting window
!     (x axis = 1.5 times y axis))
   SET WINDOW -9,9,-6,6

!--------- Plot cross at center of the Earth
   LET Rx = 0
   LET Ry = 0
   CALL CROSS

!--------- Constants and Initial Conditions
   LET G  = 19.91
   LET Me = 1
   LET M  = .001

   LET Rx = 1.1
   LET Ry = 0
   LET R  = SQR(Rx*Rx + Ry*Ry)
   CALL CROSS
   LET Vx = 0
   LET Vy = 5.5
   LET T  = 0

!--------- Time step
   LET dt = .01
   LET i = 0

!--------- Caculational loop
   DO
      LET Rx = Rx + Vx*dt
      LET Ry = Ry + Vy*dt
      LET R  = SQR(Rx*Rx + Ry*Ry)

      LET F  = G*M*Me/R^2
      LET Fx = F*(-Rx/R)
      LET Fy = F*(-Ry/R)

      LET Ax = Fx/M
      LET Ay = Fy/M

      LET Vx = Vx + Ax*dt
      LET Vy = Vy + Ay*dt
      LET V  = SQR(Vx*Vx + Vy*Vy)

      LET T  = T + dt
      LET i  = i + 1
      PLOT Rx,Ry
      IF MOD(i,40) = 0 THEN CALL CROSS
   LOOP UNTIL T > 8.45

   !--------- Subroutine "CROSS" draws
            ! a cross at Rx,Ry.
   SUB CROSS
      PLOT LINES: Rx-.2,Ry; Rx+.2,Ry
      PLOT LINES: Rx,Ry-.2; Rx,Ry+.2
   END SUB
END
```

Figure 24
Our new orbital motion program.

We also calculated an initial value of R for use in the gravitational force formula, and plotted a cross at this initial point.

We are going to fire the satellite in the +y direction, parallel to the surface of the earth. Trial and error shows us that a reasonable value for the speed of the satellite is 5.5 earth radii per hour, thus we write for our initial velocity commands

    LET Vx = 0
    LET Vy = 5.5
    LET V = SQR(Vx*Vx + Vy*Vy)

In our projectile motion program of Figure (5-23) we wanted a cross plotted every 100 time steps dt. This was done with the command

    IF MOD(i,100) = 0 THEN CALL CROSS

For our orbit program, trial and error shows that we get a good looking plot if we draw a cross every 40 time steps, each time step dt being .01 hours. Thus our new MOD line will be

    IF MOD(i,40) = 0 THEN CALL CROSS

and we will get a cross every .01 * 40 = .4 hours.

The final change is to stop plotting after one orbit. From running the program we find that one orbit takes about 9 hours, thus we can stop plotting just before one orbit with the LOOP instruction

    LOOP UNTIL T > 9

Putting all these steps together gives us the complete BASIC program shown in Figure (24).

When we run the Orbit 1 program, we get the elliptical orbit shown in Figure (25).

Exercise 10

Convert your projectile motion program to the Orbit 1 program. Use the same initial conditions so that you get the same orbit as that shown in Figure (25). (It is important to get your Orbit 1 program running correctly now, for it will be used as the basis for studying several phenomena during the rest of this chapter. If you are having problems, simply type the program in precisely as shown in Figure (24).

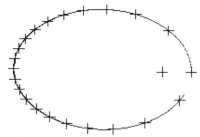

Figure 25
Output of the Orbit 1 program. The satellite is initially out at a distance x = 1.1 earth radii, and is fired in the +y direction at a speed of 5.5 earth radii per hour.

Once your program is working, it is easy to make small modifications to improve the results. To create Figure (25a) we added the command

    BOX CIRCLE -1,1,-1,1

to draw a circle to represent the earth. We also changed dt to .001 and changed the MOD command to MOD(i,539) to get an even number of crosses around the orbit. We then plotted until T = 9 hours. (With dt ten times smaller, our i counter has to be ten times bigger to get the old crosses.)

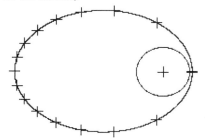

Figure 25a

Satellite Motion Laboratory

In our study of projectile motion, we could go to the laboratory and take strobe photographs in order to see how projectiles behaved. Obtaining experimental data for the study of satellite motion is somewhat more difficult. What we will do is to use the Orbit 1 program or slight modification of it to stimulate satellite motion, using it as our laboratory for the study of the behavior of satellites.

But first we wish to check that the Orbit 1 program makes predictions that are in agreement with experiment. The program is based on Newton's laws of gravity, $F_g = GMm/r^2$, Newton's law of motion $\vec{a} = \vec{F}/m$, and the procedures we developed earlier for predicting the motion of an object whose acceleration is known. Thus a verification of the results of the Orbit 1 program can be considered a verification of these laws and procedures.

Some tests of the Orbit 1 program can be made using the results of your own experience. Anyone who has listened to the launch of a low orbit satellite should be aware that the satellite takes about 90 minutes to go around the earth once. The Orbit 1 program should give the same result, which you can check in Exercise 11. Another obvious test is the prediction of the period of the moon in its orbit around the earth. It is about 4 weeks from full moon to full moon, thus the period should be approximately 4 weeks or 28 days. The fact that the apparent diameter of the moon does not change much during this time indicates that the moon is traveling in a nearly circular orbit about the earth. If you accept the astronomer's measurements that the moon orbit radius is about 60 earth radii away, then you can check the Orbit 1 program to see if it predicts a 4 week period for an earth satellite in a circular orbit of that radius (Exercise 12).

(An easy way to measure the distance to the moon was provided by the first moon landing. Because of a problem with Neil Armstrong's helmet, radio signals sent to Neil from Houston were retransmitted by Neil's microphone, giving an apparent echo. The echo was particularly noticeable while Neil was setting up a TV camera. On a tape of the mission supplied by NASA, you can hear the statement "That's good there, Neil". A short while later you hear the clear echo "That's good there, Neil". The time delay from the original statement and the echo is the time it takes a radio wave, traveling at the speed of light, to go to the moon and back. Using an inexpensive stop watch, one can easily measure the time delay as being about 2 2/5 seconds. Thus the one-way trip to the moon is 1 1/5 seconds. Since light travels 1 ft/nanosecond, or 1 billion feet per second, from this one determines that the moon is about 1.2 billion feet away. You can convert this distance to earth radii to check the astronomer's value of 60 earth radii as the average distance to the moon.)

Exercise 11

Adjust the initial conditions in your Orbit 1 program so that the satellite is in a low earth orbit, and see what the period of the orbit is.

(To adjust the initial conditions, start, for example, with $R_x = 1.01$, $R_y = 0$, $v_x = 0$ and adjust v_y until you get a circular orbit centered on the earth. As a check that the satellite did not go below the surface of the earth, you could add the line

IF R < 1 THEN PRINT "CRASHED"

Adding this line just after you have calculated R in the DO LOOP will immediately warn you if the satellite has crashed. You can then adjust the initial v_y so that you just avoid a crash. Once you have a circular orbit, you can adjust the time in the "LOOP UNTIL T > ..." command so that just one orbit is printed. This tells you how long the orbit took. You can also see how long the orbit took by adding the line in the DO LOOP

IF MOD(I, 40) = 0 THEN PRINT T, RX, RY

Looking at the values of R_x and R_y you can tell when one orbit is completed, and the value of T tells you how long it took.

Exercise 12

Put the satellite in a circular orbit whose radius is equal to the radius of the moon's orbit. (See Table 1, Planetary Units, for the value of the moon orbit radius.) See if you predict that the moon will take about 4 weeks to go around this orbit.)

KEPLER'S LAWS

A more detailed test of Newton's laws and the Orbit 1 program is provided by Kepler's laws of planetary motion.

To get a feeling for the problems involved in studying planetary motion, imagine that you were given the job of going outside, looking at the sky, and figuring out how celestial objects moved. The easiest to start with is the moon, which becomes full again every four weeks. On closer observation you would notice that the moon moved past the background of the apparently fixed stars, returning to its original position in the sky every 27.3 days. Since, as we mentioned, the diameter of the moon does not change much, you might then conclude that the moon is in a circular orbit about the earth, with a period of 27.3 days.

The time it takes the moon to return to the same point in the sky is not precisely equal to the time between full moons. A full moon occurs when the sun, earth, and moon are in alignment. If the sun itself appears to move relative to the fixed stars, the full moons will not occur at precisely the same point, and the time between full moons will not be exactly the time it takes the moon to go around once.

To study the motion of the sun past the background of the fixed stars is more difficult because the stars are not visible when the sun is up. One way to locate the position of the sun is to observe what stars are overhead at "true" midnight, half way between dusk and dawn. The sun should then be located on the opposite side of the sky. (You also have to correct for the north/south position of the sun.) After a fair amount of observation and calculations, you would find that the sun itself moves past the background of the fixed stars, returning to its starting point once a year.

From the fact that the sun takes one year to go around the sky, and the fact that its apparent diameter remains essentially constant, you might well conclude that the sun, like the moon, is traveling in a circular orbit about the earth. This was the accepted conclusion by most astronomers up to the time of Nicolaus Copernicus in the early 1500s AD.

If you start looking at the motion of the planets like Mercury, Venus, Mars, Jupiter, and Saturn, all easily visible without a telescope, the situation is more complicated. Mars, for example, moves in one direction against the background of the fixed stars, then reverses and goes backward for a while, then forward again as shown in Figure (26). None of the planets has the simple uniform motion seen in the case of the moon and the sun.

After a lot of observation and the construction of many plots, you might make a rather significant discovery. You might find what the early Greek astronomers learned, namely that if you assume that the planets Mercury, Venus, Mars, Jupiter, and Saturn travel in circular orbits about the sun, while the sun is traveling in a circular orbit about the earth, then you can explain all the peculiar motion of the planets. This is a remarkable simplification and compelling evidence that there is a simple order underlying the motion of celestial objects.

One of the features of astronomical observations is that they become more accurate as time passes. If you observe the moon for 100 orbits, you can determine the average period of the moon nearly 100 times more accurately than from the observation of a single period. You can also detect any gradual shift of the orbit 100 times more accurately.

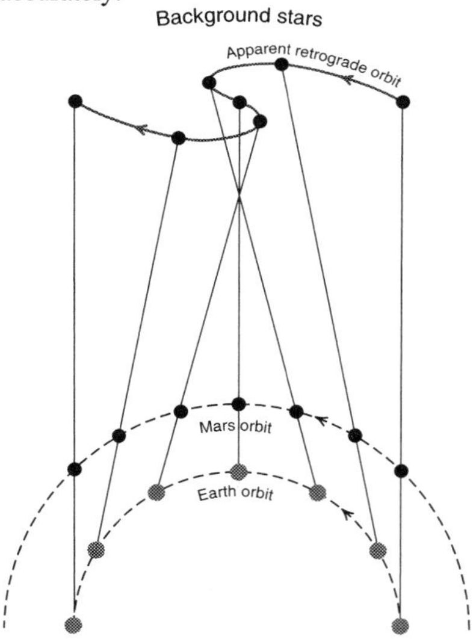

Figure 26
Retrograde motion of the planet Mars. Modern view of why Mars appears to reverse its direction of motion for a while.

Even by the time of the famous Greek astronomer Ptolemy in the second century AD, observations of the positions of the planets had been made for a sufficiently long time that it had become clear that the planets did not travel in precisely circular orbits about the sun. Some way was needed to explain the non circularity of the orbits.

The simplicity of a circular orbit was such a compelling idea that it was not abandoned. Recall that the apparently peculiar motion of Mars could be explained by assuming that Mars traveled in a circular orbit about the sun which in turn traveled in a circular orbit about the earth. By having circular orbits centered on points that are themselves in circular orbits, you can construct complex orbits. By choosing enough circles with the correct radii and periods, you can construct any kind of orbit you wish.

Ptolemy explained the slight variations in the planetary orbits by assuming that the planets traveled in circles around points which traveled in circles about the sun, which in turn traveled in a circle about the earth. The extra cycle in this scheme was called an *epicycle*. With just a few epicycles, Ptolemy was able to accurately explain all observations of planetary motion made by the second century AD.

With 1500 more years of planetary observations, Ptolemy's scheme was no longer working well. With far more accurate observations over this long span of time, it was necessary to introduce many more epicycles into Ptolemy's scheme in order to explain the positions of the planets.

Even before problems with Ptolemy's scheme became apparent, there were those who argued that the scheme would be simpler if the sun were at the center of the solar system and all the planets, including the earth, moved in circles about the sun. This view was not taken seriously in ancient times, because such a scheme would predict that the earth was moving at a tremendous speed, a motion that surely would be felt. (The principle of relativity was not understood at that time.)

For similar reasons, one did not use the rotation of the earth to explain the daily motion of sun, moon, and stars. That would imply that the surface of the earth at the equator would be moving at a speed of around a thousand miles per hour, an unimaginable speed!

In 1543, Nicolaus Copernicus put forth a detailed plan for the motion of the planets from the point of view that the sun was the center of the solar system and that all the planets moved in circular orbits about the sun. Such a theory not only conflicted with common sense about feeling the motion of the earth, but also displaced the earth and mankind from the center of the universe, two results quite unacceptable to many scholars and theologians.

Copernicus' theory was not quite as simple as it first sounds. Because of the accuracy with which planetary motion was know by 1543, it was necessary to include epicycles in the planetary orbits in Copernicus' model.

Starting around 1576, the Dutch astronomer Tycho Brahe made a series of observations of the planetary positions that were a significant improvement over previous measurements. This work was done before the invention of the telescope, using apparatus like that shown in Figure (27). Tycho Brahe did not happen to believe in the Copernican sun-centered theory, but that had little

Figure 27
Tycho Brahe's apparatus.

effect on the reason for making the more accurate observations. Both the Ptolemaic and Copernican systems relied on epicycles, and more accurate data was needed to improve the predictive power of these theories.

Johannes Kepler, a student of Tycho Brahe, started from the simplicity inherent in the Copernican system, but went one step farther than Copernicus. Abandoning the idea that planetary motion had to be described in terms of circular orbits and epicycles, Kepler used Tycho Brahe's accurate data to look for a better way to describe the planet's motion. Kepler found that the planetary orbits were accurately and simply described by ellipses, where the sun was at one of the focuses of the ellipse. (We will soon discuss the properties of ellipses.) Kepler also found a simple rule relating the speed of the planet to the area swept out by a line drawn from the planet to the sun. And thirdly, he discovered that the ratio of the cube of the orbital radius to the square of the period was the same for all planets. These three results are known as Kepler's three laws of planetary motion.

Kepler's three simple rules for planetary motion, which we will discuss in more detail shortly, replaced and improved upon the complex system of epicycles needed by all previous theories. After Kepler's discovery, it was obvious that the sun-centered system and elliptical orbits provided by far the simplest description of the motion of the heavenly objects. For Isaac Newton, half a century later, Kepler's laws served as a fundamental test of his theories of motion and gravitation. We will now use Kepler's laws in a similar way, as a test of the validity of the Orbit 1 program and our techniques for predicting motion.

Kepler's First Law

Kepler's first law states that the planets move in elliptical orbits with the sun at one focus. By analogy we should find from our Orbit 1 program that earth satellites move in elliptical orbits with the center of the earth at one focus. To check this prediction, we need to know how to construct an ellipse and determine where the focus is located.

The arch above the entrance to many of the old New England horse sheds was a section of an ellipse. The carpenters drew the curve by placing two nails on a wide board, attaching the ends of a string to each nail, and moving a pencil around while keeping the string taut as shown in Figure (28). The result is half an ellipse with a nail at each one of the focuses. (If you are in the Mormon Tabernacle's elliptical auditorium and drop a pin at one focus, the pin drop can be heard at the other focus because the sound waves bouncing off the walls all travel the same distance and add up constructively at the second focus point.)

To see if the satellite orbit from the Orbit 1 program is an ellipse, we first locate the second focus using the output shown in Figure (25a) by locating the point symmetrically across from the center of the earth as shown in Figure (29). Then at several points along the orbit we draw lines from that point to each focus as shown, and see if the total length of the lines (what would be the length of the stretched string) remains constant as we go around the orbit.

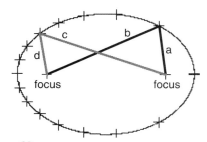

Figure 29
Checking that our satellite orbits are an ellipse. We construct a second focus, and then see if the sum of the distances from each focus to a point on the ellipse in the same for any point around the ellipse. For this diagram, we should show that a+b = c+d.

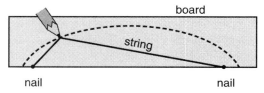

Figure 28
Ellipse constructed with two nails and a string.

Exercise 13
Using the output from your Orbit 1 program, check that the orbit is an ellipse.

Exercise 14
Slightly alter the initial conditions of your Orbit 1 program to get a different shaped orbit. (Preferably, make the orbit more stretched out.) Check that the resulting orbit is still an ellipse.

Kepler's Second Law

Kepler's second law relates the speed of the planet to the area swept out by a line connecting the sun to the planet. If we think of the sun as being at the origin of the coordinate system, then the line from the sun to the planet is what we have been calling the ***coordinate vector*** $\vec{R}$. It is also called the ***radius vector*** $\vec{R}$. Kepler's second law explicitly states *that the radius vector* $\vec{R}$ *sweeps out equal areas in equal times*.

To apply Kepler's second law to the output of our Orbit 1 program, we note that we had the computer plot a cross at equal times along the orbit. Thus the area swept out by the radius vector should be the same as $\vec{R}$ moves from one cross to the next. To check this prediction, we have in Figure (30) reproduced the output of Figure (25a), shaded the areas swept out as $\vec{R}$ moves from positions A to B, from C to D, and from E to F. These areas should look approximately equal; you will check that they are in fact equal in Exercise 15.

The most significant consequence of Kepler's second law is that in order to sweep out equal areas while the radius vector is changing length, the planet or satellite must move more rapidly when the radius vector is short, and more slowly when the radius vector is long. The planet moves more rapidly when in close to the sun, and more slowly when far away.

An extreme example of elliptical satellite orbits are the orbits of some of the comets that periodically visit the sun. Halley's comet, for example, visits the sun once every 76 years. The comet spends about 1 year in the close vicinity of the sun, where it is visible from the earth, and the other 75 years on the rest of its orbit which goes out beyond the edge of the planetary system. The comet moves rapidly past the sun, and spends the majority of the 76 year orbital period creeping around the back side of its orbit where its radius vector is very long.

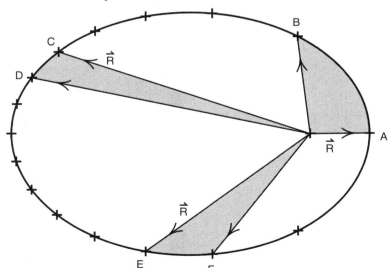

Figure 30
Kepler's Second Law. The radius vector $\vec{R}$ *should sweep out equal areas in equal time.*

Exercise 15

For both of your plots from Exercises 13 and 14, check that the satellite's radius vector sweeps out equal areas in equal times. Explicitly compare the area swept out during a time interval where the satellite is in close to the earth to an equal time interval where the satellite is far from the earth.

This exercise requires that you measure the areas of lopsided pie-shaped sections. There are a number of ways of doing this. You can, for example, draw the sections out on graph paper and count the squares, you can break the areas up into triangles and calculate the areas of the triangles, or you can cut the areas out of cardboard and weigh them.

Kepler's Third Law

Kepler's third law states that the ratio of the cube of the orbital radius R to the square of the period T is the same ratio for all the planets. We can easily use Newton's laws of gravity and motion to check this result for the case of circular orbits. The result, which you are to calculate in Exercise 16, is

$$\frac{R^3}{T^2} = \frac{GM_s}{4\pi^2} \qquad (20)$$

where M_s is the mass of the sun. In this calculation, the mass m_p of the planet, the orbital radius R, the speed v all cancelled, leaving only the sun mass M_s as a variable. Since all the planets orbit the same sun, this ratio should be the same for all the planets.

When the planet is in an elliptical orbit, the length of the radius vector $\vec{R}$ changes as the planet goes around the sun. What Kepler found was that the ratio of R^3/T^2 was constant if you used the "semi major axis" for R. The semi major axis is the half the maximum diameter of the ellipse, shown in Figure (31). As an optional Exercise (17), you can compare the ratio of R^3/T^2 for the two elliptical orbits of Exercises (13) and (14), using the semi major axis for R.

Exercise 16

Consider the example of a planet of mass m_p in a circular orbit about the sun whose mass is M_s. Using Newton's second law and Newton's law of gravity, and the fact that for circular motion the magnitude of the acceleration is v^2/R, solve for the radius R of the orbit. Then use the fact that the period T is the distance $2\pi R$ divided by the speed v, and construct the ratio R^3/T^2. All the variables except M_s should cancel and you should get the result shown in Equation 20.

Exercise 17 (optional)

A more general statement of Kepler's third law, that applies to elliptical orbits, is that R^3/T^2 is the same for all the planets, where R is the semi major axis of the ellipse (as shown in Figure (31)). Check this prediction for the two elliptical orbits used in Exercises (13) and (14). In both of those examples the satellite was orbiting the same earth, thus the ratios should be the same.

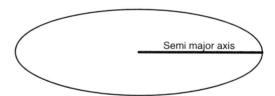

Figure 31
The semi major axis of an ellipse.

MODIFIED GRAVITY AND GENERAL RELATIVITY

After we have verified that the Orbit 1 program calculates orbits that are in agreement with Kepler's laws of motion, we should be reasonably confident that the program is ready to serve as a laboratory for the study of new phenomena we have not necessarily encountered before. To illustrate what we can do, we will begin with a question that cannot be answered in the lab. What would happen if we modified the law of gravity? What, for example, would happen if we changed the universal constant G, or altered the exponent on the r dependence of the force? With the computer program, these questions are easily answered. We simply make the change and see what happens.

These changes should not be made completely without thought. I have seen a project where a student tried to observe the effect of changing the mass of the satellite. After many plots, he concluded that the effect was not great. That is not a surprising result considering the fact that the mass m_s of the satellite cancels out when you equate the gravitational force to $m_s \vec{a}$.

One can also see that, as far as its effect on a satellite's orbit, changing the universal constant G will have an effect equivalent to changing the earth mass M_e. Since Kepler's laws did not depend particularly on what mass our sun had, one suspects that Kepler's laws should also hold when G or M_e are modified. This guess can easily be checked using the Orbit 1 program.

Changing the r dependence of the gravitational force is another matter. After developing the special theory of relativity, Einstein took a look at Newton's theory of gravity and saw that it was not consistent with the principle of relativity. For one thing, because the Newtonian gravitational force is supposed to point to the current instantaneous position of a mass, it should be possible using Newtonian gravity to send signals faster than the speed of light. (Think about how you might do that.)

From the period of time between 1905 and 1915 Einstein worked out a new theory of gravity that was consistent with special relativity and, in the limit of slowly moving, not too massive objects, gave the same results as Newtonian gravity. We will get to see how this process works when, in the latter half of this text we start with Coulomb's electric force law, include the effects of special relativity, and find that magnetism is one of the essential consequences of this combination.

Einstein's relativistic theory of gravity is more complex than the theory of electricity and magnetism, and the new predictions of the theory are much harder to test. It turns out that Newtonian gravity accurately describes almost all planetary motion we can observe in our solar system. Einstein calculated that his new theory of gravity should predict new observable effects only in the case of the orbit of Mercury and in the deflection of starlight as it passed the rim of the sun. In 1917 Sir Arthur Eddington led a famous eclipse expedition in which the deflection of starlight past the rim of the eclipsed sun could be observed. The deflection predicted by Einstein was observed, making this the first clear correction to Newtonian gravity detected in 250 years. Einstein's real fame began with the success of the Eddington expedition.

While Einstein set out to construct a theory of gravity consistent with special relativity, he was also impressed by the connection between gravity and space. Because all projectiles here on the surface of the earth have the same downward acceleration, if you were in a sealed room you could not be completely sure whether your room was on the surface of the earth, and the downward accelerations were caused by gravity, or whether you were out in space, and your room was accelerating upward with an acceleration g. These equivalent situations are shown in Figure (32).

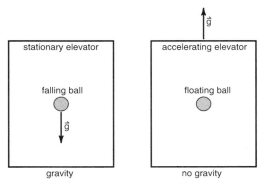

Figure 32
Equivalent situations. Explain why you would feel the same forces if you were sitting on the floor of each of the two rooms.

The equivalence between a gravitational force and an acceleration turned out to be the cornerstone of Einstein's relativistic theory of gravity. It turned out that Einstein's new theory of gravity could be interpreted as a theory of space and time, where mass caused a curvature of space, and what we call gravitational forces were a consequence of this curvature of space. This geometrical theory of gravity, Einstein's relativistic theory, is commonly called the **General Theory of Relativity**.

As they often say in textbooks, a full discussion of Einstein's relativistic theory of gravity is "beyond the scope of this text". However we can look at at least one of the predictions. As far as satellite orbit calculations are concerned, we can think of Einstein's theory as a slight modification of the Newtonian theory. We have seen that any modification of the factors G, m_s or m_e in the Newtonian gravitational force law would not have a detectable effect. The only thing we could notice is some change in the exponent of r.

With a few of quick runs of the Orbit 1 program, you will discover that the satellite orbit is very sensitive to the exponent of r. In Figure (33) we have changed the exponent from -2 to -1.9. This simply requires changing

$$G*m_s*m_e/(R \wedge 2)$$

to

$$G*m_s*m_e/(R \wedge 1.9)$$

in the formula for F_g. The result is a striking change in the orbit. When the exponent is -2, the elliptical orbit is rock steady. When we change the exponent to -1.9, the ellipse starts rotating around the earth. This rotation of the ellipse is called the **precession of the perihelion**, where the word "perihelion" describes the line connecting the two focuses of the ellipse.

A $1/r^2$ force law is unique in that only for this exponent, -2, does the perihelion, the axis of the elliptical orbit, remain steady. For any other value of the exponent, the perihelion rotates or precesses one way or another.

It turns out that a number of effects can cause the perihelion of a planet's orbit to precess. The biggest effect we have not yet discussed is the fact that there are a number of planets all orbiting the sun at the same time, and these planets all exert slight forces on each other. These slight forces cause slight perihelion precessions.

In the 250 years from the time of Newton's discovery of the law of gravity, to the early 1900s, astronomers carefully worked out the predicted orbits of the planets, including the effects of the forces between the planets themselves. This work, done before the development of computers, was an extremely laborious task. A good fraction of one's lifetime work could be spent on a single calculation.

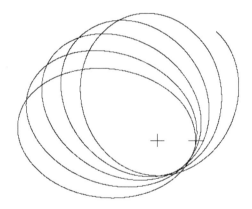

Figure 33
Planetary orbit when the gravitational force is modified to a $1/r^{1.9}$ force.

The orbit of the planet Mercury provided a good test of these calculations because its orbital ellipse is more extended than that of the other close-in planets. The more extended an ellipse, the easier it is to observe a precession. (You cannot even detect a precession for a circular orbit.) Mercury's orbit has a small but observable precession. Its orbit precesses by an angle that is slightly less than .2 degrees every *century*. This is a very small precession which you could never detect in one orbit. But the orbit of Mercury has been observed for about 3000 years, or 30 centuries. That is over a 5 degree precession which is easily detectable.

When measuring small angles, astronomers divide the degree into 60 *minutes of arc*, and for even smaller angles, divide the minute into 60 *seconds of arc*. One second of arc, 1/3600 of a degree, is a very small angle. A basketball 30 miles distant subtends an angle of about 1 second of arc. In these units, Mercury's orbit precesses about 650 seconds of arc per century.

By 1900, astronomers doing Newtonian mechanics calculations could account for all but 43 seconds of arc per century precession of Mercury's orbit as being caused by the influence of neighboring planets. The 43 seconds of arc discrepancy could not be explained. One of the important predictions of Einstein's relativistic theory of gravity is that it predicts a 43 second of arc per century precession of Mercury's orbit, a precession caused by a change in the gravitational force law and not due to neighboring planets. Einstein used this explanation of the 43 seconds of arc discrepancy as the main experimental foundation for his relativistic theory of gravity when he just presented it in 1915. The importance of the Eddington eclipse expedition in 1917 is that a completely new phenomena, predicted by Einstein's theory, was detected.

(The Eddington expedition verified more than just the fact that light is deflected by the gravitational attraction of a star. You can easily construct a theory where the energy in the light beam is related to mass via the formula $E = mc^2$, and then use Newtonian gravity to predict a deflection. Einstein's General Relativity predicts a deflection twice as large as this modified Newtonian approach. The Eddington expedition observed the larger prediction of General Relativity, providing convincing evidence that General Relativity rather than Newtonian gravity was the more correct theory of gravity.)

Exercise 18
Start with your Orbit 1 program, modify the exponent in the gravitational force law, and see what happens. Begin with a small modification so that you can see how to plot the results. (If you make a larger modification, you will have to change the plotting window to get interesting results.)

(To get the 43 seconds of arc per century precession of Mercury's orbit, using a modified gravitational force law, the force should be proportional to $1/r^{2.00000016}$ instead of $1/r^2$.)

CONSERVATION OF ANGULAR MOMENTUM

With the ability to work with realistic satellite orbits rather than just the circular orbits, we will be able to make significant tests of the laws of conservation of angular momentum and of energy, as applied to satellite motion. In this section, we will first see how Kepler's second law of planetary motion is a direct consequence of the conservation of angular momentum, and then do some calculations with the Orbit 1 program to see that a satellite's angular momentum is in fact conserved—does not change as the satellite goes around the earth. In the next section we will first take a more general look at the idea of a conservation law, and then apply this discussion to the conservation of energy for satellite orbits.

Recall that Kepler's second law of planetary motion states that a line from the sun to the planet, the radius vector, sweeps out equal areas in equal times. For this to be true when the planet is in an elliptical orbit, the planet must move faster when in close to the sun and the radius vector is short, and slower when far away and the radius vector is long.

To intuitively see that this speeding up and slowing down is a consequence of the conservation of angular momentum, one can modify the three dumbbell experiment we used to demonstrate the conservation of angular momentum. In this demonstration the instructor uses only one dumbbell. After a student assists the instructor in getting his rotation started, the instructor extends the dumbbell out to full arm's reach, for instance, when he is facing the class, and pulls his arm in when he is facing away as shown in Figure (34). Some practice is needed to maintain this pattern and not lose one's balance.

The rather expected result of this demonstration is that the instructor rotates more slowly when his arm is far out, and more rapidly when his arm is in close. If we associate the dumbbell with a satellite orbiting the earth, we see the same speeding up as the lever arm about the axis of rotation is reduced, and slowing down as the lever arm is increased.

A fairly simple geometrical construction demonstrates that the rule about the radius vector sweeping out equal areas in equal times is precisely what is required for conservation of angular momentum. In Figure (35a) we have plotted an elliptical satellite orbit showing the position of the planet for two different equal time intervals. The time intervals Δt are short enough that we can fairly accurately represent the displacement of the satellite by short, straight lines of length $v_1 \Delta t$ in the upper triangle and $v_2 \Delta t$ in the lower triangle. With this approximation we can represent the areas swept out by the radius vector by triangles as shown by the shaded areas in Figure (35a).

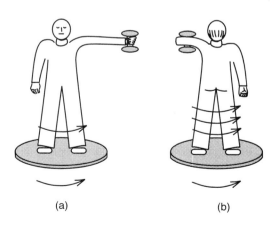

Figure 34
One dumbbell experiment.

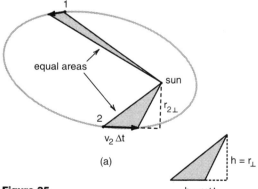

Figure 35
Calculating the area swept out by the planet during a short time interval Δt.

Now the area of a triangle is one half the base times the altitude. If you look at the lower triangle in Figure (35a), and take the side $v_2\Delta t$ as the base, then the distance labeled $r_{2\perp}$ is the altitude, as seen in the sketch in Figure (34b). Thus the area of the triangle at position 2 is

$$\left.\begin{array}{l}\text{area swept}\\\text{out at position 2}\\\text{in a time }\Delta t\end{array}\right\} = \tfrac{1}{2}\,(\text{base}) \times (\text{altitude})$$
$$= \tfrac{1}{2}\,(v_2\Delta t) \times r_{2\perp} \qquad (21)$$

When the satellite is at position 2 in Figure (35a), moving at a velocity $\vec{v}_2$, the distance of closest approach if it continued at the same velocity $\vec{v}_2$ would be the distance $r_{2\perp}$. Thus $r_{2\perp}$ is the "lever arm" for the motion of the satellite at this point in the orbit.

We get a similar formula for the area of the triangle at position 1. Using Kepler's second law which says that these areas should be equal for equal times Δt, we get

$$\tfrac{1}{2}(v_1\Delta t)r_{1\perp} = \tfrac{1}{2}(v_2\Delta t)r_{2\perp} \qquad (22)$$

Dividing Equation 22 through by Δt and multiplying both sides by 2m, where m is the mass of the satellite, gives

$$m_1 v_1 r_{1\perp} = m_2 v_2 r_{2\perp} \qquad (23)$$

Recall that the definition of a particle's angular momentum about some axis is the linear momentum $\vec{p} = m\vec{v}$ times the lever arm $r_\perp$ (see Equations 7–15, 16). Thus the left side of Equation 23 is the satellite's angular momentum at position 1, the right side at position 2. The statement that the satellite sweeps out equal areas in equal times is thus equivalent to the statement that the satellite's angular momentum $mvr_\perp$ has the same value all around the orbit. Like the dumbbell in Figure (34), the satellite moves faster when $r_\perp$ is small, and slower then $r_\perp$ is large, in order to conserve angular momentum.

As a direct check of the conservation of angular momentum in the satellite orbit program, note that if a particle is located a distance x from an axis of rotation and is moving in the y direction with a velocity v_y as shown in Figure (36a), the lever arm about the origin is simply x, and the particle's angular momentum about the origin ℓ_a is

$$\ell_a = mxv_y \qquad \begin{array}{l}\textit{particle's angular}\\\textit{momentum in Figure (36a)}\end{array} \qquad (24)$$

Using the right hand convention illustrated in Figure (7-14), we see that this particle has angular momentum directed up, out of the paper. We will call this positive angular momentum. (You can think of m as a small piece of the bicycle wheel shown in Figure 7-14.)

Now consider a particle of mass m located a distance y from the origin traveling in the – x direction as shown in Figure (36b). By the right hand convention the angular momentum is still positive (you could think of this m as another part of the same bicycle wheel), but the x velocity is now negative. Thus the formula for this particle's angular momentum is

$$\ell_b = -myv_x \qquad (25)$$

We have to put in the minus (–) sign to counteract the fact that v_x is negative but ℓ_b is positive.

It turns out that if a particle is in the xy plane at some arbitrary position $\vec{R} = (x,y)$, and has some arbitrary velocity $\vec{v} = (v_x, v_y)$ in the xy plane, then the formula for the angular momentum ℓ_o of the particle about the origin is

$$\ell_o = m\!\left(xv_y - yv_x\right) \qquad (26)$$

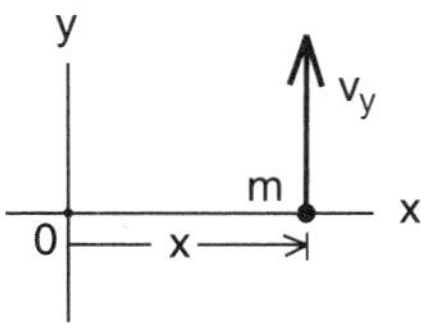

Figure 36a
Here $\ell = mxv_y$.

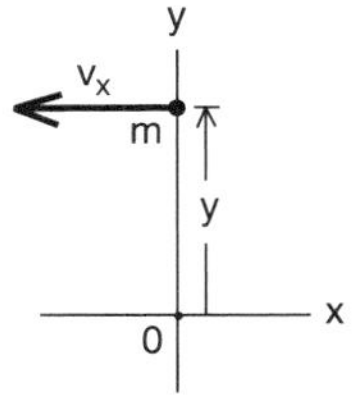

Figure 36b
Here $\ell = myv_x$.

You can see that this general result is just a combination of the two special cases we considered in Figures (36) and Equations 24 and 25. (Equation 26 also comes from the formula $\vec{\ell} = m\vec{r} \times \vec{v}$ where $\vec{r} \times \vec{v}$ is the vector cross product of $\vec{r}$ and $\vec{v}$. We will discuss vector cross products in detail later in Chapter 11. For now Equation 26 is all we need.)

With Equation 26, we can easily test whether angular momentum is in fact conserved in our satellite orbit calculations. By the end of the calculational loop, we have already calculated new values of the satellite's x and y coordinates R_x and R_y, and x and y velocity components v_x and v_y. Thus to calculate the satellite's angular momentum, all we need is the line

$$\text{LET } L_z = M * (Rx*Vy - Ry*Vx) \quad (27)$$

where we are using the name L_z because we are observing the z component of the satellite's angular momentum, as indicated in Figure (37).

To check that angular momentum is conserved, we could add a print line at the end of the calculational loop like

$$\text{IF MOD } (I, 40) = 0 \text{ THEN PRINT } Rx, Ry, Lz \quad (28)$$

By printing the values of R_x and R_y as well as L_z, we can see where the satellite is in its orbit as well as the value of the angular momentum at that point.

Exercise 19

Add lines (27) and (28) to your Orbit 1 program and check that angular momentum is conserved. Use several different initial conditions so that you can check conservation of angular momentum for different elliptical orbits. (Make sure that L_z is calculated within the calculational loop so that the latest values of R_x, R_y, V_x and V_y are used for each calculation.) Also, if you set the satellite mass m equal to 1, the values for L_z will be easier to interpret. (The value of the constant m does not matter since you are simply checking that L_z is constant during the satellite's orbital motion.)

Exercise 20

The fact that angular momentum is conserved in Exercise 19 should not be too surprising because you have already checked in earlier exercises that the elliptical orbit obeys Kepler's second law, and as we have just seen, Kepler's second law implies conservation of angular momentum. In this exercise, see if angular momentum is also conserved if we modify the gravitational force law as we did in Exercise 19. Take your program from Exercise 19, the one that prints out the values of the angular momentum, change the exponent of r in Newton's law of gravity, and see if angular momentum is conserved while the ellipse is precessing.

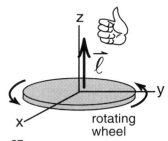

Figure 37
Angular momentum vector of a rotating wheel.

CONSERVATION OF ENERGY

In addition to angular momentum, there is another quantity that is conserved during a satellite's orbital motion. In Chapter 10, which is completely devoted to the topic of energy, we will discuss techniques for deriving formulas for various forms of energy. But it is not necessary to be able to derive energy formulas in order to be able to appreciate and use the concept.

The fundamental idea behind the concept of energy is that energy is a conserved quantity. To study the conservation of energy is often a more difficult job than studying the conservation of linear or angular momentum, because there are many forms that energy can take, and not all the forms are easy to recognize. But in certain simple examples like the motion of an earth satellite, there are only two forms of energy we have to deal with, and the conservation of energy is easy to observe.

Unlike linear and angular momentum, energy does not point anywhere. Energy is represented by a number, not a vector. You get a bill from your electric company for the amount of electrical energy you used the previous month. The electric company has a formula, based on the reading of your electric meter, for the amount of electrical energy you used. Because energy is conserved, the power company could not create the energy they sold you out of nothing, they probably got the energy either from a nuclear power plant or by burning fossil fuels. If they got the energy from fossil fuels, that energy originally came from the sun, from the combining of hydrogen nuclei to form helium nuclei. If the electricity came from a nuclear power plant, the energy came from the splitting of large uranium or plutonium nuclei into smaller nuclei. The uranium and plutonium nuclei were formed by getting their energy from a supernova explosion that must have occurred over five billion years ago.

In our discussion of energy in Chapter 10, we will see that there is a close analogy between keeping track of your checkbook balance in a bank and keeping a record of the amount of energy a system has. With a bank balance, there is a convention that if your balance is positive, the bank owes you money, and if the balance is negative, you owe the bank money. A zero balance indicates that neither owes each other anything. If the bank is not worried about your credit, it does not make much difference whether your balance is positive, negative or zero, you can still write checks, make deposits, and go about your normal business.

In the way we deal with energy, what we call the zero of energy does not make much difference either. We can think of a power company borrowing energy from a coal company just as it borrows money from a bank. In this sense the power company can have a negative energy balance just as it has a negative bank balance. The fact that energy is conserved means that the power company cannot create energy out of nothing to repay the debt. The difference between the power company and physical systems like satellites in orbit is that we let power companies pay their energy debt with cash, a physical system can increase its energy balance only by getting energy from somewhere else.

In our accounting scheme for energy, some terms are positive and some are negative. The term called *kinetic energy* is always positive. In most circumstances, kinetic energy is given by the formula $1/2\, mv^2$ where m is the mass of the object and v the object's speed. Kinetic energy is positive because neither m or $1/2\, mv^2$ can become negative.

To observe conservation of energy for satellite motion, it is necessary to account for two forms of energy. One is kinetic energy $1/2\, v^2$, the other is what is called *gravitational potential energy*. Our formula for gravitational potential energy will be $-Gm_s m_e/r$ where G is the gravitational constant, m_s and m_e the masses of the satellite and earth respectively, and r the separation between them. This formula looks much like the gravitational force formula, except that it is proportional to $1/r$ rather than $1/r^2$.

What is often upsetting to students when they first encounter the gravitational potential energy formula is the minus sign. How can energy be negative? This is essentially a result of our accounting procedure. The important feature of energy is that it is conserved. If the gravitational potential energy in some part of an orbit becomes more negative, then the kinetic energy has to become more positive so that the total is conserved, i.e., stays constant. As far as energy conservation is concerned, it does not make any difference what the total energy is, as long as it is constant.

At this point we have made no effort to explain where the formulas $1/2\, mv^2$ for kinetic energy and $-Gm_s m_e/r$ for gravitational potential energy came from. That is a subject for Chapter 10. What we are concerned with now is to see if the **Total Energy**, the sum of these two, is conserved as the satellite moves around its orbit.

$$\begin{matrix} \text{total energy} \\ \text{of a satellite} \\ \text{in orbit} \end{matrix} = \begin{matrix} \text{kinetic} \\ \text{energy} \end{matrix} + \begin{matrix} \text{gravitational} \\ \text{potential} \\ \text{energy} \end{matrix} \quad (29)$$

$$E_{tot} = \frac{1}{2} mv^2 - \frac{G\, m_s m_e}{r}$$

We will check for conservation of energy in much the same way we checked for conservation of angular momentum using our Orbit 1 program. Near the end of the calculational loop, after we have calculated the latest values of the satellite position $\vec{r}$ and velocity $\vec{v}$, and have also calculated the corresponding magnitudes r and v, we can add the line

LET Etot = Ms*V*V/2 – G*Ms*Me/R   (30)

Then we can add a print line like

IF MOD(I, 40) = 0 THEN PRINT Rx, Ry, Etot

By looking at the printed values of E_{tot} we can see whether this formula for E_{tot} is conserved as the satellite moves around.

Exercise 21
Using the steps described above, check that the satellite's total energy E_{tot} is conserved. (You will notice slight variations in the value of E_{tot}, the values are not as steady as they were in the printout of angular momentum. Exercise 22 suggests a way of improving the energy calculation and getting better results.)

As a variation, print out the values of the kinetic energy, potential energy and E_{tot}. You will see big changes in the kinetic and potential energy, while the sum E_{tot} remains nearly constant. Start the satellite with different initial conditions and check for energy conservation for different elliptical orbits.

Exercise 22
We can obtain a more accurate calculation of the satellite's total energy by slightly modifying the value of v used in the kinetic energy formula. When we put the calculation of E_{tot} at the end of the calculational loop, we are using the value of v at the end of the time step dt. It turns out that we get a more accurate energy calculation if we use a value of v that is the average of the value we had when we entered the calculational loop and the value a time dt later when we left. This averaging is easily accomplished using the following commands inserted into your calculational loop.

```
LET Vold = V              new line saving old value of v
LET Vx = ...
LET Vy = ...              your old lines calculating
                          the next new value of v
LET V = SQR ( Vx*Vx + Vy*Vy )
LET Vnew = V              saving the new value of V
LETV = ( Vold + Vnew ) /2  setting V to the average value
LET Etot = ( Ms*V*V )/2 – G*Ms*Me/R
```

The steps above using the average of V_{new} and V_{old} for V in the calculation of the kinetic energy represents the kind of specialized computer trick we have tried to avoid in this text. However, the trick works so well, the improvement in the value of the total energy is so great that it is worth the effort. This is particularly true for project work where a check for conservation of energy is the main check of the validity of the calculation. (You can usually spot computer errors by printing out the total energy, because computer errors almost never conserve energy.

Exercise 23 (optional, more like a project)

It turns out that if we modify the formula for the gravitational force, for example changing the exponent of r from + 2 to – 1.9, we also have to modify the formula for the gravitational potential energy in order to observe energy conservation. You will learn in Chapter 10 that the formula for the gravitational potential energy is the integral of the magnitude of the force. We can, for example, obtain our formula for gravitational potential energy from the gravitational force formula by the following integration

$$\int_{\infty}^{r} \frac{Gm_s m_e}{r^2} dr = -\frac{Gm_s m_e}{r} \quad (31)$$

If you modify the gravitational force formula, you can do the same kind of integration to get the corresponding potential energy formula. (In Chapter 10 we will have a lot more to say about this integration. For now you can treat the integration as a convenient device for obtaining the potential energy formula. Since the important feature of energy is that it is conserved, if you find from running your Orbit 1 program that the total energy turns out to be conserved, you know you have the correct potential energy formula no matter how it was derived.)

For this exercise, start by modifying the gravitational force law by changing the exponent of r from + 2 to – 1.9. Then run your Orbit 1 program using the formula $-Gm_s m_e/r$ for potential energy to see that this formula does not work. (Use the accurate version of the program from Exercise 22 so that you can be more confident of the results.)

Then integrate $Gm_s m_e/r^{1.9}$ to find a new potential energy formula. See if energy is conserved with your new formula. Once this is successful, try some other modification.

Chapter 9
Applications of Newton's Second Law

In the last chapter our focus was on the motion of planets and satellites, the study of which historically lead to the discovery of Newton's law of motion and gravity. In this chapter we will discuss various applications of Newton's laws as applied to objects we encounter here on earth in our daily lives. This chapter contains many of the examples and exercises that are more traditionally associated with an introductory physics course.

ADDITION OF FORCES

The main new concepts discussed in this chapter are how to deal with a situation in which several forces are acting at the same time on an object. We had a clue for how to deal with this situation in our discussion of projectile motion with air resistance, where in Figure (1) reproduced here, we saw that the acceleration $\vec{a}$ of the Styrofoam projectile was the ***vector sum*** of the acceleration $\vec{g}$ produced by gravity and the acceleration $\vec{a}_{air}$ produced by the air resistance

$$\vec{a} = \vec{g} + \vec{a}_{air} \qquad (1)$$

If we multiply Equation 1 through by m, the mass of the ball, we get

$$m\vec{a} = m\vec{g} + m\vec{a}_{air} \qquad (2)$$

We know that $m\vec{g}$ is the gravitational force acting on the ball, and it seems fairly clear that we should identify $m\vec{a}_{air}$ as the force $\vec{F}_{air}$ that the air is exerting on the ball. Thus Equation 2 can be written

$$m\vec{a} = \vec{F}_g + \vec{F}_{air}$$

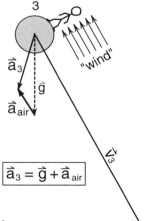

Figure 1
Vector addition of accelerations.

In other words the vector $m\vec{a}$, the ball's mass times its acceleration, is equal to the vector sum of the forces acting upon it. More formally we can write this statement in the form

$$m\vec{a} = \sum_i \vec{F}_i = \left\{ \begin{array}{l} \text{the vector sum of} \\ \text{the forces acting} \\ \text{on the object} \end{array} \right. \quad \begin{array}{l} \textit{more general} \\ \textit{form of Newton's} \\ \textit{second law} \end{array}$$

(4)

Equation 4 forms the basis of this chapter. The basic rule is that, to predict the acceleration of an object, you first identify all the forces acting on the object. You then take the vector sum of these forces, and the result is the object's mass m times its acceleration $\vec{a}$.

When we begin to apply Equation 4 in the laboratory, we will be somewhat limited in the number of different forces that we can identify. In fact there is only one force for which we have an explicit and accurate formula, and that is the gravitational force $m\vec{g}$ that acts on a mass m. Our first step will be to identify other forces such as the force exerted by a stretched spring, so that we can study situations in which more than one force is acting.

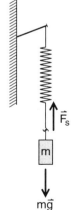

Figure 2
Spring force balanced by the gravitational force.

SPRING FORCES

The simplest way to study spring forces is to suspend a spring from one end and hang a mass on the other as shown in Figure (2). If you wait until the mass m has come to rest, the acceleration of the mass is zero and you then know that the vector sum of the forces on m is zero. In this simple case the only forces acting on m are the downward gravitational force $m\vec{g}$ and the upward spring force $\vec{F}_s$. We thus have by Newton's second law

$$\sum_i \vec{F}_i = \vec{F}_s + m\vec{g} = m\vec{a} = 0 \tag{5}$$

and we immediately get that the magnitude F_s of the spring force is equal to the magnitude mg of the gravitational force.

As we add more mass to the end of the spring, the spring stretches. The fact that the more we stretch the spring, the more mass it supports, means that the more we stretch the spring the harder it pulls back, the greater F_s becomes.

To measure the spring force, we started with a spring suspended from a nail and hung 50 gm masses on the end, as shown in Figure (3). With only one 50 gram mass, the length S of the spring, from the nail to the hook on the mass, was 45.4 cm. When we added another 50 gm mass, the spring stretched to a length of 54.8 centimeters. We added up to five 50 gram masses and plotted the results shown in Figure (4).

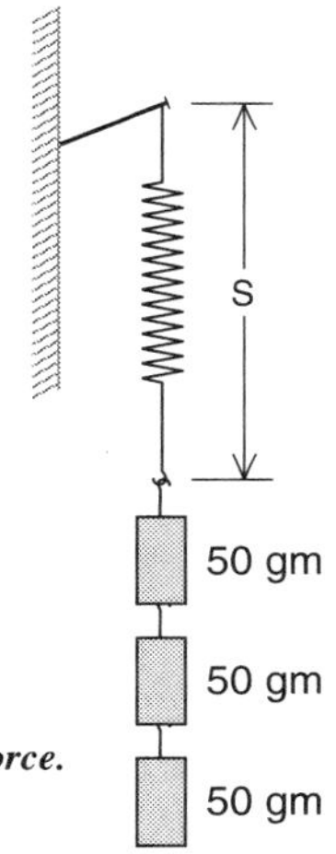

Looking at the plot in Figure (4) we see that the points lie along a straight line. This means that the spring force is linearly proportional to the distance the spring has been stretched.

To find the formula for the spring force, we first draw a line through the experimental points and note that the line crosses the zero force axis at a length of 35.9 cm. We will call this distance the unstretched length S_o.

Thus the distance the spring has been stretched is $(S - S_o)$, and the spring force should be linearly proportional to this distance. Writing the spring force formula in the form

$$F_s = k(S - S_o) \tag{6}$$

all we have left is determine the *spring constant* k.

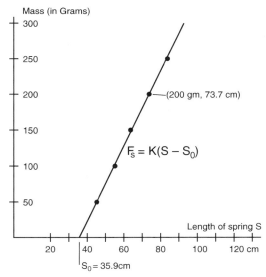

Figure 3
Calibrating the spring force.

Figure 4
Plot of the length of the spring as a function of the force it exerts.

The easy way to find the value of k is to solve Equation 6 for k and plug in a numerical value that lies on the straight line we drew through the experimental points. Using the value $F_s = 200 \text{ gm} \times 980 \text{ cm/sec} = 19.6 \times 10^4$ dynes when the spring is stretched to a distance $S = 73.7$ cm gives

$$k = \frac{F_s}{(S - S_o)} = \frac{19.6 \times 10^4 \text{ dynes}}{(73.7 - 35.9) \text{ cm}}$$

$$= 5.18 \times 10^3 \, \frac{\text{dynes}}{\text{cm}}$$

Equation 6, the statement that the force exerted by a spring is linearly proportional to the distance the spring is stretched, is known as ***Hooke's law***. Hooke was a contemporary of Isaac Newton, and was one of the first to suspect that gravitational forces decreased as $1/r^2$. There was a dispute between Hooke and Newton as to who understood this relationship first. It may be more of a consolation award that the empirical spring force "law" was named after Hooke, while Newton gets credit for the basic gravitational force law.

Hooke's law, by the way, only applies to springs if you do not stretch them too far. If you exceed the "elastic limit", i.e., stretch them so far that they do not return to the original length, you have effectively changed the spring constant k.

The Spring Pendulum

The spring pendulum experiment is one that nicely demonstrates that an object's acceleration is proportional to the vector sum of the forces acting on it. In this experiment, shown in Figure (5), we attach one end of a spring to a nail, hang a ball on the other end, pull the ball back off to one side, and let go. The ball loops around as seen in the strobe photograph of Figure (6). The orbit of the ball is improved, i.e., made more open and easier to analyze, if we insert a short section of string between the end of the spring and the nail, as indicated in Figure (5).

This experiment does not appear in conventional textbooks because it cannot be analyzed using calculus—there is no analytic solution for this motion. But the analysis is quite simple using graphical methods, and a computer can easily predict this motion. The graphical analysis most clearly illustrates the point we want to make with this experiment, namely that the ball's acceleration is proportional to the vector sum of the forces acting on the ball.

In this experiment, there are two forces simultaneously acting on the ball. They are the downward force of gravity $\vec{F}_g = m\vec{g}$, and the spring force $\vec{F}_s$. The spring force $\vec{F}_s$ always points back toward the nail from which the spring is suspended, and the magnitude of the

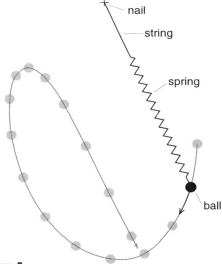

Figure 5
Experimental setup.

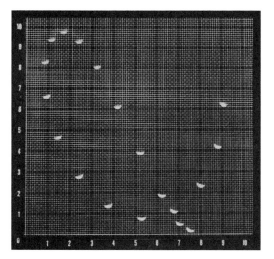

Figure 6
Strobe photograph of a spring pendulum.

spring force is given by Hooke's law $F_s = k(S - S_o)$. Since we can calibrate the spring before the experiment to determine k and S_o, and since we can measure the distance S from a strobe photograph of the motion, we can determine the spring force at each position of the ball in the photograph.

In Figure (7) we have transferred the information about the positions of the ball from the strobe photograph to graph paper and labeled the first 17 positions of the ball from –1 to 15. Consider the forces acting on the ball when it is located at the position labeled 0. The spring force $\vec{F}_s$ points from the ball up to the nail which in this photograph is located at a coordinate (50, 130). The distance S from the hook on the ball to the nail, the distance we have called the stretched length of the spring, is 93.0 cm. You can check this for yourself by marking off the distance from the edge of the ball to the nail on a piece of paper, and then measuring the separation of the marks using the graph paper (as we did back in Figure (1) of Chapter 3). We measure to the edge of the ball and not the center, because that is where the spring ends, and in calibrating the spring we measured the distance S to the end of the spring. (If we measured to the center of the ball, that would introduce

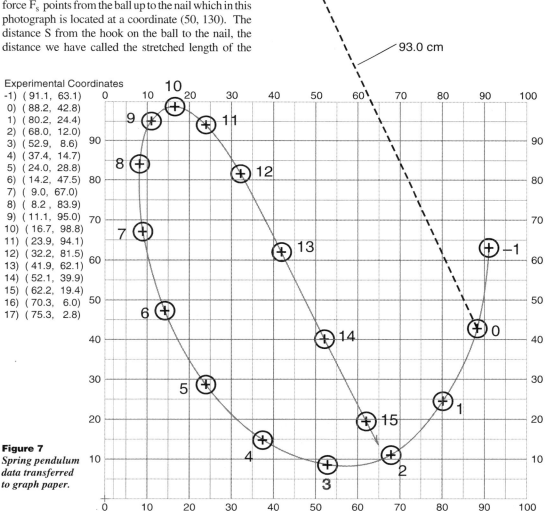

Figure 7
Spring pendulum data transferred to graph paper.

Experimental Coordinates
-1) (91.1, 63.1)
 0) (88.2, 42.8)
 1) (80.2, 24.4)
 2) (68.0, 12.0)
 3) (52.9, 8.6)
 4) (37.4, 14.7)
 5) (24.0, 28.8)
 6) (14.2, 47.5)
 7) (9.0, 67.0)
 8) (8.2, 83.9)
 9) (11.1, 95.0)
10) (16.7, 98.8)
11) (23.9, 94.1)
12) (32.2, 81.5)
13) (41.9, 62.1)
14) (52.1, 39.9)
15) (62.2, 19.4)
16) (70.3, 6.0)
17) (75.3, 2.8)

an error of about 1.3 cm, which produces a noticeable error in our results.)

It turns out that we used the same spring in our discussion of Hooke's law as we did for the strobe photograph in Figure (6). Thus the graph in Figure (4) is our calibration curve for the spring. (The length of string added to the spring is included in the unstretched length S_o.) Using Equation 6 for the spring force, we get

$$F_s = k(S - S_o)$$
$$= 5.18 \times 10^3 \frac{\text{dynes}}{\text{cm}} \times (93.0 - 35.9)\text{cm}$$
$$= 29.6 \times 10^4 \text{dynes} \quad (8)$$

The direction of the spring force is from the ball to the nail. Using a scale in which 10^4 dynes = 1 graph paper square, we can draw an arrow on the graph paper to represent this spring force. This arrow, labeled $\vec{F}_s$, starts at the center of the ball at position 0, points toward the nail, and has a length of 29.6 graph paper squares.

Throughout the motion, the ball is subject to a gravitational force $\vec{F}_g$ which points straight down and has a magnitude mg. For the strobe photograph of Figure (6), the mass of the ball was 245 grams, thus the gravitational force has a magnitude

$$F_g = mg = 245 \text{ gm} \times 980 \frac{\text{cm}}{\text{sec}^2}$$
$$= 24.0 \times 10^4 \text{ dynes} \quad (9)$$

This gravitational force can be represented by a vector labeled $m\vec{g}$ that starts from the center of the ball at position 0, and goes straight down for a distance of 24 graph paper squares (again using the scale 1 square = 10^4 dynes.)

The total force $\vec{F}_{total}$ is the vector sum of the individual forces $\vec{F}_s$ and $\vec{F}_g$

$$\vec{F}_{total} = \vec{F}_s + m\vec{g} \quad (10)$$

The vector addition is done graphically in Figure (8a), giving us the total force acting on the ball when the ball is located at position 0. On the same figure we have repeated the steps discussed above to determine the total force acting on the ball when the ball is up at position 10. Note that there is a significant shift in the total force acting on the ball as it moves around its orbit.

According to Newton's second law, it is this total force $\vec{F}_{total}$ that produces the ball's acceleration $\vec{a}$. Explicitly the vector $m\vec{a}$ should be equal to $\vec{F}_{total}$. To check Newton's second law, we can graphically find the ball's acceleration $\vec{a}$ at any position in the strobe photograph, and multiply the mass m to get the vector $m\vec{a}$.

In Figure (8b) we have used the techniques discussed in Chapter 3 to determine the ball's acceleration vector $\vec{a}\Delta t^2$ at positions 0 and 10. (Recall that for graphical work from a strobe photograph, we had $\vec{a} = (\vec{s}_2 - \vec{s}_1)/\Delta t^2$ or $\vec{a}\Delta t^2 = (\vec{s}_2 - \vec{s}_1)$, where $\vec{s}_1$ is the previous and $\vec{s}_2$ the following displacement vec-

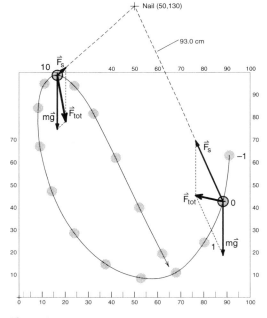

Figure 8a
Force vectors.

tors and Δt the time between images.) At position 0 the vector $\vec{a}\Delta t^2$ has a length of 5.7 cm as measured directly from the graph paper. Since $\Delta t = .1$ sec for this strobe photograph, we have, with $\Delta t^2 = .01$,

$$a\Delta t^2 = a \times .01 = 5.7 \text{ cm}$$

$$a = \frac{5.7 \text{ cm}}{.01 \text{ sec}^2} = 570 \frac{\text{cm}}{\text{sec}^2}$$

Using the fact that the mass m of the ball is 245 gm, we find that the length of the vector $m\vec{a}$ at position 0 is

$$\left|m\vec{a}\right|_{\text{at position 0}} = 245 \text{ gm} \times 570 \frac{\text{cm}}{\text{sec}^2}$$

$$= 14.0 \times 10^4 \frac{\text{gm cm}}{\text{sec}^2}$$

$$= 14.0 \times 10^4 \text{ dynes}$$

In Figure (8c) we have plotted the vector $m\vec{a}$ at position 0 using the same scale of one graph paper square equals 10^4 dynes. Since $m\vec{a}$ has a magnitude of 14.0×10^4 dynes, we drew an arrow 14.0 squares long. The direction of the arrow is in the same direction as the direction of the vector $\vec{a}\Delta t^2$ at position 0 in Figure (8b). In a similar way we have constructed the vector $m\vec{a}$ at position 10.

As a comparison between theory and experiment, we have drawn both the vectors $\vec{F}_{\text{total}}$ and $m\vec{a}$ at positions 0 and 10 in Figure (8c). While the agreement is not exact, it is the best we can expect, considering the accuracy with which we can read the strobe photographs. The important result is that the vectors $\vec{F}_{\text{total}}$ and $m\vec{a}$ can be seen to closely follow each other as the ball moves around the orbit. (In Exercise 1 we ask you to compare $\vec{F}_{\text{total}}$ and $m\vec{a}$ at a couple of more positions to see these vectors following each other.) (I once showed a figure similar to Figure (8c) to a mathematician, who observed the slight discrepancy between the vectors $\vec{F}_{\text{total}}$ and $m\vec{a}$ and said, "Gee, it's too bad the experiment didn't work." He did not have much of a feeling for experimental errors in real experiments.)

Exercise 1

Using the data for the strobe photograph of Figure (6), as we have been doing above, compare the vectors $\vec{F}_{\text{total}}$ and $m\vec{a}$ at two more locations of the ball.

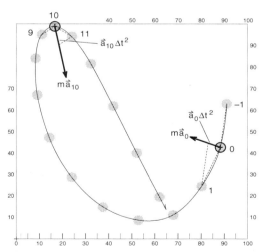

Figure 8b
Acceleration vectors.

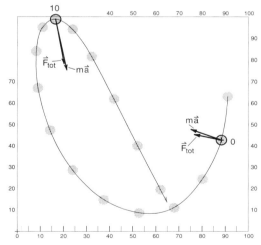

Figure 8c
Comparing $\vec{F}_{total}$ and $m\vec{a}$.

Computer Analysis of the Ball Spring Pendulum

It turns out that using the computer you can do quite a good job of predicting the motion of the ball bouncing on the end of the spring. A program for predicting the motion seen in Figure (7) is listed in the appendix of this chapter. Here all we will discuss are the essential features that you will find in the calculational loop of that program.

The main features of any program that predicts the motion of an object are the following lines, written out in English

$$
\begin{aligned}
&\text{! Calculational Loop}\\
&\text{Let } \vec{R}_{new} = \vec{R}_{old} + \vec{V}_{old} * dt\\
&\text{Let } \vec{F}_1 = \ldots \qquad \textit{find forces acting on the object}\\
&\text{Let } \vec{F}_2 = \ldots\\
&\text{Let } \vec{F}_{total} = \vec{F}_1 + \vec{F}_2 + \ldots \qquad \textit{find the vector sum of the forces}\\
&\text{Let } \vec{a} = \vec{F}_{total}/m \qquad \textit{Newton's second law}\\
&\text{Let } \vec{V}_{new} = \vec{V}_{old} + \vec{a} * dt\\
&\text{Loop Until} \ldots \text{! Repeat calculation} \qquad (11)
\end{aligned}
$$

To apply this general structure to the spring pendulum problem, we first have to be able to describe the direction of the spring force $\vec{F}_s$. This is done using the vector diagram of Figure (9). The vector $\vec{Z}$ represents the coordinate of the nail from which the spring is suspended, $\vec{S}$ the displacement from the nail to the ball, and $\vec{R}$ the coordinate of the ball. From Figure (9) we immediately get the vector equation

$$\vec{Z} + \vec{S} = \vec{R} \qquad (12)$$

which we can solve for the spring length $\vec{S}$

$$\vec{S} = \vec{R} - \vec{Z} \qquad (13)$$

From Figure (7) we see that the nail is located at the coordinate (50,130) thus

$$\vec{Z} = (50, 130)$$

Throughout the motion of the ball, the spring force points in the $-\hat{S}$ direction as indicated in Figure (10), thus the formula for the spring force can be written

$$\vec{F}_s = \left(-\hat{S}\right) k \left(S - S_o\right) \qquad (14)$$

where $k(S - S_o)$ is the magnitude of the spring force determined in Figure (4).

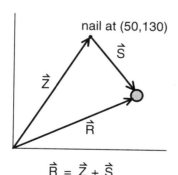

Figure 9
Vector diagram. $\vec{Z}$ is the coordinate of the nail, $\vec{R}$ the coordinate of the ball, and $\vec{S}$ the displacement of the ball from the nail.

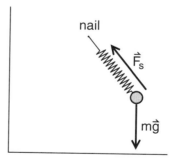

Figure 10
Force diagram, showing the two forces acting on the ball.

Using Equation 6 for the spring force, the English calculational loop for the spring pendulum becomes

! Calculational loop for spring pendulum

Let $\vec{R}_{new} = \vec{R}_{old} + \vec{V}_{old} * dt$
Let $\vec{S} = \vec{R} - \vec{Z}$
Let $\vec{F}_s = -\hat{S} * k * (S - S_o)$
Let $\vec{F}_g = mg$
Let $\vec{F}_{total} = \vec{F}_s + \vec{F}_g$
Let $\vec{A} = \vec{F}_{total}/m$
Loop Until . . .   (15)

A translation into BASIC of the lines for calculating $\vec{S}$ and $\vec{F}_s$ would be, for example,

LET Sx = Rx – Zx
LET Sy = Ry – Zy
LET S   = SQR (Sx * Sx + Sy * Sy)
LET Fsx = (– Sx / S) * k * (S – So)
LET Fsy = (– Sy / S) * k * (S – So)   (16)

The rest of the program, discussed in the Appendix, is much like our earlier projectile motion programs, with a new calculational loop. In Figure (11) we have plotted the results of the spring pendulum program, where the crosses represent the predicted positions of the ball and the squares are the experimental positions. If you slightly adjust the initial conditions for the motion of the ball, you can make almost all the crosses fall within the squares. How much adjustment of the initial conditions you have to do gives you an indication of the size of the errors involved in determining the positions of the ball from the strobe photograph.

Analytic Solution

If you pull the ball straight down and let go, the ball bounces up and down in a periodic motion that can be analyzed using calculus. The resulting motion is called a *sinusoidal oscillation* which we will discuss in considerable detail in Chapter 14. You will see that if you can use calculus to obtain an analytic solution, there are many ways to use the results. The oscillatory spring motion serves as a model for describing many phenomena in physics.

Figure 11
Output from the ball spring program. The crosses are the points predicted by the computer program, while the black squares represent the experimental data points. The program in the Appendix illustrates how the data points can be plotted on the same diagram with your computer plot.

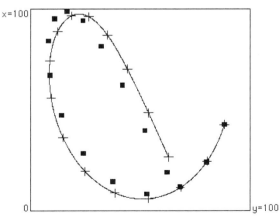

THE INCLINED PLANE

Galileo discovered the formulas for projectile motion by using an inclined plane to slow the motion down, making it easier to measure positions and velocities. He studied rolling balls, whereas we wish to study sliding objects using a frictionless inclined plane.

The frictionless inclined plane was more or less a figment of the imagination of the authors of introductory textbooks, at least until the development of the air track. And even with an air track some small effects of friction can be observed. We will discuss the inclined plane here because it illustrates a useful technique for analyzing the forces on an object, and because it leads to some interesting laboratory experiments.

As a simple experiment, place a book, a floppy disk, or some small object under one end of an air track so that the track is tilted at an angle θ as shown in Figure (12). If you keep the angle θ small, you can let the air cart bounce against the bumper at the end of the track without damaging anything.

To analyze the motion of the air cart, it helps to exaggerate the angle θ in our drawings of the forces involved as we have done this in Figure (13). The first step in handling any Newton's law problem is to identify all the forces involved. In this case there are two forces acting on the air cart; the downward force of gravity $m\vec{g}$ and the force $\vec{F}_p$ of the plane against the cart.

The main feature of a frictionless surface is that it can exert only normal forces, i.e., forces perpendicular to the surface. (Any sideways forces are the result of "friction".) Thus $\vec{F}_p$ is perpendicular to the air track, inclined at an angle θ away from the vertical direction.

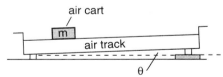

Figure 12
Tilted air cart.

What makes the analysis of this problem different from the motion of the spring pendulum discussed in the last section is the fact that the cart is constrained to move along the air track. This tells us immediately that the cart accelerates along the track, and has no acceleration perpendicular the track. If there is no perpendicular acceleration, there must be no net force perpendicular to the track. From this fact alone we can determine the magnitude of the force $\vec{F}_p$ exerted by the track.

Before we do any calculations, let us set up the problem in such a way that we can take advantage of our knowledge that the cart moves only along the track. Without thinking, we would likely take the x axis to be in the horizontal direction and the y axis in the vertical direction. But with this choice the cart has a component of velocity in both the x and y directions.

The analysis is greatly simplified if we choose one of the coordinate axes to lie along the plane. In Figure (14), we have chosen the x axis to lie along the plane, and decomposed the downward gravitational force into an x component which has a magnitude $mg\sin\theta$ and a $-y$ component of magnitude $mg\cos\theta$.

Now the analysis of the problem is easy. Starting with Newton's law in vector form, we have

$$m\vec{a} = \Sigma \vec{F}_i = m\vec{g} + \vec{F}_p \qquad (17)$$

Separating Equation 17 into its x and y components, we get

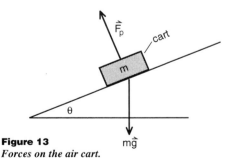

Figure 13
Forces on the air cart.

$$ma_x = \left(m\vec{g}\right)_x = mg\sin\theta \tag{18a}$$

$$\begin{aligned}ma_y &= 0 \\ &= F_p - \left(m\vec{g}\right)_y \\ &= F_p - mg\cos\theta\end{aligned} \tag{18b}$$

where we set $a_y = 0$ because the cart moves only in the x direction.

From Equation 18b we immediately get

$$F_p = mg\cos\theta \tag{19}$$

as the formula for the magnitude of the force the plane exerts on the cart.

Of more interest is the formula for a_x which we immediately get from Equation 18a

$$\boxed{a_x = g\sin\theta} \tag{20}$$

We see that the cart has a constant acceleration down the plane, an acceleration whose magnitude is equal to the acceleration due to gravity, but reduced by a factor $\sin\theta$. It is this reduction that slows down the motion, and allowed Galileo to study motion with constant acceleration using the crude timing devices available to him at that time.

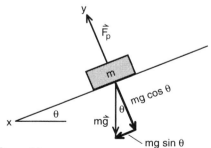

Figure 14
Choosing the x axis to lie along the plane.

Exercise 2

A one meter long air track is set at an angle of $\theta = .03$ radians. (This was done by placing a 3 millimeter thick floppy disk under one end of the track.

(a) From your knowledge of the definition of the radian, explain why, to a high degree of accuracy, the $\sin\theta$ and θ are the same for these small angles.

(b) The cart is released from rest at one end of the track. How long will it take to reach the other end. (You can consider this to be a review of the constant acceleration formulas.)

Portrait of Galileo

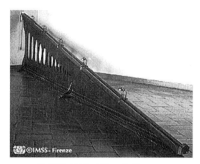

Galileo's Inclined plane

Above photos from the informative web page
http://galileo.imss.firenze.it/museo/b/egalilg.html

FRICTION

If you do the experiment suggested in Exercise 2, measuring the time it takes the cart to travel down the track when the track is tilted by a very small angle, the results are not likely to come out very close to the prediction. The reason is that for such small angles, the effects of "friction" are noticeable even on an air track.

In introductory physics texts, the word "friction" is used to cover a multitude of sins. With the air track, there is no physical contact between the cart and track. But there are air currents that support the cart and come out around the edge of the cart. These air currents usually slow the air cart down, giving rise to what we might call friction effects.

In common experience, skaters have as nearly a frictionless surface as we are likely to find. The reason that you experience little friction when skating is not because ice itself is that slippery, but because the ice melts under the blade of the skate and the skater travels along on a fine ribbon of water. The ice melts due to the pressure of the skate against the ice. Ice is a peculiar substance in that it expands when it freezes. And conversely, you can melt it by squeezing it. If, however, the temperature is very low, the ice does not melt at reasonable pressures and is therefore no longer slippery. At temperatures of $40°$ F below zero, roads on ice in Alaska are as safe to drive on as paved roads.

When two solid surfaces touch, the friction between them is caused by an interaction between the atoms in the surfaces. In general, this interaction is not understood. Only recently have computer models shed some light on what happens when clean metal surfaces interact. Most surfaces are quite "dirty" at an atomic scale, contaminated by oxides, grit and whatever. It is unlikely that one will develop a comprehensive theory of friction for real surfaces.

Friction, however, plays too important a role in our lives to be ignored. Remember the first time you tried to skate and did not have a surface with enough friction to support you. To handle friction, a number of empirical rules have been developed. One of the more useful rules is "if it squeaks, oil it". At a slightly higher level, but not much, are the formulas for friction that appear in introductory physics text books. Our lack of respect for these formulas comes from the experience of trying to verify them in the laboratory. There is some truth to them, but the more accurately one tries to verify them, the worse the results become. With this statement in mind about the friction formulas, we will state them, and provide one example. Hundreds of examples of problems involving friction formulas can be found in other introductory texts.

Inclined Plane with Friction

In our analysis of the air cart on the inclined track, we mentioned that a frictionless surface exerts only a normal force on an object. If there is any sideways force, that is supposed to be a friction force $\vec{F}_f$. In Figure (15) we show a cart on an inclined plane, with a friction force $\vec{F}_f$ included. The normal force $\vec{F}_n$ is perpendicular to the plane, the friction force $\vec{F}_f$ is parallel to the plane, and gravity still points down.

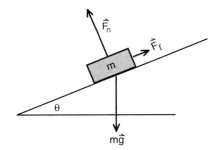

Figure 15
Friction force acting on the cart.

To analyze the motion of the cart when acted on by a friction force, we write Newton's second law in the usual form

$$m\vec{a} = \Sigma \vec{F}_i = m\vec{g} + \vec{F}_n + \vec{F}_f \quad (21)$$

The only change from Equation 13 is that we have added in the new force $\vec{F}_f$.

Since the motion of the cart is still along the plane, it is convenient to take the x axis along the plane as shown in Figure (16). Breaking Equation 21 up into x and y components now gives

$$ma_x = \Sigma F_x = mg \sin\theta - F_f \quad (22a)$$

$$ma_y = \Sigma F_y = -mg \cos\theta + F_n = 0 \quad (22b)$$

From 22b we get,

$$F_n = mg \cos\theta \quad (23)$$

which is the same result as for the frictionless plane.

The new result comes when we look at motion down the plane. Solving 22a for a_x gives

$$a_x = g\sin\theta - F_f/m \quad (24)$$

Not surprisingly, the friction force reduces the acceleration down the plane.

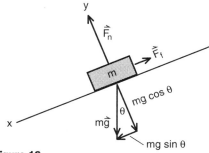

Figure 16

Coefficient of Friction

To go any further than Equation 24, we need some values for the magnitude of the friction force F_f. It is traditional to assume that F_f is proportional to the force F_n between the surfaces. Such a proportionality can be written in the form

$$F_f = \mu F_n \quad (25)$$

where the proportionality constant μ is called the *coefficient of friction*.

Equation 25 makes the explicit assumption that the friction force does not depend on the speed at which the object is moving down the plane. But it is easy to show that this is too simple a model. It is harder to start an object sliding than to keep it sliding. This is why you should not jam on the brakes when trying to stop a car suddenly. You should keep the tires rolling so that there is no sliding between the surface of the tire and the surface of the road.

The difference between non slip or static friction and sliding friction is accounted for by saying that there are two different coefficients of friction, the *static* coefficient μ_s which applies when the object is not moving, and the *kinetic* coefficient μ_k which applies when the objects are sliding. For common surfaces like a rubber tire sliding on a cement road, the static coefficient μ_s is greater than the sliding or kinetic coefficient μ_k.

Let us substitute Equation 25 into Equation 24 for the motion of an object down an inclined plane, and then see how the hypothesis that F_f is proportional to F_n can be tested in the lab. Using Equation 25 and 24 gives

$$a_x = g\sin\theta - F_f/m$$
$$ = g\sin\theta - \mu F_n/m$$

Using $F_n = mg\cos\theta$ gives

$$a_x = g\sin\theta - \mu g\cos\theta$$
$$ = g(\sin\theta - \mu\cos\theta) \quad (26)$$

Equation 26 clearly applies only if $\sin\theta$ is greater than $\mu\cos\theta$ because friction cannot pull the object back up the plane.

If we have a block on an inclined plane, and start with the plane at a very small angle, so that $\sin\theta$ is much less than $\mu\cos\theta$, the block will sit there and not slide. If you increase the angle until $\sin\theta = \mu\cos\theta$, with μ the static coefficient of friction, the block should just start to slide. Thus μ_s is determined by the condition

$$\sin\theta = \mu_s \cos\theta$$

or dividing through by $\cos\theta$

$$\mu_s = \tan\theta_s \qquad (27)$$

where θ_s is the angle at which slipping starts.

After the block starts sliding, μ is supposed to revert to the smaller coefficient μ_k and the acceleration down the plane should be

$$a_x = g\sin\theta - \mu_k g\cos\theta \qquad (26a)$$

Supposedly one can then determine the kinetic coefficient μ_k by measuring the acceleration a_x and using Equation 26a for μ_k.

If you try this experiment in the lab, you may encounter various difficulties. If you try to slide a block down a reasonably smooth board, you may get fairly consistent results and obtain values for μ_s and μ_k. But if you try to improve the experiment by cleaning and smoothing the surfaces, the results may become inconsistent because clean surfaces have a tendency to stick rather than slide.

The idea that friction forces can be described by two coefficients μ_s and μ_k allows the authors of introductory physics texts to construct all kinds of homework problems involving friction forces. While these problems may be good mental exercises, comparable to solving challenging crossword puzzles, they are not particularly appropriate for an introductory physics course. The reason is that the formula $F_f = \mu F_n$ is an over simplification of a complex phenomena. A decent treatment of friction effects belongs in a more advanced engineering oriented course where there is time to study the limitations and applicability of such a rule.

STRING FORCES

Another favorite device of the authors of introductory texts is the massless string (or rope). The idea that a string has a small mass compared to the object to which it is attached is usually a very good approximation. And strings and ropes are convenient devices for transferring a force from one object to another.

In addition, strings have the advantage that you can immediately tell the direction of the force they transmit. The force has to be along the direction of the string or rope, for a string cannot pull sideways. We used this idea when we discussed the motion of a golf ball swinging in a circle on the end of a string. The string could only pull in along the direction of the string toward the center of the circle. From this we concluded that the force acting on the ball was also toward the center of the circle, in the direction the ball was accelerating.

To see how to analyze the forces transmitted by strings and ropes, consider the example of two children pulling on a rope in a game of tug of war show in Figure (17). Let the child labeled 1 be pulling on the rope with a force $\vec{F}_1$ and child labeled 2 pulling with a force $\vec{F}_2$. Assuming that the rope is pulled straight between them, the forces $\vec{F}_1$ and $\vec{F}_2$ will be oppositely directed.

Applying Newton's second law to the rope, and assuming that the force of gravity on the rope is much smaller than either $\vec{F}_1$ or $\vec{F}_2$ and therefore can be neglected, we have

$$m_{rope}\,\vec{a}_{rope} = \left(\vec{F}_1 + \vec{F}_2\right)$$

If we now assume that the rope is effectively massless, we get

$$\vec{F}_1 + \vec{F}_2 = 0 \qquad (28)$$

Thus $\vec{F}_1$ and $\vec{F}_2$ are equal in magnitude and oppositely directed. (Note that if there were a net force on a massless rope, the rope would have an infinite acceleration.)

A convenient way to analyze the effects of a taut rope or string is to say that there is a tension T in the rope, and that this tension transmits the force along the rope. In Figure (18) we have redrawn the tug of war and included the tension T. The point where child 1 is holding the rope is subject to the left directed force $\vec{F}_1$ exerted by the child and the right directed force caused by the tension T in the rope. The total force on this point of contact is $\vec{F}_1 + \vec{T}_1$. Since the point of contact is massless, we must have $\vec{F}_1 + \vec{T}_1 = 0$ and therefore the tension T on the left side of the rope is equal to the magnitude of $\vec{F}_1$. A similar argument shows that the tension force $\vec{T}_2$ exerted on the second child is equal to the magnitude of $\vec{F}_2$. And since the magnitude of $\vec{F}_1$ and $\vec{F}_2$ are equal, the tension forces must also be equal.

Isaac Newton noted that when a force was transmitted via a massless medium, like our massless rope, or the force of gravity, the objects exerted equal and opposite forces (here $\vec{T}_1$ and $\vec{T}_2$) on each other. He called this the ***Third Law of Motion***. We will have more to say about Newton's third law in our discussion of systems of particles in Chapter 11.)

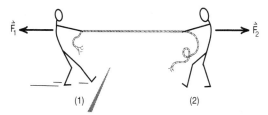

Figure 17
Tug of war.

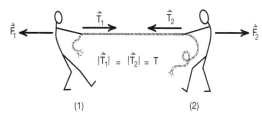

Figure 18
Tension T in the rope.

THE ATWOOD'S MACHINE

As an example of using a string to transmit forces, consider the device shown in Figure (19) which is called an *Atwood's Machine*. It simply consists of two masses at the ends of a string, where the string runs over a pulley. We will assume that the pulley is massless and the bearings in the pulley frictionless so that the only effect of the pulley is to change the direction of the string.

To predict the motion of the objects in Figure (19) we start by analyzing the forces on the two masses. Both masses are subject to the downward force of gravity, $m_1 \vec{g}$ and $m_2 \vec{g}$ respectively. Let the tension in the string be T. As a result of this tension, the string exerts an upward force $\vec{T}$ on both blocks as shown. (We saw in the last section that this force $\vec{T}$ must be the same on both masses.)

Applying Newton's second law to each of the masses, noting there is only motion in the y direction, we get

$$m_1 a_{1y} = T - m_1 g$$
$$m_2 a_{2y} = T - m_2 g \qquad (29)$$

In Equations 29, we note that there are three unknowns T, a_{1y} and a_{2y}, and only two equations. Another relationship is needed. This other relationship is supplied by the observation that the length ℓ of the string, given from Figure (19a) is

$$\ell = h_1 + h_2 \qquad (30)$$

does not change. Differentiating Equation 30 with respect to time and setting $d\ell/dt = 0$ gives

$$0 = \frac{d\ell}{dt} = \frac{dh_1}{dt} + \frac{dh_2}{dt} = v_1 + v_2$$

where $v_1 = dh_1/dt$ is the velocity of mass 1, etc. Differentiating again with respect to time gives

$$0 = \frac{dv_1}{dt} + \frac{dv_2}{dt} = a_1 + a_2 \qquad (31)$$

Thus the desired relationship is

$$a_1 = -a_2 \qquad (32)$$

(You might say that it is obvious that $a_1 = -a_2$, otherwise the string would have to stretch. But if you are dealing with more complicated pulley problems, it is particularly convenient to write down a formula for the total length of the string, and differentiate to obtain the needed extra relationship between the accelerations.)

Using Equation 32 in 29 we get

$$m_1 a_{1y} = T - m_1 g$$
$$-m_2 a_{1y} = T - m_2 g \qquad (32b)$$

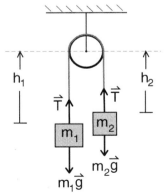

Figure 19a
An Atwood's machine consists of two masses suspended from a string looped over a pulley. The acceleration is proportional to the difference in mass of the two objects.

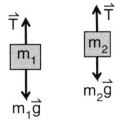

Figure 19b
Forces involved.

Solving 32a for T to get $T = m_1(a_{1y} + g)$ and using this in Equation 32b gives

$$-m_2 a_{1y} = m_1 a_{1y} + m_1 g - m_2 g$$

or

$$a_{1y} = g\left(\frac{m_1 - m_2}{m_1 + m_2}\right) \qquad (33)$$

From Equation 33 we see that the acceleration of mass m_1 is uniform, and equal to the acceleration due to gravity, modified by the factor $(m_1 - m_2)/(m_1 + m_2)$.

When you solve a new problem, see if you can check it by seeing if the limiting cases make sense. In Equation 30, if we set $m_2 = 0$, then $a_1 = g$ and we have a freely falling mass as expected. If $m_1 = m_2$, then the masses balance and $a_{1y} = 0$ as expected. When a formula checks out in its limiting cases, as this one did, there is a good chance that the result is correct.

The advantage of an of the Atwood's Machine is that by choosing m_1 close to, but not equal to m_2, you can reduce the acceleration, making the motion easier to observe, just as Galileo did by using inclined planes. If you reduce the acceleration too much by making m_1 too nearly equal to m_2, you run the risk that even small friction in the bearings of the pulley will dominate the results.

Exercise 3

In a slight complication of the Atwood's Machine, we use two pulleys instead of one as shown in Figure (20). We can treat this problem very much like the preceding example except that the length of the string is $\ell(h_1 + 2h_2)$ plus some constant length representing the part of the string that goes over the pulleys and the part that goes up to the ceiling. Calculate the accelerations of masses m_1 and m_2. For what values of m_1 and m_2 is the system balanced?

Exercise 4

If you want something a little more challenging than Exercise 3, try analyzing the setup shown in Figure (21), or construct your own setup. For Figure (21), it is enough to set up the four equations with four unknowns.

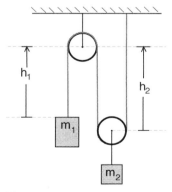

Figure 20
Pulley arrangement for Exercise 3.

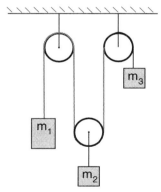

Figure 21
Pulley arrangement for Exercise 4

THE CONICAL PENDULUM

Our final example in this chapter is the conical pendulum. This is one of our favorite examples because it involves a combination of Newton's second law, circular motion, no noticeable friction, and the predictions can be checked using an old boot, shoelace and wristwatch.

For a classroom demonstration of the conical pendulum, we usually suspend a relatively heavy ball on a thin rope, with the other end of the rope attached to the ceiling as shown in Figure (22). The ball is swung in a circle so that the path of the rope forms the surface of a cone as shown. The aim is to predict the period of the ball's circular orbit.

The distances involved and the forces acting on the ball are shown in Figure (23). The ball is subject to only two forces, the downward force of gravity $m\vec{g}$, and the tension force $\vec{T}$ of the string. If the angle that the string makes with the vertical is θ, then the force T has an upward component $T_y = T\cos\theta$ and a component directed radially inward of magnitude $T_x = T\sin\theta$. (We are analyzing the motion of the ball at the instant when it is at the left side of its orbit, and choosing the x axis to point in toward the center of the circle at this instant.)

Applying Newton's second law to the motion of the ball, noting that $a_y = 0$ since the ball is not moving up and down, gives

$$ma_x = T_x$$
$$ma_y = 0 = T_y - mg \qquad (34)$$

The special feature of the conical pendulum is the fact that, because the ball is travelling in a circle, we know that it is accelerating toward the center of the circle with an acceleration of magnitude $a = v^2/r$. At the instant shown in Figure (23), the x direction points toward the center of the circle, thus $a = a_x$ and we have

$$a_x = v^2/r \qquad (35)$$

The rest of the problem simply consists of solving Equations 34 and 35 for the speed v of the ball and using that to calculate the time the ball has to go around. The easy way to solve these equations is to write them in the form

$$T_x = ma_x = \frac{mv^2}{r} \qquad (36a)$$

$$T_y = mg \qquad (36b)$$

Dividing Equation 36a by 36b and using $T_x/T_y = T\sin\theta/T\cos\theta = \tan\theta$, we get

$$\frac{T_x}{T_y} = \tan\theta = \frac{mv^2}{mgr} = \frac{v^2}{gr}$$

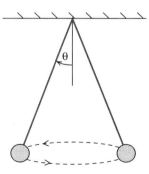

Figure 22
The conical pendulum.

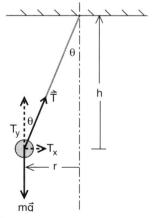

Figure 23
Forces acting on the ball.

Next use the fact that $\tan\theta = r/h$ to get

$$\tan\theta = \frac{r}{h} = \frac{v^2}{gr}$$

$$v = r\sqrt{\frac{g}{h}} \qquad (37)$$

Finally we note that the period is the distance traveled in one circuit, $2\pi r$, divided by the speed v of the ball

$$\begin{matrix}\text{period}\\ \text{of orbit}\end{matrix}\bigg\} = \frac{2\pi r}{v} = \frac{2\pi r}{r\sqrt{g/h}}$$

$$\boxed{\text{period} = 2\pi\sqrt{\frac{h}{g}}} \qquad (38)$$

The prediction of Equation 38 is easily tested, for example, by timing 10 rotations of the ball and dividing the total time by 10. Note that if the angle θ is kept small, then the height h of the ball is essentially equal to the length ℓ of the rope, and we get the formula

$$\text{period} \approx 2\pi\sqrt{\frac{\ell}{g}} \qquad (39)$$

Equation 39 is the famous formula for the period of what is called the simple pendulum, where the ball swings back and forth rather than in a circle. Equation 39 applies to a *simple pendulum* only if the angle θ is kept small. For large angles, Equation 38 is exact for a conical pendulum, but Equation 39 has to be replaced by a much more complicated formula for the simple pendulum. (We will discuss the analysis of the simple pendulum in Chapter 11 on rotations and oscillations.)

Note that the formula for the period of a simple pendulum depends only on the strength g of gravity and the length ℓ of the pendulum, and not on the mass m or the amplitude of the swing. As a result you can construct a clock using the pendulum as a timing device, where the period depends only on how long you make the pendulum.

Exercise 5 Conical Pendulum

Construct a pendulum by dangling a shoe or a boot from a shoelace.

(a) Verify that for small angles θ, you get the same period if you swing the shoe in a circle to form a conical pendulum, or back and forth to form a simple pendulum.

(b) Time 10 swings of your shoe pendulum and verify Equation 38 or 39. (You can get more accurate results using a smaller, more concentrated mass, so that you can determine the distance ℓ more accurately.) Try several values of the shoe string length ℓ to check that the period is actually proportional to $\sqrt{\ell}$.

Exercise 6

This is what we like to call a clean desk problem. Clear off your desk, leaving only a pencil and a piece of paper. Then starting from Newton's second law, derive the formula for the period of a conical pendulum.

What usually happens when you do such a clean desk problem is that since you just read the material, you think you can easily do the analysis without looking at the text. But if you are human, something will go wrong, you get stuck somewhere, and may become discouraged. If you get stuck, peek at the solution and finish the problem. Then a day or so later clean off your desk again and try to work the problem. Eventually you should be able to work the problem without peeking at the solution, and at that point you know the problem well and remember it for a long time.

When you are learning a new subject like Newton's second law, it is helpful to be fully familiar with at least one worked out example for each main topic. In that way when you encounter that topic again in your work, in a lecture, or on an exam, you can draw on that example to remember what the law is and how it is applied.

At various points in this course, we will encounter problems that serve as excellent examples of a topic in the course. The conical pendulum is a good example because it combines Newton's second law with the formula for the acceleration of a particle moving in a circle; the prediction can easily be tested by experiment, and the result is the famous law for the period of a pendulum. When we encounter similarly useful examples during the course, they will also be presented as clean desk problems.

APPENDIX
THE BALL SPRING PROGRAM

```
! --------- Plotting window
!         (x axis = 1.5 times y axis)
  SET WINDOW -40,140,-10,110

! --------- Draw & label axes
  BOX LINES 0,100,0,100
  PLOT TEXT, AT -3,0 : "0"
  PLOT TEXT, AT -13,96: "y=100"
  PLOT TEXT, AT 101,0 : "x=100"

! ---------- Experimental constants
  LET m = 245
  LET g = 980
  LET K = 5130
  LET So = 35.9
  LET Zx = 50
  LET Zy = 130

! --------- Initial conditions
  LET Rx = 88.2
  LET Ry = 42.8
  LET Vx = (80.2 - 91.1)/(2*.1)
  LET Vy = (24.4 - 63.1)/(2*.1)
  LET T  = 0
  CALL CROSS

! --------- Computer Time Step
  LET dt = .001
  LET i = 0

! --------- Calculational loop
  DO
    LET Rx = Rx + Vx*dt
    LET Ry = Ry + Vy*dt

    LET Sx = Rx - Zx
    LET Sy = Ry - Zy
    LET S  = Sqr(Sx*Sx + Sy*Sy)

    Let Fs = K*(S - So)
    LET Fx = -Fs*Sx/S
    LET Fy = -Fs*Sy/S - m*g

    LET Ax = Fx/m
    LET Ay = Fy/m

    LET Vx = Vx + Ax*dt
    LET Vy = Vy + Ay*dt

    LET T = T + dt
    LET i = i+1

    IF MOD(i,100) = 0 THEN CALL CROSS
    PLOT Rx,Ry

  LOOP UNTIL T > 1.6

! --------- Plot data
  DO
    READ Rx,Ry
    CALL BOX
  LOOP UNTIL END DATA

  DATA 88.2, 42.8
  DATA 80.2, 24.4
  DATA 68.0, 12.0
  DATA 52.9,  8.6
  DATA 37.4, 14.7
  DATA 24.0, 28.8
  DATA 14.2, 47.5
  DATA  9.0, 67.0
  DATA  8.2, 83.9
  DATA 11.1, 95.0
  DATA 16.7, 98.8
  DATA 23.9, 94.1
  DATA 32.2, 81.5
  DATA 41.9, 62.1
  DATA 52.1, 39.9
  DATA 62.2, 19.4

! --------- Subroutine "CROSS" draws
      ! a cross at Rx,Ry.
  SUB CROSS
    PLOT LINES: Rx-2,Ry;  Rx+2,Ry
    PLOT LINES: Rx,Ry-2;  Rx,Ry+2
  END SUB

! --------- Subroutine "BOX" draws
      ! a cross at Rx,Ry.
  SUB BOX
    PLOT LINES: Rx-1,Ry+1;  Rx+1,Ry+1
    PLOT LINES: Rx-1,Ry-1;  Rx+1,Ry-1
    PLOT LINES: Rx-1,Ry+1;  Rx-1,Ry-1
    PLOT LINES: Rx+1,Ry+1;  Rx+1,Ry-1
  END SUB

  END
```

The new feature is the READ statement at the top of this column. Each READ statement reads in the next values of Rx and Ry from the DATA lines below. We then call BOX which plots a box centered at Rx,Ry. The LOOP statement has this plotting continue until we run out of data. (In Figure 11, we filled in the boxes with a paint program to make them stand out.)

Chapter 10
Energy

In principle, Newton's laws relating force and acceleration can be used to solve any problem in mechanics involving particles whose size ranges from that of specks of dust to that of planets. In practice, many mechanics problems are too difficult to solve if we try to follow all the details and analyze all the forces involved. For instance $\vec{f} = m\vec{a}$ presumably applies to the motion of the objects involved in the collision of two automobiles, but it would be an enormous task to study the details of the collision by analyzing all the forces involved.

In a complicated problem, we cannot follow the motion of all the individual particles; instead we look for general principles that follow from Newton's laws and apply these principles to the system of particles as a whole. We have already discussed two such general principles: the laws of conservation of linear and angular momentum. We have found that if two cars traveling on frictionless ice collide and stick together, we can use the law of conservation of linear momentum to calculate their resulting motion. We do not have to know how they hit or any other details of the collision.

In our discussion of satellite motion, we saw that there was another quantity, which we called energy, that was conserved. Our formula for the total energy of the satellite was $E_{total} = 1/2\, mv^2 - Gmm_e/r$ where $1/2\, mv^2$ was called the **kinetic energy** and $-Gmm_e/r$ the **gravitational potential energy** of the satellite. We saw that E_{total} did not change its value as the satellite went around its orbit.

It turns out that energy is a much more complex subject than we might suspect from the discussion of satellite motion. There are many forms of energy, such as electrical energy, heat energy, light energy, nuclear energy and various forms of potential energy. Sometimes there is a simple formula for a particular form of energy, but sometime it may be hard even to figure out where the energy has gone. Despite the complexity, one simple fact remains, if we look hard enough we find that energy is conserved.

If, in fact, it were not for the conservation of energy, we would not have invented the concept in the first place. Energy is a useful concept only because it is conserved.

What we are going to do in this chapter is first take a more general look at the idea of a conservation law, and then see how we can use energy conservation to develop formulas for the various forms of energy we encounter. We will see, for example, where the formula $1/2\, mv^2$ for kinetic energy comes from, and we will show how the formula $-Gmm_e/r$ for gravitational potential energy reduces to a much simpler formula when applied to objects falling near the surface of the earth.

CONSERVATION OF ENERGY

Because energy comes in different forms, it is more difficult to state how to compute energy than how to compute linear momentum. But, as we shall see, it is not necessary to state all the formulas for all the different forms of energy. If we know the formula for some forms of energy, we can use the law of conservation of energy to deduce the other formulas as we need them. How a conservation law can be used in this way is illustrated in the following story, told by Richard Feynman in *The Feynman Lectures on Physics* (Vol. I, Addison-Wesley, Reading, Mass., 1963).

"Imagine a child, perhaps 'Dennis the Menace,' who has blocks that are absolutely indestructible, and cannot be divided into pieces. Each is the same as the other. Let us suppose that he has 28 blocks. His mother puts him with his 28 blocks into a room at the beginning of the day. At the end of the day, being curious, she counts the blocks very carefully, and discovers a phenomenal law—no matter what he does with the blocks, there are always 28 remaining! This continues for a number of days, until one day there are only 27 blocks, but a little investigating shows there is one under the rug—she must look everywhere to be sure that the number of blocks has not changed.

One day, however, the number appears to change— there are only 26 blocks. Careful investigation indicates that the window was open, and upon looking outside, the other two blocks are found. Another day careful count indicates that there are 30 blocks! This causes considerable consternation, until it is realized that Bruce came to visit, bringing his blocks with him, and he left a few at Dennis' house. After she had disposed of the extra blocks, she closes the window, does not let Bruce in, and then everything is going along all right, until one time she counts and finds only 25 blocks. However, there is a box in the room, a toy box, and the mother goes to open the toy box, but the boy says, 'No, do not open my toy box,' and screams. Mother is not allowed to open the toy box. Being extremely curious, and somewhat ingenious, she invents a scheme! She knows that a block weighs 3 ounces, so she weighs the box at a time when she sees 28 blocks, and it weighs 16 ounces. The next time she wishes to check, she weighs the box again, subtracts 16 ounces and divides by 3. She discovers the following:

$$\binom{number\ of}{blocks\ seen} + \frac{weight\ of\ box - 16\ oz}{3\ oz} = constant$$

There then appear to be some gradual deviations, but careful study indicates that the dirty water in the bathtub is changing its level. The child is throwing blocks into the water, and she cannot see them because it is so dirty, but she can find out how many blocks are in the water by adding another term to her formula. Since the original height of the water was 6 inches and each block raises the water a quarter of an inch, this new formula would be

$$\binom{number\ of}{blocks\ seen} + \frac{weight\ of\ box - 16\ oz}{3\ oz}$$
$$+ \frac{height\ of\ water - 6\ inches}{1/4\ inch} \quad (1)$$
$$= constant$$

In the gradual increase in the complexity of her world, she finds a whole series of terms representing ways of calculating how many blocks are in places where she is not allowed to look. As a result of this, she finds a complex formula, a quantity which **has to be computed**, *which always stays the same in her situation."*

Similarly, we will find a series of terms representing ways of calculating various forms of energy. Unlike the story, where some blocks are actually seen, we cannot see energy; all of the terms in our equation for energy must be computed. But if we have included enough terms and have not neglected any forms of energy, the numerical value of all the terms taken together will not change; that is, we will find that energy is conserved.

It is not necessary, however, to start with the complete energy equation. We will begin with one term. Then, as the complexity of our world increases, we will add more terms to the equation so that energy remains conserved.

MASS ENERGY

On earth, the greatest supply of useful energy ultimately comes from the sun, mainly as sunlight, which is a form of radiant energy. The energy we obtain from fossil fuel, such>é" coal and wood, and the energy we get from hydroelectric dams came originally from the sun. On a clear day, the sun delivers as much energy to half a square mile of tropical land as was released by the first atomic bomb. In about 1 millionth of a second, the sun radiates out into space an amount of energy equal to that used by all of mankind during an entire year.

The sun emits radiant energy at such an enormous rate that if it burned like a huge lump of coal, it would last about 5000 years before burning out. Yet the sun has been burning at nearly its present rate for over 5 billion years and should continue burning for another 5 billion years. How the sun could emit all of this energy was explained in 1905 when Einstein discovered that mass and energy are related through the well-known equation

$$\boxed{E = mc^2} \qquad (2)$$

where **E** is energy, **m** mass, and **c** the speed of light.

The sun's source of energy is the tiny fraction of its mass that is being converted continually to radiant energy through nuclear reactions. Similar processes occur when the hydrogen bomb is exploded. To indicate the amount of energy that is in principle available as mass energy, imagine that the mass of a 5–cent piece (5 gm) could be converted entirely into electrical energy. This electrical energy would be worth several million dollars. The problem is that we do not have the means available to convert mass completely into a useful form of energy. Even in the nuclear reactions in the sun or in the atomic or hydrogen bombs, only a few tenths of 1% of the mass is converted to energy.

Since most of the energy in the universe is in the form of mass energy, we shall begin to develop our equation for energy with Einstein's formula $E = mc^2$. As we mentioned, we will add terms to this equation as we discover formulas for other forms of energy.

Ergs and Joules

Our first step will be to use the Einstein energy formula to obtain the dimensions of energy. In the CGS system of units we have

$$E = m\,gm \times c^2 \frac{cm^2}{sec^2} = mc^2 \frac{gm\,cm^2}{sec^2}$$

The set of dimensions $gm\,cm^2/sec^2$ is called an ***erg***.

$$1 \frac{gm\,cm^2}{sec^2} = 1\,erg \quad (CGS\ units)$$

In the MKS system of units, we have

$$E = m\,kg \times c^2 \frac{m^2}{sec^2} = mc^2\,kg\,\frac{m^2}{sec^2}$$

where the set of dimensions of $kg\,m^2/sec^2$ is called a ***joule***.

$$1 \frac{kg\,m^2}{sec^2} = 1\,joule \quad (MKS\ units)$$

It turns out that for many applications the MKS joule is a far more convenient unit of energy than the CGS erg. A 100-watt light bulb uses 100 joules of energy per second, or 1 billion ergs of energy per second. The erg is too small a unit of energy for many applications, and it is primarily for this reason that the MKS system of units is more often used than the CGS system. This is particularly true when dealing with electrical phenomena.

Exercise 1

(a) Use dimensions to determine how many ergs there are in a joule. (Check your answer against the statement that a 100-watt bulb uses 100 joules or 10^9 ergs of energy per second.)

(b) As you may have guessed, a 1 watt light bulb uses 1 joule of energy per second. How many joules of energy does a 1000 watt bulb or heater use in one hour. (This amount of energy is called a **kilowatt hour** (abbreviated **kwh**) and costs a home owner about 10 cents when supplied by the local power company.)

(c) If a 5-cent piece (which has a mass of 5 grams) could be converted entirely to energy, how many kilowatt hours of energy would it produce? What would be the value of this energy at a rate of 10¢ per kilowatt hour?

KINETIC ENERGY

From the recoil definition of mass (Chapter 6), we saw that the mass of an object increases with speed, becoming very large when the speed of the object approaches the speed of light. The formula for the increase in mass with speed was simply

$$m = \frac{m_0}{\sqrt{1 - v^2/c^2}} \quad (6\text{-}14)$$

where m_0 is the mass of the particle at rest (the *rest mass*). When we combine this formula with Einstein's equation $E = mc^2$, we get as the equation for the energy of a moving particle

$$E = mc^2 = \frac{m_0 c^2}{\sqrt{1 - v^2/c^2}} \quad (3)$$

According to Equation (3), when a particle is at rest ($v = 0$), its energy is given by

$$E_0 = m_0 c^2 \quad \textit{rest energy} \quad (4)$$

This energy $m_0 c^2$ is called the *rest energy* of the particle. As a particle begins to move, its mass, and therefore its energy, increases. The *extra* energy that a particle acquires as a result of its motion is called *kinetic energy*. If mc^2 is the total energy, then the formula for the particle's kinetic energy is

$$\begin{matrix}\text{kinetic} \\ \text{energy}\end{matrix} = \begin{matrix}\text{total} \\ \text{energy}\end{matrix} - \begin{matrix}\text{rest} \\ \text{energy}\end{matrix}$$

$$KE = mc^2 - m_0 c^2 \quad (5)$$

Example 1

The muons in the motion picture *Time Dilation of the μ–Meson (Muon) Lifetime* moved at a speed of .995c. By what factor did their mass increase and what is their kinetic energy?

Solution: The first step is to calculate $\sqrt{1 - v^2/c^2}$ for the muons. An easy way to do this is as follows:

$$v = .995\,c$$

$$\frac{v}{c} = .995 = 1 - .005$$

$$\frac{v^2}{c^2} = (1 - .005)^2$$

$$= 1 - 2(.005) + (.005)^2$$

$$= 1 - .01 + \cancel{.000025}$$

We have neglected .000025 compared to .01 because it is so much smaller. We now have

$$1 - \frac{v^2}{c^2} \approx 1 - (1 - .01) = .01$$

$$\sqrt{1 - \frac{v^2}{c^2}} \approx \sqrt{.01} = .1 = \frac{1}{10}$$

(This procedure is discussed in more detail in the section on approximation formulas in Chapter 1.)

Now that we have $\sqrt{1 - v^2/c^2} = 1/10$ for these muons, we can calculate their relativistic mass

$$m = \frac{m_0}{\sqrt{1 - v^2/c^2}} = \frac{m_0}{1/10} = 10 m_0$$

Thus the mass of the muons has increased by a factor of 10. The total energy of the muons is

$$E = mc^2 = (10 m_0)c^2 = 10(m_0 c^2)$$

Hence, their total energy is also 10 times their rest energy. Their increase in energy, or their kinetic energy, is

$$KE = mc^2 - m_0 c^2 = 10 m_0 c^2 - m_0 c^2$$

$$KE = 9\, m_0 c^2$$

This kinetic energy $9 m_0 c^2$ is the amount of *additional energy* that is required to get muons moving at a speed $v = 0.995c$.

Exercise 2

Assume that an electron is traveling at a speed $v = .99995c$.

(a) What is $\sqrt{1-v^2/c^2}$ for this electron?

(b) By what factor has its mass increased over its rest mass?

(c) By what factor has its total energy increased over its rest energy?

(d) The rest mass of an electron is $m_0 = 0.911 \times 10^{-27}$ gm. What is its rest energy (in ergs)?

(e) What is the total energy (in ergs) of this electron?

(f) What is the kinetic energy of this electron in ergs?

Slowly Moving Particles

In Example 1, where the particle (muon) was moving at nearly the speed of light, we determined its increase in mass and its kinetic energy by calculating $\sqrt{1-v^2/c^2}$. However, when a particle is moving much slower than the speed of light, for instance, 1000 mi/sec or less, there is an easier way to calculate the energy of the object than by evaluating $\sqrt{1-v^2/c^2}$ directly.

In the section on approximation formulas in Chapter 1, it was shown that when v/c is much less than 1, then we can use the approximate formula

$$\frac{1}{\sqrt{1-\alpha}} \approx 1 + \frac{\alpha}{2} \qquad (1\text{-}25)$$

to get

$$\frac{1}{\sqrt{1-v^2/c^2}} \approx 1 + \frac{v^2}{2c^2} \qquad (6)$$

The approximate formula $(1 + v^2/2c^2)$ is much easier to use than $1/\sqrt{1-v^2/c^2}$. Moreover, if v/c is a small number, then the formula is quite accurate, as illustrated in Table 1. It should be noted however that when v becomes larger than about .1c, the approximation becomes less accurate. When we reach v = c, the exact formula is $1/\sqrt{1-v^2/c^2} = \infty$ but the approximate formula gives $1 + v^2/2c^2 = 1.5$. At this point the approximate formula is no good at all!

Table 1   Numerical check of the Approximation Formula $\frac{1}{\sqrt{1-v^2/c^2}} \approx 1 + \frac{v^2}{2c^2}$

v	value of exact formula $\frac{1}{\sqrt{1-v^2/c^2}}$	value of approximate formula $1 + \frac{v^2}{2c^2}$
.01c	1.000050003	1.000050000
.1c	1.005037	1.005000
.2c	1.0206	1.0200
.3c	1.048	1.045
.5c	1.148	1.125
.7c	1.41	1.25
.9c	2.30	1.40
.99c	7.1	1.49
c	∞	1.5

If we use Equation (6), the total energy of a particle becomes

$$E = mc^2$$

$$= m_0c^2 \left(\frac{1}{\sqrt{1 - v^2/c^2}}\right) \quad \text{exact formula}$$

$$\approx m_0c^2 \left(1 + \frac{v^2}{2c^2}\right) \quad \text{approximate formula}$$

$$\approx m_0c^2 + m_0c^2\frac{v^2}{2c^2}$$

The factor c^2 cancels in the second term, and we are left with the approximate formula

$$E \approx m_0c^2 + \frac{1}{2}m_0v^2 \quad \begin{array}{l}\textit{approximate formula}\\ \textit{for particles moving}\\ \textit{at speeds less than}\\ \textit{about .1c}\end{array} \quad (7)$$

Since Equation (7) contains the approximation made in Equation (6), it is not valid for particles traveling faster than about one tenth of the speed of light. For particles traveling at nearly the speed of light, we must use $E = m_0c^2/\sqrt{1 - v^2/c^2}$. But for particles traveling as slowly as a few thousand miles an hour or less, Equation (6) is so accurate that any error would be difficult to detect.

For all but the last section of this chapter, we will confine our discussion to the energy of objects traveling at slow speeds, where Equation (7) is not only accurate, but is the simplest equation to use. When we look at this equation, we can see that the mass energy $E = mc^2$ is now written in two distinct parts m_0c^2, which is the rest mass energy, and $1/2 m_0v^2$, which is the energy of motion or kinetic energy

$$\boxed{E = m_0c^2 \left(\begin{array}{c}\textit{rest}\\ \textit{energy}\end{array}\right) + \frac{1}{2}m_0v^2 \left(\begin{array}{c}\textit{kinetic}\\ \textit{energy}\end{array}\right)} \quad (7a)$$

Written in this way, our equation for total energy is beginning to resemble Equation (1), which was used to determine the number of blocks in Dennis' room. We now have two terms representing two different kinds of energy.

It is worth noting that, at one time, only the kinetic energy term $1/2 m_0v^2$ in Equation 7 was recognized as a form of energy. Before 1905, it was not known that m_0c^2 should be included in the equation for conservation of energy, because no one had ever observed the rest mass of an object to change. The first evidence that the rest energy had to be included came from the study of nuclear reactions. In these reactions enormous amounts of energy were released, producing a detectable change in the nuclear rest masses.

So long as an object is moving at a speed of .1c or less, the kinetic energy of that object will be far less than its rest mass energy. For example, let us compare the kinetic energy to the rest mass energy of a 10–gm pistol bullet that travels with a speed of about 300 m/sec. Using MKS units, we find that the bullet's kinetic energy (KE) is

$$KE = \frac{1}{2}m_0v^2$$

$$= \frac{1}{2} \times .01 \text{ kg} \times (300 \text{ m/sec})^2$$

$$= 450 \text{ joules}$$

This is enough to allow a bullet to penetrate a plank. The rest mass energy E_0 of the bullet is

$$E_0 = m_0c^2$$

$$= .01 \text{ kg} \left(3 \times 10^8 \text{ m/sec}\right)^2$$

$$= 9 \times 10^{14} \text{ joules}$$

This is the amount of energy released in a moderate-sized atomic bomb.

Exercise 3

For the preceding example of a 10 gram bullet:

a) at 10 cents per kilowatt hour, what is the value of the bullet's kinetic energy?

b) what is the value of its rest energy?

c) how fast would the bullet be traveling if it had twice as much kinetic energy?

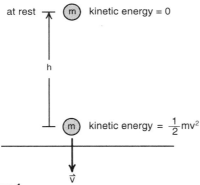

Figure 1
Falling Weight. When a weight is dropped it gains kinetic energy. This kinetic energy comes from the energy we stored in the object when we lifted it up to a height h.

GRAVITATIONAL POTENTIAL ENERGY

Let us continue our search for terms to add to our equation for energy. Suppose that a ball of mass m is dropped from a height h above the floor, as shown in Figure (1). Immediately before the ball hits the floor, it has a rest energy m_0c^2, and a kinetic energy $1/2\,m_0v^2$. Immediately before the ball was dropped, however, it had the same rest energy m_0c^2 but no kinetic energy. Where did the kinetic energy that it possessed just before it hit the floor come from?

If we were observant, we might have noted that some effort was needed to lift the ball from the floor to a height h. As the ball was lifted a new kind of energy was being stored. This new form of energy, which was released when the ball was dropped, is called *gravitational potential energy*. When it is included, our equation for energy becomes

$$E_{total} = m_0c^2 + \tfrac{1}{2}m_0v^2 + \text{gravitational potential energy} \quad (8)$$

To find the formula for the gravitational potential energy, we will assume that energy is conserved and that the total energy of the ball, immediately before it is released, is equal to the total energy of the ball immediately before it hits the ground.

When a ball is dropped from a height h, it accelerates downward with a constant acceleration g until it hits the floor. Thus we can use the constant acceleration formulas (see Appendix 1 in Chapter 4.)

$$s = v_i t + \frac{1}{2}at^2$$

$$v_f = v_i + at$$

with $v_i = 0$, $a = g$, and $s = h$ we get

$$h = \tfrac{1}{2}gt^2 \quad (12)$$

$$v_f = gt \quad (13)$$

Substituting $t = v_f/g$ from Equation 13 into Equation (12) gives

$$h = \frac{1}{2}g\left(\frac{v_f^2}{g^2}\right) = \frac{v_f^2}{2g}$$

$$\frac{1}{2}v_f^2 = gh \qquad (14)$$

Multiplying Equation 14 through by m_0 gives

$$\frac{1}{2}m_0v_f^2 = m_0gh \qquad (15)$$

Suppose that we use m_0gh as the formula for gravitational potential energy. (The greater h, the higher we have lifted the ball, the more potential energy we have stored in it.)

$$m_0gh = \left\{\begin{array}{l}\text{formula for} \\ \text{gravitational} \\ \text{potential energy}\end{array}\right. \begin{array}{l}\textit{near the} \\ \textit{surface of} \\ \textit{the earth}\end{array} \qquad (16)$$

Before the ball is released, its total energy is in the form of rest energy and gravitational potential energy

$$E_{total}\left(\begin{array}{l}\textit{before} \\ \textit{release}\end{array}\right) = m_0c^2 + m_0gh \qquad (17)$$

Just before the ball hits the floor, where it has kinetic energy but no potential energy (since h = 0), the total energy is

$$E_{total}\left(\begin{array}{l}\textit{just before} \\ \textit{hitting floor}\end{array}\right) = m_0c^2 + \frac{1}{2}m_0v_f^2 \qquad (18)$$

At first, Equations 17 and 18 for total energy look different; but since $1/2 m_0v_f^2 = m_0gh$ (Equation 15), they give the same numerical value for the ball's total energy. Thus, we conclude that we have chosen the correct formula for calculating gravitational potential energy.

Exercise 4

Call v_2 the speed of the ball when it has fallen halfway to the floor.

(a) Explain why the ball's total energy, when it has fallen halfway to the floor, is

$$E_{total}\left(\begin{array}{l}\textit{halfway} \\ \textit{down}\end{array}\right) = m_0c^2 + \frac{1}{2}m_0v_2^2 + m_0g\left(\frac{h}{2}\right)$$

(b) Calculate v_2 (just as we calculated v_f) and show that the total energy of the ball when halfway down is the same as when it was released, or just before it hit the floor.

Exercise 5

Show that the formula for gravitational potential energy has the dimensions of joules (in the MKS system) and ergs (in the CGS system).

Exercise 6

What is the gravitational potential energy (in joules and ergs) of a 100–gm ball at a height of 2 meters above the floor? (Measure h starting from the floor.)

What happens to the energy after the ball has hit the floor and is lying at rest? At this point, it no longer has kinetic energy or gravitational potential energy. Now what should we add to our equation to maintain conservation of energy? In this case, we have to look "under the rug," in the "dirty water," and "out the window" all at once. When the ball hit the floor, we heard a thump; thus, some of the ball's energy has been dissipated as sound energy. We find that there is a dent in the floor; hence we know that some of the energy has gone into rearranging the molecuies in that part of the floor. Also, because the bottom of the ball and the floor underneath became slightly warmer after the ball hit the floor, we conclude that some of the energy was converted into heat energy. (In some collisions, such as when a mining pick strikes a stone, we see what looks like a spark, which shows us that some of the kinetic energy has been changed into radiant energy, or light.)

After the ball hits the floor, the formula for total energy becomes as complicated as

$$E_{total} = m_0c^2 + \frac{1}{2}m_0v^2 + m_0gh$$

$$+ \text{ sound energy}$$
$$+ \text{ energy to cause a dent} \qquad (19)$$
$$+ \text{ heat energy } + \text{ light energy}$$

Because energy can appear in so many forms that are often difficult to detect, it was not until many years after Newton that conservation of energy was established as a general law. The law of conservation of energy is used to solve only those problems where very little energy "escapes" in a form that is difficult to detect. In a complicated collision problem we can calculate only how much energy is "lost," that is, changed to other forms of energy.

On an atomic scale, however, we do not have to think of energy as being "lost" because the various forms of energy are more easily detected. For example, we will see in Chapter 17 that the heat energy and sound energy are primarily the kinetic energy of the atoms and molecules; thus, these do not appear as separate forms of energy. It is on this small scale that the law of conservation of energy may be most accurately verified.

On the other hand, if we can neglect the effects of friction and air resistance, the law of conservation of energy can be used to solve mechanics problems that would otherwise be difficult to solve. We will illustrate this with two examples in which gravitational potential energy m_0gh is converted into kinetic energy $1/2m_0v^2$ and vice versa.

Notation

Since our discussion for the remainder of this chapter will deal with objects moving at speeds much less than the speed of light, objects whose mass m is very nearly equal to the rest mass m_0, we will stop writing the subscript 0 for the rest mass. With this notation, our formulas for kinetic energy and gravitational potential energy are simply $1/2mv^2$ and mgh. Only when we discuss objects like atomic particles whose speeds can become relativistic, will we be careful to distinguish the rest mass m_0 from the total mass m.

Example 2

Consider a simple pendulum consisting of a ball swinging on the end of a string, as shown in Figure (2). When the ball is released from a height h it has a potential energy m_0gh. As the ball swings down toward the bottom, h decreases and the ball loses potential energy but gains kinetic energy. At the bottom the original potential energy mgh has been entirely converted into kinetic energy $1/2mv^2$. Then the ball climbs again, gaining potential energy but losing kinetic energy.

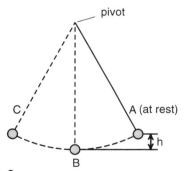

Figure 2
Application of conservation of energy to pendulum motion. The speed at B can be found by equating the kinetic energy at B $(1/2mv^2)$ to the potential energy lost in going from A to B (mgh).

Finally, at position C, the ball has swung back up to a height h and all the kinetic energy has been changed to potential energy. The ball stops momentarily at position C, and the swing is reversed. Eventually, however, the pivot becomes warm and air currents are set up by the swinging pendulum; thus, the pendulum itself gradually loses energy and finally comes to rest.

As long as we can neglect air resistance and friction in the pivot we can use the conservation of energy equation to calculate the speed of the ball at position B. Before the ball is released

$$E_{total\,A} = m_0 c^2 + mgh$$

At position B, where h = 0

$$E_{total\,B} = m_0 c^2 + \tfrac{1}{2} m v_B^2$$

If energy is conserved

$$E_{total\,A} = E_{total\,B}$$

$$\cancel{m_0 c^2} + mgh = \cancel{m_0 c^2} + \tfrac{1}{2} m v^2$$

Note that since $m_0 c^2$ did not change, it does not enter into this calculation. Here we could apply the conservation of energy equation without considering the rest energy. We now have

$$mgh = \tfrac{1}{2} m v_B^2$$

$$v_B^2 = 2gh$$

$$v_B = \sqrt{2gh}$$

Example 3

It should be noted that we are able to calculate the speed of the ball in the preceding example without an analysis of the forces involved. An even more striking example of conservation of energy that would be nearly impossible to analyze in terms of forces is that of a skier traveling down a very icy hill. If he is not an experienced skier, he may not know how to dissipate some of his kinetic energy as heat and sound by scraping the edges of his skis against the ice. If he is not able to dissipate energy, then no matter how he turns, no matter how twisted a path he takes, when he reaches bottom, all his potential energy $m_0 gh$ will have been converted to kinetic energy $1/2 m_0 v^2$, in which case his speed at the bottom of the hill will be $\sqrt{2gh}$. To see why an inexperienced skier should not try icy hills, consider that if the hill has a 500–ft rise, his speed at the bottom will be 179 ft/sec or 122 mi/hr. This result is computed not from the details of the skier's path, but from the knowledge that he was not able to dissipate energy. As we mentioned at the beginning of the chapter, the conservation of energy is one of the general principles of mechanics that can be applied successfully without knowing all the details involved in the physical situation.

Exercise 7

A car coasts along a road that leads from the top of a 300–ft-high hill, down through a valley, and up over a 200 ft high hill. Assume that the car does not dissipate energy through friction and air resistance.

(a) If the car starts at rest from atop the higher hill, how fast will it be traveling when it reaches the top of the lower hill (g = 32 ft/sec^2) ?

(b) If the car is initially moving at 80 ft/sec (55 mi/hr) when it starts coasting at the top of the higher hill, how fast will the car be moving when it reaches the top of the lower hill?

WORK

Let us take another look at the example where we dropped a ball of mass m from a height h above the floor as shown in Figure (3). At the height h, the ball had a gravitational potential energy mgh. Just before hitting the floor, all this gravitational potential energy had been converted to kinetic energy 1/2 mv². We know that the ball speeded up, accelerated, because gravity was exerting a downward force mg on the ball as it fell.

There appears to be a coincidence in this example. Gravity pulls down on the ball with a force of magnitude mg, the ball falls a distance h, and the ball gains a kinetic energy equal to (mg)×h. In this example the energy that gravity supplies to the ball by pulling down on it is equal to the gravitational force (mg) times the distance h over which the force acted. Is this a coincidence, or does this example provide a clue as to the way in which forces supply energy?

In this case, where we have a constant force mg, and the ball moves in the direction of the force for a distance h, the increase in energy is the force times the distance.

In more general examples, however, the situation can be more complex. If the object is not moving in the direction of the force, then only the component of the force in the direction of motion adds energy to the object. And if the force is not constant, we have to break the problem into many small steps, and calculate the energy gained in each step. We shall see that calculus provides powerful techniques to handle these situations.

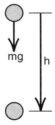

Figure 3
A ball, subject to a gravitational force mg, falling a distance h, gains a kinetic energy mgh.

We will begin the discussion with the introduction of a new term which we will call **work**. In some ways this is an unfortunate choice of a word, for everyone has their own idea of what "work" is, and it seldom coincides with the physicist's definition. In the physicist's definition, *a force does work on an object when it adds energy to the object*. More explicitly, the work a force does is equal to the energy that the force supplies. In the case of the falling ball the gravitational force supplied an amount of energy mgh, therefore that is the work that the gravitational force did as the ball fell.

$$\left.\begin{array}{l}\text{work done by the}\\\text{force of gravity}\\\text{as the ball fell}\end{array}\right\} = mgh \quad (20)$$

From Equation (20), we see that for the case where we have a constant force, and the object moves in the direction of the force, the work done is equal to the magnitude of the force times the distance moved.

$$\boxed{\text{Work} = \text{Force} \times \text{Distance}}$$

If the force is constant and the distance is in the direction of the force

(21)

Exercise 8

Show that force times distance has the same dimensions as energy. (Get the dimensions of energy from $E = mc^2$.)

As the first complication, or correction to our definition of work, suppose that the force is not in the same direction as the motion. Suppose, for example, a hockey puck slides for a distance S along frictionless ice as shown in Figure (4). During this motion a gravitational force mg is acting and the puck moves a distance S. But the puck coasts along at constant speed; it does not gain any energy at all. In this case the gravitational force does no work.

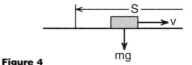

Figure 4
The force of gravity does no work on the sliding hockey puck.

The problem with the hockey puck example is that the gravitational force is down and the motion is sideways. In this case the $-y$ directed gravitational force has no component along the x directed motion of the puck. In order for the puck to gain energy, it must accelerate in the x direction, but there is no x component of force to produce that acceleration.

Now let us consider an example where the force is acting opposite to the direction of motion. If we throw a ball up in the air, the ball starts out with the kinetic energy $1/2\, mv_0^2$ that we gave it. As the ball rises, gravity acts against the motion of the ball and removes kinetic energy. When the ball has risen to a height h given by $mgh = 1/2\, mv_0^2$, all the kinetic energy is gone and the ball stops. The ball has reached the top of the trajectory. This example tells us that when the force is directed opposite to the direction of motion, the work is negative—the force removes rather than adds energy.

The Dot Product

This is where our discussion has lead so far. We have a quantity called "work" which is a form of energy. It is the energy supplied by a force acting on a moving object. Now energy, given by formulas like $E = mc^2$, is a scalar quantity; it is a number that does not point anywhere. But our formula for *work = force times distance* involves two vectors, the force $\vec{F}$ and the distance $\vec{S}$. What mathematical way can we combine the two vectors $\vec{F}$ and $\vec{S}$ to get a number for the work W? One possibility, that we discussed back in the chapter on vectors, is the scalar or dot product.

$$\boxed{W = \vec{F} \cdot \vec{S}} = |\vec{F}||\vec{S}| \cos\theta \quad (22)$$

Mathematically the dot product turns the vectors $\vec{F}$ and $\vec{S}$ into a scalar number W. Let us see if $W = \vec{F} \cdot \vec{S}$ is the correct formula for work. If $\vec{F}$ and $\vec{S}$ are in the same direction, $\theta = 0°$, $\cos\theta = 1$, and we get

$$W = \vec{F} \cdot \vec{S} = |\vec{F}||\vec{S}| \cos\theta = |\vec{F}||\vec{S}|$$

Applied to the case of a falling ball, $|\vec{F}| = mg$, $|\vec{S}| = h$ and we get $W = mgh$ which is correct.

When we throw the ball up, the angle between the downward force and upward motion is $\theta = 180°$, $\cos\theta = -1$, and we get

$$W = \vec{F} \cdot \vec{S} = |\vec{F}||\vec{S}| \cos\theta$$
$$= mgh(-1) = -mgh$$

We now predict that gravity is taking energy from the ball, which is also correct.

Finally, in the case of a hockey puck, the angle θ between the $-y$ directed force and the x directed motion is 90°. We have $\cos\theta = 0$, so that $\vec{F} \cdot \vec{S} = 0$ and the gravitational force does no work. Again the formula $W = \vec{F} \cdot \vec{S}$ works.

Exercise 9

A frictionless plane is inclined at an angle θ as shown in Figure (5). A hockey puck initially at a height h above the ground, slides down the plane. When the puck gets to the bottom, it has moved a distance $S = h/\cos\theta$ as shown. (This comes from $h = S \cos\theta$)

a) Verify the formula $S = h/\cos\theta$ for the two cases $\theta = 0$ and $\theta = 90°$. I.e., what are the values for $h/\cos\theta$ for these two cases, and are the answers correct?

b) Show that the work $W = \vec{F}_g \cdot \vec{S}$, done by the gravitational force as the puck slides down the plane, is mgh no matter what the angle θ is.

c) Explain the result of part (b) from the point of view of conservation of energy.

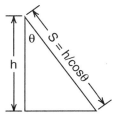

Figure 5
Diagram for Exercise 9.

Work and Potential Energy

In the discussion of energy, physicists tend to use a lot of words like *work, potential energy, kinetic energy*, etc. What we are doing is building a conceptual picture to help us organize a number of physical phenomena and related mathematical equations. You will find that when you see this picture, are familiar with the "jargon", these concepts become easy to use and powerful in their applications. Much of this chapter is to introduce the jargon and develop the picture.

The ideas of work and potential energy are closely related and play critical roles in the picture of energy. Let us discuss some examples simply from the point of view of getting used to the jargon.

Suppose I pick a ball of mass m off the floor and slowly lift it up to a height h. While lifting the ball, I have to just barely overcome the downward gravitational force mg. Therefore I exert an upward directed force of magnitude mg, and I do this for a distance h. Since my upward force and the upward displacement are in the same direction, the work I do, call it W_{me}, is my force mg times the distance h, or W_{me} = mgh. Using the ideas of potential energy discussed earlier, we can say that all the energy W_{me} = mgh that I supplied lifting the ball went into gravitational potential energy mgh.

While I was lifting the ball, gravity was pulling down. The downward gravitational force and the upward displacement were in opposite directions and therefore the work done by the gravitational force was negative. While we are storing gravitational potential energy, gravity does negative work. When we let go of the ball, gravity releases potential energy by doing positive work.

Let us consider another example where we store potential energy by doing work against a force. Suppose I tie one end of a spring to a post and pull on the other end as shown in Figure (6). As I stretch the spring, I am exerting a force $\vec{F}_{me}$ and moving the end of the spring in the same direction. Therefore I am doing positive work on the spring, and this energy is stored in what we can call the "elastic potential energy" of the stretched spring. (We know that a stretched spring has some form of potential energy, for a stretched spring can be used to launch a ball up into the air.)

Non-Constant Forces

Our example above, of storing energy in a spring by stretching it, introduces a new complication. We cannot calculate the work I do W_{me} in stretching the spring by writing $W_{me} = \vec{F}_{me} \cdot \vec{S}$. The problem is that, the farther I stretch the spring, the harder it pulls back (Hooke's law). If I slowly pull the spring out, I have to apply an increasingly stronger force. If we try to use the formula $W_{me} = \vec{F}_{me} \cdot \vec{S}$, the problem is what value of $\vec{F}_{me}$ to use. Do we use the weak $\vec{F}_{me}$ at the beginning of the pull, the strong one at the end, or some average value.

We could use an average value, but there is a more general way to calculate the work I do. Suppose I wish to pull the spring from an initial position x_i to a final position x_f. Imagine that I break this span from x_i to x_f into a bunch of small intervals of width Δx, ending at points labeled x_0, x_1, ... x_n as shown in Figure (7). During each small interval the spring force does not change by much, and I can stretch the spring through that interval by exerting a force equal to the strength of

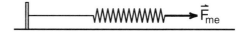

Figure 6
Doing work on a spring.

the spring force at the end of the interval. For example in stretching the spring from position x_0 to x_1, I apply a force of magnitude $F_s(x_1)$ for a distance Δx and therefore do an amount of work

$$(\Delta W_{me})_1 = F_s(x_1)\Delta x$$

To get out to position x_2, I increase my force to $F_s(x_2)$ and apply that force over another interval Δx to do an amount of work

$$(\Delta W_{me})_2 = F_s(x_2)\Delta x$$

If I keep repeating this process until I reach the final position x_f, the total amount of work I have done is

$$\begin{aligned}(W_{me})_{total} &= (\Delta W_{me})_1 + (\Delta W_{me})_2 + ... \\ &\quad + (\Delta W_{me})_n \\ &= F_s(x_1)\Delta x + F_s(x_2)\Delta x + ... \\ &\quad + F_s(x_n)\Delta x \\ &= \sum_{i=1}^{n} F_s(x_i)\Delta x \end{aligned} \quad (23)$$

In Equation 23, we still have an approximate calculation as long as the intervals Δx are of finite size. We get an exact calculation of the work I do if we take the limit as Δx goes to zero, and the number of intervals goes to infinity. In that limit, the right side of Equation 23 becomes the definite integral of $F_s(x)$ from the initial position x_i to the final position x_f:

$$(W_{me})_{total} = -\int_{x_i}^{x_f} F_s(x)\,dx \quad (24)$$

The statement of the work we did, Equation 24, can be written more formally by noting that the spring force $\vec{F}_s(x)$ is actually a vector which points opposite to the direction I pulled the spring. In addition, we should think of each Δx or dx as a small vector displacement $\vec{\Delta x}$ or $\vec{dx}$ in the direction I pulled. Since my force was directed opposite to $\vec{F}_s$, the work I did during each interval $\vec{dx}$ can be written as the dot product

$$dW_{me} = \vec{F}_{me}\cdot\vec{dx} = -\vec{F}_s\cdot\vec{dx}$$

and the formula for the total work I did becomes

$$(W_{me})_{total} = -\int_{x_i}^{x_f} \vec{F}_s\cdot\vec{dx} \quad (25)$$

Equation 25 is more general but a bit clumsier to use than 24. To use Equation 25, we would first note that I was pulling along the x axis, and thus $\vec{dx} = dx$. Then I would note that the spring force was opposite to the direction I was pulling, so that $-\vec{F}_s(x)\cdot\vec{dx} = +F_s(x)dx$

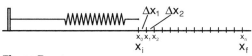

Figure 7
I can stretch the spring through a series of small intervals of length Δx. In each interval I apply a constant force that is just strong enough to get the spring to the end of the interval.

where $F_s(x)$ is the formula for the strength of the spring force. That gets me back to Equation (24) and the problem of evaluating the definite integral.

Potential Energy Stored in a Spring

Springs are useful in physics demonstrations and problems because of the simple force law (Hooke's law) which is quite accurately obeyed by real springs. In our study of the motion of a ball on the end of a spring in Chapter 9, we saw that the formula for the strength of the spring force was

$$F_s = K(S - S_0) \quad (9\text{--}6)$$

where S is the length of the spring and S_0 the unstretched length (the length at which F_s goes to zero in Figure 9–4).

We can simplify the spring force formula, get rid of the S_0, by considering a situation where an object is held in an equilibrium position by spring forces. Suppose for example we have a cart on an air track with springs

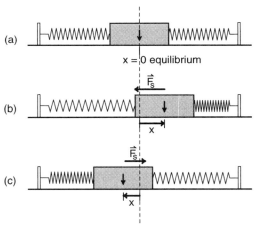

Figure 8
The spring force $\vec{F}_s$ is always opposite to the displacement x. If the spring is displaced right, $\vec{F}_s$ points left, and vice versa.

connecting the cart to each end of the track as shown in Figure (8). Mark the center of the cart with an arrow, and choose a coordinate system where x = 0 is at the equilibrium position as shown in Figure (8a).

With this setup, the spring force is always a restoring force that is pushing the cart back to the equilibrium position x = 0. If we give the cart a positive displacement as in Figure (8b), we get a left directed or negative spring force. A negative displacement shown in (8c) produces a right directed or positive spring force. And to a high degree of accuracy, the strength of the spring force is proportional to the magnitude of the displacement from equilibrium.

All of these results can be described by the formula

$$\boxed{F_s(x) = -Kx} \quad (26)$$

where the minus sign tells us that a positive displacement x produces a negative directed force and vice versa. There is no S_0 or x_0 in Equation 26 because we chose x = 0 to be the equilibrium position where $F_s = 0$. Equation 26 is what one usually finds as a statement of Hooke's law, and K is called the spring constant.

Equation 26 allows us to easily calculate the potential energy stored in the springs. If I start with the cart at rest at the equilibrium position as shown in Figure (8a), and pull the cart to the right a distance x_f, the work I do is

$$W_{me} = \int_{x=0}^{x=x_f} F_{me} \, dx = \int_{x=0}^{x=x_f} (-F_s) \, dx$$

$$= \int_{x=0}^{x=x_f} Kx \, dx \quad (27)$$

where I have to exert a force $F_{me} = -F_s$ to stretch the spring.

In Equation 27, the constant K can come outside the integral, we are left with the integral of xdx which is $x^2/2$, and we get

$$W_{me} = K\int_{x=0}^{x=x_f} x\, dx = K\frac{x^2}{2}\Big|_0^{x_f} = K\frac{x_f^2}{2}$$

Noting that all the work I do is stored as *"elastic potential energy of the spring"*, we get the formula

$$\boxed{\text{Spring potential energy} = K\frac{x^2}{2}} \qquad (28)$$

In Equation 28, we replaced x_f by x since the formula applies to any displacement x_f I choose.

Exercise 10

If you pull the cart of Figure (8) back a distance x_f from the equilibrium position and let go, all the potential energy you stored in the cart will be converted to kinetic energy when the cart crosses the equilibrium position x = 0. Use this example of conservation of energy to calculate the speed v of the cart when it crosses x = 0. (Assume that you release the cart from rest.)

Exercise 10, which you should have done by now, illustrates one of the main reasons for bothering to calculate potential energy. It is much easier to predict the speed of the ball using energy conservation than it is using Newton's second law. We can immediately find the speed of the ball by equating the kinetic energy at x = 0 to the potential energy at $x = x_f$ where we released the cart. To make the same prediction using Newton's second law, we would have to solve a differential equation and do a lot more calculation.

Exercise 11

With a little bit of cleverness, we can use energy conservation to predict the speed of the cart at any point along the air track. Suppose you release the cart from rest at a distance x_f, and want to know the cart's speed at, say, $x_f/2$. First calculate how much potential energy the cart loses in going from x_f to $x_f/2$, and then equate that to the kinetic energy $1/2\, mv^2$ that the cart has gained at $x_f/2$.

WORK ENERGY THEOREM

The reason that it is easier to apply energy conservation than Newton's second law is that when we have a formula for potential energy, we have already done much of the calculation. We can illustrate this by deriving what is called the "Work Energy Theorem" where we use Newton's second law to derive a relation between work and kinetic energy. We will first derive the theorem for one dimensional motion, and then see that it is easily extended to motion in three dimensions.

Suppose a particle is moving along the x axis as shown in Figure (9). Let a force $F_x(x)$ be acting on the particle. Then by Newton's second law

$$F_x(x) = ma_x(x) = m\frac{dv_x(x)}{dt} \qquad (29)$$

Multiplying by dx and integrating to calculate the work done by the force F_x, we get

$$\int_i^f F_x(x)dx = m\int_i^f \frac{dv_x(x)}{dt}dx \qquad (30)$$

In Equation (30), we are integrating from some initial position x_i where the object has a speed v_{xi}, to a position x_f where the speed is v_{xf}.

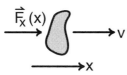

Figure 9
An x directed force acting on a particle moving in the x direction.

The next step is a standard calculus trick that you may or may not remember. We will first move things around a bit in the integral on the right side of Equation 30:

$$m\int_i^f \frac{dv_x}{dt}dx = m\int_i^f dv_x\left(\frac{dx}{dt}\right) \qquad (31)$$

Next note that $dx/dt = v_x$, the x component of the velocity of the particle. Thus the integral becomes

$$m\int_i^f dv_x\left(\frac{dx}{dt}\right) = m\int_{v_i}^{v_f} v_x dv_x \qquad (32)$$

After this transformation, we can do the integral because everything is now expressed in terms of the one variable v_x. Using the fact that the integral of $v_x dv_x$ is $v_x^2/2$, we get

$$m\int_{v_i}^{v_f} v_x dv_x = m\frac{v_x^2}{2}\bigg|_{v_i}^{v_f}$$

$$= \frac{1}{2}mv_{fx}^2 - \frac{1}{2}mv_{ix}^2 \qquad (33)$$

Using Equations (31) through (33) in Equation (30) gives

$$\boxed{\int_{x_i}^{x_f} F_x(x)dx = \frac{1}{2}mv_{fx}^2 - \frac{1}{2}mv_{ix}^2} \qquad (34)$$

The left side of Equation 34 is the work done by the force F_x as the particle moves from position x_i to position x_f. The right side is the change in the kinetic energy. Equation 34 tells us that the work done by the force F_x equals the change in the particle's kinetic energy. This is the basic idea of the work energy theorem.

To derive the three dimensional form of Equation 34, start with Newton's second law in vector form

$$\vec{F} = m\vec{a} \qquad (35)$$

Take the dot product of Equation 35 with $\vec{dx}$ and integrate from i to f to get

$$\int_i^f \vec{F}\cdot\vec{dx} = \int_i^f m\vec{a}\cdot\vec{dx} \qquad (36)$$

Writing

$$\vec{a}\cdot\vec{dx} = a_x dx + a_y dy + a_z dz \qquad (37)$$

we get

$$\int_i^f \vec{F}\cdot\vec{dx} = m\int_i^f \left(\frac{dv_x}{dt}dx + \frac{dv_y}{dt}dy + \frac{dv_z}{dt}dz\right) \qquad (38)$$

Following the same steps we used to get from Equation 31 to 33, we get

$$\int_i^f \vec{F}\cdot\vec{dx} = \frac{1}{2}mv_{fx}^2 - \frac{1}{2}mv_{ix}^2$$
$$+ \frac{1}{2}mv_{fy}^2 - \frac{1}{2}mv_{iy}^2$$
$$+ \frac{1}{2}mv_{fz}^2 - \frac{1}{2}mv_{iz}^2 \qquad (39)$$

Finally noting that by the Pythagorean theorem

$$v_i^2 = v_{ix}^2 + v_{iy}^2 + v_{iz}^2$$
$$v_f^2 = v_{fx}^2 + v_{fy}^2 + v_{fz}^2 \qquad (40)$$

we get, using (40) in (39)

$$\boxed{\int_i^f \vec{F}\cdot\vec{dx} = \frac{1}{2}mv_f^2 - \frac{1}{2}mv_i^2} \qquad (41)$$

which is the three dimensional form of the work energy theorem.

Several Forces

Suppose several forces $\vec{F}_1, \vec{F}_2, \ldots$ are acting on the particle as the particle moves from position i to position f. Then the vector $\vec{F}$ in Equations 35 through 41 is the total force $\vec{F}_{tot}$ which is the vector sum of the individual forces:

$$\vec{F} = \vec{F}_{tot} = \vec{F}_1 + \vec{F}_2 + \ldots \qquad (42)$$

Our formula for the work done by these forces becomes

$$\int_i^f \vec{F}\cdot\vec{dx} = \int_i^f (\vec{F}_1 + \vec{F}_2 + \ldots)\cdot\vec{dx}$$

$$= \int_i^f \vec{F}_1\cdot\vec{dx} + \int_i^f \vec{F}_2\cdot\vec{dx} + \ldots \qquad (43)$$

and we see that the work done by several forces is just the numerical sum of the work done by each force acting on the object. Equation 41 now has the interpretation that *the total work done by all the forces acting on a particle is equal to the change in the kinetic energy of the particle*.

Conservation of Energy

To see how the work energy theorem leads to the idea of conservation of energy, suppose we have a particle subject to one force, like the spring force $\vec{F}_s$ acting on an air cart as shown in Figure (8). If the cart moves from position i to position f, then the work energy theorem, Equation 41 gives

$$\int_i^f \vec{F} \cdot \vec{dx} = \frac{1}{2}mv_f^2 - \frac{1}{2}mv_i^2 \qquad (44)$$

In our analysis of the spring potential energy, we saw that if I slowly moved the cart from position i to position f, I had to exert a force $\vec{F}_{me}$ that just overcame the spring force $\vec{F}_s$, i.e., $\vec{F}_{me} = -\vec{F}_s$. When I moved the cart slowly, the work I did went into changing the potential energy of the cart. Thus the formula for the change in the cart's potential energy is

$$\left.\begin{array}{l}\text{change in the}\\ \text{potential energy of}\\ \text{the cart when the}\\ \text{cart moves from}\\ \text{position i to position f}\end{array}\right\} = \int_i^f \vec{F} \cdot \vec{dx}$$

$$= -\int_i^f \vec{F}_s \cdot \vec{dx} \qquad (45)$$

Equation 45 is essentially equivalent to Equation 25 which we derived in our discussion of spring forces.

Spring forces have the property that the energy stored in the spring depends only on the length of the spring, and not on how the spring was stretched. This means that the change in the spring's potential energy does not depend upon whether I moved the cart, or I let go and the spring moves the cart. We should remove $\vec{F}_{me}$ from Equation 45 and simply express the spring potential energy in terms of the spring force

$$\left.\begin{array}{l}\text{change in spring}\\ \text{potential energy}\end{array}\right\} = -\int_i^f \vec{F}_s \cdot \vec{dx} \qquad (46)$$

Equation 46 says that the change in potential energy is **minus** the work done by the force on the object as the object moves from i to f. There is a minus sign because, if the force does positive work, potential energy is released or decreases. We will see that Equation 46 is a fairly general relationship between a force and it's associated potential energy.

We are now ready to convert the work energy theorem into a statement of conservation of energy. Rewrite Equation 44 with the work term on the right hand side and we get

$$0 = \left\{-\int_i^f \vec{F}_s \cdot \vec{dx}\right\} + \left\{\frac{1}{2}mv_f^2 - \frac{1}{2}mv_i^2\right\} \qquad (47)$$

The term in the first curly brackets is the change in the particle's potential energy, the second term is the change in the particle's kinetic energy. Equation 47 says that the sum of these two changes is zero

$$0 = \left\{\begin{array}{l}\text{change in}\\ \text{potential enegy}\end{array}\right\} + \left\{\begin{array}{l}\text{change in}\\ \text{kinetic enegy}\end{array}\right\} \qquad (47a)$$

If we define the total energy of the particle as the sum of the particle's potential energy plus its kinetic energy, then the change in the particle's total energy in moving from position i to position f is the sum of the two changes on the right side of Equation 47a. Equation 47a says that this total change is zero, or that the total energy is conserved.

Conservative and Non-Conservative Forces

We mentioned that the potential energy stored in a spring depends only on the amount the spring is stretched, and not on how it was stretched. This means that the change in potential energy depends only on the initial and final lengths of the spring, and not on how we stretched it. This implies that the integral

$$-\int_i^f \vec{F}_s \cdot \vec{dx}$$

has a unique value that does not depend upon how the particle was moved from i to f.

Gravitational forces have a similar property. If I lift an object from the floor to a height h, the increase in gravitational potential energy is mgh. This is true whether I lift the object straight up, or run around the room five times while lifting it. The formula for the change in gravitational potential energy is

$$\left.\begin{array}{c}\text{change in}\\\text{gravitational}\\\text{potential energy}\end{array}\right\} = -\int_i^f \vec{F}_g \cdot \vec{dx}$$

$$= -\int_0^h F_{gy} dy$$

$$= -\int_0^h (-mg) dy$$

$$= mgh \qquad (48)$$

Again we have the change in potential energy equal to minus the work done by the force.

Not all forces, however, work like spring and gravitational forces. Suppose I grab an eraser and push it around on the table top for a while. In this case I am overcoming the friction force between the table and the eraser, and we have $\vec{F}_{me} = -\vec{F}_{friction}$. The total work done by me as I move the eraser from an initial position i to a final position f is

$$\left.\begin{array}{c}\text{work I do}\\\text{while moving}\\\text{the eraser}\end{array}\right\} = -\int_i^f \vec{F}_{me} \cdot \vec{dx}$$

$$= -\int_i^f \vec{F}_{friction} \cdot \vec{dx} \qquad (49)$$

There are two problems with this example. The integrals in Equation 49 do depend on the path I take. If I move the eraser around in circles I do a lot more work than if I move it in a straight line between the two points. And when I get to position f, there is no stored potential energy. Instead all the energy that I supplied overcoming friction has probably been dissipated in the form of heat.

Physicists divide all forces in the world into two categories. Those forces like gravity and the spring force, where the integral

$$\int_i^f \vec{F} \cdot \vec{dx}$$

depends only on the initial and final positions i and f, are called "*conservative*" forces. For these forces there is a potential energy, and the formula for the change in potential energy is minus the work the force does when the particle goes from i to f.

All the other forces, the ones for which the work integral depends upon the path, are called ***non-conservative*** forces. We cannot use the concept of potential energy for non-conservative forces because the formula for potential energy would not have a unique or meaningful value. The non-conservative forces can do work and change kinetic energy, but as we see in the case of friction, the work ends up as something else like heat rather than potential energy.

It is interesting that on an atomic scale, where energy does not disappear in subtle ways like heat, we almost always deal with conservative forces and can use the concept of potential energy.

GRAVITATIONAL POTENTIAL ENERGY ON A LARGE SCALE

In our computer analysis of satellite motion, we saw that the quantity E_{tot}, given by

$$E_{tot} = \frac{1}{2}mv^2 - \frac{GM_e m}{r} \tag{50}$$

was unchanged as the satellite moved around the earth. As shown in Figure (10), m is the mass of the satellite, $\vec{v}$ its velocity, R its distance from the center of the earth, and M_e is the mass of the earth. This was our first non trivial example of conservation of energy, where $1/2\ mv^2$ is the satellite's kinetic energy, and $-GM_e m/R$ must be the formula for the satellites's gravitational potential energy. Our discussion of the last section suggests that we should be able to obtain this formula for gravitational potential energy by integrating the gravitational force $|\vec{F}_g| = GM_e m/r^2$ from some initial to some final position.

Here on the surface of the earth, the formula for gravitational potential energy is mgh. This simple result arises from the fact that when we lift an object inside a room, the strength of the gravitational force $m\vec{g}$ acting on it is essentially constant. Thus the work I do lifting a ball a distance h is just the gravitational force mg times the height h. Since this work is stored as potential energy, the formula for gravitational potential energy is simply mgh.

In the case of satellite motion, however, the strength of the gravitational force was not constant. In our first calculation of satellite motion in Chapter 8, the satellite started 1.1 earth radii from the center of the earth and went out as far as r = 5.6 earth radii. Since the gravitational force drops off as $1/r^2$, the gravitational force was more than 25 times weaker when the satellite was far away, than when it was launched.

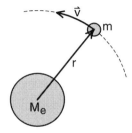

Figure 10
Earth satellite.

Zero of Potential Energy

Another difference is that the formula mgh for a ball in the room measures changes in gravitational potential energy starting from the floor where h = 0. In a rather arbitrary way, we have defined the gravitational potential energy to be zero at the floor. This is a convenient choice for people working in this room, but people working upstairs or downstairs would naturally choose their own floors rather than our floor as the zero of gravitational potential energy for objects they were studying.

Since conservation of energy deals only with changes in energy, it does not make any difference where you choose your zero of potential energy. A different choice simply adds a constant to the formula for total energy, and an unchanging or constant amount of energy cannot be detected. The most famous example of this was the fact that a particle's rest energy $m_0 c^2$ was unknown until Einstein introduced the special theory of relativity, and undetected until we saw changes in rest energy caused by nuclear reactions. In the case of the gravitational potential energy of a ball, if we use the floor downstairs as the zero of gravitational potential energy, we add the constant term $(mg)h_{floor}$ to all our formulas for E_{tot} (where h floor is the distance between floors in this building). This constant term has no detectable effect.

In finding a formula for gravitational potential energy of satellites, planets, stars, etc., we should select a convenient floor or zero of potential energy. For the motion of a satellite around the earth, we could choose gravitational potential energy to be zero at the earth's surface. Then the satellite's potential energy would be positive when its distance r from the center of the earth is greater than the earth radius r_e, and negative should r become less than r_e. Such a choice would be reasonable if we were only going to study earth satellites, but the motion of a satellite about the earth is very closely related to the motion of the planets about the sun and the motion of moons about other planets. Choosing $r = r_e$ as the distance at which gravitational potential energy is zero is neither a general or particularly convenient choice.

In describing the interaction between particles, for example an electron and a proton in a hydrogen atom, the earth and a satellite, the sun and its planets, or the stars in a galaxy, the convenient choice for the zero of potential energy is where the particles are so far apart that they do not interact. If the earth and a rock are a hundred light years apart, there is almost no gravitational force between them, and it is reasonable that they do not have any gravitational potential energy either.

Now suppose that *the earth and the rock are the only things in the universe*. Even at a hundred light years there is still some gravitational attraction, so that the rock will begin to fall toward the earth. As the rock gets closer to the earth it will pick up speed and thus gain kinetic energy. It was the gravitational force of attraction that caused this increase in speed, therefore there must be a conversion of gravitational potential energy into kinetic energy.

This gives rise to a problem. The rock starts with zero gravitational potential energy when it is very far away. As the rock approaches the earth, gravitational potential energy is converted into kinetic energy. How can we convert gravitational potential energy into kinetic energy if we started with zero potential energy?

Keeping track of energy is very much a bookkeeping scheme, like keeping track of the balance in your bank account. Suppose you begin the month with a balance of zero dollars, and start spending money by writing checks. If you have a trusting bank, this works because your bank balance simply becomes negative.

In much the same way, the rock falling toward the earth started with zero gravitational potential energy. As the rock picked up speed falling toward the earth, it gained kinetic energy at the expense of potential energy. Since it started with zero potential energy, and spent some, it must have a negative potential energy balance. From this we see that if we choose gravitational potential energy between two objects to be zero when the objects are very far apart, then the potential energy must be negative when the objects are a smaller distance apart. When we think of energy conservation as a bookkeeping scheme, then the idea of negative potential energy is no worse than the idea of a negative checking account balance.

(In the analogy between potential energy and a checking account, the discovery of rest energy m_0c^2 would be like discovering that you had inherited the bank. The checks still work the same way even though your total assets are vastly different.)

Let us now return to Equation (50) and our formula for gravitational potential energy of a satellite

$$\left.\begin{matrix}\text{gravitational}\\ \text{potential energy}\end{matrix}\right\} = -\frac{GM_e m}{r} \qquad (50a)$$

First we see that if the satellite is very far away, that as r goes to infinity, the potential energy goes to zero. Thus this formula does give zero potential energy when the earth and the satellite are so far apart that they no longer interact. In addition, the potential energy is negative, as it must be if the satellite falls in to a distance r, converting potential energy into kinetic energy.

What we have to do is to show that Equation (50a) is in fact the correct formula for gravitational potential energy. We can do that by calculating the work gravity does on the satellite as it falls in from $r = \infty$ to $r = r$. This work, which would show up as the kinetic energy of a falling satellite, must be the amount of potential energy spent. Thus the potential energy balance must be the negative of this work. Since the work is the integral of the gravitational force times the distance, we have

$$\left.\begin{array}{l}\text{gravitational}\\\text{potential energy}\\\text{at position R}\end{array}\right\} = -\left|\int_{\infty}^{R} \vec{F}_g \cdot d\vec{r}\right|$$

$$= -\left|\int_{\infty}^{R} \frac{GMm}{r^2} dr\right| \quad (51)$$

Equation 51 may look a bit peculiar in the way we have handled the signs. We have argued physically that the gravitational potential energy must be negative, and we know that it must be equal in magnitude to the integral of the gravitational force from $r = \infty$ to $r = R$. By noting ahead of time what the sign of the answer must be, we can do the integral easily without keeping track of the various minus signs that are involved. (One minus sign is in the formula for potential energy, another is the dot product since $\vec{F}_g$ points in and $d\vec{r}$ out, a third in the integral of r^{-2}, and more come in the evaluation of the limits. It is not worth the effort to get all these signs right when you know from a simple physical argument that the answer must be negative.)

Carrying out the integral in Equation 51 gives

$$\int_{\infty}^{R} \frac{GM_em}{r^2} dr = GM_em \int_{\infty}^{R} \frac{dr}{r^2}$$

$$= -\frac{GM_em}{r}\Big|_{\infty}^{R} = GM_em\left(\frac{1}{\infty} - \frac{1}{R}\right)$$

where we used the fact that the integral of $1/r^2$ is $-1/r$. Thus we get

$$\left|\int_{\infty}^{R} \frac{GM_em}{r^2} dr\right| = \frac{GM_em}{R}$$

As a result the gravitational potential energy of the satellite a distance R from the center of the earth is $-GM_em/R$ as given in Equation 50a.

Gravitational Potential Energy in a Room

Before we leave our discussion of gravitational potential energy, we should show that the formula $-GM_e m/r$ leads to the formula mgh for the potential energy of a ball in a room. To show this, let us use the formula $-GM_e m/r$ to calculate the increase in gravitational potential energy when I lift a ball from the floor, a distance R_e from the center of the earth, up to a height h, a distance $R_e + h$ from the center of the earth, as shown in Figure (11).

We have

$$PE_{\text{at height } h} = -\frac{GM_e m}{R_e + h}$$

$$PE_{\text{at floor}} = -\frac{GM_e m}{R_e}$$

$$\begin{matrix}\text{Increase} \\ \text{in PE}\end{matrix} = PE_{\text{at } h} - PE_{\text{at floor}}$$

$$= \left(-\frac{GM_e m}{R_e + h}\right) - \left(-\frac{GM_e m}{R_e}\right)$$

$$= GM_e m \left(\frac{1}{R_e} - \frac{1}{R_e + h}\right) \quad (52)$$

To evaluate the right side of Equation 52, we can write

$$\frac{1}{R_e + h} = \frac{1}{R_e}\left(\frac{1}{1 + h/R_e}\right)$$

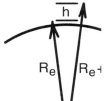

Figure 11
A height h above the surface of the earth.

Since h/R_e is a very small number compared to one, we can use our small number approximation

$$\frac{1}{1 + \alpha} \approx 1 - \alpha \quad \text{if } \alpha \ll 1$$

to write

$$\frac{1}{1 + h/R_e} \approx 1 - \frac{h}{R_e}$$

so that

$$\frac{1}{R_e + h} \approx \frac{1}{R_e}\left(1 - \frac{h}{R_e}\right) = \frac{1}{R_e} - \frac{h}{R_e^2} \quad (53)$$

Using Equation 53 in (52) gives

$$\begin{matrix}\text{Increase} \\ \text{in PE}\end{matrix} = -GM_e m \left[\frac{1}{R_e} - \left(\frac{1}{R_e} - \frac{h}{R_e^2}\right)\right]$$

$$= GM_e m \left[\frac{h}{R_e^2}\right]$$

$$= \left(\frac{GM_e}{R_e^2}\right) mh$$

Finally noting that $GM_e/R_e^2 = g$, the acceleration due to gravity at the surface of the earth, we get

$$\boxed{\begin{matrix}\text{Increase} \\ \text{in PE}\end{matrix} = mgh}$$

which is the expected result.

SATELLITE MOTION AND TOTAL ENERGY

Consider a satellite moving in a circular orbit about the earth, as shown in Figure (12). We want to calculate the kinetic energy, potential energy, and total energy (sum of the kinetic and potential energy) for the satellite. To find the kinetic energy, we analyze its motion, using Newton's laws. The only force acting on the satellite is the gravitational force $\vec{F}_g$ given by

$$|\vec{F}_g| = \frac{GMm}{r^2} \qquad \begin{array}{l}\vec{F}_g \text{ directed} \\ \text{toward the earth}\end{array}$$

where we now let M = mass of the earth and m = mass of the satellite. Since the satellite is moving at constant speed v in a circle of radius r, its acceleration is v^2/r toward the center of the circle

$$|\vec{a}| = \frac{v^2}{r} \qquad \begin{array}{l}\vec{a} \text{ directed} \\ \text{toward the earth}\end{array}$$

Since $\vec{a}$ and $\vec{F}_g$ are in the same direction, by Newton's second law

$$|\vec{F}_g| = m|\vec{a}|$$

$$\frac{GMm}{r^2} = \frac{mv^2}{r}$$

From this last equation we find that the kinetic energy $1/2\,mv^2$ of the satellite is

$$\tfrac{1}{2}mv^2 = \tfrac{1}{2}\frac{GMm}{r} \qquad \begin{array}{l}\text{kinetic} \\ \text{energy}\end{array}$$

The kinetic energy, as always, is positive.

The gravitational potential energy of the satellite is always negative. Since the satellite is a distance r from the center of the earth, its potential energy is

$$\frac{\text{potential}}{\text{energy}} = -\frac{GMm}{r}$$

The total energy of the satellite is

$$E_{total} = \frac{\text{kinetic}}{\text{energy}} + \frac{\text{potential}}{\text{energy}}$$

$$= \tfrac{1}{2}\frac{GMm}{r} + \left(-\frac{GMm}{r}\right)$$

$$E_{total} = -\tfrac{1}{2}\frac{GMm}{r} \qquad (54)$$

The total energy of a satellite in a circular orbit is negative.

Now consider a satellite in an elliptical orbit. In particular, suppose that the orbit is an extended ellipse, as shown in Figure (13). At apogee, the farthest point from the earth, the satellite is moving very slowly (explain why by using Kepler's law of equal areas). For all practical purposes, the satellite drifts out, stops at apogee, then falls back toward the earth. At apogee, the satellite has *almost no kinetic energy*; at this point its total energy is nearly equal to its negative potential energy

$$E_{total} = -\frac{GMm}{r_{apogee}}$$

Figure 13
Satellite in a very eccentric orbit. By Kepler's law of equal areas, a satellite with the above orbit would almost be at rest at apogee.

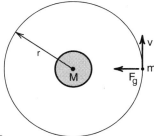

Figure 12
Satellite in a circular orbit.

Since the total energy is conserved, E_{total} remains negative throughout the orbit. If similar satellites are placed in different orbits, the one that goes out the farthest (has the greatest r_{apogee}) is the one with the least negative total energy, but all the satellites in elliptical orbits will have a negative total energy.

Suppose an extra powerful rocket is used and a satellite is launched with a positive total energy. In such a case, the positive kinetic energy must always exceed the negative potential energy. No matter how far out the satellite goes, headed for apogee, it will always have some positive kinetic energy to carry it out farther.

Even at enormous distances, where the negative potential energy $-GMm/r$ is about zero, some kinetic energy would still remain, and the satellite would escape from the earth!

By choosing potential energy to be zero when the satellite is very far out, the total energy becomes a meaningful number in itself. If the total energy is negative, the satellite will remain bound to the earth; it does not have sufficient energy to escape. If a satellite launched with positive total energy, it must escape since the negative gravitational potential energy is not sufficiently great to bind the satellite to the earth. If the satellite's total energy is zero, it barely escapes.

The orbits of comets about the sun are interesting examples of orbits of different total energies. It can be shown that when a *satellite's* total energy is positive, its orbit will be in the shape of a ***hyperbola***, which is an open-ended curve, as shown in Figure (14a). In this orbit the comet has a positive total energy and never returns.

If the total energy of the comet is zero, the orbit will be in the shape of an open curve, called a ***parabola*** (Figure 14b). A comet in this kind of orbit will not return either.

When the comet's total energy is slightly less than zero, it must return to the sun. In this situation the comet's orbit is an ellipse, even though it may be a very extended ellipse. A comparison of an extended ellipse and a parabola is shown in Figure (14c). From this figure we can see that near the sun there is not much difference in the motion of a comet with zero or slightly negative total energy. The difference can be seen at a great distance, where the zero-energy comet continues to move away from the sun, but the slightly negative-energy comet returns.

The circular, or nearly circular, motion of the planets is a limiting case of elliptical motion. The small circular orbits (Figure 14d) are occupied by planets that have large negative total energies. Thus the planets are tightly bound to the sun.

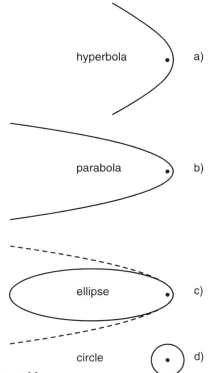

Figure 14
a) *Hyperbolic orbit of comet with positive total energy.*
b) *Parabolic orbit of comet with zero total energy.*
c) *Elliptical orbit of comet with slightly negative total energy. (Dashed lines show parabolic orbit for comparison.)*
d) *Nearly circular orbits of the tightly bound (large negative energy) planets.*

Example 4 Escape Velocity

At what speed must a shell be fired from a super cannon in order that it escapes from the earth? Does it make any difference at what angle the shell is fired, so long as it clears all obstructions? (Neglect air resistance.)

Solution: If the shell is fired at a sufficiently great initial speed so that its total energy is positive, it will eventually escape from the earth, regardless of the angle at which it is fired (so long as it clears obstructions). To calculate the minimum muzzle velocity at which the shell can escape, we will assume that the shell has zero total energy, so that it barely escapes. When $E_{total} = 0$ we have just after the shell is fired

$$0 = \frac{1}{2} mv^2 - \frac{GM_e m}{r_e}$$

which gives

$$v^2 = \frac{2GM_e}{r_e} \qquad (55)$$

Putting in numbers

$$G = 6.67 \times 10^{-8} \text{ cm}^3/\text{gm sec}^2$$
$$M_e = 5.98 \times 10^{27} \text{ gm}$$
$$r_e = 6.38 \times 10^8 \text{ cm}$$

we get

$$v^2 = \frac{2 \times (6.67 \times 10^{-8} \text{ cm}^3/\text{gm sec}^2)(5.98 \times 10^{27} \text{ gm})}{6.38 \times 10^8 \text{ cm}}$$

$$= \frac{2 \times 6.67 \times 5.98}{6.38} \frac{10^{-8} \times 10^{27}}{10^8} \text{ cm}^3\text{gm}/\text{gm cm sec}^2$$

$$= 1.25 \times 10^{12} \text{ cm}^2/\text{sec}^2$$

$$v_{escape} = 1.12 \times 10^6 \text{ cm/sec}$$

Converting this to more recognizable units, such as mi/sec, we have

$$v_{escape} = 1.12 \times 10^6 \text{ cm/sec} \times \frac{1}{1.6 \times 10^5 \text{ cm/mi}}$$

$$= 7 \text{ mi/sec} \quad (11.2 \text{ km/sec})$$

This is also equal to 25,200 mi/hr, which is far faster than the initial velocity required to put a satellite in an orbit 100 mi high.

Exercise 12
Calculate the escape velocity required to project a shell permanently away from the moon ($m_{moon} = 7.35 \times 10^{25}$ gm, $r_{moon} = 1.74 \times 10^8$ cm).

Exercise 13
Once a shell has escaped from the earth, what must its speed be to allow it to escape from our solar system?

Exercise 14
Find the escape velocities from the earth and the moon, using the planetary units given on page 8-14.

BLACK HOLES

A special feature of satellite motion we have just seen is that we can tell whether or not a satellite can escape simply by comparing kinetic energy with the gravitational potential energy. If the satellite's positive kinetic energy is greater in magnitude than the negative gravitational potential energy, then the satellite escapes, never to return on its own. This is true no matter how or from where the satellite is launched (provided it does not crash into something.)

So far we have limited our discussion to slowly moving objects where the approximate formula $1/2\,mv^2$ is adequate to describe kinetic energy. We got the formula $1/2\,mv^2$ back in Equation 7 by expanding $E = mc^2$ to get

$$E = mc^2 = \frac{m_0 c^2}{\sqrt{1 - v^2/c^2}}$$
$$\approx m_0 c^2 + 1/2\, m_0 v^2 \qquad (7)$$
$$\quad \underbrace{}_{\text{rest energy}} \;\; \underbrace{}_{\text{kinetic energy}}$$

The basic idea is that Einstein's formula $E = mc^2$ gives us a precise formula for the sum of the rest energy and the kinetic energy. In the special case the particle is moving slowly, we can use the approximate formula for $\sqrt{1 - v^2/c^2}$ to get the result shown in Equation 7.

For familiar objects like bullets, cars, airplanes, and rockets, the kinetic energy is $1/2\, m_0 v^2$, much, much, much smaller than the rest energy $m_0 c^2$. The kinetic energy of a rifle bullet, for example, is enough to allow the bullet to penetrate a few centimeters into a block of wood. The rest energy of the bullet, if converted into explosive energy, could destroy a forest. In fact, one way to tell whether or not the approximate formula $1/2\, m_0 v^2$ is reasonably accurate, is to check whether the kinetic energy is much less than the rest energy. If it is, you can use the approximate formula; if not, you can't.

We now finish our discussion of satellite motion by going to the opposite extreme, and consider the behavior of particles whose kinetic energy is much greater than their rest energy. Such a particle must be moving at a speed very close to the speed of light. We considered such a particle in Exercise 7 of Chapter 6. There we saw that electrons emerging from the Stanford linear accelerator travelled at a speed $v = .9999999999875\, c$, and had a mass 200,000 times greater than the rest mass. For such a particle, almost all the energy is kinetic energy. In the formula $E = mc^2$, only one part in 200,000 represents rest energy.

Actually we wish to go one step farther, and discuss particles with no rest energy, particles that move *at* the speed of light. The obvious example, of course, is the photon, the particle of light itself.

From one point of view there is not much difference between an electron travelling at a speed $.9999999999875\, c$ with only 1 part in 200,000 of its energy in the form of rest energy, and a photon travelling at a speed c and no rest energy. Taking this point of view, we will take as the formula for the energy of a photon $E = mc^2$, and assume that this is pure kinetic energy.

Applying the formula $E = mc^2$ to a photon implies that a photon has a mass $m_p = E_p/c^2$. We will now make the assumption that this mass m_{photon} is gravitational mass, and that photons have a gravitational potential energy $-GMm_p/r$ like other objects. Our assumption, which is slightly in error, is that Newtonian gravity, which is a non relativistic theory, applies to particles moving near to or at the speed of light. It turns out that Einstein's relativistic theory of gravity gives almost the same answers, that we are seldom off by more than a factor of 2 in our predictions.

Suppose we have a photon a distance r from a star of mass M_s. If the photon has a mass m_p, then the formula for the total energy of the photon, its kinetic energy m_pc^2 plus its gravitational potential energy $-GM_sm_p/r$ is

$$E_{tot} = m_pc^2 - \frac{GM_sm_p}{r}$$

Since m_p appears in both terms, we can factor it out (and also take out a factor of c^2) to get

$$E_{tot} = m_pc^2\left[1 - \frac{GM_s}{rc^2}\right] \quad (56)$$

Equation 56 applies only when the photon is outside the star, i.e., when the distance r is greater than the radius R of the star.

In most cases, the gravitational potential energy is much less than the kinetic energy of a photon, and gravity has little effect on the motion of the photon. For example, if a photon were grazing the surface of the sun (if r in Equation 56) were equal to the sun's radius R_{sun}) we would have

$$E_{tot} = m_pc^2\left[1 - \frac{GM_s}{R_sc^2}\right] \quad (57)$$

Putting in numbers $M_s = 1.99 \times 10^{33}$ gm, $R_s = 6.96 \times 10^{10}$ cm we have

$$\frac{GM_s}{R_sc^2} = \frac{6.67 \times 10^{-8}\frac{cm^3}{gm\ sec^2} \times 1.99 \times 10^{33}gm}{6.96 \times 10^{10}cm \times (3 \times 10^{10})^2 \frac{cm^2}{sec^2}}$$

$$= .00000212$$

Thus

$$E_{tot} = m_pc^2\left[1 - .00000212\right] \quad (58)$$

From Equation 58 we see that when a photon is as close as it can get to the surface of the sun, the gravitational potential energy contributes very little to the total energy of the photon, only 2 parts in a million.

However, suppose that the a star had *the same mass as the sun but a much, much smaller radius*. If it's radius R_s were small enough, the factor $\left[1 - GM_s/R_sc^2\right]$ in Equation 58 would become negative, and a photon grazing the surface of this star would have a negative total energy. The photon could not escape from the star. *No photons emerging from the surface of such a star could escape, and the star would cease to emit light.*

Let us see how small the sun would have to be in order that it could no longer radiate light. That would happen when the factor $\left[1 - GM_s/R_sc^2\right]$ is zero, when photons emerging from the surface of the sun have zero total energy. Putting in numbers we get

$$\frac{GM_s}{R_sc^2} = 1 \quad (59)$$

$$R_s = \frac{GM_s}{c^2}$$

$$= \frac{6.67 \times 10^{-8}\frac{cm^3}{cm\ sec^2} \times 1.99 \times 10^{33}cm}{(3 \times 10^{10})^2 cm^2/sec^2}$$

$$R_s = 1.48 \times 10^5 cm$$

$$= 1.48\ kilometer \quad (60)$$

Equation 60 tells us that an object with as much mass as the sun, confined to a sphere of radius less than 1.48 kilometers, cannot radiate light. Although we used the non relativistic Newtonian gravity in this calculation, Einstein's relativistic theory of gravity makes the same prediction.

In discussions of black holes, one often sees a reference to the radius of the black hole. What is usually meant is the radius given by Equation 59, the radius at which light can no longer escape if a mass M_s is contained within a sphere of radius R_s.

Do black holes exist? Can so much mass be concentrated in such a small sphere? The question has been difficult to answer because black holes are hard to observe since they do not emit light. They have to be detected indirectly, from the gravitational pull they exert on neighboring matter. In the sky there are many binary star systems, systems in which two stars orbit about each other. In some examples we have observed a bright star orbiting about an invisible companion. Careful analysis of the orbit of the bright star suggests that the invisible companion may be a black hole. There is recent evidence that gigantic black holes, with the mass of millions of suns, exists at the center of many galaxies, including our own.

That a black hole cannot radiate light is only one of the peculiar properties of these objects. When so much matter is concentrated in such a small volume of space, the gravitational force becomes so great that other forces cannot resist the crushing force of gravity, and as far as we know, the matter inside the black hole collapses to a point—a zero sized or very, very small sized object. At the present time we do not have a good theory for what happens to the matter inside a black hole. (We will have more to say about black holes in later chapters.)

Exercise 15
Studies of the motion of the stars in our galaxy suggests that at the center of our galaxy is a large amount of mass concentrated in a very small volume. For this problem, assume that a mass of 100 million suns is concentrated in the small volume. If this massive object is in fact a black hole, what is the radius from which light can no longer escape?

A Practical System of Units

In the CGS system of units, where we measure distance in centimeters, mass in grams and time in seconds, the unit of force is the dyne (1 dyne = 1 gm cm/sec^2) and the unit of energy is the erg (1 erg = 1 gm cm^2/sec^2). We have found the CGS system quite convenient for analyzing strobe photographs with 1 cm grids. But when we begin to talk about forces and particularly energy, the CGS system is often rather inconvenient. A force of one dyne is more on the scale of the force exerted by a fly doing push-ups than the kind of forces we deal with in the lab. A baseball pitched by Roger Clemens has a kinetic energy of over a million ergs and a 100 watt light bulb uses ten million ergs of electrical energy per second. The dyne and particularly the erg are much too small a unit for most every day situations.

In the MKS system of units, where we measure distance in meters, mass in kilograms and time in seconds, the unit of force is the newton and energy the joule. The force required to lift a 1 kilogram mass is 9.8 newtons (mg), and the energy of a Roger Clemens' pitch is over 10 joules. When working with practical electrical phenomena, the use of the MKS system is the only sensible thing to do. The unit of power, the watt, is one joule of energy per second. Thus a 100 watt light bulb consumes 100 joules of electrical energy per second. Volts and amperes are both MKS units, the corresponding CGS units are statvolts and esu, which are almost never used.

Where CGS units are far superior is in working with the basic theory of atoms, as for the case of the Bohr theory discussed in Chapter 36. This is because the electric force law has a much simpler form in CGS units. What we will do in the text from this point on is to use MKS units almost exclusively until we get through the chapters in electrical theory and applications. Then we will go back to the CGS system in most of our discussions of atomic phenomena.

Chapter 11
Systems of Particles

So far in our applications of Newton's second law, we have treated objects as individual point particles. In some problems, this appeared to be a reasonable approximation. But in others, like the motion of an apple falling from a tree, it appeared to be a remarkable result that the entire earth could be treated as a point particle located at the center of the earth. Newton supposedly invented calculus to show that one could do this.

In this chapter we will look at ways to handle systems consisting of many particles. In particular, we will see that often the concept of **center of mass** *allows us to treat a group of interacting particles as a single particle. This can lead to an enormous simplification of the analysis and a clearer understanding of the result.*

Diver - Movie
For a student project, Tobias Hays was videotaped doing a series of dives, a few of which are shown in this movie. Working frame by frame, you can see that the diver's center of mass follows a parabolic trajectory. (It was actually more instructive to use a parabolic trajectory to locate the diver's center of mass in various positions.)

CENTER OF MASS

To introduce the concept of center of mass, let us begin with some examples of mechanical systems that at first appear fairly complex, but which are greatly simplified when we focus our attention on the motion of the center of mass.

Consider the motion of the earth about the sun. To analyze the problem, we could first treat the sun as a point mass fixed at the origin and apply Newton's second law to the earth to determine the earth's elliptical orbit. On closer inspection, however, we note that the earth is not a point particle but an earth-moon system. A more accurate treatment of the problem requires us to consider two interacting particles both orbiting the sun.

With a computer program it is not too difficult to set up the earth-moon-sun system and directly solve for the motion of the earth and moon. Because the sun is so massive, it is still a good approximation to place the sun at rest at the origin of the coordinate system. We then have the earth subject to the gravitational force of the sun and the moon, and the moon experiencing the gravitational force of the sun and the earth. If we include all of these forces in the program, and start off with reasonable initial conditions, we will get the expected result that the earth goes about the sun in an elliptical orbit with a slight wobble, and the moon goes around the earth in a nearly circular orbit. If we plot the moon's orbit, exaggerating the orbit radius as in Figure (1), the orbit looks somewhat peculiar because it represents circular motion about a moving center. It looks like the drawing of epicycles from a text on ancient Greek astronomy.

For a similar but much more difficult problem, consider the globular cluster shown in Figure (2). Globular clusters are fairly common objects in our galaxy. A typical globular cluster is a swarm of several million stars all attracting each other to form a single gravitationally bound body. We can think of it as a confined gas of stars, confined not by a bottle or a rubber balloon but by the gravitational attraction between the stars.

The globular clusters in our galaxy lie outside the main body of stars of the galaxy, typically orbiting around or through the galaxy. If we wished to calculate the orbit of a globular cluster, our first approximation would be to treat it as a point particle. To do better than that with a computer program, we might try to analyze the motion of each star as we did in the earth-moon-sun problem above. But the futility of doing this would soon become obvious. Each of the millions of stars interact not only with the gravitational force of the galaxy, but with each other. Each star is subject to millions of forces, and any direct computer calculation becomes impossibly complex. One might try to simplify the problem by considering a cluster of only a few

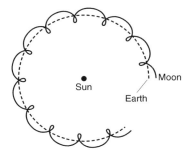

Figure 1
Motion of the moon around the sun. In this sketch, we have greatly exaggerated the size of the moon's orbit about the earth in order to show the epicycle like motion of the moon.

Figure 2
Globular cluster (NCG 5272).

hundred stars, but even then a lot of time on a super computer is needed for a meaningful prediction.

Despite all the forces involved, the motion of the cluster can easily be analyzed if we focus on the motion of the *center of mass* of the cluster. When we calculate the motion of the center of mass, all internal forces cancel, and we have to consider only the net force of the galaxy on the total mass of the stars. In the earth-moon-sun problem, the center of mass of the earth-moon system travels in an elliptical orbit around the sun. The earth and moon each orbit about this center of mass point. The calculation of the motion of the center of mass of the earth-moon is the same as calculating the motion of a point planet about the sun.

The idea of the center of mass is a familiar concept for it is a balance point. The center of mass of a long thin rod is the point where the rod balances on your finger. If you mark the balance point and throw the rod in the air, giving it a spin, the balance point follows a smooth parabolic trajectory while the rest of the rod rotates about the balance point as shown in Figure (3). The balance point or center of mass moves as if all the mass of the rod were concentrated at that point.

Although the idea of a center of mass or balance point is fairly straightforward, the formula for the center of mass of a collection of particles looks a bit peculiar at first. But when you get used to the formula, you will find that it is fairly easy to remember and leads to impressive simplifications when used in Newton's laws.

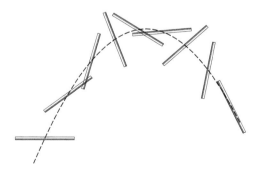

Figure 3
A meter stick tossed in the air rotates about its center of mass (balance point). The center of mass itself travels along the parabolic path of a point projectile.

Center of Mass Formula

Suppose that we have a collection of n particles $m_1, m_2, \cdots m_n$, as shown in Figure (4). The coordinate vector for particle m_1 is $\vec{R}_1$, that for m_2 is $\vec{R}_2$, etc. We define the ***total mass*** **M** of the collection of particles as simply the sum of the individual masses

$$M = m_1 + m_2 + \cdots m_i = \sum_{i=1}^{n} m_i \tag{1}$$

The ***coordinate vector*** $\vec{R}_{com}$ of the center of mass point is then defined by the formula

$$\boxed{M\vec{R}_{com} = \sum_i m_i \vec{R}_i} \quad \begin{array}{l} \textit{formula for} \\ \textit{center of mass} \\ \textit{coordinate} \end{array} \tag{2}$$

Since this is a vector equation, it can be written as the collection of the three scalar equations

$$MX_{com} = \sum_i m_i x_i \tag{2a}$$

$$MY_{com} = \sum_i m_i y_i \tag{2b}$$

$$MZ_{com} = \sum_i m_i z_i \tag{2c}$$

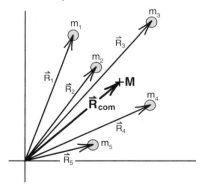

Figure 4
To calculate the center of mass of a collection of particles, you start by constructing the coordinate vector for each particle. You then use Equation 2 to calculate the center of mass coordinate vector.

where X_{com}, Y_{com} and Z_{com} are the x, y and z coordinates of the center of mass, and x_i, y_i, and z_i are the x, y, z coordinates of the i-th particle.

To see that Equation 2 does give the balance point of a collection of particles, let us consider the simple case of a horizontal massless rod of length l with two masses m_1 and m_2 at each end, as shown in Figure (5). If m_1 is placed at the origin of the coordinate system, then Equation 2a gives

$$M X_{com} = m_1 \times 0 + m_2 l$$

$$X_{com} = \frac{m_2 l}{M} = \frac{m_2 l}{m_1 + m_2}$$

If m_1 and m_2 are the masses of two children on a seesaw of length l, then the pivot should be placed a distance X_{com} from m_1 for the children to balance. (As a quick check, if $m_1 = m_2$, then $X_{com} = l/2$ which is obviously correct.)

[If you want to calculate the center of mass of a continuous object like an irregularly shaped sheet of plywood, you can mark the plywood off into small sections of mass dm_i located at $\vec{r}_i$. Then M is the mass of the whole sheet of plywood, and the x coordinate of the center of mass is

$$MX_{com} = \sum_i x_i dm_i \rightarrow \int x\, dm$$

where x_i is the x coordinate of $\vec{r}_i$. You can then replace the sum over the dm_i by an integral. This is a typical problem treated in an introductory calculus course and will not be discussed further here.]

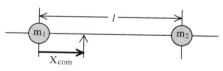

Figure 5
Calculating the center of mass for two particles.

Exercise 1
The formula for center of mass looks like it depends on where you place the origin of the coordinate system. To see that this is not true recalculate X_{com} for the two masses in Figure (5), placing the origin at m_2, and show that the pivot point comes out at the same place on the rod.

Exercise 2
Show that the center or mass of the earth-moon system is located inside the earth.

Exercise 3
An ammonia molecule consists of a nitrogen atom and three hydrogen atoms located on the corners of a tetrahedron, as shown in Figure (6). Locate the center of mass of the ammonia molecule.

Dynamics of the Center of Mass

Let us now see how the concept of center of mass can be used to handle the dynamic behavior of a system of particles. Suppose we have a system with n particles. The formula for their center of mass coordinate $\vec{R}_{com}$ is from Equation 2

$$M\vec{R}_{com} = m_1\vec{R}_1 + m_2\vec{R}_2 + \cdots + m_n\vec{R}_n \quad (2)$$

If we differentiate Equation 2 with respect to time, and note that the velocity $\vec{V}_{com}$ is given by $d\vec{R}_{com}/dt$, we get

$$M\vec{V}_{com} = m_1\vec{v}_1 + m_2\vec{v}_2 + \cdots + m_n\vec{v}_n \quad (3)$$

where $\vec{v}_i = d\vec{R}_i/dt$ is the velocity of the i-th particle.

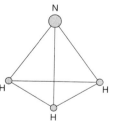

Figure 6
Structure of the ammonia molecule.

Equation 3 already has an interesting interpretation. Since $m_1\vec{v}_1$ is the linear momentum of particle 1, $m_2\vec{v}_2$ that of particle 2, etc., we see that $M\vec{V}_{com}$ is equal to the vector sum of the linear momenta of all the particles under consideration. We will come back to this point when we discuss the concept of linear momentum in more detail.

Differentiating Equation 3 with respect to time, noting that $\vec{A}_{com} = d\vec{V}_{com}/dt$ is the acceleration of the center of mass point, we get

$$M\vec{A}_{com} = m_1\vec{a}_1 + m_2\vec{a}_2 + \cdots + m_n\vec{a}_n \quad (4)$$

where $\vec{a}_i = d\vec{v}_i/dt$ is the acceleration of the i-th particle.

The final step is to use Newton's second law to replace $m\vec{a}_i$ by the vector sum of the forces acting on particle i. Calling this sum $\sum \vec{F}_i$, we get

$$M\vec{A}_{com} = \sum \vec{F}_1 + \sum \vec{F}_2 + \cdots + \sum \vec{F}_n \quad (5)$$

Equation 5 tells us that $M\vec{A}_{com}$ is equal to the sum of every force acting on all the particles.

If we wish to apply Equation 5 to the motion of something as complex as a globular cluster, it looks like we are still in trouble because, as we have mentioned, each of the millions of stars in the cluster interacts with all the other stars in the cluster. Acting on each star is the gravitational force of the galaxy plus the millions of forces exerted by the other stars.

But Newton's law of gravity provides an enormous simplification. When two objects interact gravitationally, they exert equal and opposite forces on each other as shown in Figure (7). In that case $\vec{F}_{g1} = -\vec{F}_{g2}$ or if we add these two forces of interaction we get

$$\vec{F}_{g1} + \vec{F}_{g2} = 0$$

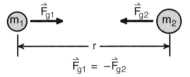

Figure 7
Two objects exert equal and opposite forces on each other.

Let us now write out Equation 5 as it might be applied to the globular cluster:

$$\begin{aligned}M\vec{A}_{com} &= \left(\vec{F}_{G1} + \vec{F}_{21} + \vec{F}_{31} + \cdots\right) \\ &+ \left(\vec{F}_{G2} + \vec{F}_{12} + \vec{F}_{32} + \cdots\right) \\ &+ \cdots\end{aligned} \quad (6)$$

where $\vec{F}_{G1}$ is the force of the galaxy on star #1, $\vec{F}_{21}$ is the force of star #2 on star #1, $\vec{F}_{31}$ the force of star #3 on star #1, etc. In the next collection of terms we have $\vec{F}_{G2}$, the force of the galaxy on star 2, $\vec{F}_{12}$ the force of star #1 on star #2, etc.

Rearranging the order in which we write the forces, we get

$$\begin{aligned}M\vec{A}_{com} &= \left(\vec{F}_{G1} + \vec{F}_{G2} + \cdots \vec{F}_{Gn}\right) \\ &+ \left(\vec{F}_{21} + \vec{F}_{12}\right) \\ &+ \left(\vec{F}_{31} + \vec{F}_{13}\right) + \cdots\end{aligned} \quad (7)$$

In Equation 7, we have separated the external forces $\vec{F}_{G1} + \vec{F}_{G2} + \cdots$ exerted by the galaxy on the individual stars from the internal forces like $\vec{F}_{12}$ and $\vec{F}_{21}$ between stars in the cluster. All the internal forces can be grouped in pairs, like $(\vec{F}_{12} + \vec{F}_{21})$, the force of star #2 on star #1 plus the force of star #1 on star #2. Because these are equal and opposite forces, all the pairs of internal forces cancel (over a trillion pairs for an average cluster), and we are simply left with the vector sum of the external forces. In our cluster example, the vector sum of all the external forces is just the net force $\vec{F}_{ext}$ the galaxy exerts on the cluster, and we are left with the fantastically simple result

$$\boxed{M\vec{A}_{com} = \vec{F}_{ext}} \quad \begin{array}{l}\textit{equation for}\\ \textit{center of}\\ \textit{mass motion}\end{array} \quad (8)$$

Equation 8 tells us that the center of mass of the globular cluster moves exactly as if the cluster were a single mass point of mass M equal to the total mass of the cluster subject to a single force $\vec{F}_{ext}$ equal to the total gravitational force exerted by the galaxy on all the stars in the cluster. This remarkable result explains why we can often represent a complex system by a single mass point in the analysis of the system's behavior.

When this result is applied to the earth-moon system shown in Figure (8), we have the following picture. When we calculate the motion of the center of mass of the earth and moon about the sun, the internal forces between the earth and moon cancel, and we are left with

$$\left(M_{earth} + M_{moon}\right)\vec{A}_{com} = \left(\vec{F}_{sun\,on\,earth} + \vec{F}_{sun\,on\,moon}\right)$$

where $\vec{F}_{sun\,on\,earth}$ is the force of the sun on the earth, and $\vec{F}_{sun\,on\,moon}$ that of the sun on the moon. The center of mass moves with the same acceleration as a point particle of mass $(M_{earth} + M_{moon})$ subject to the total force the sun exerts on the two. This results in an elliptical orbit for the center of mass.

Exercise 4
Two air carts of equal mass are connected by a spring as shown in Figure (9). A small black marker is placed at the center of the spring which remains at the center of mass of the carts. One of the carts is given a shove to the right so that the two carts move off to the right in an undulating drift. Describe the motion of the black marker.

Figure 9

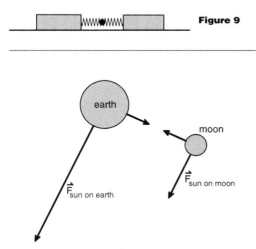

Figure 8
Forces on the earth and moon, as they go around the sun. (The force of the sun on the earth is much larger than the other three forces shown.)

NEWTON'S THIRD LAW

In our analysis of the motion of a globular cluster, the great simplification came when all the internal forces canceled in pairs in Equation 7, and we were left with the result that the acceleration of the center of mass point was determined solely by the external forces acting on the swarm of stars. The cancellation occurred because the gravitational attraction between two objects is equal in magnitude and oppositely directed.

What about other forces? What if two stars are in collision? Is the force between them still equal and opposite? If not, then we would have to take internal forces into account when predicting the motion of the center of mass? Unable to believe that this would happen, Newton proposed that *when two bodies interact, then the force between them is always equal in magnitude and oppositely directed, no matter what forces are involved.* This assumption is known as **Newton's Third Law of Mechanics**. The third law guarantees that internal forces cancel in pairs and that center of mass motion is determined only by external forces.

[Newton's first law, as we have mentioned, is that in the absence of any external forces, an object will move with uniform motion. Although there is a direct consequence of the second law $\vec{F} = m\vec{a}$, Newton explicitly stated the result, because it was not such an obvious idea in Newton's time, when a horse and buggy was the smoothest ride available.

In a traditional course in Newtonian mechanics, Newton's three laws are presented at the beginning as basic postulates and everything else is derived from them. This was a logical approach for over 200 years during which time there were no known exceptions to these laws. But with the discovery of special relativity in 1905 and quantum mechanics in 1923, we now know that Newton's laws are an approximate set of equations which apply with great accuracy to objects like stars, planets, cars and baseballs, but which have to be significantly modified when we consider objects moving at speeds near the speed of light, and which fail completely on a subatomic scale.]

CONSERVATION OF LINEAR MOMENTUM

In Chapter 6, after introducing the recoil definition of mass, we saw that the quantity $m_1\vec{v}_1 + m_2\vec{v}_2$ remained unchanged when the two objects were recoiled from each other.

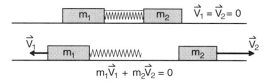

We proposed that this was one example of a more general conservation law—the conservation of linear momentum. Now, using Newton's third law, we can explicitly demonstrate how the law of conservation of momentum applies to objects obeying Newton's laws.

Our discussion begins with Equation 3, reproduced below as Equation 9, that was obtained by differentiating the formula for the center of mass of a system of particles. The result was

$$M\vec{V}_{com} = m_1\vec{v}_1 + m_2\vec{v}_2 + \cdots + m_n\vec{v}_n \quad (9)$$
$$= \vec{P}_{total}$$

where $\vec{P}_{total} = m_1\vec{v}_1 + \cdots + m_n\vec{v}_n$ is the vector sum of the momenta of all the particles in the system. We will call this the ***total momentum of the system***. We see from Equation 9 that this total momentum is simply the total mass M times the velocity of the center of mass.

Differentiating Equation 9 with respect to time, as we did in going from Equation 3 to 4 we get

$$\frac{d\vec{P}_{total}}{dt} = m_1\vec{a}_1 + m_2\vec{a}_2 + \cdots + m_n\vec{a}_n \quad (10)$$

This is essentially Equation 4, except that we are replacing $M\vec{A}_{com}$ by $d\vec{P}_{total}/dt$.

Using the fact that $m_1\vec{a}_1$ is the vector sum of all the forces acting on particle 1, $m_2\vec{a}_2$ the sum acting on particle 2, etc., we get

$$\frac{d\vec{P}_{total}}{dt} = \left\{ \begin{array}{l} \text{sum of all the forces} \\ \text{acting on all the particles} \end{array} \right. \quad (11)$$

Now use Newton's third law to cancel all the internal forces, and we are left with

$$\boxed{\frac{d\vec{P}_{total}}{dt} = \vec{F}_{ext}} \quad \begin{array}{l} \textit{Newton's second law} \\ \textit{for a system of particles} \end{array} \quad (12)$$

where $\vec{F}_{ext}$ is the vector sum of all the external forces acting on the system.

If we have an isolated system with no external forces acting on it (if our globular cluster were drifting through empty space), then

$$\boxed{\frac{d\vec{P}_{total}}{dt} = 0} \quad \begin{array}{l} \textit{for an} \\ \textit{isolated} \\ \textit{system} \end{array} \quad (13)$$

Equation 13 is our desired statement of the law of conservation of linear momentum. In words it says that the total linear momentum of an isolated system is conserved—does not change with time. (Linear momentum is also conserved if there are external forces but their vector sum is zero. For example, a cart on an air track experiences the downward force of gravity and the upward force of the air, but these forces exactly cancel.)

In deriving Equation 13, we had to use Newton's hypothesis that all internal forces cancel in pairs. And we also needed Newton's second law to relate $d(m\vec{v}_i)/dt$ to the vector sum of the forces acting on the i-th particle. However, the law of conservation of linear momentum is known to apply on a subatomic scale where the concepts of force and acceleration loose their meaning. This suggests that our derivation of momentum conservation is somehow backwards. A more logical route is to assume conservation of linear momentum as a basic principle and derive the consequences.

[To see how such a derivation would look, consider an isolated system where there are no external forces. We get from Equation 11

$$\frac{d\vec{P}_{total}}{dt} = \begin{array}{l}\text{vector sum of all the}\\ \text{internal forces acting}\\ \text{on the particles}\end{array} \quad (14)$$

But conservation of linear momentum requires that the linear momentum of an isolated system be conserved. Thus $d(P_{Total})/dt$ must be zero, and therefore the vector sum of all the internal forces must be zero. This must be true no matter what kind of forces are involved. Newton's third law is a bit more restrictive in that it requires the internal forces to cancel in pairs. The cancellation in pairs is the simplest picture, but not necessarily required for conservation of linear momentum.]

Momentum Version of Newton's Second Law

In our discussion of Newton's second law, we have consistently assumed that the mass of a particle was constant, so that we could take the m outside the derivative. For example, in going from Equation 3 to Equations 4 and 5 in the center of mass derivation, we used the following steps to relate the rate of change of momentum of the i-th particle $d(m_i\vec{v}_i)/dt$ to the total force $\vec{F}_i$ acting on the particle

$$\frac{d(m_i\vec{v}_i)}{dt} = \frac{m_i d\vec{v}_i}{dt} = m_i\vec{a}_i = \vec{F}_i \quad (15)$$

If we take $\vec{F} = m\vec{a}$ to be the basic form of Newton's second law, we have to assume m is constant in order to get $m\vec{a}$ and then $\vec{F}$.

Newton recognized that a slight generalization of the second law would make it unnecessary to assume that a particle's mass was constant. He actually expressed the second law in the form

$$\vec{F} = \frac{d\vec{p}}{dt} = \frac{d}{dt}(m\vec{v}) \quad (16)$$

where $\vec{F}$ is the total force acting on a particle of mass m, moving with a velocity $\vec{v}$. In the special case that m is constant, then Newton's second law becomes

$$\vec{F} = \frac{d(m\vec{v})}{dt} = m\frac{d\vec{v}}{dt} = m\vec{a} \quad (m = \text{constant}) \quad (17)$$

Thus we should view the Equation $\vec{F} = m\vec{a}$ as a special case of the more general law $\vec{F} = d\vec{p}/dt$.

The momentum form of Newton's second law is advantageous if we are considering problems like that in the following exercise where momentum is being transferred to an object at a known rate and we wish to determine the effective force. A more basic application is for relativistic problems where mass changes with velocity. Then we must use the momentum form of the law in order to account for the mass change. It is interesting that Newton had the insight to present the second law in a form that would handle Einstein's theory 200 years later.

Exercise 5

A boy is washing the door of his father's car by squirting a hose at the door. Assume that the water comes out of the hose at a rate of 20 kilograms (liters) per minute at a speed of 12 meters/second. When the water hits the door it dribbles down the side. What force, in newtons, does the water exert on the door?

(To solve this, simply calculate the amount of momentum per second that the water brings to the door.)

COLLISIONS

In our every day experience, collisions are something we usually try to avoid, whether it is running into a door or an automobile accident. Hitting a baseball is an obvious exception. In physics, collisions turn out to play an extremely important role, particularly in the study of elementary particles. For example, the atomic nucleus was discovered as a result of experiments involving the collision between α particles and atoms in a gold foil.

Collisions generally happen rapidly, and one is not used to observing what happens during a collision. It is usually a before and after scene, what was the situation before the collision, and what did things look like after. In most physics experiments like those involving elementary particles, that is all we can observe. However we will begin our discussion of collisions with an experiment that is explicitly designed to allow us to study the situation during the collision.

In this experiment, an air cart moving down an air track collides with a force detector mounted at the end of the track. Rather than have the metal cart bounce off the metal arm of the force detector, we slow the collision down by mounting a stretched rubber band on the end of the cart. With the rubber band colliding with the force detector, it takes several milliseconds for the collision to occur and the cart to reverse directions. During this time we can record both the force the cart exerts on the force detector and the velocity of the cart. Using the momentum form of Newton's second law, we will find that there is a particularly simple way to analyze the collision in terms of the concept of *impulse*.

When collisions are either too rapid or on too small a scale to be observed directly, we can almost always apply the law of conservation of momentum to analyze the results. In elementary particle collisions, both energy and momentum may be conserved. In some introductory physics lab experiments, momentum is conserved during the collision and energy after. We will see how this lets us make detailed predictions about the behavior of the objects involved.

Impulse

An overview of the air cart force detector experiment is shown in Figure (10). Focusing on the momentum involved, we see that the cart is initially moving down the track with a momentum $\vec{p}_i$ as shown in Figure (10a). In Figure (10b) it collides with the force detector, and in (10c) is moving back up the track with a momentum $\vec{p}_f$. The net effect of the collision is to change the cart's momentum from $\vec{p}_i$ to $\vec{p}_f$.

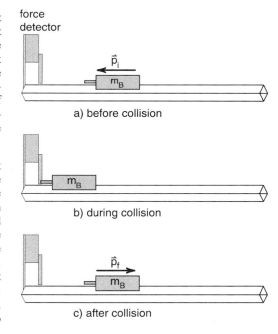

Figure 10
Collision of an aircart with the force detector. During the collision, the force detector first has to push on the cart to stop it, and then give an essentially equal push to move it out.

During the collision itself, the force detector is exerting a force $\vec{F}(t)$ on the cart. The force $\vec{F}(t)$ acts for only a short time, but can be measured in detail by the force detector. To relate this force to the observed change in the momentum of the cart, we start with Newton's law in the form

$$\vec{F}(t) = \frac{d\vec{p}}{dt} \qquad \text{(16 repeated)}$$

Multiplying through by dt gives

$$\vec{F}(t)dt = d\vec{p} \qquad (18)$$

Now integrate both sides of this equation from a time t_i before the collision, when the cart momentum was $\vec{p}_i$, to a time t_f after the collision, when the cart momentum was $\vec{p}_f$. We get

$$\int_{t_i}^{t_f} \vec{F}(t)dt = \int_{p_i}^{p_f} d\vec{p} = \vec{p}_f - \vec{p}_i \qquad (19)$$

Since $(\vec{p}_f - \vec{p}_i)$ is the change in the momentum of the cart as a result of the collision, we get

$$\begin{array}{c}\text{impulse of the}\\ \text{force } \vec{F}(t)\end{array} \int_{t_i}^{t_f} \vec{F}(t)dt = \begin{array}{c}\text{change in the}\\ \text{momentum of}\\ \text{the air cart}\end{array} \qquad (20)$$

This integral of $\vec{F}(t)$ over the time of the collision is called *the impulse* of the force $\vec{F}(t)$. The force exerted by the force detector alters the momentum of the cart, and how much it alters it is equal to the impulse of the force.

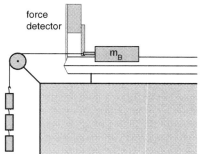

Figure 11
Calibrating the force detector.

In the air cart experiment, we will record the force $\vec{F}(t)$ from the output of the force detector, and directly compare the integral of that force with the change in the momentum of the cart.

Note that both $\vec{p}_f$ and $-\vec{p}_i$ are directed to the right. Thus the magnitude of $\vec{p}_f - \vec{p}_i$ is equal to the numerical sum $p_f + p_i$

$$\left|\vec{p}_f - \vec{p}_i\right| = \left|\vec{p}_f\right| + \left|\vec{p}_i\right|$$

As a result, the impulse supplied by the force detector has a magnitude $p_f + p_i$ or about $2p_i$ if the cart comes out at the same speed it went in. The force detector supplies $2p_i$ because one p_i is required to stop the cart, and the other is required to shove it out again.

Calibration of the Force Detector

The force detector is designed to put out a voltage that is proportional to the force exerted on the detector beam. To convert this voltage reading to a force measurement, we have to calibrate the force detector. This is easily done by running a string from the air cart, over a pulley and down to some weights as shown in Figure (11). As we add weights to the string we increase the force the cart exerts on the beam.

Figure (12) shows the output of the force detector as we added a series of 20 gm weights. Adding 3 weights changes the voltage by 42.8 millivolts (mV), thus each weight changes the output voltage by 42.8/3 = 14.3 mV. (One millivolt = 10^{-3} volts.) Each added weight corresponds to an increase of the force by $\Delta F = mg = 20 \text{ gm} \times 980 \text{ cm/s}^2 = 19600 \text{ dynes}$. Thus the factor for converting from millivolts output for this force detector, to dynes of force is

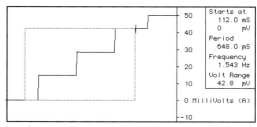

Figure 12
Voltage output when 20 gram weights are added.

$$\frac{\text{conversion}}{\text{factor}} = \frac{19600 \text{ newtons}}{14.3 \text{ millivolts}}$$
$$= 1370 \frac{\text{dynes}}{\text{millivolts}} \qquad (21)$$

The Impulse Measurement

One way to set up the collision experiment is shown in Figure (13). To soften the collision we have added a metal bracket to the cart and stretched a rubber band across the open end. Adjusting the tension in the rubber band allows us to change the length of time during which the collision occurs.

Figure (14) shows a fairly typical output of the force detector. There is no force until the rubber band reaches the detector. The force then increases and then decreases symmetrically, and becomes 0 when the cart leaves. Using our calibration factor of 1370 dynes/mV, we have graphed the force, in dynes, as a function of time in Figure (15).

The impulse of this force is the integral of the force curve from the time t_1 that the rubber band gets to the force detector, to t_2 when it leaves. Since we do not have a formula for the curve, we cannot do the integral analytically. Instead, we have to use some graphical technique to find the area under that curve. One way is to superimpose the curve on graph paper and count the squares underneath. A slightly less accurate way we will use is to construct a triangle whose area is, to our best estimate, equal to the area under the curve. We have done this with the dashed line triangle seen in Figure (15). We have adjusted the triangle so that the extra area at the top matches the area lost at the sides.

The area of a triangle is (1/2 base × altitude). In Figure (15), the base of the triangle is 3.00–2.60 = 0.40 seconds. The height is seen to be close to 56800 dynes. Thus the area is

$$\begin{array}{c}\text{area of}\\\text{triangle}\end{array} = \frac{1}{2} \times (0.40 \text{ sec}) \times (56800 \text{ dynes})$$
$$= 11360 \text{ dyne seconds} \qquad (22)$$

Equation 22 is our result for the impulse of the force F(t).

Although the force detector measures the magnitude of the force F(t) that the cart exerts on the detector, by Newton's third law, this should be equal in magnitude but oppositely directed to the force exerted by the detector on the cart. Thus the magnitude of the impulse calculated in Equation 22 should be equal to the magnitude of the change in the momentum of the cart, as a result of the collision.

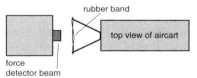

Figure 13
A rubber band is used to soften the collision.

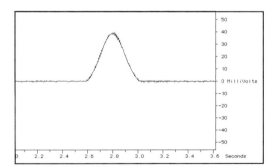

Figure 14
Output of the force detector. There is zero force before the cart arrives and after it leaves.

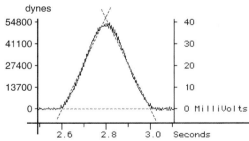

Figure 15
Force versus time graph. The area under the force curve is about equal to the area of the triangle we drew.

Exercise 6

(a) What is the direction of the force exerted by the force detector on the cart?

(b) What is the direction of the vector $\Delta \vec{p} = \vec{p}_f - \vec{p}_i$?

(c) How do these two directions compare?

Change in Momentum

To measure the momentum of the cart before and after the collision, we mounted a 10 cm long sail on the top of the cart as shown in the side view of Figure (16). Mounted above the track is a light source and a photodetector seen in the top view. When the sail on the cart interrupts the light beam, there is an abrupt change in the voltage output by the photodetector. The lower, dashed curves in Figures (14) and (17) are from the output of the photodetector. In Figure (17a) we are measuring the length of time the sail took to pass by the photocell on the way down to the force detector. We see that this time was 400 milliseconds or .400 seconds. Thus the velocity of the cart on the way down was

$$v_i = \frac{10 \text{ cm}}{.400 \text{ sec}} = 25 \frac{\text{cm}}{\text{sec}} \quad (23)$$

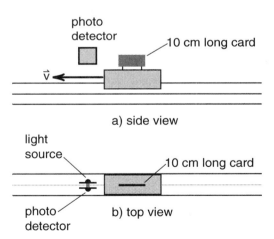

Figure 16
One way to measure the velocity of the aircart is to mount a 10 cm long sail is mounted on top of the aircart. While the sail is interrupting the light beam, there is a change in the output voltage of the photo detector.

On the way back, we find from Figure (17c) that the sail took 412 milliseconds or .412 seconds to pass the photodetector. Thus the final speed of the cart was

$$v_f = \frac{10 \text{ cm}}{.412 \text{ sec}} = 24.27 \frac{\text{cm}}{\text{sec}} \quad (24)$$

The cart, sail and rubber band apparatus had a total mass of m_{cart} which was measured to be

$$m_{cart} = 227 \text{ gm} \quad (25)$$

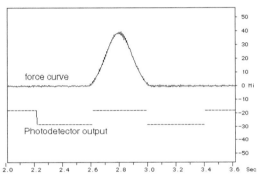

Figure 17a
Simultaneous recording of both the voltage output from the force detector (solid line) and the photo detector (dashed line).

Figure 17b,c
Measuring the length of time the sail took to go past the photodetector. The 10 cm sail took 400 milliseconds to pass on the way in (upper curve) and 412 milliseconds on the way out. We see it slowed down a bit.

(Another way to determine the speed of the aircart is to tilt the airtrack and release the cart from a known height.)

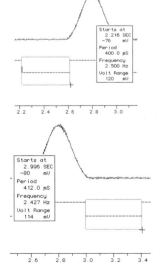

Thus the initial and final momenta have magnitudes

$$p_i = mv_i = 227 \text{ gm} \times 25.0 \frac{\text{cm}}{\text{sec}}$$
$$= 5680 \frac{\text{gm cm}}{\text{sec}}$$

$$p_f = mv_f = 227 \text{ gm} \times 24.27 \frac{\text{cm}}{\text{sec}}$$
$$= 5510 \frac{\text{gm cm}}{\text{sec}}$$

The magnitude Δp of the change in momentum is the sum of these two values

$$\Delta p = p_i + p_f = 5680 \frac{\text{gm cm}}{\text{sec}} + 5510 \frac{\text{gm cm}}{\text{sec}}$$

$$\boxed{\text{change in the linear momentum of the aircart} = \Delta p = 11200 \frac{\text{gm cm}}{\text{sec}}}$$

(26)

From Equation 22, we saw that the total impulse supplied by the force detector was

$$\boxed{\text{total impulse from the force detector} = \int_{t_i}^{t_f} \vec{F}(t)dt = 11360 \text{ dyne sec}}$$

(27)

We see that to within a quite reasonable experimental error, the total impulse supplied by the force detector equals the change in linear momentum of the aircart.

Exercise 7
Explain why Δp in Equation 26 is the sum of the magnitudes p_i and p_f.

Exercise 8
Show that the dimensions of impulse (dyne seconds) are the same as momentum (gm cm/sec).

Exercise 9
How much energy, in joules, did the cart lose in the collision with the force detector? What percentage of the cart's initial kinetic energy was this?

Momentum Conservation during Collisions

In our force detector experiment we used a rubber band to slow down the collision so that we could do a more accurate analysis of the impulse. Even so the impulsive force $\vec{F}(t)$ acted for only a very short time. The most important point of the experiment is that, no matter how short the time is, the impulse, the time integral of $\vec{F}(t)$ is equal to the total change in momentum. If we had let the metal end of the cart strike the force detector, the collision would have taken much less time, but the force would have been much greater. The integral of the larger force over the shorter time would still equal the change in the momentum of the cart.

Suppose that, instead of an air cart colliding with a force detector, we had two air carts colliding with each other. During the collision they would by Newton's third law, exert equal and opposite forces $\vec{F}(t)$ on each other. Thus they would exert upon each other equal and opposite impulses $\int \vec{F}(t)dt$. As a result, the momentum gained by one cart would precisely be equal to the momentum lost by the other. The net result is conservation of momentum during the collision.

Now consider a slightly more complex situation. Suppose I throw a red billiard ball up in the air, and you throw a blue billiard ball, and the two balls collide before landing.

During this collision, more forces are involved. There is the force of the red billiard ball on the blue one, the force of the blue billiard ball on the red one, and there are the gravitational forces acting on both. To study the change in the momenta of the balls, it appears that we must now account for all four forces at once.

However there is something special about the impulsive forces found in collisions. These forces are usually very large but **act for a very short time**. During this short time the collision forces are usually much larger than external forces like gravity. So much larger, in fact, that we can usually neglect external forces during a collision. Since the collision forces conserve linear momentum, we get the result that linear momentum is conserved during a collision even if external forces are present. The only exception would be if the collision is so slow that the external forces have time to act and change the system's momentum during the collision. This is usually not the case.

Collisions and Energy Loss

While we can use conservation of linear momentum in the analysis of a collision, often energy conservation is not applicable. If the objects are deformed or give off heat, light or sound, energy escapes in ways that are difficult to measure.

In some situations, however, we can use momentum conservation to figure out how much energy must be "*lost*" to deformation, heat, sound, etc. Suppose, for example, we hang a steel ball of mass M from a string as shown in Figure (18), and throw a putty ball of mass m at the steel ball. The putty ball is initially moving at a speed v_i, then hits and sticks to the steel ball. The two move off together at speed v_f.

Even though energy is "lost" when the putty ball squashes up against the steel ball, momentum is conserved during the collision. The initial momentum is all carried by the putty ball, and is thus

$$p_i = mv_i \quad \text{\textit{initial momentum}} \quad (28)$$

After the collision the two move off together at a speed v_f and thus have a total momentum

$$p_f = (m+M)v_f \quad (29)$$

Since momentum is conserved, we have

$$p_i = p_f$$
$$mv_i = (m+M)v_f \quad (30)$$

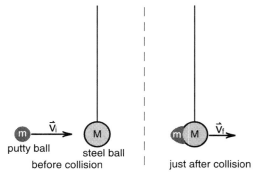

Figure 18
Collision of a putty ball with a steel ball.

Solving for the final speed v_f, we get

$$v_f = \left(\frac{m}{m+M}\right)v_i \quad (31)$$

We can now calculate the amount of energy that must be dissipated in this collision. The initial energy E_i is the kinetic energy of the putty ball

$$E_i = \tfrac{1}{2}mv_i^2 \quad (32)$$

The final energy E_f is the kinetic energy of the two together

$$E_f = \tfrac{1}{2}(m+M)v_f^2 \quad (33)$$

The energy "lost", which must have gone into deforming the putty ball, is

$$E_{lost} = E_i - E_f$$
$$= \tfrac{1}{2}mv_i^2 - \tfrac{1}{2}(m+M)v_f^2 \quad (34)$$

Using Equation 31 for v_f in Equation 34 gives

$$E_{lost} = \tfrac{1}{2}mv_i^2 - \tfrac{1}{2}(m+M)\left(\frac{mv_i}{m+M}\right)^2$$
$$= \tfrac{1}{2}mv_i^2 - \tfrac{1}{2}\left(\frac{m^2}{m+M}\right)v_i^2$$
$$= \tfrac{1}{2}mv_i^2\left(1 - \frac{m}{m+M}\right)$$

$$E_{lost} = \tfrac{1}{2}mv_i^2\left[\frac{M}{m+M}\right] \quad (35)$$

It may be somewhat surprising that even though we may not have the slightest idea how energy is lost during the deformation of the putty, we can calculate precisely how much energy this uses.

Exercise 10

Check that Equation 35 gives reasonable results for the two special cases M = 0 and M = ∞ (for m ≠ 0).

Exercise 11

A 500 gram steel ball is suspended from a string as shown in Figure (18). It is struck by a putty ball that sticks, and half the putty ball's initial kinetic energy is lost in the collision. What is the mass of the putty ball?

After the putty ball has collided with the steel ball in our preceding example, the two will rise together to a maximum height h and final angle θ_f before swinging back down again. This is illustrated in Figure (19). We can predict this height h by applying energy conservation after the collision. The kinetic energy E_f just **after the collision** is transformed into gravitational potential energy $(m + M)gh$ at the top of the swing, giving

$$\tfrac{1}{2}(m+M)v_f^2 = (m+M)gh$$

or

$$h = \frac{v_f^2}{2g} \qquad (36)$$

If you want to calculate the final angle θ_f, you use the fact that $h = \ell - (\ell \cos \theta_f)$ as can be seen from Figure (19).

This example of the steel and putty ball is essentially equivalent to the ballistic pendulum discussed in Exercise 12. In that problem a block of wood of mass M is suspended from two strings as shown in Figure (20). A bullet of mass m is fired into the block and the block with the bullet stuck inside rises to a height h as shown.

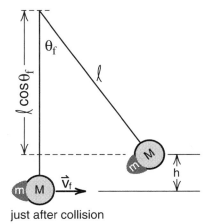

Figure 19
After the collision, energy is conserved. This allows us to calculate how high the ball rises.

You can assume that momentum is conserved during the collision, energy afterward, and from a measurement of the height h, determine the speed v_i of the bullet before it hit the block. This provides a rather simple, inexpensive way to measure the speed of bullets.

From the prospective of an introductory physics course, the ballistic pendulum experiment clearly distinguishes the use of momentum conservation and energy conservation. During the collision, momentum is conserved but energy is not. During the rise up to a height h, energy is conserved but momentum is not. To analyze such problem, you must develop an understanding of when you can apply the conservation laws and when you cannot. (For a safer ballistic pendulum demonstration, see Figure (21) on the next page.)

Exercise 12

(a) A bullet of mass m_b traveling at a speed v_b is fired into a block of wood of mass M hanging at rest as shown in Figure (20). The combined block and bullet rise to a height h. Find a formula for the speed v_b of the bullet.

(b) For part (a), suppose the bullet's mass is m_b = 10 gm, the block of wood has a mass M = 1000 gm, and the final height h is 12 cm. What was v_b in cm/sec?

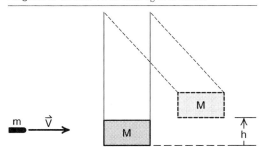

Figure 20
Ballistic pendulum experiment.

Exercise 13

In a lecture demonstration that is safer to perform than the above ballistic pendulum experiment, a wastebasket is suspended from two cords and a pillow is placed inside the wastebasket, as shown in Figure (21).

Various members of the class are selected to throw a softball into the wastebasket. A scale is constructed to indicated how fast the ball was thrown. If the mass m of the softball is 200 gm, the mass M of the wastebasket and pillow is 1000 gm, and the length of the suspension cords are 2 meters from the ceiling to the center of the wastebasket, determine the distance (x) that the end of the basket travels if the ball is thrown at 40 miles/hour.

Collisions that Conserve Momentum and Energy

When a bullet plows into a block of wood, a considerable amount of energy goes into deforming the wood and bullet. On the other hand if two hardened steel ball bearings collide, almost all the energy stays in the form of kinetic energy of the particles and very little is "lost" as heat, sound, and the deformation of the objects. When no energy is lost this way, we say that the collision is *elastic*. If energy is lost, then the collision is called *inelastic*.

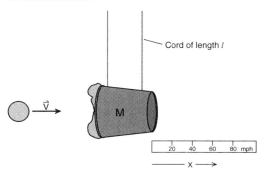

Figure 21
A safer ballistic pendulum experiment for classroom demonstrations.

On an atomic and subatomic scale we do not have the usual sound and friction, and small deformations may not be allowed. In these circumstances there is no way for energy to become "lost" and the resulting collisions are truly elastic. Thus in the study of the collisions of atomic and subatomic particles, we have examples where both energy and linear momentum are conserved.

Figure (22), shown previously in Chapter 6 as Figure (6-3), shows the track of a proton as it moves through a hydrogen bubble chamber. The incoming track ends when the proton strikes a hydrogen nucleus (another proton) that is part of the liquid hydrogen. Three dimensional stereoscopic photographs show that the two protons recoil from each other at an angle of 90°. Here we are looking at the behavior of matter on a subatomic scale where Newtonian mechanics does not apply, and we would like to find out what we can learn from this collision.

Does this photograph show us that momentum and energy are still conserved on the subatomic scale? To find out we will analyze the collision of larger objects like steel ball bearings, where momentum is conserved and energy nearly so, and see if the results explain what we see in Figure (22).

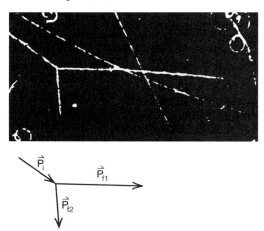

Figure 22
An incoming proton collides with a proton at rest. The two protons recoil at right angles.

Elastic Collisions

We will start with the simplest elastic collision we can think of—a ball of mass is traveling at a velocity $\vec{v}_i$ strikes an identical ball at rest, as shown in Figure (23). We will assume that the collision is straight on so that the two balls go off in the same direction at speeds v_1 and v_2 as shown. The idea is to apply conservation of momentum and energy in order to predict the final speeds v_1 and v_2.

From conservation of momentum we get

$$mv_i = mv_1 + mv_2 \quad \text{momentum conservation} \quad (27)$$

Canceling the m's we have

$$v_i = v_1 + v_2 \quad (28)$$

From conservation of energy we have

$$\frac{1}{2}mv_i^2 = \frac{1}{2}mv_1^2 + \frac{1}{2}mv_2^2 \quad \text{energy conservation} \quad (29)$$

The 1/2 m's cancel and we have

$$v_i^2 = v_1^2 + v_2^2 \quad (30)$$

If we square Equation (28) we get

$$v_i^2 = v_1^2 + 2v_1v_2 + v_2^2 \quad (31)$$

Comparing this with Equation (30) we see that

$$v_1v_2 = 0$$

I.e., either $v_2 = 0$, which means there was no collision at all, or $v_1 = 0$ which means that ball 1 stops dead in its tracks and ball 2 goes on at the initial speed v_i.

That ball 1 stops dead in its tracks explains the common toy where two or more steel balls are suspended from a string as shown in the end view—Figure (24a). One of the balls is pulled back as shown in Figure (24b) and released. At the collision, it comes to rest and the second ball goes on. Then the second ball comes back, strikes the first one and stops, and the first ball goes back up. For good hard steel balls, this process goes on for a long time before we see motion decrease.

If energy were not conserved, if some were lost in the collision, then both balls would move forward after the collision. In the extreme case the balls would stick and move off together. This is a completely inelastic collision where the maximum energy is lost (consistent with conservation of momentum). If you perform this experiment and notice that the incoming ball really comes to rest, that is experimental proof that both energy and momentum were in fact conserved in the collision.

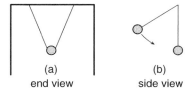

Figure 24
When the balls collide, the moving ball stops and the stationary ball moves on at the same speed.

Figure 23
Collision of two identical steel balls. We want to calculate v_1 and v_2.

In the next simplest example, suppose the two balls of Figure (23) collide but bounce off at an angle as shown in Figure (25), ball 1 coming off at a velocity $\vec{v}_1$, and ball 2 at a velocity $\vec{v}_2$ as shown. In this case we have the same formula for conservation of energy, but conservation of momentum must now be written as the vector equation

$$m\vec{v}_i = m\vec{v}_1 + m\vec{v}_2 \quad \text{momentum conservation} \quad (32)$$

Again the m's cancel and we are left with the following vector equation for the velocity vectors

$$\vec{v}_i = \vec{v}_1 + \vec{v}_2 \quad (32a)$$

The equation if pictured in Figure (26).

Recall that energy conservation gave

$$v_i^2 = v_1^2 + v_2^2 \quad \text{energy conservation} \quad (30)$$

which is simply the **Pythagorean theorem** when applied to the triangle in Figure (26). Thus the incoming speed $\vec{v}_i$ must be the hypotenuse and $\vec{v}_1$ and $\vec{v}_2$ the sides of a right triangle. Energy conservation requires that the two balls emerge from the collision at right angles.

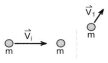

Figure 25
If the collision is not straight on, the balls come off at an angle.

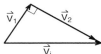

Figure 26
Equation 30 requires that the velocity vectors form a right triangle.

To state this result another way, if two equal masses collide and one is originally at rest, they always emerge at right angles (or the incoming one stops). This is experimental evidence that both energy and momentum are conserved in the collision.

It is this way that we learn that energy and momentum are both conserved during the collision of two protons in a hydrogen bubble chamber. This is a remarkable result considering that we do not have to know anything about what kind of forces were involved in the collision.

If we have elastic collisions between objects of different masses, moving at each other with different speeds, as in Figure (27), we still have the conservation of momentum

$$m_1\vec{v}_{1i} + m_2\vec{v}_{2i} = m_1\vec{v}_{1f} + m_2\vec{v}_{2f} \quad (33)$$

and conservation of energy

$$\tfrac{1}{2}m_1 v_{1i}^2 + \tfrac{1}{2}m_2 v_{2i}^2 = \tfrac{1}{2}m_1 v_{1f}^2 + \tfrac{1}{2}m_2 v_{2f}^2 \quad (34)$$

The only problem is that the algebra quickly becomes messy.

If you are in the business of working with collision problems, you will find it much easier to go to a coordinate system where the center of mass of the colliding particles is at rest. In this coordinate system the particles go in and come out symmetrically and the equations are easy to solve. Then you transform back to your original coordinate system to see what the particle velocities should be in the laboratory.

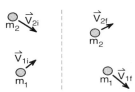

Figure 27
Arbitrary collision of two balls.

DISCOVERY OF THE ATOMIC NUCLEUS

In his early experiments with radioactivity, Ernest Rutherford found that radioactive atoms emitted three kinds of radiation which, as we have mentioned, he called α rays, β rays and γ rays. The α rays turned out to be heavy positively charged particles, later identified as helium nuclei. The β rays were beams of negatively charged particles later determined to be electrons, and the neutral γ rays turned out to be high energy particles of light (photons).

In the early 1900s, before 1912, it was not clear how these particles were emitted or what the structure of the atom was. Since J. J. Thomson's experiments with electron beams in 1895, it was known that atoms contained electrons, and it was also known that complete atoms were electrically neutral and much more massive than an electron. Thus the atom had to have mass and positive charge in some form or other, but no one knew what form. By 1912 the plum pudding model was quite popular. This was a picture in which mass and positive charge was spread throughout the atom like the pudding, and the electrons were located at various points, like the plums. A rather vague picture at best.

In 1912 Rutherford and Hans Geiger began a series of experiments using beams of radioactive particles to probe the structure of matter. These experiments could begin after Geiger had developed a tube to detect radioactive particles. This device later became known as a *Geiger counter*, and is still used through the world to monitor radiation.

In the first set of experiments, a beam of α particles were aimed at a gold foil. It was expected that some of the α particles would be slightly deflected as they passed through the positive matter in the gold atoms, or came near electrons. To the utter amazement of both Rutherford and Geiger, some of the α particles bounced straight back out of the gold foil, with essentially the same kinetic energy they had going in.

We have seen from our analysis of the elastic collision of two equal mass particles, that the incoming particle stops and the struck particle continues on. Only if the mass of the struck particle is greater than the mass of the incoming particle, will the incoming particle bounce back. And only if the mass of the struck particle is much greater than the mass of the incoming particle will the incoming particle rebound with nearly the same energy that it had coming in. Thus Rutherford and Geiger's observation that some of the α particles bounced right back out of the gold foil, indicated that they struck a solid object much more massive than an α particle.

Most α particles passed through the gold foil without much deflection, indicating that most of the volume of the gold foil was devoid of mass. The few collisions that did result in a recoil indicated that the mass in a gold foil was concentrated in incredibly small regions of space. A more detailed analysis showed that the scattering was caused by an electric force, thus they knew that both the mass and positive charge were located in a tiny region of the atom. In this way the atomic nucleus was discovered.

NEUTRINOS

The discovery of the neutrino, or at least the prediction of its existence, is another important event in the history of physics that is related to the conservation of energy and linear momentum.

After Rutherford's discovery of the nucleus, it became clear that the high energy radioactive emissions, α, β, and γ rays, must be coming from the nucleus of the atom. Thus a study of these rays should give valuable information about the nature of the nucleus itself.

After a number of years of experimentation it was determined that whenever an α particle or γ ray was emitted, the energy carried out by the α particle or γ ray was precisely equal to the energy lost by the nucleus. But decays involving β particles were different. In studying β decays, one always got a spread of energies of the β particle. Sometimes the β particle carried out almost all the energy lost by the nucleus, and sometimes only the relatively small rest energy of the electron. By the late 1920s it was clear that energy was apparently not conserved in β decay reactions. Neils Bohr proposed that the law of conservation of energy had to be modified for nuclear reactions involving β decays. The new rule was that the final energy was always less than or equal to the initial energy.

In 1930 Wolfgang Pauli, one of the founders of quantum theory, objected to the idea that energy was not conserved in β decay events. Pauli noted that the conservation of energy, linear momentum, and angular momentum are all apparently violated at the same time in β decay. Either the entire structure of physical law was being violated, or there was another explanation. Pauli's other explanation was that the energy, the linear momentum, and the angular momentum were all being carried out at the same time by an unseen particle.

If Pauli's particle existed, it would need the following properties: (1) it had to be electrically neutral because no electric charge was lost in β decays; (2) it would have to have a very small rest mass because the electron or β particle sometimes carried out almost all the available energy, leaving none for creating the new particle's rest mass; (3) the new particle had to have almost no interaction with matter, otherwise someone would have seen it.

Initially there was not much enthusiasm for Pauli's idea of an undetectable particle. At that point no one had seen an electrically neutral particle, and the fact that it did not interact with matter made it seem too speculative.

In 1932 the neutron was discovered which demonstrated that neutral particles did exist. Shortly after that Enrico Fermi developed a detailed theory of the weak interaction in which neutrinos played a significant role. (Fermi called Paul's particle the ***neutrino***, or "little neutral one" to distinguish it from the more massive neutron.) Detailed verification of Fermi's weak interaction theory convinced the physics community that neutrinos should exist.

The neutrino, actually detected in 1956 and now commonly seen in numerous experiments, is a remarkable particle in that it is subject to only the weak and gravitational interactions. All other known particles are subject to the electric and nuclear forces. For example π mesons are subject to the nuclear force, and can travel only a short distance through matter before colliding with a nucleus and being stopped.

The muon discussed in the relativity chapter, does not feel the strong nuclear force and therefore can travel much farther through matter (hundreds of meters) before being stopped. Muons are electrically charged and therefore are stopped by the weaker electric interaction. Photons also interact through the electric interaction, and therefore have a limited range traveling through matter. (An X-ray is an example of a photon passing through matter.)

The weak interaction is so weak compared to the nuclear or electric force, that a particle like the neutrino which feels only the weak interaction force, can travel incredible distances through matter before being stopped. To have a good chance of stopping a single neutrino with a stack of lead, one would need a pile of lead, light years thick.

This does not mean that neutrinos are impossible to detect. Instead of using a detector light years across to detect one neutrino, one can use a source that produces an incredible number of neutrinos and use a reasonable sized detector so that one has some chance of stopping a few neutrinos. In 1956 Cowan and Rines placed a tank car full of carbon tetrachloride cleaning fluid in front of a nuclear reactor that was estimated to emit about 10^{15} neutrinos per square centimeter per second. They observed that about two chlorine atoms per month in the tank car of carbon tetrachloride were converted by neutrino interactions into argon atoms which were counted individually.

In modern experiments carried out using high energy particle accelerators, neutrino reactions are routinely seen. The reason that more neutrinos are detected in these experiments is that the weak interaction becomes less weak as the energy of the particles involved increases. The high energy accelerators produce neutrinos with great enough energy that they are not too difficult to detect. As a result, the neutrino interactions has become an important research tool in the study of the basic interactions of matter. Neutrinos make a particularly clean tool for these studies because they have no nuclear or electric interactions. Neutrino experiments are not contaminated by effects of the nuclear and electric forces.

Neutrino Astronomy

An exciting development involving neutrinos is the birth of neutrino astronomy. In the fusion reaction that powers our sun, where four hydrogen nuclei (protons) end up as a helium 4 nucleus, the weak interaction and the β decay process comes into play in the conversion of two of the protons into the neutrons of the helium nucleus. Thus the emission of neutrinos must accompany the fusion reaction, and the neutrinos themselves must carry off a significant amount of the energy liberated by the fusion reaction.

In a star like the sun, the fusion reaction takes place down in the core of the star where the temperatures are highest. Any light emitted by the fusion reaction should take the order of about 10,000 years to work its way out. Thus if the fusion reaction in the sun were shut off today, it would be roughly 10,000 years before the sun dimmed.

Neutrinos, however, escape from the core of the sun without delay. If the fusion reaction stopped and we were monitoring the neutrinos from the sun, we would know about it within 8 minutes. As a result there is considerable incentive to observe the solar neutrinos, for that gives us a picture of what is happening in the sun's core now.

To study solar neutrinos, and do other experiments like look for decay of the proton, several large neutrino detectors have been set up around the world. Solar neutrinos have been monitored fairly carefully for over a decade, and there is an unexplained, perhaps disturbing result. Only about one third as many neutrinos are being emitted by the sun as we expect from what we think the fusion reaction should produce. Perhaps we are not detecting all we should, but the detectors are getting better and the number remains at 1/3. This is one of the major puzzles of astronomy.

That neutrino astronomy is really here was dramatically illustrated with the supernova explosion of 1987. On the average, supernovas occur about once per century per galaxy. Kepler saw the last supernova explosion in our galaxy 400 years ago. In 1987 a graduate student spotted the sudden appearance of a bright star in the large Magellanic cloud, a close small neighboring galaxy. This was the first supernova explosion in the local region of our galaxy in 400 years.

In a supernova explosion, huge quantities of neutrinos should be emitted. In fact a fair fraction of the energy of the explosion should be carried out by neutrinos.

Theoretical models of supernova explosions suggest that light should take about three hours to work its way out through the expanding envelope of gas before it starts its trek through space at the speed c. Neutrinos, on the other hand, should escape without being slowed down, and have a three hour head start on the light. If neutrinos have no rest mass, and therefore travel at the speed of light, they should have reached the earth about three hours before the light. Two of the major neutrino detectors, one in the US and one in a tunnel in the Alps, detected significant pulses of neutrinos about three hours before the flare-up of the star was seen. (This was determined by a later analysis of the neutrino data.) That event marks the birth of neutrino astronomy on a galactic scale.

Chapter 12
Rotational Motion

Our discussion of rotational motion begins with a review of the measurement of angles using the concept of radians. We will refer to an angle measured in radians as an **angular distance**. *If we are discussing an object that is rotating, we will describe the rotation in terms of the increase in angular distance, namely an* **angular velocity**. *And if the speed of rotation is changing, we will describe the change in terms of an* **angular acceleration**.

In Chapter 7, linear momentum and angular momentum were treated as distinctly separate topics. The main point of this chapter is to develop a close analogy between the two concepts. The linear momentum *of an object is its mass m times its linear velocity* $\vec{v}$. *We will see that* angular momentum *can be expressed as an angular mass times an angular velocity. (Angular mass is more commonly known as moment of inertia). Then, using the formalism of the vector cross product (mentioned in Chapter 2), we will see that angular momentum can be treated as a vector quantity, which explains the bicycle wheel experiments we discussed in Chapter 7.*

The fundamental concept of Newtonian mechanics is that the total force $\vec{F}$ *acting on an object is equal to the time rate of change of the object's linear momentum;* $\vec{F} = d\vec{p}/dt$. *Using the vector cross product formalism, we will obtain a complete angular analogy to this equation. We will find that a quantity we call an angular force is equal to the time rate of change of angular momentum. (The angular force is more commonly known as* torque*).*

The angular analogy to Newton's second law looks a bit peculiar at first. It involves lever arms and vectors that point in funny directions. After some demonstrations to show that the equation appears to give reasonable results, we apply the equation to predict the motion of a gyroscope. The prediction appears to be absurd, but we find that that is the way a gyroscope behaves.

Our focus in this chapter is on angular momentum because that concept will play such an important role in our later discussions of atomic physics and electrons and nuclear magnetic resonance. There are other important and interesting topics such as rotational kinetic energy and the calculation of moments of inertia which we discuss in more detail in the appendix. These topics are not difficult and lead to some good lecture demonstrations and laboratory experiments. We put them in an appendix because they do not play the essential role that angular momentum does in our later discussions.

RADIAN MEASURE

From the point of view of doing calculations, it is more convenient to measure an angle in radians than the more familiar degrees. In radian measure the angle θ shown in Figure (1) is the ratio of the arc length s to the radius r of the circle

$$\theta \equiv \frac{s}{r} \text{ radians} \qquad (1)$$

Since s and r are both distances, the ratio s/r is a dimensionless quantity. However we will find it convenient for the angular analogy to keep the name radians as if it were the actual dimension of the angle. For example we will measure angular velocities in **radians per second**, which is analogous to linear velocities measured in meters per second.

Since the circumference of a circle is $2\pi r$, the number of radians in a complete circle is

$$\theta \binom{\text{complete}}{\text{circle}} = \frac{2\pi r}{r} = 2\pi$$

In discussing rotation, we will often refer to going around one complete time as one complete **cycle**. In one cycle, the angle θ increases by 2π. Thus 2π is the number of **radians per cycle**. We will find it convenient to assign these dimensions to the number 2π:

$$2\pi \; \frac{\text{radians}}{\text{cycle}} \qquad (2)$$

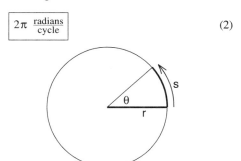

Figure 1
The angle θ in radians is defined as the ratio of the arc length s to the radius r: θ = s/r.

To relate radians to degrees, we use the fact that there are 360 degrees/cycle and dimensional analysis to find the number of degrees/radian

$$360 \, \frac{\text{degrees}}{\text{cycle}} \times \frac{1}{2\pi} \, \frac{\text{cycles}}{\text{radian}}$$

$$= \frac{360}{2\pi} \, \frac{\text{degrees}}{\text{radian}} = 57.3 \, \frac{\text{degrees}}{\text{radian}} \qquad (3)$$

Fifty seven degrees is a fairly awkward unit angle for purposes of drafting and navigation; no one in his or her right mind would mark a compass in radians. However, in working with the dynamics of rotational motion, radian measure is the only reasonable choice.

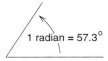

Angular Velocity

The typical measure of angular velocity you may be familiar with is revolutions per minute (RPM). The tachometer in a sports car is calibrated in RPM; a typical sports car engine gives its maximum power around 5000 RPM. Engine manufacturers in Europe are beginning to change over to revolutions per second (RPS), but somehow revving an engine up to 83 RPS doesn't sound as impressive as 5000 RPM. (Tachometers will probably be calibrated in RPM for a while.) In physics texts, angular velocity is measured in radians per second. Since there are 2π radians/cycle, 83 revolutions or cycles per second corresponds to $2\pi \times 83 = 524$ radians/second. Few people would know what you were talking about if you said that you should shift gears when the engine got up to 524 radians per second.

Exercise 1

What is the angular velocity, in radians per second, of the hour hand on a clock?

Our formal definition of angular velocity is the time rate of change of an angle. We almost always use the Greek letter ω (omega) to designate angular velocity

$$\begin{array}{c}\text{angular}\\ \text{velocity}\end{array} \omega \equiv \frac{d\theta}{dt} \frac{\text{radians}}{\text{second}} \quad (4)$$

When thinking of angular velocity ω picture a line marked on the end of a rotating shaft. The angle θ is the angle that the line makes with the horizontal as shown in Figure (2). As the shaft rotates, the angle θ(t) increases with time, increasing by 2π every time the shaft goes all the way around.

Angular Acceleration

When we start a motor, the angular velocity of the shaft starts at ω = 0 and increases until the motor gets up to its normal speed. During this start-up, ω(t) changes with time, and we have an angular acceleration α defined by

$$\begin{array}{c}\text{angular}\\ \text{acceleration}\end{array} \alpha \equiv \frac{d\omega}{dt} \frac{\text{radians}}{\text{second}^2} \quad (5)$$

The angular acceleration α has the dimensions of radians/sec^2 since the derivative gives us another factor of time in the denominator. Combining Equation 4 and 5 relates α to θ by

$$\alpha = \frac{d^2\theta}{dt^2} \quad (6)$$

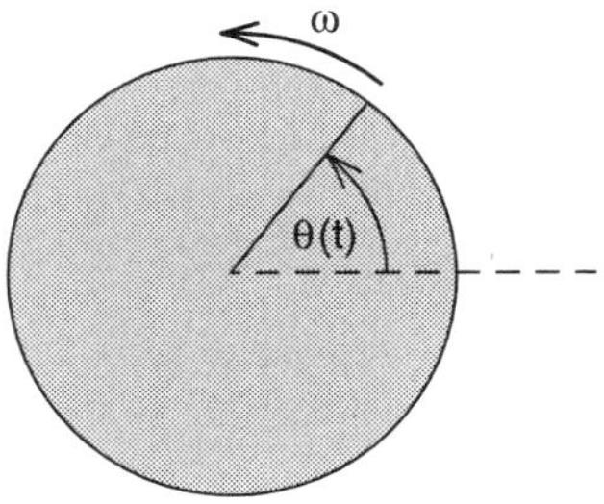

Figure 2
End of a shaft rotating at an angular velocity ω.

Angular Analogy

At this point we have a complete analogy between the rotation of a motor shaft and one dimensional linear motion. This analogy becomes clear when we write out the definitions of position, velocity, and acceleration:

	Linear motion	Angular motion
Distance	x meters	θ radians
Velocity	$v = \frac{dx}{dt} \frac{\text{meters}}{\text{second}}$	$\omega = \frac{d\theta}{dt} \frac{\text{radians}}{\text{second}}$
Acceleration	$a = \frac{dv}{dt} \frac{\text{meters}}{\text{second}^2}$	$\alpha = \frac{d\omega}{dt} \frac{\text{radians}}{\text{second}^2}$
	$= \frac{d^2x}{dt^2}$	$= \frac{d^2\theta}{dt^2}$ (7)

As far as these equations go, the analogy is precise. Therefore any formulas that we derived for linear motion in one dimension must also apply to angular motion. In particular the constant acceleration formulas, derived in Chapter 3, must apply. If the linear and angular accelerations a and α are constant, then we get

Constant Acceleration Formulas

Linear motion (a = const)   Angular motion (α = const)

$$x = v_0 t + \frac{1}{2}at^2 \qquad \theta = \omega_0 t + \frac{1}{2}\alpha t^2 \quad (8)$$

$$v = v_0 + at \qquad \omega = \omega_0 + \alpha t \quad (9)$$

Exercise 2

An electric motor, that turns at 3600 rpm (revolutions per minute) gets up to speed in 1/2 second. Assume that the angular acceleration α was constant while the motor was getting up to speed.

a) What was α (in radians/sec^2)?

b) How many radians, and how many complete cycles, did the shaft turn while getting up to speed?

Tangential Distance, Velocity and Acceleration

So far we have used the model of a rotating shaft to illustrate the concepts of angular distance, velocity and acceleration. We now wish to shift the focus of our discussion to the dynamics of a particle traveling along a circular path. For this we will use the model of a small mass m on the end of a massless stick of length r shown in Figure (3). The other end of the stick is attached to and is free to rotate about a fixed axis at the origin of our coordinate system. The presence of the stick ensures that the mass m travels only along a circular path of radius r. The quantity $\theta(t)$ is the angular distance travelled and $\omega(t)$ the angular velocity of the particle.

When we are discussing the motion of a particle in a circular orbit, we often want to know how far the particle has travelled, or how fast it is moving. The distance s along the path (we could call the tangential distance) travelled is given by Equation 1 as

$$\boxed{s = r\theta} \quad \text{tangential distance} \tag{10}$$

The speed of the particle along the path, which we can call the tangential speed v_t, is the time derivative of the tangential distance s(t)

$$v_t = \frac{ds(t)}{dt} = \frac{d}{dt}[r\theta(t)] = r\frac{d\theta(t)}{dt}$$

where r comes outside the derivative since it is constant. Since $d\theta(t)/dt$ is the angular velocity ω, we get

$$\boxed{v_t = r\omega} \quad \text{tangential velocity} \tag{11}$$

The tangential acceleration a_t, the acceleration of the particle along its path, is the time derivative of the tangential velocity

$$a_t = \frac{dv_t(t)}{dt} = \frac{d[r\omega(t)]}{dt} = r\frac{d\omega(t)}{dt} = r\alpha$$

$$\boxed{a_t = r\alpha} \quad \text{tangential acceleration} \tag{12}$$

where again we took the constant r outside the derivative, and used $\alpha = d\omega/dt$.

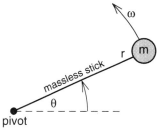

Figure 3
Mass rotating on the end of a massless stick.

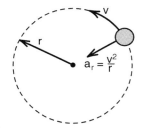

Figure 4
Particle moving at a constant speed in a circle of radius r accelerates toward the center of the circle with an acceleration of magnitude $a_r = v^2/r$.

Radial Acceleration

If the angular velocity ω is constant, if we have a particle traveling at constant speed in a circle, then $\alpha = d\omega/dt = 0$ and there is no tangential acceleration a_t. However, we have known from almost the beginning of the course that a particle traveling at constant speed v in a circle of radius r has an acceleration directed toward the center of the circle, of magnitude v^2/r, as shown in Figure (4). We will now call this center directed acceleration the *radial acceleration* a_r

$$a_r = \frac{v_t^2}{r} \qquad \textit{radial acceleration} \qquad (13)$$

Exercise 3

Express the radial acceleration a_r in terms of the orbital radius r and the particle's angular velocity ω.

If a particle is traveling in a circular orbit, but its speed v_t is not constant, then it has both a radial acceleration $a_r = v_t^2/r$, and a tangential acceleration $a_t = r\alpha$. The radial acceleration is always directed toward the center of the circle and always has a magnitude v^2/r. The tangential acceleration, if it exists, is tangential to the circle, pointing forward (counterclockwise) if α is positive and backward if α is negative. These accelerations are shown in Figure (5).

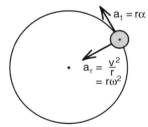

Figure 5
Motion with radial and tangential acceleration.

Bicycle Wheel

For much of the remainder of the chapter, we will use a bicycle wheel, often weighted with wire wound around the rim, to illustrate various phenomena of rotational motion. Conceptually we can think of the bicycle wheel as a collection of masses on the ends of massless rods as shown in Figure (6). The massless rods form the spokes of the wheel, and we can think of the masses m as fusing together to form the wheel. When forming a wheel, all the masses have the same radius r, same angular velocity ω and same angular acceleration α. If we choose one point on the wheel from which to measure the angular distance θ, then as far as angular motion is concerned, it does not make any difference whether we are discussing the mass on the end of a rod shown in Figure (3) or the bicycle wheel shown in Figure (6). Which model we use depends upon which provides a clearer insight into the phenomena being discussed.

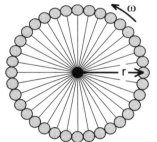

Figure 6
Bicycle wheel as a collection of masses on the end of massless rods.

ANGULAR MOMENTUM

In Chapter 7, we defined the angular momentum ℓ of a mass m traveling at a speed v in a circle of radius r as

$$\ell = mvr \tag{7-11}$$

As we saw, in Figure (7-9) reproduced here, the quantity $\ell = mvr$ did not change when we had a ball moving in a circle on the end of a string, and we pulled in on the string. The radius of the circle decreased, but the speed increased to keep the product (vr) constant. This was our introduction to the concept of the conservation of angular momentum.

Figure 7–9
Ball on the end of a string, swinging in a circle.

After that, we went on to consider some rather interesting experiments where we held a rotating bicycle wheel while standing on a freely turning platform. We found that these experiments could be explained qualitatively if we thought of the angular momentum of the bicycle wheel as being a vector quantity $\vec{\ell}$ which pointed along the axis of the wheel, as shown in Figure (7-15) reproduced below. What we will do now is develop the formalism which treats angular momentum as a vector.

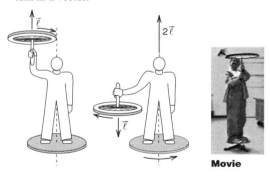

Figure 7-15
When the bicycle wheel is turned over and its angular momentum points down, the person starts rotating with twice as much angular momentum, pointing up.

Angular Momentum of a Bicycle Wheel

We will begin our discussion of the angular momentum of a bicycle wheel using the picture of a bicycle wheel shown in Figure (6), i.e., a collection of balls on the end of massless rods or spokes. If the wheel is rotating with an angular velocity ω, then each ball has a tangential velocity v_t given by Equation 11a

$$v_t = r\omega \qquad (11 \text{ repeated})$$

If the i-th ball in the wheel (identified in Figure 7) has a mass m_i, then its angular momentum ℓ_i will be given by

$$\ell_i = m_i v_t r = m_i(r\omega)r$$

$$\ell_i = (m_i r^2)\omega \tag{14}$$

Assuming that the total angular momentum L of the bicycle wheel is the sum of the angular momenta of each ball (we will discuss this assumption in more detail shortly) we get

$$L = \sum_i \ell_i = \sum_i \left(m_i r^2 \omega\right) \tag{15}$$

Since each mass m_i is at the same radius r and is traveling with the same angular velocity ω, we get

$$L = \left(\sum_i m_i\right) r^2 \omega$$

Noting that $M = \left(\sum_i m_i\right)$ is the total mass of the bicycle wheel, we get

$$\boxed{L = Mr^2\omega} \qquad \begin{array}{l}\textit{angular momentum}\\ \textit{of a bicycle wheel}\end{array} \tag{16}$$

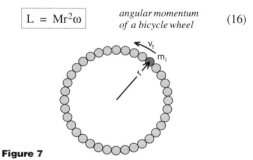

Figure 7
The angular momentum of the i'th ball is $m_i v_i r_i$.

Angular Velocity as a Vector

To explain the bicycle wheel experiment discussed in Chapter 7, we assumed that the angular momentum $\vec{L}$ was a vector pointing along the axis of the wheel as shown in Figure (8a). We can obtain this vector concept of angular momentum by first defining a vector angular velocity $\vec{\omega}$ as shown in Figure (8b). We will say that if a wheel is rotating with an angular velocity $\omega_{rad/sec}$, the vector $\vec{\omega}$ has a magnitude of $\omega_{rad/sec}$, and points along the axis of rotation as shown in Figure (8b). Since the axis has two directions, we use a right hand convention to select among them. Curl the fingers of your right hand in the direction of the direction of the rotation, and the thumb of your right hand will point in the direction of the vector $\vec{\omega}$.

Angular Momentum as a Vector

Since the vector $\vec{\omega}$ points in the direction we want the angular momentum vector $\vec{L}$ to point, we can obtain a vector formula for $\vec{L}$ by simply replacing ω by $\vec{\omega}$ in Equation 16 for the angular momentum of the bicycle wheel

$$\vec{L} = (Mr^2)\vec{\omega}$$
vector formula for the angular momentum of a bicycle wheel (17)

ANGULAR MASS OR MOMENT OF INERTIA

Equation 17 expresses the angular momentum $\vec{L}$ of a bicycle wheel as a numerical quantity (Mr^2) times the vector angular velocity $\vec{\omega}$. This is not very different from linear momentum $\vec{p}$ which is the mass (M) times the linear velocity vector $\vec{v}$

$$\vec{p} = M\vec{v} \qquad (18)$$

We obtain an analogy between linear and angular momentum if we call the quantity (Mr^2) the ***angular mass*** of the bicycle wheel. Designating the angular mass by the letter I, we get

$$\vec{L} = I\vec{\omega} \qquad (19)$$

$$\boxed{I = Mr^2} \qquad \begin{array}{l}\textit{angular mass}\\ \textit{(moment of inertia)}\\ \textit{of a bicycle wheel}\end{array} \qquad (20)$$

The quantity I is usually called ***moment of inertia*** rather than angular mass, but angular mass provides a better description of what we are dealing with. We will use either name, depending upon which seems more appropriate.

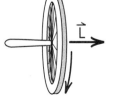

Figure 8a
The angular momentum vector.

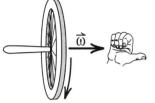

Figure 8b
The angular velocity vector.

Calculating Moments of Inertia

Equation 20 is not the most general formula for calculating moments of inertia. The bicycle wheel is special in that all the mass is essentially out at a single radius r. If, instead, we had a solid wheel where the mass was spread out over different radii, we would have to conceptually break the wheel into a number of separate rim-like wheels of radii r_i and mass m_i, calculate the moment of inertia of each rim, and add the results together to get the total moment of inertia.

In Appendix A we have relatively complete discussion of how to calculate moments of inertia, and how moment of inertia is related to rotational kinetic energy. There you will see that rotational kinetic energy is $1/2\, I\omega^2$, which is analogous to the linear kinetic energy $1/2\, Mv^2$. This material is placed in an appendix, not because it is difficult, but because we do not wish to digress from our discussion of the analogy between linear and angular momentum. At this point, one example and one exercise should be a sufficient introduction to the concept of moment of inertia.

Example 1

Calculate the moment of inertia, about its axis, of a cylinder of mass M and outside radius R. Assume that the cylinder has uniform density.

Solution: We conceptually break the cylinder into a series of concentric cylinders of radius r and thickness dr as shown in Figure (9). Each hollow cylinder has a mass given by

$$dm = M \times \frac{\text{end area of hollow cylinder}}{\text{total end area}}$$

$$= M \times \frac{2\pi r\, dr}{\pi R^2} = M \times \frac{2r\, dr}{R^2} \quad (21)$$

Since all the mass in the hollow cylinder is out at a radius r, just as it is for a bicycle wheel, the hollow cylinder has a moment of inertia dI given by

$$dI = (dm) \times r^2$$

$$= M \times \frac{2r\, dr}{R^2} \times r^2 = \frac{2Mr^3 dr}{R^2} \quad (22)$$

The total moment of inertia of the cylinder is the sum of the moments of inertia of all the hollow cylinders. This addition is done by integrating the formula for dI from r = 0 out to r = R.

$$I\binom{\text{solid}}{\text{cylinder}} = \int_{r=0}^{r=R} dI$$

$$= \int_{r=0}^{r=R} \frac{2Mr^3 dr}{R^2}$$

$$= \frac{2M}{R^2} \int_{r=0}^{r=R} r^3 dr$$

$$= \frac{2M}{R^2} \left.\frac{r^4}{4}\right|_0^R = \frac{2MR^4}{4R^2}$$

$$\boxed{I\binom{\text{solid}}{\text{cylinder}} = \frac{1}{2} MR^2} \quad (23)$$

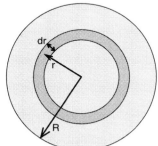

Figure 9
Calculating the moment of inertia of a cylinder about its axis of rotation.

Two points are made in Example 1, The first is that calculating the moment of inertia of an object usually requires an integration, because different parts of the object are out at different distances r from the axis of rotation. Secondly we see that the moment of inertia of a solid cylinder is less than the moment of inertia of a bicycle wheel of the same mass and outer radius ($1/2\,MR^2$ for the cylinder versus MR^2 for the bicycle wheel). This is because all the mass of the bicycle wheel is out at the maximum radius R, while most of the mass of the solid cylinder is in at smaller radii.

A considerable amount of time can be spent discussing the calculation of moments of inertia of various shaped objects. Rather than do that here, we will simply present a table of the moments of inertia of common objects of mass M and outer radius R, about an axis that passes through the center.

Object	Moment of Inertia
cylindrical shell	$1\,MR^2$
solid cylinder	$1/2\,MR^2$
spherical shell	$2/3\,MR^2$
solid sphere	$2/5\,MR^2$

Exercise 4

As shown in Figure (10) we have a thick-walled hollow brass cylinder of mass M, with an inner radius R_i and outer radius R_o. Calculate its moment of inertia about its axis of symmetry. Check your answer for the case $R_i = 0$ (a solid cylinder) and for $R_i = R_o$ (which corresponds to the bicycle wheel).

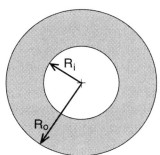

Figure 10
Thick-walled hollow cylinder.

VECTOR CROSS PRODUCT

The idea of having the angular velocity $\vec{\omega}$ being a vector pointing along the axis of rotation gave us a nice analogy between linear momentum $\vec{p} = M\vec{v}$ and angular momentum $\vec{L} = I\vec{\omega}$. But to obtain the dynamical equation for angular momentum, the one analogous to Newton's second law for linear momentum, we need the mathematical formalism of the vector cross product defined back in Chapter 2. Since we have not used the vector cross product before now, we will briefly review the topic here.

If we have two vectors $\vec{A}$ and $\vec{B}$ like those shown in Figure (11), the vector cross product $\vec{A} \times \vec{B}$ is defined to have a magnitude

$$\left|\vec{A} \times \vec{B}\right| = \left|\vec{A}\right|\left|\vec{B}\right|\sin\theta \tag{24}$$

where $\left|\vec{A}\right|$ and $\left|\vec{B}\right|$ are the magnitudes of the vectors $\vec{A}$ and $\vec{B}$, and θ is the small angle between them. Note that when the vectors are parallel, $\sin\theta = 0$ and the cross product is zero. The cross product is a maximum when the vectors are perpendicular. This is just the opposite from the scalar dot product which is a maximum when the vectors are parallel and zero when perpendicular. Conceptually you can think of the dot product as measuring parallelism while the cross product measures perpendicularity.

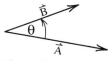

Figure 11
The vectors $\vec{A}$ and $\vec{B}$.

The other major difference between the dot and cross product is that with the dot product we end up with a number (a scalar), while with the cross product, we end up with a vector. The direction of $\vec{A} \times \vec{B}$ is the most peculiar feature of the cross product; it is ***perpendicular*** to the plane defined by the vectors $\vec{A}$ and $\vec{B}$. If we draw $\vec{A}$ and $\vec{B}$ on a sheet of paper as we did in Figure (11), then the directions perpendicular to both $\vec{A}$ and $\vec{B}$ are either up out of the paper or down into the paper. To decide which of these two directions to choose, we use the following right hand rule. (This is an arbitrary convention, but if you use it consistently in all of your calculations, everything works out OK).

Right Hand Rule for Cross Products

To find the direction of the vector $\vec{A} \times \vec{B}$, point the fingers of your right hand in the direction of the first vector in the product (namely $\vec{A}$). Then, without breaking your knuckles, curl the finger of your right hand toward the second vector $\vec{B}$. Curl them in the direction of the small angle θ. If you do this correctly, the thumb of your right hand will point in the direction of the cross product $\vec{A} \times \vec{B}$. Applying this to the vectors in Figure (11), we find that the vector $\vec{A} \times \vec{B}$ points up out of the paper as shown in Figure (12).

Exercise 5

(a) Follow the steps we just mentioned to show that $\vec{A} \times \vec{B}$ from Figure (11) does point up out of the paper.

(b) Show that the vector $\vec{B} \times \vec{A}$ points down into the paper.

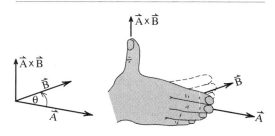

Figure 12
Right hand rule for vector cross product $\vec{A} \times \vec{B}$. Point the fingers of your right hand in the direction of the first vector $\vec{A}$ and then curl them in the direction of the second vector $\vec{B}$ (without breaking your knuckles). Your thumb will then point in the direction of the cross product $\vec{A} \times \vec{B}$.

If you did the exercise (5b) correctly, you found that $\vec{B} \times \vec{A}$ points in the opposite direction from the vector $\vec{A} \times \vec{B}$. In all previous examples of multiplication you have likely to have encountered, the order in which you did the multiplication made no difference. For example, both 3 x 5 and 5 x 3 give the same answer 15. But now we find that $\vec{A} \times \vec{B} = -\vec{B} \times \vec{A}$ and the order of the multiplication does make a difference. Mathematicians say that cross product multiplication ***does not commute***.

There is one other special feature of the cross product worth noting. If $\vec{A}$ and $\vec{B}$ are parallel, or anti parallel, then they do not define a unique plane and there is no unique direction perpendicular to both of them. Various possibilities are indicated in Figure (13). But when the vectors are parallel or anti parallel, $\sin \theta = 0$ and the cross product is zero. The special case where the cross product does not have a unique direction is when the cross product has zero magnitude with the result that the lack of uniqueness does not cause a problem.

Figure 13
If the vectors $\vec{A}$ and $\vec{B}$ are either parallel or antiparallel, then as shown above, there is a whole plane of vectors perpendicular to both $\vec{A}$ and $\vec{B}$.

CROSS PRODUCT DEFINITION OF ANGULAR MOMENTUM

Let us now see how we can use the idea of a vector cross product to obtain a definition of angular momentum vectors. To explain the bicycle wheel experiments, we wanted the angular momentum to point along the axis of the wheel as shown in Figure (14a). Since there are two directions along the axis, we have arbitrarily chosen the direction defined by the right hand convention shown. (Curl the fingers of your right hand in the direction of the rotation and your thumb will point in the direction of $\vec{L}$).

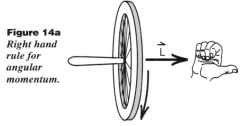

Figure 14a
Right hand rule for angular momentum.

In Figure (14b) we went to the masses and spoke model of the bicycle wheel, and selected one particular mass which we called m_i. This mass is located at a coordinate vector $\vec{r}_i$ from the center of the wheel, and is traveling with a velocity $\vec{v}_i$. According to our definition of angular momentum in Chapter 7, using the formula $\ell = mvr$, the ball's angular momentum should be

$$\ell = m_i r_i v_i \qquad \text{(7-11 again)}$$

What we want to do now is to turn this definition of angular momentum into a vector that points down the axis of the wheel. This we can do with the vector cross product of $\vec{r}_i$ and $\vec{v}_i$. We will try the definition of the vector $\vec{\ell}_i$ as

$$\vec{\ell}_i = m_i \left(\vec{r}_i \times \vec{v}_i \right) \qquad (25)$$

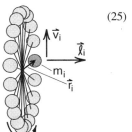

Figure 14b
Angular momentum of one of the balls in the ball-spoke model of a bicycle wheel.

Exercise 6

a) Look at Figure (14c) showing the vectors $\vec{r}_i$ (which point into the paper) and $\vec{v}_i$. Point the fingers of your right hand in the direction of $\vec{r}_i$ and then curl them toward the vector $\vec{v}_i$. Does your thumb point in the direction of the vector $\vec{\ell}_i$ shown? (If it does not, you have peculiar knuckle joints or are not following instructions).

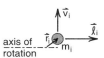

Figure 14c
The three vectors $\vec{r}_i$, $\vec{v}_i$ and $\vec{\ell}_i$

b) Choose any other mass that forms the bicycle wheel shown in Figure (14b). Call that the mass m_j. Show that the vector $\vec{\ell}_j = m_j \left(\vec{r}_j \times \vec{v}_j \right)$ also points down the axis, parallel to $\vec{\ell}_i$. Try this for several different masses, say one at the top, one at the front, and one at the bottom of the wheel.

If you did Exercise 6 correctly, you found that all the angular momentum vectors $\vec{\ell}_i = m_i \left(\vec{r}_i \times \vec{v}_i \right)$ were parallel to each other, all pointing down the axis of the wheel. We will define the total angular momentum of the wheel as the vector sum of the individual angular momentum vectors $\vec{\ell}_i$

$$\vec{L} \begin{pmatrix} \text{total angular} \\ \text{momentum} \\ \text{of wheel} \end{pmatrix} = \sum_i \vec{\ell}_i = \sum_i m_i \vec{r}_i \times \vec{v}_i \qquad (26)$$

It is easy to add the vectors $\vec{\ell}_i$ because they all point in the same direction, as shown in Figure (15). Thus we can add their magnitudes numerically. (It is just the numerical sum we did back in Equation 15).

Figure 15
Since all the angular momentum vectors $\vec{\ell}_i$ point in the same direction, we can add them up numerically.

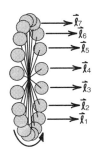

12-12   Rotational Motion

To do the sum starting from Equation (26) we note that for each mass m_i, the vectors $\vec{r}_i$ and $\vec{v}_i$ are perpendicular, thus

$$|\vec{r}_i \times \vec{v}_i| = r v \sin \theta$$
$$= r v \quad (\text{for } \theta = 90°)$$

Then note that for a rotating wheel, the speed v of the rim is related to the angular velocity ω by

$$v = r\omega \qquad \text{(11 repeated)}$$

so that

$$|\vec{r}_i \times \vec{v}_i| = rv = r(r\omega) = r^2\omega \qquad (27)$$

Finally note that the vector $\vec{\omega}$ points in the same direction as $\vec{r}_i \times \vec{v}_i$, so that Equation 27 can be written as the vector equation

$$\vec{r}_i \times \vec{v}_i = r^2\vec{\omega} \quad \begin{array}{l}\textit{for all}\\ \textit{mass } m_i\end{array} \qquad (28)$$

Using Equation 28 in Equation 26 gives

$$\vec{L} = \sum_i \vec{\ell}_i = \sum_i m_i \vec{r}_i \times \vec{v}_i$$
$$= \left(\sum_i m_i\right) r^2 \vec{\omega}$$
$$= Mr^2\vec{\omega}$$

$$\boxed{\vec{L} = Mr^2\vec{\omega}} \quad \begin{array}{l}\textit{angular momentum}\\ \textit{of a rotating}\\ \textit{bicycle wheel}\end{array} \qquad (29)$$

where M is the sum of the individual masses m_i. Equation 29 is the desired vector version of our original Equation 16.

The important point to get from the above discussion is that by using the vector cross product definition of angular momentum $\vec{\ell}_i = m_i \vec{r}_i \times \vec{v}_i$, all the $\vec{\ell}_i$ for each mass in the wheel pointed down the axis of the wheel, and we could thus calculate the total angular momentum by numerically adding up the individual $\vec{\ell}_i$.

The $\vec{r} \times \vec{p}$ Definition of Angular Momentum

A slight rewriting of our definition of angular momentum, Equation 25, gives us a more compact, easily remembered result. Noting that the linear momentum $\vec{p}$ of a particle is $\vec{p} = m\vec{v}$, then a particle's angular momentum $\vec{\ell}$ can be written

$$\vec{\ell} = m\vec{r} \times \vec{v}$$
$$= \vec{r} \times (m\vec{v})$$

$$\boxed{\vec{\ell} = \vec{r} \times \vec{p}} \qquad (30)$$

In Chapter 7, we saw that the magnitude of the angular momentum ℓ if a particle was given by the formula

$$\ell = r_\perp p \qquad (7\text{-}15)$$

where $r_\perp$ was the lever arm or perpendicular distance from the path of the particle to the point O about which we were measuring the angular momentum. This was illustrated in Figure (7-10) (reproduced here), where a ball of momentum $\vec{p}$, passing by an axis O, is caught by a hook and starts rotating in a circle.

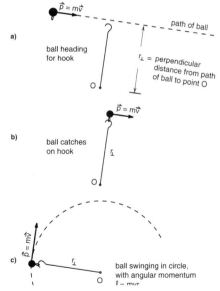

Figure 7-10
As the ball is caught by the hook, its angular momentum, about the point O, remains unchanged. It is equal to $(r_\perp p)$.

After the ball is caught it is traveling in a circle with an angular momentum $\ell = r_\perp mv = r_\perp p$. By defining the angular momentum as $r_\perp p$ even before the ball was caught, we could say that the ball had the same angular momentum $r_\perp p$ before it was caught by the hook as it did afterward; that the angular momentum was unchanged when the ball was grabbed by the hook.

The idea that the angular momentum is the linear momentum times the perpendicular lever arm $r_\perp$ follows automatically from the cross product definition of angular momentum $\vec{\ell} = \vec{r} \times \vec{p}$. To see this, consider a ball with momentum $\vec{p}$ moving past an axis O as shown in Figure (16a). At the instant of time shown, the ball is located at a coordinate vector $\vec{r}$ from the axis. The angle between the vectors $\vec{r}$ and $\vec{p}$ is the angle θ shown in Figure (16b). The vector cross product $\vec{r} \times \vec{p}$ is given

$$\left|\vec{\ell}\right| = \left|\vec{r} \times \vec{p}\right| = rp \sin\theta \tag{31}$$

However we note that the lever arm or perpendicular distance $r_\perp$ is given from Figure (16a)

$$r_\perp = r \sin\theta \tag{32}$$

Combining Equations 31 and 32 gives

$$\left|\vec{\ell}\right| = \left|\vec{r} \times \vec{p}\right| = (r \sin\theta)p = r_\perp p \tag{33}$$

which is the result we used back in Chapter 7.

Figure 16a
The coordinate vector $\vec{r}$ and the lever arm $r_\perp$ are related by $r_\perp = r \sin\theta$.

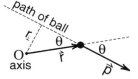

Figure 16b
The angle between $\vec{r}$ and $\vec{p}$ is θ.

Exercise 7

Using the vectors $\vec{r}$ and $\vec{p}$ in Figure (16), does the vector $\vec{\ell} = \vec{r} \times \vec{p}$ point up out of the paper or down into the paper?

The intuitive point you should get from this discussion is that the magnitude of the vector cross product $\vec{r} \times \vec{p}$ is equal to the magnitude of p times the perpendicular lever arm $r_\perp$. We will shortly encounter the cross product $\vec{r} \times \vec{F}$ where $\vec{F}$ is a force vector. We will immediately know that the magnitude of $\vec{r} \times \vec{F}$ is $r_\perp F$ where again $r_\perp$ is a perpendicular lever arm.

ANGULAR ANALOGY TO NEWTON'S SECOND LAW

We now have the mathematical machinery we need to formulate a complete angular analogy to Newton's second law. We do this by noting that to go from linear momentum $\vec{p}$ to angular momentum $\vec{\ell}$, we took the cross product with the coordinate vector $\vec{r}$

$$\vec{\ell} = \vec{r} \times \vec{p} \qquad (30\ \text{repeated})$$

The origin of the coordinate vector $\vec{r}$ is the point about which we wish to calculate the angular momentum.

To obtain a dynamical equation for angular momentum $\vec{\ell}$, we start with Newton's second law which is a dynamical equation for linear momentum $\vec{p}$

$$\vec{F} = \frac{d\vec{p}}{dt} \qquad (11\text{-}16)$$

where $\vec{F}$ is the vector sum of the forces acting on the particle.

With one mathematical trick, we can reexpress Newton's second law in terms of angular momentum. The mathematical trick involves evaluating the expression

$$\frac{d}{dt}(\vec{r} \times \vec{p}) \qquad (34)$$

In the ordinary differentiation of the product of two functions a(t) and b(t), we would have

$$\frac{d}{dt}(ab) = \left(\frac{da}{dt}\right)b + a\left(\frac{db}{dt}\right) \qquad (35)$$

The same rules apply if we differentiate a vector cross product. Thus

$$\frac{d}{dt}(\vec{r} \times \vec{p}) = \left(\frac{d\vec{r}}{dt}\right) \times \vec{p} + \vec{r} \times \left(\frac{d\vec{p}}{dt}\right) \qquad (36)$$

Equation 36 can be simplified by noting that

$$\vec{v} = \frac{d\vec{r}}{dt}$$

so that

$$\left(\frac{d\vec{r}}{dt}\right) \times \vec{p} = \vec{v} \times \vec{p} = \vec{v} \times (m\vec{v}) = 0 \qquad (37)$$

This product is zero because the vectors $\vec{v}$ and $\vec{p} = m\vec{v}$ are parallel to each other, and the cross product of parallel vectors is zero. Thus Equation 36 becomes

$$\frac{d}{dt}(\vec{r} \times \vec{p}) = \vec{r} \times \frac{d\vec{p}}{dt} \qquad (38)$$

With this result, let us return to Newton's law for linear momentum

$$\vec{F} = \frac{d\vec{p}}{dt} \qquad (39)$$

As long as we do the same thing to both sides of an equation, it is still a correct equation. Taking the vector cross product $\vec{r} \times$ on both sides gives

$$\vec{r} \times \vec{F} = \vec{r} \times \frac{d\vec{p}}{dt} \qquad (40)$$

Using Equation 38 in Equation 40 gives

$$\vec{r} \times \vec{F} = \frac{d}{dt}(\vec{r} \times \vec{p}) \qquad (41)$$

Finally note that $\vec{r} \times \vec{p}$ is the particle's angular momentum $\vec{\ell}$, thus

$$\boxed{\vec{r} \times \vec{F} = \frac{d\vec{\ell}}{dt}} \qquad (42)$$

Equation (39) told us that the net linear force is equal to the time rate of change of linear momentum. Equation 42 tells us that something, $\vec{r} \times \vec{F}$, is equal to the time rate of change of angular momentum. What should we call this quantity $\vec{r} \times \vec{F}$? The obvious name, from an angular analogy would be an ***angular force***. Then we could say that the angular force is the time rate of change of angular momentum, just as the linear force is the time rate of change of linear momentum.

The world does not use the name angular force for $\vec{r} \times \vec{F}$. Instead it uses the name ***torque***, and usually designates it by the Greek letter τ ("tau")

$$\boxed{\text{torque}\ \vec{\tau} \equiv \vec{r} \times \vec{F}} \qquad \begin{array}{l}\textit{definition}\\ \textit{of torque}\end{array} \qquad (43)$$

With this naming, the angular analogy to Newton's second law is

$$\boxed{\vec{\tau} = \frac{d\vec{\ell}}{dt}} \qquad \begin{array}{l}\textit{torque = rate}\\ \textit{of change of}\\ \textit{angular momentum}\end{array} \qquad (44)$$

ABOUT TORQUE

To gain an intuitive picture of the concept of torque $\vec{\tau} = \vec{r} \times \vec{F}$, imagine that we have a bicycle wheel with a fixed axis, and push on the rim of the wheel with a force $\vec{F}$ as shown in Figure (17). In (17a) the force $\vec{F}$ is directed through the axis of the wheel, in this case the force has no lever arm $r_\perp$. In (17b), the force is applied above the axis, while in (17c) the force is applied below the axis.

Intuitively, you can see that the wheel will not start turning if you push right toward the axis. When you push above the axis as in (17b), the wheel will start to rotate counter clockwise. By our right hand convention this corresponds to an angular momentum directed up out of the paper.

In (17c), where we push below the axis, the wheel will start to rotate clockwise, giving it an angular momentum directed down into the paper.

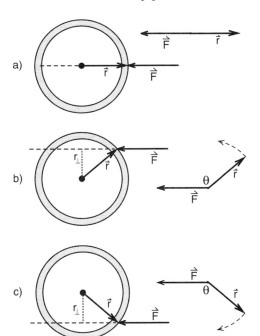

Figure 17
Both a force $\vec{F}$ and a lever arm $r_\perp$ are needed to turn the bicycle wheel. The product $r_\perp F$ is the magnitude of the torque τ acting on the wheel.

Exercise 8
In Figure (17) we have separately drawn the vectors $\vec{F}$ and $\vec{r}$ for each diagram. Using the right hand rule for cross products, find the direction of $\vec{\tau} = \vec{r} \times \vec{F}$ for each of these three diagrams.

If you did Exercise 8 correctly, you found that $\vec{r} \times \vec{F} = 0$ for Figure (17a), that $\vec{\tau} = \vec{r} \times \vec{F}$ pointed up out of the paper in (17b), and down into the paper in (17c). Thus we find that when we apply a zero torque as in (17a), we get zero change in angular momentum. In (17b) we applied an upward directed torque, and saw that the wheel would start to turn to produce an upward directed angular momentum. In (17c), the downward directed torque produces a downward directed angular momentum. These are all results we would expect from the equation $\vec{\tau} = d\vec{\ell}/dt$.

In our discussion of angular momentum, we saw that $\vec{\ell} = \vec{r} \times \vec{p}$ had a magnitude $|\vec{\ell}| = r_\perp p$ where $r_\perp$ was the perpendicular lever arm. A similar result applies to torque. By the same mathematics we find that the magnitude of the torque $\vec{\tau}$ produced by a force $\vec{F}$ is

$$|\vec{\tau}| = r_\perp F \tag{45}$$

where $r_\perp$ is the perpendicular lever arm seen in Figures (17b,c).

Intuitively, the best way to remember torque is to think of it as a force times a lever arm. To turn an object, you need both a force and a lever arm. In Figure (17a), we had a force but no lever arm. The line of action of the force went directly through the axis, with the result that the wheel did not start turning. In both cases (17b) and (17c), there was both a force and a lever arm $r_\perp$, and the wheel started turning.

To get the direction of the torque, to determine whether $\vec{\tau}$ points up or down (and thus gives rise to an up or down angular momentum), use the right hand rule applied to the vector cross product $\vec{\tau} = \vec{r} \times \vec{F}$. A convention, which we will use in the next chapter on *Equilibrium*, is to say that a torque that points up out of the paper is a positive torque, while a torque pointing down into the paper is a negative one. With this convention, we see that the force in Figure (17b) is exerting a positive torque (and creating positive angular momentum), while the force in Figure (17c) is producing a negative torque (and creating negative angular momentum).

CONSERVATION OF ANGULAR MOMENTUM

In our discussion of a system of particles in Chapter 11, we saw that if we had a system of many interacting particles, with internal forces $\vec{F}_{i\,internal}$ between the particles, as well as various external forces $\vec{F}_{i\,external}$, we obtained the equation

$$\vec{F}_{external} = \frac{d\vec{P}}{dt} \quad (11\text{-}12)$$

where $\vec{F}_{external}$ is the vector sum of all the external forces acting on the system, and $\vec{P}$ is the vector sum of all the momenta $\vec{p}_i$ of the individual particles. This result was obtained using Newton's third law and noting that all the internal forces cancel in pairs. In the case where there is no net external force acting on the system, then $d\vec{P}/dt = 0$ and the total linear momentum of the system is conserved.

We can obtain a similar result for angular momentum by starting with the definition of the total angular momentum $\vec{L}$ of a system as being the vector sum of the angular momentum of the individual particles $\vec{\ell}_i$

$$\boxed{\vec{L} \equiv \sum_i \vec{\ell}_i} \quad \begin{array}{l}\textit{definition of the}\\ \textit{total angular momenta}\\ \textit{of a system of particles}\end{array} \quad (46)$$

Differentiating Equation (46) with respect to time gives

$$\frac{d\vec{L}}{dt} = \sum_i \frac{d\vec{\ell}_i}{dt} \quad (47)$$

For an individual particle i, we have

$$\frac{d\vec{\ell}_i}{dt} = \vec{\tau}_i = \vec{r}_i \times \vec{F}_i \quad \begin{array}{l}\textit{Equation 44}\\ \textit{applied to}\\ \textit{particle i}\end{array} \quad (48)$$

where $\vec{F}_i$ is the vector sum of the forces acting on the particle i. As shown in Figure (18), we can take $\vec{r}_i$ to be the coordinate vector of the i-th particle. For this discussion, we can locate the origin of the coordinate system anywhere we want.

Substituting Equation (48) into Equation (47) gives

$$\frac{d\vec{L}}{dt} = \sum_i \frac{d\vec{\ell}_i}{dt} = \sum_i \vec{r}_i \times \vec{F}_i$$

Now break the net force $\vec{F}_i$ into the sum of the external forces $\vec{F}_{i\,external}$ and the sum of the internal forces $\vec{F}_{i\,internal}$. This gives

$$\begin{aligned}\frac{d\vec{L}}{dt} &= \sum_i \vec{r}_i \times \vec{F}_{i\,external} + \sum_i \vec{r}_i \times \vec{F}_{i\,internal} \\ &= \sum_i \vec{\tau}_{i\,external} + \sum \vec{\tau}_{i\,internal}\end{aligned} \quad (49)$$

Next assume that all the internal forces are equal and opposite as required by Newton's third law, **and** are directed toward or away from each other. In Figure (19) we consider a pair of such internal forces and note that both coordinate vectors $\vec{r}_1$ and $\vec{r}_2$ have the same perpendicular lever arm $\vec{r}_\perp$. Thus the equal and opposite forces $\vec{F}_{1,2\,external}$ and $\vec{F}_{2,1\,internal}$ create equal and opposite torques which cancel each other in Equation (49). The result is that all torques produced by internal forces cancel in pairs, and we are left with the general result

$$\boxed{\vec{\tau}_{external} = \frac{d\vec{L}}{dt}} \quad (50)$$

where $\vec{\tau}_{external}$ is the vector sum of all the external torques acting on the system of particles, and $\vec{L}$ is the vector sum of the angular momentum of all of the particles.

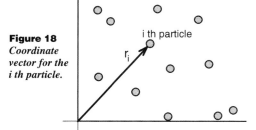

Figure 18
Coordinate vector for the i th particle.

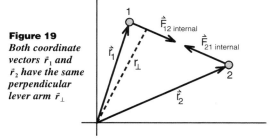

Figure 19
Both coordinate vectors $\vec{r}_1$ and $\vec{r}_2$ have the same perpendicular lever arm $\vec{r}_\perp$.

In order to define torque or angular momentum, we have to choose an axis or origin for the coordinate vectors $\vec{r}_i$. (Both torque and angular momentum involve the lever arm $\vec{r}_\perp$ about that axis.) Equation 50 is remarkably general in that it applies no matter what origin or axis we choose. In general, choosing a different axis will give us different sums of torques and a different total angular momentum, but the new torques and angular momenta will still obey Equation 50.

In some cases, there is a special axis about which there is no external torque. In the bicycle wheel demonstrations where we stood on a rotating platform, the freely rotating platform did not contribute any external torques about it own axis, which we called the z axis. As long as we did not touch another person or some furniture, then the z component of the external torques were zero. Since Equation 50 is a vector equation, that implies

$$\tau_{z\,external} = \frac{dL_z}{dt} = 0 \qquad (51)$$

and we predict that the z component of the total angular momentum (us and the bicycle wheel) should be unchanged, remain constant, no matter how we turned the bicycle wheel. This is just what we saw.

Another consequence of Equation 50 is that if we have an isolated system of particles with no net external torque acting on it, then the total angular momentum will be unchanging, will be conserved. This is one statement of the law of conservation of angular momentum. Our derivation of this result relied on the assumption of Newton's third law that all internal forces are equal and opposite and directed toward each other. Since angular momentum is conserved on an atomic, nuclear and subnuclear scale of distance, where Newtonian mechanics no longer applies, our derivation is in some sense backwards. We should start with the law of conservation of angular momentum as a fundamental law, and show for large objects which obey Newtonian mechanics, the sum of the internal torques must cancel. This is the kind of argument we applied to the conservation of linear momentum in Chapter 11 (see Equation 11-14).

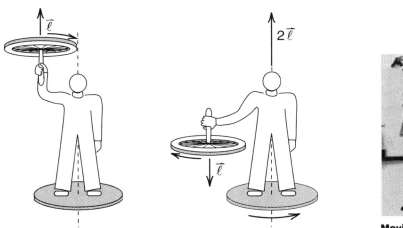

Figure 7–15 repeated
Since the platform is completely free to rotate about the z axis, there are no z directed external torques acting on the system consisting of the platform, person and bicycle wheel. As a result the z component of angular momentum is conserved when the bicycle wheel is turned over. (Note: when the wheel is being held up, we are looking at the under side.)

GYROSCOPES

The gyroscope provides an excellent demonstration of the predictive power of the equation $\vec{\tau} = d\vec{L}/dt$. Gyroscopes behave in peculiar, non intuitive ways. The fact that a relatively straightforward application of the equation $\vec{\tau} = d\vec{L}/dt$ predicts this bizarre behavior, provides a graphic demonstration of the applicability of Newton's laws from which the equation is derived.

Start-up

For this discussion, a bicycle wheel with a weighted rim will serve as our example of a gyroscope. To weight the rim, remove the tire and wrap copper wire around the rim to replace the tire. The axle needs to be extended as shown in Figure (20).

As an introduction to the gyroscope problem, start with the bicycle wheel at rest, hold the axle fixed, and apply a force $\vec{F}$ to the rim as shown in Figure (20). The force shown will cause the wheel to start spinning in a direction so that the angular momentum $\vec{L}$ points to the right as shown. (Curl the fingers of your right hand in the direction of rotation and your thumb points in the direction of $\vec{L}$.)

The force $\vec{F}$, in Figure (20), produces a torque $\vec{\tau} = \vec{r} \times \vec{F}$ that also points to the right as shown. (The right hand convention used here is to point your fingers in the direction of the first vector $\vec{r}$, curl them in the direction of the second vector $\vec{F}$, and your thumb points in the direction of the cross product $\vec{r} \times \vec{F} = \vec{\tau}$.)

When we start with the bicycle wheel at rest, and apply the right directed torque shown in Figure (20), we get a right directed angular momentum $\vec{L}$. Thus the torque $\vec{\tau}$ and the resulting angular momentum $\vec{L}$ point in the same direction. In addition, the longer we apply the torque, the faster the wheel spins, and the greater the angular momentum $\vec{L}$. Thus both the direction and magnitude of $\vec{L}$ are consistent with the equation $\vec{\tau} = d\vec{L}/dt$.

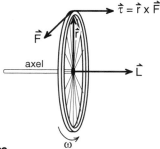

Figure 20
Spinning up the bicycle wheel. Note that the resulting angular momentun $\vec{L}$ points in the same direction as the applied torque $\vec{\tau}$.

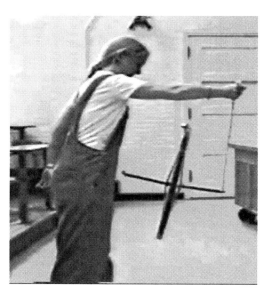

Figure 25 Movie
The gyroscope really works!

Precession

When we apply the equation $\vec{\tau} = d\vec{L}/dt$ to a gyroscope that is already spinning, and apply the torque in a direction that is not parallel to $\vec{L}$, the results are not so obvious.

Suppose we get the bicycle wheel spinning rapidly so that it has a big angular momentum vector $\vec{L}$, and then suspend the bicycle wheel by a rope attached to the end of the axle as shown in Figure (21).

To predict the motion of the spinning wheel, the first step is to analyze all the external forces acting on it. There is the gravitational force $m\vec{g}$ which points straight down, and can be considered to be acting at the center of mass of the bicycle wheel, which is the center of the wheel as shown. Then there is the force of the rope which acts along the rope as shown. No other detectable external forces are acting on our system of the spinning wheel.

One thing we know about the force $\vec{F}_{rope}$ is that it acts at the point labeled O where the rope is tied to the axle. If we take the sum of the torques acting on the bicycle wheel about the suspension point O, then $\vec{F}_{rope}$ has no lever arm about this point and therefore contributes no torque. The only torque about the suspension point O is produced by the gravitational force $m\vec{g}$ whose lever arm is $\vec{r}$, the vector going from point O down the axle to the center of the bicycle wheel as shown in Figure (21).

The formula for this gravitational torque $\vec{\tau}_g$ is

$$\tau_g = \vec{r} \times m\vec{g} \qquad \begin{array}{l}\textit{torque about point O}\\ \textit{produced by the}\\ \textit{gravitational force}\\ \textit{on the bicycle wheel}\end{array} \qquad (52)$$

The new feature of the gyroscope problem, which we have not encountered before, is that the torque $\vec{\tau}$ does not point in the same direction as the angular momentum $\vec{L}$ of the bicycle wheel. If we look at Figure (21), point the fingers of our right hand in the direction of the vector $\vec{r}$, and curl our fingers in the direction of the vector $m\vec{g}$, then our thumb points down into the paper. This is the definition of the direction of the vector cross product $\vec{r} \times m\vec{g}$. But the angular momentum $\vec{L}$ of the bicycle wheel points along the axis of the wheel to the right in the plane of the paper. In order to view both the angular momentum vector $\vec{L}$ and the torque vector $\vec{\tau}$ in the same diagram, we can look down on the bicycle wheel from the ceiling as shown in Figure (22).

When we started the wheel spinning, back in Figure (20), the torque $\vec{\tau}$ and angular momentum $\vec{L}$ pointed in the same direction, and we had the simple result that the longer we applied the torque, the more angular momentum we got. Now, with the torque and angular momentum pointing in different directions as shown in Figure (22), we expect that the torque will cause a change in the direction of the angular momentum.

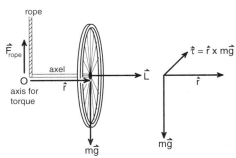

Figure 21
Suspend the spinning bicycle wheel by a rope attached to the axle. The gravitational force $m\vec{g}$ has a lever arm $\vec{r}$ about the axis O. This creates a torque $\vec{\tau} = \vec{r} \times m\vec{g}$ pointing into the paper.

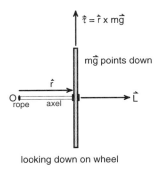

Figure 22
Looking down from the ceiling, the vector $m\vec{g}$ points down into the paper and $\vec{\tau} = \vec{r} \times m\vec{g}$ points to the top of the page. In this view we can see both the vectors $\vec{L}$ and $\vec{\tau}$.

To predict the change in $\vec{L}$, we start with the angular form of Newton's second law

$$\vec{\tau} = \frac{d\vec{L}}{dt}$$

and multiply through by the short (but finite) time interval dt to get

$$d\vec{L} = \vec{\tau}\, dt \qquad (53)$$

Equation 53 gives us $d\vec{L}$, which is the change in the bicycle wheel's angular momentum as a result of applying the torque $\vec{\tau}$ for a short time dt.

To see the effect of this change $d\vec{L}$, we will use some of the terminology we used in the computer prediction of motion. Let us call $\vec{L}_{old}$ the old value of the angular momentum that the bicycle wheel had before the time interval dt, and $\vec{L}_{new}$ the new value at the end of the time interval dt. Then $\vec{L}_{new}$ will be related to $\vec{L}_{old}$ by the equation

$$\vec{L}_{new} = \vec{L}_{old} + d\vec{L} \qquad (54)$$

Using Equation 53 for $d\vec{L}$ gives

$$\vec{L}_{new} = \vec{L}_{old} + \vec{\tau}\, dt \qquad (55)$$

A graph of the vectors $\vec{L}_{old}$, $\vec{L}_{new}$, and $\vec{\tau}\, dt$ is shown in Figure (23). In this figure the perspective is looking down on the bicycle wheel, as in Figure (22).

Since the torque $\vec{\tau}$ is in the horizontal plane, the vector $\vec{L}_{new} = \vec{L}_{old} + \vec{\tau}\, dt$ is also in the horizontal plane. And since $\vec{\tau}$ and $\vec{L}_{old}$ are perpendicular to each other, $\vec{L}_{new}$ has essentially the same length as $\vec{L}_{old}$. What is happening is that the vector $\vec{L}$ is starting to rotate counter clockwise (as seen from above) in the horizontal plane.

One final, important point. For this experiment we were careful to spin up the bicycle wheel so that before we suspended the wheel from the rope, the wheel had a big angular momentum pointing along its axis of rotation. When we apply a torque to change the direction of $\vec{L}$, the axis of the wheel and the angular momentum vector $\vec{L}$ move together. As a result the axis of the bicycle wheel also starts to rotate counter clockwise in the horizontal plane. The bicycle wheel, instead of falling as expected, starts to rotate sideways.

Once the bicycle wheel has turned an angle dθ sideways, the axis of rotation and the torque $\vec{\tau}$ also rotate by an angle dθ, so that the torque $\vec{\tau}$ is still perpendicular to $\vec{L}$ as shown in Figure (24). Since $\vec{\tau}$ always remains perpendicular to $\vec{L}$, the vector $\vec{\tau}\, dt$ cannot change the length of $\vec{L}$. Thus the angular momentum vector $\vec{L}$ remains constant in magnitude and rotates or "precesses" in the horizontal plane. This is the famous precession of a gyroscope which is nicely demonstrated using the bicycle wheel apparatus of Figure (21).

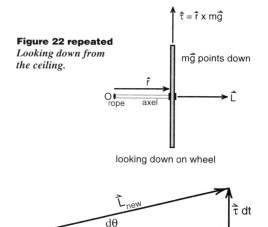

Figure 22 repeated
Looking down from the ceiling.

Figure 23
The vectors $\vec{L}_{old}$, $\vec{L}_{new}$ and $\vec{\tau}\, dt$ as seen from the top view of Figure 22.

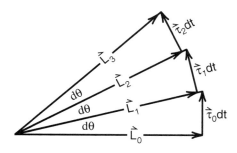

Figure 24
After each time step dt, the angular momentum vector $\vec{L}$ (and the bicycle wheel axis) rotates by another angle dθ.

To calculate the rate of precession we note from Figures (23) or (24) that the angle $d\theta$ is given by

$$d\theta = \frac{\tau dt}{L} \quad (56)$$

where we use the fact that τdt is a very short length, and thus $\sin(d\theta)$ and $d\theta$ are equivalent. Dividing both sides of Equation 56 through by dt, we get

$$\frac{d\theta}{dt} = \frac{\tau}{L} \quad (57)$$

But $d\theta/dt$ is just the angular velocity of precession, measured in radians per second. Calling this precessional velocity $\Omega_{precession}$ (Ω is just a capital ω "omega"), we get

$$\Omega_{precession} = \frac{\tau}{L} \quad \begin{array}{l}\text{precessional}\\\text{angular velocity}\\\text{of a gyroscope}\end{array} \quad (58)$$

Exercise 9

A bicycle wheel of mass m, radius r, is spun up to an angular velocity ω. It is then suspended on an axle of length h as shown in Figure (21). Calculate

(a) the angular momentum L of the bicycle wheel.

(b) the angular velocity of precession.

(c) the time it takes the wheel to precess around once (the period of precession). [You should be able to obtain the period of precession from the angular velocity of precession by dimensional analysis.]

(d) A bicycle wheel of total mass 1kg and radius 40cm, is spun up yo a frequency $f = 2\pi\omega = 10$ cycles/sec. The handle is 30cm long. What is the period of precession in seconds? Does the result depend on rhe mass of the bicycle wheel?

If you try the bicycle wheel demonstration that we discussed, the results come out close to the prediction. Instead of falling as one might expect, the wheel precesses horizontally as predicted. There is a slight drop when you let go of the wheel, which can be compensated for by releasing the wheel at a slight upward angle.

If you look at the motion of the wheel carefully, or study the motion of other gyroscopes (particularly the air bearing gyroscope often used in physics lectures) you will observe that the axis of the wheel bobs up and down slightly as it goes around. This bobbing, or epicycle like motion, is called *nutation*. We did not predict this nutation because we made the approximation that the axis of the wheel exactly follows the angular momentum vector. This approximation is very good if the gyroscope is spinning rapidly but not very good if $\vec{L}$ is small. Suppose, for example we release the wheel without spinning it. Then it simply falls. It starts to rotate, but along a different axis. As it starts to fall it gains angular momentum in the direction of $\vec{\tau}$. A more accurate analysis of the motion of the gyroscope can become fairly complex. But as long as the gyroscope is spinning fast enough so that the axis moves with $\vec{L}$, we get the simple and important results discussed above.

APPENDIX
Moment of Inertia and Rotational Kinetic Energy

In the main part of the text, we briefly discussed moment of inertia as the angular analogy to mass in the formula for angular momentum. As linear momentum $\vec{p}$ of an object is its mass m times its linear velocity $\vec{v}$

$$\vec{p} = m\vec{v} \quad \textit{linear momentum} \tag{A1}$$

the angular momentum $\vec{\ell}$ is the angular mass or moment of inertia I time the angular velocity $\vec{\omega}$

$$\vec{\ell} = I\vec{\omega} \quad \textit{angular momentum} \tag{A2}$$

In the simple case of a bicycle wheel, where all the mass is essentially out at a distance (r) from the axis of the wheel, the moment of inertia about the axis is

$$I = Mr^2 \quad \begin{array}{l}\textit{moment of inertia}\\ \textit{of a bicycle wheel}\end{array} \tag{A3}$$

where M is the mass of the wheel.

When the mass of an object is not all concentrated out at a single distance (r) from the axis, then we have to calculate the moment of inertia of individual parts of the object that are at different radii r, and tie together the various pieces to get the total moment of inertia. This usually involves an integration, like the one we did in Equations 21 through 23 to calculate the moment of inertia of a solid cylinder.

For topics to be discussed later in the text, the earlier discussion of moment of inertia is all we need. But there are topics, such as rotational kinetic energy and its connection to moment of inertia, which are both interesting, and can be easily tested in both lecture demonstrations and laboratory exercises. We will discuss these topics here.

ROTATIONAL KINETIC ENERGY

Let us go back to our example, shown in Figure (3) repeated here, of a ball of mass m, on the end of a massless stick of length r, rotating with an angular velocity ω. The speed v of the ball is given by Equation 11 as

$$v = r\omega \tag{11 repeated}$$

and the ball's kinetic energy will be

$$\begin{aligned}\text{kinetic energy} &= \tfrac{1}{2}mv^2 = \tfrac{1}{2}m(r\omega)^2 \\ &= \tfrac{1}{2}(mr^2)\omega^2\end{aligned} \tag{A4}$$

Since the ball's moment of inertia I about the axis of rotation is mr^2, we get as the formula for the ball's kinetic energy

$$\text{kinetic energy} = \tfrac{1}{2}I\omega^2 \quad \begin{array}{l}\textit{analogous}\\ \textit{to } 1/2\, mv^2\end{array} \tag{A5}$$

We see the angular analogy working again. The ball has a kinetic energy, due to its rotation, which is analogous to $1/2\, mv^2$, with the linear mass m replaced by the angular mass I and the linear velocity v replaced by the angular velocity ω.

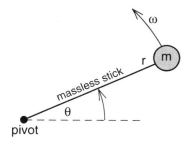

Figure 3 repeated
Mass rotating on the end of a massless stick.

If we have a bicycle wheel of mass M and radius r rotating at an angular velocity ω, we can think of the wheel as being made up of a collection of masses on the ends of rods as shown in Figure (6) repeated here. For each individual mass m_i, the kinetic energy is $1/2\, m_i v^2$ where $v = r\omega$ is the same for all the masses. Thus the total kinetic energy is

$$\text{kinetic energy of bicycle wheel} = \sum_i \tfrac{1}{2} m_i r^2 \omega^2$$

$$= \tfrac{1}{2} r^2 \omega^2 \sum_i m_i$$

$$= \tfrac{1}{2} r^2 \omega^2 M$$

where the sum of the masses $\sum m_i$ is just the mass M of the wheel. The result can now be written

$$\text{kinetic energy of bicycle wheel} = \tfrac{1}{2}(Mr^2)\omega^2$$

$$= \tfrac{1}{2} I \omega^2 \qquad \text{(A6)}$$

If we call Mr^2 the angular mass, or moment of inertia I of the bicycle wheel, we again get the formula $1/2\, I\omega^2$ for kinetic energy of the wheel. Thus we see that, in calculating this angular mass or moment of inertia, it does not make any difference whether the mass is concentrated at one point as in Figure (3), or spread out as in Figure (6). The only criterion is that the mass or masses all be out at the same distance r from the axis of rotation.

In most of our examples we will consider objects like bicycle wheels or hollow cylinders where the mass is essentially all at a distance r from the axis of rotation, and we can use the formula Mr^2 for the moment of inertia. But often the mass is spread out over different radii and we have to calculate the angular mass. An example is a rotating shaft shown back in Figure (9), where the mass extends from the center where $r = 0$ out to the outside radius $r = R$.

Suppose we have an arbitrarily shaped object rotating an angular velocity ω about some axis, as shown in Figure (A1). To find the moment of inertia, we will calculate the kinetic energy of rotation and equate that to $1/2 I\omega^2$ to obtain the formula for I. To do this we conceptually break the object into many small masses dm_i located a distance r_i from the axis of rotation as shown. Each dm_i will have a speed $v_i = r_i \omega$, and thus a kinetic energy

$$\text{kinetic energy of object} = \sum_i \left(\tfrac{1}{2} m_i v_i^2\right)$$

$$= \sum_i \left(\tfrac{1}{2} m_i r_i^2 \omega^2\right)$$

$$= \tfrac{1}{2} \omega^2 \sum_i \left(m_i r_i^2\right)$$

$$= \tfrac{1}{2} \omega^2 I \qquad \text{(A7)}$$

From Equation A7, we see that the general formula for moment of inertia is

$$\boxed{I = \sum_i m_i r_i^2} \qquad \text{(A8)}$$

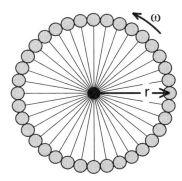

Figure 6 repeated
Bicycle wheel as a collection of masses on the end of massless rods.

Figure A1
Calculating the moment of inertia of an object about the axis of rotation.

In example 1, Equations 21 through 23, we showed you how to calculate the moment of inertia of a solid cylinder about its axis of symmetry. In that example we broke the cylinder up into a series of concentric shells of radius r_i and mass dm_i, calculated the moment of inertia of each shell $(dm_i \, r_i^2)$, and summed the results as required by Equation A7. As in most cases where we calculate a moment of inertia, the sum is turned into an integral.

In Exercise 3 which followed Example 1, we had you calculate the moment of inertia, about its axis of symmetry, of a hollow thick-walled cylinder. The calculation was essentially the same as the one we did in Example 1, except that you had to change the limits of integration. The following exercise gives you more practice calculating moments of inertia, and shows you what happens when you change the axis about which the moment of inertia is calculated.

Exercise A1
Consider a uniform rod of mass M and length L as shown in Figure (A2).

a) Calculate the moment of inertia of the rod about the center axis, labeled axis 1 in Figure (A2).

b) Calculate the moment of inertia of the rod about an axis that goes through the end of the rod, axis 2 in Figure (A2). About which axis is the moment of inertia greater? Explain why.

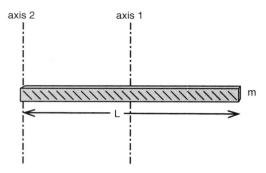

Figure A2
Calculating the moment of inertia of a long thin rod.

COMBINED TRANSLATION AND ROTATION

In our discussion of the motion of a system of particles, we saw that the motion was much easier to understand if we focused our attention on the motion of the center of mass of the system. The simple feature of the motion of the center of mass, was that the effects of all internal forces cancelled. The center of mass moved as if it were a *point particle of mass M*, equal to the total mass of the system, subject to a force $\vec{F}$ equal to the vector sum of all the *external forces* acting on the object.

When the system is a rigid object, we have a further simplification. The motion can then be described as the motion of the center of mass, plus rotation about the center of mass. To see that you can do this, imagine that you go to a coordinate system that moves with the object's center of mass. In that coordinate system, the object's center of mass point is at rest, and the only thing a rigid solid object can do is rotate about that point.

A key advantage of viewing the motion of a rigid object this way is that the kinetic energy of a moving, rotating, solid object is simply the kinetic energy of the center of mass motion plus the kinetic energy of rotation. Explicitly, if an object has a total mass M, and a moment of inertia I_{com} about the center of mass (parallel to the axis of rotation of the object) then the formula for the kinetic energy of the object is

$$\begin{matrix}\text{kinetic energy} \\ \text{of moving and} \\ \text{rotating object}\end{matrix} = \tfrac{1}{2}MV_{com}^2 + \tfrac{1}{2}I_{com}\,\omega^2 \quad (A9)$$

where V_{com} is the velocity of the center of mass and ω the angular velocity of rotation about the center of mass.

More important is the idea that motion can be separated into the motion of the center of mass plus rotation about the center of mass. To emphasize the usefulness of this concept, we will first consider an example that can easily be studied in the laboratory or at home, and then go through the proof of the equation.

Example—Objects Rolling Down an Inclined Plane

Suppose we start with a cylindrical object at the top of an inclined plane as shown in Figure (A3), and measure the time the cylinder takes to roll down the plane. Since we do not have to worry about friction for a rolling object, we can use conservation of energy to analyze the motion.

If the cylinder rolls down so that its height decreases by h as shown, then the loss of gravitational potential energy is mgh. Equating this to the kinetic energy gained gives

$$mgh = \frac{1}{2} m v_{com}^2 + \frac{1}{2} I \omega^2 \qquad (A10)$$

where m is the mass of the cylinder, v_{com} the speed of the axis of the cylinder, I the moment of inertia about the axis and ω the angular velocity.

If the cylinder rolls without slipping, there is a simple relationship between v_{com} and ω. We are picturing the rolling cylinder as having two kinds of motion—translation and rotation. The velocity of any part of the cylinder is the vector sum of $\vec{v}_{com}$ plus the velocity due to rotation.

At the point where the cylinder touches the inclined plane, the rotational velocity has a magnitude $v_{rot} = \omega r$, and is directed back up the plane as shown in Figure (15). If the cylinder is rolling without slipping, the velocity of the cylinder at the point of contact must be zero, thus we have

$$\vec{v}_{com} + \vec{\omega} r = 0 \quad \text{rolling without slipping} \qquad (A11)$$

Thus we get for magnitudes

$$\omega r = v_{com} ; \quad \omega = v_{com}/r \qquad (A12)$$

Using Equation A12 in A10 gives

$$\begin{aligned} mgh &= \frac{1}{2} m v_{com}^2 + \frac{1}{2} I \frac{v_{com}^2}{r^2} \\ &= \frac{1}{2} \left(m + \frac{I}{r^2} \right) v_{com}^2 \end{aligned} \qquad (A13)$$

Let us take a look at what is happening physically as the cylinder rolls down the plane. In our earlier analysis of a block sliding without friction down the plane, all the gravitational potential energy mgh went into kinetic energy $1/2\, m v_{com}^2$. Now for a rolling object, the gravitational potential has to be shared between the kinetic energy of translation $1/2\, m v_{com}^2$ and the kinetic energy of rotation $1/2\, I \omega^2$. The greater the moment of inertia I, the more energy that goes into rotation, the less available for translation, and the slower the object rolls down the plane.

In our discussion of moments of inertia, we saw that for two cylinders of equal mass, the hollow thin-walled cylinder had twice the moment of inertia as the solid one. Thus if you roll a hollow and a solid cylinder down the plane, the solid cylinder will travel faster because less gravitational potential energy goes into the kinetic energy of rotation. You get to figure out how much faster in Exercise A2.

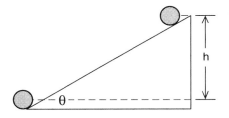

Figure A3
Calculating the speed of an object rolling down a plane.

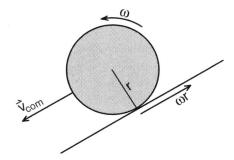

Figure A4
The velocity at the point of contact is the sum of the center of mass velocity and the rotational velocity. This sum must be zero if there is no slipping.

Before you work Exercise A2, think about this question. The technician who sets up our lecture demonstrations has a metal sphere, and does not know for sure whether the sphere is solid or hollow. (It could be a solid sphere made of a light metal, or a hollow sphere made from a more dense metal.) How could you find out if the sphere is solid or hollow?

Exercise A2

You roll various objects down the inclined plane shown in Figure (A3).

(a) a thin walled hollow cylinder

(b) a solid cylinder

(c) a thin walled sphere

(d) a solid sphere

and for comparison, you also slide a frictionless block down the plane:

(e) a frictionless block

For each of these, calculate the speed v_{com} after the object has descended a distance h. (It is easy to do all cases of this problem by writing the object's moment of inertia in the form $I = \alpha MR^2$, where $\alpha = 1$ for the hollow cylinder, 1/2 for the solid cylinder, etc.) What value of α should you use for the sliding block?

Writing your results in the form $\vec{v}_{com} = \beta\sqrt{2gh}$ summarize your results in a table giving the value of β in each case. ($\beta = 1$ for the sliding block, and is less than 1 for all other examples.)

Exercise A3   A Potential Lab Experiment

In Exercise A2 you calculated the speed v_{com} of various objects after they had descended a distance h. A block sliding without friction has a speed v given by $mgh = 1/2\,mv^2$, or $v = \sqrt{2gh}$. The rolling objects were moving slower when they got to the bottom. For all heights, however, the speed of a rolling object is slower than the speed of the sliding block by the same constant factor. Thus the rolling objects moved down the plane with constant acceleration, but less acceleration than the sliding block. It is as if the acceleration due to gravity were reduced from the usual value g. Using this idea, and the results of Exercise A2, predict how long each of the rolling objects take to travel down the plane. This prediction can be tested with a stop watch.

PROOF OF THE KINETIC ENERGY THEOREM

We are now ready to prove the kinetic energy theorem for rotational motion. If we have an object that is rotating while it moves through space, its total kinetic energy is the sum of the kinetic energy of the center of mass motion plus the kinetic energy of rotational motion about the center of mass. The proof is a bit formal, but shows what you can do by working with vector equations.

Consider a solid object, shown in Figure (A5), that is moving and rotating. Let $\vec{R}_{com}$ be the coordinate vector of the center of mass of the object. We will think of the object as being composed of many small masses m_i which are located at $\vec{R}_i$ in our coordinate system, and a displacement $\vec{r}_i$ from the center of mass as shown. As we can see from Figure (A5), the vectors $\vec{R}_{com}$, $\vec{R}_i$ and $\vec{r}_i$ are related by the vector equation

$$\vec{R}_i = \vec{R}_{com} + \vec{r}_i \qquad (A14)$$

We can obtain an equation for the velocity of the small mass m_i by differentiating Equation A14 with respect to time

$$\frac{d\vec{R}_i}{dt} = \frac{d\vec{R}_{com}}{dt} + \frac{d\vec{r}_i}{dt} \qquad (A15)$$

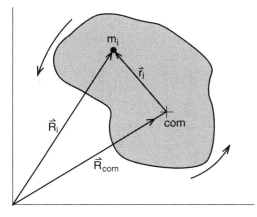

Figure A5
Analyzing the motion of a small piece of an object.

which can be written in the form

$$\vec{V}_i = \vec{V}_{com} + \vec{v}_i \quad (A16)$$

where $\vec{V}_i = d\vec{R}_i/dt$ is the velocity of m_i in our coordinate system, $\vec{V}_{com} = d\vec{R}_{com}/dt$ is the velocity of the center of mass of the object, and $\vec{v}_i = d\vec{r}_i/dt$ is the velocity of m_i in a coordinate system that is moving with the center of mass of the object.

The kinetic energy of the small mass m_i is

$$\begin{aligned}\tfrac{1}{2}m_i V_i^2 &= \tfrac{1}{2}m_i \vec{V}_i \cdot \vec{V}_i \\ &= \tfrac{1}{2}m_i\left(\vec{V}_{com} + \vec{v}_i\right)\cdot\left(\vec{V}_{com} + \vec{v}_i\right) \\ &= \tfrac{1}{2}m_i\left(V_{com}^2 + 2\vec{V}_{com}\cdot\vec{v}_i + v_i^2\right) \\ &= \tfrac{1}{2}m_i V_{com}^2 + \tfrac{1}{2}m_i v_i^2 + m_i\vec{V}_{com}\cdot\vec{v}_i\end{aligned}$$
(A17)

The total kinetic energy of the object is the sum of the kinetic energy of all the small pieces m_i

$$\begin{aligned}\left.\begin{array}{l}\text{total}\\ \text{kinetic}\\ \text{energy}\end{array}\right\} &= \sum_i \tfrac{1}{2} m_i V_i^2 \\ &= \tfrac{1}{2}V_{com}^2 \sum_i m_i + \sum_i \tfrac{1}{2} m_i v_i^2 + \vec{V}_{com}\cdot\sum_i m_i\vec{v}_i\end{aligned}$$
(A18)

In two of the terms, we could take the common factor V_{com} outside the sum.

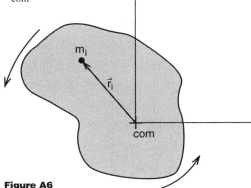

Figure A6
Here we moved the origin of the coordinate system to the center of mass.

Now the quantity $m_i\vec{v}_i$ that appears in the last term of Equation A18 is the linear momentum of m_i as seen in a coordinate system where the center of mass is at rest. To evaluate the sum of these terms, let us choose a new coordinate system whose origin is at the center of mass of the object as shown in Figure (A6). In this coordinate system the formula for the center of mass of the small masses m_i is

$$\vec{r}_{com} = \sum_i m_i\vec{r}_i = 0 \quad (A19)$$

Differentiating Equation 19 with respect to time gives

$$\sum_i m_i\frac{d\vec{r}_i}{dt} = \sum_i m_i\vec{v}_i = 0 \quad (A20)$$

Equation A20 tells us that when we are moving along with the center of mass of a system of particles, the total linear momentum of the system, the sum of all the $m_i\vec{v}_i$, is zero.

Using Equation A20 in A18 gets rid of the last term. If we let $M = \sum_i m_i$ be the total mass of the object, we get

$$\left.\begin{array}{l}\text{total}\\ \text{kinetic}\\ \text{energy}\end{array}\right\} = \tfrac{1}{2}MV_{com}^2 + \sum_i \tfrac{1}{2}m_i v_i^2 \quad (A21)$$

Equation A21 applies to any system of particles, whether the particles make up a rigid object or not. The first term, $1/2 MV_{com}^2$ is the kinetic energy of center of mass motion, and $\sum 1/2\, m_i v_i^2$ is the kinetic energy as seen by someone moving along with the center of mass. If the object is solid, then in a coordinate system where the center of mass is at rest, the only thing the object can do is rotate about the center of mass. As a result the kinetic energy in that coordinate system is the kinetic energy of rotation. If the moment of inertia about the axis of rotation is I_{com}, then the total kinetic energy is $1/2 MV_{com}^2 + 1/2\, I_{com}\omega^2$ where ω is the angular velocity of rotation. This is the result we stated in Equation A9.

Chapter 13
Equilibrium

When does a structure fall over, when does a bridge collapse, how do you lift a weight in a way that prevents serious injury to your back? We begin to answer such questions by applying Newton's laws to an object that has neither linear nor angular acceleration. The most interesting special case is when an object is at rest and will stay that way, when it is not about to tip over or collapse.

EQUATIONS FOR EQUILIBRIUM

If the center of mass of an object is not accelerating, then we know that the vector sum of the external forces acting on it is zero. If the object has no angular acceleration, then the sum of the torques about any axis must be zero. These two conditions

sum of external forces
$$\sum_i \vec{F}_{i\,external} = 0 \quad (1)$$

sum of torques about any axis
$$\sum_i \vec{\tau}_{i\,external} = 0 \quad (2)$$

are what we will consider to be the required conditions for an object to be in *equilibrium*.

Equations 1 and 2 are a complete statement of the basic physics to be discussed in this chapter. Everything else will be examples to show you how to effectively apply these equations in order to understand and predict when an object will be in equilibrium. In particular we wish to show you some techniques that make it quite easy to apply these equations.

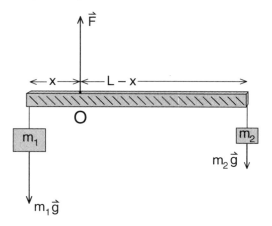

Figure 1
Masses m_1 and m_2 suspended from a massless rod. At what position x do we suspend the rod in order for the rod to balance?

Example 1 *Balancing Weights*

As our first example, suppose we have a massless rod of length L and suspend two masses m_1 and m_2 from the ends of the rod as shown in Figure (1). The rod is then suspended from a string located a distance x from he left end of the rod. What is the distance x and how strong a force $\vec{F}$ must be exerted by the string?

Solution: The first step is to sketch the situation and draw the forces involved, as we did in Figure (1). Our system will be the rod and the two masses. The external forces acting on this system are the two gravitational forces $m_1\vec{g}$ and $m_2\vec{g}$, and the force of the string $\vec{F}$. Since all the forces are y directed, when we set the vector sum of these external forces to zero we have

$$\Sigma F_y = F - m_1 g - m_2 g = 0 \quad (3)$$

Thus we get for F

$$F = (m_1 + m_2)g \quad (4)$$

and we see that F must support the weight of the two masses.

To figure out where to suspend the rod, we use the condition that *the net torque produced by the three external forces must be zero*. Since a torque is a force times a lever arm about some axis, you have to choose an axis before you can calculate any torques. The important point in equilibrium problems is that *you can choose the axis you want*. We will see that by intelligently selecting an axis, we can simplify the problem to a great extent.

Our definition of a torque $\vec{\tau}$ caused by a force $\vec{F}$ is

$$\vec{\tau} = \vec{r} \times \vec{F} \quad (5)$$

where $\vec{r}$ is a vector from the axis O to the point of application of the force $\vec{F}$ as shown in Figure (2).

In this chapter we do not need the full vector formalism for torque that we used in the discussion of the gyroscope. Here we will use the simpler picture that the magnitude of a torque caused by a force $\vec{F}$ is equal to the magnitude of F times the lever arm $r_\perp$, which is the distance of closest approach between the axis and the line of action of the force $\vec{F}$ as shown in Figure (2). If the torque tends to cause a counter clockwise rotation, as it is in Figure (2), we will call this a positive torque. If it tends to cause a clockwise rotation, we will call that a negative torque.

(In Figure (2), the vector $\vec{r} \times \vec{F}$ points up out of the page. If the vector $\vec{F}$ were directed to cause a clockwise rotation, then $\vec{r} \times \vec{F}$ would point down into the paper. [It is good practice to check this for yourself.] Thus we are using the convention that torques pointing up are positive, and those pointing down are negative.)

Returning to our problem of the rod and weights shown in Figure (1), let us take as our axis for calculating torques, the point of suspension of the rod, labeled point O in Figure (1). With this choice, the force $\vec{F}$ which passes though the point of suspension, has no lever arm about point O and therefore produces no torque about that point.

The gravitational force $m_1 \vec{g}$ has a lever arm x about point O and is tending to rotate the rod counter clockwise. Thus $m_1 \vec{g}$ produces a positive torque of magnitude $m_1 g x$. The other gravitational force $m_2 \vec{g}$ has a lever arm $(L - x)$ and is tending to rotate the rod clockwise. Thus $m_2 \vec{g}$ is producing a negative torque magnitude $-m_2 g(L-x)$. Setting the sum of the torques about point O equal to zero gives

$$m_1 g x - m_2 g (L - x) = 0 \qquad (6)$$

$$\boxed{x = \left(\frac{m_2}{m_1 + m_2}\right) L} \qquad (7)$$

Let us check to see that Equation 7 is a reasonable result. If $m_1 = m_2$, then we get $x = L/2$ which says that with equal weights, the rod balances in the center. If, in the extreme, $m_1 = 0$, then we get $x = L$, which tells us that we must suspend the rod directly over m_2, also a reasonable result. And if $m_2 = 0$ we get $x = 0$ as expected.

We obtained Equation 7 by setting the torques about the balance point equal to zero. This choice had the advantage that the suspending force $\vec{F}$ had no lever arm and therefore did not appear in our equations. We mentioned earlier that the condition for equilibrium was that the ***sum of the torques be zero about any axis***. In Exercises 1 and 2 we have you select different axes about which to set the torques equal to zero. With these other choices, you will still get the same answer for x, namely Equation 7, but you will have two unknowns, F and x, and have to solve two simultaneous equations. You will see that we simplified the work by choosing the suspension point as the axis and thereby eliminating F from our equation.

Exercise 1

If we choose the left end of the rod as our axis, as shown in Figure (3), then only the forces $\vec{F}$ and $m_2 \vec{g}$ produce a torque

(a) Is the torque produced by $\vec{F}$ positive or negative?

(b) Is the torque produced by $m_2 \vec{g}$ positive or negative?

(c) Write the equation setting the sum of the torques about the left end equal to zero. Then combine that equation with Equation 4 for F and solve for x. You should get Equation 7 as a result.

Exercise 2

Obtain two equations for x and F in Figure (1) by first setting the torques about the left end to zero, then by setting the torques about the right end equal to zero. Then solve these two equations for x and see that you get the same result as Equation 7

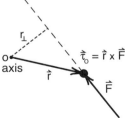

Figure 2
The torque $\vec{\tau} = \vec{r} \times \vec{F}$ has a magnitude $\tau = r_\perp \times F$.

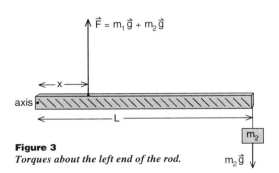

Figure 3
Torques about the left end of the rod.

GRAVITATIONAL FORCE ACTING AT THE CENTER OF MASS

When we are analyzing the torques acting on an extended object, we can picture the gravitational force on the whole object as acting on the center of mass point. To prove this very convenient result, let us conceptually break up a large object of mass M into many small masses m_i as shown in Figure (4), and calculate the total gravitational torque about some arbitrary axis O. An individual particle m_i located a distance x_i down the x axis from our origin O produces a gravitational torque τ_i given by

$$\tau_i = (m_i g) x_i \tag{8}$$

where $m_i g$ is the gravitational force and x_i the lever arm. Adding up the individual torques τ_i to obtain the total gravitational torque τ_O gives

$$\begin{aligned} \tau_O &= \sum_i \tau_i = \sum_i m_i g x_i \\ &= g \sum_i m_i x_i \end{aligned} \tag{9}$$

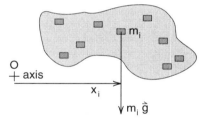

Figure 4
Conceptually break the large object of mass M into many small pieces of mass m_i, located a distance x_i down the x axis from our arbitrary origin O.

But the sum $\sum m_i x_i$ is by definition equal to M times the x coordinate of the center of mass of the object

$$MX_{com} \equiv \sum_i m_i x_i \tag{10}$$

Using Equation 10 in 9 gives

$$\tau_O = MgX_{com} \tag{11}$$

Equation 11 says that the gravitational torque about any axis O is equal to the total gravitational force Mg times the horizontal coordinate of the center of mass of the object. Thus the gravitational torque is just the same as if all of the mass of the object were concentrated at the center of mass point.

Exercise 3

A wheel and a plank each have a mass M. The center of the wheel is attached to one end of a uniform beam of length L. A nail is driven through the center of mass of the plank and nailed into the other end of the beam as shown in Figure (5). Where do you attach a rope around the beam so that the beam will balance? Explain how you got your answer.

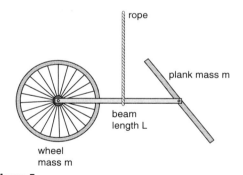

Figure 5
A wheel and a plank are attached to the ends of a uniform beam.

TECHNIQUE OF SOLVING EQUILIBRIUM PROBLEMS

In our discussion of the balance problem shown in Figure (1), we saw that there were several ways to solve the problem. We always have the condition that for equilibrium the vector sum of the forces is zero $(\Sigma_i \vec{F}_i = 0)$ and sum of the torques $\vec{\tau}_i$ about any axis is zero $(\Sigma_i \vec{\tau}_i = 0)$. By choosing various axes we can easily get enough, or more than enough equations to solve the problem. If we are not careful about the way we do this however, we can end up with a lot of simultaneous equations that are messy to solve.

Our first solution of the equilibrium condition for Figure (1) suggests a technique for simplifying the solution of equilibrium problems. In Equation 6 we set to zero the sum of the torques about the balance point O shown in Figure (1) reproduced here. We wanted to calculate the position x of the balance point, and were not particularly interested in the magnitude of the force $\vec{F}$. By taking the torques about the point O where $\vec{F}$ has no lever arm, F does not appear in our equation. As a result the only variable in Equation 6 is x, which can be immediately solved to give the result in Equation 7. As we saw in Exercises 1 and 2, if we chose the torques about any other point, both variables x and F appear in our equations, and we have to solve two simultaneous equations.

We will now consider some examples and exercises that look hard to solve, but turn out to be easy if you take the torques about the correct point. The trick is to find a point that eliminates the unknown forces you do not want to know about.

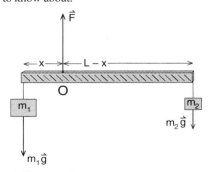

Figure 1 (Repeated)
We can eliminate any force by a proper choice of axis.

Example 3 *Wheel and Curb*

A boy is trying to push a wheel up over a curb by applying a horizontal force $\vec{F}_{boy}$ as shown in Figure (6a). The wheel has a mass m, radius r, and the curb a height h as shown. How strong a force does the boy have to apply?

Solution: We will consider the wheel to be the object in equilibrium, and as a first step sketch all the forces acting on the wheel as shown in Figure (6b). We can treat this as an equilibrium problem by noting that as the wheel is just about to go up over the curb, there is no force between the bottom of the wheel and the road.

There is, however, the force of the curb on the wheel, labeled $\vec{F}_{curb}$ in Figure (6b). We know the point at which $\vec{F}_{curb}$ acts but we do not know off hand either the magnitude or direction of $\vec{F}_{curb}$, nor are we asked to find $\vec{F}_{curb}$.

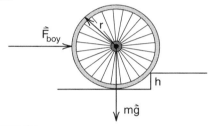

Figure 6a
A boy, exerting a horizontal force on the axle of a wheel, is trying to push the wheel up over a curb. How strong a force must the boy exert?

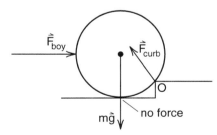

Figure 6b
Forces on the wheel as the wheel is just about to go up over the curb.

We can eliminate the unknown force $\vec{F}_{curb}$ by setting to zero the torques about the point O where the curb touches the wheel. Since $\vec{F}_{curb}$ has no lever arm about this point, it will not appear in the resulting equation.

In Figure (6c) we have sketched the geometry of the problem. About the point O the force $\vec{F}_{boy}$ has a lever arm $(r-h)$ and is tending to cause a clockwise rotation about point O. Thus $\vec{F}_{boy}$ is producing (by our convention) a negative torque, of magnitude $F_{boy}(r-h)$. The gravitational force $m\vec{g}$ has a lever arm ℓ shown in Figure (6c), and is tending to produce a counter clockwise rotation. Thus it is producing a positive torque of magnitude $+mg\ell$ about point O. Since there are no other torques about point O, setting the sum of the torques equal to zero gives

$$-F_{boy}(r-h) + mg\ell = 0 \qquad (12)$$

$$F_{boy} = mg\left(\frac{\ell}{r-h}\right)$$

We immediately see that if the curb is as high as the axle, if $r = h$, there is no finite force that will get the wheel over the curb.

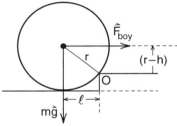

Figure 6c
Geometry of the problem.

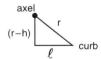

Figure 6d

The final slip in solving this problem is to relate the distance ℓ to the wheel radius r and curb height h. As shown in Figure (6d), the right triangle from the axle to the curb has sides ℓ, $(r - h)$ and hypotenuse r. By the Pythagorean theorem we get

$$r^2 = \ell^2 + (r-h)^2$$

or

$$\ell = \sqrt{2rh - h^2} \qquad (13)$$

which finishes the problem.

Exercise 4

The direction of $\vec{F}_{curb}$ is slightly off in Figure (6b). Explain what would happen to the wheel if $\vec{F}_{curb}$ pointed as shown.

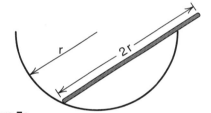

Figure 7a
A frictionless rod is placed in a hemispherical frictionless bowl. What is the equilibrium position of the rod?

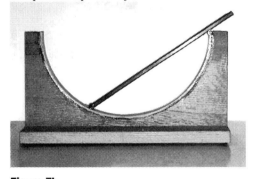

Figure 7b
We simulated a frictionless rod in a hemispherical frictionless bowl by placing ball bearing rollers at one end of the rod and the edge of the "bowl". The rod always comes to rest at this angle.

Example 4 Rod in a Frictionless Bowl

We include the problem here, first because it gives some practice with what we mean by a frictionless surface, but more importantly it is an example where we can gain considerable insight without solving any equations.

You place a frictionless rod of length 2r in a frictionless hemispherical bowl of radius r. Where does the rod come to rest? (Put in just enough friction to have it come to rest.) The situation is diagramed in Figure (7a). In Figure (7b), we have made a reasonably accurate simulation of the problem by using a semi circular piece of plastic for the bowl and placing small rollers on one end of the rod and one rim of the bowl to mimic the frictionless surfaces. In Figure (7c) we have sketched the forces acting on the rod. There is the downward force of gravity $m\vec{g}$ that acts at the center of mass of the rod, the force $\vec{F}_b$ exerted by the bowl on the end of the rod, and the force $\vec{F}_r$ exerted by the rim.

The idealization that we have a frictionless surface is equivalent to the statement that the surface can only exert normal forces, forces perpendicular to the surface. Thus the force $\vec{F}_b$ exerted by the frictionless surface of the bowl is normal to the bowl and points toward the center of the circle defining the bowl. The force $\vec{F}_r$ between the rim of the bowl and the frictionless rod must be perpendicular to the rod as shown.

Off hand we know nothing about the magnitude of the forces $\vec{F}_b$ and $\vec{F}_r$, only their directions. If we extend the lines of action of $\vec{F}_b$ and $\vec{F}_r$ they will intersect at some point above the rod as shown in Figure (7c). If we set to zero the sum of the torques about this intersection point, where neither $\vec{F}_b$ or $\vec{F}_r$ has a lever arm, then neither $\vec{F}_b$ or $\vec{F}_r$ will contribute. The only remaining torque is that produced by the gravitational force $m\vec{g}$. If the rod is in equilibrium, then the torque produced by $m\vec{g}$ about the intersection point must also be zero, with the result that the line of action of $m\vec{g}$ must also pass through the intersection point as shown in Figure (7d). Thus the rod will come to rest when the center of mass of the rod lies directly below the intersection point of $\vec{F}_b$ and $\vec{F}_r$. This result is nicely demonstrated by comparing the prediction, Figure (7d), with the experiment, Figure (7b).

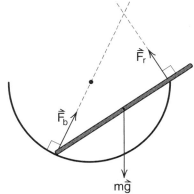

Figure 7c
Forces acting on the rod. Because the bowl is frictionless, $\vec{F}_b$ is perpendicular to the surface of the bowl. Because the rod is frictionless $\vec{F}_r$ is perpendicular to the rod.

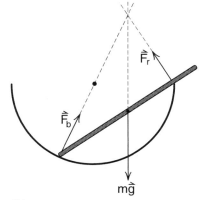

Figure 7d
For equilibrium, the center of mass must lie directly below the intersection point of $\vec{F}_b$ and $\vec{F}_r$.

Exercise 5

A spherical ball of mass m, radius r, is suspended by a string of length ℓ attached to a frictionless wall as shown in Figure (8).

(a) show that the line of action of the tension force T (the line of the string) passes through the center of the ball.

(b) find the tension T.

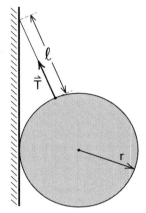

Figure 8
Ball suspended from a frictionless wall.

Exercise 6 Ladder Problem

A ladder is leaning against a frictionless wall at an angle θ as shown in Figure (9). Assume that the ladder is massless, and that a person of mass m is on the ladder.

The force between the ground and the bottom of the ladder can be decomposed into a normal component $\vec{F}_n$, and a horizontal component $\vec{F}_f$ that can exist only if there is friction between the ladder and the ground.

It is traditional in introductory texts to say that the ladder will start to slip if the friction force $\vec{F}_f$ exceeds a value of μF_n where μ is called the "coefficient of static friction". This idea is reasonable in that as the normal force $\vec{F}_n$ increases, so does the gripping or friction force $\vec{F}_f$. However the coefficient μ depends so much upon the circumstances of the particular situation, that the theory is not particularly useful. What, for example, should you use for the value of μ if the ends of the ladder sink down into the ground?

However for the sake of this problem, assume that the ladder will just start to slip when $F_f = \mu F_n$. Assume that μ has the value $\mu = 1/\sqrt{3} = .557$.

(a) at what angle θ would you place the ladder so that it will not start to slip until the person climbing it just reaches the top?

(b) at what angle θ would you place the ladder so that it will not start to slip until the person has gone half way up?

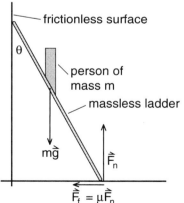

Figure 9
Ladder leaning against a frictionless wall.

Example 5 A Bridge Problem

A bridge is constructed from massless rigid beams of length ℓ. The ends of the beams are connected by a single large bolt that acts more or less like a big hinge. As a result the only forces you can have in each beam is either tension or compression (i.e. each beam either pulls or pushes along the length of the beam.) The idea is to be able to calculate the tension or compression force in any of the beams when a load is placed on the bridge.

In this example, we will place a mass m in the center of the right most span as shown in Figure (10a). To illustrate the process of calculating tension or compression in the beams, we will calculate the force in the upper left hand beam. For now we will assume that the beam is under tension and exerts a force $\vec{T}_d$ on joint d as shown. If it turns out that the beam is under compression, then the magnitude of T_d will turn out to be negative. Thus we do not have to know ahead of time whether the beam is under tension or compression.

Solution: When you have a statics problem involving an object with a lot of pieces, and you want to calculate the force in one of the pieces, the first step is to isolate part of the object and consider that as a separate system with external forces acting on it. In Figure (10b) we have chosen as our isolated system the part of the bridge made from the girders that have been drawn in heavy lines. The external forces acting on this isolated system are the gravitational force $m\vec{g}$ acting on the mass m, the supporting force $\vec{F}_2$ that holds up the right end of the bridge, (we will assume that the ends of the bridge are free to slide back and forth, so that the supporting forces $\vec{F}_1$ and $\vec{F}_2$ point straight up). In addition the tension (or compression) forces we have labeled $\vec{T}_d$, $\vec{T}_{c1}$ and $\vec{T}_{c2}$ are also acting on our isolated section of the bridge.

Looking at Figure (10b), it is immediately clear that the forces we do not want to know anything about are the tension forces $\vec{T}_{c1}$ and $\vec{T}_{c2}$. We can eliminate these forces by setting to zero the torques about the joint labeled c. Using our convention that counter clockwise torques are positive and clockwise ones are negative, we get

$$\Sigma T_{about\,c} = F_2(2\ell) - mg\left(\frac{3}{2}\ell\right) + T_d\left(\frac{\sqrt{3}}{2}\ell\right) = 0 \quad (13)$$

where we used the fact that T_d's lever arm h is the altitude of an equilateral triangle.

In Equation 13 we have two unknowns, F_2 and T_d. Thus we need another equation. If we take the bridge as a whole, and calculate the torques on about the point a (to eliminate the force $\vec{F}_1$), we get

$$\Sigma T_{about\,a} = F_2(3\ell) - mg\left(\frac{5}{2}\ell\right) = 0 \quad (14)$$

Solving Equation 14 for F_2 gives

$$F_2 = \frac{5}{6}mg \quad (15)$$

Using this value in Equation 13 gives

$$\boxed{T_d = \frac{-mg}{3\sqrt{3}}} \quad (18)$$

The minus (–) sign indicates that the beam is under compression.

Exercise 7

Find the tension (or compression in the beam that goes from point (d) to point (e) in the bridge problem of Figure (10a).

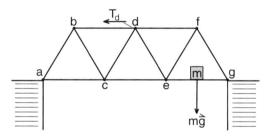

Figure 10a
Bridge with a truck on the last span.

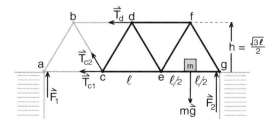

Figure 10b
Finding the tension in the span from b to d.

Exercise 8 Working with Rope

As most sailors know, if you use rope correctly, you can create very large tension forces without exerting that strong a force yourself. Suppose, for example, you wish to make a raft out of two long logs with two short spacer planks between them as shown in Figure (11a). You wish to hold the raft together with a rope around the center as shown.

The first step is to tie, as tightly as you can, the logs together as shown in the end view of Figure (11b). Then take another piece of rope and tie it as tightly as you can as shown in Figure (11c). If you do a reasonably good job, you can create a large tension in the rope holding the logs together.

To analyze the problem, let T_1 be the tension in the rope holding the logs together, and T_2 the tension in the line between the ropes as shown in Figure (11d). For this problem, assume that angle θ in Figure (11d) is 5 degrees, and the tension T_2 that you could supply in winding the line around the ropes was 200 newtons (enough force to lift a 20 kilogram mass). What is the tension T_1 in the ropes holding the logs together?

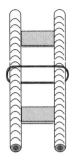

Figure 11a
Constructing a raft by tying two logs together, with wood spacers.

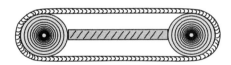

Figure 11b
End view of raft.

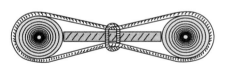

Figure 11c
Tightening the rope.

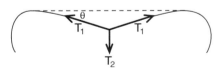

Figure 11d
Tensions in the ropes.

LIFTING WEIGHTS AND MUSCLE INJURIES

The previous exercise on tying a raft together illustrates the fact that with some leverage, you can create large tensions in a rope. Similar large forces can exist in your muscles when you lift weights, particularly if you do not lift the weights properly.

To illustrate the importance of lifting heavy objects correctly, consider the sketch of Figure (12) showing a shopper holding a funny looking 10 kg shopping bag out at arms length. We wish to determine the forces that must be exerted on the backbone and by the back muscles in order to support this extra weight.

To analyze the forces, think of the upper body and arm as essentially a rigid object supported by the backbone and back muscles as shown. Since we are interested in the extra forces required to lift the weight, we will ignore the weight of the upper body itself.

The external forces acting on the upper body are the upward compressional force $\vec{F}_b$ acting on the backbone, and the downward force $\vec{F}_m$ acting at the point where the back muscle is attached to the thighbone. (This is in reaction to the upward force exerted on the thigh bone by the contacting back muscle.) There is also the weight $m\vec{g}$ of the shopping bag. We are letting L be the horizontal distance from the backbone to the shopping bag, and ℓ the separation between the point where the thigh bone pushes up on the backbone and pulls down on the back muscle.

If we want to solve for the force $\vec{F}_m$ exerted by the back muscle we can eliminate $\vec{F}_b$ from our equation by setting to zero the sum of the torques about point b, the point where the thigh bone contacts the backbone. We get

$$\Sigma \tau_{about\ b} = F_m \ell - MgL = 0$$

$$F_m = mg\left(\frac{L}{\ell}\right) \qquad (9)$$

We see that the back muscle has to pull down with a force that is a factor of L/ℓ times greater than the weight mg of the shopping bag. With $\ell = 2$ cm, you see that if you hold the shopping bag out at arms length, say L = 80 cm, then F_m is $L/\ell = 40$ times as great as the weight of the shopping bag. For you to lift a 10 kg bag at arms length, your back is essentially lifting a 10 kg × 40 = 400 kg mass, which has a weight of almost half a ton! If instead you pulled your arm in so that L was only 20 cm, then F_m drops 1/4 of its original value. Do your back a favor and do not lift heavy objects out at arms length.

Exercise 9

Write a single equation that allows you to solve for the compressional force $\vec{F}_b$ exerted on the backbone, as shown in Figure (12).

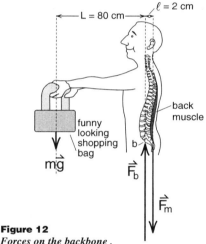

Figure 12
Forces on the backbone.

Exercise 10

Figure (13) shows the structure of the lower leg and foot. Assume that this person is raising her heel a bit off the ground, so that her foot is touching the floor at only one place, namely the point labeled P in the diagram. Also assume that the calf muscle is attached to the foot bones at the point labeled (a), and that the leg bone acts at the pivot point (b). If her mass is 60 kilograms, what must be the forces exerted by the calf muscle and the leg bone? Compare the strength of these forces with her weight. (This and the next problem adapted from Halliday & Resnick.)

Exercise 11

The arm in Figure (14) is holding a 20 kilogram mass. The arm pivots around the points marked with a small black circle. What is the compressional force on the bones in this joint? (Neglect the mass of the arm.)

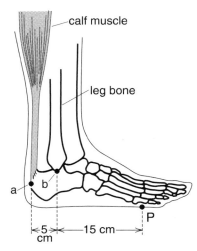

Figure 13
Foot

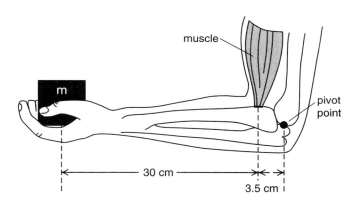

Figure 14
Arm

Chapter 14
Oscillations and Resonance

Oscillations and vibrations play a more significant role in our lives than we realize. When you strike a bell, the metal vibrates, creating a sound wave. All musical instruments are based on some method to force the air around the instrument to oscillate. Oscillations from the swing of a pendulum in a grandfather's clock to the vibrations of a quartz crystal are used as timing devices. When you heat a substance, some of the energy you supply goes into oscillations of the atoms. Most forms of wave motion involve the oscillatory motion of the substance through which the wave is moving.

Despite the enormous variety of systems that oscillate, they have many features in common, features exhibited by the simple system of a mass on a spring. As a result we will focus our attention on the analysis of the motion of a mass on a spring, describing ways in which other forms of oscillation are similar.

OSCILLATORY MOTION

Suspend a mass on the end of a spring as shown in Figure (1), gently pull the mass down and let go. The mass will oscillate up and down about the equilibrium position. How do we describe the kind of motion we are looking at?

One of the best ways to see what kind of motion we are dealing with is to perform the demonstration illustrated in Figure (2a). In that demonstration, we place a rotating wheel beside an oscillating mass, and view the two objects via a TV camera set off to the side as shown. A white tape is placed around the mass, and a short stick is mounted on the rotating wheel as seen in the edge view, Figure (2b). This edge view is the one displayed by the TV camera.

The wheel is mounted on a variable speed motor, which allows us to adjust the angular velocity of the ω of the wheel so that the wheel goes around once in precisely the same length of time it takes the mass to bob up and down once. The height of the wheel is adjusted so that when the mass is at rest at its equilibrium position, the white stripe on the mass lines up with the axis of the wheel. As a result if the stick is in a horizontal position (3 o'clock or 9 o'clock) and the mass is at rest, the stick and the white stripe line up on the television image.

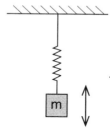

Figure 1
Mass suspended from a spring. If you gently pull the mass down and release it, it will oscillate up and down about the equilibrium position.

Now pull the mass down so that the white stripe lines up with the lowest position of the stick the same height as the stick when the stick is at the bottom position (6 o'clock). Start the motor rotating at the correct frequency and release the mass when the stick is at the bottom. If you do this just right, some practice may be required, you will see in the television picture that the white stripe and the stick move up and down together as if they were a single object.

From this demonstration we conclude that *the up and down oscillatory motion of the mass is the same as circular motion viewed sideways*. As a result we can use what we know about circular motion to understand oscillatory motion. As a start, we will say that the oscillatory motion has an angular frequency ω that is the same as the angular velocity ω of the rotating wheel when the mass and the stick go up and down together.

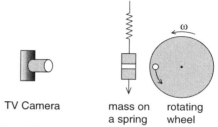

Figure 2a
Lecture setup for comparing the oscillation of a mass on a spring with a rotating wheel. A stick is mounted on the rotating wheel, and a TV camera off to the side provides a side view of the oscillating mass and rotating wheel.

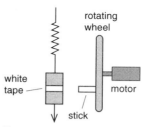

Figure 2b
Side view of the oscillating mass and rotating wheel, as seen by the TV camera. When the motor is adjusted to the correct frequency, the mass and the stick are observed to move up and down together.

THE SINE WAVE

There is another way to picture the sideways view of circular motion. As more or less a thought experiment, suppose that we take our rotating wheel with a stick shown in Figure (2a) and shine a parallel beam of light at it, sideways, as shown in Figure (3). Now picture a truck with a big billboard mounted on the back, driving away from the light source as shown on the right side of Figure (3). The image of the stick will move up and down on the billboard as the truck moves forward.

Finally picture the line traced out in space by the image of the stick on the moving billboard. The image is going up and down with a frequency ω and moves forward at a speed v, the speed of the truck. The result is an undulating curve we call a *sine wave*.

To make this definition of the sine wave more specific, assume the truck crosses the point x = 0 just when the angle θ of the stick is zero as shown in Figure (3). Let us imagine that the truck drives at a speed $v = \omega$, so that the distance $x = vt = \omega t$ that the truck has travelled is the same as the angular distance $\theta = \omega t$ that the stick has travelled. Finally let the radius of the circle around which the stick is travelling be r = 1, so that the undulating curve goes up to a maximum value y = + 1 and down to a minimum value y = – 1. With these conditions, the curve traced out is the mathematical function

$$y = \sin \theta = \sin (\omega t) \tag{1}$$

Let us remove the truck and billboard and look at the sine curve itself more carefully as shown in Figure (4). The horizontal axis of the sine curve is the angular distance $\theta = \omega t$ that the rotating stick has travelled. Starting at 0 when $\theta = 0$, the sine curve completes one full cycle or undulation just when the wheel has gone around once and $\theta = 2\pi$. Thus one cycle of a sine wave goes from 0 to 2π as shown in Figure (4). The sine wave reaches a maximum height at $\theta = \pi/2$, goes back to zero again at $\theta = \pi$, has a minimum value at $3\pi/2$ and completes the cycle at $\theta = \omega t = 2\pi$.

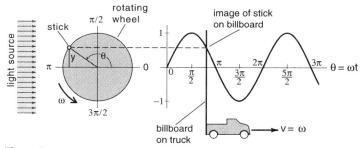

Figure 3
Project the image of the stick onto the back of a truck moving at a speed $v = \omega t$, and the image traces out a sine wave.

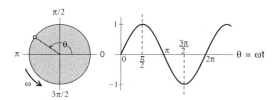

Figure 4
The sine curve $y = \sin \theta = \sin \omega t$.

Exercise 1

Somewhere back in the dim past, you learned that $\sin\theta$ was the ratio of the opposite side to the hypotenuse in a right triangle. Applied to Figure (5), this is

$$\sin\theta \equiv y/r \tag{2}$$

Show that this older definition of $\sin\theta$, at least for angles θ between 0 and $\pi/2$, is the same as the definition of $\sin\theta$ we are using in Figure (3) and (4).

As you can see from Exercise 1, our rotating stick picture of the sine wave is mathematically equivalent to the definition of $\sin\theta$ you learned in your first trigonometry class. What may be new conceptually is the dynamic aspect of the definition. Figures (3) and (4) connect rotational motion to oscillatory motion and to the shape of a sine wave. The relationship between the static picture and the dynamic one is that the angular distance θ is equal to the angular velocity ω times the elapsed time t.

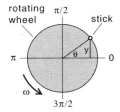

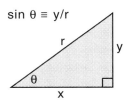

Figure 5
Our old definition of $\sin\theta$.

The basic question for the dynamic picture is how long does one oscillation take. The time for one oscillation is called the period T of the oscillation. We can therefore ask what is the period T of an oscillation whose angular velocity, or angular frequency is ω.

The solution to this problem is to note that the sine wave completes one cycle when $\theta = 2\pi$. But θ is just the angular distance ωt. Thus, if t = one period T, we have

$$\theta = \omega T = 2\pi$$

$$\boxed{T = \frac{2\pi}{\omega}} \quad \begin{array}{l}\textit{period of}\\ \textit{a sinusoidal}\\ \textit{oscillation}\end{array} \tag{3}$$

To remember formulas like Equation 3, we can use the same set of dimensions we used in our discussion of angular motion. If we remember the dimensions

$\omega \dfrac{\text{radians}}{\text{second}}$ *angular frequency*

$2\pi \dfrac{\text{radians}}{\text{cycle}}$

$T \dfrac{\text{seconds}}{\text{cycle}}$ *period*

$f \dfrac{\text{cycles}}{\text{second}}$ *frequency*

then we can go back and forth between the quantities ω, T and f simply by making the dimensions come out right.

For example

$$T \frac{\sec}{\text{cycle}} = \frac{2\pi \frac{\text{radians}}{\text{cycle}}}{\omega \frac{\text{radians}}{\sec}} = \frac{2\pi}{\omega} \frac{\sec}{\text{cycle}}$$

$$f \frac{\text{cycles}}{\sec} = \frac{\omega \frac{\text{radians}}{\sec}}{2\pi \frac{\text{radians}}{\text{cycle}}} = \frac{\omega}{2\pi} \frac{\text{cycles}}{\sec}$$

$$T \frac{\sec}{\text{cycle}} = \frac{1}{f \frac{\text{cycles}}{\sec}} = \frac{1}{f} \frac{\sec}{\text{cycle}}$$

We repeated these dimensional exercises, because it is essential that you be able to easily go back and forth between quantities like frequency, angular frequency, and period.

Exercise 2
(a) A spring is vibrating at a rate of 2 seconds per cycle. What is the angular velocity ω of this oscillation?

(b) What is the period of oscillation, in seconds, of an oscillation where $\omega = 1$?

Exercise 3
As shown in Figure (6), an air cart sitting on an air track has springs attached to the ends of the track as shown. We are taking x = 0 to be the equilibrium position of the cart. It turns out that the car oscillates back and forth with the same kind of sinusoidal motion as the mass on the end of a spring shown in Figures (1) and (2).

Assume that the mathematical formula for the coordinate x of the cart is

$$x = x_0 \sin \omega t$$

where

$$x_0 = 4 \text{ cm} ; \quad \omega = 3 \frac{\text{radians}}{\sec}$$

(a) Figure (7) is an x–t graph of the position of the air cart in Figure (6). We have drawn in the sine curve, and drawn tick marks at important points along the x and t axis. On the x axis, the tick marks are at the maximum and minimum values of x. On the t axis, the tick marks are at 1/4, 1/2, 3/4 and 1 complete cycle. Insert on the graph, the numerical values that should be associated with these tick marks.

(b) Where will the cart be located at the time $t = 300\pi$ seconds?

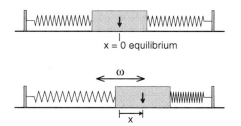

Figure 6
Mass with springs on an air cart. We take the equilibrium position to be x = 0.

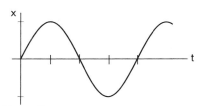

Figure 7
The x-t graph for the motion of the air cart in Figure 6.

Phase of an Oscillation

Our sine waves, by definition, begin at 0 when $\theta = 0$. This is equivalent to saying that in Figure (2), the truck crossed $x = 0$ at the same instant that the rotating wheel crossed $\theta = 0$. We did not have to make this choice, the rotating wheel could have been at any angle ϕ when the truck crossed $t = 0$ as shown in Figure (8). If the stick were up at an angle ϕ when the truck crossed the zero of the horizontal axis, we say that the resulting sine wave has a phase ϕ. The formula for the resulting sine curve is

$$y = \sin(\omega t + \phi) \quad \substack{\phi \text{ is the} \\ \text{phase angle}} \quad (4)$$

You can see from this equation that at $t = 0$, the angle of the sine wave is $\theta = \omega t + \phi = \phi$

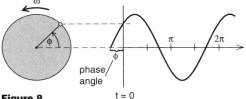

Figure 8
The phase angle ϕ.

In Figure (9) we have sketched the sine wave for several different phases. At a phase $\phi = \pi/2$, the wave starts at 1 for $\theta = 0$ and goes down to 0 at $\theta = \pi/2 = 90°$. This is just what $\cos\theta$ does, and we have what is called a cosine wave. We have

$$\cos(\omega t) = \sin(\omega t + \pi/2) \quad (6)$$

When the phase angle gets up to π or $180°$, the sine wave is reversed into a $-\sin(\omega t)$ wave. At $\phi = 2\pi$ we are back to the sine wave we started with.

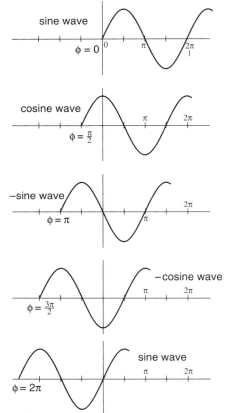

Figure 9
Various phases of the sine wave. When $\phi = \pi/2$, the wave is called a cosine wave. (It matches the definition of a cosine, which starts out at 1 for $\theta = 0$. At a phase angle $\phi = \pi$ or $180°$, the sine wave has reversed and become $-\sin(\omega t)$. At $\phi = 2\pi$, we are back to a sine wave again.

Exercise 4

In trigonometry class, or somewhere perhaps, you were given the trigonometric identity

$$\sin(a+b) = \sin a \cos b + \sin b \cos a$$

Use this result to show that

$$\sin(\omega t + \pi/2) = \cos \omega t \qquad (7a)$$

$$\sin(\omega t + \pi) = -\sin \omega t \qquad (7b)$$

$$\sin(\omega t + 3\pi/2) = -\cos \omega t \qquad (7c)$$

$$\sin(\omega t + 2\pi) = \sin \omega t \qquad (7d)$$

These are the results graphed in Figure (9).

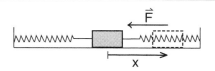

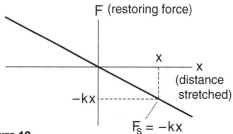

Figure 10
If the cart is displaced a distance x from equilibrium, there is a restoring force F = - kx.

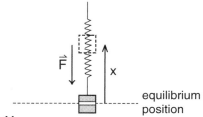

Figure 11
Restoring force for a mass on a spring.

MASS ON A SPRING; ANALYTIC SOLUTION

Let us now apply Newton's laws to the motion of a mass on a spring and see how well the results compare with the sinusoidal motion we observed in the demonstration of Figure (2). In our analysis of the spring pendulum in Chapter 9 (see Figure 9-4) we saw that the spring exerted a force whose strength was linearly proportional to the amount the spring was stretched, a result known as Hooke's law.

When we have the one dimensional motion of a mass oscillating about its equilibrium position, as illustrated in Figure (10), then we get a very simple formula for the net force on the object. If we displace the cart of Figure (10) by a distance x from equilibrium, there is a restoring force whose magnitude is proportional to x pushing the cart back toward the equilibrium position.

We can completely describe this restoring force by the formula

$$\boxed{F = -kx} \qquad \begin{array}{l}\textit{Hooke's law}\\ \textit{restoring force}\end{array} \qquad (8)$$

If $x = 0$, the cart is at its equilibrium position and there is no force. If x is positive, as in Figure (10), the restoring force is negative, pointing back to the equilibrium position. And if x is negative, the restoring force points in the positive direction. All these cases are handled by the formula $F = -kx$.

For a mass bobbing up and down on a spring, shown in Figure (11), there is both a gravitational and a spring force acting on the mass. But if you measure the net force starting from the equilibrium position, you still get a linear restoring force. The net force is always directed back toward the equilibrium position and has a strength proportional to the distance x the mass is away from equilibrium. Thus Equation 8 still describes the net force on the mass.

Exercise 5

Describe experiments you could carry out in the laboratory to measure the force constant k for the air cart setup of Figure (10). To do this you are given the air cart setup, a string, a pulley, and some small weights.

Sketch the setup you would use to make the measurements, and include some simulated data to show how you would obtain a numerical value of k from your data. (This is the kind of exercise you would do ahead of time if you planned to do a project studying the oscillatory motion of the air cart and spring system.)

To apply Newton's laws to the problem of the oscillating mass, let x(t) be the displacement from equilibrium of either the air cart of Figure (10) or the mass of Figure (11). The velocity v_x and the acceleration a_x of either the cart or mass is

$$v_x \equiv \frac{dx(t)}{dt} \qquad (9)$$

$$a_x \equiv \frac{dv_x}{dt} = \frac{d^2x(t)}{dt^2} \qquad (10)$$

The x component of Newton's second law becomes

$$F_x = ma_x$$

$$-kx = m\frac{d^2x}{dt^2} \qquad (11a)$$

where we used Hooke's law, Equation 8 for F_x.

The result, Equation 11a, involves both the variable x(t) and its second derivative d^2x/dt^2. An equation involving derivatives is called a ***differential equation***, and one like Equation 11a, where the highest derivative is the second derivative, is called a ***second order differential equation***. Differential equations are harder to solve than algebraic equations like $x^2 = 4$, because the answer is a function or a curve, rather than simple numbers like ± 2.

When working with differential equations, there is a traditional form in which to write the equation. The highest derivative is written to the far left, all terms with the unknown variable are put on the left side of the equation, and the coefficient of the highest derivative is set to one. (A reason for this tradition is that only a few differential equations have been solved. If you write them all in a standard form, you may recognize the one you are working with.) Putting Equation 11a in the standard form by dividing through by m and rearranging terms gives

$$\boxed{\frac{d^2x}{dt^2} + \frac{k}{m}x = 0} \qquad (11)$$

A standard way to solve a differential equation is to guess the answer, and then plug your guess into the equation to see if the guess works. A course in differential equations basically teaches you how to make good guesses. In the absence of such a course, we have to use whatever knowledge we have about the system in order to make as good a guess as we can. That is why we did the demonstration of Figure (2). In that demonstration we saw the oscillating mass moved the same way as a stick on a rotating wheel, when the wheel is viewed sideways. We then saw that this sideways view of rotating motion is described by the mathematical function $\sin(\omega t)$. From this we suspect that a good guess for the function x(t) may be

$$x(t) = \sin(\omega t) \qquad \text{initial guess} \qquad (12)$$

In order to see if this guess is any good, we need to substitute values of x and d^2x/dt^2 into Equation 11. To do this, we need derivatives of $x = \sin(\omega t)$.

From your calculus course you learned that

$$\frac{d}{dt}\sin(\omega t) = \omega\cos(\omega t) \qquad (13)$$

$$\frac{d}{dt}\cos(\omega t) = -\omega\sin(\omega t) \qquad (14)$$

Thus if we start with
$$x = \sin(\omega t) \quad (12)$$
we have
$$\frac{dx}{dt} = \omega \cos(\omega t) \quad (15)$$

$$\frac{d^2x}{dt^2} = \frac{d}{dt}\left(\frac{dx}{dt}\right) = \frac{d}{dt}\left(\omega \cos \omega t\right)$$

$$= \omega\left(-\omega \sin \omega t\right)$$

$$\frac{d^2x}{dt^2} = -\omega^2 \sin(\omega t) \quad (16)$$

To check our guess $x = \sin(\omega t)$, we substitute the values of x from Equation 12 and d^2x/dt^2 from Equation 16 into the differential Equation 11. We get

$$\frac{d^2x}{dt^2} + \frac{k}{m}x = 0$$

$$\left[-\omega^2 \sin(\omega t)\right] + \frac{k}{m}\left[\sin(\omega t)\right] \stackrel{?}{=} 0 \quad (17)$$

We put the question mark over the equal sign, because the question we want to answer is whether $x = \sin \omega t$ can be a solution to our differential equation. Can the sum of these two terms be made equal to zero as required by Equation 11?

The first thing we note in Equation 17 is that the function (sin ωt) cancels. This is encouraging, because if we ended up with two different functions of time, say a sin ωt in one term and a cos ωt in the other, there would be no way to make the sum of the two terms to be zero for all time, and we would not have solved the equation. However because the (sin ωt) cancels, we are left with

$$-\omega^2 + \frac{k}{m} \stackrel{?}{=} 0 \quad (18)$$

Equation 18 is easily solved with the choice

$$\boxed{\omega = \sqrt{\frac{k}{m}}} \quad \begin{array}{l} \textit{angular frequency} \\ \textit{of oscillating cart} \end{array} \quad (19)$$

Thus we have shown not only that $(x = \sin \omega t)$ is a solution of Newton's second law, but we have also solved for the frequency of oscillation ω. Newton's second law predicts that the air cart will oscillate at a frequency $\omega = \sqrt{k/m}$. This result is easily tested by experiment.

Exercise 6

(a) In the formula $(x = \sin \omega t)$, ω is the angular frequency of oscillation, measured in radians per second. Using the formula $\omega = \sqrt{k/m}$ and dimensional analysis, find the predicted formula for the period T of the oscillation, the number of *seconds per cycle*.

(b) A mass m = 245 gms is suspended from a spring as shown in Figure (12). The mass is observed to oscillate up and down with a period of 1.37 seconds. From this determine the spring constant k. How could this result have been used in the spring pendulum experiment discussed in Chapter 9 (page 9-3)?

(You can check your answer, since the ball and spring of Figure (12) are the same ones we used in the spring pendulum experiment.)

Figure 12
The spring constant can be determined by measuring the period of oscillation.

The guess we made in Equation 12, ($x = \sin \omega t$) is not the only possible solution to our differential Equation 11. In the following exercises, you show that

$$x = A \sin (\omega t) \quad (12a)$$

is also a solution, where A is an arbitrary constant. Since the function $\sin (\omega t)$ oscillates back and forth between the values $+1$ and -1, the function $A \sin (\omega t)$ oscillates back and forth between $+A$ and $-A$. Thus A represents the *amplitude* of the oscillation. The fact that Equation 12a, with arbitrary A, is also a solution to Newton's second law, means that a sine wave with any amplitude is a solution. (This is true as long as you do not stretch the spring too much. If you pull a spring out too far, if you exceed what is called the *elastic limit*, the spring does not return to its original shape and its spring constant changes.)

Exercise 7

As a guess, try Equation 12a as a solution to the differential Equation 11. Follow the same kind of steps we used in checking the guess $x = \sin (\omega t)$, and see why Equation 12a is a solution for any value of A.

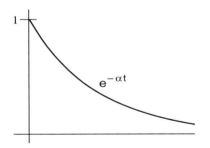

Figure 13
The exponential decay curve $e^{-\alpha t}$.

Exercise 8

(a) Show that the guess

$$x = A \cos (\omega t) \quad (12b)$$

is also a solution to our differential Equation 11. This should be an expected result, because the only difference between a sine wave and a cosines wave is the choice of the time $t = 0$ when we start measuring the oscillation.

(b) The sine and cosine waves are only special cases of the more general solution

$$x = A \sin (\omega t + \phi) \quad (12c)$$

where ϕ is the phase of the oscillation discussed in Figure (9). Show that Equation 12c is also a solution of our differential Equation 11. [Hint: the derivative of $\sin (\omega t + \phi)$ is $-\omega \cos (\omega t + \phi)$. You can, if you want, prove this result using Equations 13 and 14 and the trigonometric identities

$\sin (a + b) = \sin a \cos b + \cos a \sin b$

$\cos (a + b) = \cos a \cos b - \sin a \sin b$

Remember that ϕ is a constant.]

Exercise 9

We do not want you to think every function is a solution to Equation 11. Try as a guess

$$x = e^{-\alpha t} \quad (20)$$

which represents an exponentially decaying curve shown in Figure (13). To do this you need to know that

$$\frac{d}{dt} e^{-\alpha t} = -\alpha e^{-\alpha t} \quad (21)$$

When you try Equation 20 as a guess, what goes wrong? Why can't this be a solution to our differential equation? [Or, by what crazy way could you make it a solution?]

Conservation of Energy

Back in Chapter 10 we calculated the formula for the potential energy stored in the springs when we pulled the cart of Figure (10) a distance x from equilibrium. The result was

$$\text{spring potential energy} \quad U_{spring} = \tfrac{1}{2} k x^2 \quad (10\text{-}28)$$

We then used the law of conservation of energy to predict how fast the cart would be moving when it crossed the $x = 0$ equilibrium line if it were released from rest at a position $x = x_0$. The idea was that the potential energy $\tfrac{1}{2} k x_0^2$ the springs have when the cart is released, is converted to kinetic energy $\tfrac{1}{2} m v_0^2$ the cart has when it is at $x = 0$ and its speed is $v = v_0$.

Exercise 10

See if you can derive Equation 10-28 without looking back at Chapter 10. If you cannot, review the derivation now.

Using conservation of energy to predict the speed of the air cart was particularly useful back in Chapter 10 because at that time we did not have the analytic solution for the motion of the cart. Now that we have solved Newton's second law to predict the motion of the cart, we can turn the problem around, and see if energy is conserved by the analytic solution.

An analytic solution for the position x(t) and velocity v(t) of the cart is

$$\begin{aligned} x(t) &= \sin(\omega t) \\ v(t) &= \frac{dx}{dt} = \omega \cos(\omega t) \end{aligned} \quad (22)$$

For this solution, the kinetic and potential energies are

$$\text{kinetic energy} \quad \tfrac{1}{2} m v^2 = \tfrac{1}{2} m \omega^2 \cos^2(\omega t) \quad (23)$$

$$\text{potential energy} \quad \tfrac{1}{2} k x^2 = \tfrac{1}{2} k \sin^2(\omega t) \quad (24)$$

The total energy E_{tot} of the cart at any time t is

$$\text{total energy} = \text{kinetic energy} + \text{potential energy}$$

$$E_{tot} = \tfrac{1}{2} m \omega^2 \cos^2(\omega t) + \tfrac{1}{2} k \sin^2(\omega t) \quad (25)$$

At first sight, Equation 25 does not look too promising. It seems that E_{tot} is some rather complex function of time, hardly what we expect if energy is conserved. However remember that the frequency ω is related to the spring constant k by $\omega = \sqrt{k/m}$, thus we have

$$\begin{aligned} \tfrac{1}{2} m \omega^2 &= \tfrac{1}{2} m \left(\sqrt{\tfrac{k}{m}} \right)^2 \\ &= \tfrac{1}{2} m \tfrac{k}{m} = \tfrac{1}{2} k \end{aligned} \quad (26)$$

Thus the two terms in our formula for E_{tot} have the same coefficient $\tfrac{1}{2} k$, and E_{tot} becomes (using Equation 26 in 25)

$$E_{tot} = \tfrac{1}{2} k \left[\cos^2(\omega t) + \sin^2(\omega t) \right] \quad (27)$$

Equation 27 can be simplified further using the trigonometric identity

$$\cos^2(a) + \sin^2(a) = 1 \quad (28)$$

for any value of a. Thus the term in square brackets in Equation 27 has the value 1, and we are left with

$$\boxed{E_{tot} = \frac{k}{2}} \quad (29)$$

The total energy of the mass and spring system is constant as the oscillator moves back and forth. Energy is conserved after all!

Exercise 11

What is the total energy of an oscillating mass whose amplitude of oscillation is A? [Start with the solution $x(t) = A \sin(\omega t)$, calculate v(t), and then calculate

$$E_{tot} = \tfrac{1}{2} k x^2 + \tfrac{1}{2} m v^2.$$

THE HARMONIC OSCILLATOR

The sinusoidal motion we have been discussing, which results when an object is subject to a linear restoring force $F = -kx$, is called *simple harmonic motion* and the oscillating system is often called a *harmonic oscillator*. These general names are used because there are many examples in physics of simple harmonic motion. In some cases the sine wave solution $\sin(\omega t)$ is an exact solution of a Hooke's law problem. In many other cases, the solution is approximate, valid only for small amplitude oscillations where the displacements x do not become too big. In the following sections we will consider examples of both kinds of problems.

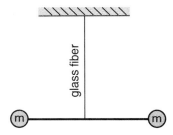

Figure 14a
Side view of the torsion pendulum used in the Cavendish experiment.

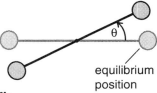

Figure 14b
Top view of the torsion pendulum. The light drawing shows the equilibrium position of the pendulum, the dark drawing shows the pendulum displaced by an angle θ. In this displaced position, the glass fiber exerts a restoring torque $\tau_{restoring} = -k\theta$.

The Torsion Pendulum

One example of simple harmonic motion is provided by part of the apparatus used by Cavendish to detect the gravitational force between two lead balls. The apparatus, illustrated back in Figure (8-8), contains two small lead balls mounted on a light rod, which in turn is suspended from a glass fiber as shown in Figure (14a). (Such glass fibers are easy to make. Heat the center of a glass rod in a Bunsen burner until the glass is about to melt, and then pull the ends of the rod apart. The soft glass stretches out into a long thin fiber.)

If you let the rod with two balls come to rest at its equilibrium position, then rotate then rod by an angle θ in the horizontal plane as shown in the top view of Figure (14b), the glass fiber exerts a torque tending to rotate the rod back to its equilibrium position. Careful experiments have shown that the restoring torque exerted by the glass fiber is proportional to the angular distance θ that the rod has been rotated from equilibrium. The rod is acting like an angular spring, producing a restoring torque τ_r given by an angular version of Hooke's law

$$\tau_r = -k\theta \quad \text{angular version of Hooke's law} \quad (30)$$

In the Cavendish experiment two large balls of mass M are placed near the small balls as shown in Figure (15).

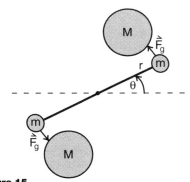

Figure 15
In the Cavendish experiment, a torque is exerted on the torsion pendulum by the gravitational force of the large lead spheres. In the new equilibrium position, the gravitational torque just balances the restoring torque of the torsion pendulum.

The gravitational forces $\vec{F}_g$ between the large and small balls produce a net torque τ_g on the suspended rod of magnitude

$$\tau_g = 2(rF_g) \tag{31}$$

where r is the distance from the center of the rod to the small mass (the lever arm), and the factor of 2 is from the fact that we have a gravitational force acting on each pair of the balls.

The suspended rod will finally come to rest at an angle θ_0 where the gravitational torque τ_g just balances the glass fiber restoring torque τ_r, so that there is no net torque the rod. Equating the magnitudes of τ_r and τ_g, Equations 30 and 31 gives us

$$|\tau_g| = |\tau_r|$$
$$2rF_g = k\theta_0 \tag{32}$$

Equation 32 can then be solved for the gravitational force τ_g in terms of the rod length 2r, the restoring constant k, and the rest angle θ_0.

The problem the Cavendish experiment has to overcome is the fact that the gravitational force between the two lead balls is extremely weak. You need an apparatus where the tiny gravitational torque τ_g produces an observable deflection θ_0. That means that the restoring torque τ_r must also be very small. That was why the long glass fiber was used to suspend the rod, for it produces an almost immeasurably small restoring torque.

In order to carry out the experiment and measure the gravitational force F_g, you need to know the restoring torque constant k that appears in Equation 32. But the feature of the glass fiber that makes it good for the experiment, the small value of k, makes it hard to directly measure the value of k. To determine k by direct measurement would mean applying known forces of magnitude F_g, but the only forces around that are sufficiently weak are the gravitational forces you are trying to measure.

Fortunately there is an easy way to obtain an accurate value of the restoring constant k. Remove the large lead balls, displace the rod from equilibrium by some reasonable angle θ as shown in Figure (14b), and let go. You will observe the rod to swing back and forth in an oscillatory motion. The rod, two balls, and glass fiber of Figure (14a) form what is called a ***torsion pendulum***, and the oscillation is caused by the restoring torque of the glass fiber. The glass fiber is acting like an angular spring, creating an angular harmonic motion in strict analogy to the linear harmonic motion of a mass suspended from a spring.

The analogy applies directly to the equations of motion of the two systems. For a linear one dimensional system like a mass on a spring, Newton's second law is

$$F_x = ma_x = m\frac{d^2x}{dt^2}$$

The angular version of Newton's second law, applied to the simple case of an object rotating about a fixed axis, is from Equation 30 of Chapter 12

$$\tau = I\alpha = I\frac{d^2\theta}{dt^2} \tag{12-20}$$

where τ is the net torque, I the angular mass or moment of inertia, and α the angular acceleration of the object about its axis of rotation.

For the linear harmonic oscillator (mass on a spring), the force F_x is a linear restoring force $F_x = -kx$, which gives rise to the equation of motion and differential equation

$$F_x = -kx = m\frac{d^2x}{dt^2} \tag{11a}$$

$$\frac{d^2x}{dt^2} + \frac{k}{m}x = 0 \tag{11}$$

For our torsion pendulum, the restoring torque is $\tau_r = -k\theta$ which gives rise to the equation of motion and differential equation

$$\tau_r = -k\theta = I\frac{d^2\theta}{dt^2} \tag{33a}$$

$$\frac{d^2\theta}{dt^2} + \frac{k}{I}\theta = 0 \tag{33}$$

Equations 11 and 33 are the same if we substitute the angular distance θ for the linear distance x, and the angular mass I for the linear mass m. For the linear motion, we saw that the spring oscillated back and forth at an oscillation frequency ω_0 and period T given by

$$\omega_0 = \sqrt{\frac{k}{m}}$$

$$T = \frac{2\pi}{\omega_0} = 2\pi\sqrt{\frac{m}{k}} \tag{19}$$

By strict analogy, we expect the torsion pendulum to oscillate with a frequency ω_0 and period T given by

$$\omega_0 = \sqrt{\frac{k}{I}}$$

$$T = \frac{2\pi}{\omega_0} = 2\pi\sqrt{\frac{I}{k}} \tag{34}$$

where I is the moment of inertia of the rod and two balls about the axis defined by the glass fiber (as shown in Figure 14).

As a result, by observing the period of oscillation of the rod and two balls (with the big masses M removed), you can determine the restoring constant k of the glass fiber, and use that result in Equation 32 to solve for the gravitational force F_g. Because you can measure periods accurately by timing many swings, k can be measured accurately, and the Cavendish experiment allows you to do a reasonably good job of measuring the gravitational force F_g.

Exercise 12

Solve the differential Equation 33 by starting with the guess

$$\theta(t) = A\sin(\omega_0 t) \tag{35}$$

Check that Equation 35 is in fact a solution of Equation 33, and find the formula for the frequency ω_0 of the oscillation. Also use dimensional analysis find the period of oscillation.

Exercise 13

In the commercial Cavendish experiment apparatus shown in Figure 8-8, the small balls each have a mass of 170 gms, the distance between the small balls is 12 cm, and the observed period of oscillation is 24 minutes.

(a) Calculate the value of the restoring constant k of the glass fiber.

(b) How big a torque, measured in dyne centimeters, is required to rotate the glass fiber by an angle of one degree. (Remember to convert degrees to radians.)

The Simple Pendulum

Perhaps the most well-known example of oscillatory motion is the simple pendulum which consists of a mass swinging back and forth on the end of a string or rod. The regular swings of this pendulum serve as the basic timing device of the grandfather's clock.

When we begin to analyze the simple pendulum, we will find that it is not quite so simple after all. The restoring force is not strictly a linear restoring force and we end up with a differential equation whose solution is more complex than the sinusoidal oscillations we have been discussing. What allows us to include this example in our discussion of simple harmonic motion is the fact that, *for small amplitude oscillations*, the restoring force is approximately linear, and the resulting motion is approximately sinusoidal.

Figure (16) is a sketch of a simple pendulum consisting of a small mass m swinging on the end of a string of length ℓ. The downward gravitational force $m\vec{g}$ has a component of magnitude $(mg\sin\theta)$ directed along the circular path of the ball.

Since the ball is constrained to move along the circular arc, we can analyze the motion of the ball by equating the tangential forces acting along the arc to the mass times the tangential acceleration. The tangential component of the gravitational force is always directed toward the bottom equilibrium position, thus it is a restoring force of the form

$$F_{tangential} = -mg\sin\theta \qquad (36)$$

As the mass moves along the arc, the speed of the ball is related to the angle θ by

$$v_{tangential} = \ell\frac{d\theta}{dt} \qquad (37)$$

a result from the beginning of our discussion of circular motion in Chapter 12. (See the discussion before Equation 12-11.) Differentiating Equation 37, we get for the tangential acceleration

$$a_{tangential} = \frac{dv_{tangential}}{dt} = \ell\frac{d^2\theta}{dt^2} \qquad (38)$$

Thus Newton's second law gives

$$F_{tangential} = ma_{tangential}$$

$$-mg\sin\theta = m\ell\frac{d^2\theta}{dt^2} \qquad (39)$$

Dividing through by $m\ell$ and rearranging terms gives us the differential equation

$$\boxed{\frac{d^2\theta}{dt^2} + \frac{g}{\ell}\sin\theta = 0} \quad \begin{array}{l}\textit{equation for a}\\ \textit{simple pendulum}\end{array} \quad (40)$$

Equation 40, the differential equation for the simple pendulum, is more complex than the equations we have been discussing that lead to simple harmonic motion. If you try as a guess that the motion is sinusoidal and try the solution $\theta = \sin(\omega t)$, it does not work. You are asked to see why in the following exercise.

Exercise 14
Try substituting the guess

$$\theta = \sin(\omega t)$$

into Equation 40 and see what goes wrong. Why can't you make the left side zero with this guess?

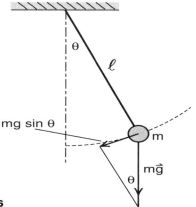

Figure 16
Simple pendulum consisting of a mass swinging on the end of a string. The gravitational force has a component $mg\sin\theta$ in the direction of motion of the mass.

There is a solution to Equation 40, it is just not the sine curve we have been discussing. The solution is a curve called an *elliptic integral*, a curve generated much as we generated the sine curve in Figure (3), except that the stick whose shadow generates the curve has to move around an elliptical path rather than around the circular path used in Figure (3). Elliptic integrals carry us farther into the theory of functions than we want to go in this text, thus we will not discuss the exact solution of the differential Equation 40.

Small Oscillations

The problem with Equation 40 is the appearance of the function $\sin \theta$ in the second term on the left hand side. It is this term that seems to keep us from using the oscillatory solution.

In Figure (17) we look again at the geometry of the simple pendulum. In that figure we have a right triangle whose small angle is θ, hypotenuse the string length ℓ, and opposite side x. The definition of the sine of the angle θ is

$$\sin \theta \equiv \frac{\text{opposite side}}{\text{hypotenuse}} = \frac{x}{\ell} \qquad (41)$$

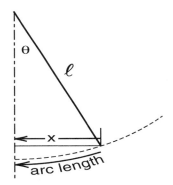

Figure 17

$\sin \theta = \frac{x}{\ell}$ $\theta = \frac{arc\,length}{\ell}$

The definition of the angle θ, in radian measure, is the arc length divided by the radius ℓ of the circular arc

$$\theta \equiv \frac{\text{arc length}}{\text{radius}} = \frac{\text{arc length}}{\ell} \qquad (42)$$

From Figure (17) we see that for small angles θ the opposite side x and the arc length ℓ are about the same. The smaller the angle θ, the more nearly equal they are. If we restrict our analysis to small amplitude swings, we can replace $\sin \theta$ by θ in Equation 40, giving us the differential equation

$$\boxed{\frac{d^2\theta}{dt^2} + \frac{g}{\ell}\theta = 0} \quad \begin{array}{l}\textit{equation for small}\\ \textit{oscillations of a}\\ \textit{simple pendulum}\end{array} \qquad (43)$$

Equation 43 is an equation for simple harmonic motion. If we try the guess $\theta = \sin(\omega_0 t)$, and plug the guess into Equation 43, we can solve the equation provided the frequency ω_0 and the period of oscillation T have the values

$$\omega_0 = \sqrt{\frac{g}{\ell}}$$

$$T = \frac{2\pi}{\omega_0} = 2\pi\sqrt{\frac{\ell}{g}} \quad \begin{array}{l}\textit{period of a}\\ \textit{simple pendulum}\end{array} \qquad (44)$$

Exercise 15

Substitute the guess $\theta = A \sin(\omega_0 t)$ into Equation 43 and show that you get a solution provided $\omega_0^2 = g/\ell$. Then use dimensional analysis to derive a formula for the period of the oscillation.

From Equation 44, we see that the period of the oscillation of a simple pendulum depends only on the gravitational acceleration g and the length ℓ of the pendulum. It does not depend on!tXe mass m of the swinging object, nor on the amplitude of the oscillation, provided that the amplitude is kept small. For these reasons the simple pendulum makes a good timing device.

Exercise 16

How long should a simple pendulum be so that it's period of oscillation is one second?

Simple and Conical Pendulums

In Chapter 9 we analyzed the motion of a conical pendulum. The conical pendulum also consists of a mass on a string, but the mass is swung around in a circle as shown in Figure (18), rather than back and forth along an arc as for a simple pendulum.

From our analysis of the conical pendulum, we found that the period of rotation was given by the formula

$$T_{cp} = 2\pi \sqrt{\frac{h}{g}} \quad \begin{array}{l}\textit{period of a}\\ \textit{conical pendulum}\end{array} \quad (9\text{-}34)$$

where h is the height shown in Figure (18). Considering the trouble we went through to get an approximate solution to the simple pendulum, it seems surprising that Equation 9-34 is an exact solution to Newton's second law for any achievable radius x of the circle.

For small circles, where $x \ll \ell$, the height h and the string length ℓ are approximately the same and we have

$$T_{cp} = 2\pi \sqrt{\frac{g}{h}} \approx 2\pi \sqrt{\frac{g}{\ell}} \quad (45)$$

But this is just the period of a simple pendulum if the oscillations are kept small. Since the two pendulums have the same period for small oscillations, it makes no difference, as far as the period is concerned, whether we swing the balls back and forth or around in a circle. This prediction is easily checked by experiment.

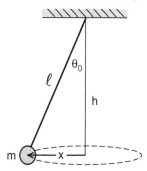

Figure 18
The conical pendulum.

Exercise 17

You can do your own experiments to show that as you increase the amplitude of a simple pendulum, the period of oscillation starts to get longer. In contrast, when you increase the radius of the circle for a conical pendulum, the height h and the period T_{cp} become shorter.

(a) From your own experiments estimate how much longer the period of a simple pendulum is when the maximum angle θ_{max} is 90° than when θ_{max} is small. (Is it 20% longer, 30% longer? Do the experiment and find out. Does this percentage depend on the length ℓ of the string?

(b) For a conical pendulum, at what angle θ_0 (shown in Figure 18) is the period half as long as it is for small angles θ_0? Give your answer for θ_0 in degrees.

Exercise 18

Another way to analyze the simple pendulum is to treat the mass and the string as a rigid object that can rotate about an axis through the top end of the string as shown in Figure (19). Then use the angular version of Newton's second law in the form

$$\tau = I \frac{d^2\theta}{dt^2} \qquad (12\text{-}20)$$

where τ is the net torque, and (I) the moment of inertia, of the mass and string about the axis 0.

(a) When the string is at an θ angle as shown in Figure (19), what is the torque τ about the axis 0, exerted by the gravitational force $m\vec{g}$?

(b) What is the moment of inertia I of the mass and string about the axis 0?

(c) Show that when you use the above values for τ and (I) in Equation 12-20 you get the same differential Equation 40 that we got earlier for the simple pendulum.

Exercise 19 A Physical Pendulum

A uniform rod of length ℓ is pivoted at one end as shown in Figure (20). It is free to swing back and forth about this axis, forming what is called a physical pendulum. A *simple pendulum* is one where the mass is all concentrated at the end as in Figure (19). In a *physical pendulum* the mass is distributed in some other way, in this case uniformly along the rod.

(a) What is the torque τ about the axis 0 exerted by the gravitational force on the rod? (In Chapter 13 near Equation 13-11, we showed that when calculating the torque exerted by a gravitational force, you may assume that all the mass is concentrated at the center of gravity of the object.)

(b) What is the moment of inertia (I) of the rod about the axis at the end of the rod? (See exercise 5 in Chapter 12.)

(c) Write the differential equation for the motion of the rod. (Use the procedure outlined in Exercise 18.)

(d) Find the period of small oscillations of the rod.

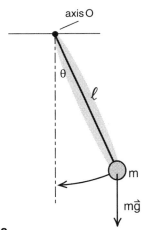

Figure 19
The simple pendulum treated as a rigid object.

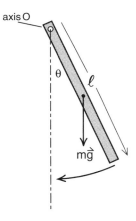

Figure 20
A physical pendulum.

NON LINEAR RESTORING FORCES

The simple pendulum is an example of an oscillator with a non linear restoring force. In Figure (21), we show the actual restoring force (mg sin θ) and the linear approximation (mg θ) that we used in order to solve the differential equation for the pendulum's motion. You can see that if the angle θ always remains small, much less than $\pi/2$ in magnitude, then the linear force (mg θ) is a good approximation to the non linear force (mg sin θ). Since the linear force gives rise to sinusoidal simple harmonic motion, we expect sinusoidal motion for small oscillations of the simple pendulum. What we are seeing is that a linear restoring force is described by a straight line, and that the non linear restoring force can be approximated by a straight line in the region of small oscillations.

In physics, there are many examples of complex, non linear restoring forces which for small amplitudes can be approximated by a linear restoring force, and which therefore lead to small amplitude sinusoidal oscillations. A rather wild example which we will discuss shortly, is the collapse of the Tacoma Narrows bridge. The bridge undoubtedly started oscillating with small amplitude sinusoidal oscillations. What happened was that these oscillations were continually driven by the shedding wind vortices until the amplitude of oscillation became large and the restoring force was no longer linear. (There was still a more or less sinusoidal motion almost up to the point when the bridge collapsed.)

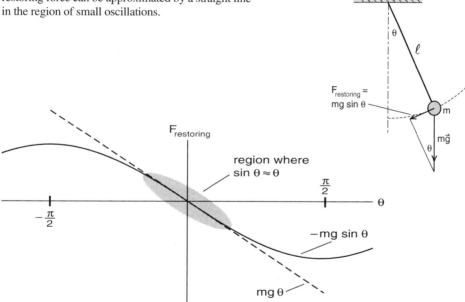

Figure 21
The non linear restoring force mg sin θ can be approximated by the straight line (linear term) mg θ if we keep the angle θ small.

MOLECULAR FORCES

One of the most important examples of a non linear restoring force is the molecular force between atoms. Consider, for example, the hydrogen molecule which consists of two hydrogen atoms held together by a molecular force. (We will discuss the origin of the molecular force in Chapter 17.)

In the hydrogen molecule, the hydrogen atoms have an equilibrium separation, and the molecular force provides a restoring force to this equilibrium separation. The restoring force, however, is quite non linear. If you try to squeeze the atoms together, you quickly build up a large repulsive force that keeps the atoms from penetrating far into each other.

If you try to pull the atoms apart, there is an attractive force that pulls the atoms back together. The attractive force never gets too big, and then dies out when the separation gets much larger than an atomic diameter.

In Figure (22) we have sketched the molecular force as a function of the separation of the atoms, the origin being at the equilibrium position. This graph is not too unlike Figure (21) where we have the force curve for the simple pendulum. For the pendulum, the equilibrium position is at $\theta = 0$, thus the origin of both curves represents the equilibrium position.

While the overall shape of the force curves for the simple pendulum and the molecular force are quite different, right in close to the origin both curves can be approximated by a straight line, a linear restoring force. As long as the amplitudes of the oscillation remain small, we effectively have a linear restoring force and any oscillations should be simple harmonic motion.

In Chemistry texts one often sees molecular forces as being represented by springs as shown in Figure (23). The spring force, given by Hooke's law, is our ideal example of a linear restoring force. We can now see that, while the molecular force in Figure (22) does not look like a linear spring force, if the amplitude of oscillation remains small, the spring force provides a reasonably good approximation to the actual molecular force. The chemist's diagrams are not so bad after all.

In a crystal, like quartz, where you have many atoms held together by molecular forces, it is possible to get all the atoms oscillating together. Each atom only oscillates a very small distance about its equilibrium position, but all the oscillations can add up to produce a fairly large, quite detectable oscillation of the crystal as a whole. An advantage of a quartz crystal is that these oscillations can be both driven and detected by electric fields. This vibration or simple harmonic motion of a small quartz crystal is used as the basic timing device for digital watches, computers, and almost all forms of modern electronics.

In Galileo's time we used small oscillations of a non linear harmonic oscillator, the simple pendulum, as a basic time device. Now we use the small oscillations of a non linear harmonic oscillator, the atoms in a quartz crystal, as our most convenient timing device. The main thing we have changed in the last 300 years is not the basic physics, but the size and frequency of the device.

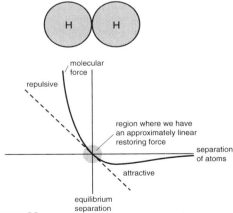

Figure 22
Sketch of the molecular force between two hydrogen atoms. As long as the atoms stay close to the equilibrium position, the force can be represented by a straight line—a linear restoring force.

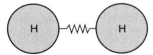

Figure 23
Representation of the molecular force by a spring force.

DAMPED HARMONIC MOTION

If you start a pendulum swinging or a quartz crystal oscillating, and do not keep the oscillation going with some kind of external force, the oscillation will eventually die out due to friction forces. Such a dying oscillation is called *damped harmonic motion*.

The analysis of damped harmonic motion starts out quite easily. In Newton's second law add a damping force like the air resistance term we added to our analysis of projectile motion. (See Chapter 3, Figure 31.) We could write, for example

$$F_{tot} = F_{restoring} + F_{damping}$$

$$F_{tot} = -kx - bv_x \quad (46)$$

where $x(t)$ is the coordinate of the oscillator, $v_x(t) = dx/dt$ its velocity, and we are assuming a simple linear damping proportional to $(-v)$ with a strength b.

Using Equation 46 in Newton's second law gives

$$F_{tot} = -kx - bv_x = ma_x \quad (47)$$

With $a_x = d^2x/dt^2$, this becomes, after dividing through by m and rearranging terms

$$\boxed{\frac{d^2x}{dt^2} + \frac{b}{m}\frac{dx}{dt} + \frac{k}{m}x = 0} \quad (48)$$

Equation 48 is our new differential equation for damped harmonic motion. It is like our old differential Equation 11a for undamped oscillation, except that it has the additional term $\frac{b}{m}\frac{dx}{dt}$ representing the damping.

If we were doing a computer solution of harmonic motion, adding the damping term represents hardly any extra effort at all. In the appendix to this chapter we discuss a short computer program to handle harmonic motion. Starting with the first version of the program that has no damping, you can include damping by changing the line

LET F = – k ∗ x

to the new line

LET F = – k ∗ x – b ∗ v (49)

and placing your choice for the damping constant b in the initial conditions.

In contrast, when working with differential equations analytically you find that a very small change in the equation can make a great deal of difference in the effort required to obtain a solution. Adding a bit of damping to a harmonic oscillator changes the curve from a pure sinusoidal motion to a dying sine wave. If you try using a pure sine wave as a guess for the solution to the differential Equation 48 for damped harmonic motion, the guess does not work because the pure sine wave has the wrong shape. The decay of the sine wave has to be built into your guess before the guess stands a chance of working.

The difficult part about solving differential equations is that you essentially have to know the answer before you can solve the equation. You only have to know general features like the fact that in working with Equation 11a you are dealing with a sinusoidal oscillation. You can then use the differential equation to determine explicit features like the frequency of the oscillation. It is helpful to have a physical example to tell you what the general features of the motion are, so that you can begin the process of solving the equation. That is why we begin this chapter with the demonstration in Figure (2) that the motion of a mass on a spring is similar to circular motion seen sideways, namely sinusoidal motion.

To set up a physical model for damped harmonic motion is not too difficult. One way to add damping to the air cart and springs oscillator is to run a string from the air cart over a pulley to a small weight as shown in Figure (24). The idea was to have the weight move up and down in a glass of water to give us fluid damping. But it turned out that there was enough friction in the pulley itself to give us considerable damping.

To record the motion of the cart, we used the air cart velocity detector that we used in Chapter 8 to study the momentum of air carts during collisions. Figure (25a) shows the velocity of the air cart damped only by the friction in the pulley. In Figure (25b) water was added to the glass so that the weight on the string was moving up and down in water. The result was considerably more damping with the curve almost dying out before any oscillations take place.

It turns out that mechanical oscillators like a pendulum or a mass on a spring are not particularly convenient devices for studying damped harmonic motion, or forced harmonic motion which is the subject of the next section. It is hard to control the damping, just adding the pulley in Figure (24) gave us almost too much damping. Worse yet, the damping that we get from friction in a pulley, or a mass moving up and down in water, is not a simple linear damping force of the form $-bv$. What is remarkable about these systems is that much more complex forms of damping give us results similar to what we would get with linear damping.

In Chapter 27 we will study the behavior of basic electric circuits made from electrical components called capacitors, inductors, and resistors. It turns out that the amplitude of the currents in these circuits obey differential equations that are exactly like our oscillator Equations (11) and (45). The damping is caused by the resistor in the circuit, the damping is accurately given by a linear damping term proportional to the amount of resistance in the circuit. (The resistance can be changed simply by turning a knob on a resistance pot.)

Figure (26), taken from Chapter 27, is an example of damped harmonic motion in an electric circuit. Here we have a curve with enough oscillations so that we can see how the wave is damped. In Chapter 27 we will see that the amplitude of the oscillation dies exponentially, following a mathematical curve of the form $e^{-\alpha t}$. As a result, the wave in Figure (26) has the form

$$x = \left(Ae^{-\alpha t}\right) \sin \omega t$$

decaying   sine wave (50)
amplitude oscillation

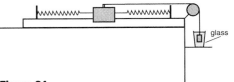

Figure 24
Adding damping to the air cart oscillator.

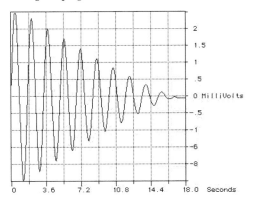

Figure 25a
Damping caused by the pulley and weight alone.

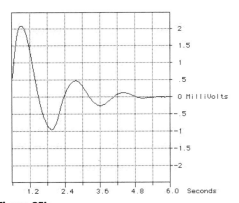

Figure 25b
Resulting motion when water was added to the glass.

It turns out that if we use Equation 50 as our guess for a solution to our differential Equation 48 for damped harmonic motion, the guess works, and we can determine both the frequency ω and decay rate α in terms of the constants that appear in Equation 48.

When we study electric circuits, you will get much more experience with the exponential function $e^{-\alpha t}$, and you will have a better laboratory setup for studying damped and forced harmonic motion. In other words, now, with our somewhat crude mechanical experiments and lack of familiarity with exponential damping, is not the best part of the course to go deeply into the mathematical analysis of these motions. What we will do instead is discuss the motions more or less qualitatively and leave the more detailed analysis for later.

Exercise 20 Damped Harmonic Motion

We won't let you completely off the hook for doing mathematical analysis of damped harmonic motion. Start with Equation 5a as a guess for the form for the displacement x(t) for a damped harmonic oscillator

$$x(t) = Ae^{-\alpha t}\sin(\omega t)$$

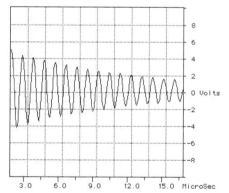

Figure 26
Damped harmonic motion seen in an electric circuit. Note the difference in time scales. The electrical oscillations we will study are usually of much higher frequency than the mechanical ones.

use the following rules of differentiation to calculate dx/dt and d^2x/dt^2

$$\frac{d}{dt}e^{-\alpha t} = -\alpha e^{-\alpha t}$$

$$\frac{d}{dt}[a(t)b(t)] = \frac{da}{dt}b + a\frac{db}{dt}$$

and show that when you try this guess in the differential Equation 48, you do in fact get a solution, and that ω and α are given by

$$\omega = \sqrt{\frac{k}{m} - \frac{b^2}{4m^2}} \qquad \alpha = \frac{b}{2m} \qquad (51)$$

You can see that in the absence of damping, when b = 0, we get back to our old result $\omega = \sqrt{k/m}$.

Critical Damping

In Figure (25b) the damping was so great that the motion damped out almost before the curve had a chance to oscillate. It turns out that there is a critical amount of damping that just kills all oscillations. Any further increase in damping and the mass just coasts to rest.

The idea of critical damping can be seen in our analytic solution for damped harmonic motion obtained Exercise 20. Equation 51 gives us a formula for the frequency of oscillation ω in terms of the constants k and b. We can see as the frequency of oscillation goes to zero, i.e., the period of oscillation becomes infinite when

$$\frac{k}{m} = \frac{b^2}{4m^2} \; ; \quad b = 2\sqrt{mk}$$

$$b = \sqrt{4mk} \qquad \begin{array}{l}\textit{critical}\\ \textit{damping}\end{array} \qquad (52)$$

Equation 52 is the condition for critical damping because if the period of oscillation is infinite there are no oscillations.

RESONANCE

When you are pushing a child on a swing, you time your pushes to coincide with the motion of the child. Usually you give a shove just after the child has swung back and is starting forward again.

If you push forward just as the child is swinging forward, your force $\vec{F}$ and the child's velocity $\vec{v}$ are in the same direction, the dot product $\vec{F} \cdot \vec{v}$ is positive, and you are adding energy to the child's motion. Initially the energy you add goes into increasing the amplitude of the swing. After a while friction effects become large enough that the energy you add in each push is dissipated by friction in each swing. (If there is not enough friction, or you push too hard, the child will end up going over the top.)

The key to getting the child swinging was to time your shoves so that $\vec{F} \cdot \vec{v}$ was always positive. If you pushed the child at random intervals, so that $\vec{F} \cdot \vec{v}$ was sometimes positive, sometimes negative, you would be sometimes adding energy and sometimes removing it. The net result would be that your shoves would not be particularly effective in helping the child to swing.

To make sure that you are always adding energy to the child's swing, you want to time your shoves with the natural frequency of oscillation of the child. When you do this, we say that your shoves are in *resonance* with the oscillation of the child.

The striking feature of resonance is that a small repeated force can produce a large oscillation. If the damping is small then by adding just a little energy with each shove, the energy accumulates until you end up with a very energetic oscillation. A rather dramatic consequence of this effect is shown in Figure (27) where we see the Tacoma Narrows bridge oscillating wildly and then collapsing.

The new bridge was dedicated in April of 1940. Three months later a reasonably stiff breeze started the bridge oscillating, an oscillation that finally destroyed the bridge's integrity.

The brute force of the wind itself did not destroy the bridge. The bridge was designed to handle far stronger winds. What happened was that as the wind was blowing over the bridge, vortices began to peel off the bridge. Whenever fluid flows past a cylindrical object at the right speed, vortices began to peel off, first on one side of the cylinder, then the other, and are carried downstream, forming a wake of vortices seen in the wind tunnel photograph of Figure (28). This vortex structure is called a **Karmen vortex street** after the hydrodynamicist Theodore Von Karmen.

In the case of the Tacoma Narrows bridge, vortices alternately peeled off the top and bottom of the down wind side of the bridge, rocking the bridge at its natural frequency of oscillation. While no separate jolt by any one vortex would have much effect on the bridge, the

Figure 27a
Tacoma Narrows bridge oscillating in the winds of a mild gale on July 1, 1940.

Figure 27b
After a couple of hours the bridge collapsed.

resonance between the peeling of vortices and the oscillation of the bridge caused the oscillations to grow to destructive proportions.

The example of the Tacoma Narrows bridge illustrates how widely the ideas of simple harmonic motion and resonance apply to physical systems. The bridge is far more complex than a mass on a spring, and the *vortex street* (line of vortices) exerts a rather complex driving force. However the bridge had a natural frequency, the vortices provided a small driving force at that frequency, and we got a resonant amplification of the oscillation.

To apply Newton's second law to resonant motion, we have to add an oscillating driving force to the system under study. As we have seen from the Tacoma Narrows bridge discussion, we do not need to know the exact form of the driving force, all we need is a repetitive force that can be timed with the natural oscillation. For the theoretical analysis we can use the simplest mathematical form we can find for the driving force, which turns out to be a sine wave.

To write a formula for the driving force, let ω_0 be the natural frequency of oscillation ($\omega_0 = \sqrt{k/m}$ if the damping is small), and let ω be the frequency of the driving force. Then the total force, acting on the oscillating system like a mass on a spring, can be written

$$F_{x\,tot} = -kx - bv + F_d \sin(\omega t) \qquad (52)$$

Figure 28
Karman vortex street in the flow of water past a circular cylinder. The vortices peel off of alternate sides of the cylinder and flow downstream forming a double line of vortices. (Reynolds number = 140.) Photograph by Sadatoshi Taneda.

where in the driving term, F_d, represents the amplitude or strength of the sinusoidal driving force. Using Equation 52 in Newton's second law $F_{x\,tot} = ma_x$ gives

$$-kx - b\frac{dx}{dt} + F_d \sin(\omega t) = m\frac{d^2x}{dt^2} \qquad (53)$$

Dividing through by m and rearranging terms gives

$$\boxed{\frac{d^2x}{dt^2} + \frac{b}{m}\frac{dx}{dt} + \frac{kx}{m} = F_d \sin(\omega t)} \qquad (54)$$

Equation 54 is the standard form for the differential equation representing forced or resonant harmonic motion. It is the simplest equation we can write whose solution has the features we associate with the phenomena of resonance.

In our study of electric circuits, we can easily create a circuit whose behavior is accurately described by Equation 54. We saw that we could use resistors to add linear damping of the form $-bv$. It is not hard to add a purely sinusoidal driving force of the form $F_d \sin(\omega t)$, where we can adjust the driving frequency by turning a knob. In other words, with electric circuits we can accurately study the predictions of Equation 54.

With mechanical systems like a mass on a spring, it is hard to get linear damping, and the sinusoidal driving force is usually simulated by some trick such as wiggling the supported end of the spring at a frequency ω. Despite the crudeness of the experiment, the equation gives a surprisingly good prediction of what we see.

Since we will later have a laboratory setup that accurately matches Equation 54, we will postpone (until Appendix 1) the mathematical solution of Equation 54. Instead we will investigate the resonance phenomena qualitatively, using the simple setup of a mass on a spring, where we hold the other end of the spring in our hand and move our hand up and down at a frequency ω as shown in Figure (29).

Resonance Phenomena

We have already seen that if we hold our hand still, pull the mass down, and let go, the mass oscillates up and down at the natural frequency $\omega_0 = \sqrt{k/m}$. You can observe some damping, because the mass finally stops oscillating. But the damping is small and has no noticeable effect on the resonant frequency. (We can neglect the term $b^2/4m^2$ in Equation 51.)

Now try a very different experiment. Stop the mass from oscillating, and slowly move your hand up and down a small distance. If you do this slowly enough and carefully enough, the mass will move up and down with your hand (just as if the spring were not there). In this case the formula for the motion of the mass is

$$x = x_0 \sin(\omega t) \quad \omega \ll \omega_0 \quad (55)$$

where the frequency ω of the oscillation of the mass is the frequency of ω the oscillation of your hand. This only happens if you oscillate your hand at a frequency much much lower than the natural oscillation frequency ω_0.

In the next experiment, keep the oscillations of your hand small in amplitude, but start moving your hand up and down rapidly, at a frequency ω considerably greater than the natural frequency ω_0. Now, what happens is that the mass oscillates at the same frequency as your hand, but out of phase. When your hand is going down, the mass is coming up, and vice versa. Now the formula for the displacement x of the mass is

$$x = -x_0 \sin(\omega t) \quad \omega \gg \omega_0 \quad (56)$$

where the minus sign tells us that the mass is oscillating out of phase with our hand.

A way we can write both Equations 55 and 56 is in the form

$$x = x_0 \sin(\omega t + \phi) \quad (57)$$

where ϕ is the phase angle of the oscillation (see Figures 8 and 9 and Equation 4 at the beginning of this chapter for a discussion of phase angle.) In Equation 56, where $\omega \ll \omega_0$ and there was no phase difference, the phase angle ϕ is zero. In Equation 56 where $\omega \gg \omega_0$, and the motion is completely out of phase, the phase angle ϕ is π or 180°.

Equations 55 and 56 represent the two extremes of driven harmonic motion. The mass moves with a small amplitude at the same frequency ω as the driving force. When the driving frequency is much less than the natural frequency ω_0, the difference in phase between the driving force and the response of the mass is zero degrees. When $\omega \gg \omega_0$ the phase difference increases to π or 180°.

As the third experiment, start at the low frequency where the mass is following your hand, and slowly increase the frequency ω of oscillation of your hand, keeping the amplitude of oscillation constant. As ω approaches ω_0, the amplitude of oscillation of the mass increases. When you get close to the natural frequency ω_0, the oscillation becomes so large that the mass will most likely jump off the spring. This is the phenomena of resonance, the phenomenon that destroyed the Tacoma Narrows bridge.

How big the oscillation of the mass becomes depends mainly how close you are to resonance, how close ω is to ω_0, and how big the damping force is relative to the driving force. The formula for the amplitude x_0 of the motion of the mass obtained by solving Equation 54 is

$$x_0 = \frac{F_d/m}{\sqrt{(\omega^2 - \omega_0^2)^2 + \omega^2 b^2/m^2}} \quad (58)$$

where F_d is the strength of the driving force, ω the driving frequency, $\omega_0 = \sqrt{k/m}$ the natural frequency and b the damping constant. In the absence of damping (b=0), Equation 58 predicts an infinite amplitude at the resonant frequency $\omega = \omega_0$. Such an infinite amplitude is prevented either by damping or by the destruction of the system. (A damping mechanism could have saved the Tacoma Narrows bridge.)

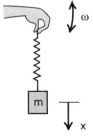

Figure 29
Experimental setup for a qualitative study of resonance phenomena.

Exercise 21

To derive Equation 21, you start with Equation 57 as a guess

$$x = x_0 \sin(\omega t + \phi) \quad (57)$$

and substitute that into the differential Equation 54. It turns out that you can get a solution provided that the amplitude x_0 has the value given by Equation 58, and the phase angle ϕ is given by

$$\boxed{\tan(\phi) = \frac{(b/m)\omega}{\omega^2 - \omega_0^2}} \quad (59)$$

Doing the work, actually substituting Equation 57 into 54 and getting Equations 58 and 59 for x_0 and ϕ is a somewhat messy job which we leave to Appendix 1 on the next page. Here we would rather have you develop an intuitive feeling for the solutions in Equations 58 and 59 by working the following exercises.

(a) Write the formula for x_0 in the case b = 0. Sketch the resulting curve using the axes shown in Figure (30). Explain what happens as $\omega \to \omega_0$.

(b) When the damping is not zero, find the formula for the amplitude of oscillation x_0 at the resonant frequency $\omega \to \omega_0$. Check the dimensions of your answer.

(c) What is the phase angle ϕ at resonance? How does the phase angle change as we go from $\omega \ll \omega_0$ to $\omega \gg \omega_0$?

In Figure (31) we have graphed the amplitude x_0 for a fixed driving force F_d, as a function of ω for several values of the damping constant b. The main point to get from this diagram is that the smaller the damping, the sharper the resonance.

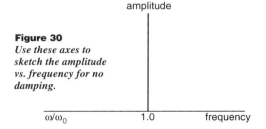

Figure 30
Use these axes to sketch the amplitude vs. frequency for no damping.

Transients

There is one more qualitative experiment we want to do with our simple apparatus of the hand held spring and mass of Figure (29). Instead of gently starting the mass moving as we had you do in the earlier experiments, let the mass fall from some small height and move your hand up and down at the same time.

If you just let the mass drop from some small height, it will oscillate up and down at the resonant frequency ω_0. If you just start moving your hand slowly at a frequency ω, the mass will move at the same frequency as your hand, building up to an amplitude given more or less by Equation 58 and shown in Figure (31), the driven oscillation we have been discussing.

If you drop the weight and move your hand at the same time, you get both kinds of motion at once. You get the natural oscillation at a frequency ω_0 that eventually dies out due to damping, and the driven oscillation at the frequency ω that eventually builds up to an amplitude x_0. For a while, before the natural oscillation has died out, the resulting motion is a mixture of two frequencies of oscillation and can look quite complex. The natural oscillation is called a ***transient*** because it eventually dies out. But until the transients do die out, forced harmonic motion can be fairly complicated to analyze. In the next chapter we will study a powerful technique called *Fourier analysis* that allows us to study complex motions that involve such a mixture of oscillations.

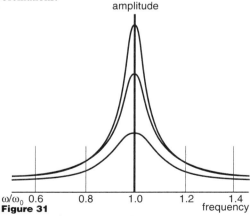

Figure 31
Amplitude of the oscillation for various values of the damping constant. The amplitude of the driving force F_d is the same for all curves.

APPENDIX 14–1
SOLUTION OF THE DIFFERENTIAL EQUATION FOR FORCED HARMONIC MOTION

The equation we wish to solve is Equation 54

$$\frac{d^2x}{dt^2} + \frac{b}{m}\frac{dx}{dt} + \frac{k}{m}x = \frac{F_d}{m}\sin(\omega t) \quad (54)$$

where our guess for a solution is Equation 57

$$x = x_0 \sin(\omega t + \phi) \quad (57)$$

The quantities x_0 and ϕ are the unknown amplitude and phase of the oscillation that we wish to determine.

To simply plug our guess into Equation 54 and grind away leads to a sufficiently big mess that we could easily make a mistake. We will instead simplify things as much as possible to make the calculation easier. The first step is to define the constants

$$\omega_0^2 = k/m \,; \quad b' = b/m \,; \quad F' = F_d/m \quad (60)$$

Next we wish to get the phase angle into the forcing term so that it appears only once in our equation. We can do this by using a time scale t' where

$$\omega t' = \omega t + \phi \implies \omega t = \omega t' - \phi \quad (61)$$

In terms of the new constants and t' our differential equation becomes

$$\frac{d^2x}{dt^2} + b'\frac{dx}{dt} + \omega_0^2 x = F' \sin(\omega t' - \phi) \quad (62)$$

Our guess, and its first and second derivative are

$$x = x_0 \sin(\omega t') \quad (63a)$$

$$\frac{dx}{dt} = \omega x_0 \cos(\omega t') \quad (63b)$$

$$\frac{d^2x}{dt^2} = -\omega^2 x_0 \sin(\omega t') \quad (63c)$$

In deriving dx/dt we used the fact that

$$\frac{dx}{dt} = \frac{dx}{dt'}\frac{dt'}{dt} = \frac{dx}{dt'}$$

where $t' = t - \phi/\omega$ so that $dt'/dt = 1$. We will also use the trigonometric identity

$$\sin(a+b) = \sin(a)\cos(b) + \cos(a)\sin(b)$$

to write

$$F' \sin(\omega t' - \phi)$$
$$= F' \sin(\omega t')\cos(-\phi) + F' \sin(-\phi)\cos(\omega t') \quad (64)$$
$$= F' \sin(\omega t')\cos(\phi) - F' \sin(\phi)\cos(\omega t')$$

where we used

$$\cos(-\phi) = \cos(\phi), \quad \sin(-\phi) = -\sin(\phi)$$

Substituting Equation 63 and 64 into 62, and separately collecting terms with $\sin(\omega t')$ and $\cos(\omega t')$, we get

$$\sin(\omega t')\Big[-\omega^2 x_0 + \omega_0^2 x_0 - F'\cos(\phi)\Big] +$$
$$\cos(\omega t')\Big[\omega b' x_0 + F' \sin(\phi)\Big]$$
$$= 0 \quad (65)$$

Because there are both $\cos(\omega t')$ terms and $\sin(\omega t')$ terms in Equation 65, there is no way to make everything add up to zero for all times unless the coefficients of both $\cos(\omega t')$ and of $\sin(\omega t')$ are separately equal to zero. This gives us the two equations

$$F'\sin(\phi) = -\omega b' x_0 \quad (66)$$

$$F'\cos(\phi) = (\omega_0^2 - \omega^2) x_0 \quad (67)$$

If we divide Equation 66 by Equation 67, the F' and x_0 cancel and we get

$$\frac{\sin(\phi)}{\cos(\phi)} = \frac{-\omega b'}{\omega_0^2 - \omega^2}$$

$$\boxed{\tan(\phi) = \frac{(b/m)\omega}{\omega^2 - \omega_0^2}} \quad (59)$$

This is the result we stated earlier, namely Equation 59.

To solve for the amplitude x_0 of the oscillation, we can use Equation 66 to get

$$x_0 = \frac{F'}{\omega b'}\sin(\phi) \quad (68)$$

To find the $\sin(\phi)$ from the $\tan(\phi)$ which we already know, construct the right triangle shown in Figure (32). We have made the opposite and adjacent sides so that the ratio comes out as $\tan(\phi)$, and the hypotenuse is given by the Pythagorean theorem. Thus $\sin(\phi)$ is

$$\sin(\phi) = \frac{\omega b'}{\sqrt{(\omega^2 - \omega_0^2)^2 + \omega^2 b'^2}} \quad (69)$$

Substituting 69 into 68, and setting $b' = b/m$, $F' = F_d/m$, we get

$$\boxed{x_0 = \frac{F_d/m}{\sqrt{(\omega^2 - \omega_0^2)^2 + \omega^2 b^2/m^2}}} \quad (58)$$

which is our earlier Equation 58 for x_0.

Transients

In our qualitative discussion of forced harmonic motion, we saw that in addition to the driven oscillation $x_{driven} = x_0 \sin(\omega t + \phi)$ we have just studied, we could also have transient motion at the natural frequency ω_0. In controlled experiments, you observe that any transient motion present initially finally dies out and you are eventually left with just the driven motion.

The transient motion $x_{tr}(t)$ is just damped harmonic motion that satisfies the equation of motion

$$\frac{d^2x}{dt^2} + \frac{b}{m}\frac{dx}{dt} + \frac{k}{m}x = 0 \quad (48)$$

The question we wish to answer now is whether the forced harmonic motion equation

$$\frac{d^2x}{dt^2} + \frac{b}{m}\frac{dx}{dt} + \frac{k}{m}x = F_d \sin(\omega t) \quad (54)$$

allows us to have both driven and transient motion at the same time. In other words, is a guess of the form

$$x = x_0 \sin(\omega t + \phi) + x_{tr} \quad (70)$$

where x is the sum of the driven motion and an arbitrary amount of transient motion, is this sum also a solution of Equation 54?

If you substitute our new guess 70 into 54, the driven term satisfies the whole equation and the transient terms add up to zero because of Equation 48, thus we do get a solution. Transient motions are allowed by Newton's second law.

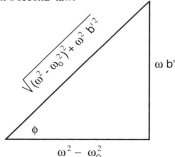

Figure 32
Triangle to go from tan ϕ to sin ϕ.

APPENDIX 14-2
COMPUTER ANALYSIS
OF OSCILLATORY MOTION

In this appendix, we will use the computer to analyze the motion of the oscillating air cart shown in Figure (33). For this problem, the computer solution does not have the elegance of the calculus solution we have been discussing. The calculus approach gives us a single solution valid for all values of the experimental parameters. With the computer we have to alter the program and rerun it any time we want to change a parameter such as the mass of the cart, the spring constant, or the initial position or velocity. The calculus approach gives us a single formula valid for all values of the experimental parameters.

However, the advantage of using the computer is that we can easily modify the program to include new physical phenomena. For example, to add damping, all we have to do is change the command

$$\text{LET } F = -K*X$$

to the command

$$\text{LET } F = -K*X - b*V$$

and rerun the program. To add damping to the calculus solution, we had to work with a differential equation (48) that was much more difficult to solve than the equation for undamped motion (11).

The computer opens up a number of possibilities for student project work. For example, in our discussion of the simple pendulum shown in Figure (21), we had to limit our analysis to small amplitude swings of the pendulum. For large amplitude swings, the restoring force became non linear which led to a differential equation that is difficult to solve. As the amplitude increases, there is a lengthening of the period that is easy to measure but difficult to predict using calculus. However with the computer, it is as easy to use the exact force $M_g \sin \theta$ as it is to use the approximate linear force $M_g \theta$. Thus with the computer you can predict the lengthening of the period and compare your results with experiment.

In Appendix I, we made a considerable effort to predict the effects of adding a time dependent driving force to a harmonic oscillator. The work paid off in that we got Equations 58 and 59 which provide a general description of resonance phenomena. With the computer you do not get these elegant formulas, but it is much easier to add a time dependent force and see what happens. In effect the computer solutions can be used as a laboratory to test the predictions of Equations 58 and 59. This provides an opportunity for a lot of project work.

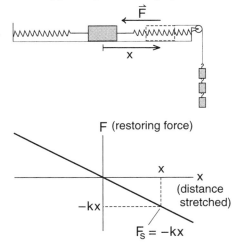

Figure 33
Reproduction of Figure (10), showing an oscillating air cart. If the cart is displaced a distance x from equilibrium, there is a restoring force F = - kx. The force is measured by adding weights as shown.

English Program

In Chapters 5 and 8 our usual approach for solving a new mechanics problem on the computer was to modify an old working program. But because the harmonic oscillator is an easy one dimensional problem, we will start over with a new program. Our general procedure has been to first write an English program that described the steps using familiar notation. Once we checked the steps to see that the program did what we wanted, we then translated the program into an actual computer language such as BASIC.

The English program for the oscillating air cart is shown in Figure 34. In the first section, we state the experimental constants, namely the mass M of the air cart and the spring restoring constant K. For this particular experiment, the cart has a mass M of 191 grams, and the spring constant K was 3947 dynes/cm. As indicated in Figure (33), the spring constant K was determined by tying a string to the mass, running the string over a pulley, and hanging weights on the other end. We got a linear force verses the distance curve like the one in Figure (9-4), and used the same method to find K.

In the next section of the program, we choose an explicit set of initial conditions. For this problem we start the cart from rest $(V_0 = 0)$ at a distance 10 cm to the right of the equilibrium position $(X_0 = 0)$. The cart is released at time $(T_0 = 0)$.

In the lab we observed that the period of oscillation was about 1.5 seconds. Thus a calculational time step $dt = .01$ seconds gives us about 150 points for one oscillation, enough points for a smooth plot.

The calculational loop is similar to the one in the projectile motion program of Figure (5-18), page 5-16, except that for one dimensional motion we do not need vectors, and the old command

    LET $\vec{A} = \vec{g}$

is replaced by

    LET F = –K∗X
    LET A = F/M

On the next page we repeat this English program and show its translation into the computer language BASIC.

English Program

! --------- Experimental constants
    LET M = 191 grams   (cart mass)
    LET K = 3947 dynes/cm (spring constant)

! --------- Initial conditions
    LET X_0 = 10 cm
    LET V_0 = 0   (release from rest)
    LET T_0 = 0   (start clock)

! --------- Computer Time Step
    LET dt = .01

! --------- Calculational loop
    DO
        LET $X_{new} = X_{old} + V_{old}*dt$
        LET F = –K∗X (spring force)
        LET A = F/M
        LET $V_{new} = V_{old} + A_{old}*dt$
        LET $T_{new} = T_{old} + dt$
        PLOT X vs T
    LOOP UNTIL T > 15

END

Figure 34
English program for the motion of an oscillating cart on an air track.

English Program

```
! --------- Experimental constants
    LET M = 191 grams      (cart mass)
    LET K = 3947 dynes/cm  (spring constant)

! --------- Initial conditions
    LET X₀ = 10 cm
    LET V₀ = 0       (release from rest)
    LET T₀ = 0       (start clock)

! --------- Computer Time Step
    LET dt = .01

! --------- Calculational loop
    DO
        LET Xnew = Xold + Vold*dt
        LET F = –K*X    (spring force)
        LET A = F/M
        LET Vnew = Vold + Aold*dt
        LET Tnew = Told + dt
        PLOT  X vs T
    LOOP UNTIL T > 15

END
```

Figure 34 repeated
English program for the motion of an oscillating cart on an air track.

The BASIC Program

Because no vectors are involved in the harmonic oscillator program, the translation into BASIC is almost automatic. Drop the subscripts "new" and "old", fix up the PLOT statement, add the plotting window commands, and you have the result shown in Figure (35a). Select **RUN** and you get the plot of oscillating motion shown in Figure (35b).

BASIC Program

```
! --------- Plotting window
    SET WINDOW -5,15,-15,15
    PLOT LINES: -5,0; 15,0  ! T axis
    PLOT LINES: 0,-15; 0,15 ! X axis

! --------- Experimental constants
    LET M = 191
    LET K = 3947

! --------- Initial conditions
    LET X = 10
    LET V = 0
    LET T = 0

! --------- Computer Time Step
    LET dt = .01

! --------- Calculational loop
    DO
        LET X = X + V*dt
        LET F = -K*X
        LET A = F/M
        LET V = V + A*dt
        LET T = T + dt
        PLOT T,X;
    LOOP UNTIL T > 15

END
```

Figure 35a
BASIC program for the motion of an oscillating cart on an air track.

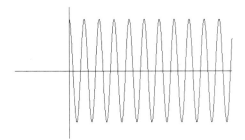

Figure 35b
Output of the BASIC program, showing the oscillation of the cart.

The plot of Figure (35b) nicely shows the sinusoidal oscillation, but does not tell us the numerical value of the period of oscillation. To determine the period, we modified the program as shown in Figure (36a). The main change is to replace the **PLOT** statement by a **PRINT** statement. To reduce the output, we included **MOD** statement (as described in Exercise 5-5, page 5-9) so that every tenth calculated point would be printed. From the output shown in Figure (36b), we see that the period is close to 1.4 seconds. A more accurate value of the period can be obtained by not using the MOD statement and printing every value as shown in Figure (36c). From this section of data we see that the period is closer to 1.39 seconds.

Exercise 22

Show that the frequency of oscillation seen in the computer output of Figure (36) is consistent with the calculus derived equation

$$\omega = \sqrt{k/M}$$

T	X	U
.1	9.08409	-19.9673
.2	6.32275	-35.8782
.3	2.27695	-44.5005
.4	-2.23141	-44.0825
.5	-6.28646	-34.7092
.6	-9.06442	-18.2848
.7	-10.001	1.8542
.8	-8.9058	21.6165
.9	-6.00143	36.9875
1.	-1.87788	44.8444
1.1	2.62716	43.5913
1.2	6.5985	33.4826
1.3	9.22935	16.5719
1.4	9.98527	-3.70531
1.5	8.71269	-23.2298
1.6	5.67013	-38.0352
1.7	1.47568	-45.1137
1.8	-3.01854	-43.0275
1.9	-6.89955	-32.2002
2.	-9.37893	-14.8315
2.1	-9.95297	5.55026
2.2	-8.50508	24.8045
2.3	-5.32939	39.0196
2.4	-1.07103	45.308
2.5	3.4049	42.3921
2.6	7.18913	30.8642
2.7	9.5129	13.0663
2.8	9.90412	-7.38598
2.9	8.28333	-26.3378
3.	4.97979	-39.9392

Figure 36b
Numerical output.

```
! --------- Print labels
    PRINT "T","X","U"

! --------- Experimental constants
    LET M = 191
    LET K = 3947

! --------- Initial conditions
    LET X = 10
    LET U = 0
    LET T = 0

! --------- Computer Time Step
    LET dt = .01
    LET i = 0

! --------- Calculational loop
    DO
        LET X = X + U*dt
        LET F = -K*X
        LET A = F/M
        LET U = U + A*dt
        LET T = T + dt
        LET i = i+1
        IF MOD(i,10) = 0 THEN PRINT T,X,U
    LOOP UNTIL T > 3

    END
```

Figure 36a
Program for numerical output.

1.36	9.92701	4.55175
1.37	9.97253	2.49094
1.38	9.99744	.424976
1.39	10.0017	-1.64187
1.4	9.98527	-3.70531
1.41	9.94822	-5.7611
1.42	9.89061	-7.80499

Figure 36c
Detailed numerical output. By printing every calculated numerical value, we can more accurately determine the period of oscillation.

Damped Harmonic Motion

In Figure (37a) we modified the projectile motion program of Figure (35a) to include damping The only change, shown in boxes in Figure (37a) is to replace

LET F = −K∗X

by

LET F = −K∗X −b∗V

where we gave b the numerical value of 100 to get the result shown in Figure (35b).

Exercise 23

(This is more of an introduction to project work)

In our analysis of damped harmonic motion in Exercise 20, we predicted that the frequency for damped harmonic motion would be

$$\omega = \sqrt{\frac{k}{M} - \frac{b^2}{4m^2}} \quad (51)$$

In the special case that

$$\frac{k}{M} = \frac{b^2}{4m^2} \;;\quad b = \sqrt{4mk} \quad (51a)$$

we get $\omega = 0$ which is the case of critical damping, where oscillations cease.

Run the damped harmonic oscillator program of Figure (35a) for values of b near $\sqrt{4mk}$ and show that oscillations cease when you get to this critical value.

```
!      Damped oscillation

! --------- Plotting window
  SET WINDOW -5,15,-15,15
  PLOT LINES: -5,0; 15,0   ! T axis
  PLOT LINES: 0,-15; 0,15  ! X axis

! --------- Experimental constants
  LET M = 191
  LET K = 3947
  LET b = 100

! --------- Initial conditions
  LET X = 10
  LET V = 0
  LET T = 0

! --------- Computer Time Step
  LET dt = .01

! --------- Calculational loop
  DO
     LET X = X + V*dt
     LET F = -K*X - b*V
     LET A = F/M
     LET V = V + A*dt
     LET T = T + dt
     PLOT T,X;
  LOOP UNTIL T > 15

END
```

Figure 37a
BASIC program for the damped harmonic motion.

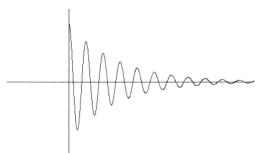

Figure 37b
Plot of damped harmonic motion.

Chapter 15
One Dimensional Wave Motion

In the last few chapters we have followed the straightforward procedure of identifying the forces on an object, setting the vector sum of the forces equal to the mass times acceleration, and solving the resulting equation. We began with problems like those involving circular motion where we knew the acceleration and could solve the equation immediately, the conical pendulum being an example. With oscillatory motion we ended up with a differential equation whose solution had to be guessed. The observation that oscillatory motion looks like circular motion viewed sideways helped greatly in this guess.

For damped and forced harmonic motion, it was not hard to write the differential equations, but the solutions involved mathematical functions and techniques that may not have been not familiar to the reader. In this chapter we are dealing with the subject of wave motion, where it turns out that the differential equation describing the motion has derivatives in both time and space. Setting up and solving such an equation requires mathematical discussions that are best left to a more advanced level course. Fortunately we can study the physics of wave motion without working with differential equations.

If we went through the effort to derive the differential equation for wave motion, we would end up with what is called a **wave equation**. Once you have a wave equation, you can guess a solution and plug in your guess just as we did for the simpler equations for oscillatory motion. For oscillatory motion, when we plugged in our guess $\sin(\omega t)$, we ended up with a simple equation $\omega = \sqrt{k/m}$ for the frequency of the oscillation. For wave equations, if you plug in a guess representing a wave traveling through the medium, you end up with a simple equation for the speed of the wave.

There are some famous wave equations in physics. In 1860 James Clerk Maxwell combined the equations for electricity with those for magnetism and, to his surprise, ended up with a wave equation. He initially had no idea what the wave was, but he could calculate the speed of the wave. Whatever wave he was dealing with travelled at a speed of 3×10^8 meters per second or 1 foot per nanosecond. As he knew of only one thing that travelled at that speed—light—he concluded that he had an equation for light waves and that the theory leading to this equation was the theory of light. He had discovered that light was an electric and magnetic phenomena.

In 1925, Louis De Broglie explained some baffling phenomena in atomic physics by proposing that electrons have a wave nature. Erwin Schrödinger then went further and derived a wave equation for the electron, an equation known as **Schrödinger's equation** that serves as the theoretical foundation for almost all of chemistry.

In 1929 Paul Dirac constructed a relativistic generalization of Schrödinger's equation. The problem with Dirac's equation was that it had solutions for two different kinds of wave, one representing the electron and the other an unknown particle of the opposite charge. A particle similar to the electron but opposite in charge, the positron, was observed in a cloud chamber experiment carried out by Carl Anderson in 1933.

It turns out that the relativistic wave equation for all elementary particles has two solutions, one solution like the electron representing matter, the other, like the positron, representing **antimatter**. And the wave equations predict that if a matter particle encounters its corresponding antimatter particle, the two particles can annihilate each other. There is an entire world of antimatter, the existence of which was predicted by Dirac's wave equation.

With wave equations playing such an important role in physics, one might think it is unfortunate that we are not prepared to derive and solve wave equations. Actually that is a blessing. There are certain general, simple principles that apply to all forms of wave motion, principles that allow you to understand and predict many features of the behavior of waves. These principles apply not only to waves like water and sound waves whose behavior can be deduced from Newtonian mechanics, but to light and electron waves where Newtonian mechanics does not apply. Thus by learning these general principles of wave motion, you are developing a foundation in physics that goes beyond Newtonian mechanics.

The two basic principles of wave motion we will discuss in this text are the **principle of superposition** and the **Huygens principle**. The principle of superposition is a fancy way of saying that waves add. If two waves are moving through each other, they produce a total wave that is the sum of the two waves. Since waves can have negative amplitudes (troughs) this addition of waves can produce cancellation. Two waves running into each other can, under the right circumstances, cancel each other out. This cancellation is clearly a wave phenomena, particles are not expected to do that.

The other general principle of wave motion is Huygens principle, which tells us how waves spread out in space. In this and the next chapter we will focus our attention on one dimensional waves which do not spread out. Thus we do not need Huygens principle at this point. Later in Chapter 33, we discuss two and three dimensional waves and phenomena such as interference and diffraction. In that chapter we do everything using the principle of superposition and the Huygens principle. In that chapter no calculus is used and we obtain results that apply to a broad spectrum of phenomena, even to sub atomic particles where concepts like velocity and acceleration no longer have meaning.

WAVE PULSES

We begin our discussion of wave motion with the wave pulses we described in Chapter 1 on Special Relativity. To create a wave pulse on a stretched rope, you flick the end of the rope and a pulse travels down the rope as shown in Figure (1-4) reproduced here. This is called a ***transverse wave*** because the particles in the rope move perpendicular or *transverse* to the direction of motion of the wave pulse.

With a stretched Slinky we were able to observe two different kinds of wave motion, the transverse wave seen in Figure (1-5) and a compressional wave seen in Figure (1-6). The compressional wave is also called a ***longitudinal wave*** because the particles in the spring are moving longitudinally or parallel to the direction of motion of the wave pulse.

Sound waves are usually compressional waves traveling through matter. A sound wave pulse in air can be viewed as a region of compressed gas where the molecules are closer together as shown in Figure (1). It is the region of compression that moves through the gas in much the same way as the region of compressed coils moves along the Slinky as seen in Figure (1-6).

To create the compressional wave on the Slinky we pulled back on the end of the Slinky and let go. This gives a small impulse directed down the Slinky. In much the same way we can use a loudspeaker cone to create the pressure pulse in the air column of Figure (1). Here the impulse can be provided by applying a voltage pulse to the speaker causing the speaker cone to suddenly jump forward. (If the speaker cone suddenly jumps back, you get a pulse consisting of a region of low pressure traveling down the tube.)

A transverse or sideways force in the medium tends to restore the medium to its original shape. For a transverse wave on a stretched rope, the tension on the rope provides the restoring force. For waves on the surface of a liquid, gravity or surface tension supplies the restoring force. But for waves passing through the bulk of a liquid or a gas, there are no transverse restoring forces and the only kind of waves we get are the compressional sound waves.

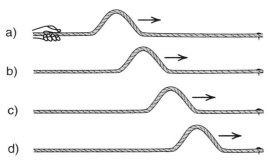

Figure 1-4
Wave traveling down a rope.

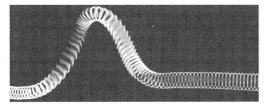

Figure 1-5
Transverse wave on a Slinky

Figure 1-6
Compressional wave on a Slinky.

Figure 1
A sound wave pulse traveling down through a tube of air. The pulse consists of a region of compressed air where the air molecules are closer together. This region of compression moves through the gas much as the region of compressed coils moves along the Slinky in Figure 1-6.

The main difference between a liquid and a solid is that in a liquid the molecules can slide past each other, while in a solid the molecules are held in place by molecular forces. These forces which prevent molecules from sliding past each other can also supply a transverse restoring force allowing a solid to transmit both transverse and compressional waves. An earthquake, for example, is a sudden disruption of the earth that produces both transverse waves called *S waves* and longitudinal or compressional waves called *P waves*. These waves can easily be detected using a device called a seismograph which monitors the vibration of the earth. It turns out that the S and P waves from an earthquake travel at different speeds, and will thus arrive at a seismometer at different times. By measuring the difference in arrival time and knowing the speed of the waves, you can determine how far away the earthquake was.

Exercise 1
The typical speed of a transverse S wave through the earth is about 4.5 kilometers per second, while the compressional P wave travels nearly twice as fast, about 8.0 kilometers per second. On your seismograph, you detect two sharp pulses indicating the occurrence of an earthquake. The first pulse is from the P wave, the second from the S wave. The pulses arrive three minutes apart. How far away did the earthquake occur?

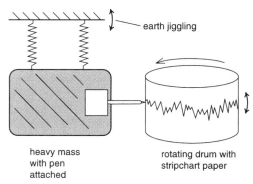

heavy mass with pen attached

rotating drum with stripchart paper

Figure 2
Sketch of a simple seismograph for detecting earthquake waves. When the earth shakes, the mass tends to remain at rest, thus the pen records the relative motion of the stationary mass and shaking earth.

(Building a seismograph is a favorite high school science fair project. Basically you suspend a large mass from springs and have a pen which is attached to the mass draw a line on moving stripchart paper as shown in Figure (2). When the earth shakes, the stripchart shakes with the earth, but the mass remains more or less stationary. The result is a squiggly line on the stripchart whose amplitude is the amplitude of vibration of the earth.

SPEED OF A WAVE PULSE

One way to predict the speed of a wave is to set up the differential equation for the wave, plug in a traveling wave solution and let the equation tell you the speed. Without the wave equation we can in some cases deduce the speed of the wave using clever tricks. One example is the transverse wave on a rope, whose speed we will calculate now. Another is the speed of Maxwell's wave of electric and magnetic forces which we will discuss in Chapter 32.

To calculate the speed of a transverse wave on a rope, consider a wave pulse moving down a rope at velocity $\vec{v}$ as shown in Figure (3a). To analyze the pulse, imagine that you are running along with the pulse at the same velocity $\vec{v}$. From your point of view, shown in Figure (3b), the pulse is at rest and the rope is moving back through the pulse at a speed v.

Now look at the top of the wave pulse. For any reasonably shaped pulse, the top of the pulse will be circular, fitting around a circle of radius r as shown in Figure (3c). This radius r is also called the **radius of curvature** of the rope at the top of the pulse.

Finally consider a short piece of rope of length ℓ at the top of the pulse as shown in Figure (3d). If this piece of rope subtends an angle 2θ on the circle, as shown, then $\ell = 2r\theta$ and the mass m of this section of rope is

$$m = \mu\ell = \mu 2r\theta \quad \begin{array}{l}\textit{mass of short}\\ \textit{section of rope}\end{array} \quad (1)$$

where μ is the mass per unit length of the rope.

The net force on this piece of rope is caused by the tension T in the rope. As seen in Figure (3d), the ends of the piece of rope point down at an angle θ. Thus the tension at each end has a downward component $T \sin(\theta)$ for a total downward force F_y of magnitude

$$F_y = 2T \sin(\theta) \approx 2T\theta \quad \text{downward component of tension force} \quad (2)$$

If we keep the angle θ small, just look at a very small section of the rope, then we can approximate $\sin(\theta)$ by θ as we did in Equation 2.

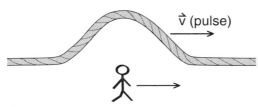

Figure 3a
Wave pulse, and an observer, moving to the right at a speed v.

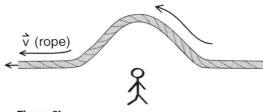

Figure 3b
From the moving observer's point of view, the pulse is stationary and the rope is moving through the pulse at a speed v.

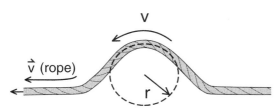

Figure 3c
Assume that the top of the pulse fits over a circle of radius r.

The final step is to note that this section of rope is moving at a speed v around a circle of radius r. Thus we know its acceleration; it is accelerating downward, *toward the center of the circle*, with a magnitude v^2/r.

$$a_y = \frac{v^2}{r} \quad \text{downward acceleration of section of rope} \quad (3)$$

Applying Newton's second law to the downward component of the motion of the section of rope, we get using Equations 1, 2 and 3

$$F_y = ma_y$$
$$2T\theta = (\mu 2r\theta)\frac{v^2}{r} \quad (4)$$

Both the variables r and θ cancel, and we are left with

$$T = \mu v^2$$

$$\boxed{v = \sqrt{\frac{T}{\mu}}} \quad \begin{array}{l}\text{speed of a wave pulse} \\ \text{on a rope with tension } T, \\ \text{mass per unit length } \mu\end{array} \quad (5)$$

A result we stated back in Chapter 1.

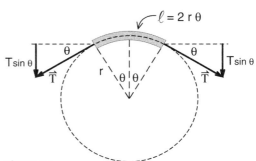

Figure 3d
The ends of the rope point down at an angle θ, giving a net restoring force $F_y = 2T \sin \theta$.

DIMENSIONAL ANALYSIS

In the above derivation of the speed of a transverse pulse on a rope, we avoided solving a differential equation by observing that the rope at the top of the pulse, from the moving observer's point of view, was moving with circular motion whose acceleration we know. For other kinds of wave pulses, particularly the compressional pulse seen in Figure (1-6), we do not have a simple circular motion, and a non calculus derivation of the wave speed becomes even more convoluted than the derivation we just went through. We could do it, but it is not worth the effort, especially since there are more straightforward ways of predicting wave speeds when one has the differential equation for the wave motion.

What we will do instead is use a technique called ***dimensional analysis*** to predict the speed of the wave. With dimensional analysis, you do not work out equations. Instead you determine what the relevant variables are, and then combine those variables in such a way that the dimensions are correct. If you have selected the correct variables, you get an answer that is correct to within a constant factor, and sometimes the correct answer.

To see how dimensional analysis works, let us first apply it to the example we just worked out—to find the speed of a transverse wave pulse on a rope. *(For clarity, we will italicize the variable names. We will also use MKS units.)* The first step is to do some experiments to find out what variables the speed depends upon. You choose a rope, stretch it, and soon discover that the speed of the pulse depends upon the tension T. Thus T is one of the variables. Then you try two ropes of the same length but different mass m, and discover that you get different wave speeds for the same tension. Thus the mass m is one of the relevant variables. Another experiment with 2 ropes of the same mass but different lengths, gives different wave speeds. Thus the rope length L is also important. Further experiments indicate that the speed of the pulse does not depend upon such variables as the color of the rope, the material from which it is constructed, or the time of day. Thus you conclude that the relevant variables and their dimensions are

$$T \frac{\text{kg m(meter)}}{\text{sec}^2}, \quad m \text{ kg}, \quad L \text{ m} \qquad (6)$$

From these variables we have to construct the velocity.

$$v \frac{\text{m}}{\text{sec}} \qquad (7)$$

The only variable with the dimensions of seconds in it is the tension T, thus T must be included in our formula for v. To get rid of kilograms, we must divide T by m to give

$$T \frac{\text{kg m}}{\text{sec}^2} * \frac{1}{m \text{ kg}} = \frac{T}{m} \frac{\text{m}}{\text{sec}^2}$$

We are getting there, but we must have the same power of meters and sec in order to get a velocity. If we multiply T/m by L meters, we get

$$\frac{T}{m} \frac{\text{m}}{\text{sec}^2} * L \text{ m} = \frac{TL}{m} \frac{\text{m}^2}{\text{sec}^2}$$

Finally we get the correct dimensions by taking the square root, giving

$$v = \sqrt{\frac{TL}{m}} \frac{\text{m}}{\text{sec}} = \sqrt{\frac{T}{\mu}} \frac{\text{m}}{\text{sec}} \qquad (8)$$

where we noted that $\mu = m/L$ is the mass per unit length.

Equation 8 tells us that no matter what the theory is, if the only relevant variables are T, m and L, the speed of the wave must be proportional to $\sqrt{T/\mu}$ for the dimensions to work out. We may have missed a factor of $1/2$ or 2π, but the functional dependence must be right.

Let us now use dimensional analysis to predict the speed of the compressional Slinky pulse shown in Figure (1-6), or any compressional pulse on a stretched spring. Since a stretched spring has a tension T and a mass per unit length μ, one might guess that $\sqrt{T/\mu}$ could also be the formula for the compressional wave. However compressional and transverse waves do not have the same speed. Even more important, you can get different wave speeds for the same value of T/μ, by using different springs. It turns out that the tension T is not a relevant variable.

Compressional waves depend upon the stiffness of a material, not the tension. For example a compressional sound pulse will travel down a steel rod whether or not the rod is under tension. Pulling on the ends of the steel rod does not noticeably change the speed of the sound pulse. Increasing the tension in a spring stretches the spring and therefore changes the mass per unit length.

It is the change in mass per unit length, not the change in tension, which affects the speed of the compressional pulse.

What variable is related to the inherent stiffness of a spring? The one that comes to mind is the spring constant k that appears in Hooke's law

$$F = -kx;$$
$$|k| = \frac{F}{x} \frac{\text{newtons}}{\text{meter}} \qquad \textit{Hooke's law} \qquad (9)$$

The stiffer the spring, the greater the spring constant k.

Suppose we decide, after enough experimentation, that the relevant variables for the compressional pulse on a spring are the spring constant k, spring mass m and spring length L. We obtain the dimensions of k from Hooke's law,

$$k \frac{\text{newtons}}{\text{meter}} = k \frac{\text{kg} \cdot \text{m/sec}^2}{\text{m}} = k \frac{\text{kg}}{\text{sec}^2}$$

thus we have to construct a quantity with dimensions m/sec from the variables

$$k \frac{\text{kg}}{\text{sec}^2}, \qquad M \text{ kg}, \qquad L \text{ m}$$

The only way we can do it is to divide k by m to get rid of kilograms and multiply by L^2 to get

$$\frac{kL^2}{m} \frac{\text{m}^2}{\text{sec}^2}$$

Taking the square root gives a quantity with the dimensions of a velocity

$$v = \sqrt{\frac{kL^2}{m}} = \sqrt{\frac{kL}{\mu}} \qquad \begin{array}{l}\textit{speed of a}\\ \textit{compressional}\\ \textit{wave on a spring}\end{array} \qquad (10)$$

where again $\mu = m/L$ is the mass per unit length. This is our prediction for the speed of a compressional wave on a spring. The actual speed could differ by a constant factor like 2, but it must have this functional dependence if we are correct in our assumption that the only relevant variables are k, m and L.

In the formula $v = \sqrt{kL/\mu}$, the appearance of the product kL, rather than k alone may at first seem surprising. But it turns out that the inherent stiffness of a spring is proportional to kL and not just k, with the result that the speed of the pulse is related to the stiffness, as we suspected.

To see why the inherent stiffness is related to kL, imagine that we wind a long spring and cut it in half to create two identical springs of length L_1. As shown in Figure (4a), if we apply a force F to one of the springs, and measure the distance Δx that the spring stretches, we can use Hooke's law to calculate that the spring constant k_1 is given by

$$k_1 = \frac{F}{\Delta x} \qquad (11)$$

Now attach the two springs back together and stretch the combination with the same force F as shown in Figure (4b). Since each spring feels the same force F, each stretches a distance Δx, and the pair stretch a distance $2\Delta x$. Thus from Hooke's law the k_2 of the combination is given by

$$k_2 = \frac{F}{2\Delta x} = \frac{1}{2}\left(\frac{F}{\Delta x}\right) = \frac{k_1}{2} \qquad (12)$$

where we used $k_1 = F/\Delta x$ from Equation 11.

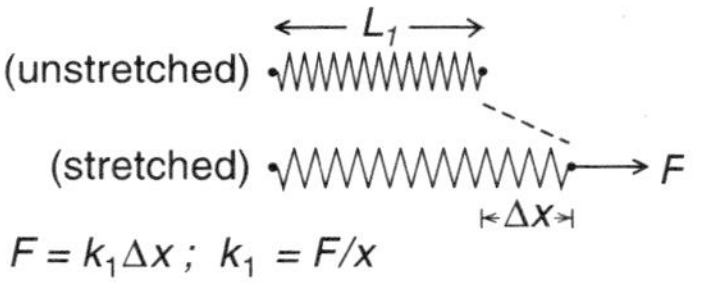

(unstretched) ⟵ L_1 ⟶ ⋯⋯⋯⋯⋯

(stretched) ⋯⋯⋯⋯⋯⋯⋯⟶ F
⊢Δx⊣

$F = k_1 \Delta x ; \quad k_1 = F/x$

Figure 4a
Measuring the spring constant of a spring.

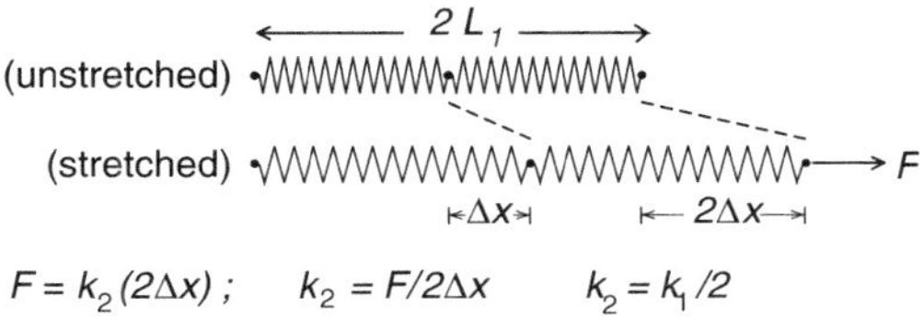

(unstretched) ⟵ $2L_1$ ⟶ ⋯⋯⋯⋯⋯⋯⋯⋯

(stretched) ⋯⋯⋯⋯⋯⋯⋯⋯⋯⋯⟶ F
⊢Δx⊣ ⊢— $2\Delta x$ —⊣

$F = k_2(2\Delta x) ; \quad k_2 = F/2\Delta x \quad k_2 = k_1/2$

Figure 4b
Measuring the spring constant of two connected springs.

When we attach two identical springs together, we end up with a longer spring but we are not changing the inherent stiffness of the spring. More importantly we do not change the speed of the wave pulse. Connecting the two springs and keeping the tension F the same merely gives the pulse a longer distance to travel.

Note, however, that when we attach the two springs, the spring constant is cut in half, but the length is doubled, with the result that the product kL is unchanged. Explicitly we have

$$k_2 L_2 = \left(\frac{k_1}{2}\right)(2L_1) = k_1 L_1 \quad (13)$$

It should now appear more reasonable that the speed of the wave pulse should be given by $\sqrt{kL/\mu}$. Both the quantity kL, and the mass per unit length μ are inherent properties of the spring that do not depend on the length of the spring. It is thus reasonable that the speed of the wave pulse should also involve only these variables.

Project Suggestion

We have spent some time discussing the speed of pulses on a stretched spring for two reasons. One is that we used these pulses as our main example of wave motion in our introduction to special relativity in Chapter 1. The second is that measuring the speed of pulses on a spring makes a nice project, not much equipment is needed, and you can fairly easily measure the variables needed to test Equation 10.

An additional advantage is that Equation 10 might or might not be right. Since it was derived by dimensional analysis, it could be off by a constant factor like 1/2 or 2π. Therefore you have the challenge of determining whether or not there are some missing constant factors.

We expect that the wave speed should be proportional to $\sqrt{kL/\mu}$, and if this does not turn out to be correct, we have made some mistake in our analysis of what variables are important. For example, in our analysis, we said nothing about the unstretched length L_0 of the spring. Should L_0 also appear in the formula for v_{wave}? The way to find out is to do some experiments.

The experiments are made a bit easier by noting that

$$v = \sqrt{\frac{kL}{\mu}} = \sqrt{\frac{kL^2}{m}} = L\sqrt{\frac{k}{m}}$$

where we used $\mu = m/L$.

SPEED OF SOUND WAVES

The quantity kL that appeared in our formula for the speed of a wave pulse is essentially the stiffness of a unit length spring. By stiffness we mean the ratio of the force applied to stretch a unit length of spring, to the amount of stretch Δx that we get. The same ideas also apply to stretching a steel rod or any one dimensional object that has an elasticity and obeys Hooke's law.

In engineering texts, the force applied to a unit length, area, or volume is given the generic name **stress**, and the resulting displacement that the stress causes is called a **strain**. The ratio of the *stress to the strain*, is called the **modulus**. For our spring, the stress is the tension force F, and the strain is the change in length per unit length, or $\Delta x/L$. The ratio of the stress to the strain, $F/(\Delta x/L) = FL/\Delta x$ is called **Young's modulus**. From Hooke's law, $F/\Delta x = k$, thus Young's modulus is $FL/\Delta x = kL$, the quantity we have been discussing.

When we have a compressional wave in a gas, we can think of the compression as being caused by a pressure pulse that travels through the gas. In the region where the gas is compressed, there is a slight excess pressure. The speed of the wave pulse depends upon the response of the gas to this excess pressure.

Using the engineering terminology, the excess pressure ΔP represents the stress and the fractional change in volume, $\Delta V/V$ the corresponding strain. In this case the ratio of the stress ΔP to the strain $\Delta V/V$ is called the **bulk modulus B**. The formula for B is thus

$$B = \Delta P/(\Delta V/V) \quad \text{bulk modulus} \quad (15)$$

In Chapter 17 we will discuss the concept of a pressure in a gas, and see how changes in pressure are related to changes in volume. Until we get to that chapter, any detailed discussion of the concept of bulk modulus is premature. What we will do now is assume that it is the bulk modulus B essentially represents the "stiffness" of the gas and should appear in the formula for the speed of a sound wave. We will then use dimensional analysis to figure out what the formula should be.

In the ratio $B = \Delta P/(\Delta V/V)$, the denominator $\Delta V/V$ is dimensionless, thus B has the dimensions of pressure which is a force per unit area.

$$B \frac{\text{newtons}}{\text{meter}^2} = B \frac{\text{kg m}}{\text{sec}^2 \text{m}^2} = B \frac{\text{kg}}{\text{sec}^2 \text{m}} \quad (15)$$

To construct a quantity involving B that has the dimensions of a velocity, we have to get rid of the kilograms by dividing by some quantity related to the mass of the gas. The only reasonable choice is the gas density ρ kg/m^3, thus we now have

$$\frac{B \text{ kg/(sec}^2 \text{ m)}}{\rho \text{ kg/m}^3} = \frac{B}{\rho} \frac{\text{m}^2}{\text{sec}^2} \quad (16)$$

which is the square of a velocity. Taking the square root, we get

$$\boxed{v = \sqrt{\frac{B}{\rho}}} \quad \text{speed of a sound wave} \quad (17)$$

a result we stated back in Chapter 1.

Equation 17 holds not only for a gas, but also for compressional sound waves in a liquid and a solid. Liquids and solids, being far more incompressible than a gas, have a much greater bulk modulus B and therefore higher speeds of sound. The speed of sound in various substances is given in Table 15-1.

Substance	Speed of Sound in meters/sec
Gases (at atmospheric pressure)	
Air at 0° C	331
Air at 20° C	343
Helium at 20° C	965
Hydrogen at 20° C	1284
Liquids	
Water at 0° C	1402
Water at 20° C	1482
Sea water at 20° C	1522
Solids	
Aluminum	6420
Steel	5941
Granite	6000
Nuclear matter	near c

Table 15-1

From this table we see that the speed of sound is considerably higher in the light gases helium and hydrogen than in the more dense gas air. (The difference in density is why helium and hydrogen filled balloons float in air.) The compressibility or bulk modulus is usually the same for all gases at a given pressure, thus the higher speed in hydrogen or helium is due to the lower density, as we would expect from the formula $v = \sqrt{B/\rho}$.

Exercise 2
Steel is much stiffer than aluminum. (You make much better springs from steel than aluminum.) Yet the speed of sound is greater in aluminum than steel. Why?

Exercise 3
You tap the end of a 10 meter long steel rod with a hammer. How long before the tap can be detected at the other end of the rod?

Exercise 4
A dimensional analysis problem that you should attempt now.

When working with the theory of electric and magnetic phenomena, using the MKS system of units, one encounters two rather mysterious constants labeled ε_0 (epsilon naught) and μ_0 (mu naught). The constant ε_0 appears in the formula for electric forces, and μ_0 in the formula for magnetic forces. These constants have the following dimensions

$$\varepsilon_0 \frac{(\text{coulomb})^2 (\text{seconds})^2}{\text{kilogram}(\text{meter})^3} \quad (18)$$

$$\mu_0 \frac{\text{kilogram meter}}{(\text{coulomb})^2} \quad (19)$$

where a coulomb is a unit of electrical charge. The numerical values of ε_0 and μ_0 are to be found on the inside cover of this text along with other important physical constants.

(a) what combination of the constants ε_0 and μ_0 have the dimensions of a velocity?

(b) from the numerical value of this velocity, what do you think it is the velocity of?

LINEAR AND NONLINEAR WAVE MOTION

Few sights are more awesome than the crashing of ocean rollers on a rocky beach during a storm. The waves seen in Figure (3) of Chapter 1, produced by hurricane Bertha hundreds of miles out to sea, were crashing against the rocky shores of Mt. Desert Island, Maine, in July 1990. Hundreds of tourists and local television station reporters were at the beach to observe the event. At one spot, called 'Thunder Hole', the crashing waves created a loud boom and a geyser of water that went 40 or 50 feet in the air.

A very different sight are the circular ripples emerging from where raindrops have hit a puddle of water, seen in Figure (1-2), reproduced here. The special feature of these ripples is that they maintain their identity as they move through each other. They are still circular waves even after moving through other waves.

Figure 1-3
This ocean wave from Hurricane Bertha (July 31, 1990).

Conceptually we can separate all wave motion into two classes. There are the relatively smooth waves that can pass through each other like the circular ripples of Figure (1-2), and the relatively wild waves that crest, crash and change their shapes. The relatively smooth waves all obey what is called a ***linear*** wave equation. The properties of these waves are well understood and their behavior easy to predict. The wild waves obey ***nonlinear*** wave equations. We know very little about the behavior of nonlinear waves, and in most cases find it very difficult to make predictions about their behavior. (Ocean waves, for example, are linear until they start to crest. When you see whitecaps, the waves have become nonlinear.)

In this text we will restrict our discussion to the smooth, linear waves that behave like the circular ripples. Fortunately, most kinds of wave motion we encounter in nature, including almost all examples of light waves and the probability waves of quantum mechanics, are linear and therefore relatively easy to analyze. But there are growing applications for nonlinear waves, particularly in the field of laser optics.

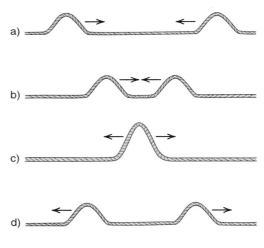

Figure 5
Two wave crests running into each other add up to produce a bigger crest.

Figure 1-2
Rain drops creating circular waves on a puddle.

THE PRINCIPLE OF SUPERPOSITION

Figure (1-2) illustrates one aspect of linear wave motion. The waves can move through each other and emerge undisturbed. The waves are still circular and unbroken after they have crossed.

There is another simple feature of this wave motion that is a bit harder to see from that picture. While the waves are crossing, they produce a wave whose height is the sum of the heights of the individual waves. If two crests are moving through each other, the crests add to produce a higher crest. Two troughs produce a deeper trough, and a crest and a trough will tend to cancel as they move through each other.

This adding of the heights of crossing waves is more easily illustrated for the case of one dimensional wave pulses traveling down a rope. In Figure (5), two similar crests add together to produce a doubly high crest for an instant. In Figure (6) we see that a similar shaped crest and trough will cancel at the instant they are together. Figure (7) illustrates the idea that as any two wave shapes move through each other, they produce a wave shape whose height at any point along the rope is the sum of the heights of the individual waves moving through each other.

The concept that waves can maintain their identity as they move through each other, and that they produce a resultant wave whose height or *amplitude* is the sum of the heights or amplitudes of the individual waves, this concept is known as the *principle of superposition*. In more colloquial language, the principle of superposition says that *waves add*. The principle of superposition is one of the key concepts of linear wave motion. It distinguishes linear from nonlinear wave motion. When nonlinear waves interact, you get something different than the simple sum of the two waves.

Before leaving our discussion of the principle of superposition, we wish to take one further look at part (c) of Figure (6). That is the point where an equal shaped crest and trough are right on top of each other, precisely cancelling each other out. This kind of cancellation of waves is a common feature of wave motion. In fact, it is what distinguishes *wave motion* from what we have been calling *particle motion*. If two particles run into each other, they do not cancel like the waves of Figure (6). They bounce or crash but not cancel.

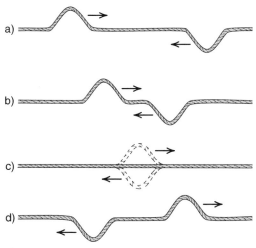

Figure 6
When a crest meets a trough, there is a short time when the waves cancel.

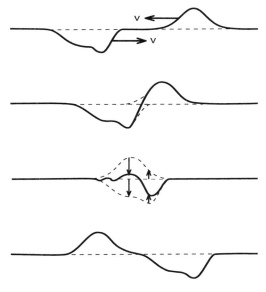

Figure 7
In general, for linear wave motion. We obtain the shape of the resulting wave by adding the amplitudes of the individual waves.

SINUSOIDAL WAVES

If you ask someone to describe wave motion, they are likely to picture water waves and sketch a curve that looks like a sine wave. A sine wave represents just one of many possible shapes for a wave. But it is an important shape because it is often seen in nature and it is easy to handle mathematically. We will see shortly that any arbitrary wave shape can be constructed from sine waves, thus the sine wave can be thought of as a basic building block of wave motion.

To relate the mathematical sine function to wave motion, recall our definition of sine function shown back in Figure (14-4). The point of that figure is that the sine function is the sideways projection of circular motion. As the arrow rotates at an angular velocity ω, the angle θ that the arrow has rotated increases as $\theta = \omega t$. On the right we have graphed the height of the rotating arrow as a function of the angle $\theta = \omega t$ to obtain a sine curve.

To actually create the sine wave shape seen in Figure (14-4), you can start shaking one end of a long rope as shown in Figure (8). If you move your hand up and down with a sinusoidal oscillation, a sinusoidal shaped wave will start traveling down the rope, at a speed $v_{wave} = \sqrt{T/\mu}$. This creates an example of what is called a ***traveling sine wave***.

The problem with creating traveling sine waves on a rope, is that the wave reaches the end of the rope, reflects, and moves back through the incoming wave, complicating the situation. A better example of traveling sine waves can be seen on the surface of a lake or the ocean where there is plenty of room for the waves to move before they strike an object or a shore.

There are two distinct ways to view a traveling sine wave. One is to move along with the wave. Then all you see is a stationary sinusoidal shape. The other is to stand still and let the wave pass by you. Then you will see the wave oscillate up and down as successive crests and troughs pass by you. This is illustrated in Figure (9) where we have sketched a traveling sinusoidal water wave passing a fixed post in the water. If you move along with the wave, then the shape of the wave does not change. But if you look at the post, the level of the water is moving up and down with a sinusoidal oscillation.

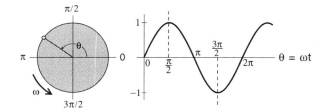

Figure 14-4
Definition of the function $\sin\theta = \sin(\omega t)$.

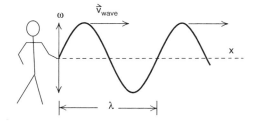

Figure 8
Sine wave created on a stretched rope.

Wavelength, Period, and Frequency

We usually describe a wave in terms of its wavelength λ, frequency f or period T. The easy way to remember how to go back and forth between these quantities, is to use dimensions.

When we view the shape of a traveling sine wave, the predominant feature is the wavelength, the distance λ between crests shown in Figure (8). Considering that one full cycle of the wave fits between the crests, we can assign the dimensions of *meters per cycle* to λ.

$$\lambda \frac{\text{meters}}{\text{cycle}} \qquad \textit{wavelength}$$

When we let the wave pass by us and view the up and down motion of the surface, we see an oscillation whose period is T *seconds per cycle* and frequency is f *cycles per second*.

Since the period T is the length of time it takes one wavelength λ of the wave to pass by at a speed v_{wave}, we have (*distance = speed times time*)

$$\lambda \frac{\text{meters}}{\text{cycle}} = v_{wave}\left(\frac{\text{meters}}{\text{second}}\right) T\left(\frac{\text{second}}{\text{cycle}}\right) \qquad (20)$$

By assigning the dimensions meter/cycle to λ and sec/cycle to T, we can get the relationship $\lambda = v_{wave}T$ from dimensions without having to memorize formulas, or even having to think very much.

Figure 9
Traveling sine wave on the surface of water. The sine wave shape moves as a unit along the surface at a speed v_{wave}. But if we look at a fixed post in the water, the water level at the post oscillates up and down with a sinusoidal oscillation.

As an example of using dimensions to derive a formula, let us see if we can get a formula for the frequency f of a wave of wavelength λ. The idea of using dimensions is to try something, then see if the dimensions match. If they don't match, change the formula until they do. As a guess, let us try the formula

$$f = v_{wave}\lambda \qquad \textit{guess}$$

Putting in dimensions, we have

$$f\left(\frac{\text{cycles}}{\text{sec}}\right) = v_{wave}\left(\frac{\text{meters}}{\text{sec}}\right) \times \lambda\left(\frac{\text{meters}}{\text{cycle}}\right)$$

$$= v_{wave}\lambda \frac{\text{meters}^2}{\text{sec cycle}}$$

Clearly the dimensions do not match. We have to change the formula so that meters cancel and we get cycles upstairs. This can be done if we move λ downstairs, giving

$$f\left(\frac{\text{cycles}}{\text{sec}}\right) = \frac{v_{wave}(\text{meters/sec})}{\lambda(\text{meters/cycle})}$$

$$= \frac{v_{wave}}{\lambda}\left(\frac{\text{cycles}}{\text{sec}}\right) \qquad (21)$$

which works. Thus the correct formula is $f = v_{wave}/\lambda$, a result that can be a bit tricky to figure out other ways.

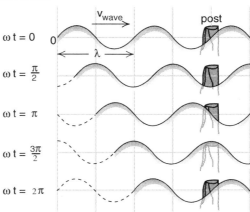

Angular Frequency ω

In Figure (14-4), we reminded ourselves that a sinusoidal oscillation is equivalent to the sideways view of circular motion. If the vector on the left side of Figure (14-4) is rotating with an angular velocity ω **radians per cycle**, we get one rotation, one period T of the oscillation, when the angle $\theta = \omega t$ goes from 0 at t = 0, to 2π at t = T. Thus at the end of one period, $\theta = \omega T = 2\pi$.

Again we can avoid memorizing new formulas by using dimensions. We note that 2π is the number of radians in a full circle or cycle. Thus we will assign to it the dimensions

$$\boxed{2\pi \frac{\text{radians}}{\text{cycle}}} \quad (23)$$

Then to find the formula, for example, for the wave's angular frequency ω in terms of the wave's period T, we have

$$\omega\left(\frac{\text{radians}}{\text{sec}}\right) = \frac{2\pi(\text{radians / cycle})}{T(\text{sec / cycle})}$$

$$= \frac{2\pi}{T}\left(\frac{\text{radians}}{\text{sec}}\right) \quad (23)$$

Exercise 5 (Do this one now.)

For a traveling sine wave moving at a speed v_{wave}, use dimensions to find

(a) λ in terms of v_{wave} and ω

(b) v_{wave} in terms of λ and T

(c) T in terms of λ and ω

(d) f in terms of ω

Spacial Frequency k

When we let a traveling wave pass by us, we observe a sinusoidal oscillation in *time*. This oscillation can be described in terms of the number of seconds in each cycle (T seconds/cycle), in terms of the frequency (f cycles/second) or the angular frequency (ω radians/second).

If instead we look at the whole wave at one instant of time, or move along with the wave, we see a sinusoidal oscillation in *space*. We have described this spacial oscillation in terms of the wavelength, the number of meters in each cycle (λ meters/cycle). What we are missing is a spacial analogy to frequency, the *number of cycles or radians per meter*.

By dimensions we immediately see that

$$\frac{1}{\lambda \frac{\text{meters}}{\text{cycle}}} = \frac{1}{\lambda}\frac{\text{cycles}}{\text{meter}}$$

is the special analogy to the time frequency f, and that

$$\frac{2\pi(\text{radians/cycle})}{\lambda(\text{meters/cycle})} = \frac{2\pi}{\lambda}\left(\frac{\text{radians}}{\text{meter}}\right)$$

is the spacial analogy to the angular frequency ω.

In physics texts, it is not common to use a special symbol to designate the spacial frequency $1/\lambda$ (cycles/meter), but it is standard practice to designate the angular spacial frequency $2\pi/\lambda$ (radians/meter) by the letter k

$$k\frac{\text{radians}}{\text{meter}} \equiv \frac{2\pi}{\lambda}\frac{\text{radians}}{\text{meter}} \quad \begin{array}{l}\textit{spacial}\\ \textit{frequency}\end{array} \quad (24)$$

The standard name for k is the lackluster expression *wave number,* which says very little about the quantity. Instead we will refer to k as the *spacial frequency* of the wave. The higher the spacial frequency k, the more radians we get in a meter, just as the higher the time frequency ω, the more radians we get in one second.

When you first study wave motion, it may be an irritating complication to have two kinds of frequency, f cycles/sec and ω radians/sec (or two spacial frequencies $1/\lambda$ cycles/meter and k radians/meter). Why not stick with cycles which are much easier to visualize than radians? The answer is that in the formulas for sine waves, the sine function basically has to be expressed in terms of an angle, as in $\sin \theta$, and radians are an angle. To convert time t to an angle, we multiply by ω as in

$$\theta \text{ radians} = \omega \frac{\text{radians}}{\text{sec}} \times t \text{ sec}$$
$$= \omega t \text{ radians} \quad (25a)$$

while to convert the distance x to an angle we multiply by k as in

$$\theta \text{ radians} = k \frac{\text{radians}}{\text{meter}} \times x \text{ meters}$$
$$= kx \text{ radians} \quad (25b)$$

Using Equations 25a or 25b, we can express the single function $\sin \theta$ either as $\sin \omega t$, a sine wave in time shown in Figure (10a), or as $\sin kx$, a sine wave in space shown in Figure (10b). From these graphs we can see that when ωt gets up to 2π, we have one period T, and when kx gets up to 2π we have one wavelength λ.

Exercise 6 (Try this now.)
You have a traveling sine wave moving at a speed v_{wave}. Using dimensions find the formula for v_{wave} in terms of the wave's time frequency ω and spacial frequency k.

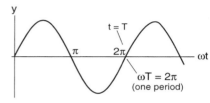
a) Time: $\sin(\theta) = \sin(\omega t)$

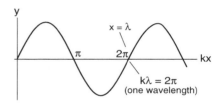
b) Space: $\sin(\theta) = \sin(kx)$

Figure 10
Sine waves in time and space.

Traveling Wave Formula

Thus far we have formulas for a time varying sine wave $\sin \omega t$, and a space varying wave $\sin kx$. Now we want a formula for a traveling sine wave whose amplitude varies in both space and time. The answer turns out to be

$$y = \sin\theta = \sin(kx - \omega t) \quad \text{traveling sine wave} \quad (26)$$

What we will do is show that this formula represents a sine wave moving down the x axis.

Figure (11) shows a sinusoidal shape that is moving down the x axis at a speed v_{wave}. If we describe the wave by the function $\sin\theta$, then it is the origin $\sin\theta = 0$ that moves down the axis at a speed v_{wave}. Thus what we need is a formula for θ so that when we set $\theta = 0$, that point does move down the x axis at the desired speed.

The answer we gave in Equation 26 suggests that the correct formula for θ is

$$\theta = kx - \omega t \quad (27)$$

Setting $\theta = 0$ we get

$$\theta = 0 = kx - \omega t \, ; \quad kx = \omega t$$

$$x = \left(\frac{\omega}{k}\right)t \quad (28)$$

But if the $\theta = 0$ point travels at a speed v_{wave}, then after a time t, it has traveled a distance x given by

$$x = v_{wave}t \quad (29)$$

Comparing Equations (28) and (29), we see that the point $\theta = 0$ moves along the x axis at a speed

$$v_{wave} = \frac{\omega}{k} \quad (30)$$

If you did Exercise 6, you recognize that the quantity ω/k has the dimensions of a velocity

$$\frac{\omega}{k} \frac{\text{radian/sec}}{\text{radian/meter}} = \frac{\omega}{k} \frac{\text{meter}}{\text{sec}} = v_{wave} \quad (31)$$

Thus the origin does move down the x axis at a speed v_{wave}, and the formula $(kx - \omega t)$, is our desired traveling wave formula.

Exercise 7

Explain what kind of a wave is represented by the formula $y = \sin(kx + \omega t)$.

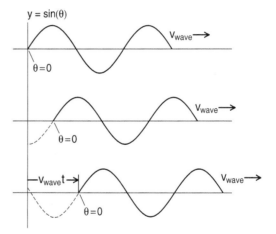

Figure 11
The cycle begins at $\theta = 0$.

Phase and Amplitude

The Equation 26 for a traveling wave can be generalized by noting that the wave can have an arbitrary amplitude A, and an arbitrary constant phase angle φ to give

$$y = A \sin(kx - \omega t + \phi) \quad (32)$$

The amplitude A just makes the sine wave bigger or smaller, and the phase angle φ shifts the sine wave to the left or right.

To see precisely how the phase angle φ shifts the sine wave, we have in Figure (12) compared $\sin \theta$ and $\sin(\theta + \phi)$. The function $\sin(\theta + \phi)$ crosses zero when the angle $(\theta + \phi) = 0$ or at $\theta = -\phi$. Thus adding a phase angle φ shifts the sine wave back a distance $-\phi$ radians. If, for example, we set $\phi = \pi/2$, the wave is shifted back 1/4 of a wavelength, and we have converted a sine wave into a cosine wave.

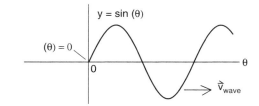

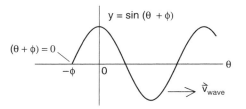

Figure 12
Adding a phase angle φ shifts the wave back a distance φ.

STANDING WAVES

In addition to the waves traveling down a rope, another kind of wave pattern that is easy to achieve are those shown in Figure (13). All you have to do is shake the end of the rope at the right frequency and one of these waves will appear. Change the frequency and you can change to one of the other patterns.

The waves in Figure (13) are called *standing waves* because the pattern does not move along the rope. The points of zero amplitude, the points called *nodes* of the wave, stay at fixed positions while the rope between nodes oscillates back and forth.

The difference between the wave patterns is characterized by the number of nodes. In Figure (13) all the waves have nodes at the ends, and there are zero, one, two and three nodes in between as we go from the left to the right pattern.

The two kinds of waves on a rope, the traveling wave of Figures (8) and (9) and the standing wave of Figure (13) are closely related to each other. A careful demonstration shows how traveling waves can turn into standing waves.

If you start shaking a rope a traveling wave starts down the rope as shown in Figure (8). After a while the wave reaches the other end of the rope, is reflected, and starts moving back the other way. This reflection is most easily seen if you send a single pulse down the rope so that you can see it bounce off the fixed end and come back to you.

If you send a series of pulses down the rope, if you create a traveling sine wave, then the reflected pulses have to move back through the pulses that are still coming in. You now have the superposition of two traveling waves moving through each other in opposite directions.

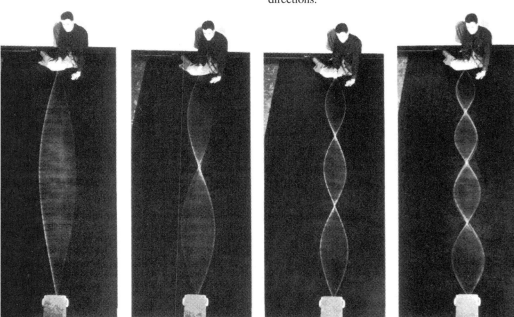

Figure 13
Standing waves on a rope.

You might think that the sum of two traveling waves moving through each other could lead to a complex pattern, and when you try it in a demonstration it often looks that way. The problem with a demonstration is that the reflected wave reaching your hand can reflect again and you begin to build up a mixture of many waves.

If you do the demonstration carefully, however, you can observe a simple result. *The sum of the traveling wave and the reflected wave, moving through each other, is a standing wave.* The addition of two traveling waves of equal amplitude and wavelength moving in opposite directions through each other is illustrated in Figure (14). The two traveling waves are shown on lines (a) and (b) at five different times t = 0, T/4, T/2, 3T/4 and T, where T is the period of the waves. On the bottom line (c) we have added the amplitudes of the two traveling waves to get a picture of the amplitude of the resulting wave.

In the first frame t = 0, the two waves match exactly, producing a sum that has twice the amplitude of either traveling wave. A quarter of a period later, at t = T/4, the traveling waves are precisely opposite each other. And the sum is zero all along the wave. This never happens in a traveling wave. A traveling wave is never completely flat, there is always a crest moving along.

At time t = T/2, half a period later, the traveling waves again line up producing a wave of twice the amplitude. Note now that the points in the sum wave that were below the axis at t = 0 are now above the axis at t = T/2, and vice versa. At time t = 3T/4 the traveling waves are again out of phase and add up to zero. At t = T, we are back to where we started.

From line (c) of Figure (14), we see that the nodes of the sum wave remain stationary and the rope between the nodes oscillates up and down. This is exactly what we see in the photographs of the standing rope waves in Figure (13).

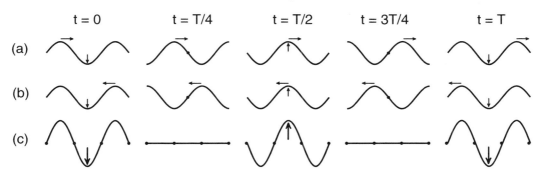

Figure 14
Making a standing wave out of two traveling waves on a rope. In the top line (a) we show a traveling wave moving to the right. A section of rope is shown at times t = 0, T/4, T/2, 3T/4 and T, where T is the period of the wave. In (b) we have, in the same section of rope, five views of a traveling wave moving to the left. In (c), we have added the two traveling waves and get a standing wave with stationary nodes. To demonstrate the addition, at the center of each time frame we have drawn arrows to show the height of the wave at that point. The length of the bottom arrow is the sum of the lengths of the upper two arrows. To add two waves, you add up the heights at each point along the wave.

Using a trigonometric identity, we can show mathematically that the sum of two traveling waves moving through each other creates a standing wave. The formula for a sine wave traveling to the right is from Equation 26.

$$y_{\text{moving right}} = A \sin(kx - \omega t) \quad (26)$$

If you worked Exercise 6, you found that the formula for a similar wave moving left is

$$y_{\text{moving left}} = A \sin(kx + \omega t) \quad (33)$$

The formula for the sum of the two waves is

$$y_{\text{sum wave}} = A \sin(kx - \omega t) + A \sin(kx + \omega t)$$

The trigonometric identity we will use is

$$\sin(a \pm b) = \sin(a)\cos(b) \pm \cos(a)\sin(b) \quad (34)$$

This gives

$$\sin(kx - \omega t) = \sin(kx)\cos(\omega t) - \cos(kx)\sin(\omega t)$$

$$\sin(kx + \omega t) = \sin(kx)\cos(\omega t) + \cos(kx)\sin(\omega t)$$

Add these two sine waves, the $\cos(kx)\sin(\omega t)$ terms cancel and we are left with

$$y_{\text{sum wave}} = 2A \sin kx \cos \omega t \quad (35)$$

To interpret Equation 35, write it in the form

$$y_{\text{sum wave}} = A(x) \cos \omega t \quad (36a)$$

where the x dependent amplitude $A(x)$ is

$$A(x) = 2A \sin kx \quad (36b)$$

Equation 36a tells us that the entire wave is oscillating in time as $\cos(\omega t)$. However the amplitude of the oscillation depends upon the position x long the wave. Equation (36b) tells us how the amplitude varies with position. It varies sinusoidally as sin kx, with nodes permanently located at the points where sin kx = 0. This sinusoidal variation in the amplitude along the wave is clearly seen in the photographs of the standing waves on the rope, Figure (13).

WAVES ON A GUITAR STRING

Perhaps the clearest example of standing waves are the waves on the strings of a stringed instrument such as the guitar. The advantage of working with these waves is that you get to both see the shape of the wave and hear its frequency.

The shape of guitar string waves are the same as the standing rope waves in Figure (13). In Figure (15), we have sketched the allowed standing wave patterns on a string of length L. Because the string is fixed at the ends, we can only have waves with nodes at the ends.

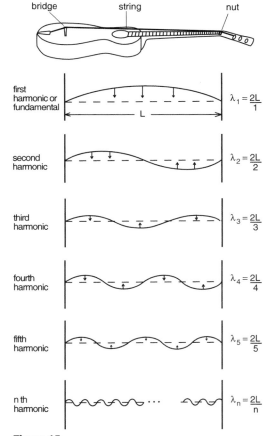

Figure 15
Allowed standing waves on a guitar string. The formula for the wavelength of the nth harmonic is seen to be $\lambda_n = 2L/n$.

The wave with no nodes between the ends is called the *fundamental* or *first harmonic*. Its wavelength is 2L, twice the length of the string. In the second harmonic, with one node in the middle, a full wave fits on the string at one time and we have $\lambda_2 = L$. Each time we add a node we go up to one higher harmonic. *(The second harmonic is also called the first overtone, etc.)*

The general formula for the wavelength of the nth harmonic can be seen is we write the progression of wavelengths in the form

$$\lambda_1 = \frac{2L}{1}; \; \lambda_2 = \frac{2L}{2}; \; \lambda_3 = \frac{2L}{3}, \; \text{etc.}$$

It is clear that the formula for λ_n is simply

$$\boxed{\lambda_n = \frac{2L}{n}} \quad \begin{array}{l}\textit{wavelength of}\\ \textit{the nth harmonic}\end{array} \quad (37)$$

The easiest way to remember Equation 36 is draw a sketch of the allowed standing waves and write down the progression $\lambda_1 = 2L/1$, $\lambda_2 = 2L/2$, etc.

Frequency of Guitar String Waves

What notes do you hear when you pluck a guitar string? That depends very much upon how you pluck it. Usually when you pluck the string, you create a number of the standing wave patterns at one time, and the note you hear is a rich mixture of the frequencies of the individual waves.

With care, however, you can pluck the string so that most of the vibration is in one of the harmonics. A gentle pluck at the center of the string will excite mostly the fundamental or first harmonic. Pluck the string 1/4 of the way from one end and briefly place your finger at the center of the string to create a node there. This way you can excite mostly the second harmonic. You will then notice that the sound of the note is one octave above the sound of the fundamental. Accomplished guitar players can selectively excite still higher harmonics.

What are the frequencies of oscillations of these various standing wave patterns? We can answer this question because of our knowledge that standing waves can be made from two traveling waves moving through each other. The resulting standing wave has the same frequency and wavelength as the traveling wave, thus we can use the traveling wave formulas to determine the frequency of the standing wave.

Using dimensions, we see that a traveling wave of wavelength λ cm/cycle, traveling at a speed v cm/sec, has a frequency f cycles/second given by

$$f \, \frac{\text{cycles}}{\text{sec}} = v \, \frac{\text{meter}}{\text{sec}} \times \frac{1}{\lambda \, \text{meter/cycle}}$$

$$= \frac{v}{\lambda} \frac{\text{cycles}}{\text{sec}} \quad (38)$$

The speed v of a transverse wave on a string, that has a mass per unit length μ, and tension T, is from Equation 5

$$v_{\text{wave}} = \sqrt{\frac{T}{\mu}} \quad (5)$$

Using Equation 5 in Equation 38 gives us as the formula for the frequency of the traveling wave

$$f = \frac{1}{\lambda}\sqrt{\frac{T}{\mu}} \quad (39)$$

The same Equation 39 must also apply to the standing waves on the guitar string. We get for the frequency f_n of the nth harmonic, which has a wavelength f_n,

$$\boxed{f_n = \frac{1}{\lambda_n}\sqrt{\frac{T}{\mu}}} \quad \begin{array}{l}\textit{frequency of the}\\ \textit{nth harmonic}\end{array} \quad (40)$$

For anyone who has tuned a guitar, Equation 40 makes a lot of sense. First note that when you go from the first harmonic $\lambda_1 = 2L$, to the second harmonic $\lambda_2 = L$, the wavelength is cut in half and the frequency doubles. A doubling of the frequency of a note corresponds to going up one octave.

When you are tuning the guitar, you raise the frequency of a string by tightening it and increasing the tension. That is also predicted by Equation 40.

On a guitar or most stringed instruments the low notes are played on fat wires that have a greater mass per unit length μ than the skinny wires used for the high notes. The reason for using the fat wire is that you can increase the tension and still keep the frequency down. The more tension in the wire and the more mass in the wire, the more energy you can store in the wire and the louder the sound you can produce. It is hard to get as much sound out of the low frequency strings and you need all the help you can get.

Exercise 8

You have a guitar string of length L, with a tension T and a mass per unit length μ. (L is the distance from the nut to the bridge.)

(a) What is the f frequency of the fundamental mode of vibration? Express your answer in terms of L, T, and μ.

(b) Show that the nth harmonic has a frequency n times as great as the fundamental.

Exercise 9

One end of a wire is attached to a post as shown in Figure (16). The wire is then run over a pulley where a mass m is hung on the other end. The distance d from the post to the pulley is 1 meter and the mass of one meter length of the wire is 5 grams.

(a) How big a mass m must be hung on the wire in order to get the wire to vibrate in its fundamental mode at a frequency of 440 cycles/second, which is middle A? (Answer: 395.10 kg).

(b) Describe four distinct ways one could double the frequency of oscillation of the wire.

(c) How much mass would you have hung on the wire in Figure (16) to get the wire to oscillate in its fundamental mode at a frequency two octaves above middle A? (Answer 6321.63 kg.)

Sound Produced by a Guitar String

When you pluck a guitar string, the standing wave on the string produces a traveling sound wave in the air. This is analogous to plucking the end of a rope to produce a traveling wave along the rope as illustrated in Figure (8b). The wavelength of the sound wave is determined by the frequency of oscillation of the string and the speed of sound. (It is not the same as the wavelength of standing waves on the string.)

Exercise 10

A guitar string is tuned to oscillate at a frequency of 440 cycles/second in its fundamental mode. What are the wavelengths of the sound waves produced by the first three harmonics

(a) in air at 20° C

(b) in helium at 20° C

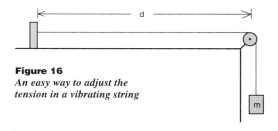

Figure 16
An easy way to adjust the tension in a vibrating string

Chapter 16
Fourier Analysis, Normal Modes and Sound

In Chapter 15 we discussed the principle of superposition—the idea that waves add, producing a composite wave that is the sum of the component waves. As a result, quite complex wave structures can be built from relatively simple wave forms. In this chapter our focus will be on the analysis of complex wave forms, finding ways to determine what simple waves went into constructing a complex wave.

As an example of this process, consider what happens when white sunlight passes through a prism. White light is a mixture of all the colors, all the wavelengths of the visible spectrum. When white light passes through a prism, a rainbow of colors appears on the other side. The prism separates the individual wavelengths so that we can study the composition of the white light. If you look carefully at the spectrum of sunlight, you will observe certain dark lines; some very specific wavelengths of light are missing in light from the sun. These wavelengths were absorbed by elements in the outer atmosphere of the sun. By noticing what wavelengths are missing, one can determine what chemical elements are in the sun's atmosphere. This is how the element Helium (named after Helios, Greek for sun) was discovered.

This example demonstrates how the ability to separate a complex waveform (in this case white sunlight) into its component wavelengths or frequencies, can be a powerful research tool. The sounds we hear, like those produced by an orchestra, are also a complex mixture of waves. Even individual instruments produce complex wave forms. Our ears are very sensitive to these wave forms. We can distinguish between a note played on a Stradivarius violin and the same note played by the same person on an inexpensive violin. The only difference between the two sounds is a slight difference in the mixture of the component waves, the harmonics present in the sound. You could not tell which was the better violin by looking at the waveform on an oscilloscope, but your ear can easily tell.

You can hear these subtle differences because the ear is designed in such a way that it separates the complex incoming sound wave into its component frequencies. The information your brain receives is not what the shape of the complex sound wave is, but how much of each component wave is present. In effect, your ear is acting like a prism for sound waves.

When we study sound in the laboratory, the usual technique is to record the sound wave amplitude using a microphone, and display the resulting waveform on an oscilloscope or computer screen. If you want to, you can generate more or less pure tones that look like sine waves on the screen. Whistling is one of the best ways to do this. But if you record the sound of almost any instrument, you will not get a sine wave shape. The sound from virtually all instruments is some mixture of different frequency waves. To understand the subtle differences in the quality of sound of different instruments, and to begin to understand why these differences occur, you need to be able to decompose the complex waves you see on the oscilloscope screen into the individual component waves. You need something like an ear or a prism for these waves.

A way to analyze complex waveforms was discovered by the French mathematician and physicist Jean Baptiste Fourier, who lived from 1786 to 1830. Fourier was studying the way heat was transmitted through solids and in the process discovered a remarkable mathematical result. He discovered that any continuous, repetitive wave shape could be built up out of harmonic sine waves. His discovery included a mathematical technique for determining how much of each harmonic was present in any given repetitive wave. This decomposition of an arbitrary repetitive wave shape into its component harmonics is known as **Fourier analysis**. We can think of Fourier analysis effectively serving as a mathematical prism.

The techniques of Fourier analysis are not difficult to understand. Appendix A of this chapter is a lecture on Fourier analysis developed for high school students with no calculus background (explicitly for my daughter's high school physics class). To apply Fourier analysis you have to be able to determine the area under a curve, a process known in calculus as integration. While the idea of measuring the area under a curve is not a difficult concept to grasp, the actual process of doing this, particularly for complex wave shapes, can be difficult. Everyone who takes a calculus course knows that integration can be hard. The integrals involved in Fourier analysis, particularly the analysis of experimental data are much too hard to do by hand or by analytical means.

People find integration hard to do, but computers don't. With a computer one can integrate any experimental wave shape accurately and rapidly. As a result, Fourier analysis using a computer is very easy to do. A particularly fast way of doing Fourier analysis on the computer was discovered by Cooley and Tukey in the 1950s. Their computer technique or algorithm is known as the **Fast Fourier Transform** or **FFT** for short. This algorithm is so commonly used that one often refers to a Fourier transform as an FFT.

The ability to analyze data using a computer, to do things like Fourier analysis, has become such an important part of experimental work that older techniques of acquiring data with devices like strip chart recorders and stand alone oscilloscopes have become obsolete. With modern computer interfacing techniques, important data is best recorded in a computer for display and analysis. We have developed the **MacScope™** program, which will be used often in this text, for recording and displaying experimental data. The main reason for writing the program was to make it a simple and intuitive process to apply Fourier analysis. In this chapter you will be shown how to use this program. With the computer doing all the work of the analysis, it is not necessary to know the mathematical processes behind the analysis, the steps are discussed in the appendix. But a quick reading of the appendix should give you a feeling for how the process works.

HARMONIC SERIES

We begin our discussion with a review of the standing waves on a guitar string, shown in Figure (15-15) reproduced here. We saw that the wavelengths λ_n of the allowed standing waves are given by the formula

$$\lambda_n = \frac{2L}{n} \quad (15\text{-}37)$$

Where L is the length of the string, and n takes on integer values n = 1, 2, 3, ...

Each of these standing waves has a definite frequency of oscillation that was given in Equation 15-38 as

$$f_n \frac{cycles}{sec} = \frac{v}{\lambda_n} \frac{meters/sec}{meters/cycle}$$

$$= \frac{v}{\lambda_n} \frac{cycles}{sec} \quad (15\text{-}38)$$

If we substitute the value of λ_n from Equation 15-37 into Equation 15-38, we get as the formula for the corresponding frequency of vibration

$$f_n = \frac{v_{wave}}{\lambda_n} = \frac{v_{wave}}{2L/n} = n\left(\frac{v_{wave}}{2L}\right) \quad (1)$$

For n = 1, we get

$$f_1 = \frac{v_{wave}}{2L} \quad (2)$$

All the other frequencies are given by

$$\boxed{f_n = nf_1} \quad \text{harmonic series} \quad (3)$$

This set of frequencies is called a **harmonic series**. The **fundamental frequency** or **first harmonic** is the frequency f_1. The **second harmonic** f_2 has twice the frequency of the first. The third harmonic f_3 has a frequency three times that of the first, etc. Note also that the fundamental has the longest wavelength, the second harmonic has half the wavelength of the fundamental, the third harmonic one third the wavelength, etc.

It was Fourier's discovery that any continuous repetitive wave could be built up by adding together waves from a harmonic series. The correct harmonic series is the one where the fundamental wavelength λ_1 is equal to the period over which the waveform repeats.

To begin our discussion of Fourier analysis and the building up of waveforms from a harmonic series, we will first study the motion of two air carts connected by springs and riding on an air track. We will see that these *coupled air carts* have several distinct *modes of motion*. Two of the modes of motion are purely sinusoidal, with precise frequencies. But any other kind of motion appears quite complex.

However, when we record the complex motion, we discover that the velocity of either cart is repetitive. A graph of the velocity as a function of time gives us a continuous repetitive wave. According to Fourier's theorem, this waveform can be built up from sinusoidal waves of the harmonic series whose fundamental frequency is equal to the repetition frequency of the wave. When we use Fourier analysis to see what harmonics are involved in the motion, we will see that the apparent complex motion of the carts is not so complex after all.

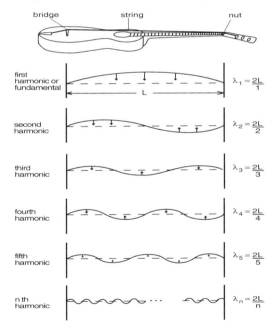

Figure 15-15
Standing waves on a guitar string.

NORMAL MODES OF OSCILLATION

The reason musical instruments generally produce complex sound waves containing various frequency components is that the instrument has various ways to undergo a resonant oscillation. Which resonant oscillations are excited with what amplitudes depends upon how the instrument is played.

In Chapter 14 we studied the resonant oscillation of a mass suspended from a spring, or equivalently, of a cart on an air track with springs attached to the end of the cart, as shown in Figure (1). This turns out to be a very simple system—there is only one resonant frequency, given by $\omega = \sqrt{k/m}$. The only natural motion of the mass is purely sinusoidal at the resonant frequency. This system does not have the complexity found in most musical instruments.

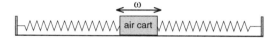

Figure 1
Cart and springs on an air track.

Figure 2
System of coupled air carts.

Things become more interesting if we place two carts on the air track, connected by springs as shown in Figure (2). We will call this a system of two *coupled* air carts.

When we analyzed the one cart system, we found that the force on the cart was simply $F = -kx$, where x is the displacement of the cart from its equilibrium position. With two carts, the force on one cart depends not only on the position of that cart, but also on how far away the other cart is. A full analysis of this coupled cart system, using Newton's second law, leads to a pair of coupled differential equations whose solution involves matrices and eigenvalues. In this text we do not want to get into that particular branch of mathematics. Instead we will study the motion of the carts experimentally, and find that the motion, which at first appears complex, can be explained in simple terms.

In order to record the motion of the aircarts, we have mounted the velocity detector apparatus shown in Figure (3). The apparatus consists of a 10 turn wire coil mounted on top of the cart, that moves through the magnetic field of the iron bars suspended above the coil. The operation of the velocity detector apparatus depends upon Faraday's law of induction which will be discussed in detail in Chapter 30 on Faraday's law. For now all we need to know is that a voltage is induced in the wire coil, a voltage whose magnitude is proportional to the velocity of the cart. This voltage signal

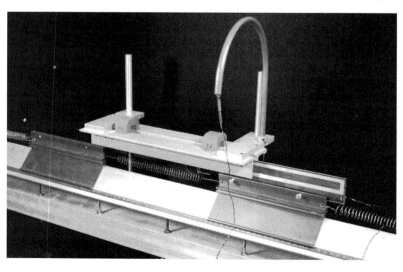

Figure 3
Recording the velocity of one of the aircarts. A 10 turn coil is mounted on top of one of the carts. The coil moves through the magnetic field between the angle irons, and produces a voltage proportional to the velocity of the cart. This voltage is then recorded by the Macintosh oscilloscope.

from the wire coil is carried by a cable to the Macintosh oscilloscope where it is displayed on the computer screen.

When you first start observing the motion of the coupled aircarts, it appears chaotic. One cart will stop and reverse direction while the other is moving toward or away from it, and there is no obvious pattern. But after a while, you may discover a simple pattern. If you pull both carts apart and let go in just the right way, the carts come together and go apart as if one cart were the mirror image of the other. This motion of the carts is illustrated in Figure (4a). We will call this the *vibrational mode* of motion.

In Figure (4b) we have used the velocity detector to record the motion of one of the coupled air carts when the carts are moving in the vibrational mode. You can see that the curve closely resembles a sine wave.

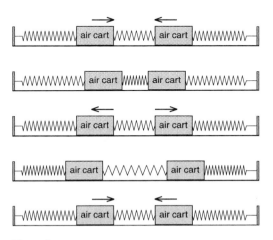

Figure 4a
Vibrational motion of the coupled air carts.

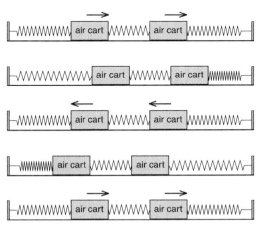

Figure 5a
Sloshing mode of motion of the coupled aircarts.

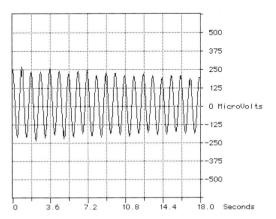

Figure 4b
Vibrational mode of oscillation of the coupled aircarts. The voltage signal is proportional to the velocity of the cart that has the coil on top.

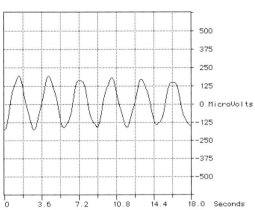

Figure 5b
A pure sloshing mode is harder to get. Here we came close, but it is not quite a pure sine wave.

If you play around with the carts for a while longer, you will discover another way to get a simple sinusoidal motion. If you pull both carts to one side, get the positions just right, and let go, the carts will move back and forth together as illustrated in Figure (5a). We will call this the ***sloshing mode*** of motion of the coupled air carts. In Figure (5b) we have recorded the velocity of one of the carts in the sloshing mode, and see that the curve is almost sinusoidal.

In general, the motion of the two carts is not sinusoidal. For example, if you pull one cart back and let go, you get a velocity curve like that shown in Figure (6). If you start the carts moving in slightly different ways you get differently shaped curves like the one seen in Figure (7). Only the vibrational and sloshing modes result in sinusoidal motion, all other motions are more complex. To study the complex motion of the carts, we will use the techniques of ***Fourier analysis***.

FOURIER ANALYSIS

As we mentioned in the introduction, Fourier analysis is essentially a mathematical prism that allows us to decompose a complex waveform into its constituent pure frequencies, much as a prism separates sunlight into beams of pure color or wavelength. We have just studied the motion of coupled air carts, which gave us an explicit example of a relatively complex waveform to analyze. While the two carts can oscillate with simple sinusoidal motion in the vibrational and sloshing modes of Figures (4) and (5), in general we get complex patterns like those in Figures (6) and (7). What we will see is that, by using Fourier analysis, the waveforms in Figures (6) and (7) are not so complex after all.

The MacScope program was designed to make it easy to perform Fourier analysis on experimental data. The MacScope tutorial gives you considerable practice using MacScope for Fourier analysis. What we will do here is discuss a few examples to see how the program, and how Fourier analysis works. We will then apply Fourier analysis to the curves of Figures (6) and (7) to see what we can learn. But first we will see how MacScope handles the analysis of more standard curves like a sine wave or square wave.

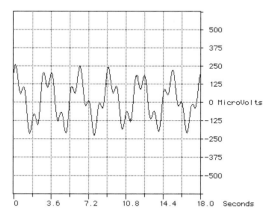

Figure 6
Complex motion of the coupled air carts.

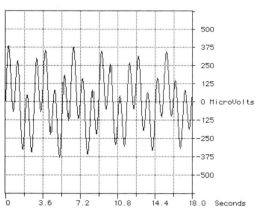

Figure 7
Another example of the complex motion of the coupled air carts.

Analysis of a Sine Wave

In Figure (8a) we attached MacScope to a sine wave generator and recorded the resulting waveform. The sine wave generator, which is usually called a **signal generator**, is an electronic device that we will use extensively in laboratory work on the electricity part of the course. Typically the device has a dial that allows you to select the frequency of the wave, a knob that allows you to adjust the wave amplitude, and some buttons by which you can select the shape of the wave. Typically you can choose between a sine wave shape, ∧∨∧∨∧∨, a square wave shape ⊓⊔⊓⊔⊓⊔, and a triangular wave ∧∨∧∨∧∨ .

In Figure (8a) we have selected one cycle of the wave and see that the frequency of the wave is 1515 Hz (1.515 KHz), which is about where we set the frequency dial at on the signal generator. To get Figure (8b) we pressed the **Expand** button that appears once a section of curve has been selected. This causes the selected section of the curve to fill the whole display rectangle.

The selection rectangle is obtained by holding down the mouse bottom and dragging across the desired section of the curve. The starting point of the selection rectangle can be moved by holding down the shift key while moving the mouse. The data box shows the period **T** *of the selected rectangle, and the corresponding frequency* **f**. *If you wish the data box to remain after the selection is made, hold down the option key when you release the mouse button. This immediately gives you the* **ImageGrabber™**, *which allows you to select any section of the screen to save as a PICT file for use in a report or publication.*

In Figure (9), we went up to the **Analyze** menu of MacScope and selected ***Fourier Analysis***. As a result we get the window shown in Figure (10). At the top we see the selected one cycle of a sine wave. Beneath, we see a rectangle with one vertical bar and a scale labeled **Harmonics**. The vertical bar is in the first position, indicating that the section of the wave which we selected has only a first harmonic. Beneath the ex-

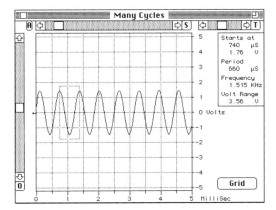

Figure 8a
Sine wave from a signal generator. Selecting one cycle of the wave, we see that the frequency of the wave is about 1.5 kilocycles (1.515 KHz). (Here we also see some of the controls that allow you to study the experimental data. The scrollbar labeled "S" lets you move the curve sideways. The "T" scrollbar changes the time scale, and the "O" scrollbar moves the curve up and down. In the text, we will usually not show controls unless they are important to the discussion.)

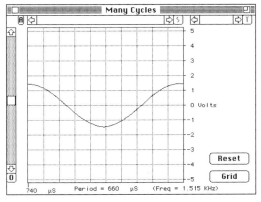

Figure 8b
Once we have selected a section of curve, a new control labeled **Expand** *appears. When we press the* **Expand** *button, the selected section of the curve fills the entire display rectangle as seen above. (The control then becomes a* **Reset** *button which takes us back to the full curve.)*

panded curve, there is a printout of the period and frequency of the selected section. Here we see again that the frequency is 1.515 KHz.

For Figure (11), we pressed **Reset** to see the full curve, selected 4 cycles of the sine wave, and expanded that. In the lower rectangle we now see one vertical bar over the 4th position, indicating that for the selected wave we have a pure fourth harmonic. In the MacScope tutorial we give you a MacScope data file for a section of a sine wave. Working with this data file, you should find that if you select one cycle of the wave, anywhere along the wave, you get an indication of a first harmonic. Select two cycles, and you get an indication of a pure second harmonic, etc.

The basic function of the Fourier analysis program in MacScope is to determine how you can construct the selected section of the wave from **harmonic sine waves**. The first harmonic has the frequency of the selected section. For example, suppose that you had a 10 cycle per second sine wave and selected one cycle. That selection would have a frequency of 10 Hz and a period of 0.1 seconds. The second harmonic would have a frequency of 20 Hz, and the third harmonic 30 Hz, etc. The nth harmonic frequency is n times greater than the first harmonic. Fourier's discovery was that *any continuous repeating wave form can be constructed from the harmonic sine waves.* So far we have considered only the obvious examples of sections of a pure sine wave. We will now go on to more complex examples to see how a waveform can be constructed by adding up the various harmonics.

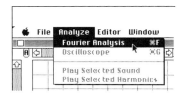

Figure 9
Choosing Fourier Analysis.

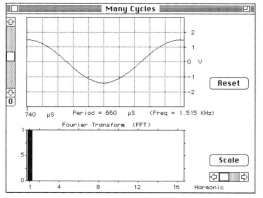

Figure 10
In the MacScope program, **Fourier Analysis** *acts only on the selected section of the curve. Since we selected only one cycle, we see only a first harmonic.*

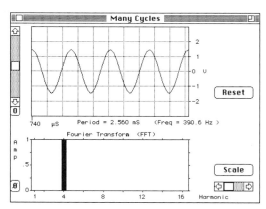

Figure 11
When 4 cycles are selected, the wave form consists of a pure 4th harmonic. We see the period of the selected section of wave, which is 4 times longer than one cycle. This makes the frequency 1/4 as high. (The difference between 1515 Hz/4 and 390.6 Hz indicates the accuracy of graphical selection.)

Analysis of a Square Wave

The so called square wave ⊓⊔⊓⊔⊓⊔, whose shape is shown in Figure (12), is commonly used in electronics labs to study the response of various electronic circuits. The waveform regularly jumps back and forth between two levels, giving it a repeated rectangular shape. The ideal mathematical square wave jumps instantaneously from one level to another. The square waves we study in the lab are not ideal; some time is always required for the transition.

It is traditional to use the square wave as the first example to show students how a complex wave form can be constructed from harmonic sine waves. This is a bit ironic, because the ideal square wave has discontinuous jumps from one level to another, and therefore does not satisfy Fourier's theorem that any *continuous wave shape* can be made from harmonic sine waves. The result is that if you try to construct an ideal square wave from sine waves, you end up with a small blip at the discontinuity (called the Gibb's effect). Since our focus is experimental data where there is no true discontinuity, we will not encounter this problem.

In Figure (12) we have selected one cycle of the square wave. Selecting *Fourier Analysis*, we get the result shown in Figure (13). We have clicked on the *Expand* button so that only the selected section of the wave shows in the upper rectangle. In the lower rectangle, which we will now call the FFT window, we see that this section of the square wave is made up from various harmonics. The MacScope program calculates 128 harmonics, but we have clicked three times on the *Scale* button to expand the harmonics scale so that we can study the first 16 harmonics in more detail.

In Figure (14) we have clicked on the first bar in the FFT window, the bar that represents the amplitude of the first harmonic. In the upper window you see one cycle of a sine wave superimposed upon the square wave. This is a picture of the first harmonic. It represents the best possible fit of the square wave by a single sine wave. If you want a better fit, you have to add in more sine waves.

In Figure (15) we clicked on the bar in the 3rd harmonic position, the bar representing the amplitude of the 3rd harmonic in the square wave. In the upper window you see a sine wave with a smaller amplitude and three times the frequency of the first harmonic. If you select a single harmonic, as we have just done in Figure (15), MacScope prints the frequency of both the first harmonic and the selected harmonic above the FFT window. Here you can see that the frequency of the first harmonic is 201.6 Hz and the selected harmonic frequency is 604.9 Hz as expected.

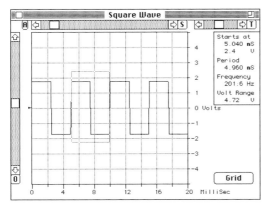

Figure 12
The square wave. The wave shape goes back and forth periodically between two levels. Here we have selected one cycle of the square wave.

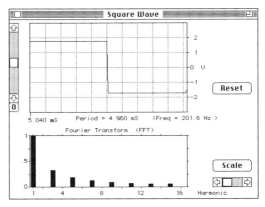

Figure 13
Expanding the one cycle selected, and choosing **Fourier Analysis***, we see that this wave form has a number of harmonics. The computer program calculates the first 128 harmonics. We used the* **Scale** *button to display only the first 16.*

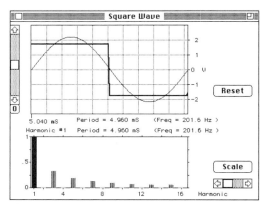

Figure 14
Select the first harmonic by clicking on the first bar.

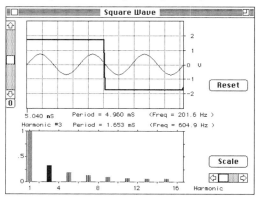

Figure 15
Select the second harmonic by clicking on the second bar.

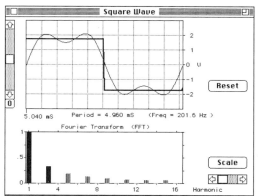

Figure 16
Sum of the first and second harmonic is obtained by selecting both.

If we select both the first and third harmonics together (either by dragging a rectangle over both bars, or by holding down the shift key while selecting them individually), the upper window displays the sum of the first and third harmonic, as shown in Figure (16). You can see that the sum of these two harmonics gives us a waveform that is closer to the shape of the square wave than either harmonic alone. We are beginning to build up the square wave from sine waves.

In Figures (17, 18 and 19), we have added in the 5th, 7th and 9th harmonics. You can see that the more harmonics we add, the closer we get to the square wave.

One of the special features of a square wave is that it contains only odd harmonics—all the even harmonics are absent. Another is that the amplitude of the nth harmonic is 1/n times as large as that of the first harmonic. For example, the third harmonic has an amplitude only 1/3 as great as the first. This is represented in the FFT window by drawing a bar only 1/3 as high as that of the first bar. In the MacScope program, the harmonic with the greatest amplitude is represented by a bar of height 1. All other harmonics are represented by proportionally shorter bars.

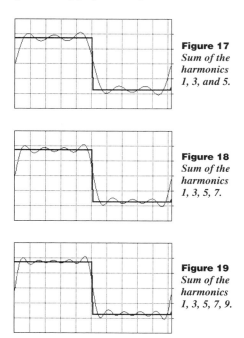

Figure 17
Sum of the harmonics 1, 3, and 5.

Figure 18
Sum of the harmonics 1, 3, 5, 7.

Figure 19
Sum of the harmonics 1, 3, 5, 7, 9.

Repeated Wave Forms

Before we apply the Fourier transform capability of MacScope to the analysis of experimental data, there is one more feature of the analysis we need to discuss. What we are doing with the program is reconstructing a selected section of a waveform from harmonic sine waves. Anything we build from harmonic sine waves *exactly repeats at the period of the first harmonic*. Thus our reconstructed wave will always be a repeating wave, beginning again at the same height as the beginning of the previous cycle.

You can most easily see what we mean if you select a non repeating section of a wave. In Figure (20) we have gone back to a sine wave, but selected one and a half cycles. In the FFT window you see a slew of harmonics. To see why these extra harmonics are present, we have in Figure (21) selected the first 9 of them and in the upper window see what they add up to. It is immediately clear what has gone wrong. The selected harmonics are trying to reconstruct a repeating version of our 1.5 cycle of the sine wave. The extra spurious harmonics are there to force the reconstructed wave to start and stop at the same height as required by a repeating wave.

If you are analyzing a repeating wave form and select a section that repeats, then your harmonic reconstruction will be accurate, with no spurious harmonics. If your data is not repeating then you have to deal one way or another with this problem. One technique often used by engineers is to select a long section of data and smoothly force the ends of the data to zero so that the selected data can be treated as repeating data. Hopefully, forcing the ends of the data to zero does not destroy the information you are interested in. You can often accomplish the same thing by throwing away higher harmonics, assuming that the lower harmonics contain the interesting features of the data. You can see that neither of these techniques would work well for our one and a half cycles of a sine wave selected in Figure (21).

In this text, our use of Fourier analysis will essentially be limited to the analysis of repeating waveforms. As long as we select a section that repeats, we do not have to worry about the spurious harmonics.

(Most programs for the acquisition and analysis of experimental data have an option for doing Fourier analysis. Unfortunately, few of them allow you to select a precise section of the experimental data for analysis. As a result the analysis is usually done on a non repeating section of the data, which distorts the resulting plot of the harmonic amplitudes. For a careful analysis of data, the ability to precisely select the data to be analyzed is an essential capability.)

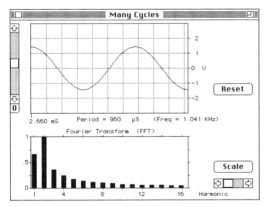

Figure 20
When we select one and a half cycles of a sine wave, we get a whole bunch of spurious harmonics.

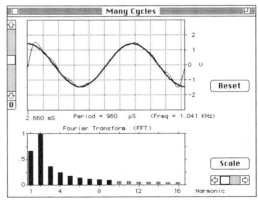

Figure 21
The spurious harmonics result from the fact that the reconstructed wave must be repeating—must start and stop at the same height.

ANALYSIS OF THE COUPLED AIR CART SYSTEM

We are now ready to apply Fourier analysis to our system of coupled air carts. Recall that there were two modes of motion that resulted in a sinusoidal oscillation of the carts, the vibrational motions shown in Figure (4) and the sloshing mode shown in Figure (5). In Figures (22) and (23) we expanded the time scales so that we could accurately measure the period and frequency of these oscillations. From the data rectangles, we see that the frequencies were 1.11Hz and 0.336 Hz for the vibrational and sloshing modes respectively.

When the carts were released in an arbitrary way, we generally get the complex motion seen in Figures (6) and (7). What we wish to do now is apply Fourier analysis to these waveforms to see if any simple features underlie this complex motion.

In Figure (24), which is the waveform of Figure (6), we see that there is a repeating pattern. The fact that the pattern repeats means that it can be reconstructed from harmonic sine waves, and we can use our Fourier analysis program to find out what the component sine waves are.

In Figure (24) we have selected precisely one cycle of the repeating pattern. This is the crucial step in this experiment—finding the repeating pattern and selecting one cycle of it. How far you have to look for the pattern to repeat depends upon the mass of the carts and the strength of the springs.

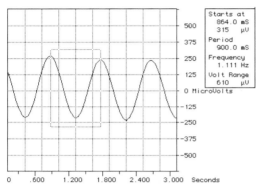

Figure 22
Vibrational mode of Figure (4). We have expanded the time scale so that we could accurately measure the period and frequency of the oscillation.

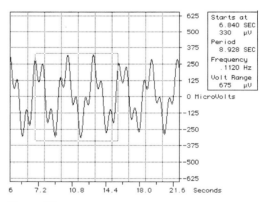

Figure 24
Complex mode of the coupled aircarts. We see that the waveform repeats, and have selected one cycle of the repeating wave.

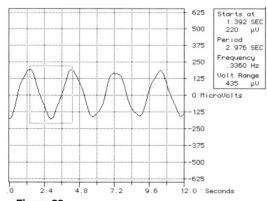

Figure 23
Sloshing mode of Figure (5). Less of an expansion of the time scale was needed to measure the period here.

Expanding the repeating section of the complex waveform, and choosing **Fourier Analysis** gives us the results shown in Figure (25). What we observe from the Fourier analysis is that the complex waveform is a mixture of two harmonics, in this case the third and tenth harmonic. If we click on the bar showing the amplitude of the third harmonic, we see that harmonic drawn in the display window of Figure (26), and we find that the frequency of this harmonic is 0.336 Hz. This is the frequency of the sloshing mode of Figure (23). Clicking on the bar above the tenth harmonic, we get the harmonic drawn in the display window of Figure (27), and see that the frequency of this mode is 1.12 Hz, within a fraction of a percent of the 1.11 Hz frequency of the vibrational mode of Figure (22).

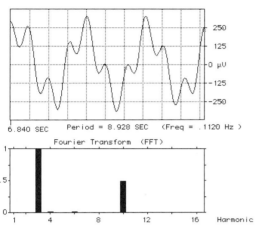

Figure 25
Fourier analysis of one cycle of the complex waveform. The FFT rectangle shows us that the wave consists of only two harmonics.

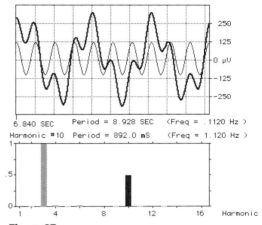

Figure 27
Selecting the tenth harmonic, we see that its frequency is essentially equal to the frequency of the vibrational mode of motion.

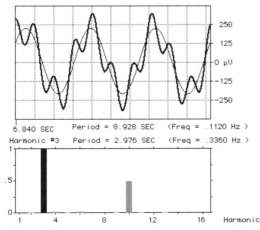

Figure 26
When we click on the third harmonic bar, we see that the frequency of the third harmonic is precisely the frequency of the sloshing mode of oscillation.

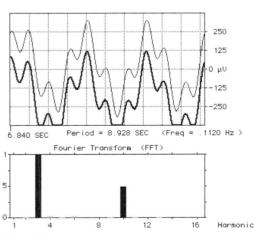

Figure 28
Selecting both modes shows us that the complex motion is simply the sum of the two sinusoidal modes of motion.

If we select both the third and tenth harmonics, the sum of these two harmonics is shown in Figure (28). These two harmonics together so closely match the experimental data that we had to move the experimental curve down in order to see both curves. What we have learned from this experiment is that *the complex motion of Figure (6) is a mixture of the two simple, sinusoidal modes of motion.*

Back in Figures (6) and (7), we displayed two waveforms, representing different complex motions of the same two carts. Starting with the second waveform of Figure (7), we selected one cycle of the motion, expanded the selected section, and chose Fourier Analysis. The result is shown in Figure (29). What we see is that the second complex waveform is also a mixture of the third and tenth harmonics. The first waveform in Figure (25) had more of the third harmonic, more of the sloshing mode, while the second waveform of Figure (29) has more of the tenth harmonic, the vibrational mode. Both complex waveforms are simply mixtures of the vibrational and sloshing modes. *They have different shapes because they are different mixtures.*

This experiment is beginning to demonstrate that for the two coupled aircarts, there is a strict limitation to the kind of motion the carts can have. They can either move in the vibrational mode, or in the sloshing mode, or in some combination of the two modes. *No other kinds of motion are allowed! The various complex motions are just different combinations of the two modes.*

Adding another aircart so that we have three coupled aircarts, the motion becomes still more complex. However if we look carefully, we find that the waveform eventually repeats. Selecting one repeating cycle and choosing Fourier Analysis, we got the results shown in Figure (30). We observe that this complex motion is made up of three harmonics.

The sinusoidal modes of motion of the coupled air carts are called *normal modes*. The general rule is that if you have n coupled objects, like n carts on an air track connected by springs, and they are confined to move in 1 dimension, there will be n normal modes. (With 2 carts, we saw 2 normal modes. With 3 carts, 3 normal modes, etc.) This result, which will play an important role in our discussion of the specific heat of molecules, can be extended to motion in 2 and 3 dimensions. For example, if you have n coupled particles that can move in 3 dimensions, as in the case of a molecule with n atoms, then the system should have 3n normal modes of motion. Such a molecule should have 3n independent ways to vibrate or move. We will have more to say about this subject in the next chapter.

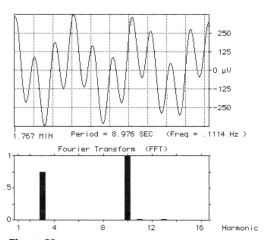

Figure 29
The second complex mode of motion, from Figure (7), is simply a different mixture of the same vibrational and sloshing modes of motion.

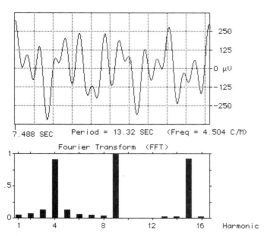

Figure 30
One cycle of the waveform for three coupled aircarts. With three carts, we get three normal modes of motion.

THE HUMAN EAR

The human ear performs a frequency analysis of sound waves that is not unlike the Fourier analysis of wave motion which we just studied. In the ear, the initial analysis is done mechanically, and then improved and sharpened by a sophisticated data analysis network of nerves. We will focus our attention on the mechanical aspects of the ear's frequency analysis.

Figure (31) is a sketch of the outer and inner parts of the human ear. Sound waves, which consist of pressure variations in the air, are funneled into the auditory canal by the external ear and impinge on the eardrum, a large membrane at the end of the auditory canal. The eardrum (tympanic) membrane vibrates in response to the pressure variations in the air. This vibrational motion is then transferred via a lever system of three bones (the malleus, incus, and stapes) to a small membrane covering the oval window of the snail shaped cochlea.

The cochlea, shown unwound in Figure (32), is a fluid filled cavity surrounded by bone, that contains two main channels separated by a membrane called the ***basilar membrane***. The upper channel (scala vestibuli) which starts at the oval window, is connected at the far end to the lower channel (scala tympani) through a hole called the helicotrema. The lower channel returns to the round window which is also covered by a membrane. If the stapes pushes in on the membrane at the oval window, fluid flows around the helicotrema and causes a bulge at the round window.

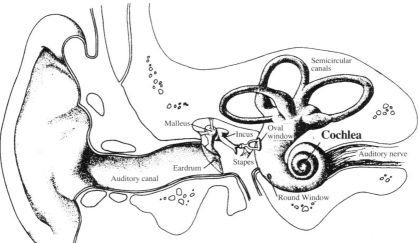

Figure 31
The human ear. Sound, entering the auditory canal, causes vibrations of the eardrum. The vibrations are transferred by a bone lever system to the membrane covering the oval window. Vibrations of the oval window membrane then cause wave motion in the fluid in the cochlea.
(Adapted from Lindsey and Norman, Human Information Processing.)

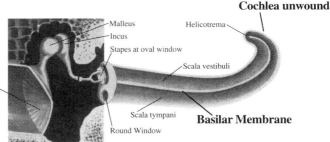

Figure 32
Lever system of the inner ear and an unwound view of the cochlea. The basilar membrane separates the two main fluid channels in the cochlea. Vibrations of the basilar membrane are detected by hair cells. (Adapted from **Principles of Neural Science** Edited by E. R. Kandel and J. H. Schwartz, Elsevier/North-Holland, p260.)

The purpose of the lever system between the eardrum and the cochlea is to efficiently transfer sound energy to the cochlea. The eardrum membrane is about 25 times larger in area than the membrane across the oval window. The lever system transfers the total force on the eardrum to an almost equal force on the oval window membrane. Since *force* equals *pressure* times *area*, a small pressure variation acting on the large area of the eardrum membrane results in a large pressure variation at the small area at the oval window. The higher pressures are needed to drive a sound wave through the fluid filled cochlea.

If the oval window membrane is struck by a pulse, a pressure wave travels down the cochlea. The basilar membrane, which separates the two main fluid channels, moves in response to the pressure wave, and a series of hair cells along the basilar membrane detect the motion. It is the way in which the basilar membrane responds to the pressure wave that allows for the frequency analysis of the wave.

Figure (33) is an idealized sketch of a straightened out cochlea. (See Appendix B for more realistic sketches.) At the front end, by the oval window, the basilar membrane is narrow and stiff, while at the far end it is about 5 times as wide and much more floppy. To see why the basilar membrane has this structure, we have in Figure (34) sketched a mechanical model that has a similar function as the membrane. In this model we have a series of masses mounted on a flexible steel band and attached by springs to fixed rods as shown. The masses are small and the springs stiff at the front end. If we shake these small masses, they resonate at a high frequency $\omega = \sqrt{k/m}$. Down the membrane model, the masses get larger and the springs weaken with the result that the resonant frequency becomes lower.

If you gently shake the steel band at some frequency ω_0 a small amplitude wave will travel down the band and soon build up a standing wave of that frequency. If ω_0 is near the resonant frequency of one of the masses, that mass will oscillate with a greater amplitude than the others. Because the masses are connected by the steel band, the neighboring masses will be carried into a slightly larger amplitude of motion, and we end up with a peak in the amplitude of oscillation centered around the mass whose frequency ω is equal to ω_0. (In the sketch, we are shaking the band at the resonant frequency ω_7 of the seventh mass.)

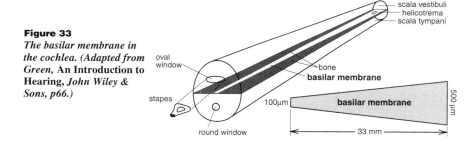

Figure 33
*The basilar membrane in the cochlea. (Adapted from Green, **An Introduction to Hearing**, John Wiley & Sons, p66.)*

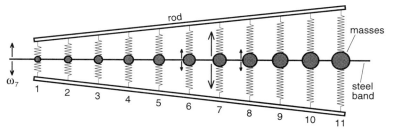

Figure 34
Spring model of the basilar membrane. As we go down the steel band, the masses become larger, the springs weaker, and the resonant frequency drops. If we vibrate the end of the band at some frequency, the mass which resonates at that frequency will have the biggest amplitude of oscillation.

Because the masses m get larger and the spring constants k get smaller toward the far end, there is a continuous decrease in the resonant frequency as we go down the band. If we start shaking at a high frequency, the resonant peak will occur up near the front end. As we lower the frequency, the peak of the oscillation will move down the band, until it finally gets down to the lowest resonant frequency mass at the far end. We can thus measure the frequency of the wave by observing where along the band the maximum amplitude of oscillation occurs.

The basilar membrane functions similarly. The stiff narrow membrane at the front end resonates at a high frequency around 20,000 Hz, while the wide floppy back end has a resonant frequency in the range of 20 to 30 Hz. Figure (35) shows the amplitude of the oscillation of the membrane in response to driving the fluid in the cochlea at different frequencies. We can see that as the frequency increases, the location of the maximum amplitude moves toward the front of the membrane, near the stapes and oval window. Figure (36) depicts the shape of the membrane at an instant of maximum amplitude when driven at a frequency of a few hundred Hz. The amplitude is greatly exaggerated; the basilar membrane is about 33 millimeters long and the amplitude of oscillation is less than .003 mm.

Although the amplitude of oscillation is small, it is accurately detected by a system of about 30,000 hair cells. How the hair cells transform the oscillation of the membrane into nerve impulse signals is discussed in Appendix B at the end of this chapter.

The human ear is capable of detecting tiny changes in frequency and very subtle mixtures of harmonics in a sound. Looking at the curves in Figure (35) (which were determined from a cadaver and may not be quite as sharp as the response curves from a live membrane), it is clear that it would not be possible to make the ear's fine frequency measurements simply by looking for the peak in the amplitude of the oscillation of the membrane. But the ear does not do that. Instead, measurements are continuously made all along the membrane, and these results are fed into a sophisticated data analysis network before the results are sent to the brain. The active area of current research is to figure out how this data analysis network operates.

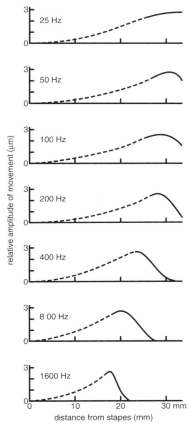

Figure 35
Amplitude of the motion of the basilar membrane at different frequencies. (Adapted from **Principles of Neural Science** *Edited by E. R. Kandel and J. H. Schwartz, Elsevier/North-Holland, p 263.)*

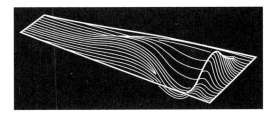

Figure 36
Response of the basilar membrane to a moderately low frequency driving force. (From Vander,A; Sherman,J; and Luciano,D. **Human Physiology,** *4th edition, 1985, P662. McGraw Hill Publishing Co., NY.)*

STRINGED INSTRUMENTS

The stringed instruments provide the clearest example of how musical instruments function. The only possible modes of oscillation of the string are those with nodes at the ends, and we have seen that the frequencies of these modes form a harmonic series $f_n = nf_1$. This suggests that if we record the sound produced by a stringed instrument and take a Fourier transform to see what harmonics are present in the sound, we can tell from that what modes of oscillation were present in the vibrating string.

This is essentially correct for the electric stringed instruments like the electric guitar and electric violin. Both of these instruments have a magnetic pickup that detects the velocity of the string at the pickup, using the same principle as the velocity detector we used in the air cart experiments discussed earlier in this chapter. The voltage signal from the magnetic pickup is then amplified electronically and sent to a loudspeaker. Thus the sound we hear is a fairly accurate representation of the motion of the string, and an analysis of that sound should give us a good idea of which modes of oscillation of the strings were excited.

The situation is different for the acoustic stringed instruments, like the acoustic guitar used by folk singers, and the violin, viola, cello and base, found in symphony orchestras. In these instruments the vibration of the string does not produce that much sound itself. Instead, the vibrating string excites resonances in the sound box of the instrument, and it is the sound produced by the resonating sound box that we hear. As a result the quality of the sound from an acoustic string instrument depends upon how the sound box was constructed. Subtle differences in the shape of the sound box and the stiffness of the wood used in its construction can lead to subtle differences in the harmonics excited by the vibrating string. The human ear is so sensitive to these subtle differences that it can easily tell the difference between a great instrument like a 280 year old Stadivarius violin, and even the best of the good instruments being made today. (It may be that it takes a couple of hundred years of aging for a very good violin to become a great one.)

To demonstrate the difference between electric and acoustic stringed instruments, and to illustrate how Fourier analysis can be used to study these differences, my daughter played the same note, using the same bowing technique, on the open E string of both her electric and her acoustic violins. Using the same microphone in the same setting to record both, we obtained the results shown in Figure (37).

From Figure (37) we immediately see why acoustic stringed instruments sound differently from their electric counterparts. With the acoustic instruments you get a far richer mixture of harmonics. In the first trial, labeled *E(1) Electric Violin*, the string was bowed so that it produced a nearly pure 4th harmonic. The sound you hear corresponds to a pure tone of frequency $657.8/4 = 264$ Hz. The corresponding sound produced by the acoustic violin has predominately the same 4th harmonic, but a lot of the sound is spread through the first 8 harmonics.

A number of recordings were made, so that we could see how the sounds varied from one playing to the next. The examples shown in Figure (37) are typical. It is clear in all cases that for this careful bowing a single harmonic predominated in the electric violin while the acoustic violin produced a mixture of the first eight. It is rather surprising that the ear hears all of these sounds as representing the same note, but with a different quality of sound.

We chose the violin for this comparison because by using a bow, one can come much closer to exciting a pure mode of vibration of the string. We see this explicitly in the *E(1) Electric Violin* example. When you pluck or strum a guitar, even if you pluck only one string, you get a far more complex sound than you do for the violin. If you pluck a chord on an acoustic guitar, you get a very complex sound. It is the complexity of the sound that gives the acoustic guitar a richness that makes it so effective for accompanying the human voice.

Figure 37
Comparison of the sound of an electric and an acoustic violin. In each case the open E string was bowed as similarly as possible. The electric violin produces relatively pure tones. The interaction between the string and the sound box of the acoustic violin gives a much richer mixture of harmonics.

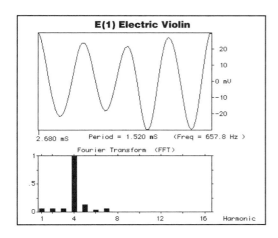
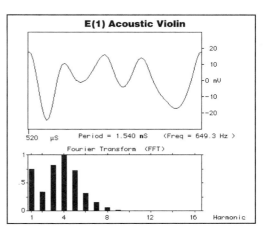
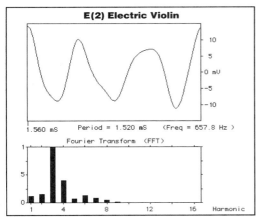
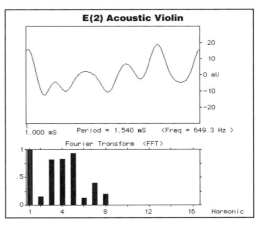
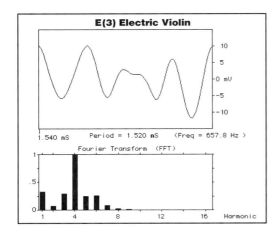
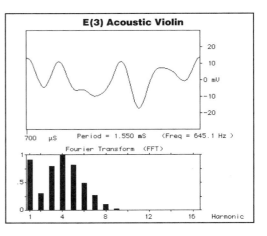

Recording the sound of an instrument and using Fourier analysis can be an effective tool for studying musical instruments, but care is required. For example, in comparing two instruments, start by choosing a single note, and try to play the note the same way on both instruments. Make several recordings so that you can tell whether any differences seen are due to the way the note was played or due to the differences in the instruments themselves. With careful work, you can learn a lot about the nature of the instrument.

In the 1970s, before we had personal computers, students doing project work would analyze the sound produced by instruments or the voice, using a time sharing mainframe computer system to do the Fourier analysis. Hours of work were required to analyze a single sound, but the results were so interesting that they served as the incentive to develop the MacScope™ program when the Macintosh computer became available

I particularly remember an early project in which two students compared the spinet piano, the upright piano and the grand piano. They recorded middle C played on each of these pianos. Middle C on the spinet consisted of a wide band of harmonics. From the upright there were still a lot of harmonics, but the first, third, and fifth began to predominate. The grand piano was very clean with essentially only the first, third, and fifth present. You could clearly see the effect of the increase in the size of the musical instrument.

The same year, another student, Kelly White, took a whale sound from the Judy Collins record *Sound of the Humpback Whales*. Listening to the record, the whale sounds are kind of squeaky. But when the sound was analyzed, the results were strikingly similar to those of the grand piano. The analysis suggested that the whale sounds were by an instrument as large as, or larger, than a grand piano. (The whale's blowhole acts as an organ pipe when the whale makes the sound.)

WIND INSTRUMENTS

While the string instruments are all based on the oscillation of a string, the wind instruments, like the organ, flute, trumpet, clarinet, saxophone, and glass bottle, are all based on the oscillations of an air column. Of these, the bottle is the most available for studying the nature of the oscillations of an air column.

When you blow carefully across the top of a bottle, you hear a sound with a very definite frequency. Add a little water to the bottle and the frequency of the note rises. When you shortened the length of a string, the pitch went up, thus it is not surprising that the pitch also goes up when you shorten the length of the air column.

You might guess that the mode of vibration you set up by blowing across the top of the bottle has a node at the bottom of the bottle and an anti node at the top where you are blowing. For a sine wave the distance from a node to the next anti node is 1/4 of a wavelength, thus you might predict that the sound has a wavelength 4 times the height of the air column, and a frequency

$$f = \frac{v_{sound}}{\lambda} = \frac{v_{sound}}{4d} \quad (\text{for } \lambda = 4d) \tag{4}$$

This prediction is not quite right as you can quickly find out by experimenting with various shaped bottles. Add water to different shaped bottles, adjusting the levels of the water so that all the bottles have the same height air columns. When you blow across the top of the different bottles, you will hear distinctly different notes, the fatter bottles generally having lower frequencies than the skinny ones. Unlike the vibrating string, it is not just the length of the column and the speed of the wave that determine the frequency of oscillation, the shape of the container also has a noticeable effect.

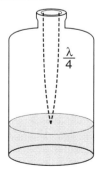

Figure 38
Blowing across the top of a bottle produces a note whose wavelength is approximately 4 times the height of the air column.

Despite this additional complexity, there is one common feature to the air columns encountered in musical instruments. They all have unique frequencies of oscillation. Even an organ pipe that is open at both ends—a situation where you might think that the length of the column is not well defined—the air column has a precise set of frequencies. The fussiness in defining the length of the column does not fuzz out the sharpness of the resonance of the column.

Organ pipes, with their straight sides, come closest to the simple standing waves we have seen for a stretched instrument string. In all cases, the wave is excited at one end by air passing over a sharp edge creating a turbulent flow behind the edge. This turbulence excites the air column in much the same way that dragging a sticky bow across a violin string excites the oscillation of the string.

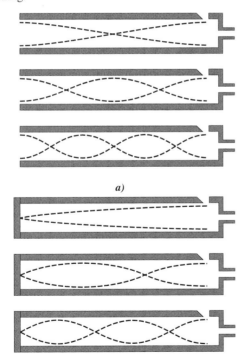

Figure 39
Modes of oscillation of an air column in open and closed organ pipes.

The various modes of oscillation of an air column in an organ pipe are shown in Figure (39). All have an anti node at the end with sharp edges where the turbulence excites the oscillation. The open ended pipes shown in Figure (39a) also have anti nodes at the open end, while the closed pipes of Figure (39b) have a node at the far end. The pictures in Figure (39a) are a bit idealized, but give a reasonably accurate picture of the shape of the standing wave. We can compensate for a lack of accuracy of these pictures by saying, for example, that the anti node of the open ended pipes lies somewhat beyond the end of the pipe.

Exercise 1

(a) Find the formula for the wavelength λ_n of the nth harmonic of the open ended pipes of Figure (39).

(b) Assuming that the frequencies of vibration are given by $f(cycle/sec) = v(meter/sec) / \lambda(meter/cycle)$ where v is the speed of sound, what is the formula for the allowed frequencies of the open organ pipe of length L?

(c) What should be the length L of an open ended organ pipe to produce a fundamental frequency of 440 cycles/second, middle A?

(d) Repeat the calculations of parts a, b, and c for the closed organ pipes of Figure (39).

(e) If you have the opportunity, find a real organ and check the predictions you have made (or try the experiment with bottles).

If you start with an organ pipe, and drill holes in the side, essentially converting it into a flute or one of the other wind instruments like a clarinet, you considerably alter the shape and frequency of the modes of oscillation of the air column. As a first approximation you could say that you create an anti node at the first open hole. But then when you play these instruments you can make more subtle changes in the pitch by opening some holes and closing others. The actual patterns of oscillation can become quite complex when there are open holes, but the simple fact remains that, no matter how complex the wave pattern, there is a precise set of resonant frequencies of oscillation. It is up to the maker of the instrument to locate the holes in such a way that the resonances have the desired frequencies.

PERCUSSION INSTRUMENTS

We all know that the string and wind instruments produce sound whose frequency we adjust to produce melodies and chords. But what about drums? They seem to just make noise. Surprisingly, drumheads have specific modes of oscillation with definite frequencies, just as do vibrating strings and air columns. But one does not usually adjust the fundamental frequency of oscillation of the drumhead, and the frequencies of the higher modes of oscillation do not follow the harmonic patterns of string and wind instruments.

To observe the standing wave patterns corresponding to modes of vibration of a drumhead, we can drive the drumhead at the resonant frequency of the oscillation we wish to study. It turns out to be a lot easier to drive a drumhead at a precise frequency than it is to find the normal modes of the coupled air cart system.

The experiment is illustrated in Figure (40). The apparatus consists of a hollow cardboard cylinder with a rubber sheet stretched across one end to act as a drumhead. At the other end is a loudspeaker attached to a signal generator. When the frequency of the signal generator is adjusted to the resonant frequency of one of the normal modes of the drumhead, the drumhead will start to vibrate in that mode of oscillation.

To observe the shape and motion of the drumhead in one of its vibrational modes, we place a strobe light to one side of the drumhead as shown. If you adjust the strobe to the same frequency as the normal mode vibration, you can stop the motion and see the pattern.

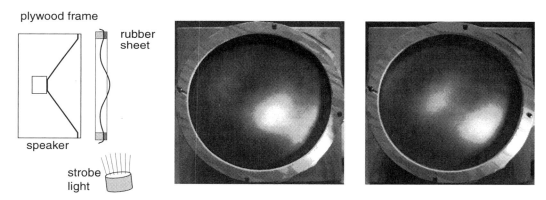

Figure 40
Studying the modes of oscillation of a drumhead.

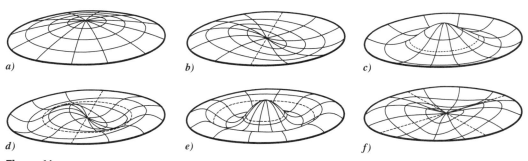

Figure 41
Modes of oscillation of a drumhead. (Adapted from **Vibration and Sound** *by Phillip M. Morse, 2nd ed., McGraw-Hill, New York, 1948.)*

Turn the frequency a bit off resonance, and you get a slow motion moving picture of the motion of that mode.

Some of the low frequency normal mode or standing wave patterns of the drumhead are illustrated in Figure (41). In the lowest frequency mode, Figure (41a), the entire center part of the drumhead moves up and down, much like a guitar string in its lowest frequency mode. In this pattern there are no nodes except at the rim of the drumhead.

In the next lowest frequency mode, shown in Figure (41b), one half the drumhead goes up while the other half goes down, again much like the second harmonic mode of the guitar string. The full two dimensional nature of the drumhead standing waves begins to appear in the next mode of Figure (41c) where the center goes up while the outside goes down. Now we have a circular node about half way out on the radius of the drumhead.

As we go up in frequency, we observe more complex patterns for the higher modes. In Figure (41d) we see a pattern that has a straight node like (41b) and a circular node like (41c). This divides the drumhead into 4 separate regions which oscillate opposite to each other. Finer division of the drumhead into smaller regions can be seen in Figures (41e) and (41f). The frequencies of the various modes are listed with each diagram. You can see that there is no obvious progression of frequencies like the harmonic progression for the modes of a stretched string.

When you strike a drumhead you excite a number of modes at once and get a complex mixture of frequencies. However, you do have some control over the modes you excite. Bongo drum players, for example, get different sounds depending upon where the drum is struck. Hitting the drum in the center tends to excite the lowest mode of vibration and produces a lower frequency sound. Striking the drum near the edges excites the higher harmonics, giving the drum a higher frequency sound.

Even more complex than the modes of vibration of a drumhead are those of the components of a violin. To construct a successful violin, the front and back plates of a violin must be tuned before assembly. Figure (42) shows a violin backplate under construction, while Figure (43) shows the first 6 modes of oscillation of a completed backplate. Note again that the resonant frequencies do not form a harmonic series.

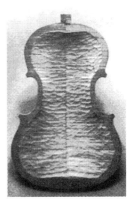

Figure 42
Back plate of a violin under construction. The resonant frequencies are tuned by carving away wood from different sections of the plate.

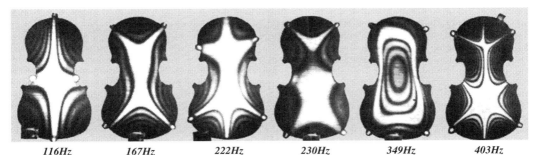

116Hz   167Hz   222Hz   230Hz   349Hz   403Hz

Figure 43
Modes of oscillation of the backplate made visible by holographic techniques. Quality violins are made by tuning the frequencies of the various modes. (Figures 42 and 43 from "The Acoustics of Violin Plates," by Carleen Maley Hutchins, **Scientific American,** *October 1981.)*

SOUND INTENSITY

One of the amazing features of the human eardrum is its ability to handle an extreme range of intensities of sound waves. We define the ***intensity*** of a sound wave as the ***amount of energy per second being carried by a sound wave through a unit area***. In the MKS system of units, this would be the number of joules per second passing through an area of one square meter. Since one joule per second is a unit of power called a ***watt***, the MKS unit for sound intensity is ***watts per square meter***. The human ear is capable of detecting sound intensities as faint as 10^{-12} watts/m^2, but can also handle intensities as great as 1 watt/m^2 for a short time. This is an astounding range, a factor of 10^{12} in relative intensity.

The ear and brain handle this large range of intensities by essentially using a logarithmic scale. Imagine, for example, you are to sit in front of a hi fi set playing a pure tone, and you are told to mark off the volume control in equal steps of loudness. The first mark is where you just barely hear the sound, and the final mark is where the sound just begins to get painful. Suppose you are asked to divide this range of loudness into what you perceive as 12 equal steps. If you then measured the intensity of the sound you would find that the intensity of the sound increased by approximately a factor of 10 after each step. Using the faintest sound you can hear as a standard, you would measure that the sound was 10 times as intense at the end of the first step, 100 times as intense after the second step, 1000 times at the third, and 10^{12} times as intense at the final step.

The idea that the intensity increases by a factor of 10 for each equal step in loudness is what we mean by the statement that the loudness is based on a logarithmic scale. We take the faintest sound we can hear, an intensity $I_0 = 10^{-12}$ watts/m^2 as a basis. At the first setting $I = I_0$, at the second setting $I_1 = 10\, I_0$, at the 3rd setting $I_2 = 100\, I_0$, etc. The factor I/I_0 by which the intensity has increased is thus

$$I_0/I_0 = 1 \qquad (I_0 = 10^{-12} \text{ watts/m}^2)$$
$$I_1/I_0 = 10$$
$$I_2/I_0 = 100$$
$$\cdots\cdots$$
$$I_{12}/I_0 = 10^{12} \qquad\qquad (5)$$

Taking the logarithm to the base 10 of these ratios gives

$$\text{Log}_{10}(I_0/I_0) = \text{Log}_{10}(1) = 0$$
$$\text{Log}_{10}(I_1/I_0) = \text{Log}_{10}(10) = 1$$
$$\text{Log}_{10}(I_2/I_0) = \text{Log}_{10}(100) = 2$$
$$\cdots\cdots$$
$$\text{Log}_{10}(I_{12}/I_0) = \text{Log}_{10}(10^{12}) = 12 \qquad (6)$$

Bells and Decibels

The scale of loudness defined as $\text{Log}_{10}(I/I_0)$ with $I_0 = 10^{-12}$ watts/m^2 is measured in bells, named after Alexander Graham Bell, the inventor of the telephone.

$$\left.\begin{array}{l}\text{loudness of a sound}\\ \text{measured in bells}\end{array}\right\} \equiv \text{Log}_{10}\!\left(\frac{I}{I_0}\right)$$
$$I_0 = 10^{-12} \text{watts/m}^2 \qquad (7)$$

From this equation, we see that the faintest sound we can hear, at $I = I_0$, has a loudness of zero bells. The most intense one we can stand for a short while has a loudness of 12 bells. All other audible sounds fall in the range from 0 to 12 bells.

It turns out that the bell is too large a unit to be convenient for engineering applications. Instead one usually uses a unit called the ***decibel (db)*** which is 1/10 of a bell. Since there are 10 decibels in a bell, the formula for the ***loudness β***, in decibels, is

$$\beta(\text{decibels}) = (10\text{ db})\text{Log}_{10}\!\left(\frac{I}{I_0}\right) \qquad (8)$$

On this scale, the loudness of sounds range from 0 decibels for the faintest sound we can hear, up to 120 decibels (10×12 bells) for the loudest sounds we can tolerate.

The average loudness of some of the common or well known sounds is given in Table 1.

Table 1   Various Sound Levels in db

threshold of hearing	0
rustling leaves	10
whisper at 1 meter	20
city street, no traffic	30
quiet office	40
office, classroom	50
normal conversation at 1 meter	60
busy traffic	70
average factory	80
jack hammer at 1 meter	90
old subway train	100
rock band	120
jet engine at 50 meters	130
Saturn rocket at 50 meters	200

An increase in loudness of 10 db corresponds to an increase of 1 bell, or an increase of intensity by a factor of 10. A rock band at 110 db is some 100 times as intense (2 factors of 10) as a jack hammer at 90 db. A Saturn rocket is about 10^{20} times as intense as the faintest sound we can hear.

Our sensitivity to sound depends not only to the intensity of the sound, but also to the frequency. About the lowest frequency note one can hear, and still perceive as being sound, is about 20 cycles/second. As you get older, the highest frequencies you can hear decreases from around 20,000 cycles/sec for children, to 15,000 Hz for young adults to under 10,000 Hz for older people. If you listen to too much, too loud rock music, you can also decrease your ability to hear high frequency sounds.

Figure (44) is a graph of the average range of sound levels for the human ear. The faintest sounds we can detect are in the vicinity of 4000 Hz, while any sound over 120 db is almost uniformly painful. The frequency ranges and sound levels usually encountered in music are also shown.

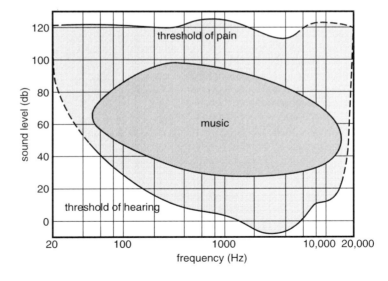

Figure 44
Average range of sound levels for the human ear. Only the very young can hear sound frequencies up to 20,000 Hz. (Adapted from **Fundamental Physics** *by Halliday and Resnick, John Wiley & Sons.)*

Sound Meters

Laboratory experiments involving the intensity or loudness of sound are far more difficult to carry out than those involving frequencies like the Fourier analysis experiments already discussed. From the output of any reasonably good microphone, you can obtain a relatively good picture of the frequencies involved in a sound wave. But how would you go about determining the intensity of a sound from the microphone output? (There are commercial sound meters which have a scale that shows the ambient sound intensity in decibels. Such devices are often owned by zoning boards for checking that some factory or other noise source does not exceed the level set by the local zoning ordinance, often around 45 db. The point of our question is, how would you calibrate such a device if you were to build one?)

The energy in a wave is generally proportional to the square of the amplitude of the wave. A sound wave can be viewed as oscillating pressure variations in the air, and the energy in a sound wave turns out to be proportional to the square of the amplitude of the pressure variations. The output of a microphone is more or less proportional to the amplitude of the pressure variations, thus we expect that the intensity of a sound wave should be more or less proportional to the square of the voltage output of the microphone. However, there is a great variation in the sensitivity of different microphones, and in the amplifier circuits used to produce reasonable signals. Thus any microphone that you wish to use for measuring sound intensities has to be calibrated in some way.

Perhaps the easiest way to begin to calibrate a microphone for measuring sound intensities is to use the fact that very little sound energy is lost as sound travels out through space. Suppose you had a speaker radiating 100 watts of sound energy, and for simplicity let us assume that the speaker radiates uniformly in all directions and that there are no nearby walls.

If we are 1 meter from this speaker, all the sound energy is passing out through a 1 meter radius sphere centered on the speaker. Since the area of a sphere is $4\pi r^2$, this 1 meter radius sphere has an area of 4π meters^2, and the average intensity of sound at this 1 meter distance must be

$$\begin{pmatrix} \text{average intensity of} \\ \text{sound 1 meter from} \\ \text{a 100 watt speaker} \end{pmatrix} = \frac{100 \text{ watts}}{4\pi \text{ meters}^2}$$

$$= 8.0 \frac{\text{watts}}{\text{meters}^2} \quad (9)$$

If we wish to convert this number to decibels, we get

$$\beta \begin{pmatrix} \text{sound intensity 1 meter} \\ \text{from a 100 watt speaker} \end{pmatrix}$$

$$= (10 \text{ db}) \text{Log} \frac{I}{I_0}$$

$$= (10 \text{ db}) \text{Log}_{10} \left(\frac{8.0 \text{ watts}/\text{m}^2}{10^{-12} \text{ watts}/\text{m}^2} \right)$$

$$= (10 \text{ db}) \times \text{Log}_{10} (8 \times 10^{12})$$

$$= (10 \text{ db}) \times 12.9$$

$$= 129 \text{ db} \quad (10)$$

From our earlier discussion we see that this exceeds the threshold of pain. One meter from a 100 watt speaker is too close for our ears. But we could place a microphone there and measure the amplitude of the signal output for our first calibration point.

Move the speaker back to a distance of 10 meters and the area that the sound energy has to pass through increases by a factor of 100 since the area of a sphere is proportional to r^2. Thus as the same 100 watts passes through this 100 times larger area, the intensity drops to 1/100 of its value at 1 meter. At a distance of 10 meters the intensity is thus $8/100 = .08$ watts/m^2 and the loudness level is

$$\beta \begin{pmatrix} \text{10 meters from a} \\ \text{100 watt speaker} \end{pmatrix}$$

$$= (10 \text{ db}) \text{Log}_{10} \left(\frac{.08 \text{ watts}/\text{m}^2}{10^{-12} \text{ watts}/\text{m}^2} \right)$$

$$= 109 \text{ db} \qquad (11)$$

We see that when the intensity drops by a factor of 100, it drops by 20 db or 2 bells.

To calibrate your sound meter, record the amplitude of the signal on your microphone at this 10 meter distance, then set the microphone back to a distance of 1 meter, and cut the power to the speaker until the microphone reads the same value as it did when you recorded 100 watts at 10 meters. Now you know that the speaker is emitting only 1/100th as much power, or 1 watt. Repeating this process, you should be able to calibrate a fair range of intensities for the microphone signal. If you get down to the point where you can just hear the sound, you could take that as your value of I_0, which should presumably be close to $I_0 = 10^{-12}$ watts/m^2. Then calibrate everything in db and you have built a loudness meter. (The zoning board, however, might not accept your meter as a standard for legal purposes.)

Exercise 2
What is the loudness, in db, 5 meters from a 20 watt speaker? (Assume that the sound is radiated uniformly in all directions).

Exercise 3
You are playing a monophonic record on your stereo system when one of your speakers cuts out. How many db did the loudness drop? (Assume that the intensity dropped in half when the speaker died. Surprisingly you can answer this question without knowing how loud the stereo was in the first place. The answer is that the loudness dropped by 3 db).

Speaker Curves

When you buy a hi fi loudspeaker, you may be given a frequency response curve like that in Figure (45), for your new speaker. What the curve measures is the intensity of sound, at a standard distance, for a standard amount of power input at different frequencies. It is a fairly common industry standard to say that the frequency response is "flat" over the frequency range where the intensity does not fall more than 3 db from its average high value. In Figure (45), the response of that speaker, with the woofer turned on, is more or less "flat" from 62 Hz up to 30,000 Hz.

Why the 3 db cutoff was chosen, can be seen in the result of Exercise 3. There you saw that *if you reduce the intensity of the sound by half, the loudness drops by 3 db.* This is only 3/120 (or 1/36) of our total hearing range, not too disturbing a variation in what is supposed to be a flat response of the speaker.

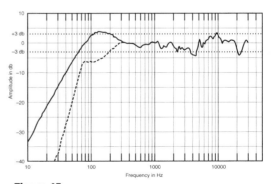

Figure 45
Speaker response curve from a recent audio magazine. The dashed line shows the response when the woofer is turned off. (We added the dotted lines at + and – 3 db.)

APPENDIX A
FOURIER ANALYSIS LECTURE

In our discussion of Fourier Analysis, we saw that any wave form can be constructed by adding together a series of sine and cosine waves. You can think of the Fourier transform as a mathematical prism which breaks up a sound wave into its various wavelengths or frequencies, just as a light prism breaks up a beam of white light into its various colors or wavelengths. In MacScope, the computer does the calculations for us, figuring out how much of each component sine wave is contained in the sound wave. The point of this lecture is to give you a feeling for how these calculations are done. The basic ideas are easy, only the detailed calculations that the computer does would be hard for us to do.

Square Wave

In Figure A-1 we show a MacScope window for a square wave produced by a Hewlett Packard oscillator.

We have selected precisely one cycle of the wave, and see that the even harmonics are missing. A careful investigation shows that the amplitude of the Nth odd harmonic is 1/N as big as the first (e.g., the 3rd harmonic is 1/3 as big as the first, etc.). Thus the mathematical formula for a square wave F(t) can be written:

$$F(t) = (1)\sin(t) + (1/3)\sin(3t) + (1/5)\sin(5t) + (1/7)\sin(7t) \ldots$$

where, for now, we are assuming that the period of the wave is precisely 2π seconds. The coefficients (1), (1/3), (1/5), (1/7), which tell us how much of each sine wave is present, are called the ***Fourier coefficients***. Our goal is to calculate these coefficients.

Calculating Fourier Coefficients

In general we cannot construct an arbitrary wave out of just sine waves, because sine waves, $\sin(t)$, $\sin(2t)$, etc., all have a value 0 at $t = 0$ and at $t = 2\pi$. If our wave is not zero at the beginning ($t = 0$) of our selected period, or not zero at the end ($t = 2\pi$), then we must also include cosine waves which have a value 1 at those points. Thus the general formula for breaking an arbitrary repetitive wave into sine and cosine waves is:

$$\begin{aligned}F(t) = A_0 &+ A_1\cos(1t) + A_2\cos(2t) \\ &+ A_3\cos(3t) + \ldots \\ &+ B_1\sin(1t) + B_2\sin(2t) \\ &+ B_3\sin(3t) + \ldots \quad \text{(A-1)}\end{aligned}$$

The question is: How do we find the coefficients A_0, A_1, A_2, B_1, B_2 etc. in Equation (A1)? (These are the ***Fourier coefficients***.)

To see how we can determine the Fourier coefficients, let us take an explicit example. Suppose we wish to find the coefficient B_3, representing the amount of $\sin(3t)$ present in the wave. We can find B_3 by first multiplying Equation (1) through by $\sin(3t)$ to get:

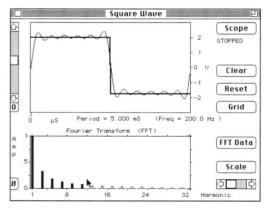

Figure A-1
The square wave has only odd harmonics.

$$\begin{aligned}
F(t)\sin(3t) = &\ A_0 \sin(3t) \\
&+ A_1 \cos(1t)\sin(3t) \\
&+ A_2 \cos(2t)\sin(3t) \\
&+ A_3 \cos(3t)\sin(3t) \\
&+ ... \\
&+ B_1 \sin(1t)\sin(3t) \\
&+ B_2 \sin(2t)\sin(3t) \\
&+ B_3 \sin(3t)\sin(3t) \\
&+ ...
\end{aligned} \quad (A-2)$$

At first it looks like we have created a real mess. We have a lot of products like cos(t)sin(3t), sin(t)sin(3t), sin(3t)sin(3t), etc. To see what these products look like, we plotted them using **True BASIC™** and obtained the results shown in Figure (A2) (on the next page).

Notice that in all of the plots involving sin(3t), the product $\sin(3t)\sin(3t) = \sin^2(3t)$ is special; it is the only one that is always positive. (It has to be since it is a square.) A careful investigation shows that, in all the other non square terms, there is as much negative area as positive area, as indicated by the two different shadings in Figure (A3). If we define the "net area" under a curve as *the positive area minus the negative area*, then only the sin(3t)sin(3t) term on the right side of the Equation A-2 has a net area.

The mathematical symbol for finding the net area under a curve (in the interval t = 0 to t = 2π) is:

$$\left.\begin{aligned}\text{Area under} \\ \sin(3t)\sin(3t) \\ \text{from } t = 0 \\ \text{to } t = 2\pi\end{aligned}\right\} = \int_0^{2\pi} \sin(3t)\sin(3t)\,dt \quad (A-3)$$

(Those who have had calculus say we are taking the **integral** of the term sin(3t)sin(3t).)

A basic rule learned in algebra is that if we do the same thing to both sides of an equation, the sides will still be equal. This is also true if we do something as peculiar as evaluating the net area under the curves on both sides of an equation.

If we take the net area under the curves on the right side of Equation A-2, only the $\sin^2(3t)$ term survives and we get:

$$\int_0^{2\pi} F(t)\sin(3t)\,dt = B_3 \int_0^{2\pi} \sin^2(3t)\,dt \quad (A-4)$$

In Figure (A4), we have replotted the curve $\sin^2(3t)$, and drawn a line at height y = .5. We see that the peaks above the y = .5 line could be flipped over to fill in the valleys below the y = .5 line. Thus $\sin^2(3t)$ in the interval t = 0 to t = 2π has a net area equal to that of a rectangle of height .5 and length 2π. I.e., the net area is π:

$$\int_0^{2\pi} \sin^2(3t)\,dt = \pi \quad (A-5)$$

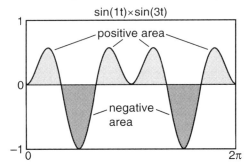

Figure A-3
Only the square terms such as sin(3t)sin(3t) have a net area. This curve has no net area.

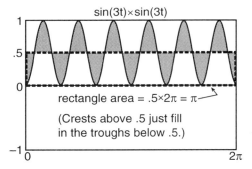

Figure A-4
The area under the curve $\sin^2(3t)$ is equal to the area of a rectangle .5 high by 2π long.

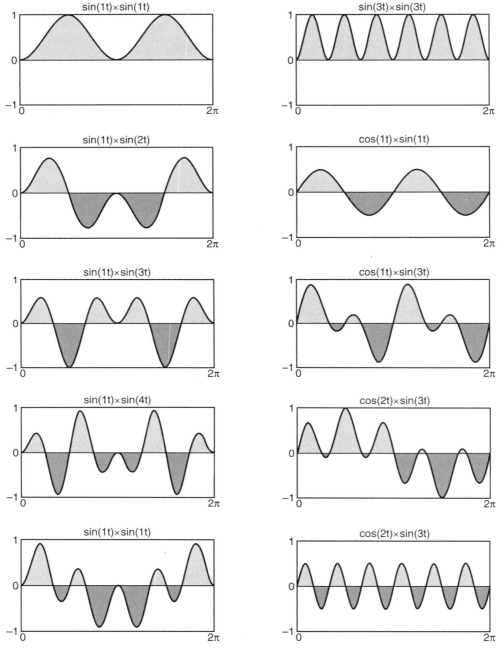

Figure A-2
Product wave patterns.

Substituting Equation A-5 in Equation A-4 and solving for the Fourier coefficient B_3 gives:

$$B_3 = \frac{1}{\pi}\int_0^{2\pi} F(t)\sin(3t)\,dt \quad (A\text{-}6)$$

Similar arguments show that the general formulas for the Fourier coefficients A_n and B_n are:

$$\boxed{A_n = \frac{1}{\pi}\int_0^{2\pi} F(t)\cos(nt)\,dt} \quad (A\text{-}7)$$

$$\boxed{B_n = \frac{1}{\pi}\int_0^{2\pi} F(t)\sin(nt)\,dt} \quad (A\text{-}8)$$

These integrals, which were nearly impossible to do before computers, are now easily performed even on small personal computers. Thus the computer has made Fourier analysis a practical experimental tool.

Amplitude and Phase

Instead of writing the Fourier series as a sum of separate sine and cosine waves, it is often more convenient to use amplitudes and phases. The basic formula we use is **

$$A\cos(t) + B\sin(t) = C\cos(t-\phi) \quad (A\text{-}9)$$

where

C = amplitude

$C^2 = A^2 + B^2$

ϕ = phase

$\tan(\phi) = B/A$

The function $\cos(t-\phi)$ is illustrated in Figure (A-5). We see that C is the amplitude of the wave, and the phase angle ϕ is the amount the wave has been moved to the right. (When $t = 0$, $\cos(t-\phi) = \cos(-\phi)$.) With Equation A-9, we can rewrite equation A-1 in the form:

$$\begin{aligned}F(t) &= C_0 + C_1\cos(t-\phi_1)\\ &+ C_2\cos(2t-\phi_2)\\ &+ C_3\cos(3t-\phi_3)\\ &+ \ldots \end{aligned} \quad (A\text{-}10)$$

The advantage of Equation A-10 is that the coefficients C represent how much of each wave is present, and sometimes we do not care about the phase angle ϕ. For example our ears are not particularly sensitive to the phase of the harmonics in a musical note, thus the tonal quality of a musical instrument is determined almost entirely by the amplitudes C of the harmonics the instrument produces.

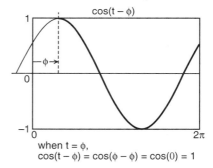

when $t = \phi$,
$\cos(t-\phi) = \cos(\phi-\phi) = \cos(0) = 1$

Figure A-5
The function $\cos(t-\phi)$. When $\phi = 90°$ you get a sine wave.

** Start with
$\cos(x-y) = \cos(x)\times\cos(y) + \sin(x)\times\sin(y)$
Let $x = t$, $y = \phi$, and multiply through by C to get
$C\cos(t-\phi) = \{C\cos(\phi)\}\cos(t) + \{C\sin(\phi)\}\sin(t)$
This is Equation A–9 if we set
$A = \{C\cos(\phi)\}$; $B = \{C\sin(\phi)\}$
Thus $\tan\phi = \sin\phi/\cos\phi = B/A$

In the Fourier transform plots we have shown so far, the graph of the harmonics has been representing the amplitudes C. If you wish to see a plot of the phases φ, then press the button labeled ø as shown, and you get the result seen in Figure(A-6). In that figure we are looking at the phases of the odd harmonic sine waves that make up a square wave. Since

$$\sin(t) = \cos(t - 90°)$$

all the sine waves should have a phase shift of 90°.

If for any reason, you need accurate values of the Fourier coefficients, they become available if you press the **FFT Data** button to get the results shown in Figure (A-7). When you do this, the **Editor** window is filled with a text file containing the A, B coefficients accurate to 3 or 4 significant figures.

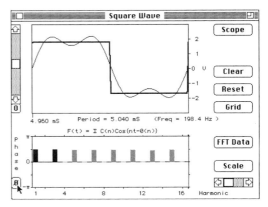

Figure A-6
When you press the ø button, the Fourier Transform display shifts from amplitudes to phases. Since the square wave is made up of pure odd harmonic sine waves, each odd harmonic should have a 90 degree or π/2 phase shift.

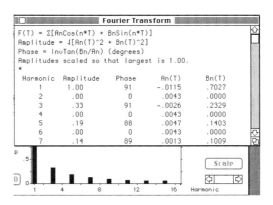

Figure A-7
For greater numerical accuracy, you can press the FFT Data button. This gives you a text file with the A, B coefficients given to four places.

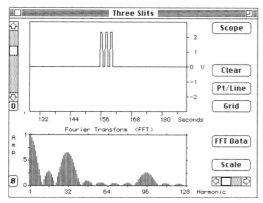

Figure A-8
Amplitude. The Fourier transform of a 3-slit pattern gives the amplitude of the diffraction pattern that would be produced by a laser beam passing through these slits. (Selecting the data to give wider slits would correspond to using a different wavelength laser beam.)

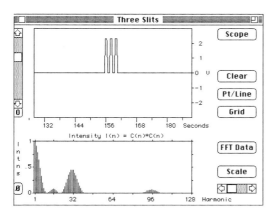

Figure A-9
Intensity. In the lab, you see the intensity of the diffraction pattern. MacScope will display the intensity of the Fourier transform if you click one more time on the ø button.

Amplitude and Intensity

An experiment that has become possible with MacScope, is to have students compare the Fourier transform of a multiple slit grating with the diffraction pattern produced by a laser beam passing through that grating. For example, in Figure (A-8) we have taken the transform of a 3-slit grating. In this case, the 3-slit "pattern" was made simply by turning a 2 volt power supply on and off. We are now working on ways for students to record the slit pattern directly.

The problem with Figure (A-8) is that the Fourier transform of the slits gives the amplitude of the diffraction pattern, while in the lab one measures the intensity of the diffraction pattern. The intensity of a light wave is proportional to the square of its amplitude.

In order that students can compare a slit Fourier transform with an experimental diffraction pattern, we have designed MacScope so that one more press on the ø button takes us from a display of phases to intensities. Explicitly, the ø button cycles from *amplitudes* to *phases* to *intensities*. In Figure (A-9) we have clicked over to intensities, and this pattern may be directly compared with the intensity of the 3-slit diffraction pattern seen in lab.

(The Fourier transform of the diffraction pattern amplitude should give the slit pattern. Unfortunately, if you take the Fourier transform of the experimental diffraction pattern, you are taking the transform of the intensity, or square, of the amplitude. What you get, as Chris Levey of our department demonstrated, is the *convolution* of the slit pattern with itself.)

APPENDIX B
INSIDE THE COCHLEA

In Figure (32), our simplified unwound view of the cochlea, we show only the basilar membrane separated by two fluid channels (the scala vestibuli which starts at the oval window, and the scala tympani which ends at the round window). That there is much more structure in the cochlea is seen in the cochlea cross section of Figure (B-1). The purpose of this additional structure is to detect the motion of the basilar membrane in a way that is sensitive to the harmonic content of the incoming sound wave.

Recall that when the basilar membrane is excited by a sinusoidal oscillation, the maximum amplitude of the response of the basilar membrane is located at a position that depends upon the frequency of the oscillation.

As seen in Figure (35), the lower the frequency, the farther down the membrane the maximum amplitude occurs. Along the top of the basilar membrane is a system of hair cells that detects the motion of the membrane and sends the needed information to the brain.

Figure (B-2a) is a close up view of the hair cells that sit atop of the basilar membrane. (There are about 30,000 hair cells in the human ear.) Above the hair cells is another membrane called the tectorial membrane which is hinged on the left hand side of that figure. Fine hairs go from the top of each hair cell up to the tectorial membrane as shown. When the basilar membrane is deflected by an incoming sound wave the hairs are bent as shown in Figure (B-2b). It is the bending of the hairs that triggers an electrical impulse in the hair cell.

Figure (B-3) is a mechanical model of how the bending of the hairs creates the electrical impulse. The fluid in the cochlea duct surrounding the hair cells has a high concentration of positive potassium ions (k^+). The

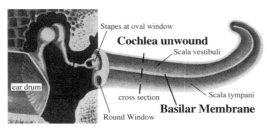

Figure 32 (repeated)
The cochlea unwound.

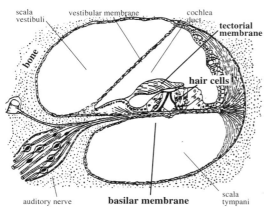

Figure B-1
Cross section of the cochlea. (From Vander,A; Sherman,J; and Luciano,D. **Human Physiology**, *4th edition, 1985, P662. McGraw Hill Publishing Co., NY.)*

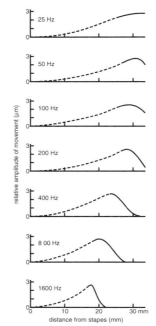

Figure 35 (repeated)
Amplitude of the motion of the basilar membrane at different frequencies, as we go down the basilar membrane.

bending of the hair cell opens small channels allowing potassium ions to flow into the hair cell. This flow of positive charge into the cell changes the electrical potential of the cell, triggering reactions that will eventually result in an electrical impulse in the nerve fiber that is connected to the hair cell.

After the channel at the top of the hair cell closes, the excess potassium is pumped out of the hair cell, and the cell returns to its normal resting voltage, ready to fire again.

There are various ways that a hair cell can transmit frequency information to the nervous system. One is by its location down the basilar membrane. The lower the frequency of the sound wave, the farther down the membrane an oscillation of the membrane takes place. Thus high frequency waves excite cells at the front of the basilar membrane, while low frequency oscillations excite cells at the back end.

Secondly, hair calls in a given area show special sensitivities to different frequencies. Figure (B-4) shows the amplitude, in db, of the sound wave required to excite a nerve fiber connected to that particular region of hair cells. You can see that the nerve is most sensitive to a 2 kilocycle (2kHz) frequency. At 2 kHz, that nerve fires when excited by a 15 db sound wave. It is not excited by a 4 kHz wave until the sound intensity rises to 80 db.

Ultimately the exquisite sensitivity of the human ear to different frequency components in a sound wave results from the fact that there are about 30,000 hair cells continuously monitoring the motion of the basilar membrane. Effective processing of this vast amount of information leads to the needed sensitivity. Much of this processing of information occurs in the nervous system in the ear, before the information is sent to the brain.

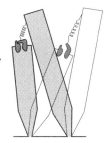

Figure B-3
*Model of the valves at the base of the hair cells.
(From Shepard, G.M., Neurobiology, 3rd Edition, 1994, P316. Oxford Univ. Press.)*

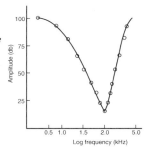

Figure B-4
Frequency dependence. A much lower amplitude sound will excite this nerve fiber at 2kHz than any other frequency. Different nerve fibers connected to the hair cells have different frequency dependence.

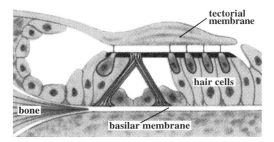
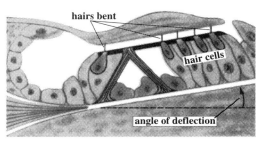

Figure B-2 a,b
When the basilar membrane is deflected by a sound wave, the hair cells are bent. This opens a valve at the base of the hair call. (Figures 2 & 4 adapted from Kandel, E; Schwartz, J; and Jessell, T; Principles of Neural Science, 3rd Edition, 1991; pages 486 and 489.)

Chapter 17
Atoms, Molecules and Atomic Processes

To extract the basic laws of mechanics from the variety and confusion of the world around us required looking at matters on a large scale, looking out at the moon and planets whose motion is regular, periodic, and easier to understand. In this and the next two chapters we take a similarly large leap to the small scale of distance where simplicity and periodic behavior again allow us to gain insight into the working of nature. Here we find the world of atoms and their constituent particles, a world in which we observe the basic forces and particles ultimately responsible for the variety about us.

The jump down to the small scale of atoms is comparable to the jump out from the study of projectile motion in the lab to the analysis of satellite orbits. Imagine, for example, that we could enlarge the golf balls used in our strobe labs to the size of the earth. The same enlargement of a hydrogen atom would give us an object about the size of a golf ball.

Only with the development of the new generation of microscopes in the late 1980s has it become possible to see and work with individual atoms. Figure (1) is the first atomic sized logo consisting of xenon atoms on a background of nickel, made by scientists at the IBM Research Laboratories in 1990. But despite great improvements in seeing and working with individual atoms, the images we now get are still fuzzy and we are restricted to looking at atoms in solid structures, atoms that do not move around.

Our knowledge of atoms comes not from looking through microscopes, but instead from the study of chemical reactions, the measurement of the physical properties of substances, and the bombardment of materials with x-rays and other particles. This study essentially began with John Dalton's construction of the first periodic table in 1808. Other milestones were Thomson's discovery of the electron in 1895, Rutherford's discovery of the atomic nucleus in 1912, Neils Bohr's model of the hydrogen atom in 1913, and the discovery of the rules of quantum mechanics in the mid 1920s.

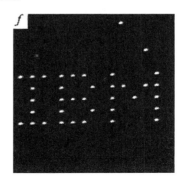

Figure 1
Thirty five xenon atoms were dragged across a nickel surface to form the letters IBM. (D. M. Eigler & E. K. Schweizer, Nature, *5 April 1990.)*

MOLECULES

Atoms attract each other to form molecules, like the water molecule H_2O sketched in Figure (2). It was from x-ray studies of ice, the crystalline form of water, that we know the distance from the center of the oxygen atom to the center of a hydrogen atom is $.958 \times 10^{-8}$ cm and that the hydrogen atoms are spread out at an angle of 104.5 degrees as shown.

X-ray studies of large biological molecules began in the late 1950s. For example, myoglobin is a substance found in muscle tissue. The myoglobin molecule contains over 2500 atoms, mostly carbon, hydrogen, oxygen, nitrogen, and one iron atom. For determining the precise structure of the myoglobin molecule from x-rays of crystals of myoglobin, John Kendrew and Max Peritz received the 1963 Nobel Prize in chemistry. Their model of the molecule is shown in Figure (3).

Recent advances in computer modeling now provide detailed views of numerous kinds of molecules. An example is Figure (4) showing the cholera toxin B-subunit.

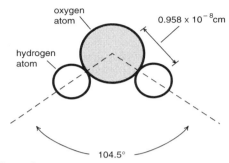

Figure 2
The water molecule H_2O. We know the precise location of the centers of the three atoms.

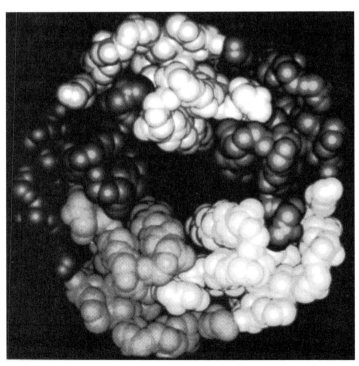

Figure 4
Computer model of the cholera toxin B-subunit. (Courtesy of Argonne National Laboratory.)

Figure 3
Model of the myoglobin molecule. (Photograph courtesy of J.C. Kendrew and H.C. Watson.)

ATOMIC PROCESSES

The myoglobin molecule provides a hint of the complex structures that can be formed from atoms, a complexity we wish to avoid in this chapter by concentrating on basic, simpler atomic processes. To help illustrate these processes, we have illustrated in Figures (5) through (11) a set of sketches drawn on the blackboard by Richard Feynman and copied to his book of introductory physics lectures. Such simple sketches, full of information, were characteristic of Feynman's style.

In the first three Figures (5, 6, and 7) we have views of three forms of matter made from water molecules, namely ice, water and steam. In the form of ice, the water molecules fit into a hexagonal structure with a hole in the center, as seen in Figures (5a,b). When water freezes to form snowflakes, the hexagonal structure repeats and we get a six sided symmetry seen in all snowflakes, examples of which is shown in Figure (5c).

When ice melts to form water, shown in Figure (6), the rigid structure of ice disappears and the water molecules can now slide past each other. In addition the holes in the hexagonal structure fill in with the result that water is more dense than ice. That is why ice floats in water.

The third form of water is steam, shown in Figure (7). In the gaseous state the water molecules move freely about, interacting only when they collide with each other. The picture of steam is more or less what we would see if we could look at the steam emerging from a teakettle on an atomic scale. The separation of the molecules is on the average about 10 times the diameter of a water molecule. As a result, the steam is about 1000 times less dense than liquid water. The transition from water to steam, either by evaporation or by boiling, involves a competition between molecular forces and thermal forces. We will discuss that competition shortly.

Figure 5a
Ball and stick model of an ice crystal.

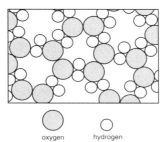

Figure 5b
Sketch of the arrangement of the water molecules in an ice crystal.

Figure 5c
Snowflakes reflect the 6-sided structure of the ice crystal.

Some atomic processes are illustrated in Figures (8) through (10). Figure (8) is what you might see in a snapshot of the surface of a glass of water. On our scale of distance, such a surface looks quiet, but on an atomic scale it is an active place with molecules continually entering and leaving the water. Evaporation occurs if more water molecules leave the water than return. If you put a cover over the glass, the concentration of water molecules in the air above the water builds up to the point where just as many water molecules return as leave, and the level of the water stops dropping. We would then say that the evaporation has ceased.

In Figure (9) we see what happens to a block of carbon when it is heated in an atmosphere of oxygen. By themselves oxygen atoms combine in pairs to form O_2 or oxygen molecules, and carbon atoms attract each other to form solids like diamond, graphite, soot, or Buckeyballs (soccerball shaped structures of carbon). But there is a greater attraction between a carbon and an oxygen atom than between two carbon or two oxygen atoms. If the carbon and oxygen are heated, the various atoms bounce into each other at high speeds, the old structures break apart and molecules of carbon monoxide ●○ and carbon dioxide ○●○ are formed. Energy is released in the process in the form of heat and light, and we say that the carbon is burning.

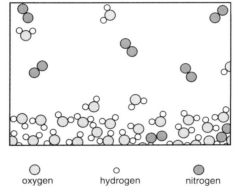

Figure 8
Water evaporating in air.

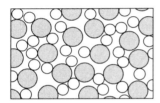

Figure 6
Water magnified a billion times.

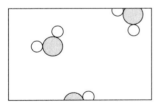

Figure 7
Steam.

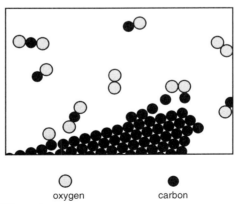

Figure 9
Carbon burning in oxygen.

The process of salt dissolving in water is illustrated in Figure (10). Table salt is a stable crystal structure made from sodium and chlorine atoms. Strong electric forces hold these atoms together in the crystal. But let water molecules come into contact with the salt crystal, and the water molecules work their way in between the sodium and chlorine atoms, allowing these atoms to move freely and independently throughout the water.

One of our favorite sketches is Figure (11), the odor of violets. Many of our common experiences have a simple origin on an atomic scale.

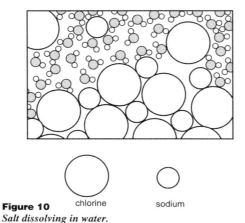

Figure 10
Salt dissolving in water.

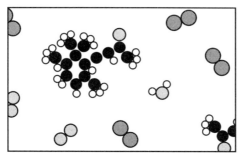

Figure 11
Odor of violets.

THERMAL MOTION

What we have not been able to display in the sketches of atomic processes is the constant juggling of the atoms and molecules. This juggling, which we will call **thermal motion**, becomes more intense as a substance becomes warmer and can cause major changes in the structure of matter. When ice is warmed to above the melting point, the juggling or thermal motion breaks up the rigid structure of the ice crystal, allowing the water molecules to slide past each other to form liquid water. With more heating, the thermal motion can increase to the point that the water molecules fly apart.

Surprisingly this thermal motion can be seen on a considerably larger scale than the scale of atoms. In 1827, the botanist Robert Brown observed that tiny pollen particles in water, when seen through a microscope, moved around in a juggling, random fashion. Wondering whether these particles were alive and swimming, Brown studied these and other small particles in circumstances where nothing should be alive, and concluded that this random motion, now called **Brownian motion**, had nothing to do with life, but was related to the motion of the molecules.

With a laser, microscope and TV camera, it is easy to set up a demonstration of the Brownian motion of cigarette smoke particles in air. The apparatus, shown in Figure (12a), consists of a small cavity between two microscopic slides, into which we inject smoke through a small tube.

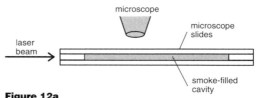

Figure 12a
Brownian Motion. A small cavity, made from microscope slides is filled with cigarette smoke, is illuminated from the side by a laser beam and viewed from above by a microscope and TV camera.

Once inside the cavity the smoke is illuminated from the side by a laser beam. Through the microscope, whose image can be displayed using a TV camera, we clearly see the illuminated smoke particles moving around in a relatively slow jagged motion. Figure (12b) is one minute movie showing the motion of the smoke particles.

In Figure (12c), a student recorded the motion of a single smoke particle for 38 TV frames. Although we may think of smoke particles as being small, they are huge compared to the air molecules in which they are immersed. Smoke particles have a mass many orders of magnitude larger than that of the air molecules. Yet the motion we see is caused by the constant bombardment of these huge particles by the air molecules.

At first you might believe that if a large particle were constantly bombarded on all sides by billions of tiny particles the effect of the collisions would cancel out and the big particle would just sit there. But it turns out that if the collisions are random, then fluctuations in the collisions will cause the large particle to move around with the jerky motion we see in the Brownian motion demonstration. What is more remarkable, *the average kinetic energy of the smoke particles, as they wander about, is the same as the average kinetic energy of the air molecules bombarding them.*

The air molecules themselves are not all the same; air is mostly nitrogen and oxygen, some carbon dioxide and water, and smaller amounts of other gases and pollutants. It turns out that each species of molecule in the air has precisely the same average kinetic energy due to thermal motion. As a result, the oxygen molecules, for example, having a slightly greater mass than the nitrogen molecules, must have a slightly smaller average speed in order that the average kinetic energy $1/2\ mv^2$ be the same. The smoke particles, with their huge masses, have a much slower average speed than the air molecules. The air molecules move at roughly the speed of sound in air, while the smoke particles move slowly enough for us to see and follow them on the TV screen.

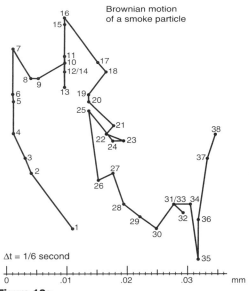

Figure 12c
Brownian motion of a smoke particle. This is the result of a student project by Lisa Stigler, where the motion of a smoke particle was recorded by a TV camera using the apparatus of Figure (12a). (We cannot use this plot to estimate the average speed of the smoke particles, because the smoke particle undergoes many collisions between TV frames. However this plot does illustrate the random walk nature of the motion of the particle. One feature of a random walk, that may be observable from plots like this, is that the average distance from the starting point should be proportional to the square root of the elapsed time.)

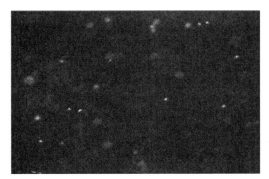

Figure 12b
First frame of a one minute movie showing the brownian motion of smoke particles. The frames are 1/15 of a second apart in this movie. (Click on the image to see the movie. Press the "esc" button to close the movie.)

Exercise 1

If we know the relative masses of the molecules in air, and know the average speed of one of the species, we can use the fact that all species of particles have the same average kinetic energy, in order to calculate the average speed of the other particles. It turns out that if we have a sample of air at room temperature (27°C), the average speed of the nitrogen molecule is 483 meters/sec. (We will calculate this number shortly.)

Using the mass of a hydrogen atom as a standard mass of 1, the mass of a nitrogen molecule is 28, and an oxygen molecule 32. (These are often called the **molecular weights** of the molecules). In Table 1 we have listed various constituents of the cigarette smoke, the relative mass of the particles, and the average thermal speed of two of the species. Use the fact that all the species have the same average kinetic energy to fill in the table of average speeds. (We gave you the speed of helium so that you could check your calculations.)

Particle	Symbol	Mass*	Speed**
Hydrogen molecule	H_2	2.0	...
Helium atom	He	4.0	1370 m/sec
Water molecule	H_2O	18.0	...
Nitrogen molecule	N_2	28.0	518 m/sec
Oxygen molecule	O_2	32.0	...
Carbon Dioxide	CO_2	44.0	...
Smoke particle		10^{10}	

* Mass relative to Hydrogen atom
** Average speed at room temperature

Table 1
Particles involved in the Brownian motion demonstration. Fill in the column for average speed, using the fact that these particles have the same average kinetic energy.

THERMAL EQUILIBRIUM

If you place a hot cup of coffee and a cold glass of milk on the table and leave the room for several hours, when you return the coffee and the milk are both at room temperature. We say that the coffee, the milk, and the air in the room are in thermal equilibrium. A faster way to reach thermal equilibrium, at least between the coffee and the milk, is to pour the milk into the coffee and stir.

On an atomic scale, what does it mean to say that the molecules of the coffee and those of the milk are in thermal equilibrium? We obtain a hint from our discussion of Brownian motion. The cigarette smoke represents a well stirred mixture of smoke particles and air molecules. These should therefore be in thermal equilibrium, just as the well stirred molecules in the coffee and milk. In the case of Brownian motion, the smoke particles and the air molecules had the same average thermal kinetic energy. We expect that the same may be true for the molecules in the mixture of coffee and milk.

It is an almost general rule that when two objects are in thermal equilibrium, the molecules that make up these objects have the same average thermal kinetic energy. When we first placed a hot cup of coffee and a cold cup of milk on the table, the molecules in the coffee had a greater average kinetic energy, and the molecules in the milk a lesser average kinetic energy, than the molecules in the air. But after a few hours, the coffee molecules slowed down and the milk molecules speeded up until all three sets of molecules, coffee, milk, and air attained the same average kinetic energy.

The process of reaching thermal equilibrium is usually a result of random collisions between molecules. If a fast molecule collides with a slow one, chances are that the slow one will speed up and the fast one will slow down. It requires a detailed analysis, which we will not attempt, to show that if the collisions are random, they tend to equalize the kinetic energy of the particles.

TEMPERATURE

If there is any scientific concept familiar to everybody, it is the concept of temperature. All of our lives we have been poked with thermometers and listened to weather forecasts about tomorrow's temperature. How is that quantity, measured by various kinds of thermometers, related to the atomic processes we have been discussing?

Most thermometers are a black art which depends upon such properties as the thermal expansion of mercury or alcohol, the stiffness of a spring, or the color changes of a material, etc. There is, however, one kind of a thermometer whose function can be understood from a simple molecular picture. That is the *ideal gas thermometer* which we will discuss shortly. We will see that for an ideal gas thermometer, the temperature reading is proportional to the average thermal kinetic energy of the gas molecules in the thermometer.

When you measure the temperature of an object, you have to wait until the thermometer and the object are in thermal equilibrium. (You wait until the reading on the thermometer in your mouth stops changing.) When in thermal equilibrium, the molecules of the object and those of the thermometer have the same average thermal kinetic energy. If we are using an ideal gas thermometer, the reading is proportional to this average kinetic energy. Thus if we use an ideal gas thermometer as an experimental definition of temperature, we are effectively defining temperature as being proportional to the average thermal kinetic energy of the molecules.

Absolute Zero

An immediate consequence of temperature being related to thermal kinetic energy, is that there must be a lowest possible temperature. When the thermal kinetic energy is gone, you cannot go any lower in temperature. It thus seems reasonable to define an absolute zero of temperature as the state where the molecules have no thermal kinetic energy, and choose a temperature scale that starts at this absolute zero and goes up proportionally to the thermal kinetic energy.

However, as you approach absolute zero, as you try to remove the last vestiges of thermal kinetic energy, nature has a surprise in store. No matter what you do, there is some unremovable kinetic energy left. One of the basic predictions of quantum mechanics is that a confined particle cannot have zero kinetic energy, and the closer the confinement the more kinetic energy it has to have. A molecule in a liquid or a solid is confined to the small volume bounded by its neighbors, and therefore cannot have a kinetic energy less than that required for that volume.

The unremovable kinetic energy is called *zero point energy*. This energy is so small that for most substances it is not noticeable unless you carry out specially designed experiments to detect it. However, zero point energy shows up clearly in the case of liquid helium. All substances except helium freeze when cooled to a sufficiently low temperature. We can remove enough kinetic energy from the molecules so that they settle into a solid structure. But the molecular force between helium atoms is so weak that the zero point kinetic energy alone is enough to keep helium a liquid. You cannot freeze helium by cooling alone, you must also subject it to high pressures.

The existence of zero point energy suggests that we will encounter problems with the definition of temperature as we approach absolute zero. Suppose, for example, we have two substances with different zero point energies in thermal equilibrium. If the temperature is so low that any thermal kinetic energy is much less than the zero point energies, then we have a situation in which molecules with different vibrational kinetic energies are in thermal equilibrium. If we insist that two substances in thermal equilibrium are at the same temperature, then we can no longer say that temperature is proportional to the vibrational kinetic energy of the molecules.

The ideal gas thermometer does not get us out of this problem because it does not work at very low temperatures. Before the zero point energies become important, any gas we use in an ideal gas thermometer becomes liquid or solid and we no longer have an ideal gas as a working substance.

In the next chapter on entropy and the second law of thermodynamics, we will discuss the consequences of the basic idea that *order does not naturally arise from disorder*. In that discussion we will describe a method of defining temperature that applies to all temperature ranges. This thermodynamic definition of temperature is consistent with the ideal gas thermometer over the range that ideal gas thermometers operate, but also correctly describes temperatures near absolute zero where we have to deal with zero point energy.

Temperature Scales

For the rest of this chapter, we will put aside any worries about zero point energy, and simply assume that the temperature of an object is proportional to the average thermal kinetic energy of the molecules in the object, and that absolute zero is where no thermal kinetic energy remains.

From this point of view, the simplest way to define a temperature scale is to equate the temperature with the average thermal kinetic energy, and measure temperature in energy units such as ergs as shown in Figure (13). But you probably have not heard anyone describe temperature in ergs, and for good reason. Telling your doctor that you are running a fever of 6.4423×10^{-14}, an increase of 23×10^{-18} over normal could be a bit hard to explain when you are sick. It is much easier to say that you have a temperature of 100° F or about 38° C. Ergs are too awkward a unit for most purposes.

Historically, thermometers were invented and temperature scales established long before the relation between temperature and the average kinetic energy of molecules became known. Throughout the world the most widely used temperature scale is the Centigrade scale, where the temperature of melting ice is arbitrarily set at 0° C (zero degrees Centigrade), and the

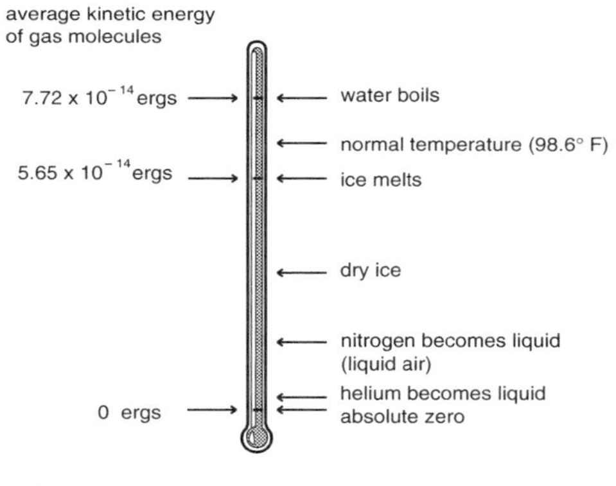

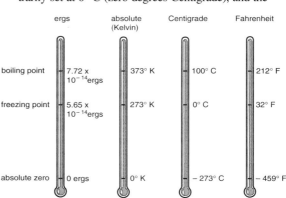

Figure 13a
Temperature scale in ergs.

Figure 13b
Comparison of various temperature scales.

boiling of water at 100° C. Commonly, changes in temperature are measured with a mercury thermometer. This device registers temperature changes when the mercury in a thin glass column expands or contacts. On the Centigrade scale, the distance between 0° C and 100° C is marked into 100 equally spaced smaller intervals which we call degrees.

A less arbitrary scale is the **Kelvin** or **absolute** scale, which measures temperature in Centigrade size degrees beginning at absolute zero. Using the absolute scale, we find that helium boils at 4 degrees Kelvin, ice melts at 273 degrees Kelvin and water boils at 373 degrees Kelvin. A comparison of various temperature scales (ergs, degrees Kelvin, degrees Centigrade, and degrees Fahrenheit) is shown in Figure (13b).

Those who define standard nomenclature for physical quantities have decided, in their great wisdom, that the word "degrees" shall be omitted when talking about temperature in degrees Kelvin. Thus we should say that helium boils at 4 kelvins or 4K, ice melts at 273 kelvins or 273K, and the temperature difference between melting ice and boiling water is 100 kelvins or 100K. At least this nomenclature is easy to say and should not be confusing when you get used to it. We do not feel the same way about all recent changes in nomenclature.

The conversion from one temperature scale to another is a relatively straightforward process. If you went to an American school, somewhere along the way you were taught how to convert from Fahrenheit to Centigrade degrees. You do not need to worry about that because we will not be using the obsolete Fahrenheit scale. But we will often want to convert from the absolute scale to the energy units, ergs or joules. The conversion is written in the somewhat peculiar form

$$\left. \begin{array}{l} \text{average kinetic energy} \\ \text{of gas molecules} \\ \text{in ergs or joules} \end{array} \right\} = \frac{3}{2} kT \quad (1)$$

where T is the temperature in kelvins, and the conversion factor k, known as **Boltzman's constant** has the numerical value

$$\left. \begin{array}{l} \text{Boltzman's} \\ \text{constant k} \end{array} \right\} = 1.38 \times 10^{-16} \frac{\text{ergs}}{\text{kelvin}} \quad (2)$$

If you are using MKS units, the value of k is 1.38×10^{-23} joules/kelvin.

The important feature of Equation 1 is that the average kinetic energy of the molecules is proportional to the absolute temperature measured in kelvins. We have written the proportionality constant as 3/2 k, putting in the numerical factor of 3/2 to get rid of a factor 2/3, as you will see shortly. Basically, think of Boltzman's constant as the conversion factor to go from temperature units to energy units or vice versa.

Exercise 2
Use Equation 1 to calculate the temperature of melting ice in ergs. Compare your answer with the result in Figure (13).

Exercise 3
What would be the temperature in Kelvins of a gas if the particles in the gas had an average kinetic energy of 1 erg?

Exercise 4
In Exercise 1 we said that the average speed of nitrogen molecules at room temperature was 518 meters/sec, and asked you to use that result to calculate the average speed of the other molecules and particles in the cigarette smoke. Now you are to calculate the speed of the nitrogen molecules using the fact that their average kinetic energy is 3/2 kT.

It is traditional to take room temperature as 300K = 27° C. Assume that a nitrogen molecule is 28 times as massive as a hydrogen atom, whose mass is essentially the same as a proton, or 1.67×10^{-24} grams. See if you get the answer of 518 meters/sec.

MOLECULAR FORCES

Much of the behavior of matter we see as we look around us is the result of a competition between molecular forces holding atoms together and thermal motion tending to pull them apart. Molecular forces can be subtle enough to form objects as complex as the myoglobin molecule. Yet knowing just some of the basic features of molecular forces is enough to provide an insight into processes like evaporation, osmotic pressure, elasticity of rubber, and the behavior of an ideal gas. At the end of Chapter 19 we will discuss the so-called **Leonard Jones potential** as a model for molecular forces. That model is far more detailed than we need for our current discussion. All we need to know now is reviewed in Figure (14).

In Figure (14a), where the atoms are about an atomic diameter apart, the attractive molecular force between the two atoms is less than one percent of its maximum value. The point is that unless the atoms are very close together, within an atomic diameter of each other, molecular forces are negligible. This is why atoms in a gas often act as independent free particles. In the air we breath, the average spacing of atoms is about ten molecular diameters, so that molecular forces play no role except when molecules collide.

When atoms get closer than an atomic diameter, the attractive molecular force increases rapidly, reaching a maximum at a separation at about one tenth of an atomic diameter as shown in Figure (14c). Then the force rapidly drops to zero at the spacing shown in Figure (14d). When the force is zero, this is the equilibrium distance which determines the size of the atom in a chunk of matter. Effectively we can say that when the atoms are at their equilibrium separation, they are just touching, as we drew them in Figure (14d).

Try to shove the atoms closer together than the equilibrium position, and you encounter a repulsive force that builds very rapidly, much faster than the attractive force increases as you pull the atoms apart. This repulsion makes atoms behave as hard, nearly incompressible spherical objects. This repulsive force is often referred to as the repulsive core of the atom.

In Figure (14f) we have sketched the potential energy corresponding to the molecular force. As you can see the potential energy forms a well with the bottom at the equilibrium position. When two atoms form a molecule, like hydrogen (H_2), oxygen (O_2) or nitrogen (N_2), you can picture one of the atoms as sitting in the potential well created by the other, and vice versa. We only have to think about one of the atoms, for the same thing is happening to the other.

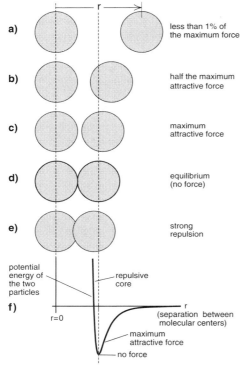

Figure 14
Interaction of two atoms via a Leonard Jones potential (f). When the atoms have an equilibrium separation (d), their potential energy of interaction is a minimum, and we can visualize one of the atoms as sitting at the bottom of the potential energy well. If the separation either increases or decreases, there is a force back toward equilibrium. The repulsion quickly builds up if you try to shove the atoms together, and the attraction dies rapidly after the atoms become separated by about one atomic diameter.

If the atom is sitting at rest at the bottom of its potential well, it is at its equilibrium position shown in Figure (14d). If the atom moves either way, in or out, it is subject to a restoring force pushing it back to the equilibrium position. If it does not move too far from its equilibrium position, the restoring force is very similar to the restoring force of a spring at equilibrium, as indicated in Figure (15). That is why one can often quite accurately picture molecular forces as spring forces between atoms.

The advantage of the potential energy diagram is that it allows us to think of atomic processes in terms of the energy involved. The distance from the zero of potential energy down to the bottom of the well is the ***binding energy*** of the molecule as indicated in Figure (16). This is the energy required to pull a molecule apart, starting with the atoms in their equilibrium position.

We have seen that the average thermal kinetic energy of an atom or molecule is $3/2\,kT$ where k is Boltzman's constant and T the temperature in Kelvins. If the thermal kinetic energy is much less than the binding energy of the molecule, as we have shown in Figure (16), then the atom can move back and forth around the bottom of the potential well but not climb out.

From the depth and shape of the potential well one can deduce general features of the behavior of matter, such as why solids and liquids expand when heated. But before we look at such fine details, there is much to understand about atomic processes just from the fact that there is an attractive molecular force with a repulsive core. We will look at these more general features first and then return to the details we see in Figure (16).

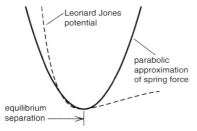

a) Comparison of the Leonard Jones potential and the parabolic potential of a spring force.

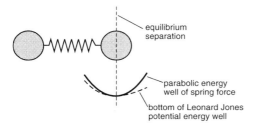

b) Modeling the molecular force as a spring force. As long as the atom stays near the equilibrium position, the Leonard Jones force and the spring force are equivalent.

Figure 15
Physics and chemistry texts often picture molecules with spring forces between the atoms. At first this may seem to be a highly unrealistic picture. But when you carefully compare the spring potential energy and the Leonard Jones potential energy right near the equilibrium position, the two curves have the same shape. Thus the spring force is a good model as long as the atoms stay near their equilibrium positions.

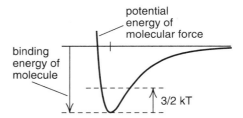

Figure 16
Binding energy. If an atom is in its equilibrium position, we can think of it as sitting at the bottom of its potential energy well. To remove the atom from the molecule, we have to lift it out of the well. Thus the binding energy is the depth of the well. If the atom has a thermal kinetic energy $3/2\,kT$, and this thermal energy is less than the binding energy, the molecule should stay together.

Evaporation

Simple features of molecular forces lead to a reasonable understanding of the transition from a liquid to a gaseous state, the process of evaporation. We start with a picture of a liquid as a collection of molecules that all attract each other, can move around past each other, but are nearly incompressible because the repulsive core in the molecular force prevents atoms from being squeezed into each other. The incompressibility of water can be seen from the fact that water in the deepest parts of the ocean, where the pressures are some 800 times atmospheric pressure, is only about 3% denser than the water at the surface.

A molecule in a liquid is free to move around because of its thermal kinetic energy and because there is essentially no net force on it. Although attracted to all of its neighbors, the neighbors surround the molecule as shown in Figure (17), and the net force is zero.

The situation is different for a molecule on the surface as shown in Figure (18). Such a molecule has neighbors only to the sides and below. If we try to lift such a molecule out of the surface, there will be a net force exerted by all of the molecules beneath it, pulling the molecule back in. To extract a molecule from the surface requires that you do work against these attractive forces. The amount of work required to extract a molecule from the surface depends upon the type of liquid and the temperature of the liquid, but some energy is required as long as the surface exists.

Example 2

To estimate the amount of energy required to extract a water molecule from the surface of water, we note that to boil 1 gram of water requires 2.25×10^{10} ergs of energy. Since there are 3.3×10^{22} molecules in 1 gram of water, this represents an energy of 7×10^{-13} ergs per molecule. Some of the energy you supply goes into displacing the air above the water to make room for the steam, but most of it goes into supplying the energy each molecule needs to escape water at 100°C (373K).

Exercise 5

(a) What is the average kinetic energy of a molecule at a temperature of 373K?

(b) Is this enough energy for an average water molecule to escape through the surface of the water?

(c) At what temperature does the thermal kinetic energy equal to the 7×10^{-13} ergs needed to escape?

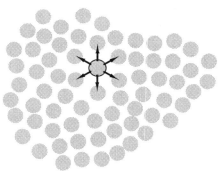

Figure 17
A molecule in the interior of a liquid is attracted by all its neighbors which surround it. As a result the net force is zero and the molecule is free to move about through the liquid.

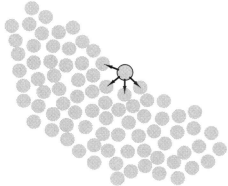

Figure 18
A molecule on the surface is attracted to its neighbors beneath it. To pull a molecule out of the surface, you have to overcome these forces. As a result it takes energy to remove a molecule from the liquid. This surface force is often referred to as **surface tension.**

If you worked Exercise 6, you realize that the average molecule, even in boiling water, does not have nearly enough thermal kinetic energy to escape through the surface. Yet even at room temperature water evaporates; even at these lower temperatures some molecules escape through the surface. The reason is that, while 3/2 kT is the average thermal kinetic energy of the molecules, some molecules have more kinetic energy than average, some less. Some have so much more kinetic energy than average that they can escape.

The rate of evaporation depends very much on the distribution of thermal kinetic energies. At a given temperature T what fraction of the molecules have a kinetic energy sufficiently far above average to be able to escape? It turns out that for a substance in thermal equilibrium, there is a precise formula for the distribution of thermal kinetic energies, a formula known as the ***Boltzman distribution*** which we discuss in Chapter 22. For now we will not go into that much detail. Instead, we will simply recognize that some molecules are hotter than average, some colder than average, and that it is the very hottest ones that have enough energy to escape.

If it is the hot molecules that escape during evaporation, then the cooler ones must be left behind and ***evaporation must be a cooling process***. There must, however, be a net loss of molecules from the surface for cooling to occur. As we noted at the beginning of the chapter, the surface of water is a dynamic place where water molecules are continually leaving and returning. A returning water molecule, even if relatively cool when in the air above the water, gains as much kinetic energy when it reenters the water as the hot molecule lost when escaping. Thus reentering molecules become hot when they get back in the water, and thus the returning or condensation of water molecules is a warming process.

Whether you get evaporation or condensation depends upon the number of water molecules in the air above the water. If you cover a glass of water with a dish, soon the number of water molecules in the air in the glass builds up to the point that there is a balance between molecules leaving and molecules entering the liquid surface. When this balance is achieved, evaporation ceases and we say that the air above the water is at 100% relative humidity. In order to get cooling from evaporation, the relative humidity of the air must be less than 100%.

The human body uses evaporation for cooling which is effective on a hot, dry day but not on a humid one. When the relative humidity approaches 100% there is no net loss of water molecules and no cooling. Incidentally, you blow on soup to cool it, not necessarily because your breath is cooler than the soup, but because you are replacing the moist air over the soup with drier air so that more evaporation can take place.

PRESSURE

When you try to compress a liquid, you are trying to shove the atoms into each other which is very difficult to do because of the repulsive core of the molecular force. It is also hard to compress a gas—try blowing air into a soda bottle. But when you compress air you are not squeezing air molecules together, you are not trying to overcome a molecular force at all. The air that you breathe is mostly empty space, the separation of air molecules being about 10 molecular diameters. There is a completely different explanation for why it is difficult to compress a gas, why a gas exerts a pressure that you must overcome to compress it.

One of the simplest demonstrations of the pressure exerted by a gas is provided by a rubber balloon. When you blow up a balloon, the rubber of the balloon is trying to compress the gas, force it down to a smaller volume. The molecules of the gas exert an outward force on the rubber, preventing it from collapsing. This outward force is caused by the collisions of the gas molecules with the rubber, as illustrated in Figure (19). Whenever an air molecule strikes and bounces off of the rubber, there is a net transfer of outward directed linear momentum to the rubber. Since the collisions are occurring continually, there is a continual transfer of momentum to the rubber, which, according to Newton's second law $\vec{F} = d\vec{p}/dt$, means that the molecules exert a force on the rubber. On the average, the direction of momentum transfer is outward, perpendicular to the surface of the rubber. Thus the force exerted by the gas molecules is also perpendicular to the surface.

The total force exerted on some part of the balloon depends upon the area of the balloon we are talking about. We can simplify the discussion by talking about the *force on a unit area* of the balloon surface, and give that force the special name ***pressure***. We know that force is a vector quantity, but the force that a gas exerts on a surface is always directed perpendicular to the surface, no matter what orientation the surface has.

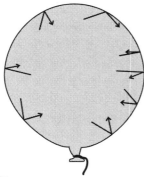

Figure 19a
The air molecules bouncing off the inside surface of the balloon, transfer an outward directed momentum to the rubber. The average momentum per second transferred by these collisions is the average force the molecules exert on a section of the balloon surface.

Figure 19b
Balloon placed on liquid nitrogen. If you cool the air molecules inside the balloon, they do not strike the rubber as hard, exert less of a force, and the balloon collapses. In the final picture, there is only a puddle of liquid air inside.

Thus we can let the word "pressure" stand for the magnitude of the force on a unit area, and determine the direction of the force from the orientation of the surface.

$$\boxed{\begin{array}{l}\text{pressure} \\ \text{of a gas}\end{array} = \begin{array}{l}\text{magnitude of the force} \\ \text{exerted by the gas on a} \\ \text{unit area of surface}\end{array}} \quad (3)$$

In the case of a gas inside a balloon, it is clear that the behavior of the gas molecules is more or less the same throughout the balloon. As a result, the pressure should be essentially the same on all surface areas of the balloon. Instead of using the word "pressure" to merely describe the force on surfaces of the balloon, we can say that *the pressure is in the gas itself*. Once you know the pressure of the gas, you can then calculate the force the gas exerts on some particular surface by multiplying the pressure times the area of the surface, and noting that the force is directed perpendicular to the surface.

The pressure of the gas, the force the gas molecules exert on a surface, depends upon how fast the molecules are moving when they hit the surface. The faster their average speed, the greater the force and pressure. Since the motion of the molecules is thermal motion, whose average kinetic energy is 3/2 kT, the average speed and pressure must increase with temperature. If you heat the balloon, the gas pressure increases and the balloon expands. If you cool it, it contracts.

An excellent demonstration of the dependence of air pressure on temperature is to place a balloon in a bucket of liquid nitrogen, as shown in Figure (19b). A common Styrofoam ice bucket makes an excellent container for liquid nitrogen for this demonstration. When you place the balloon on the surface of the liquid nitrogen, the balloon sits there for a while, and then begins to shrink as the air molecules cool down. The shrinking continues until the balloon collapses and all you have inside is a puddle of liquid air. Now any further contraction of the balloon would require squeezing the air molecules themselves together which is opposed by the repulsive core of the molecular forces. In a sense, the collapse of the cooling gas in the balloon was halted by molecular forces.

If you take the balloon out of the liquid nitrogen, the air inside warms up and the balloon expands again to its original volume. Remarkably, a typical balloon can undergo this cycle a number of times without breaking.

Stellar Evolution

The balloon has features in common with our sun. The sun is not the solid object it appears. A better model is that the sun is a bag of gas, somewhat like a balloon, but with the constraining force of the rubber replaced by gravity. The sun looks like it has a distinct surface, but that is an illusion created by the fact that the gas molecules in the sun, which are mostly hydrogen, become ionized and opaque at temperatures in excess of 3000 Kelvins. As you go down into the sun, the temperature of the gas increases. The distance at which it reaches 3000 Kelvins, the gas changes from transparent to opaque and that is what we see as the surface of the sun.

For the past five billion years, the sun has gotten energy from the conversion of hydrogen nuclei to helium nuclei. And there is another five billion years worth of hydrogen left before the sun runs out of fuel. At that point the sun will do something spectacular. It will expand so that the earth will be orbiting near the sun's surface. (We discuss this process in more detail in Chapter 20.) But eventually the sun will settle down and begin to cool off.

As the sun cools, it will contract very much like the balloon in our demonstration. And like the balloon, the collapse will be halted when the atoms are so close together that the repulsive core of the molecular forces prevents further contraction. At that point the sun will have become what is called a **white dwarf**, an object about the size of the earth slowly cooling until it becomes a dark ember.

If the sun were just a bit bigger, about 1.4 times its current mass, the gravitational collapse would not be halted by the molecular forces between atoms. That is the mass at which gravity is strong enough to crush the atoms together, to overcome the atomic repulsive cores, with the result that you end up with a neutron star. That is also a topic we discuss in Chapter 20.

THE IDEAL GAS LAW

We have discussed the picture of how the repeated collisions of the gas molecules inside a balloon exert an outward force on the rubber, and how, if we raise the temperature, the molecules travel faster, strike harder, and exert a greater force. We will now, with a fairly simple derivation, obtain an explicit relation between the temperature of the molecules and the force they exert, a relationship known as the *ideal gas law*.

There are many ways to derive the ideal gas law, depending upon what assumptions you are willing to make about averaging over molecular speeds. The less you are willing to assume, the harder the derivation is. But there is one rather surprising feature of all the derivations. They all give the same correct answer. The usual procedure in textbooks is to make the derivation as complex as students will tolerate, apologize for or hide the approximations, and announce that the answer is correct. What we will do is present the simplest derivation we can find that gives the right answer. When an argument looks too simple to be true, but gives the right answer, that means that you may have extracted an important basic feature from a complex situation.

The Ideal Piston

While a balloon is a very practical container for gas, the curved surfaces make it a bit difficult to use for theoretical analysis. Instead we will, more or less as a thought experiment, use an idealized device called the frictionless piston. Figure (20) is a diagram of a frictionless piston in a cylinder of cross-sectional area A. In the cylinder is a gas -- like air at room temperature. The gas molecules are bouncing around, colliding with the walls of the cylinder and the face of the piston. Because of the collisions, the gas molecules exert a force on the piston, and because the piston is frictionless, we must exert an oppositely directed force $\vec{F}$, as shown, to keep the cylinder from expanding. We are assuming that there is no gas behind the piston, so that only the force $\vec{F}$ keeps the gas from expanding.

We know that the force F exerted by the gas increases with temperature because the balloon expanded when we heated it. We also know that the average thermal kinetic energy of the gas molecules increases as $3/2\, kT$ as the temperature rises. What we wish to do now is relate these observations to obtain a formula for how the force F depends upon the temperature T.

To relate F to T we start with the simplest possible model of the gas in the cylinder, namely a gas consisting of one molecule, bouncing back and forth at a speed v, as shown in Figure (21). Each time the molecule strikes and bounces off the piston, its momentum changes by 2mv. Since linear momentum is conserved during the collision, the piston picks up an outward, x-directed, linear momentum of magnitude 2mv as a result of the collision. (Remember problem 7-5, where two skaters on frictionless ice were tossing a ball back and forth. When one of the skaters caught the ball, she picked up the ball's momentum mv. When she threw the ball back, she recoiled, picking up an additional mv of momentum. As a result she picked up 2mv of momentum with each catch and toss.)

If we designate by Δp the magnitude of the x-directed linear momentum that the piston gains from each collision we have

$$\begin{matrix}\text{momentum}\\ \text{transferred} \\ \text{per collision}\end{matrix} \quad \Delta p = 2mv \qquad (4)$$

The momentum $\Delta p = 2mv$ is transferred to the piston each time the molecule comes back and strikes the piston. If Δt is the time between collisions, then the amount of momentum transferred per second is $\Delta p/\Delta t$.

Figure 20
Frictionless cylinder in a piston. Picture the force that the gas molecules exert on the piston as being counterbalanced by an external force $\vec{F}$ as shown.

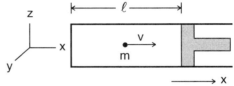

Figure 21
Analysis of a one molecule gas.

Using Newton's second law, in the form $\vec{F} = d\vec{p}/dt$, we see that this rate of transfer of linear momentum $\Delta p/\Delta t$ is just the average force F that the molecule is exerting on the piston

$$\begin{array}{c}\text{average force}\\ \text{exerted by}\\ \text{molecule}\\ \text{on piston}\end{array} \quad F = \frac{\Delta p}{\Delta t} \quad \begin{array}{c}\text{average rate}\\ \text{of momentum}\\ \text{transferred}\\ \text{to the piston}\end{array} \quad (5)$$

To calculate the time Δt between collisions, note that if the distance from the end of the cylinder to the piston is ℓ, and the molecule is traveling at a speed v, it covers the distance down and back, 2ℓ, in a time

$$\Delta t = \frac{2\ell \text{ cm}}{v \text{ cm/sec}} = \frac{2\ell}{v} \text{ sec} \quad (6)$$

With Equation (4) for Δp and (6) for Δt in (5), we get, for the average force F exerted by this one molecule of gas

$$F = \frac{\Delta p}{\Delta t} = \frac{2mv}{(2\ell/v)} = \frac{mv^2}{\ell} \quad (7)$$

Note the appearance of mv^2 in Equation 7. We are already beginning to see a connection between the molecule's kinetic energy and the force it exerts.

If you could actually set up a one molecule, one dimensional, gas like that shown in Figure (21), Equation 7 would accurately describe the average force of that gas on the piston. No approximations have been made yet. The approximations enter when we go to a gas of N molecules, moving in three dimensions, at various speeds. The simplest, most outrageous approximation we can make is that all of the molecules have the same speed v, and that 1/3 of them are bouncing back and forth in the x-direction, 1/3 in and out in the y-direction, and 1/3 up and down in the z-direction. Such a gas with N/3 molecules bouncing back and forth, would exert a force N/3 times as great as our one molecule gas in Figure (21).

$$F\left[\begin{array}{c}N \text{ molecule}\\ \text{gas}\end{array}\right] = \frac{N}{3} F\left[\begin{array}{c}1 \text{ molecule}\\ \text{gas}\end{array}\right]$$

$$= \frac{N}{3}\left(\frac{mv^2}{\ell}\right)$$

$$= \frac{2}{3\ell} \times N \times \left(\frac{1}{2}mv^2\right) \quad (8)$$

Since, in this approximation, all the molecules have the same speed v, then the factor $1/2\,mv^2$ in Equation 8 must be the average thermal kinetic energy of the molecules. Replacing $1/2\,mv^2$ by $3/2\,kT$ gives

$$F\left[\begin{array}{c}N \text{ molecule}\\ \text{gas}\end{array}\right] = \frac{2}{3\ell} N \times \frac{3}{2} kT \quad (9)$$

Now you see why the factor of 3/2 was inserted into the formula $3/2\,kT$ for the average thermal kinetic energy. The 3/2 cancels the 2/3 that appeared in Equation 9, and we get

$$F\left[\begin{array}{c}N \text{ molecule}\\ \text{gas}\end{array}\right] = \frac{N}{\ell} kT \quad (10)$$

We are almost finished. Equation 10, despite our approximations, is the correct formula for the force exerted by a gas of N molecules at a temperature T. The only problem with Equation 10 is the explicit dependence on the length ℓ of the cylinder. We can remove this explicit dependence by expressing the force on the cylinder in terms of the pressure P of the gas.

In our earlier discussion, we said that the pressure of a gas inside a balloon was equal to the force per unit area exerted by the gas on the surface of the balloon (Equation 3). If we have a gas at pressure P in a cylinder of cross-sectional area A, as shown in Figures (20) and (21), then the force exerted on the piston, whose area is A, must be

$$F = PA \quad \begin{array}{c}\text{pressure}\\ \text{times area}\end{array} \quad (11)$$

Substituting Equation 11 for F in Equation 10, and multiplying through by the cylinder length ℓ, gives

$$P(A\ell) = NkT \quad (12)$$

The final step is to note that $(A\ell)$, the area of the cylinder times its length, is the volume V of the cylinder. Thus we get

$$\boxed{PV = NkT} \quad \text{ideal gas law} \quad (13)$$

Equation 13 is known as the ***ideal gas law***. Despite the approximations we used to derive it, it is accurate as long as the particles in the gas are separated enough that one can neglect the molecular forces between particles. To express this another way, any gas that obeys Equation 13 is known as an ***ideal gas***. To a very high degree of accuracy, the air around us behaves as an ideal gas.

Ideal Gas Thermometer

The ideal gas law, PV = NkT, incorporates such laws as Boyle's law and Charle's law which you may have encountered in an introductory chemistry course. One can construct numerous examples and homework problems applying this law. What we will do for our first application is to describe the ideal gas thermometer which we will use, at least for now, as our experimental definition of temperature.

An example of an ideal gas thermometer is shown in Figure (22). The glass tube and the plug of mercury come about as close as you can get to a cylinder with a frictionless piston. You can make one of these devices by sealing a glass tube at the bottom, pouring some mercury in, evacuating most of the air above the mercury, and sealing the top of the tube. If the tube is fairly small, when you turn the tube over, the mercury plug will slide down until it sits on the remaining air. There will be a vacuum above the plug. How high the plug rides depends upon the length ℓ of the mercury plug and how much gas you left in the tube before sealing it.

We can use the ideal gas law to predict how the height of the plug varies with the temperature of the gas in the tube. When we do this, we obtain a messy looking formula with factors like the density ρ of the mercury, the number N of air molecules in the tube, the area A of the tube, the acceleration g due to gravity and Boltzman's constant k. But when we take another look at the result we see that most of the factors are constants, and the height h of the air column turns out to be strictly proportional to the temperature T of the gas in the tube. Let us see how this all works out.

Rewriting the ideal gas law as an equation for the temperature T of the gas molecules, we get

$$T = \frac{PV}{Nk} = \frac{PhA}{Nk} \qquad (14)$$

where V = hA is the volume occupied by the air which is in a column of height h and area A. The mercury plug of length ℓ riding on top of the gas, exerts a gravitational force mg on the gas, where the mass m of the mercury is equal to the mercury's density ρ times its volume ℓA. Thus

$$\left.\begin{array}{l}\text{weight of}\\ \text{mercury}\\ \text{column}\end{array}\right\} = mg = \rho\ell Ag \qquad (15)$$

The force mg is the total force exerted by the mercury column on the air. The force per unit area, which must equal the pressure of the gas if the plug is balanced on the gas is

$$\begin{array}{l}\textit{pressure of gas}\\ \textit{beneath a plug}\\ \textit{of mercury}\\ \textit{of length } \ell\end{array} \quad P = \frac{mg}{A} = \frac{\rho\ell Ag}{A} = \rho\ell g \qquad (16)$$

Equation 16 will turn out to be useful in other experiments, for it tells us how to measure the pressure of a gas in terms of the height ℓ of a mercury plug that the gas can support.

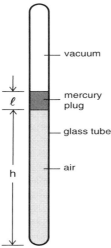

Figure 22
Ideal gas thermometer. If we heat the gas beneath the mercury plug, the gas expands raising the plug. If we have an ideal gas, then the height of the plug depends only on the temperature of the gas and not the kind of gas we used.

Using Equation 16 in 14, we get the desired result

$$T = \frac{PhA}{Nk} = \left(\frac{\rho \ell g A}{Nk}\right)h = \chi h \qquad (17)$$

where $\chi = \rho \ell g A/Nk$ is a collection of constants. The basic result is that the gas temperature T is strictly proportional to the height h of the air column.

To use an ideal gas thermometer, we do not need to evaluate the constants in the formula for χ. Instead immerse the thermometer in ice water and mark the bottom of the plug 0°C. Then put the thermometer in boiling water and mark that 100°C. Mark off the distance between 0°C and 100°C in 100 equally spaced intervals and you have a centigrade thermometer.

The fascinating feature of an ideal gas thermometer is that you can quickly determine the temperature at which the gas volume should go to zero, the temperature we have called *absolute zero*. On a sheet of graph paper, mark off a temperature scale on the bottom that runs backwards from 100° C to 0° C and goes on out quite away into negative temperatures. On the vertical axis plot the height h of the air column. For this plot, you have only 2 experimental points, the height at 0° C and at 100° C. Connect these two points by a straight line (that is what the formula T = χh says you should do), and you find that h goes to zero at a temperature of – 273° C. That is all there is to it!

From our discussion of molecular forces, you can see that any ideal gas thermometer you actually build has to fail before you get to absolute zero. At some point as you cool the air in the thermometer, you end up with a puddle of liquid air as we did in the balloon demonstration. Even before the air becomes liquid, the spacing between the air molecules is reduced to the point where the molecular forces between air molecules becomes important. The attractive molecular forces reduce the pressure of the gas, the gas no longer obeys the ideal gas law, and we cannot believe the readings of the thermometer. This problem can be put off by using helium gas that remains a gas down to a temperature of 4 kelvins, but that's the limit. To work at temperatures closer to absolute zero you need a different experimental definition of temperature, like the thermodynamic definition we discuss in Chapter 18.

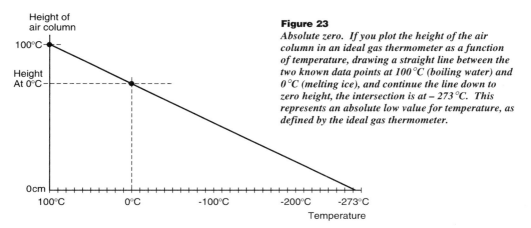

Figure 23
Absolute zero. If you plot the height of the air column in an ideal gas thermometer as a function of temperature, drawing a straight line between the two known data points at 100 °C (boiling water) and 0 °C (melting ice), and continue the line down to zero height, the intersection is at – 273 °C. This represents an absolute low value for temperature, as defined by the ideal gas thermometer.

The Mercury Barometer and Pressure Measurements

Columns of mercury are useful not only for making thermometers, but also for making devices to measure pressure. We will begin with a discussion of the mercury barometer whose construction is shown in Figure (24).

As shown in Figure (24a), start with a u shaped glass tube about a meter long, sealed at one end, and work mercury into it until the sealed section is nearly full. Then invert the tube as shown in Figure (24b). The mercury in the sealed section will slide down, leaving a vacuum behind it, until the difference in the heights of the mercury columns is about 76 cm as shown.

The height difference of the two columns tells us the pressure of the atmosphere. To see why, conceptually break the mercury column up into two parts as shown in Figure (24c). The bottom part is the loop of mercury that goes from point 1 (at the open end of the mercury) to point 2 (at the equal height in the closed section). The upper part, goes from point (2) up to the vacuum, a section whose height we designate by the letter h. This column sits over the bottom loop and exerts a downward force equal to the weight mg of a column of mercury of height h.

The mercury in the bottom section between points (1) and (2), is completely free to move up one side or the other. Since it does not move, the weight mg of the mercury column pushing down on the left side at point (2) must be balanced by the force of the atmosphere pushing down at the open end, point (1). As indicated in Figure (25), the molecules of the air are colliding with the surface of the mercury, exerting a force in the same way that the air molecules in a balloon push out on the rubber. If the air molecules are at a pressure P_a (pressure of the atmosphere) and the glass tube has an area A, the force exerted by the air is the pressure times the area.

$$\left. \begin{array}{l} \text{force exerted} \\ \text{by atmosphere} \\ \text{on air column} \end{array} \right\} = P_a A \qquad (18)$$

The weight mg of the mercury column pushing down at point (2) is

$$mg = \rho(Ah)g \qquad (19)$$

where ρ is the density of mercury and Ah is the volume of mercury in the column of area A and height h.

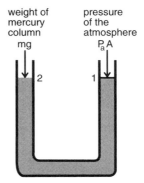

Figure 25
The weight of the mercury column above point (2) must be balanced by the force exerted by the atmosphere at point (1).

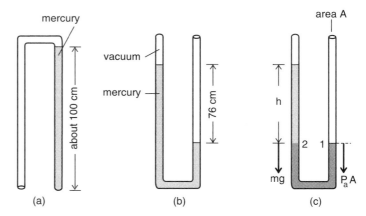

Figure 24
Construction of a mercury barometer. When you turn the tube over, going from (a) to (b), the mercury slides down the sealed leg, leaving a vacuum behind. The difference in heights h of the two columns (c) is a measure of atmospheric pressure.

Equating the forces on the two sides of the mercury in the bottom section gives

$$P_a A = mg = \rho h A g$$

$$\boxed{P_a A = \rho g h} \qquad (20)$$

The result is that the atmospheric pressure is proportional to the height difference h in the two columns of mercury. As we have mentioned, the dimensions of pressure in CGS units is dynes per square centimeter, while in MKS units it is Newton's per square meter, a set of dimensions given the name *pascal*. Neither set of units is particularly convenient. Using $\rho = 13.6$ gm/cm^3 for the density of mercury, and using the value h = 76 cm for an average value for the height of the mercury column, we get

$$P_a A = 13.6 \frac{\text{gm}}{\text{cm}^3} \times 980 \frac{\text{cm}}{\text{sec}^2} \times 76 \text{ cm}$$

$$= 1.01 \times 10^6 \frac{\text{gm cm/sec}^2}{\text{cm}^2}$$

$$P_a A = 1.01 \times 10^6 \text{ dynes /cm}^2 \qquad (21a)$$

Converting to MKS units, where one newton equals 10^5 dynes and 1 m$^2 = 10^4$ cm^2 we have

$$P_a A = 1.01 \times 10^6 \frac{\text{dynes}}{\text{cm}^2} \times \frac{10^4 \text{ cm}^2/\text{m}^2}{10^5 \text{ dynes/newton}}$$

$$= 1.01 \times 10^5 \frac{\text{newtons}}{\text{m}^2}$$

$$= 1.01 \times 10^5 \text{ pascals} \qquad (21b)$$

Just as it was inconvenient to measure temperature in ergs, it is rather inconvenient for the weatherman to announce today's barometric pressure in either dynes per square centimeter or pascals. Numbers in the range of 10^6 or 10^5 do not go over well with the listening audience.

What is much easier is to simply express the pressure in terms of the height of the mercury column that the atmospheric pressure will support. "A low pressure system moved over the area today, and the barometer reading dropped to 752 millimeters of mercury" sounds much better than saying that the pressure dropped to 1.0023×10^5 pascals.

The millimeter of mercury, as a unit of pressure, has been given the name *torr*, one more name inflicted upon students by those who decide what the standard names shall be. For comparison's sake, we can express atmospheric pressure as

$$\begin{aligned} P_a &= 1.01 \times 10^6 \text{ dynes/cm}^2 \\ &= 1.01 \times 10^5 \text{ pascals} \\ &= 101 \text{ kilo pascals} \\ &= 76 \text{ cm Hg} \qquad (22) \\ &= 760 \text{ mm Hg} \\ &= 760 \text{ torr} \\ &= 14.7 \text{ lbs/in}^2 \end{aligned}$$

At different times you may encounter any of these units.

When you are working with vacuum pumps and vacuum gauges, even the torr, 1 millimeter of mercury, is too large a unit to be convenient. Many gauges are calibrated in *microns*, which is the pressure exerted by one micron or one millionth of a meter of mercury.

$$\begin{aligned} 1 \text{ "micron"} &= 10^{-6} \text{ meters Hg} \\ &= 10^{-4} \text{ cm Hg} \\ &= 10^{-3} \text{ mm Hg} \qquad (23) \\ &= 10^{-3} \text{ torr} \end{aligned}$$

In the electron gun experiments we discuss in Chapter 28, the glass tube containing the electron beam is evacuated to a pressure of around one micron. Current technology allows you to work with much better vacuums in the range of 10^{-6} to 10^{-7} microns. Such vacuums are needed to maintain clean surfaces when studying the atomic structure of surfaces or creating complex electronic chips.

Exercise 6

What is the pressure P in dynes/cm^2 and pascals, inside an apparatus where the pressure gauge reads

a) 1 cm Hg

b) 1 micron

c) 1 kilo pascal

d) 1 torr

e) 10^{-9} torr

f) 50 microns

g) 30 lbs/in^2 (U.S. tire pressure gauge)

h) 200 kilo pascals (European tire pressure gauge)

AVOGADRO'S LAW

Speaking of inconvenient units like measuring temperature in ergs or pressure in dynes per cm^2, we have something particularly inconvenient in the form of the ideal gas law PV = NkT. To use this equation, we have to know the number N of the molecules in the gas. To actually count the molecules is essentially an impossible requirement.

Instead of counting individual molecules, we can lump them in large units, and count the number of units. The standard unit for counting molecules is the *mole*. If you have a mole of molecules or any other object, you have 6×10^{23} of them.

$$\boxed{1 \text{ mole of objects} = 6 \times 10^{23} \text{ objects}} \quad (23)$$

The idea of a mole is that it is a convenient counting device for handling large numbers. You might, for example, hear an astronomer say that there is about a mole of stars in the visible universe. By that the astronomer would mean that he thinks that the visible universe contains about 6×10^{23} stars. (That estimate may not be too many orders of magnitude off.) As another example, it would take about a mole of baseballs to fill the volume of the earth with baseballs.

The number 6×10^{23} (more accurately 6.02×10^{23}), which is known as *Avogadro's number* or constant, is essentially the number of hydrogen atoms in one gram of hydrogen. Since hydrogen atoms and protons have essentially the same mass, a mole of protons also has a mass of 1 gram.

When you have a bottle of hydrogen gas, the hydrogen atoms combine in pairs to form hydrogen molecules. Thus a mole of hydrogen molecules has a mass of 2 grams. An oxygen atom is 16 times as massive as a hydrogen atom, thus a mole of oxygen molecules, with 2 atoms in each molecule, has a mass of $2 \times 16 = 32$ grams. The mass of a mole of a given kind of molecule is usually called the *molecular weight*, but more properly the *molecular mass*, of that kind of molecule.

(The integer numbers that appear in the mass of atoms arises from the fact that protons and neutrons which make up the atomic nucleus have about the same mass, and the mass of the electrons is much much smaller. Most oxygen atoms for example have a nucleus with 8 protons and 8 neutrons, and that is why an oxygen atom is 16 times as massive as a hydrogen atom.)

We will use the symbol N_A to designate Avogadro's number

$$N_A = 6 \times 10^{23} \frac{\text{particles}}{\text{mole}} \quad \text{Avogadro's number} \quad (24)$$

We can now rewrite the ideal gas law in the form

$$PV = NkT = \frac{N}{N_A} \times (kN_A)T \quad (25)$$

We do this because N/N_A is the number of moles of the substance rather than the number of molecules. Designating this by the symbol n, we have

$$n \equiv \frac{N \text{ molecules}}{N_A \text{ molecules/mole}} = \frac{N}{N_A} \text{ moles} \quad (26)$$

In addition the product of Boltzman's constant k times Avogadro's number N_A is called the gas constant R

$$R \equiv kN_A \quad \text{(gas constant)}$$

$$= 1.38 \times 10^{-23} \frac{\text{joules}}{\text{kelvin}} \times \frac{6.02 \times 10^{23}}{\text{mole}} \quad (27)$$

$$R = 8.31 \frac{\text{joules}}{\text{mole kelvin}} \quad \text{MKS units}$$

In this case the MKS units are much more convenient. The gas constant R in the CGS system is 10^{-7} times smaller.

Expressing the ideal gas law in terms of the number of moles n and the gas constant R gives

$$\boxed{PV = nRT} \quad (28)$$

which is the alternate form of the ideal gas law. If your perspective is from an atomic point of view, you would use the form PV = NkT. But if you were a chemist and had to actually measure the quantities involved, you would use the form PV = nRT.

Our first example of the use of Equation 28, will be to determine the volume of one mole of molecules at 0° C (273 K) and atmospheric pressure (1.01×10^5 pascals). We have

PV = nRT

$1.01 * 10^5 \frac{\text{newtons}}{\text{meter}^2} \times V$

$= 1 \text{ mole} \times 8.31 \frac{\text{joules}}{\text{mole K}} \times 273 \text{ K}$

First let us check the dimensions. The moles and the kelvins cancel on the right side, and we get

$$V \sim \text{joules} \times \frac{\text{m}^2}{\text{newton}} = \text{kg} \frac{\text{m}^2}{\text{sec}^2} \times \frac{\text{m}^2}{\text{kg m/sec}^2}$$

$$= \text{meter}^3$$

The numerical value is

$$V = \frac{8.31 \times 273}{1.01 \times 10^5} \text{ m}^3$$

$$= 22.4 \times 10^{-3} \text{ m}^3$$

Noting that 10^{-3} m^3 is one liter, we get

$$\boxed{V = 22.4 \text{ liters}} \quad \begin{array}{l} \textit{volume of 1 mole} \\ \textit{of any gas at 0° C} \\ \textit{and atmospheric} \\ \textit{pressure} \end{array} \quad (29)$$

The important point is that a mole of any kind of gas has a volume of 22.4 liters at the standard conditions of 0° C and atmospheric pressure. (One often uses the notation **STP** for this **standard temperature and pressure**.) At STP, 22.4 liters of hydrogen have a mass of 2 grams, nitrogen 28 grams, and oxygen 32 grams.

Exercise 7
A helium nucleus contains 2 protons and 2 neutrons. The mass of 22.4 liters of helium gas at STP is 4 grams. What does that say about the molecular force between helium atoms?

The calculation of the volume of a mole of a gas emphasizes an important point about the behavior of an ideal gas. Namely, if we have equal volumes of gas at the same temperature and pressure, the volumes will contain the **same number of molecules**. This was first suggested by the Italian scientist Amedeo Avogadro (1776--1856), and is known as **Avogadro's law**.

HEAT CAPACITY

From our discussion of temperature and other processes from an atomic and molecular point of view, it is obvious that the way to raise the temperature of an object is to add energy. At higher temperatures the average thermal kinetic energy of the molecules increases, and that energy must come from somewhere. Historically the relationship between heat and energy was not so clear. As late as 1798, 71 years after Newton's death, there were accepted theories that treated heat as a substance called *caloric* that flowed from hot substances to cooler ones.

It was Benjamin Thomson, later known as Count Rumford, who proposed that heat was, in fact, a form of energy. Rumford was boring cannons for Prince Maximilian of Bavaria, and was quite aware that when the drills were dull, the cannons became hot. Thomson proposed that the mechanical work he put into turning the drills was converted to heat energy that raised the temperature of the cannons. Forty years later, Joule accurately measured the amount of work required to raise the temperature of various substances.

Traditionally heat energy was defined as the amount of heat required to raise the temperature of one gram of water one degree centigrade. This unit of heat is called the *calorie*. In terms of mechanical energy, the conversion factor is

$$1 \text{ calorie} = 4.186 \text{ joules} \tag{30}$$

This is the relationship between mechanical work and heat that Joule studied.

Exercise 8

A 1 kilogram mass is dropped into a bucket containing 1 liter (10^3 cm^3) of water. Assume that all of the kinetic energy of the mass ends up as heat energy, raising the temperature of the water.

(a) From what height would you have to drop the mass to raise the temperature one degree centigrade?

(b) (More realistic question.) How much would the temperature rise if you dropped the mass from a height of one meter?

Specific Heat

The amount of heat energy required to raise the temperature of a unit mass of a substance one degree is called the ***specific heat capacity*** or specific heat of the substance. For example, since it requires one calorie to heat one gram of water one degree centigrade, we can say that the specific heat of water is 1 calorie/gm °C, or 4.186 joules/gm °C.

Molar Heat Capacity

If, instead of measuring the heat capacity of a unit mass, we measure the heat capacity of a mole of a substance, we call the result the ***molar heat capacity***. For example a water molecule H$_2$O with 2 hydrogen and 1 oxygen atom is 18 times as massive as a hydrogen atom. Its molecular weight is 18, and thus a mole of water has a mass of 18 grams. As a result it takes 18 calories to raise the temperature of a mole of water 1 degree centigrade, and thus the molar heat capacity of water is 18 calories/mole °C or $18 \times 4.186 = 75.3$ joules/mole °C. For the units, instead of degrees centigrade, we can use kelvins, which are the same size. Thus we can write

$$\left.\begin{array}{l}\text{molar specific}\\ \text{heat of water}\end{array}\right\} = 75.3 \, \frac{\text{joules}}{\text{mole K}} \tag{31}$$

as an example of a molar specific heat.

Predicting the specific heat of a substance, even with an understanding of the atomic and molecular processes involved, turned out to be a much more difficult subject than expected. The first time a failure of Newtonian mechanics was detected was during the efforts to predict the specific heats of various gases. This failure was due to quantum mechanics being necessary to fully understand what happened to the added heat energy.

There is one example, however, where the simple picture of atoms we have been discussing gives the correct answer. That is for the specific heat of helium gas. We will discuss that example here, and leave all other discussions of specific heat to Chapter 20, an entire chapter devoted to the subject.

Molar Specific Heat of Helium Gas

A gas of helium atoms is about the simplest substance you can picture. Since helium does not form molecules, the gas simply consists of individual atoms moving around and bouncing off of each other. If the temperature of the gas is T, then the average thermal kinetic energy of the atoms is 3/2 kT.

If you have a mole of helium gas at a temperature T, then the thermal energy of the atoms should be the average energy of 1 atom, 3/2 kT, times the number N_A atoms in a mole. Thus we easily estimate that the thermal energy E_{He} of a mole of helium atoms is

$$E_{He} = N_A \times \left(\frac{3}{2} kT\right)$$

$$= \frac{3}{2}(N_A k)T$$

Using the fact that $N_A k = R$, the gas constant, we get

$$E_{He} = \frac{3}{2} RT \quad \begin{array}{l} \textit{thermal energy} \\ \textit{of a mole of} \\ \textit{helium atoms} \end{array} \quad (32)$$

If we raise the temperature one degree, from T to (T + 1), the thermal energy goes from 3/2 RT to 3/2 R(T + 1), an increase of 3/2 R. Thus the molar specific heat, which we will call C_V, is

$$C_V = \frac{3}{2}R = \frac{3}{2} \times 8.31 \; \frac{\text{joules}}{\text{mole K}}$$

$$\boxed{C_V \text{(helium)} = 12.5 \; \frac{\text{joules}}{\text{mole K}} = \frac{3}{2}R} \quad (33)$$

As we mentioned, we get the right answer. Equation 33 is in agreement with experiment.

The subscript V on the symbol C_V is there to remind us to measure the specific heat at constant volume. If you add heat to a gas, and at the same time allow the gas to expand, some of the energy goes into the work required to expand the volume, pushing the surrounding gas aside. This is a complication that we will discuss in Chapter 18. For now we will leave the subscript V on C_V to remind us not to let the volume increase.

Other Gases

It took almost no effort to correctly predict the specific heat of helium gas. What complications do we face when we try to predict the specific heat of other gases?

The problem is that other gases form molecules. From one point of view the molecules themselves are the gas particles, so that their average thermal kinetic energy must be 3/2 kT just like the helium atoms. So far so good, but molecules have an internal structure. An oxygen molecule, for example, consists of two oxygen atoms held together by the molecular force we discussed back in Figures (14, 15, and 16). As we saw in Figure (15), we can fairly accurately picture the molecule as two atoms held together by a spring force as shown here in Figure (26).

If an oxygen molecule collides with another molecule in the gas, one would expect that the molecule would start to vibrate, and perhaps rotate. This vibration and rotation represent forms of internal motion of the molecule that are quite distinct from the motion of the molecule as a whole, distinct from what we would call the center of mass motion.

If the center of mass motion has an average thermal kinetic energy 3/2 kT just like helium atoms, but the molecules can have internal motions and internal energy, you would expect that it would require more energy to heat a mole of oxygen than a mole of helium. For with the oxygen you not only have to supply the kinetic energy of the center of mass motion, but also the internal energy of the molecules. And that is correct. The molar specific heat of oxygen is 20.8 joules/(mole K), as compared to 12.5 joules/(mole K) for helium.

However it is when we try to calculate how much the internal energy of the molecules contribute to the specific heat, we run into trouble. As far back as 1858, James Clerk Maxwell, who was working on these calculations, repeatedly failed, and suspected that the failure was due to a problem with Newtonian mechanics.

Figure 26
Model of an oxygen molecule.

EQUIPARTITION OF ENERGY

The success of our calculation of the specific heat of helium gas lies in the fact that in our discussion of thermal processes, the helium atom can be treated as a rigid, undeformable sphere. Striking a helium atom is more analogous to striking a golf ball than hitting whiffle ball. When you hit a golf ball, most of the energy of the impact goes into the kinetic energy of the ball, and the motion of the ball is quite predictable. Hit a whiffle ball and most of the energy goes into mushing the ball; predicting where the whiffle ball will go is difficult.

All gas atoms except helium form molecules. While the individual atoms can usually be treated as hard spheres, for thermal calculations, the molecule as a whole is generally not rigid. Strike a molecule and some of the energy goes into center of mass motion of the molecule, but some goes into internal motions of the individual atoms. It seems that it would be rather hard to say much about the energy of an object that is vibrating, rotating, and flying through space.

However, if you have a gas of molecules in thermal equilibrium, the laws of Newtonian mechanics combined with the mathematical laws of probability, make a surprisingly simple prediction of where the energy goes. This prediction is called the *equipartition of energy theorem* which we will now describe.

As a background for the concepts involved in the equipartition of energy theorem, let us go back to the normal modes experiment of Chapter 16 where we had two air carts on an air track, connected by springs as shown in Figure (16-3). We found that the air carts had two distinct kinds of motion. There was the high frequency mode of motion where the two air carts oscillated against each other, moving together and then apart in a sinusoidal motion. Then there was the low frequency sloshing mode where the carts went back and forth along the track more or less together.

When we started the carts moving in a random way, recorded the motion, and did a Fourier analysis, we found that the apparently complex motion was merely a combination of the two simple sinusoidal modes of motion, the so-called *normal modes*. The carts were not free to move in an arbitrary way, their motion had to either be all of the vibrational mode, or all of the sloshing mode, or some combination of the two. The only thing that was arbitrary about the motion of the carts was how much of each of the two normal modes was present.

In our earlier discussion of center of mass motion in Chapter 11, we considered the example of two air carts, joined to each other by a spring, but free to move down the track as shown in Figure (11-9). We saw that when we gave one of the carts a shove, the center of mass of the two carts moved at a uniform speed down the track, while the carts themselves oscillated about the center of mass. In this example we again have two normal modes of motion. One is the motion of the center of mass, and the other is the oscillation about the center of mass. If we shove the carts just right we can have pure center of mass motion. Or we can have the carts oscillate with no center of mass motion. Or we can have some combination of the two kinds of motion.

These examples begin to show a pattern. If you have two masses connected by springs, that are constrained to move on a one dimensional track, the objects will have precisely two normal modes of motion. What the actual modes are depends upon the way the springs are connected. When the springs were connected to the ends of the air track, there was no center of mass motion, but we had two vibrational modes. When the carts were free to move down the track, we had the mode representing center of mass motion, but only one vibrational mode.

Another term often used to describe the way these carts are moving is the expression *degrees of freedom*. Two carts moving in one dimension have 2 degrees of freedom of motion. The degrees of freedom are the center of mass motion and the vibrational mode shown in Figure (11-9), or the two vibrational modes that we get from the setup in Figure (16-3).

Figure 11-9
Oscillating carts

If one connects three air carts with springs and starts them moving, the resulting motion can be quite complex. But if you record the motion, select one cycle of the repeating pattern and do a Fourier analysis, you get the simple result that there are three normal modes, as seen in the results from a student project shown in Figure (16-30). What is not so easy is to find out what the individual normal modes are. And for our discussion now, it is not important to find them. The important result is that if you have three carts connected in some way by springs, and they are constrained to move in one dimension there will be three normal modes or degrees of freedom. (If you have no springs, you still have 3 degrees of freedom, namely the center of mass motion of each of the 3 carts.

The counting of normal modes or degrees of freedom generalizes to more than one dimension. If you have one particle moving in three dimensions, it has three degrees of freedom, one for center of mass motion in the x direction, one for center of mass motion in the y direction and one for center of mass motion in the z direction.

If we have two particles in 3 dimensions, there are 6 degrees of freedom. If they are independent particles, then each has three degrees of freedom of motion of the center of mass. If they are connected by a spring, which is a good model of a diatomic molecule, there are still 6 degrees of freedom but we count them in a different way. There are the 3 degrees of freedom of the center of mass motion, and one degree of freedom for the kind of vibrational motion we saw in Figure (11-9) where we had two air carts connected by a spring. That accounts for 4 degrees of freedom; what are the other two?

When two connected particles move in three dimensions, what they can do that they could not do in one dimension is rotate about the center of mass. One can envision independent rotations about the x, the y, and the z axis, but one of these rotations does not count. In the picture we are developing, we will view the atoms themselves as perfectly smooth spheres, so that you cannot tell whether the atom itself is rotating or not. From this point of view, if the separation of two atoms in a diatomic molecule is along the z axis as shown in Figure (27), then rotation about the z axis cannot be detected and does not count as one of the degrees of freedom. Only rotations about the x and y axis contribute. Thus for a diatomic molecule moving in 3 dimensions, the 6 available degrees of freedom are 3 for center of mass motion, 2 for rotation, and one for vibration.

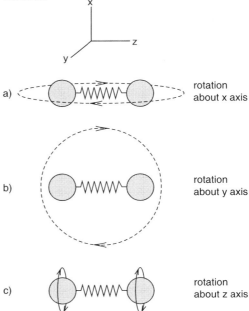

Figure 27
The three independent rotations of a molecule. For a diatomic molecule, the rotation about the z axis does not count. (Picture the atoms as perfectly smooth spheres. Then a collision could not start the z axis rotation, and you could not tell that it was rotating this way.)

If we have a system of n small spherical particles connected by spring-like forces, a reasonable model for many molecules, there should be 3n degrees of freedom. For example, the ammonia molecule with one nitrogen and three hydrogen atoms shown in Figure (28) should have 12 degrees of freedom. Three are center of mass motion, and now all 3 degrees of rotation should be counted. The remaining 6 must be vibrational normal modes, one for each spring like force. If you could kick an ammonia (or, let us say, a large scale model of one), record the motion of one of the molecules, and Fourier analyze a repeated pattern of the motion, you should be able to detect up to 6 normal mode frequencies of vibration.

We are now ready to state the equipartition of energy theorem that was first derived by James Clark Maxwell in 1858. Using Newtonian mechanics and the mathematical laws of probability, Maxwell showed that if a gas of molecules is in thermal equilibrium, then the average thermal energy of each molecule is 1/2 kT times the number of degrees of freedom possessed by the molecule. The theorem implies that, as you add thermal energy to a system of molecules, the energy is shared equally, on the average, between the available degrees of freedom.

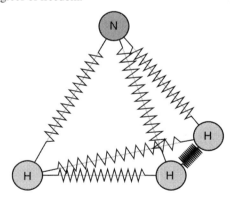

Figure 28
The ammonia molecule is a tetrahedral structure with one nitrogen atom and three hydrogen atoms. Here we are modeling the forces between atoms as spring forces.

The theorem is particularly easy to understand for the case of a gas of monatomic particles. If we have a single particle moving in 3 dimensions, we can write its kinetic energy $1/2\, mv^2$ in the form $1/2\, m(v_x^2 + v_y^2 + v_z^2)$ where we used the Pythagorean theorem to express v^2 in terms of its components. Thus the kinetic energy breaks up into 3 distinct terms

$$\text{kinetic energy} = \tfrac{1}{2}mv_x^2 + \tfrac{1}{2}mv_y^2 + \tfrac{1}{2}mv_z^2$$

which we can call the kinetic energies of x motion, y motion, and z motion respectively.

If the particle is in thermal equilibrium, then on the average v_x^2 should be the same as v_y^2 and v_z^2. Thus the kinetic energies associated with each of the degrees of freedom of the molecule (x motion, y motion, and z motion) should be the same, and the sum should be the total average kinetic energy 3/2 kT. If 3/2 kT is shared 3 ways, each degree of freedom should get 1/2 kT kinetic energy on the average, as required by the equipartition of energy theorem.

Real Molecules

Applying Maxwell's equipartition of energy theorem, we can make definite predictions about the specific heat of various kinds of molecules. We will begin with a brief review of the calculation of the specific heat of a monatomic gas like helium. A monatomic gas atom has 3 degrees of freedom, thus on the average its kinetic energy is

$$E(1 \text{ molecule}) = 3 \times \left(\tfrac{1}{2}kT\right) = \tfrac{3}{2}kT \quad (34)$$

Since the specific heat C_V deals with one mole of a substance, we multiply Equation 34 through by Avagadro's number N_A (number of particles in a mole) to get

$$E(1 \text{ mole}) = N_A\left(\tfrac{3}{2}kT\right) = \tfrac{3}{2}(N_A k)T$$

$$= \tfrac{3}{2}RT \quad (35)$$

where $N_A k = R$ is the gas constant.

Finally C_V is defined as the change in energy ΔE for a small change in temperature ΔT. Differentiating Equation (35) gives $\Delta E = (3/2)R\Delta T$, so that

$$\frac{\Delta E}{\Delta T} \equiv C_V = \frac{3}{2}R \qquad (36)$$

where the subscript V reminds us to keep the volume V of the gas constant so that none of the gas energy goes into doing the work of expanding the gas.

In the above derivation, we immediately see that the factor of 3 in the formula $C_V = 3/2\ R$ came from our assumption that the molecule has 3 degrees of freedom. If we had a molecule with n degrees of freedom, then the equipartition of energy theorem predicts that the specific heat should be

for a molecule   *prediction of the*
with n degrees $C_V = \frac{n}{2}R$ *equipartition of*   (37)
of freedom   *energy theorem*

To see how good the predictions of the equipartition of energy theorem are, we have in Table 2 listed the specific heats of some common gasses, and compared the results with the predicted values.

FAILURE OF CLASSICAL PHYSICS

If you worked out a complex theory that made detailed predictions, and when you compared the predictions with experiment, you got the results shown in Table 1, you should be disappointed. The agreement is simply terrible. The predictions work only for the monatomic gases (gases that remain individula atoms and do not form molecules). There is an increase in specific heat when we go to larger molecules, but not the predicted increase. In going from carbon dioxide to methane, where there is a considerable increase in the number of degrees of freedom, there is actually a decrease in the specific heat.

Maxwell worked on this problem for a number of years, carefully checking that he had correctly applied the mathematical laws of statistics to Newtonian mechanics, but he could find no error in his work. By 1879 he became convinced that Newtonian mechanics was flawed, and that eventually some new theory would have to be developed to replace it. The new theory, of course, was quantum mechanics, discovered nearly 50 years later. Maxwell's work in the 1860s and 1870s provided the first real evidence that Newtonian mechanics was not correct in all applications.

Molecule	Number of particles	Expected number degrees of freedom	Expected C_V (joules/mole)	Experimental C_V
helium	1	3	3/2 R = 12.5	12.5
argon	1	3	3/2 R = 12.5	12.6
nitrogen N_2	2	6	6/2 R = 25	20.7
oxygen O_2	2	6	6/2 R = 25	20.8
carbon dioxide CO_2	3	9	9/2 R = 37.5	29.7
methane NH_4	5	15	15/2R = 62.5	29.0

Table 2
Specific heats of various molecules. Theory and experiment agree only for the monatomic gases.

Freezing Out of Degrees of Freedom

While a straightforward application of the equipartition of energy theorem fails miserably, the ideas involved are not completely useless. A look at the specific heat of hydrogen gas as a function of temperature, Figure (29) gives us a clue as to what is happening at low temperatures below 100 K. The specific heat is 3/2 R, which is what we would expect for a monatomic gas with only 3 degrees of freedom. At these temperatures none of the thermal energy is going into exciting the internal motion of the atoms. At these low temperatures, the hydrogen molecules are acting like incompressible hard spheres.

Up at room temperature, the specific heat of hydrogen has jumped to 5/2 R. It appears that two additional degrees of freedom have appeared, and some of the thermal energy is now going into internal motions of the molecule. At still higher temperatures, just as the molecules are being torn apart, their specific heat reaches 7/2 R, indicating 7 degrees of freedom, one more than we expected. We can explain the 7 degrees of freedom by assuming that 1/2 kT of thermal energy goes, on the average, into the spring potential energy.

Going back down in temperature, we have the following picture. At very high temperature all the degrees of freedom are active and energy is shared equally among them as required by the equipartition of energy theorem. As we go down in temperature some of the degrees of freedom appear to freeze out. By room temperature we have lost two degrees of freedom, and down at 100 K, only the 3 translational degrees of freedom are left. Why the degrees of freedom freeze out is what is not explained by Newtonian mechanics. This is purely a quantum mechanical effect.

In Maxwell's time, the idea that matter consisted of atoms was a hypothesis rather than an experimentally proved fact. What atoms consisted of, whether they were indivisible hard spheres or had an internal structure was unknown. By applying Newtonian mechanics to models of atoms and molecules, he was trying to learn about the nature of these objects. The fact that monatomic gases have a specific heat $C_V = 3/2R$ was evidence that the atoms were in fact acting like hard, indivisible objects. The failure to predict molecular specific heats turned out to be evidence that Newtonian mechanics was failing.

We know that atoms themselves consist of many particles—a nucleus surrounded by electrons. If we applied Newtonian mechanics to this structure, we would assume that each atom should have many degrees of freedom and that the nucleus and electrons should individually pick up thermal energy as the atom is heated. This simply does not happen. Applying the language we have used above, we can say that temperatures at which we ordinarily study atoms, the internal degrees of freedom of the atom are frozen out.

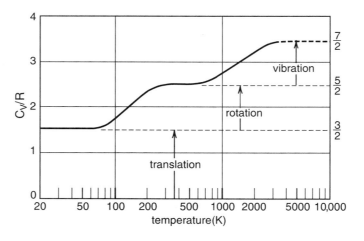

Figure 29
Specific heat of the hydrogen molecule. If each degree of freedom contributes 1/2R to the specific heat, then as the temperature drops, we see that various degrees of freedom freeze out. (Diagram adapted from Halliday and Resnick.)

THERMAL EXPANSION

When you heat a substance, in most cases the substance expands. For example, the mercury (or alcohol) in the bulb at the bottom of a thermometer expands when heated, forcing more mercury up the small tube above the bulb. If the column is marked off in degrees centigrade, you have a typical mercury thermometer as illustrated in Figure (30). As another example, you are aware of the cracks left between sections of cement in sidewalks and cement highways. These cracks are there so that on a hot day when the cement expands, the sidewalk or road will not buckle.

The reason for thermal expansion can be understood at an atomic level in terms of the shape of the molecular force potential well. Figure (31) is a redrawing of the molecular force potential well of Figure (16), with some added information. Let $2r$ be the separation of the two atoms as indicated at the top of the diagram. If the molecule is at a very low temperature, the atom will essentially sit at the bottom of the potential well and the separation will be $2r_0$ as shown.

If we raise the temperature of the molecule, the atoms gain thermal kinetic energy whose average value is $3/2\,kT$. As a result they will move back and forth at a higher level in the potential well, a height we have indicated as level 1 in the diagram. Due to the shape of the potential energy well, due to the fact that the repulsive core rises faster than the attractive side, the average separation $2r_T$ of the atoms at a temperature T is greater than the average separation $2r_0$ at low temperatures.

Although our discussion of molecular forces focused on two atom molecules, the general shape of the molecular force potential well is the same when you have many atoms forming a liquid or a solid. Thus when you heat a liquid or a solid, the atoms gain an average thermal kinetic energy $3/2\,kT$, effectively rise up in the molecular force potential well, and due to the shape of the well, have a slightly greater average separation. The substance expands.

You can see that the amount of expansion depends upon the detailed shape of the potential well which varies from one substance to another. Thus a thermometer based on the expansion properties of mercury does not have to give precisely the same reading as a thermometer using alcohol, except at the calibration points $0°C$ and $100°C$. And neither of these thermometers has to agree with the ideal gas thermometer. Since the ideal gas thermometer is based on the universal ideal gas law, one should use the ideal gas thermometer as a standard against which you calibrate mercury, alcohol, and other thermometers.

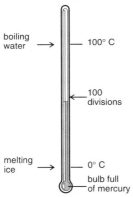

Figure 30
The typical mercury thermometer is based on the thermal expansion of mercury. There is no guarantee that a mercury thermometer and an ideal gas thermometer will agree at any temperatures except $0°C$ and $100°C$.

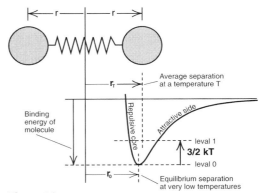

Figure 31
As you raise the temperature of a substance, the average thermal kinetic energy $3/2\,kT$ rises and the molecule sits higher in its potential well. Because the well is lopsided, the average separation of the atom becomes greater and the substance expands.

OSMOTIC PRESSURE

We would like to conclude this chapter, this brief view of atoms, molecules and thermal processes, with a discussion of two familiar phenomena that can be understood qualitatively from a molecular point of view. One is the elasticity of rubber, and the other is the process of osmosis, which is essential for biological systems.

Osmosis is a rather peculiar but important effect that is easily explained with an atomic model. Ordinarily, when a liquid can flow between two vessels at the same height, the liquid will tend to seek the same height in both vessels. But this does not always happen. Suppose we have a tank separated by a membrane, as shown in Figure (32). On the right side (side 2) of the membrane we place pure water, indicated by the small molecules. On the left (side 1) we place a solution of water and some other substance consisting of large molecules. The membrane has a special characteristic: the small water molecules can pass through it easily, whereas the big molecules are prevented from passing through because the holes in the membrane are too small. Initially, the two compartments are filled to the same level on each side of the membrane.

Assume that some definite fraction, for instance 50%, of all water molecules that strike the membrane pass through it. Initially, more water molecules strike the membrane from side 2 than from side 1 simply because there are more water molecules on side 2. If more water molecules strike from side 2 than side 1, and if 50% of **all** the water molecules striking the membrane pass through it, there must be a net flow of water from side 2 to side 1.

As the flow continues, the level of the liquid on side 1 rises and the solution becomes diluted. As the solution becomes further diluted, the number of water molecules on side 1 facing each cm^2 of the membrane increases, the more flow back to side 2, so that the net flow into side 1 decreases. From this description alone, however, we would not expect the flow to stop, since we never get pure water on side 1.

The flow does stop eventually though, because the level in side 1 rises to such an extent that the pressure at the bottom of side 1 becomes considerably greater than the pressure at the bottom of side 2 (a result of the increased weight of the column of water). This additional pressure, known as *osmotic pressure*, finally stops the flow of water from side 2 to side 1. The flow of the small molecules through the membrane is called *osmosis*; thus osmotic pressure is the pressure that finally stops osmosis.

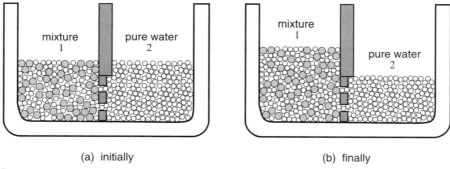

Figure 32
Osmosis. The two sides of the container are separated by a membrane that allows the water (small molecules) but not the large molecules to pass through. If the liquid levels are the same initially, as in (a), some of the pure water will flow through the membrane, raising the level on the side with the large molecules as shown in (b). This process is called osmosis.

Osmosis and osmotic pressure are crucial in biological processes. Osmosis is involved in the separation of nutrients and wastes in our own cells, the flow of fluids in our bodies, the flow of sap in plant life, and a number of other important processes.

The function and composition of blood is critically dependent on osmosis and osmotic pressure. Blood consists of red cells, white cells, and a fluid called *plasma*. The red cells are membrane sacs containing about 60% water and 40% hemoglobin molecules, molecules closely related to, but about four times as large as, the giant myoglobin molecule described in at the beginning of this chapter. We may think of the red blood cell as representing side 1 in Figure (32) where the big molecules are the hemoglobin molecules. If red blood cells are removed from blood and placed in pure water, they absorb so much water by osmosis that they burst. The red hemoglobin flows away, leaving an empty, pale misshapen sac.

The function of the red blood cell and its hemoglobin is to carry oxygen to the other cells in the body. The plasma, which consists of 90% water, 9% protein molecules, and 1% salts, serves as a fluid in which to dissolve needed proteins and salts to be carried to the cells, and to make the blood fluid enough to flow through the minute capillaries.

The capillary walls through which blood flows are porous membranes that permit water and salts to pass freely through, but that restrict the passage of proteins. Pure water could not be pumped through the bloodstream because it would leak out through the capillary walls. You may wonder how blood plasma, which is 90% water, can be pumped through the porous capillaries. The reason is that the 9% protein molecules in the plasma is sufficient to draw just enough water back into the capillary by osmosis to replace the water molecules that do leak out. Just as many water molecules are drawn back in as leak out, even though the pressure of the plasma inside the capillary is greater than the pressure of the fluids outside the leaky walls. Thus, side 1 in Figure (32) behaves in the same way as the capillary with the blood plasma inside it.

ELASTICITY OF RUBBER

A model for the elasticity of rubber was presented by Richard Feynman in a lecture to freshmen at Caltech in 1960. We select this model, not so much for its accuracy in describing the detailed behavior of molecules in rubber, but for developing an intuition for thermal processes. The mechanisms underlying the model and the behavior of rubber are fundamentally the same. The beauty of the model is that it is so outrageous that you are forced to think differently about thermal processes.

A Model of Rubber

Imagine that you enter a large room where there are a number of heavy chains loosely suspended from one end of the room to the other, as shown in Figure (33). These are massive chains, like the anchor chains used on old sailing ships, but they are hanging loosely, so that except for their weight, they are not exerting any force pulling the walls together.

On the floor are hundreds of cannonballs, lying there a couple of layers deep. This is our room at "absolute zero".

Now turn up the temperature in the room. The cannonballs start to jiggle and vibrate with an average thermal kinetic energy $3/2\ kT$. In this model, nothing melts. Instead, as we turn up the temperature the jiggling becomes stronger and stronger. When the average thermal kinetic energy $3/2\ kT$ becomes as large as the gravitational potential energy mgh of a cannonball near the ceiling, then we will have cannonballs flying all around the room. We will have a gas of cannonballs.

As the cannonballs fly around, they strike the chains, kinking them up as indicated in Figure (34). The kinked-up chains are no longer hanging loose, instead they are taut and pulling the side walls of the room in.

If we raise the temperature of the gas of cannonballs, the cannonballs strike the chains harder and the chains pull harder on the walls.

Here is an experiment that stretches the imagination even more. Suppose we start with the room with a gas of cannonballs at a temperature T, and chains kinked by the colliding cannonballs, and suddenly pull the sides of the room apart so that the chains are straight and tight. When we suddenly straighten out the kinked chains, the chains will slap against the cannonballs transforming the work we do pulling the chains straight into increased thermal kinetic energy of the cannonballs. As a result by suddenly stretching the chains we raise the temperature of the cannonballs.

If we let the walls go back suddenly, the chains initially go slack, and it takes some of the thermal kinetic energy of the cannonballs to kink the chains up again. As a result of unstretching the chains, the temperature of the cannonballs drops.

One's lips are a good detector of small temperature changes. Place a loose rubber band between your lips and suddenly stretch it. You will notice that the rubber band becomes distinctly warmer. Now quickly release the rubber band by bringing your hands together. The rubber band becomes distinctly cool. Rubber consists of a long chain of molecules that are kinked up by thermal motion. When you stretch the rubber band, you increase the thermal kinetic energy of the molecules and raise their temperature. Releasing the band reduces the thermal motion and drops the temperature. The elastic restoring force you felt when you stretched the band is caused by thermal motions kinking the long chain molecules.

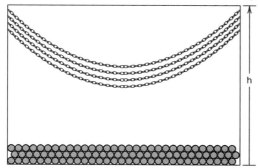

Figure 33
Room with suspended chains and cannonballs.

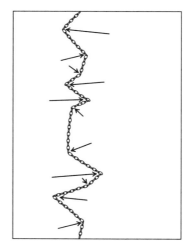

Figure 34
Top view of chains being struck by cannonballs.

Chapter 18
Entropy

Movie Undive

The focus of the last chapter was the thermal energy of the atoms and molecules around us. While the thermal energy of an individual molecule is not large, the thermal energy in a reasonable collection of molecules, like a mole, is a noticeable amount. Suppose, for example, you could extract all the thermal energy in a mole of helium atoms at room temperature. How much would that energy be worth at a rate of 10 cents per kilowatt hour?

This is an easy calculation to do. The atoms in helium gas at room temperature have an average kinetic energy of $3/2\, kT$ per molecule, thus the energy of a mole is $N_A(3/2\, KT) = 3/2\, RT$, where N_A is Avagadro's number and $R = N_A K$. Since ice melts at 273 K and water boils at 373 K, a reasonable value for room temperature is 300 K, about one quarter of the way up from freezing to boiling. Thus the total thermal energy in a mole of room temperature helium gas is

$$\left.\begin{array}{l}\text{thermal energy}\\\text{in a mole of}\\\text{room temperature}\\\text{helium gas}\end{array}\right\} = \frac{3}{2}RT$$

$$= \frac{3}{2} \times 8 \frac{joules}{mole\ K} \times 300K$$

$$= 3600\ joules$$

This is enough energy to lift 1 kilogram to a height of 360 feet!

To calculate the monetary value of this energy, we note that a kilowatt hour is 1000 watts of electric power for 3600 seconds, or 1000×3600 joules. Thus at a rate of 10 cents per kilowatt hour, the thermal energy of a mole of helium gas is worth only .01 cents.

The value .01 cents does not sound like much, but that was the value of the energy in only one mole of helium. Most substances have a greater molar heat capacity than helium due to the fact that energy is stored in internal motions of the molecule. Water at room temperature, for example, has a molar heat capacity six times greater than that of helium. Thus we would expect that the thermal energy in a mole of water should be of the order of 6 times greater than that of helium, or worth about .06 cents.

A mole of water is only 18 grams. A kilogram of water, 1000 grams of it, is 55 moles, thus the thermal energy in a kilogram or liter of water should have a value in the neighborhood of .06 cents/mole $\times$ 55 moles = 3.3 cents. Now think about the amount of water in a swimming pool that is 25 meters long, 10 meters wide, and 2 meters deep. This is $500\, m^3$ or 500×10^3 liters, their being 1000 liters/m^3. Thus the commercial value of the heat energy in a swimming pool of water at room temperature is $5 \times 10^5 \times 3.3$ cents or over $16,000.

The point of this discussion is that there is a lot of thermal energy in the matter around us, energy that would have enormous value if we could get at it. The question is why don't we use this thermal energy rather than getting energy by burning oil and polluting the atmosphere in the process?

INTRODUCTION

A simple lecture demonstration helps provide insight into why we cannot easily get at, and use, the $16,000 of thermal energy in the swimming pool. In this demonstration, illustrated in Figure (1), water is pumped up through a hose and squirted down onto a flat plate placed in a small bucket as shown. If you use a vibrator pump, then the water comes out as a series of droplets rather than a continuous stream.

To make the individual water droplets visible, and to slow down the apparent motion, the water is illuminated by a strobe. If the time between strobe flashes is just a bit longer than the time interval which the drops are ejected, the drops will appear to move very slowly. This allows you to follow what appears to be an individual drop as it moves down toward the flat plate.

For our discussion, we want to focus on what happens to the drop as it strikes the plate. As seen in the series of pictures in Figure (2), when the drop hits it flattens out, creating a wave that spreads out from where the drop hits. The wave then moves down the plate into the pool of water in the bucket. [In Figure (1) we see several waves flowing down the plate. Each was produced by a separate drop.]

Let us look at this process from the point of view of the energy involved. Before the drop hits, it has kinetic energy due to falling. When it hits, this kinetic energy goes into the kinetic energy of wave motion. The waves then flow into the bucket, eventually dissipate, and all the kinetic energy becomes thermal energy of the water molecules in the bucket. This causes a slight, almost undetectable, increase in the temperature of the water in the bucket (if the water drop had the same temperature as the water in the bucket, and we neglect cooling from evaporation).

We have selected this demonstration for discussion, because with a slight twist of the knob on the strobe, we can make the process appear to run backwards. If the time interval between flashes is just a bit shorter than the pulse interval of the pump, the drops appear to rise from the plate and go back into the hose. The situation looks funny, but it makes a good ending to the demonstration. Everyone knows that what they see couldn't possibly happen—or could it?

Does this reverse flow violate any laws of physics? Once a drop has left the plate it moves like a ball thrown up in the air. From the point of view of the laws of physics, nothing is peculiar about the motion of the drop from the time it leaves the plate until it enters the hose.

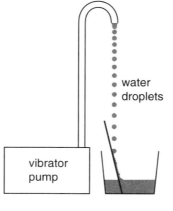

Figure 1
Water droplets are created by a vibrator pump. If you illuminate the drops with a strobe light, you can make them appear to fall or rise.

Where the situation looks funny is in the launching of the drop from the plate. But it turns out that none of the laws of physics we have discussed so far is violated there either. Let us look at this launching from the point of view of the energy involved. Initially the water in the bucket is a bit warm. This excess thermal energy becomes organized into a wave that flows up the plate. As seen in Figure (3) the wave coalesces into a drop that is launched up into the air. No violation of the law of conservation of energy is needed to describe this process.

While the launching of the drop in this reversed picture may not violate the laws of physics we have studied, it still looks funny, and we do not see such things happen in the real world. There has to be some reason why we don't. The answer lies in the fact that in the reversed process we have converted thermal energy, the disorganized kinetic energy of individual molecules, into the organized energy of the waves, and finally into the more concentrated kinetic energy of the upward travelling drop. We have converted a disorganized form of energy into an organized form in a way that nature does not seem to allow.

It is not impossible to convert thermal energy into organized kinetic energy or what we call useful work. Steam engines do it all the time. In a modern electric power plant, steam is heated to a high temperature by burning some kind of fuel, and the steam is sent through turbines to produce electricity. A certain fraction of the thermal energy obtained from burning the fuel ends up as electrical energy produced by the electric generators attached to the turbines. This electric energy can then be used to do useful work running motors.

The important point is that power stations cannot simply suck thermal energy out of a reservoir like the ocean and turn it into electrical energy. That would correspond to our water drop being launched by the thermal energy of the water in the bucket.

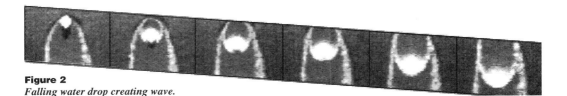

Figure 2
Falling water drop creating wave.

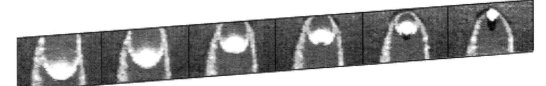

Figure 3
Rising wave launching water drop.

Even more discouraging is the fact that power plants do not even use all the energy they get from burning fuel. A typical high efficiency power plant ends up discarding, into the atmosphere or the ocean, over 2/3 of the energy it gets from burning fuel. Less than 1/3 of the energy from the fuel is converted into useful electrical energy. Car engines are even worse. Less than 1/5 of the energy from the gasoline burned goes into powering the car; most of the rest comes out the exhaust pipe.

Why do we tolerate these low efficiency power plants and even lower efficiency car engines? The answer lies in the problem of *converting a disorganized form of energy into an organized one*. Or to state the problem more generally, of trying to create order from chaos.

The basic idea is that a disorganized situation does not naturally organize itself—in nature, things go the other way. For example, if you have a box of gas, and initially the atoms are all nicely localized on one side of the box, a short time later they will be flying around throughout the whole volume of the box. On their own, there is almost no chance that they will all move over to that one side again. If you want them over on one side, you have to do some work, like pushing on a piston, to get them over there. It takes work to create order from disorder.

At first, it seems that the concepts of order and disorder, and the related problems of converting thermal energy into useful work should be a difficult subject to deal with. If you wished to formulate a physical law, how do you go about even defining the concepts. What, for example, should you use as an experimental definition of disorder? It turns out, surprisingly, that there is a precise definition of a quantity called *entropy* which represents the amount of disorder contained in a system. Even more surprising, the concept of entropy was discovered before the true nature of heat was understood.

The basic ideas related to entropy were discovered in 1824 by the engineer Sadi Carnot who was trying to figure out how to improve the efficiency of steam engines. Carnot was aware that heat was wasted in the operation of a steam engine, and was studying the problem in an attempt to reduce the waste of heat. In his studies Carnot found that there was a *theoretically maximum efficient engine whose efficiency depended upon the temperature of the boiler relative to the temperature of the boiler's surroundings*.

To make his analysis, Carnot had to introduce a new assumption not contained in Newton's law of mechanics. Carnot's assumption is equivalent to the idea that *you cannot convert thermal energy into useful work in a process involving only one temperature*. This is why you cannot sell the $16,000 worth of thermal energy in the swimming pool—you cannot get it out.

This law is known as the *Second Law of Thermodynamics*. (The first law is the law of conservation of energy itself.) The second law can also be expressed in terms of entropy which we now know represents the disorder of a system. The second law states that in any process, the total entropy (disorder) of a system either stays the same or increases. Put another way, it states that in any process, the total order of a system cannot increase; it can only stay the same, or the system can become more disordered.

To develop his formulas for the maximum efficiency of engines, Carnot invented the concept known as a *Carnot engine*, based on what is called the *Carnot cycle*. The Carnot engine is not a real engine, no one has ever built one. Instead, you should think of it as a *thought experiment*, like the ones we used in Chapter 1 to figure out what happened to moving clocks if the principle of relativity is correct.

The question we wish to answer is, how efficient can you make an engine or a power plant, if the second law of thermodynamics is correct? If you cannot get useful work from thermal energy at one temperature, how much work can you get if you have more than one temperature? It turns out that there is a surprisingly simple answer, but we are going to have to do quite a bit of analysis of Carnot's thought experiment before we get the answer. During the discussion of the Carnot engine, one should keep in mind that we are making this effort to answer one basic question—what are the consequences of the second law of thermodynamics— what are the consequences of the idea that order does not naturally arise from disorder.

WORK DONE BY AN EXPANDING GAS

The Carnot thought experiment is based on an analysis of several processes involving the ideal piston and cylinder we discussed in the last chapter. We will discuss each of these processes separately, and then put them together to complete the thought experiment.

The ideal piston and cylinder is shown in Figure (4). A gas, at a pressure p, is contained in the cylinder by a frictionless piston of cross-sectional area A. (Since no one has yet built a piston that can seal the gas inside the cylinder and still move frictionlessly, we are now already into the realm of a thought experiment.) A force F is applied to the outside of the piston as shown to keep the piston from moving. The gas, at a pressure p, exerts an outward force

$$p\left(\frac{newtons}{meter^2}\right) \times A\left(meter^2\right) = pA(newtons)$$

on the cylinder, thus F must be given by

$$F = pA \qquad (1)$$

to keep the cylinder from moving.

If we decrease the force F just a bit to allow the gas in the cylinder to expand, the expanding gas will do work on the piston. This is because the gas is exerting a force pA on the cylinder, while the cylinder is moving in the direction of the force exerted by the gas. If the piston moves out a distance Δx as shown in Figure (5), the work ΔW done by the gas is the force pA it exerts times the distance Δx

$$\Delta W = (pA)\Delta x \qquad (2)$$

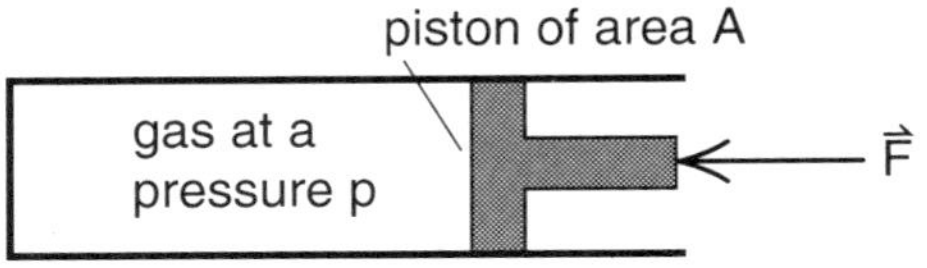

Figure 4
In the ideal piston and cylinder, the piston confines the gas and moves frictionlessly.

After this expansion, the volume of the gas has increased by an amount $\Delta V = A\Delta x$. Thus Equation 2 can be written in the form $\Delta W = pA\Delta x$ or

$$\boxed{\Delta W = p\Delta V} \qquad (3)$$

Equation 3 is more general than our derivation indicates. Any time a gas expands its volume by an amount ΔV, the work done by the gas is $(p\Delta V)$ no matter what the shape of the container. For example, if you heat the gas in a balloon and the balloon expands a bit, the work done by the gas is $(p\Delta V)$ *where ΔV is the increase in the volume of the balloon.*

Exercise 1

In our introduction to the concept of pressure, we dipped a balloon in liquid nitrogen until the air inside became a puddle of liquid air (see Figure 17-19). When we took the balloon out of the liquid nitrogen, the air slowly expanded until the balloon returned to its original size. During the expansion, the rubber of the balloon was relatively loose, which means that the air inside the balloon remained at or very near to atmospheric pressure during the entire time the balloon was expanding.

(a) If the final radius of the balloon is 30 cm, how much work did the gas inside the balloon do as the balloon expanded? (You may neglect the volume of the liquid air present when the expansion started.) (Answer: 1.1×10^4 joules.)

(b) Where did the gas inside the balloon get the energy required to do this work?

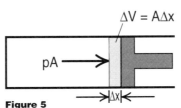

Figure 5
The work done by an expanding gas is equal to the force pA it exerts, times the distance Δx the piston moves.

SPECIFIC HEATS C_V AND C_p

In our earlier discussion of specific heat, we dealt exclusively with the "molar" specific heat at constant volume C_V. We always assumed that we kept the gas at constant volume so that all the energy we added would go into the internal energy of the gas. If we had allowed the gas to expand, then some of the energy would have gone into the work the gas did to expand its volume, and we would not have had an accurate measure of the amount of energy that went into the gas itself.

Sometimes it is convenient to heat a gas while keeping the gas *pressure*, rather than *volume*, constant. This is more or less the case when we heat the gas in a balloon. The balloon expands, but the pressure does not change very much if the expansion is small.

Earlier we defined the molar heat capacity C_V as the amount of energy required to heat one mole of a substance one kelvin, if the volume of the substance is kept constant. Let us now define the molar heat capacity C_p as the amount of energy required to heat one mole of a substance one kelvin if the pressure is kept constant. For gases C_p is always larger than C_V. This is because, when we heat the gas at constant pressure, the energy goes both into heating and expanding the gas. When we heat the gas at constant volume, the energy goes only into heating the gas. We can write this out as an equation as follows

energy required to heat 1 mole of a gas 1K at constant pressure $\quad C_p = \begin{pmatrix} \text{increase in thermal} \\ \text{energy of the gas} \\ \text{when the temperature} \\ \text{increases 1K} \end{pmatrix}$

$\qquad\qquad + \begin{pmatrix} \text{work done} \\ \text{by the} \\ \text{expanding} \\ \text{gas} \end{pmatrix}$ (4)

Noting that since C_V is equal to the increase in thermal energy of the gas, and that the work done is $p\Delta V$, we get

$$C_p = C_V + p\Delta V \qquad (5)$$

In the special case of an ideal gas, we can use the ideal gas law $pV = nRT$, setting $n = 1$ for 1 mole

$$pV = RT \quad \text{(1 mole of gas)} \qquad (6)$$

If we let the gas expand a bit at constant pressure, we get differentiating Equation 6, keeping p constant**

$$p\Delta V = R\Delta T \quad \text{(if p is constant)} \qquad (7)$$

** By differentiating the equation $(pV = RT)$, we mean that we wish to equate the *change* in (pV) to the *change* in (RT). To determine the change in (pV), for example, we let (p) go to $(p + \Delta p)$ and (V) go to $(V + \Delta V)$, so that the product (pV) becomes

$$pV \rightarrow (p + \Delta p)(V + \Delta V)$$
$$= pV + (\Delta p)V + p\Delta V + (\Delta p)\Delta V$$

If we neglect the second order term $(\Delta p)\Delta V$ then

$$pV \rightarrow pV + (\Delta p)V + p\Delta V$$

Then if we hold the pressure constant $(\Delta p = 0)$, we see that the change in (pV) is simply $(p\Delta V)$. Since R is constant, the change in RT is simply $R\Delta T$.

If the temperature increase is $\Delta T = 1$ kelvin, then Equation 7 becomes

$$p\Delta V = R \quad (1 \text{ mole, p constant}, \Delta T = 1K) \quad (8)$$

Using Equation 8 in Equation 5 we get the simple result

$$\boxed{C_p = C_V + R} \quad (9)$$

The derivation of Equation 9 illustrates the kind of steps we have to carry out to calculate what happens to the heat we add to substances. For example, in going from Equation 6 to Equation 7, we looked at the change in volume when the temperature but not the pressure was varied. When we make infinitesimal changes of some quantities in an equation while holding the quantities constant, the process is called *partial differentiation*. In this text we will not go into a formal discussion of the ideas of partial differentiation. When we encounter the process, the steps should be fairly obvious as they were in Equation 7.

*(The general subject that deals with changes produced by adding or removing heat from substances is called **thermodynamics**. The full theory of thermodynamics relies heavily on the mathematics of partial derivatives. For our discussion of Carnot's thought experiments, we need only a small part of thermodynamics theory.)*

Exercise 2

(a) Back in Table 2 on page 31 of Chapter 17, we listed the values of the molar specific heats for a number of gases. While the experimental values of did not agree in most cases with the values predicted by the equipartition of energy, you can use the experimental values of C_V to accurately predict the values of C_p for these gases. Do that now.

(b) Later in this chapter, in our discussion of what is called the *adiabatic* expansion of a gas (an expansion that allows no heat to flow in), we will see that the ratio of C_p/C_V plays an important role in the theory. It is common practice to designate this ratio by the Greek letter γ

$$\boxed{\gamma \equiv C_p/C_V} \quad (10)$$

(i) Explain why, for an ideal gas, γ is always greater than 1.

(ii) Calculate the value of γ for the gases, listed in Table 2 of Chapter 17.

Answers:

gas	γ
helium	1.66
argon	1.66
nitrogen	1.40
oxygen	1.40
CO_2	1.28
NH_4	1.29

ISOTHERMAL EXPANSION AND PV DIAGRAMS

In the introduction, we pointed out that while a swimming pool of water may contain $16,000 of thermal energy, we could not extract this energy to do useful work. To get useful energy, we have to burn fuel to get heat, and convert the heat to useful work. What seemed like an insult is that even the most efficient power plants turn only about 1/3 of the heat from the fuel into useful work, the rest being thrown away, expelled either into the atmosphere or the ocean.

We are now going to discuss a process in which heat is converted to useful work with 100% efficiency. This involves letting the gas in a piston expand at constant temperature, in a process called an *isothermal expansion*. (The prefix "iso" is from the Greek meaning "equal," thus, isothermal means equal or constant temperature.) This process cannot be used by power plants to make them 100% efficient, because the process is not repetitive. Some work is required to get the piston back so that the expansion can be done over again.

Suppose we start with a gas in a cylinder of volume V_1 and let the gas slowly expand to a volume V_2 as shown in Figure (6). We control the expansion by adjusting the force $\vec{F}$ exerted on the back side of the piston.

While the gas is expanding, it is doing work on the piston. For each ΔV by which the volume of the gas increases, the amount of work done by the gas is $p\Delta V$. The energy required to do this work must come from somewhere. If we did not let any heat into the cylinder, the energy would have to come from thermal energy, and the temperature would drop. (This is one way to get work out of thermal energy.)

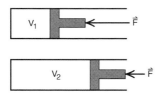

Figure 6
Isothermal expansion of the gas in a cylinder. The force $\vec{F}$ on the cylinder is continually adjusted so that the gas expands slowly at constant temperature.

However, we wish to study the process in which the gas expands at constant temperature. To keep the temperature from dropping, we have to let heat flow into the gas. Since the temperature of the gas is constant, there is no change in the thermal energy of the gas. Thus *all the heat that flows in goes directly into the work done by the gas*.

To calculate the amount of work done, we have to add up all the $p\Delta V$'s as the gas goes from a volume V_1 to a volume V_2. If we graph pressure as a function of volume, in what is called a *pV diagram*, we can easily visualize these increments of work $p\Delta V$ as shown in Figure (7).

Suppose the pressure of the gas is initially p_1 when the volume of the cylinder is V_1. As the cylinder moves out and the gas expands, its pressure will drop as shown in Figure (7), reaching the lower value p_2 when the cylinder volume reaches V_2. At each step ΔV_i, when the pressure is p_i, the amount of work done by the gas is $p_i \Delta V_i$. The total work, the sum of all the $p_i \Delta V_i$, is just the total area under the pressure curve, as seen in Figure (7).

The nice feature of a graph of pressure versus volume like that shown in Figure (7), is *the work done by the gas is always the area under the pressure curve*, no matter what the conditions of the expansion are. If we had allowed the temperature to change, the shape of the pressure curve would have been different, but the work done by the gas would still be the area under the pressure curve.

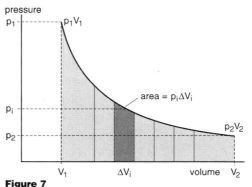

Figure 7
The work done by an expanding gas is equal to the sum of all $p\Delta V$'s, which is the area under the pressure curve.

Isothermal Compression

If we shoved the piston back in, from a volume V_2 to a volume V_1 in Figure (7), we would have to do work on the gas. If we kept the temperature constant, then the pressure would increase along the curve shown in Figure (7) and the work we did would be precisely equal to the area under the curve. In this case work is done on the gas (we could say that during the compression the gas does negative work). When work is done on the gas, the temperature of the gas will rise unless we let heat flow out of the cylinder. Thus if we have an isothermal compression, where there is no increase in the thermal energy of the gas, then we have the pure conversion of useful work into the heat expelled by the piston. This is the opposite of what we want for a power plant.

Isothermal Expansion of an Ideal Gas

If we have one mole of an ideal gas in our cylinder, and keep the temperature constant at a temperature T_1, then the gas will obey the ideal gas equation.

$$pV = RT_1 = \text{constant} \tag{11}$$

Thus the equation for the pressure of an ideal gas during an isothermal expansion is

$$p = \frac{\text{constant}}{V} \tag{11a}$$

and we see that the pressure decreases as $1/V$. This decrease is shown in the pV diagram of Figure (8).

ADIABATIC EXPANSION

We have seen that we can get useful work from heat during an isothermal expansion of a gas in a cylinder. As the gas expands, it does work, getting energy for the work from heat that flows into the cylinder. This represents the conversion of heat energy at one temperature into useful work. The problem is that there is a limited amount of work we can get this way. If we shove the piston back in so that we can repeat the process and get more work, it takes just as much work to shove the piston back as the amount of work we got out during the expansion. The end result is that we have gotten nowhere. We need something besides isothermal expansions and compressions if we are to end up with a net conversion of heat into work.

Another kind of expansion is to let the cylinder expand without letting any heat in. This is called an ***adiabatic expansion***, where *adiabatic* is from the Greek *(a-*not + *dia-through* + *bainein*-to go). If the gas does work during the expansion, and we let no heat energy in, then *all the work must come from the thermal energy* of the gas. The result is that the gas will cool during the expansion. In an adiabatic expansion, we are converting the heat energy *contained in the gas* into useful work. If we could keep this expansion going we could suck all the thermal energy out of the gas and turn it into useful work. The problem, of course, getting the piston back to start the process over again.

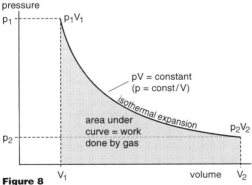

Figure 8
In the isothermal expansion of an ideal gas, we have $pV = \text{constant}$. Thus the pressure decreases as $1/V$.

It is instructive to compare an isothermal expansion to an adiabatic expansion of a gas. In either case the pressure drops. But in the adiabatic expansion, the pressure drops faster because the gas cools. In Figure (9), we compare the isothermal and adiabatic expansion curves for an ideal gas. Because the adiabatic curve drops faster in the pV diagram, there is less area under the adiabatic curve, and the gas does less work. This is not too surprising, because less energy was available for the adiabatic expansion since no heat flowed in.

For an ideal gas, the equation for an adiabatic expansion is

$$pV^\gamma = \text{constant}; \quad \gamma = \frac{C_p}{C_V} \qquad (12)$$

a result we derive in the appendix. (You calculated the value of γ for various gases in Exercise 2.) The important point now is not so much this formula, as the fact that the adiabatic curve drops faster than the isothermal curve.

If we compress a gas adiabatically, all the work we do goes into the thermal energy of the gas, and the temperature rises. Thus with an adiabatic expansion we can lower the temperature of the gas, and with an adiabatic compression raise it.

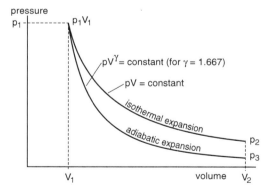

Figure 9
Comparison of isothermal and adiabatic expansions. In an adiabatic expansion the gas cools, and thus the pressure drops faster.

Exercise 3

In the next section, we will discuss a way of connecting adiabatic and isothermal expansions and compressions in such a way that we form a complete cycle (get back to the starting point), and get a net amount of work out of the process. Before reading the next section, it is a good exercise to see if you can do this on your own.

In order to see whether or not you are getting work out or putting it in, it is useful to graph the process in a pV diagram, where the work is simply the area under the curve.

To get you started in this exercise, suppose you begin with an ideal gas at a pressure p_1, volume V_1, and temperature T_1, and expand it isothermally to p_2, V_2, T_1 as shown in Figure (10a). The work you get out is the area under the curve.

If you then compressed the gas isothermally back to p_1, V_1, T_1, this would complete the cycle (get you back to where you started), but it would take just as much work to compress the gas as you got from the expansion. Thus there is no net work gained from this cycle. A more complex cycle is needed to get work out.

If we add an adiabatic expansion to the isothermal expansion as shown in Figure (10b) we have the start of something more complex. See if you can complete this cycle, i.e., get back to p_1, V_1, T_1, using adiabatic and isothermal expansions or compressions, and get some net work in the process. See if you can get the answer before we give it to you in the next section. Also show graphically, on Figure (10b) how much work you do get out.

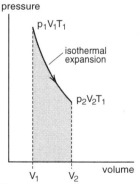

Figure 10a
pV diagram for an isothermal expansion from volume V_1 to volume V_2.

THE CARNOT CYCLE

With the isothermal and adiabatic expansion and compression of an ideal gas in a frictionless cylinder, we now have the pieces necessary to construct a Carnot cycle, the key part of our thought experiment to study the second law of thermodynamics.

The goal is to construct a device that continually converts heat energy into work. Such a device is called an *engine*. Both the isothermal and adiabatic expansions of the gas converted heat energy into work, but the expansions alone could not be used as an engine because the piston was left expanded. Carnot's requirement for an engine was that after a complete cycle all the working parts had to be back in their original condition ready for another cycle. Somehow the gas in the cylinder has to be compressed again to get the piston back to its original position. And the compression cannot use up all the work we got from the expansion, in order that we get some net useful work from the cycle.

The idea for Carnot's cycle that does give a net amount of useful work is the following. Start off with the gas in the cylinder at a high temperature and let the gas expand isothermally. We will get a certain amount of work from the gas. Then rather than trying to compress the hot gas, which would use up all the work we got, cool the gas to reduce its pressure. Then isothermally compress the cool gas. It should take less work to compress the low pressure cool gas than the work we got from the high pressure hot gas. Then finish the cycle by heating the cool gas back up to its original temperature. In this way you get back to the original volume and temperature (and therefore pressure) of the cylinder; you have a complete cycle, and hopefully you have gotten some useful work from the cycle.

To cool the gas, and then later heat it up again, Carnot used an adiabatic expansion and then an adiabatic compression. We can follow the steps of the Carnot cycle on the pV diagram shown in Figure (11). The gas starts out at the upper left hand corner at a high temperature T_1, volume V_1, pressure p_1. It then goes through an isothermal expansion from a volume V_1 to a volume V_2, remaining at the initial temperature T_1. The hot gas is then cooled down to a low temperature T_3 by an adiabatic expansion to a volume V_3. The cool, low pressure gas is then compressed isothermally to a volume V_4, where it is then heated back to a higher temperature T_1 by an adiabatic compression. The volume V_4 is chosen just so that the adiabatic compression will bring the temperature back to T_1 when the volume gets back to V_1.

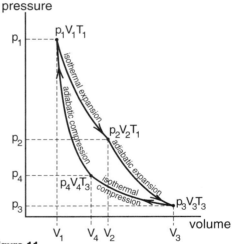

Figure 11
The Carnot Cycle. The gas first expands at a high temperature T_1. It is then cooled to a lower temperature T_3 by an adiabatic expansion. Then it is compressed at this lower temperature, and finally heated back to the original temperature T_1 by an adiabatic compression. We get a net amount of work from the process because it takes less work to compress the cool low pressure gas than we got from the expansion of the hot high pressure gas.

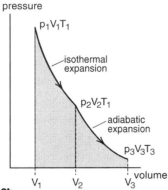

Figure 10b
pV diagram for an isothermal expansion followed by an adiabatic expansion.

In this set of 4 processes, we get work out of the two expansions, but put work back in during the two compressions. Did we really get some net work out? We can get the answer immediately from the pV diagram. In Figure (12a), we see the amount of work we got out of the two expansions. It is the total area under the expansion curves. In Figure (12b) we see how much work went back in during the two compressions. It is the total area under the two compression curves. Since there is more area under the expansion curves than the compression curves, we got a net amount of work out. The net work out is, in fact, just equal to the 4 sided area between the curves, seen in Figure (13).

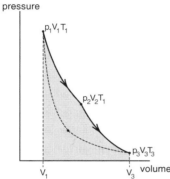

Figure 12a
The work we get out of the two expansions is equal to the area under the expansion curves.

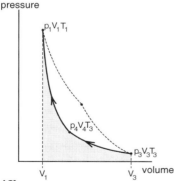

Figure 12b
The work required to compress the gas back to its original volume is equal to the area under the compression curves.

Thermal Efficiency of the Carnot Cycle

The net effect of the Carnot cycle is the following. During the isothermal expansion while the cylinder is at the high temperature T_H, a certain amount of thermal or heat energy, call it Q_H, flows into the cylinder. Q_H must be equal to the work the gas is doing during the isothermal expansion *since the gas' own thermal energy does not change at the constant temperature*. (Here, all the heat in becomes useful work.)

During the isothermal compression, while the cylinder is at the lower temperature T_L, (the gas having been cooled by the adiabatic expansion), an amount of heat Q_L is expelled from the cylinder. Heat must be expelled because we are doing work on the gas by compressing it, and *none of the energy we supply can go into the thermal energy of the gas* because its temperature is constant. (Here all the work done becomes expelled heat.)

Since no heat enters or leaves the cylinder during the adiabatic expansion or compression, all flows of heat have to take place during the isothermal processes. Thus the net effect of the process is that an amount of thermal energy or heat Q_H flows into the cylinder at the high temperature T_H, and an amount of heat Q_L flows out at the low temperature T_L, and we get a net amount of useful work W out equal to the 4-sided area seen in Figure (13). By the law of conservation of energy, the

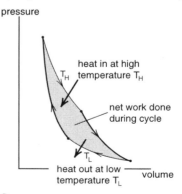

Figure 13
The net work we get out of one complete cycle is equal to the area bounded by the four sided shape that lies between the expansion and compression curves.

work W must be equal to the difference between Q_H in and Q_L out

$$W = Q_H - Q_L \tag{13}$$

We see that the Carnot engine suffers from the same problem experienced by power plants and automobile engines. They take in heat Q_H at a high temperature (produced by burning fuel) and do some useful work W, but they expel heat Q_L out into the environment. To be 100% efficient, the engine should use all of Q_H to produce work, and not expel any heat Q_L. But the Carnot cycle does not appear to work that way.

One of the advantages of the Carnot cycle is that we can calculate Q_H and Q_L, and see just how efficient the cycle is. It takes a couple of pages of calculations, which we do in the appendix, but we obtain a remarkably simple result. The ratio of the heat in, Q_H, to the heat out, Q_L, is simply equal to the ratio of the high temperature T_H to the low temperature T_L.

$$\boxed{\frac{Q_H}{Q_L} = \frac{T_H}{T_L}} \quad \begin{array}{l}\textit{for a carnot} \\ \textit{cycle based on} \\ \textit{an ideal gas}\end{array} \tag{14}$$

One suspects that if you do a lot of calculation involving integration, logarithms, and quantities like the specific heat ratio, and almost everything cancels to leave such a simple result as Equation 14, then there might be a deeper significance to the result than expected. Equation 14 was derived for a Carnot cycle operating with an ideal gas. It turns out that *the result is far more general* and has broad applications.

Exercise 4
A particular Carnot engine has an efficiency of 26.8%. That means that only 26.8% of Q_H comes out as useful work W and the rest, 73.2% is expelled at the low temperature T_L. The difference between the high and low temperature is 100 K ($T_H - T_L = 100$ K). What are the values of T_H and T_L? First express your answer in kelvins, then in degrees centigrade. (The answer should be familiar temperatures.)

Exercise 5
If you have a 100% efficient Carnot engine, what can you say about T_H and T_L?

Reversible Engines

In our discussion of the principle of relativity, it was immediately clear why we developed the light pulse clock thought experiment. You could immediately see that moving clocks should run slow, and why that was a consequence of the principle of relativity.

We now have a new thought experiment, the Carnot engine, which is about as idealized as our light pulse clock. We have been able to calculate the efficiency of a Carnot engine, but it is not yet obvious what that has to do either with real engines, or more importantly with the second law of thermodynamics which we are studying. It is not obvious because we have not yet discussed one crucial feature of the Carnot engine.

The Carnot engine is explicitly designed to be reversible. As shown in Figure (14), we could start at point 1 and go to point 4 by an adiabatic expansion of the gas. During this expansion the gas would do work but no heat is allowed to flow in. Thus the work energy would come from thermal energy and the gas would cool from T_H to T_L.

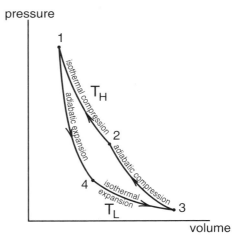

Figure 14
The Carnot cycle run backward.

The next step of the reverse Carnot cycle is an isothermal expansion from a volume V_4 to a volume V_3. During this expansion, the gas does an amount of work equal to the area under the curve as shown in Figure (15a). Since there is no change in the internal energy of a gas when the temperature of the gas remains constant, the heat flowing in equals the work done by the gas. This is the same amount of heat Q_L that flowed out when the engine ran forward.

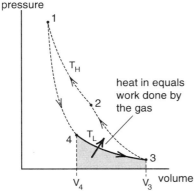

a) *During the isothermal expansion, some heat flows into the gas to supply the energy needed for the work done by the gas.*

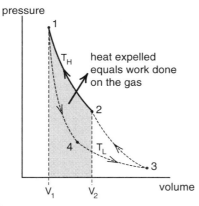

b) *A lot more work is required, and a lot more heat is expelled when we compress the hot gas isothermally.*

Figure 15
Heat flow when the Carnot cycle runs backward. Since more heat flows out than in, some work W is required for the cycle. The net effect is that the work W pumps heat out of the gas, giving us a refrigerator.

In going from point 3 to point 2, we adiabatically compress the gas to heat it from the lower temperature T_L to the higher temperature T_H. Since the compression is adiabatic, no heat flows in or out.

In the final step from point 2 to point 1, we isothermally compress the gas back to its original volume V_1. Since the gas temperature remains constant at T_H, there is no change in thermal energy and all the work we do, shown as the area under the curve in Figure (15b), must be expelled in the form of heat flowing out of the cylinder. The amount of heat expelled is just Q_H, the amount that previously flowed in when the engine was run forward.

We have gone through the reverse cycle in detail to emphasize the fact that the engine should run equally well both ways. In the forward direction the engine takes in a larger amount of heat Q_H at the high temperature T_H, expels a smaller amount of heat Q_L at the lower temperature T_L, and produces an amount of useful work W equal to the difference $Q_H - Q_L$.

In the reverse process, the engine takes in a smaller amount of heat Q_L at the low temperature T_L, and expels a larger amount Q_H at the higher temperature T_H. Since more heat energy is expelled than taken in, an amount of work $W = Q_H - Q_L$ must now be supplied to run the engine. When we have to supply work to pump out heat, we do not usually call the device an engine. The common name is a **refrigerator**. In a refrigerator, the refrigerator motor supplies the work W, a heat Q_L is sucked out of the freezer box, and a total amount of energy $Q_L + W = Q_H$ is expelled into the higher room temperature of the kitchen. If we have a Carnot refrigerator running on an ideal gas, then the heats Q_L and Q_H are still given by Equation 14

$$\frac{Q_L}{Q_H} = \frac{T_L}{T_H} \qquad \text{(14 repeated)}$$

where T_L and T_H are the temperatures on a scale starting from absolute zero such as in the kelvin scale.

Exercise 6

How much work must a Carnot refrigerator do to remove 1000 joules of energy from its ice chest at 0° C and expel the heat into a kitchen at 27° C?

ENERGY FLOW DIAGRAMS

Because of energy conservation, we can view the flow of energy in much the same way as the flow of some kind of a fluid. In particular we can construct flow diagrams for energy that look much like plumbing diagrams for water. Figure (16) is the energy flow diagram for a Carnot engine running forward. At the top and the bottom are what are called ***thermal reservoirs***—large sources of heat at constant temperature (like swimming pools full of water). At the top is a thermal reservoir at the high temperature T_H (it could be kept at the high temperature by burning fuel) and at the bottom is a thermal reservoir at the low temperature T_L. For power plants, the low temperature reservoir is often the ocean or the cool water in a river. Or it may be the cooling towers like the ones pictured in photographs of the nuclear power plants at Three Mile Island.

In the energy flow diagram for the forward running Carnot engine, an amount of heat Q_H flows out of the high temperature reservoir, a smaller amount Q_L is expelled into the low temperature reservoir, and the difference comes out as useful work W. If the Carnot engine is run on an ideal gas, Q_H and Q_L are always related by $Q_H/Q_L = T_H/T_L$.

Figure (17) is the energy flow diagram for a Carnot refrigerator. A heat Q_L is sucked out of the low temperature reservoir, an amount of work W is supplied (by some motor), and the total energy $Q_H = Q_L + W$ is expelled into the high temperature reservoir.

Maximally Efficient Engines

We are now ready to relate our discussion of the Carnot cycles to the second law of thermodynamics. The statement of the second law we will use is that you cannot extract useful work from thermal energy at one temperature. (The colloquial statement of the first law of thermodynamics—conservation of energy—is that you can't get something for nothing. The second law says that you can't break even.)

Up until now we have had to point out that our formula for the efficiency of a Carnot engine was based on the assumption that we had an ideal gas in the cylinder. If we use the second law of thermodynamics, we can show that it is impossible to construct any engine, by any means, that is more efficient than the Carnot engine we have been discussing. This will be the main result of our thought experiment.

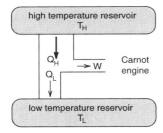

Figure 16
Energy flow diagram for a Carnot engine. Since energy is conserved, we can construct a flow diagram for energy that resembles a plumbing diagram for water. In a Carnot cycle, Q_H flows out of a "thermal reservoir" at a temperature T_H. Some of this energy goes out as useful work W and the rest, Q_L, flows into the low temperature thermal reservoir at a temperature T_L.

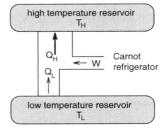

Figure 17
The Carnot refrigerator is a Carnot engine run backwards. The work W plus the heat Q_L equals the heat Q_H pumped up into the high temperature reservoir.

18-16 Entropy

Let us suppose that you have constructed a Super engine that takes in more heat Q_H^* from the high temperature reservoir, and does more work W^*, while rejecting the same amount of heat Q_L as a Carnot engine. In the comparison of the two engines in Figure (18) you can immediately see that your Super engine is more efficient than the Carnot engine because you get more work out for the same amount of heat lost to the low temperature reservoir.

Now let us run the Carnot engine backwards as a refrigerator as shown in Figure (19). The Carnot refrigerator requires an amount of work W to suck the heat Q_L out of the low temperature reservoir and expel the total energy $Q_H = W + Q_L$ into the high temperature reservoir.

You do not have to look at Figure (19) too long before you see that you can use some of the work W^* that your Super engine produces to run the Carnot refrigerator. Since your engine is more efficient than the Carnot cycle, $W^* > W$ and you have some work left over.

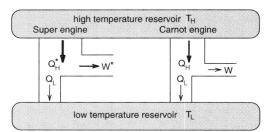

Figure 18
Comparison of the Super engine with the Carnot engine.

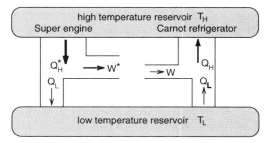

Figure 19
Now run the Carnot engine backward as a refrigerator.

The next thing you notice is that you do not need the low temperature reservoir. All the heat expelled by your Super engine is taken in by the Carnot refrigerator. The low temperature reservoir can be replaced by a pipe and the new plumbing diagram for the combined Super engine and Carnot refrigerator is shown in Figure (20). The overall result of combining the super engine and Carnot refrigerator is that a net amount of work $W_{net} = W^* - W$ is extracted from the high temperature reservoir. The net effect of this combination is to produce useful work from thermal energy at a single temperature, **which is a violation of the second law of thermodynamics**.

Exercise 7

Suppose you build a Super engine that takes the same amount of heat from the high temperature reservoir as a Carnot engine, but rejects less heat $Q_L^* < Q_L$ than a Carnot engine into the low temperature. Using energy flow diagrams show what would happen if this Super engine were connected to a Carnot refrigerator. (You would still be getting useful work from thermal energy at some temperature. From what temperature reservoir would you be getting this work?)

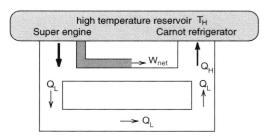

Figure 20
If you connect the Super engine to the Carnot refrigerator, you can eliminate the low temperature reservoir and still get some work W_{net} out. This machine extracts work from a single temperature, in violation of the second law of thermodynamics.

Reversibility

We have just derived the rather sweeping result that if the second law of thermodynamics is correct, you cannot construct an engine that is more efficient than a Carnot engine based on an ideal gas. You may wonder why the cycles based on an ideal gas are so special. It turns out that they are not special. What was special about the Carnot engine is that it was reversible, that it could be run backwards as a refrigerator. You can use precisely the same kind of arguments we just used to show that *all reversible engines must have precisely the same efficiency as a Carnot engine*. It is a requirement of the second law of thermodynamics.

There were two reasons we went through the detailed steps of constructing a Carnot engine using an ideal gas in a frictionless piston. The first was to provide one example of how an engine can be constructed. It is not a very practical example, commercial engines are based on different kinds of cycles. But the Carnot engine illustrates the basic features of all engines. In all engines the process must be repetitive, at least two temperatures must be involved, and only some of the heat extracted from the high temperature reservoir can be converted to useful work. Some heat must be expelled at a lower temperature.

While all reversible engines have the same efficiency, we have to work out at least one example to find out what that efficiency is. You might as well choose the simplest possible example, and the Carnot cycle using an ideal gas is about as simple as they get. Because of the second law of thermodynamics, you know that even though you are working out a very special example, the answer $Q_H/Q_L = T_H/T_L$ applies to all reversible engines operating between two temperature reservoirs. This is quite a powerful result from the few pages of calculations in the appendix.

APPLICATIONS OF THE SECOND LAW

During the oil embargo in the middle 1970s, there was a sudden appreciation of the consequences of the second law of thermodynamics, for it finally became clear that we had to use energy efficiently. Since that time there has been a growing awareness that there is a cost to producing energy that considerably exceeds what we pay for it. Burning oil and coal depletes natural limited resources and adds carbon dioxide to the atmosphere which may contribute to global climate changes. Nuclear reactors, which were so promising in the 1950s, pose unexpected safety problems, both now as in the example of Chernobyl, and in the very distant future when we try to deal with the storage of spent reactor parts. Hydroelectric power floods land that may have other important uses, and can damage the agricultural resources of an area as in the case of the Aswan Dam on the Nile River. More efficient use of energy from the sun is a promising idea, but technology has not evolved to the point where solar energy can supply much of our needs. What we have learned is that, for now, the first step is to use energy as efficiently as possible, and in doing this, the second law of thermodynamics has to be our guide.

During the 1950s and 60s, one of the buzz words for modern living was the *all electric house*. These houses were heated electrically, electric heaters being easy and inexpensive to install and convenient to use. And it also represents one of the most stupid ways possible to use energy. In terms of a heat cycle, it represents the 100% conversion of work energy into thermal energy, what we would have called in the last section, a 0% efficient engine. There are better ways of using electric power than converting it all into heat.

You can see where the waste of energy comes in when you think of the processes involved in producing electric power. In an electric power plant, the first step is to heat some liquid or gas to a high temperature by burning fuel. In a common type of coal or oil fired power plant, mercury vapor is heated to temperatures of 600 to 700 degrees centigrade. The mercury vapor is then used to run a mercury vapor turbine which cools the mercury vapor to around 200° C. This cooler mercury vapor then heats steam which goes through a steam turbine to a steam condenser at temperatures around 100° C. In a nuclear reactor, the first step is often to heat liquid sodium by having it flow through pipes that pass through the reactor. The hot sodium can then be used to heat mercury vapor which runs turbines similar to those in a coal fired plant. The turbines are attached to generators which produce the electric power.

Even though there are many stages, and dangerous and exotic materials used in power stations, we can estimate the maximum possible efficiency of a power plant simply by knowing the highest temperature T_H of the boiler, and the lowest temperature T_L of the condenser. If the power plant were a reversible cycle running between these two temperatures, it would take in an amount of heat Q_H at the high temperature and reject an amount of heat Q_L at the low temperature, where Q_H and Q_L are related by $Q_H/Q_L = T_H/T_L$ (Eq. 14). The work we got out would be

$$W = Q_H - Q_L \quad \text{amount of work from a reversible cycle} \quad (15)$$

We would naturally define the efficiency of the cycle as the ratio of the work out to the heat energy in

$$\text{efficiency} = \frac{W}{Q_H} = \frac{Q_H - Q_L}{Q_H} \quad (16)$$

If we solve Equation 14 for Q_L

$$Q_L = Q_H \frac{T_L}{T_H}$$

and use this in Equation 16, we get

$$\text{efficiency} = \frac{Q_H - Q_L}{Q_H} = \frac{Q_H(1 - T_L/T_H)}{Q_H}$$

$$\boxed{\text{efficiency} = \frac{T_H - T_L}{T_H}} \quad \text{efficiency of a reversible cycle} \quad (17)$$

Since by the second law of thermodynamics no process can be more efficient than a reversible cycle, Equation 17 represents the maximum possible efficiency of a power plant.

The important thing to remember about Equation 17 is that the temperatures T_H and T_L start from absolute zero. The only way we could get a completely efficient engine or power plant would be to have the low temperature at absolute zero, which is not only impossible to achieve but even difficult to approach. You can see from this equation why many power plants are located on the shore of an ocean or on the bank of a large river. These bodies are capable of soaking up large quantities of heat at relatively low temperatures. If an ocean or river is not available, the power plant will have large cooling towers to condense steam. Condensing steam at atmospheric pressure provides a low temperature of $T_L = 100°$ C or 373 K.

Equation 15 also tells you why power plants run their boilers as hot as possible, using exotic substances like mercury vapor or liquid sodium. Here one of the limiting factors is how high a temperature turbine blades can handle without weakening. Temperatures as high as 450° C or around 720 K are about the limit of current technology. Thus we can estimate the maximum efficiency of power plants simply by knowing how high a temperature turbine blades can withstand, and that the plant uses water for cooling. You do not have to know the details of what kind of fuel is used, what kind of exotic materials are involved, or how turbines and electric generators work, as long as they are efficient. Using the numbers $T_H = 720$ k, $T_L = 373$ k we find that the maximum efficiency is about

$$\begin{aligned}\text{maximum efficiency} &= \frac{T_H - T_L}{T_H} = \frac{720 - 373}{720} \\ &= .48\end{aligned} \quad (18)$$

Thus about 50% represents a theoretical upper limit to the efficiency of power plants using current technology. In practice, well designed power plants reach only about 33% efficiency due to small inefficiencies in the many steps involved.

You can now see why the all electric house was such a bad idea. An electric power plant consumes three times as much fuel energy as it produces electric energy. Then the electric heater in the all electric house turns this electric energy back into thermal energy. If the house had a modern oil furnace, somewhere in the order of 85% of the full energy can go into heating the house and hot water. This is far better than the 33% efficiency from heating directly by electricity.

Electric Cars

One of the hot items in the news these days is the electric car. It is often touted as the pollution free solution to our transportation problems. There are advantages to electric cars, but not as great an advantage as some new stories indicate.

When you plug in your electric car to charge batteries, you are not eliminating the pollution associated with producing useful energy. You are just moving the pollution from the car to the power plant, which may, however, be a good thing to do. A gasoline car may produce more harmful pollutants than a power station, and car pollutants tend to concentrate in places where people live creating smog in most major cities on the earth. (Some power plants also create obnoxious smog, like the coal fired plants near the Grand Canyon that are harming some of the most beautiful scenery in the country.)

In addition to moving and perhaps improving the nature of pollution, power plants have an additional advantage over car engines—they are more efficient. Car engines cannot handle as high a temperature as a power plant, and the temperature of the exhaust from a car is not as low as condensing steam or ocean water. Car engines seldom have an efficiency as high as 20%; in general, they are less than half as efficient as a power plant. Thus there will be a gain in efficiency in the use of fuel when electric cars come into more common use.

(One way electric cars have for increasing their efficiency is to replace brakes with generators. When going down a hill, instead of breaking and dissipating energy by heating the brake shoes, the gravitational potential energy being released is turned into electric energy by the generators attached to the wheels. This energy is then stored as chemical energy in the batteries as the batteries are recharged.)

The Heat Pump

There is an intelligent way to heat a house electrically, and that is by using a heat pump. The idea is to use the electric energy to pump heat from the colder outside temperature to the warmer inside temperature. Pumping heat from a cooler temperature to a warmer temperature is precisely what a refrigerator does, while taking heat from the freezer chest and exhausting it into the kitchen. The heat pump takes heat from the cooler outside and exhausts it into the house.

As we saw in our discussion, it takes work to pump heat from a cooler to a higher temperature. The ratio of the heat Q_L taken in at the low temperature, to the heat Q_H expelled at the higher temperature, is $Q_H/Q_L = T_H/T_L$ for a maximally efficient refrigerator. The amount of work W required is $W = Q_H - Q_L$. The efficiency of this process is the ratio of the amount of heat delivered to the work required.

$$\begin{aligned}\text{efficiency of} \\ \text{heat pump}\end{aligned} = \frac{Q_H}{W} = \frac{Q_H}{Q_H - Q_L} \qquad (19)$$

$$= \frac{T_H}{T_H - T_L}$$

where the last step in Equation 19 used $Q_L = Q_H T_L/T_H$.

Exercise 8

Derive the last formula in Equation 19.

When the temperature difference $T_H - T_L$ is small, we can get very high efficiencies, i.e., we can pump a lot of heat using little work. In the worst case, where $T_L = 0$ and we are trying to suck heat from absolute zero, the efficiency of a heat pump is 1—heat delivered equals the work put in—and the heat pump is acting like a resistance heater.

To illustrate the use of a heat pump let us assume that it is freezing outside ($T_L = 0°\text{C} = 273\,\text{K}$) and you want the inside temperature to be $27°\text{C} = 300\,\text{K}$. Then a heat pump could have an efficiency of

$$\text{efficiency of heat pump running from } 0°\text{C to } 27°\text{C} = \frac{T_H}{T_H - T_L}$$

$$= \frac{300\,\text{k}}{300\,\text{k} - 273\,\text{k}}$$

$$= 11.1$$

In other words, as far as the second law of thermodynamics is concerned, we should be able to pump eleven times as much heat into a house, when it is just freezing outside, as the amount of electrical energy required to pump the heat. Even if the electrical energy is produced at only 30% efficiency, we should still get $.30 \times 11.1 = 3.3$ times as much heat into the house as by burning the fuel in the house at 100% conversion of fuel energy into heat.

Exercise 9

This so-called "heat of fusion" of water is 333kJ/kg. What that means is that when a kilogram (1 liter) of water freezes (going from 0°C water to 0° ice), 333 kilojoules of heat are released. Thus to freeze a liter of 0°C water in your refrigerator, the refrigerator motor has to pump 333×10^3 joules of heat energy out of the refrigerator into the kitchen. The point of the problem is to estimate how powerful a refrigerator motor is required if you want to be able to freeze a liter of water in 10 minutes.

Assume that the heat is being removed at a temperature of 0°C and being expelled into a kitchen whose temperature is 30°C, and that the refrigerator equipment is 100% efficient. (We will account for a lack of efficiency at the end of this problem.)

In the United Sates, the power of motors is generally given in "horsepower", a familiar but archaic unit. The conversion factor is 1 horsepower = 746 watts, and a power of 1 watt is 1 joule per second.

Calculate the horsepower required, then double the answer to account for lack of efficiency. (Answer: 0.16 horsepower.)

Exercise 10

Here is a problem that should give you some practice with the concepts of efficiency. You have the choice of buying a furnace that converts heat energy of oil into heat in the house with 85% efficiency. I.e., 85% of the heat energy of the oil goes into the house, and 15% goes up the chimney. Or you can buy a heat pump which is half as efficient as a Carnot refrigerator. (This is a more realistic estimate of the current technology of refrigeration equipment.) At very low temperatures outside, heat pumps are not as efficient, and burning oil in your own furnace is more efficient. But if it does not get too cold outside, heat pumps are more efficient. At what outside temperature will the heat pump and the oil furnace have the same efficiency? Assume that the electric energy you use is produced by a power plant that is 30% efficient. (Answer: -26° C.)

The Internal Combustion Engine

We finish this section on practical applications with a brief discussion of the internal combustion engine. The main point is to give an example of an engine that runs on a cycle that is different from a Carnot cycle. It is more difficult to apply the second law of thermodynamics to an internal combustion engine because it does not take heat in or expel heat at constant temperatures like the Carnot engine, but we can still analyze the work we get out using a pV diagram.

The pV diagram for an internal combustion engine is shown in Figure (21). At position 1, a fuel and air mixture have been compressed to a small volume V_1 by the piston which is at the top of the cylinder. If it is a gasoline engine, the fuel air mixture is ignited by a spark from a sparkplug. If it is a diesel engine, the mixture of diesel fuel and air have been heated to the point of combustion by the adiabatic compression from point 4 to point 1 that has just taken place. One of the advantages of a diesel engine is that an electrical system to produce the spark is not needed. This is particularly important for boat engines where electric systems give all sorts of problems. (We said this was a section on practical applications.)

After ignition, the pressure and temperature of the gas rise rapidly to p_2, T_2 before the piston has had a chance to move. Thus the volume remains at V_1 and the pV curve goes straight up to point 2. The heated gas then expands adiabatically, and cools some, driving the piston down to the bottom of the cylinder. This is the stroke from which we get work from the engine.

We now have a cylinder full of hot burned exhaust gases. In a 4 cycle engine, a valve at the top of the cylinder is opened, and a piston is allowed to rise, pushing the hot exhaust gases out into the exhaust pipes. Not much work is required to do this. This is the part of the cycle where (relatively) low temperature thermal energy is exhausted to the environment.

While the piston goes back down, the valves are set so that a mixture of air and fuel are sucked into the piston. When the cylinder is at the bottom of the piston, we have a cool, low pressure fuel air mixture filling the full volume V_4. We are now at the position labeled (4) in Figure (21). It took two strokes (up and down) of the piston to go from position 3 to position 4.

In the final stroke, the valves are shut and the rising piston adiabatically compresses the gas back to the starting point p_1, V_1, T_1. While the increase in temperature during this compression is what is needed to ignite the diesel fuel, you do not want the temperature to rise enough to ignite the air gasoline mixture in a gasoline engine. This can sometimes happen in a gasoline engine, causing a knock in the engine, or sometimes allowing the engine to run for a while after you have shut off the ignition key and stopped the spark plug from functioning.

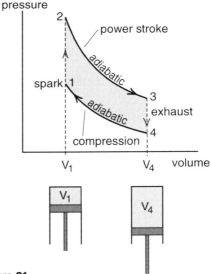

Figure 21
PV diagram for an internal combustion engine. When the piston is all the way up in the cylinder the volume is V_1. When it is all the way down, the volume has increased to V_2.

ENTROPY

The second law of thermodynamics provided us with the remarkable result that the efficiency of all reversible engines is the same. Detailed calculation of this efficiency using a Carnot engine based on an ideal gas gave us a surprisingly simple formula for this efficiency, namely $Q_H/Q_L = T_H/T_L$. Our preceding examples involving car engines, power plants, refrigerators and heat pumps illustrate how important this simple relationship is to mankind.

When you do a calculation and a lot of stuff cancels out, it suggests that your result may have a simpler interpretation than you originally expected. This turns out to be true for our calculations of the heat flow in a Carnot engine. To get a new perspective on our equation for heat flow, let us write the equation in the form

$$\frac{Q_H}{T_H} = \frac{Q_L}{T_L} \quad (20)$$

In this form the equation for heat flow is beginning to look like a conservation law for the quantity Q/T. During the isothermal expansion, an amount Q_H/T_H of this quantity flowed into the piston. During the isothermal compression, Q_L/T_L flowed out. We find that if the engine is reversible, the amount of Q/T that flowed in is equal to the amount of Q/T that flowed out. The net effect is that there was no change in Q/T during the cycle.

To get a better insight into what this quantity Q/T may be, consider a nonreversible engine operating between T_H and T_L, an engine that would be less efficient than the Carnot engine. Assume that the less efficient engine and the Carnot engine both take in the same amount of heat Q_H at the high temperature T_H. Then the less efficient engine must do less work and expel more heat at the low temperature T_L. Thus Q_L for the less efficient engine is bigger than Q_L for the Carnot engine. Since $Q_L/T_L = Q_H/T_H$ for the Carnot engine, Q_L/T_L must be greater than Q_H/T_H for the less efficient engine

$$\frac{Q_L}{T_L} > \frac{Q_H}{T_H} \quad \text{\textit{for an engine that is less efficient than a Carnot engine}} \quad (21)$$

We see that the inefficient engine expelled more Q/T than it took in. The inefficient, non reversible, engine creates Q/T while reversible engines do not.

As we have done throughout the course, whenever we encounter a quantity that is conserved, or sometimes conserved, we give it a name. We did this for linear momentum, angular momentum, and energy. Now we have a quantity Q/T that is unchanged by reversible engines, but created or increased by irreversible inefficient ones. We are on the verge of defining the quantity physicists call *entropy*. We say on the verge of defining entropy, because Q/T is not entropy itself; it represents the *change in entropy*. We can say that when the gas expanded, the entropy of the gas *increased* by Q_H/T_H. And when the gas was compressed, the entropy decreased by Q_L/T_L.

For a reversible engine there is no net change in entropy as we go around the cycle. But for an irreversible, inefficient engine, more entropy comes out than goes in during each cycle. The net effect of an inefficient engine is to create entropy.

What is this thing called entropy that is created by inefficient irreversible engine? Consider the most inefficient process we can imagine—the electric heater which converts useful work in the form of electrical energy into heat. From one point of view, the device does nothing but create entropy. If the heater is at a temperature T, and the electric power into the heater is W watts, all this energy is converted to heat and entropy is produced at a rate of W/T in units of entropy per second. (Surprisingly, there is no standard name for a unit of entropy. The units of Q/T are of course joules/kelvin, the same as Boltzman's constant.)

The process of converting energy in the form of useful work into the random thermal energy of molecules can be viewed as the process of turning order into disorder. Creating entropy seems to be related to creating disorder. But the surprising thing is that we have an explicit formula Q/T for changes in entropy. How could it be possible to measure disorder, to have an explicit formula for changes in disorder? This question baffled physicists for many generations.

Ludwig Boltzman proposed that entropy was related to the number of ways that a system could be arranged. Suppose, for example, you go into a woodworking shop and there are a lot of nails on the wall with tools hanging on them. In one particular woodworking shop you find that the carpenter has drawn an outline of the tool on the wall behind the nail. You enter her shop you find that the tool hanging from each nail exactly matches the outline behind it. Here we have perfect order, every tool is in its place and there is one and only one way the tools can be arranged. We would say that, as far as locating tools is concerned the shop is in perfect order, it has no disorder or entropy.

On closer inspection, we find that the carpenter has two saws with identical outlines, a crosscut saw and a rip saw. We also see that the nails are numbered, and see the cross cut saw on nail 23 and the rip saw on nail 24. A week later when we come back, the cross cut is on nail 24 and the rip saw on 23. Thus we find that her system is not completely orderly, for there are two different ways the saws can be placed. This way of organizing tools has some entropy.

A month later we come back to the shop and find that another carpenter has taken over and painted the walls. We find that there are still 25 nails and 25 tools, but now there is no way to tell which tool belongs on which nail. Now there are many, many ways to hang up the tools and the system is quite disordered. We have the feeling that this organization, or lack of organization, of the tools has quite high entropy.

To put a numerical value on how disorganized the carpenter shop is, we go to a mathematician who tells us that there are N! ways to hang N tools on N nails. Thus there are $25! = 1.55 \times 10^{25}$ different ways the 25 tools can be hung on the 25 nails. We could use this number as a measure of the disorder of the system, but the number is very large and increases very rapidly with the number of tools. If, for example, there were 50 tools hung on 50 nails, there would be 3.04×10^{64} different ways of hanging them. Such large numbers are not convenient to work with.

When working with large numbers, it is easier to deal with the logarithm of the number than the number itself. There are approximately 10^{51} protons in the earth. The log to the base 10 of this number is 51 and the natural logarithm, $\ln(10^{51})$, is 2.3 times bigger or 117. In discussing the number of protons in the earth, the number 117 is easier to work with than 10^{51}, particularly if you have to write out all the zeros.

If we describe the disorder of our tool hanging system in terms of the logarithm of the number of ways the tools can be hung, we get a much more reasonable set of numbers, as shown in Table 1.

Setup	Number of ways to arrange tools	Logarithm of the number of ways
all tools have unique positions	1	0
two saws can be interchanged	2	.7
25 tools on unmarked nails	1.5×10^{25}	58
50 tools on unmarked nails	3.0×10^{65}	148

TABLE 1

The table starts off well. If there is a unique arrangement of the tools, only one way to arrange them, the logarithm of the number of ways is 0. This is consistent with our idea that there is no disorder. As the number of tools on unmarked nails increases, the number of ways they can be arranged increases at an enormous pace, but the logarithm increases at a reasonable rate, approximately as fast as the number of tools and nails. This logarithm provides a reasonable measure of the disorder of the system.

We could define the entropy of the tool hanging system as the logarithm of the number of ways the tools could be hung. One problem, however, is that this definition of entropy would have different dimensions than the definition introduced earlier in our discussions of engines. There changes in entropy, for example Q_H/T_H, had dimensions of joules/kelvin, while our logarithm is dimensionless. However, this problem could be fixed by multiplying our dimensionless logarithm by some fundamental constant that has the dimensions of joules/kelvin. That constant, of course, is Boltzman's constant k, where $k = 1.38 \times 10^{-23}$ joules/kelvin. We could therefore take as the formula for the entropy (call it S) of our tool hanging system as

$$\boxed{S = k \ln(n)} \quad \begin{array}{l}\textit{entropy of our tool}\\ \textit{hanging system}\end{array} \quad (22)$$

where n is the number of ways the tools can be hung. Multiplying our logarithm by k gives us the correct dimensions, but very small values when applied to as few items as 25 or 50 tools.

Equation 22 appears on Boltzman's tombstone as a memorial to his main accomplishment in life. Boltzman believed that Equation 22 should be true in general. That, for example, it should apply to the atoms of the gas inside the cylinder of our heat engine. When heat flows into the cylinder and the entropy increases by an amount Q_H/T_H, the number of ways that the atoms could be arranged should also increase, by an amount we can easily calculate using Equation 22. Explicitly, if before the heat flowed in there were n_{old} ways the atoms could be arranged, and after the heat flowed in n_{new} ways, then Equation 22 gives

$$\begin{array}{l}\text{change in}\\ \text{entropy}\end{array} = k \ln(n_{new}) - k \ln(n_{old}) = \frac{Q_H}{T_H}$$

Since $\ln(n_{new}) - \ln(n_{old}) = \ln(n_{new}/n_{old})$, we get

$$\ln\left(\frac{n_{new}}{n_{old}}\right) = \frac{Q_H}{kT_H} \quad (23)$$

Taking the exponent of both sides of Equation 16, using the fact that $e^{\ln(x)} = x$, we get

$$\frac{n_{new}}{n_{old}} = e^{Q_H/kT_H} \quad (24)$$

Thus Boltzman's equation gives us an explicit formula for the fractional increase in the number of ways the atoms in the gas atoms in the cylinder can be arranged.

Boltzman committed suicide in 1906, despondent over the lack of acceptance of his work on the statistical theory of matter, of which Equation 22 is the cornerstone. And in 1906 it is not too surprising that physicists would have difficulty dealing with Boltzman's equation. What is the meaning of the number of ways you can arrange gas atoms in a cylinder? From a Newtonian perspective, there are an infinite number of ways to place just one atom in a cylinder. You can count them by moving the atoms an infinitesimal distance in any direction. So how could it be that 10^{24} atoms in a cylinder have only a finite way in which they can be arranged?

This question could not be satisfactorily answered in 1906, the answer did not come until 1925 with the discovery of quantum mechanics. In a quantum picture, an atom in a cylinder has only certain energy levels, an idea we will discuss later in Chapter 35. Even when you have 10^{24} atoms in the cylinder, the whole system has only certain allowed energy levels. At low temperatures the gas does not have enough thermal energy to occupy very many of the levels. As a result the number of ways the atoms can be arranged is limited and the entropy is low. As the temperature is increased, the gas atoms can occupy more levels, can be arranged in a greater number of ways, and therefore have a greater entropy.

The concept of entropy provides a new definition of absolute zero. A system of particles is at absolute zero when it has zero entropy, when it has one uniquely defined state. We mentioned earlier that quantum mechanics requires that a confined particle has some kinetic energy. All the kinetic energy cannot be removed by cooling. This gives rise to the so-called *zero point energy* that keeps helium a liquid even at absolute zero. However, a bucket of liquid helium can be at absolute zero as long as it is in a single unique quantum state, even though the atoms have zero point kinetic energy.

In our discussion of temperature in the last chapter, we used the ideal gas thermometer for our experimental definition of temperature. We pointed out, however, that this definition would begin to fail as we approached very low temperatures near absolute zero. At these temperatures we need a new definition which agrees with the ideal gas thermometer definition at higher temperatures. The new definition which is used by the physics community, is based on the efficiency of a reversible engine or heat cycle. You can measure the ratio of two temperatures T_H and T_C, by measuring the heats Q_H and Q_C that enter and leave the cycle, and use the formula $T_H/T_C = Q_H/Q_C$. Since this formula is based on the idea that a reversible cycle creates no entropy ($Q_H/T_H = Q_C/T_C$), we can see that the concept of entropy forms the basis for the definition of temperature.

The Direction of Time

We began the chapter with a discussion of a demonstration that looked funny. We set the strobe so that the water drops appeared to rise from the plate in the bucket and enter the hose. Before our discussion of the second law of thermodynamics, we could not find any law of physics that this backward process appeared to violate. Now we can see that the launching of the drops from the plate is a direct contradiction of the second law. In that process, heat energy in the bucket converts itself at one temperature into pure useful work that launches the drop.

When you run a moving picture of some action backwards, effectively reversing the direction of time, in most cases the only law of physics that is violated is the second law of thermodynamics. The only thing that appears to go wrong is that disordered systems appear to organize themselves on their own. Scrambled eggs turn into an egg with a whole yoke just by the flick of a fork. Divers pop out of swimming pools propelled, like the drops in our demonstration, by the heat energy in the pool (see movie). All these funny looking things require remarkable coincidences which in real life do not happen.

For a while there was a debate among physicists as to whether the second law of thermodynamics was the only law in nature that could be used to distinguish between time running forward and time running backward.

When you study processes like the decay of one kind of elementary particle into another, the situation is so simple that the concepts of entropy and the second law of thermodynamics do not enter into the analysis. In these cases you can truly study whether nature is symmetric with the respect to the reversal of time. If you take a moving picture of a particle decay, and run the movie backwards, will you see a process that can actually happen? For example if a muon decays into an electron and a neutrino, as happened in our muon lifetime experiment, running that moving picture backward would have neutrinos coming in, colliding with an electron, creating a muon. Thus, if the basic laws of physics are truly symmetric to the reversal of time, it should be possible for a neutrino and an electron to collide and create a muon. This process is observed.

In 1964 Val Fitch and James Cronin discovered an elementary particle process which indicated that nature was not symmetric in time. Fitch found a violation of this symmetry in the decay of a particle called the neutral k meson. For this discovery, Fitch was awarded the Nobel prize in 1980. Since the so-called "weak interaction" is responsible for the decay of k mesons, the weak interaction is not fully symmetric to the reversal of time. The second law of thermodynamics is not the only law of physics that knows which way time goes.

Movie
Time reversed motion picture of dive

APPENDIX: CALCULATION OF THE EFFICIENCY OF A CARNOT CYCLE

The second law of thermodynamics tells us that the efficiency of all reversible heat engines is the same. Thus if we can calculate the efficiency of any one engine, we have the results for all. Since we have based so much of our discussion on the Carnot engine running on an ideal gas, we will calculate the efficiency of that engine.

To calculate the efficiency of the ideal gas Carnot engine, we need to calculate the amount of work we get out of (or put into) isothermal and adiabatic expansions. With these results, we can then calculate the net amount of work we get out of one cycle and then the efficiency of the engine. To simplify the formulas, we will assume that our engine is running on one mole of an ideal gas.

Isothermal Expansion

Suppose we have a gas at an initial volume V_1, pressure p_1, temperature T, and expand it isothermally to a volume V_2, pressure p_2, and of course the same temperature T.

The P-V diagram for the process is shown in Figure (A-1). The curve is determined by the ideal gas law, which for 1 mole of an ideal gas is

$$pV = RT \qquad (A-1)$$

The work we get out of the expansion is the shaded area under the curve, which is the integral of the pressure curve from V_1 to V_2.

Using Equation 1, we get

$$\begin{aligned} W &= \int_{V_1}^{V_2} p\,dV = \int_{V_1}^{V_2} \frac{RT}{V} dV \\ &= RT \int_{V_1}^{V_2} \frac{dV}{V} = RT \ln V \Big|_{V_1}^{V_2} \\ &= RT \Big[\ln(V_2) - \ln(V_1) \Big] \qquad (A-2) \\ &= RT \ln\left(\frac{V_2}{V_1}\right) \end{aligned}$$

Thus the work we get out is RT times the logarithm of the ratio of the volumes.

Adiabatic Expansion

It is a bit trickier to calculate the amount of work we get out of an adiabatic expansion. If we start with a mole of ideal gas at a volume V_1, pressure p_1, and temperature T_H, the gas will cool as it expands because the gas does work and we are not letting any heat in. Thus when the gas gets to the volume V_2, at a pressure p_2, its temperature T_C will be cooler than its initial temperature T_H.

The pV diagram for the adiabatic expansion is shown in Figure (A-2). To get an equation for the adiabatic expansion curve shown, let us assume that we change the volume of the gas by an infinitesimal amount ΔV. With this volume change, there will be an infinitesimal pressure drop Δp, and an infinitesimal temperature

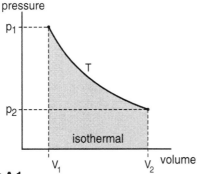
Figure A-1
Isothermal expansion.

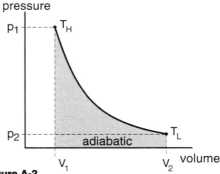
Figure A-2
Adiabatic expansion.

drop ΔT. We can find the relationship between these small changes by differentiating the ideal gas equations. Starting with

$$pv = RT$$

and differentiating we get

$$p\Delta V + (\Delta p)V = R\Delta T \quad \text{(A-3)}$$

We now have to introduce the idea that the expansion is taking place adiabatically, i.e., that no heat is entering. That means that the work $p\Delta V$ done by the gas during the infinitesimal expansion ΔV must all have come from thermal energy. But the decrease in thermal energy is $C_V \Delta T$. Thus we have from conservation of energy

$$p\Delta V + C_V \Delta T = 0$$

or

$$\Delta T = \frac{-p\Delta V}{C_V} \quad \text{(A-4)}$$

The – (minus) sign tells us that the temperature drops as work energy is removed.

Using Equation A-4 for ΔT in Equation A-3 gives

$$p\Delta V + \Delta pV = R\left(\frac{-p\Delta V}{C_V}\right)$$

Combining the $p\Delta V$ terms gives

$$p\Delta V\left(1 + \frac{R}{C_V}\right) + \Delta pV = 0$$

$$p\Delta V\left(\frac{C_V + R}{C_V}\right) + \Delta pV = 0 \quad \text{(A-5)}$$

Earlier in the chapter, in Equation 9, we found that for an ideal gas, C_V and C_p were related by $C_p = C_V + R$. Thus Equation A-5 simplifies to

$$p\Delta V\left(\frac{C_p}{C_V}\right) + \Delta pV = 0 \quad \text{(A-6)}$$

It is standard notation to define the ratio of specific heats by the constant γ

$$\frac{C_p}{C_V} \equiv \gamma \quad \text{(A-7)}$$

thus Equation A-6 can be written in the more compact form

$$\gamma p\Delta V + \Delta pV = 0 \quad \text{(A-8)}$$

The next few steps will look like they were extracted from a calculus text. They may or may not be too familiar, but you should be able to follow them step-by-step.

First we will replace ΔV and Δp by dV and dp to indicate that we are working with calculus differentials. Then dividing through by the product pV gives

$$\gamma \frac{dV}{V} + \frac{dp}{p} = 0 \quad \text{(A-9)}$$

Doing an indefinite integration of this equation gives

$$\gamma \ln(V) + \ln(p) = \text{const} \quad \text{(A-10)}$$

The γ can be taken inside the logarithm to give

$$\ln(V^\gamma) + \ln(p) = \text{const} \quad \text{(A-11)}$$

Next exponentiate both sides of Equation A-11 to get

$$e^{\ln(V^\gamma) + \ln(p)} = e^{\text{const}} = \text{another const} \quad \text{(A-12)}$$

where e^{const} is itself a constant. Now use the fact that $e^{a+b} = e^a e^b$ to get

$$e^{\ln(V^\gamma)} e^{\ln(p)} = \text{const} \quad \text{(A-13)}$$

Finally use $e^{\ln(x)} = x$ to get the final result

$$\boxed{pV^\gamma = \text{const}} \quad \begin{array}{l}\textit{adiabatic}\\ \textit{expansion}\end{array} \quad \text{(A-14)}$$

Equation A-14 is the formula for the adiabatic curve seen in Figure (A-2). During an isothermal expansion, we have pV= RT where T is a constant. Thus if we compare the formulas for isothermal and adiabatic expansions, we have for any ideal gas

$$\begin{aligned} pV &= \text{const} & \textit{isothermal expansion} \\ pV^\gamma &= \text{const} & \textit{adiabatic expansion} \\ \gamma &= C_p/C_V & \textit{ratio of specific heats} \\ C_p &= C_V + R \end{aligned} \quad \text{(A-15)}$$

The Carnot Cycle

We now have the pieces in place to calculate the efficiency of a Carnot cycle running on one mole of an ideal gas. The cycle is shown in Figure (11) repeated here as Figure (A-3).

During the isothermal expansion from point 1 to 2, the amount of heat that flows into our mole of gas is equal to the work one by the gas. By Equation A-2, this work is

$$W_{12} = Q_H = RT_H \ln(V_2/V_1) \quad \text{(A-16)}$$

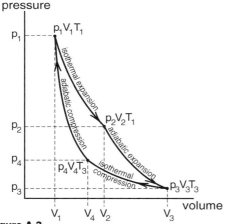

Figure A-3
The Carnot cycle.

The heat Q_L expelled at the low temperature T_L is equal to the work we do compressing the gas isothermally in going from point 3 to 4. This work is

$$W_{34} = Q_L = RT_C \ln(V_4/V_3) \quad \text{(A-17)}$$

Taking the ratio of Equations A-16 to A-17 we get

$$\frac{Q_H}{Q_L} = \frac{T_H \ln(V_2/V_1)}{T_C \ln(V_4/V_3)} \quad \text{(A-18)}$$

The next step is to calculate the ratio of the logarithms of the volumes using the adiabatic expansion formula $pV^\gamma = \text{constant}$.

In going adiabatically from 2 to 3 we have

$$p_2 V_2^\gamma = p_3 V_3^\gamma \quad \text{(A-19)}$$

and in going from 4 to 1 adiabatically we have

$$p_4 V_4^\gamma = p_1 V_1^\gamma \quad \text{(A-20)}$$

Finally, use the ideal gas law pV = RT to express the pressure p in terms of volume and temperature in Equations A-19 and A-20. Explicitly use

$$\begin{aligned} p_1 &= RT_H/V_1; & p_2 &= RT_H/V_2 \\ p_3 &= RT_H/V_3; & p_4 &= RT_H/V_4 \end{aligned} \quad \text{(A-21)}$$

to get for Equation (19)

$$\frac{RT_H}{V_2} V_2^\gamma = \frac{RT_C}{V_3} V_3^\gamma$$

or

$$T_H V_2^{\gamma-1} = T_C V_3^{\gamma-1} \quad \text{(A-22)}$$

and similarly for Equation (20)

$$T_H V_1^{\gamma-1} = T_C V_4^{\gamma-1} \quad \text{(A-23)}$$

as you can check for yourself.

If we divide Equation A-22 by A-23 the temperatures T_H and T_C cancel, and we get

$$\frac{V_2^{\gamma-1}}{V_1^{\gamma-1}} = \frac{V_3^{\gamma-1}}{V_4^{\gamma-1}}$$

or

$$\left(\frac{V_2}{V_1}\right)^{\gamma-1} = \left(\frac{V_3}{V_4}\right)^{\gamma-1} \qquad \text{(A-24)}$$

Taking the $(\gamma-1)$th root of both sides of Equation A-24 gives simply

$$\left(\frac{V_2}{V_1}\right) = \left(\frac{V_3}{V_4}\right) \qquad \text{(A-25)}$$

Since $V_2/V_1 = V_3/V_4$, the logarithms in Equation A-18 cancel, and we are left with the surprisingly simple result

$$\frac{Q_H}{Q_L} = \frac{T_H}{T_L} \qquad \text{(14 repeated)}$$

which is our Equation 14 for the efficiency of a Carnot cycle. As we mentioned, when you are doing a calculation and a lot of stuff cancels to give a simple result, there is a chance that your result is more general, or has more significance than you expected. In this case, Equation 14 is the formula for the efficiency of any reversible engine, no matter how it is constructed. We happened to get at this formula by calculating the efficiency of a Carnot engine running on one mole of an ideal gas.

Chapter 19
The Electric Interaction Atomic & Molecular Forces

THE FOUR BASIC INTERACTIONS

The world around us is a complex place with enormous variety of a myriad of interactions. But if you look in the right places, from the right point of view, you may find great simplicity. Planetary motion is one example. If you look at the sun and planets alone, ignoring things on a larger scale like other stars and galaxies, and on a smaller scale like the makeup of the planets and the atmosphere of the sun, you have a system of 10 objects whose behavior is accurately determined by a single force law. The system is simple enough that mankind learned about physics by studying it.

*In this and the next chapter we take a first look at several of the basic patterns and laws of physics. Perhaps the most important discovery in the twentieth century, actually more of a gradual realization made during the first half of the twentieth century, was that all of the phenomena of nature, everything we see around us, can be explained in terms of **four basic forces or interactions**. In some circumstances, as in the case of planetary motion, a single force dominates and the structures it creates are obvious. In most cases however, the structures are complex and it is no easy task to uncover the underlying forces.*

*What we will do in these two chapters is to describe the four basic interactions, focusing our attention on examples where the action of the force is most clearly seen. For the **gravitational force**, a planetary system* *provides the most clear and detailed example of a structure created by gravity. Using this example as a guide we can see that larger structures like globular clusters and galaxies are an extension of the gravitational interaction to more complex situations. Although we do not attempt to calculate in detail the motion of the millions or billions of stars in such objects, we gain an intuitive feeling for the kind of structures the gravitational interaction creates.*

The **electric interaction***, the second of the four basic interactions to be discovered, is most clearly seen at an atomic level. To introduce the electric interaction, we will take the simple point of view that atoms consist of a tiny nucleus made from protons and neutrons, surrounded by electrons held to the nucleus by the electric force. The model should be familiar, it is essentially a scaled down version of the solar system, with an electric force replacing the gravitational force. After a brief discussion of the kinds of nuclei that can be made from protons and neutrons, we will look at the properties of the electric interaction that give rise to complete atoms. This involves the concept of electric charge, the fact that electricity, like gravity, is a $1/r^2$ force law, and the fact that, on an atomic scale, the electric force is much, much stronger than gravity.*

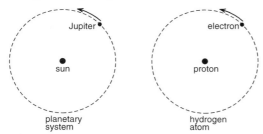

Figure 0
The classical model of the hydrogen atom closely resembles a planetary system

When you bring complete atoms close to each other, there are weak residual electric forces which result from small distortions of the atomic structure. These weak residual forces are the **molecular forces** *that hold atoms together to form molecules, crystals and most of the variety we see in the world about us. In this chapter we will consider only the simplest example of a molecular force to see how such residual forces arise when atoms interact.*

In the next chapter we leave the scale of atoms and molecules and look down inside the atomic nucleus, where we find two more forces at work. There is an attractive force, called the **nuclear interaction***, that is even stronger than electricity. It holds the nucleus together against the electric repulsion between the proton in the nucleus. There is also another force, called the* **weak interaction***, which allows neutrons to decay into protons and protons into neutrons (the β decay reaction discussed in Chapter 6). The nature of nuclear reactions, the stability of atomic nuclei, and the abundance of the elements depend upon a delicate interplay of the nuclear, the weak, and the electric interactions.*

In the past 30 years we have been able to sharpen our view of nuclear and subnuclear matter. We have been able to look inside the proton and neutron and discover that they are not elementary particles, but instead, composite objects made from **quarks***. We have also found that the basic nuclear force is the one between quarks. The force that holds protons together in the nucleus is a residual of the quark force, just as molecular forces that hold atoms together are a residual of the electric force. Because of a mathematical analogy to the theory of color, the force between quarks is called the* **color force***. The color force now replaces the residual nuclear force as one of the four basic interactions.*

Once we see how nature can be explained in terms of just four basic forces, we cannot help wondering why the number is four. Are there more basic forces, some of them yet undetected? Or are there fewer than four basic forces, some of the four being equivalent on a more fundamental level?

Einstein spent the latter half of his life trying to find a unified way of viewing the gravitational and electric interactions. He tried to find a single more fundamental theory from which both electricity and gravity emerged. He did not succeed, and no one else has yet done so either. However from the knowledge gained by looking inside the proton and neutron, the knowledge that led to the discovery of quarks, Steven Weinberg and Sheldon Glashow were able to construct a theory that unified the electric and weak interactions. These two forces which appeared so different in their effects on matter, turn out to be two components of a more fundamental interaction. Thus we are now down to three basic interactions. Why three? Can these be unified? We do not know yet.

We end this discussion with another look at the gravitational interaction. On an atomic scale, gravity is so weak compared to electricity that only recently has it been experimentally determined that electrons fall down rather than up in the earth's gravitational field. The only reason we human beings personally know about gravity is the fact that we are standing on a huge chunk of matter, the earth. It takes a lot of matter to create a big enough gravitational force for us to notice.

But there is a lot of matter in the universe. Sometimes, as in the case of a neutron star, so much matter is packed in such a small space that the gravitational force becomes stronger than the electric force. In a neutron star, gravity has forced the electrons back into the nucleus, to form pure nuclear matter. If the neutron star gets too big, if gravity gets a bit stronger, it can overwhelm the nuclear force and crush the star to form a black hole. Gravity, a force so weak that we could barely detect it using the Cavandish experiment, can become the strongest of all forces.

ATOMIC STRUCTURE

To set the stage for our discussion of the electric interaction, we will first construct a brief overview of atomic and nuclear structure. The components of atoms and nuclei that are of interest are the proton and neutron found in the nucleus, and the electron which orbits outside. Protons and neutrons are each about 1836 times as massive as the electron, thus most of the mass of an atom is located in the nucleus.

The simplest of all atoms is hydrogen with one proton for a nucleus and one electron outside. The electron is attracted to the proton by a $1/r^2$ electric force, just as the earth is attracted to the sun by the $1/r^2$ gravitational force. And because the proton is much more massive than the electron, the proton sits nearly at rest at the center of the atom while the electron orbits outside, much as the earth orbits the sun.

Also like the solar system, the atom is mostly empty space. A proton has a diameter of about 10^{-13} cm, while a hydrogen atom is one hundred thousand times bigger. If the hydrogen atom were enlarged to the point where the proton nucleus were the size of the sun, the electron would be orbiting out at a distance over 10 times the radius of the pluto's orbit. In this sense there is more empty space in an atom than in the solar system.

In Newton's law of gravity, the gravitational force on an object is proportional to an object's mass. The fact that all gravitational forces are attractive can be viewed as a consequence of the fact that there is only positive mass. In the electric interaction, there are both attractive and repulsive electric forces. We will see that for the electric interaction, the concept of electric charge plays a role similar to that of mass for the gravitational interaction. The existence of both attractive and repulsive electric forces leads to having both positive and negative charge.

One new feature of having both attractive and repulsive electric forces is that the net electric force between two objects can be zero, due to the cancellation of attractive and repulsive components. This cancellation of electric force can be represented by a cancellation of electric charge, giving us an object which we say is *electrically neutral*. Because there are no repulsive gravitational forces and no negative mass, there is no such thing as a gravitationally neutral object.

When an atom has the same number of electrons in orbit as the number of protons in its nucleus, the atom is electrically neutral. If you have two electrically neutral atoms separated by a reasonable distance, like the atoms in a gas, the electric forces between the atoms cancel. Thus neutral atoms in a gas move by each other with almost no interaction. There is an interaction only when the atoms get too close in a collision and the electric forces no longer cancel.

Atoms are classified by the number of protons in the nucleus. If the nucleus has one proton, the atom belongs to the element hydrogen. If there are 2 protons in the nucleus, it is a helium atom. Three protons gives us lithium, on up through the periodic table. The largest naturally occurring atom, on the earth at least, is uranium with 92 protons. Atoms with over 100 protons in the nucleus have been created artificially.

The periodic table, and the classification of the elements, were developed by chemists studying the chemical properties of matter. However chemical reactions, with the possible exception of cold fusion, have virtually no effect on the atomic nucleus. If you have a lead nucleus with its 82 protons, there is no set of chemical reactions that can change it to a gold nucleus with 79 protons. This is what doomed the alchemists of the middle ages to failure.

The chemistry of an atom depends upon the behavior of the electrons in an atom, and the electron behavior depends significantly on the number of electrons. Since an electrically neutral atom has the same number of electrons as protons, different elements with different numbers of protons have different numbers of electrons and thus different chemical properties. For example, hydrogen with one electron is an excellent fuel, helium with 2 electrons is chemically inert, and lithium with 3 electrons is a highly reactive alkali metal.

The periodic table is not merely a list of atoms according to the number of protons in the nucleus. The table exhibits many striking patterns or regularities in the chemical behavior of the elements. For example, helium with 2 electrons, neon with 10, argon with 18, krypton with 36, xenon with 54 and radon with 86 electrons are all chemically inert gases. These so-called noble gases enter into few if any chemical reactions. Now add one electron to each of these atoms (and one proton to the nucleus), and you get lithium (3 electrons), sodium (11 electrons), potassium (19 electrons), etc., all reactive alkali metals.

The patterns in the chemical properties of the elements exhibited by the periodic table are a consequence of the electric interaction, but they cannot be explained using Newtonian mechanics. Scientists had to wait until the discovery of quantum mechanics before a detailed explanation of the periodic table unfolded. After we have discussed some of the basic ideas of quantum mechanics in later chapters, we will see that there are fairly simple explanations for the main features of the periodic table, like the difference between noble gasses and alkali metals mentioned above. For now, however, where we have only developed a background of Newtonian mechanics, we will go no further than treating the periodic table as a list of the elements according to the number of protons in the nucleus.

Element	Chemical symbol	No. of protons	No. of neutrons	Element	Chemical symbol	No. of protons	No. of neutrons
Hydrogen	H	1	0	Iodine	I	53	74
Helium	He	2	2	Xenon	Xe	54	78
Lithium	Li	3	4	Cesium	Cs	55	78
Beryllium	Be	4	5	Barium	Ba	56	82
Boron	B	5	6	Lanthanum	La	57	82
Carbon	C	6	6	Cerium	Ce	58	82
Nitrogen	N	7	7	Praseodymium	Pr	59	82
Oxygen	O	8	8	Neodymium	Nd	60	82
Fluorine	F	9	10	Promethium	Pm	61	86
Neon	Ne	10	10	Samarium	Sm	62	90
Sodium	Na	11	12	Europium	Eu	63	90
Magnesium	Mg	12	12	Gadolinium	Gd	64	94
Aluminum	Al	13	14	Terbium	Tb	65	94
Silicon	Si	14	14	Dysprosium	Dy	66	98
Phosphorus	P	15	16	Holmium	Ho	67	98
Sulfur	S	16	16	Erbium	Er	68	98
Chlorine	Cl	17	18	Thulium	Tm	69	100
Argon	A	18	22	Ytterbium	Yb	70	104
Potassium	K	19	20	Lutetium	Lu	71	104
Calcium	Ca	20	20	Hafnium	Hf	72	108
Scandium	Sc	21	24	Tantalum	Ta	73	108
Titanium	Ti	22	26	Tungsten	W	74	110
Vanadium	V	23	28	Rhenium	Re	75	112
Chromium	Cr	24	28	Osmium	Os	76	116
Manganese	Mn	25	30	Iridium	Ir	77	116
Iron	Fe	26	30	Platinum	Pt	78	117
Cobalt	Co	27	32	Gold	Au	79	122
Nickel	Ni	28	30	Mercury	Hg	80	122
Copper	Cu	29	34	Thallium	Tl	81	124
Zinc	Zn	30	34	Lead	Pb	82	126
Gallium	Ga	31	38	Bismuth	Bi	83	126
Germanium	Ge	32	42	Polonium	Po	84	124 (3 yr)
Arsenic	As	33	42	Astatine	At	85	125 (8 hr)
Selenium	Se	34	46	Radon	Rn	86	136 (3 days)
Bromide	Br	35	44	Francium	Fr	87	136 (21 min)
Krypton	Kr	36	48	Radium	Ra	88	138 (1622 yr)
Rubidium	Rb	37	48	Actinium	Ac	89	138 (22 hr)
Strontium	Sr	38	50	Thorium	Th	90	140 (80,000 yr)
Yttrium	Y	39	50	Protactinium	Pa	91	140 (34,000 yr)
Zirconium	Zr	40	50	Uranium	U	92	146 (4.5 billion yr)
Niobium	Nb	41	52	Neptunium	Np	93	144 (2.2 million yr)
Molybdenum	Mo	42	56	Plutonium	Pu	94	145 (24,000 yr)
Technetium	Tc	43	54 (> 100 yr)	Americium	Am	95	144 (490 yr)
Ruthenium	Ru	44	58	Curium	Cm	96	146 (150 day)
Rhodium	Rh	45	58	Berkelium	Bk	97	150 (1000 yr)
Palladium	Pd	46	60	Californium	Cf	98	153 (800 yr)
Silver	Ag	47	60	Einsteinium	Es	99	155 (480 days)
Cadmium	Cd	48	66	Fermium	Fm	100	153 (23 hr)
Indium	In	49	66	Mendelevium	Md	101	155 (1.5 hr)
Tin	Sn	50	70	Nobelium	No	102	152 (3 sec)
Antimony	Sb	51	70	Lawrencium	Lw	103	154 (8 sec)
Tellurium	Te	52	78				

Table 1
The most commonly found (most abundant in nature) isotope of each element is listed. In cases where an element has no stable isotopes, the isotope with the longest lifetime is listed.

Isotopes

The atomic nucleus contains not only the protons we have been discussing, but also neutrons. The nuclear force, unlike the electric force, is the same between protons and neutrons, and ignores electrons. The nuclear force between *nucleons* (protons or neutrons), is attractive if the nucleons are close but not too close, and repulsive if you try to shove nucleons into each other. As a result, due to the nuclear force, protons and neutrons in a nucleus stick to each other forming a ball of nuclear matter as indicated in Figure (1). The nuclear force is strong enough to hold the nucleus together despite the electrical repulsion between the protons. In the nucleus of a given element, there is no precisely fixed number of neutrons. The hydrogen nucleus which has one proton, can have zero, one, or two neutrons. Helium nuclei, which have 2 protons usually also have 2 neutrons, but can be found with only one neutron. Generally the light elements have roughly equal numbers of protons and neutrons while the heavy elements like uranium have a considerable excess of neutrons.

Atoms of the same element with different numbers of neutrons in the nucleus are called different **isotopes** of the element. We distinguish different isotopes of an element by appending to the name of the element a number equal to the total number of protons and neutrons in the nucleus. The hydrogen atom with just a single proton for a nucleus is hydrogen-1. If there is one neutron in addition to the proton, the atom is called hydrogen-2, and with 2 neutrons and one proton, we have hydrogen-3. These isotopes and the isotopes of helium are indicated schematically in Figure (2).

The naturally occurring isotope of uranium, an unstable element, but one with such a long half life (4.5 billion years) that it has survived since the formation of the earth, is uranium-238 or U-238. Since uranium has 92 protons, the number of neutrons in U-238 must be 238 - 92 = 144. Another uranium isotope, U-235 with 3 fewer neutrons, is a more highly radioactive material from which atomic bombs can be constructed. (U-238 is so long-lived, so stable that it is quite safe to handle. There was some discussion of using U-238 for the keels of America's

Figure 1a
Picture the nucleus as a spherical ball of protons and neutrons.

Figure 1b
Model of the uranium nucleus constructed from styrofoam balls. The dark balls represent protons.

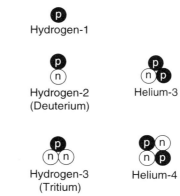

Figure 2
Isotopes of hydrogen and helium.

cup yachts because of its very high density, but this was disallowed as being too high tech.) A list of the most common or longest lived isotopes for each element is given in Table 1.

For historical reasons, the isotopes of hydrogen are given special names. Hydrogen-2, with one proton and one neutron, is known as *deuterium*. Just over one in ten thousand hydrogen atoms in naturally occurring hydrogen are the deuterium isotope. Water molecules (H_2O), in which one of the hydrogen atoms is the deuterium isotope, are called *heavy water*. Heavy water played an important role in the unsuccessful German effort to build a nuclear bomb during World War II. Hydrogen–3, with one proton and 2 neutrons, is called *tritium*. Tritium is unstable with a half life of 12.5 years. Along with deuterium, tritium plays an important role in mankind's attempt to build a nuclear fusion reactor.

Why are some isotopes stable while others are not? Why are there roughly equal numbers of neutrons as protons in the small stable isotopes and an excess of neutrons in the large ones? Why are some elements more abundant than others—for example, why does the earth have an iron core? These are questions whose answers depend upon an interplay inside the nucleus between the nuclear, the electric, and the weak interaction. We reserve a discussion of these questions for the next chapter where we discuss the nuclear and weak interactions in more detail.

THE ELECTRIC FORCE LAW

Since electrons, protons, and neutrons make up almost everything we see around us (except for photons or light itself), a description of the electric force between these three particles provides a fairly complete picture of the electric interaction, insofar as it affects our lives. For electrons, protons, and neutrons at rest, this interaction is completely summarized in Figure (3).

As we see, protons repel each other, electrons repel each other, and a proton and an electron attract each other. There is no electric force on a neutron. The strength of the electric force between these particles drops off as $1/r^2$ and has a magnitude shown in Equation 1. We know, to extremely high precision, that the attractive force between an electron and proton has the same strength as the repulsive force between two protons or two electrons, when the particles have the same separation r.

It is surprising how complete a summary of the electric interaction Figure (3) represents. We have only shown the forces between the particles at rest. But if you combine these results with the special theory of relativity, you can deduce the existence of magnetism and derive the formulas for magnetic forces. We will do this in Chapter 28.

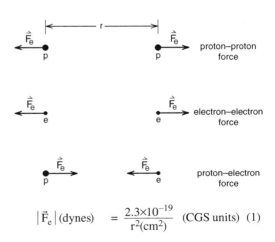

Figure 3
The electric interaction between protons and electrons at rest. There is no electric force on a neutron.

Strength of the Electric Interaction

If two electrons or two protons are separated by a distance of one centimeter, then according to Equation 1 there is a repulsive force between them—whose strength is 2.3×10^{-19} dynes. Since a dyne is the weight of one milligram of mass, 2.3×10^{-19} dynes is a very small force. But, of course, protons and electrons are very small particles.

To get a better idea of how strong the electric force is, let us compare it with the gravitational force. If we have 2 protons any distance r apart, then the ratio of the electric repulsion to the gravitational attraction is

$$\frac{\text{electric repulsion}}{\text{gravitational attraction}} = \frac{|\vec{F}_e|}{|\vec{F}_g|} \quad (2)$$

$$= \frac{2.3 \ast 10^{-19}/r^2}{Gm_p m_p / r^2}$$

Since both electricity and gravity are $1/r^2$ forces, the r^2 cancel out in Equation 2 and we are left with

$$\frac{|\vec{F}_e|}{|\vec{F}_g|} = \frac{2.3 \times 10^{-19}}{Gm_p^2}$$

$$= \frac{2.3 \times 10^{-19}}{6.67 \times 10^{-8} \times (1.67 \times 10^{-24})^2}$$

$$= 1240000000000000000000000000000000000 \quad (3)$$

The electrical force is some 10^{36} times stronger than gravity. This is true no matter how far apart the protons are. The only reason that electric forces do not completely swamp gravitational forces is that there are both attractive and repulsive electric forces which tend to cancel on a large scale, when many electrons and protons are involved.

ELECTRIC CHARGE

From a historical perspective, the electric interaction was carefully studied and the electric force law well known long before the discovery of electrons and nuclei, even before there was much evidence for the existence of atoms. The simple summary of the electric force law given by Equation 1 could only be written after the 1930s, when we finally began to understand what was going on inside an atom. Prior to that, the electric force law was expressed in terms of electric charge, a concept invented by Benjamin Franklin. What we want to do in this section is to show how the concept of electric charge evolves from the forces pictured in Figure (3), and why electric charge is such a useful concept.

To convert Equation 1 into the more standard form of the electric force law, we will begin by writing the numerical constant 2.3×10^{-19} dyne cm^2, or 2.3×10^{-20} newton meter^2, in the form Ke^2 to give

$$|\vec{F}_e|_{\text{electron}} = \frac{Ke^2}{r^2} \quad \begin{array}{l} \textit{electric force} \\ \textit{between two} \\ \textit{electrons} \end{array} \quad (4)$$

where e is called the ***charge on an electron*** and k is a numerical constant whose size depends on the system of units we are using.

The form of Equation 4 is chosen to make the electric force law look like the gravitational force law. To see this explicitly, compare the formulas for the magnitude of the electric and the gravitational forces between two electrons

$$|\vec{F}_{\text{gravitational}}| = \frac{Gm_e m_e}{r^2}$$

$$|\vec{F}_{\text{electric}}| = \frac{Kee}{r^2} \quad (5)$$

In words, we said that the gravitational force between two electrons was proportional to the product of the masses m_e, and inversely proportional to the square of the separation $1/r^2$. Now we say that the electric force is proportional to the product of the charges (e), and inversely proportional to the square of the distance $1/r^2$. By introducing the constant (***e***) as the ***charge on the electron***, we have electric charge playing nearly the same role for the electric force law as mass does for the gravitational force law.

To get the numerical value for the charge (e) on an electron, we note that in the CGS system of units it is traditional to set the proportionality constant (K) equal to one, giving

$$\left|\vec{F}_e\right|(\text{CGS}) = \frac{e^2}{r^2} \quad \substack{\text{electric force}\\ \text{law in CGS}\\ \text{units } (K=1)} \tag{6}$$

Comparing Equations 1 and 6 we get

$$e^2 = 2.3 \times 10^{-19} \text{ dynes cm}^2$$

$$e = 4.8 \times 10^{-10} \sqrt{\text{dynes cm}^2} \quad \substack{\text{charge on}\\ \text{electron in}\\ \text{CGS units}} \tag{7}$$

In the MKS system, the proportionality constant K is not 1, therefore Equation 6 does not apply to that system. Calculations involving electric forces on an atomic scale are simpler in the CGS system because of the choice $K = 1$. The MKS system turns out, however, to be far more convenient when working with practical or engineering applications of electrical theory. As a result we will use the MKS system throughout the chapters on electric fields and their application, and restrict our use of the CGS system to discussions of atomic phenomena.

You will notice that the dimensions of e, displayed in Equation 4 are fairly messy. To avoid writing $\sqrt{\text{dynes cm}^2}$ all the time, this set of units is given the name ***esu*** which stands for electric charge as measured in the ***electrostatic system of units***. Thus we can rewrite Equation 7 as

$$e = 4.8 \times 10^{-10} \text{ esu} \tag{7a}$$

as the formula for the amount of charge on an electron. (In the MKS system, electric charge is measured in ***coulombs*** rather than esu. The difference between a coulomb and an esu arises not only from the different set of units (newtons vs dynes) but also from the different choice of K in the MKS system. Any further discussion of the MKS system will be reserved for later chapters.

Exercise 1

What would be the value of the electric force constant K in a system of units where distance was measured in centimeters and the charge e on the electron was set equal to 1?

Positive and Negative Charge

It was Ben Franklin who introduced the concept of two kinds of electric charge. Franklin noticed that you get opposite electrical effects when you rub a glass rod with silk, or rub a rubber rod with cat fur. He decided to call the charge left on the glass rod positive charge, and the charge left on the rubber rod negative charge. What we will see in this section is how this choice of positive and negative charge leads to the electron having a negative charge $-e$, and a proton a positive charge $+e$.

$$\text{charge on an electron} = -e \quad (8a)$$

$$\text{charge on a proton} = +e \quad (8b)$$

Despite the fact that the electron's charge turns out to be negative, e is still called the ***charge on an electron***.

The basis for saying we have two kinds of charge is the fact that with the electric interaction we have both attractive and repulsive forces. With the choice that electrons are negative and protons are positive, then the rules shown in Figure 3 can be summarized as follows: ***like charges (2 protons or 2 electrons) repel, opposite charges (proton and electron) attract***. We can explain the lack of any force on the neutron by saying that the neutron has no charge—that it is ***neutral***.

Choosing one charge as positive and one as negative automatically gives us a reversal in the direction of the force when we switch from like to opposite charges

$$\left|\vec{F}_e\right|_{\text{proton proton}} = \frac{(+e)(+e)}{r^2} = \frac{e^2}{r^2}$$

$$\left|\vec{F}_e\right|_{\text{electron electron}} = \frac{(-e)(-e)}{r^2} = \frac{e^2}{r^2}$$

$$\left|\vec{F}_e\right|_{\text{electron proton}} = \frac{(-e)(+e)}{r^2} = \frac{-e^2}{r^2}$$

(9)

Addition of Charge

The concept of charge is particularly useful when we have to deal with complex structures involving many particles. To see why, let us start with the simplest electrical structure, the hydrogen atom, and gradually add more particles. We will quickly see that the electric force law, in the form $\left|\vec{F}_e\right| = e^2/r^2$ becomes difficult to use.

In Figure (4) we have a model of the hydrogen atom consisting of a proton at the center and an electron orbiting about it. The proton sits nearly at rest at the center because it is 1836 times as massive as the electron, much as our sun sits at the center of our solar system because it is so much more massive than the planets.

The proton and electron attract each other with a force of magnitude (in CGS units) of $\left|\vec{F}_e\right| = e^2/r^2$. Since this is similar in form to the gravitational force between the earth and the moon, we expect the electron to travel in elliptical orbits around the proton obeying Kepler's laws. This would be exactly true if Newton's laws worked on the small scale of the hydrogen atom as they do on the larger scale of the earth-moon system.

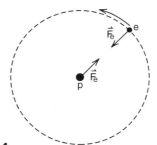

Figure 4
The hydrogen atom consists of a proton at the center with an electron moving about it. The particles are held together by the attractive electric force between them.

Exercise 2 (Do this now)

Hydrogen atoms are approximately 10^{-8} cm in diameter. Assume that the electron in the hydrogen atom in Figure (4) is traveling in a circular orbit about the proton. Use Newton's law $\vec{F} = m\vec{a}$ and your knowledge about the acceleration $\vec{a}$ of a particle moving in a circular orbit to predict the speed, in cm/sec, of the electron in its orbit. How does the electron's speed compare with the speed of light? (It had better be less.)

An analysis of the hydrogen atom is easy and straight forward using the force law $|\vec{F}_e| = e^2/r^2$. But the analysis gets more difficult as the complexity of the problem increases. Suppose, for example, we have two hydrogen atoms separated by a distance r. Let r be quite a bit larger than the diameter of a hydrogen atom, as shown in Figure (5a).

Even though r is much larger than the size of the individual hydrogen atoms, there are still electric forces between the protons and electrons in the two atoms. The two protons repel each other with a force $\vec{F}_{pp}$, the electrons repel each other with a force $\vec{F}_{ee}$, the proton in the left atom attracts the electron in the right atom with a force $\vec{F}_{ep}$. Sa you can see, eight separate forces are involved, as shown in Figure (5b).

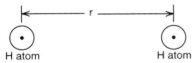

(a) Two hydrogen atoms separated by a distance r

(b) Forces between the particles in the two atoms
$|\vec{F}_{pp}| = |\vec{F}_{ee}| = |\vec{F}_{pe}| = e^2/r^2$

Figure 5
When we have two hydrogen atoms fairly far apart, then there is essentially no net force between the atoms. The reason is that the repulsive forces between like particles in the two atoms are cancelled by the attractive force between oppositely charged particles.

If the separation r of the atoms is large compared to the diameter of each hydrogen atom, then all these eight forces have essentially the same magnitude e^2/r^2. Since half are attractive and half are repulsive, they cancel and we are left with no net force between the atoms.

Exercise 3

A complete carbon atom has a nucleus with 6 protons, surrounded by 6 orbiting electrons. If you have two complete separate carbon atoms, how many forces are there between the particles in the two different atoms?

With just two simple hydrogen atoms we have to deal with 8 forces in order to calculate the total force between the atoms. If we have to deal with something as complex as calculating the force between two carbon atoms, we have, as you found by doing Exercise 3, to deal with 72 forces. Yet the answer is still zero net force. There must be an easier way to get this simple result.

The easier way is to use the concept of **net charge** Q which is the **sum of the charges in the object**. A hydrogen atom has a net charge

$$Q_{hydrogen} = (+e)_{proton} + (-e)_{electron}$$
$$= 0 \qquad (10)$$

The **net force** between two objects with net charges Q_1 and Q_2 is simply

$$\boxed{|\vec{F}_{net}| = \frac{KQ_1Q_2}{r^2} \quad \begin{array}{l}\textit{Coulomb's law, where}\\ \textit{K = 1 for CGS units}\end{array}} \quad (11)$$

Equation 1, which looks very much like Newton's law of gravity, except that charge replaces mass, is known as **Coulomb's law**. The proportionality constant K is 1 for CGS units.

Applying Coulomb's law to the force between two complete carbon atoms, we see immediately that the complete atoms have zero net charge, and therefore by Coulomb's law there is no net force between them. The cancellation of individual forces seen in Figure (4) is accounted for by the cancellation of charge in Coulomb's law.

To see that the addition of charge and Coulomb's law work in situations where charge does not cancel, suppose we had two helium nuclei separated by a distance r. (A bare helium nucleus, which is a helium atom missing both its electrons, would be called a *doubly ionized* helium atom.) In Figure (6) we have sketched the forces between the two protons in each nucleus. Both protons in nucleus number 1 are repelled by both protons in nucleus number 2, giving rise to a net repulsive force four times as strong as the force between individual protons, or a force of magnitude $4e^2/r^2$.

Applying Coulomb's law to these two nuclei, we see that the charge on each nucleus is 2e, giving for the charges Q_1 and Q_2

$Q_1 = 2e$ = total charge on nucleus #1

$Q_2 = 2e$ = total charge on nucleus #2

Thus Coulomb's law gives (with K = 1 for CGS units)

$$|\vec{F}_e| = \frac{Q_1 Q_2}{r^2} = \frac{(2e)(2e)}{r^2} = \frac{4e^2}{r^2} \qquad (12)$$

which is in agreement with Figure (6).

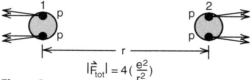

Figure 6
We see that the repulsive force between 2 helium nuclei is 4 times as great as the repulsion e^2/r^2 between 2 protons. Using Coulomb's law $|\vec{F}| = Q_1 Q_2 / r^2$ with $Q_1 = Q_2 = 2e$ for the helium nuclei gives the same result.

Exercise 4

This exercise is designed to give you a more intuitive feeling for the enormous magnitude of the electric force, and how complete the cancellation between attractive and repulsive forces is in ordinary matter.

Imagine that you could strip all the electrons from two garden peas, leaving behind two small balls of pure positive charge. Assume that there is about one mole (6×10^{23}) protons in each ball.

(a) What is the total charge Q on each of these two balls of positive charge? Give the answer in esu.

(b) The two positively charged peas are placed one meter (100 cm) apart as shown. Use Coulomb's law

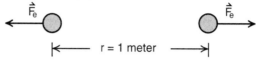

to calculate the magnitude of the repulsive force between them. Give your answer in dynes and in metric tons. (1 metric ton = 10^9 dynes ≈ 1 english ton.)

If you worked Exercise 4 correctly, you found that two garden peas, stripped of all electrons and placed one meter apart, would repel each other with a force of nearly 10^{16} tons! Yet when you actually place two garden peas a meter apart, or only a centimeter apart, there is no observable force between them. The 10^{16} ton repulsive forces are so precisely cancelled by 10^{16} ton attractive forces between electrons and protons that not even a dyne force remains.

Exercise 5

What would be the repulsive force between the peas if only one in a billion (one in 10^9) electrons were removed from each pea, and the peas were placed one meter apart?

CONSERVATION OF CHARGE

Up to this point, we have used the concept of electric charge to simplify the calculation of the electric force between two objects containing many electrons and protons. But the fact that electrons and protons have precisely opposite charges suggests that in nature electric charge has a deeper significance. That deeper significance is the conservation of electric charge. Like the conservation of energy, linear momentum, and angular momentum, the *conservation of electric charge* appears to be a basic law with no known exceptions.

When we look beyond the familiar electrons and protons, into the world of subnuclear particles, we find a bewildering array of hundreds of different kinds of particles. In the chaos of such an array of particles, two features stand out. Almost all of the particles are unstable, and when the unstable particles decay, electric charge is conserved. In looking at the particle decays, it becomes clear that there really is something we call electric charge that is passed from one particle to another, and not lost when a particle decays.

We will illustrate this with a few examples. We have already discussed several unstable particles, the muon introduced in the muon lifetime experiment, the π mesons, created for cancer research, and the neutron which, by itself outside a nucleus, has a half life of nine minutes. The muon decays into an electron and a neutrino, and the neutron decays into a proton, electron and an antineutrino (the antiparticle of the neutrino). There are three separate π mesons. The negative charged one decays into an electron and an antineutrino, the positive one into a positron (antielectron) and neutrino, and the neutral one into two photons.

We can shorten our description of these decays by introducing shorthand notation for the particles and their properties. We will use the Greek letter μ (mu) for the muon, π for the π mesons, ν (nu) for the neutrino and γ (gamma) for protons. We designate the charge of the particle by the superscript + for a positive charge, – for a negative, and 0 for uncharged. Thus the three π mesons are designated π^+, π^0, and π^- for the positive, neutral and negative ones respectively. In later discussions, it will be useful to know whether we are dealing with a particle or an antiparticle. We denote antiparticles by putting a bar over the symbol, thus ν represents a neutrino, and $\bar{\nu}$ an antineutrino. Since a particle and an antiparticle can annihilate each other, a particle and an antiparticle must have opposite electric charges if they carry charge at all, so that charge will not be lost in the annihilation. As a result the antiparticle of the electron e^- is the positively charged positron which we designate $\bar{e}^+$.

Using these conventions, we have the following notation for the particles under discussion (photons and neutrinos are uncharged):

Notation	Particle		Particle
p^+	proton	$\bar{\nu}^0$	antineutrino
n^0	neutron	μ^-	muon
e^-	electron	π^+	pi plus
$\bar{e}^+$	positron	π^0	pi naught
γ^0	photon	π^-	pi minus
ν^0	neutrino		

The particle decays we just described can now be written as the following reactions.

$$\mu^- \rightarrow e^- + \bar{\nu}^0 \quad \text{muon decay} \quad (a)$$
$$n \rightarrow p^+ + e^- + \bar{\nu}^0 \quad \text{neutron decay} \quad (b)$$
$$\pi^- \rightarrow \bar{e}^+ + \nu^0 \quad \text{pi minus decay} \quad (c)$$
$$\pi^0 \rightarrow \gamma^0 + \gamma^0 \quad \text{pi naught decay} \quad (d)$$
$$\pi^- \rightarrow e^- + \bar{\nu}^0 \quad \text{pi minus decay} \quad (e) \quad (14)$$

Note that in all of these decays, the particles change but the charge does not. If we start with a negative charge, like the negative muon, we end up with a negative particle, the electron. If we start with a neutral particle like the π^0, we end up with no net charge, in this case two photons.

Among the hundreds of elementary particle decays that have been studied, no one has found an example where the total charge changed during the process. It is rather impressive that the concept of positive and negative charge, introduced by Ben Franklin to explain experiments involving rubber rods and cat fur, would gain even deeper significance at the subnuclear level.

Stability of Matter

The conservation of electric charge may be related to the stability of matter. The decay of elementary particles is not an exceptional occurrence, it is the general rule. Of the hundreds of particles that have been observed, only four are stable, the proton, the electron, the photon and the neutrino. (Neutrons are also stable if buried inside a nucleus, for reasons we will discuss in the next chapter.) All the other particles eventually, and often very quickly, decay into these four.

The question we should ask is not why particles decay, but instead why these four particles do not. We know the answer in the case of two of them. Photons, and perhaps, neutrinos, have zero rest mass. As a result they travel at the speed of light, and time does not pass for them. If a photon had a half life, that half life would become infinite due to time dilation.

Why is the electron stable? It appears that the stability of the electron is due to the conservation of energy and electric charge. The electron is the least massive charged particle. There is nothing for it to decay into and still conserve charge and energy.

That leaves the proton. Why is it stable? We do not know for sure. There are a couple of possibilities which are currently under study. One is that perhaps the proton has some property beyond electric charge that is conserved, and that the proton is the least massive particle with this property. This was the firm belief back in the 1950s.

In the 1960s, with the discovery of quarks and the combining of the electric and weak interaction theories, it was no longer obvious that the proton was stable. Several theories were proposed, theories that attempted to unify the electric, weak, and nuclear force. These so-called **Grand Unified Theories** or **GUT** for short, predicted that protons should eventually decay, with a half life of about 10^{31} years. Since the universe is only 10^{10} years old, that is an incredibly long time.

It is not impossible to measure a half life of 10^{31} years. You do not have to wait that long. Instead you look at 10^{31} or 10^{32} particles, and see if a few decay in one year. Since a mole of particles is 6×10^{23} particles, you need about a billion moles of protons for such an experiment. A mole of protons (hydrogen) weighs one gram, a billion moles is a million kilograms or a thousand metric tons. You get that much mass in a cube of water 10 meters on a side, or in a large swimming pool. For this reason, experiments designed to detect the decay of the proton had to be able to distinguish a few proton decays per year in a swimming pool sized container of water.

So far none of these detectors has yet succeeded in detecting a proton decay (but they did detect the neutrinos from the 1987 supernova explosion). We now know that the proton half-life is in excess of 10^{32} years, and as a result the Grand Unified Theories are in trouble. We still do not know whether the proton is stable, or just very long lived.

Quantization of Electric Charge

Every elementary particle that has been detected individually by particle detectors has an electric charge that is an integer multiple of the charge on the electron. Almost all of the particles have a charge $+ e$ or $- e$, but since the 1960s, a few particles with charge 2e have been observed. Until the early 1960s it was firmly believed that this *quantization* of charge in units of $+ e$ or $- e$ was a basic property of electric charge.

In 1961, Murray Gell-Mann, who for many years had been trying to understand the bewildering array of elementary particles, discovered a symmetry in the masses of many of the particles. This symmetry, based on the rather abstract mathematical group called SU3, predicted that particles with certain properties, could be grouped into categories of 8 or 10 particles. This grouping was not unlike Mendeleev's earlier grouping of the elements in the periodic table.

When the periodic table was first constructed, there were gaps that indicated missing, as yet undiscovered elements. In Gell-Mann's SU3 symmetry there were also gaps, indicating missing or as yet undetected elementary particles. In one particular case, Gell-Mann accurately predicted the existence and the properties of a particle that was later discovered and named the ω^- (omega minus). The discovery of the ω^- verified the importance of Gell-Mann's symmetry scheme.

In 1964 Gell-Mann, and independently George Zweig also from Caltech, found an exceedingly simple model that would explain the SU3 symmetry. They found that if there existed three different kinds of particles which Gell-Mann called *quarks*, then you could make up *all* the known heavy elementary particles out of these three quarks, and the particles you make up would have just the right SU3 symmetry properties. It was an enormous simplification to explain hundreds of "elementary particles" in terms of 3 kinds of quarks.

Our discussion of quarks will be mainly reserved for the next chapter. But there is one property of quarks that fits into our current discussion of electric charge. The charge on a quark or anti quark can be $+1/3\,e$, $-1/3\,e$, $+2/3\,e$ or $-2/3\,e$. The charge e on the electron turns out not to be the fundamental unit of charge.

An even stranger property of quarks is the fact that they exist only inside elementary particles. For example, a proton or neutron is made up of three quarks and a π meson of two. All particles made from quarks have just the right number of quarks, in just the right combination, so that the total charge of the particle is an integer multiple of the electron charge e. Although the quarks themselves have a fractional charge $\pm 1/3\,e$ or $\pm 2/3\,e$, they are always found in combinations that have an integer net charge.

You might ask, why not just tear a proton apart and look at the individual quarks? Then you would see a particle with a fractional charge. It now appears that, due to an unusual property of the so-called *color force* between quarks, you cannot simply pull quarks out of a proton. The reason is that the color force, unlike gravity and electricity, becomes stronger, not weaker, as the separation of the particles increases. We will see later how this bizarre feature of the color force makes it impossible to extract an individual quark from a proton.

MOLECULAR FORCES

A naïve application of Coulomb's law would say that complete atoms do not interact. A complete atom has as many electrons outside as protons in the nucleus, and thus zero total charge. Thus by Coulomb's law, which says that the electric force between two objects is proportional to the product of the charges on them, one would predict that there is no electric force between two complete atoms. Tell that to the two hydrogen atoms that bind to form hydrogen molecules, oxygen atoms that bind to form to O_2 molecules we breathe, or the hydrogen and oxygen atoms that combine to form the water molecules we drink. These are all complete atoms that have combined together through electric forces to form molecular structures.

The reason that neutral atoms attract electrically to form molecules is the fact that the negative charge in an atom is contained in the electrons which are moving about the nucleus, and their motion can be affected by the presence of other atoms.

When trying to understand molecular forces, the planetary picture of an atom, with electrons in orbits like the planets moving around the sun, is not a particularly useful or accurate model. A more useful picture, which has its origin both in quantum mechanics and Newtonian mechanics is to picture the electrons as forming a cloud of negative charge surrounding the nucleus. You can imagine the electrons as moving around so fast that, as far as neighboring atoms are concerned, the electrons in an atom simply fill up a region around the nucleus with negative electric charge. When doing accurate calculations with quantum mechanics, one finds that the electron clouds have definite shapes, shapes which the chemists call *orbitals*.

One does not need quantum mechanics in order to get a rough understanding of the origin of molecular forces. Simple arguments about the behavior of the electron clouds gives a fairly good picture of what the chemists call *covalent bonding*. We will illustrate this with a discussion of the hydrogen molecule.

Hydrogen Molecule

To construct a hydrogen molecule, imagine that we start with a single proton and a complete hydrogen atom as indicated in Figure (7a). Here we are representing the hydrogen atom by a proton with the electron moving around to more or less fill a spherical region around the proton. In this case the external proton is attracted to the sphere of negative charge by a force that is as strong as the repulsion from the hydrogen nucleus. As a result there is very little force between the external proton and the neutral atom. Here, Coulomb's law works.

Now bring the external proton closer to the hydrogen atom, as shown in Figure (7b). Picture the hydrogen nucleus as fixed, nailed down, and look at what happens to the hydrogen electron cloud. The electron is now beginning to feel the attraction of the external proton as well as its own proton. The result is that the electron cloud is distorted, sucked over a bit toward the external proton. Now the center of the electron cloud is a bit closer to the external proton than the hydrogen nucleus, and the attractive force between the electron cloud and the proton is slightly stronger than the repulsive force between the protons. The external proton now feels a net attractive force to the neutral hydrogen atom because of the distortion of the electron cloud. A naïve application of Coulomb's law ignores the distortion of the electron cloud and therefore fails to predict this attractive force.

Since the external proton in Figure (7b) is attracted to the hydrogen atom, if we let go of the proton, it will be sucked into the hydrogen atom. Soon the electron will start orbiting about both protons, and the external proton will be sucked in until the repulsion between the protons just balances the electrical attraction.

Since the two protons are identical, the electron will have no reason to prefer one proton over the other, and will form a symmetric electron cloud about both protons as shown in Figure (7c). The result is a complete and stable object called a *hydrogen molecule ion*.

(a) A proton far from a complete hydrogen atom.

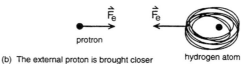

(b) The external proton is brought closer distorting the electron cloud

(c) Electron orbits both protons   Hydrogen molecule ion

Figure 7a
Formation of a hydrogen molecule ion. To visualize how a hydrogen molecule ion can be formed, imagine that you bring a proton up to a neutral hydrogen atom.

Figure 7b
When the proton gets close, it distorts the hydrogen electron cloud. Since the distorted cloud is closer to the external proton, there is a net attractive force between the proton and the distorted hydrogen atom.

Figure 7c
If the protons get too close, they repel each other. As a result there must be some separation where there is neither attraction or repulsion. This equilibrium separation for the protons in a hydrogen molecule ion is 1.07 Angstroms. (1 Angstrom = 10^{-8} cm.)

The final step in forming a hydrogen molecule is to note that the hydrogen molecule ion of Figure (7c) has a net charge +e, and therefore will attract another electron. If we drop in another electron, the two electrons form a new symmetric cloud about both protons and we end up with a stable H_2 molecule, shown in Figure (8).

Although the discussion related to Figures (7, 8) is qualitative in nature, it is sufficient to give a good picture of the difference in character between atomic and molecular forces. Atomic forces, the pure e^2/r^2 Coulomb force that binds electrons to the nucleus, is very strong and fairly simple to understand. Molecular forces, which are also electrical in origin but which depend on subtle distortion of the electron clouds, are weaker and more complex. Molecular forces are so subtle that you can make very complex objects from them, for example, objects that can read and understand this page. The sciences of chemistry and biology are devoted primarily to understanding this complexity.

Figure 8
The hydrogen molecule ion of Figure (7c) has a net positive charge +e, and therefore can attract and hold one more electron. In that case both electrons orbit both protons and we have a complete hydrogen molecule. The equilibrium separation expands to 1.48 Angstroms.

Molecular Forces—A More Quantitative Look

It is commonly believed that quantum mechanics, which can be used to predict the detailed shape of electron clouds, is needed for any quantitative understanding of molecular forces. This is only partly true. We can get a fair understanding of molecular forces from Newtonian mechanics, as was demonstrated by the student Bob Piela in a project for an introductory physics course. This section will closely follow the approach presented in Piela's project.

In this section we will discuss only the simplest of all molecules, the hydrogen molecule ion consisting of two protons and one electron and depicted in Figure (7c). We will use Newtonian mechanics to get a better picture of how the electron holds the molecule together, and to see why the lower, the more negative the energy of the electron, the more tightly the protons are bound together.

If you do a straightforward Newtonian mechanics calculation of the hydrogen molecule ion, letting all three particles move under the influence of the Coulomb forces between them, the system eventually flies apart. As a number of student projects using computer calculations have shown, eventually the electron gets captured by one of the protons and the other proton gets kicked out of the system. With Newtonian mechanics we cannot explain the stability of the hydrogen molecule ion, quantum mechanics is required for that.

Piela avoided the stability problem by assuming that the two protons were fixed at their experimentally known separation of 1.07×10^{-8} cm (1.07 angstroms) as shown in Figure (9a), and let the computer calculate the orbit of the electron about the two fixed protons, as seen in Figure (9b). By letting the calculation run for a long time and plotting the position of the electron at equal intervals of time, as in a strobe photograph of the electron's motion, one obtains the dot pattern seen in Figure (9c). This dot pattern can be thought of as the classical electron cloud pattern for the electron.

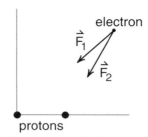

a) Electric force acting on the electron.

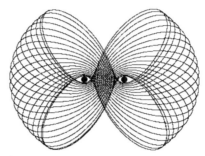

b) Line drawing plot of the orbit of the electron about the two protons

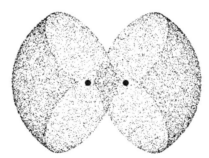

c) Dots showing the position of the electron at equal time intervals (effectively a strobe photograph).

Figure 9
Orbit of an electron about two fixed protons.

The Bonding Region

Of the dots we have drawn in Figure (9c) some are more effective than others in holding the molecule together. When the electron is between the protons, it pulls in on both protons providing a net bonding force. But when the electron is outside to the left or right, it tends to pull the protons apart. We can call the region where the electron gives rise to a net attractive force the bonding region, while the rest of space, where the electron tends to pull the protons apart, can be called the anti-bonding region.

To see how we can distinguish the bonding from the anti-bonding region, consider Figure (10) where we show the forces the electron exerts on the protons for several positions of the electron. In (10a), the electron is between but above the protons, giving rise to the forces $\vec{F}_1$ and $\vec{F}_2$ shown. We also show the x components which are $\vec{F}_{1x}$ and $\vec{F}_{2x}$. These components are of more interest to us than $\vec{F}_1$ and $\vec{F}_2$ because the electron, while in orbit, will spend an equal time above and below the protons. Thus on the average the y components cancel, and the net effect of the electron is described by the x components $\vec{F}_{1x}$ and $\vec{F}_{2x}$ alone. We can see from Figure (10a) that $\vec{F}_{1x}$ and $\vec{F}_{2x}$ are pulling the protons together. This electron is clearly in the bonding region.

It is a little bit harder to see the anti-bonding forces. In Figure (10b), we show the electron first to the left of the protons, then to the right. Again we concentrate on the x components $\vec{F}_{1x}$ and $\vec{F}_{2x}$ because the y components will, on the average, cancel. However, for orbits like that shown in Figure (9b), the electron spends the same amount of time on the left as the right, and thus we should average $\vec{F}_{1x}$ and $\vec{F}_{2x}$ for these two cases. When we do this, we see that the average $\vec{F}_{1x}$ points left, the average $\vec{F}_{2x}$ points right, and these two average forces are pulling the protons apart.

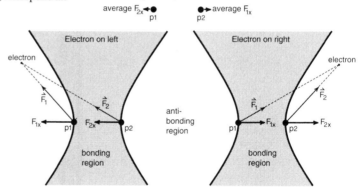

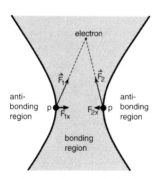

a) When the electron is in the bonding region, the x component of its electric forces pulls the protons together. (The y components average out since the electron spends as much time above as below the protons.)

b) When the electron is in the anti-bonding region, the average x component of its electric forces pulls the protons apart.

Figure 10
Bonding and anti bonding regions. When the electron is in the bonding region, the electric force exerted by the electron on the protons pulls the protons together. When in the anti-bonding region, the electric force pulls the protons apart.

If you are calculating an orbit and want to test whether the electron is in the bonding or anti-bonding region, simply compare $\vec{F}_{1x}$ and $\vec{F}_{2x}$. If the electron is to the right of the protons and $\vec{F}_{1x}$ is bigger than $\vec{F}_{2x}$, the protons are being pulled apart. If you are to the left of the protons, and $\vec{F}_{2x}$ is bigger than $\vec{F}_{1x}$, then again the protons are being pulled apart. Otherwise the protons are being pulled together and the electron is in the bonding region.

In Figure (11), we replotted the electron dot pattern of Figure (9c), but before plotting each point, checked whether the electron was in the bonding or anti-bonding region. If it were in the bonding region, we plotted a dot, and if it were in the anti-bonding region we drew a cross. After the program ran for a while, it became very clear where to draw the lines separating the bonding from the anti-bonding regions.

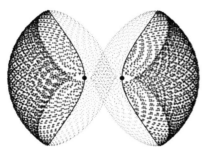

a) Electron cloud for a –10eV electron. Points in the bonding region are plotted as dots, outside in the anti-bonding region as crosses.

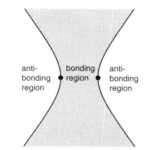

b) The area of dots in a) show us where the bonding region is.

Figure 12
Determining the bonding region from the computer plot.

Electron Binding Energy

One of the features of Figure (12) is that there are many dots out in the anti-bonding region. It looks as if there are more dots out there pulling the protons apart than inside holding them together. In this case is the electron actually helping to hold the molecule together?

One way to tell whether or not a system will stay together or fall apart is to look at the total energy of the system. If it costs energy to pull a system apart, it will stay together. But if energy is released when a system comes apart, it will fall apart.

In our earlier discussion in Chapter 8 of the motion of a satellite around the earth, we saw that if the total energy of the satellite were negative, the satellite would be bound to the earth and could not escape. On the other hand, if the total energy of the satellite were positive, the satellite would eventually escape no matter what direction it was heading (assuming it did not crash). These predictions about total energy applied if we did not include rest energy, and assumed that the gravitational potential energy was zero when the satellite and earth were infinitely far apart. This gave us as the formula for gravitational potential energy

$$E_{tot}\begin{bmatrix} planet \\ and \\ satellite \end{bmatrix} = \frac{1}{2}m_s v^2 - \frac{Gm_s M_e}{r} \qquad (8\text{-}29)$$

where m_s and M_e are the masses of the satellite and earth, and v the speed of the satellite.

In describing the motion of an electron in an atom or molecule, we can use the convention that the electron's electric potential energy is zero when it is infinitely far away from the protons. With this convention, the formula for the electric potential energy between an electron and a proton a distance r apart is $-Ke^2/r$, which is analogous to the gravitational potential energy $-Gm_s M_e/r$. (Simply replace $Gm_s M_e$ by Ke^2 to go from a discussion of gravitational forces in satellite motion to electric forces in atoms.)

Thus the formula for the total energy of an electron in orbit about a single stationary proton should be (in analogy to Equation 8-29)

$$E_{tot}\begin{bmatrix}electron\\ and\\ proton\end{bmatrix} = \frac{1}{2}m_e v^2 - \frac{Ke^2}{r} \quad (15)$$

where m_e is the mass of the electron, v its speed, and r the separation between the electron and proton.

Just as in the case of satellite motion, the total energy tells you whether the electron is bound or will eventually escape. If the total energy is negative, the electron cannot escape, while if the total energy is positive, it must escape.

Another way of describing the electron's behavior is to say that if the electron's total energy is negative, it is down in some kind of a well and needs outside help, outside energy, in order to escape. The more negative the electron's total energy, the deeper it is in the well, and the more tightly bound it is. We can call the amount by which the electron's total energy is negative the **binding energy** of the electron. The binding energy is the amount of energy that must be supplied to free the electron.

Electron Volt as a Unit of Energy

In discussing the motion of an electron in an atom, quantities like meters and kilograms and joules are awkwardly large. There is, however, a unit of energy that is particularly convenient for discussing many applications, including the motion of electrons in atoms. This unit of energy, called the **electron volt** (abbreviated eV), is the amount of energy an electron would gain if it hopped from the negative to the positive terminal of a 1 volt battery. The numerical value is

$$1 \text{ electron volt (eV)} = 1.6 \times 10^{-19} \text{joules}$$
$$= 1.6 \times 10^{-12} \text{ergs} \quad (16)$$

As an example, an electron in a cold (unexcited) hydrogen atom has a total energy of –13.6 eV. The fact that the electron's energy is negative means that the electron is bound to the proton–cannot escape.

The value –13.6 eV means that, in order to pull the electron out of the hydrogen atom, we would have to supply 13.6 eV of energy. In other words, the binding energy of the electron in a cold hydrogen atom is 13.6 eV. The number is 13.6 eV is much easier to discuss and remember than 2.16×10^{-18} joules.

Another unit that is convenient for discussing atoms is the **angstrom** (abbreviated Å) which is 10^{-8} cm or 10^{-10} meters.

$$1 \text{ Angstrom} \left(\text{Å}\right) = 10^{-8} \text{cm}$$
$$= 10^{-10} \text{m} \quad (17)$$

A hydrogen atom has a diameter of 1 Å and all atoms are approximately the same size. Even the largest atom, Uranium, has a diameter of only a few angstroms. In the hydrogen molecule ion, the separation of the protons is 1.07 Å.

Electron Energy in the Hydrogen Molecule Ion

We have seen that the strength of the binding of an electron in an atom is related to the total energy of the electron. The more negative the energy of the electron, the more tightly it is bound.

Let us now return to our discussion of the hydrogen molecule ion to see if the total energy of the electron in its orbit about the two protons is in any way related to the effectiveness which the electron binds the proton together.

When the electron is orbiting about two protons, there are two electric potential energy terms, one for each proton. Thus the formula for the electron's total energy is

$$E_{tot}\begin{bmatrix}electron\\ in\ H_2^+\\ molecule\end{bmatrix} = \frac{1}{2}m_e v^2 - \frac{Ke^2}{r_1} - \frac{Ke^2}{r_2} \quad (18)$$

which is the same as Equation 15 except for the additional potential energy term. In this equation, r_1 is the distance from the electron to proton #1, and r_2 to proton #2.

When we wrote computer programs for satellite motion, we found that it was much easier to work in a system of units where the earth mass, earth radius and hour were set to 1. In these units the gravitational constant G was simply 20, and we never had to work with awkwardly large numbers.

Similarly, we can simplify electron orbit calculations by choosing a set of units that are convenient for these calculations. In what we will call *atomic units*, we will set the mass m_e of the electron, the electric charge e, the angstrom, and the electron volt all to 1. When we do this, the electric force constant K has the simple value of 14.40. These choices are summarized in Table 2.

Using atomic units, the formula for the total energy of the electron in the hydrogen molecule ion (Equation 18) reduces to

$$E_{tot}(H_2^+) = \frac{v^2}{2} - \frac{14.40}{r_1} - \frac{14.40}{r_2} \quad (18a)$$

since $m_e = e = 1$. The real advantage of this formula is that it directly gives the electron's total energy in electron volts, no conversion is required. Because energy is conserved, because the electron's total energy does not change as the electron goes around in its orbit, we can name the orbit by E_{tot}. For example, the orbit shown back in Figures (9) and (11), had a total energy of –10 eV. We can say that this was a "–10 eV orbit".

When Bob Piela did his project on the hydrogen molecule ion, his main contribution was to show how the energy of the electron in orbit was related to the bonding force exerted by the electron on the protons. Piela's results are easily seen in Figure (13). In (13a), we show the –10 eV orbit superimposed upon a sketch of the bonding region. In (13b), the same orbit is shown as a strobe photograph. As we mentioned earlier, there appear to be a lot more dots outside the bonding region than inside, and it does not look like a –10 eV electron does a very good job of binding the protons in the H_2^+ molecule.

In Figure (13c) we have plotted a –20 eV orbit. The striking feature is that as the electron energy is reduced, made more negative, the electron spends more time in the bonding region doing a better job of holding the molecule together. In Figure (13d) the electron energy is dropped to –30 eV and the majority of the electron cloud is now in the bonding region. It is easy to see that at –30 eV the electron does a good job of binding the protons. With Piela's diagrams it is easy to see how the electron bonds more strongly when its energy is lowered.

Atomic Units

Constant	Symbol	Atomic Units	MKS Units
electron volt	eV	1	1.6×10^{-19} joules
angstrom	Å	1	10^{-10} m
electron mass	m_e	1	9.1×10^{-31} kg
electric charge	e	1	1.6×10^{-19} coulombs
electric force constant	K	14.40	9×10^9 m/farad
Bohr radius	r_b	.51 Å	
separation of protons in H_2^+		1.07 Å	

Table 2

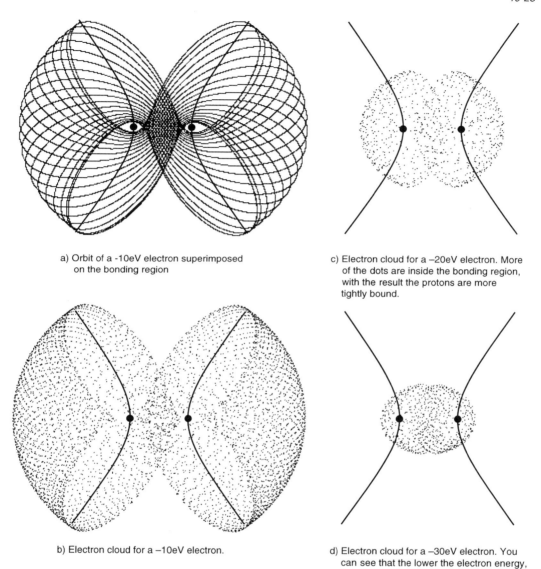

a) Orbit of a -10eV electron superimposed on the bonding region

b) Electron cloud for a -10eV electron.

c) Electron cloud for a -20eV electron. More of the dots are inside the bonding region, with the result the protons are more tightly bound.

d) Electron cloud for a -30eV electron. You can see that the lower the electron energy, the stronger the bonding.

Figure 13
As we decrease the electron's total energy, the electron spends more time in the bonding region, with the result that the protons are more tightly bound. Thus we see that the lower the electron energy, the stronger the binding.

HYDROGEN MOLECULE ION

```
! --------- Plotting window
!    (x axis = 1.5 times y axis)
   SET WINDOW -3,6,-3,3


! --------- Experimental constants in Atomic Units
   LET K = 14.40      !Electric force constant
   LET Kr = K/3       !Ficticious repulsive force
   LET Qe = 1         !Charge on electron (magnitude)
   LET Qp = 1         !Charge on Proton
   LET Me = 1         !Electron mass
   LET Mp = 1836.1    !Proton mass
   LET Rbohr = .5292  !Bohr radius
   LET Dion = 1.07    !Proton separation

! --------- Position of proton #2
   LET Zx = Dion
   LET Zy = 0
   LET Z  = SQR(Zx*Zx + Zy*Zy)

! --------- Plot crosses at protons
   LET Rx = 0
   LET Ry = 0
   CALL BigCROSS
   LET Rx = Zx
   LET Ry = Zy
   CALL BigCROSS

! --------- Initial conditions
   LET Rx = 1.5
   LET Ry = 1.6
   LET R  = SQR(Rx*Rx + Ry*Ry)

   LET Sx = Rx - Zx    !Vector equation (S = R - Z)
   LET Sy = Ry - Zy
   LET S  = SQR(Sx*Sx + Sy*Sy)

   LET Vx = -1
   LET Vy = 0
   LET V  = SQR(Vx*Vx + Vy*Vy)

   LET T = 0
   LET i = 0

! --------- Print total energy
   CALL ENERGY

! --------- Computer time step
   LET dt = .001

! --------- Calculational loop
   DO
     LET Rx = Rx + Vx*dt
     LET RY = Ry + Vy*dt
     LET R  = SQR(Rx*Rx + Ry*Ry)

     LET Sx = Rx - Zx
     LET Sy = Ry - Zy
     LET S  = SQR(Sx*Sx + Sy*Sy)

     !Calculate force
     LET F1 = K*Qp*Qe/R^2 - Kr*Qe*Qp/R^3   !Force by proton 1
     LET F2 = K*Qp*Qe/S^2 - Kr*Qe*Qp/S^3   !Force by proton 2

     LET F1x = -(Rx/R)*F1    !points in -R direction
     LET F1y = -(Ry/R)*F1

     LET F2x = -(Sx/S)*F2    !points in -S direction
     LET F2y = -(Sy/S)*F2

     LET Fx = F1x + F2x      !Vector sum of forces
     LET Fy = F1y + F2y

     !Newton's Second law
     LET Ax = Fx/Me
     LET Ay = Fy/Me

     LET Vold = SQR(Vx*Vx + Vy*Vy)
     LET Vx = Vx + Ax*dt
     LET VY = Vy + Ay*dt
     LET Vnew = SQR(Vx*Vx + Vy*Vy)
     LET V = (Vold + Vnew)/2

     LET T = T + dt
     LET i = i + 1

     IF MOD(i,50) = 0 THEN

       PLOT Rx,Ry

       IF Rx > Zx THEN
          IF -F2x > -F1x THEN CALL CROSS
       END IF

       IF Rx < 0 THEN
          IF -F1x < -F2x THEN CALL CROSS
       END IF

     END IF

   LOOP UNTIL T > 100

! --------- Subroutine ENERGY prints out total energy.
SUB ENERGY
   LET Etot = Me*V*V/2 - K*Qe*Qp/R - K*Qe*Qp/S
   !Add potential energy of repulsive core
   LET Etot = Etot + (1/2)*Kr*Qe*Qp/R^2 +  (1/2)*Kr*Qe*Qp/S^2
   PRINT T,Etot
END SUB

! --------- Subroutine CROSS draws a cross at Rx,Ry.
SUB CROSS
   PLOT LINES: Rx-.01,Ry; Rx+.01,Ry
   PLOT LINES: Rx,Ry-.01; Rx,Ry+.01
END SUB

! --------- Subroutine BigCROSS draws a cross at Rx,Ry.
SUB BigCROSS
   PLOT LINES: Rx-.04,Ry; Rx+.04,Ry
   PLOT LINES: Rx,Ry-.04; Rx,Ry+.04
END SUB

END
```

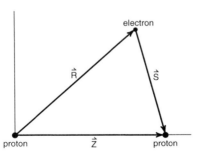
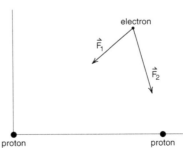

Figure 14
Computer program for the hydrogen molecule ion.

Chapter 20
Nuclear Matter

In the last chapter our focus was on what one might call electronic matter—the structures that result from the interaction of the electrons in atoms. Now we look at nuclear matter, found both in the nuclei of atoms and in neutron stars. The structures we see result from an interplay of the basic forces of nature. In the atomic nucleus, the nuclear, electric, and weak interactions are involved. In neutron stars and black holes, gravity also plays a major role.

NUCLEAR FORCE

In 1912 Ernest Rutherford discovered that all the positive charge of an atom was located in a tiny dense object at the center of the atom. By the 1930s, it was known that this object was a ball of positively charged protons and electrically neutral neutrons packed closely together as illustrated in Figure (19-1) reproduced here. Protons and neutrons are each about 1.4×10^{-13} cm in diameter, and the size of a nucleus is essentially the size of a ball of these particles. For example, iron 56, with its 26 protons and 30 neutrons, has a diameter of about 4 proton diameters. Uranium 235 is just over 6 proton diameters across. (One can check, for example, that a bag containing 235 similar marbles is about six marble diameters across.)

That the nucleus exists means that there is some force other than electricity or gravity which holds it together. The protons are all repelling each other electrically, the neutrons are electrically neutral, and the attractive gravitational force between protons is some 10^{-38} times weaker than the electric repulsive force. The force that holds the nucleus together must be attractive and even stronger than the electric repulsion. This attractive force is called the *nuclear force*.

The nuclear force treats protons and neutrons equally. In a real sense, the nuclear force cannot tell the difference between a proton and a neutron. For this reason, we can use the word nucleon to describe either a proton or neutron, and talk about the nuclear force between nucleons. Another feature of the nuclear force is that it ignores electrons. We could say that electrons have no *nuclear charge*.

The properties of the nuclear force can be deduced from the properties of the structures it creates—namely atomic nuclei. The fact that protons and neutrons maintain their size while inside a nucleus means that the nuclear force is both attractive and repulsive. Try to pull two nucleons apart and the attractive nuclear force holds them together, next to each other. But try to squeeze two nucleons into each other and you encounter a very strong repulsion, giving the nucleons essentially a solid core.

We have seen this kind of behavior before in the case of molecular forces. Molecular forces are attractive, holding atoms together to form molecules, liquids and crystals. But if you try to push atoms into each other, try to compress solid matter, the molecular force becomes repulsive. It is the repulsive part of the molecular force that makes solid matter hard to compress, and the repulsive part of the nuclear force that makes nuclear matter nearly incompressible.

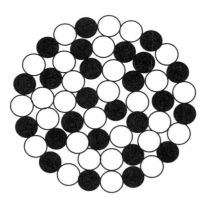

Figure 19-1a
Sketch of an atomic nucleus, showing it as a ball of protons and neutrons.

Figure 19-1b
Styrofoam model of a Uranium nucleus. (The dark balls represent protons.)

Range of the Nuclear Force

While the attractive nuclear force must be stronger than the electric force to hold the protons together in the nucleus, it is not a long range $1/r^2$ force like electricity and gravity. It drops off much more rapidly than $1/r^2$, with the result that if two protons are separated by more than a few proton diameters, the electric repulsion becomes stronger than the nuclear attraction. The separation R_0 at which the electric repulsion becomes stronger than the nuclear attraction, is about 4 proton diameters. This distance R_0, which we will call the ***range of the nuclear force***, can be determined by looking at the stability of atomic nuclei.

If we start with a small nucleus, and keep adding nucleons, for a while the nucleus becomes more stable if you add the right mix of protons and neutrons. By more stable, we mean more tightly bound. To be explicit, the more stable, the more tightly bound a nucleus, the more energy that is required, per nucleon, to pull the nucleus apart. This stability, this tight binding, is caused by the attractive nuclear force between nucleons.

Iron 56 is the most stable nucleus. It takes more energy per nucleon to take an Iron 56 nucleus apart than any other nucleus. If the nucleus gets bigger than Iron 56, it becomes less stable, less tightly bound. If a nucleus gets too big, bigger than a Lead 208 or Bismuth 209 nucleus, it becomes unstable and decays by itself.

The stability of Iron 56 results from the fact that an Iron 56 nucleus has a diameter about equal to the range of the nuclear force. In an Iron 56 nucleus every nucleon is attracting every other nucleon. If we go to a nucleus larger than Iron 56, then neighboring nucleons still attract each other, but protons on opposite sides of the nucleus now repel each other. This repulsion between distant protons leads to less binding energy per particle, and instability.

NUCLEAR FISSION

One way the instability of large nuclei shows up is in the process of nuclear fission, a process that is explained by the liquid drop model of the nucleus developed by Neils Bohr and John Wheeler in 1939.

In this model, we picture nuclear matter as being essentially an incompressible liquid. The nucleons cannot be pressed into each other, or pulled apart, but they are free to slide around each other like the water molecules in a drop of water. As a result of the liquid nature of nuclear matter, we can learn something about the behavior of nuclei by studying the behavior of drops of water.

In our discussion of entropy at the beginning of Chapter 18, we discussed a demonstration in which a stream of water is broken into a series of droplets by vibrating the hose leading to the stream. If you put a strobe light on the stream, you can stop the apparent motion of the individual droplets. The result is a strobe photograph of the projectile motion of the droplets.

If you use a closely focused television camera, you can follow the motion of individual drops. Adjust the strobe so that the drop appears to fall slowly, and you can watch an individual drop oscillate as it falls. As shown in Figure (1), the oscillation is from a rounded pancake shape (images 3 & 4) to a vertical jelly bean shape (images 6 & 7). Bohr and Wheeler proposed that similar oscillations should take place in a large nucleus like Uranium, particularly if the nucleus were struck by some outside particle, like an errant neutron.

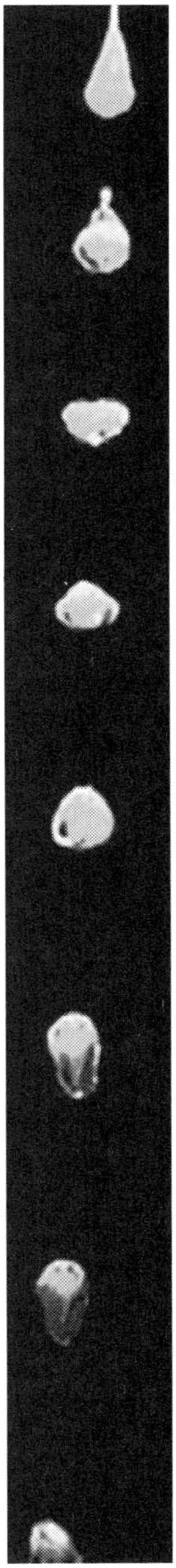

Figure 1
Oscillations of a liquid drop.

Suppose we have an oscillating Uranium nucleus, and at the present time it has the dumbbell shape shown in Figure (2a,b). In this shape we have two nascent spheres (shown by the dotted circles) connected by a neck of nuclear matter. The nascent spheres are far enough apart that they are beyond the range R_0 of the nuclear force, so that the electrical repulsion is stronger than the nuclear attraction. The only thing that holds this nucleus together is the neck of nuclear matter between the spheres.

If the Uranium nucleus is struck too vigorously, if the neck is stretched too far, the electric force will cause the two ends to fly apart, releasing a huge quantity of electrical potential energy. This process, shown in Figure (3) is called *nuclear fission*.

In the fission of Uranium 235, the large Uranium nucleus breaks up into two moderate sized nuclei, for example, Cesium 140 and Zirconium 94. Because larger nuclei have a higher percentage of neutrons than smaller ones, when Uranium breaks up into smaller, less neutron rich nuclei, some free neutrons are also emitted as indicated in Figure (3). These free neutrons may go out and strike other Uranium nuclei, causing further fission reactions.

If you have a small block of Uranium, and one of the Uranium nuclei fissions spontaneously (it happens once in a while), the extra free neutrons are likely to pass out through the edges of the block and nothing happens. If, however, the block is big enough, (if it exceeds a *critical mass* of about 13 pounds for a sphere), then neutrons from one fissioning nucleus are more likely to strike other Uranium nuclei than to escape. The result is that several other nuclei fission, and each of these cause several others to fission. Quickly you have a large number of fissioning nuclei in a process called a *chain reaction*. This is the process that occurs in an uncontrolled way in an atomic bomb and in a controlled way in a nuclear reactor.

The energy we get from nuclear fission, the energy from all commercial nuclear reactors, is electrical potential energy released when the two nuclear fragments fly apart. The fragments shown in Figure (3) are at that point well beyond the range R_0 of the attractive nuclear force, and essentially feel only the repulsive electric force between the protons. These two balls of positive charge have a large positive electric potential energy which is converted to kinetic energy as the fragments fly apart.

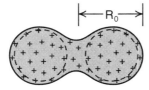

Figure 2a
Uranium nucleus in a dumbbell shape.

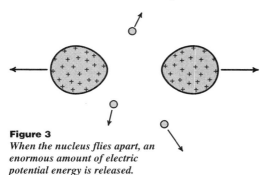

Figure 3
When the nucleus flies apart, an enormous amount of electric potential energy is released.

Figure 2b
Styrofoam model of a Uranium nucleus in a dumbbell shape.

To get a feeling for the amount of energy released in a fission reaction, let us calculate the electric potential energy of two fragments, say a Cesium and a Zirconium nucleus when separated by a distance $2R_0$, twice the range of the nuclear force.

In CGS units, the formula for the electric potential energy of 2 particles with charges Q_1 and Q_2 separated by a distance R is

$$\text{electric potential energy} \quad U_{electric} = \frac{Q_1 Q_2}{r} \quad \text{CGS units} \quad (1)$$

For our problem, let Q_1 be the charge on a Cesium nucleus (55 protons) and Q_2 the charge on a Zirconium nucleus (40 protons).

$$Q_{Cesium} = 55e$$
$$Q_{Zirconium} = 40e$$
$$r = 2R_0$$

and we get

$$U_{electric} = \frac{(55\,e) \times (40\,e)}{2R_0}$$
$$= 1.1 \times 10^3 \frac{e^2}{R_0} \quad (2)$$

We would like to compare the energy released in nuclear fission reactions with the energies typically involved in chemical reactions. It takes a fairly violent chemical reaction to rip the electron completely out of a hydrogen atom. The amount of energy to do that, to ionize a hydrogen atom is e^2/r_b where r_b is the Bohr radius of 5×10^{-9} cm.

$$\text{energy to ionize a hydrogen atom} = \frac{e^2}{r_b} = 13.6\,eV \quad (3)$$

We evaluated the number e^2/r_b earlier and found it to have a numerical value of 13.6 electron volts. This is a large amount of energy for a chemical reaction, more typical chemical reactions, arising from molecular forces, have involved energies in the 1 to 2 electron volt range.

To compare the strength of nuclear fission reactions to chemical reactions, we can compare the electric potential energies in Equations 2 and 3. If we take the range R_0 of the nuclear force to be 4 proton diameters then

$$R_0 = 4 \times \left(1.4 \times 10^{-13}\,cm\right) = 5.6 \times 10^{-13}\,cm$$

Since the Bohr radius is 5×10^{-9} cm, we see that R_0 is essentially $10^{-4} R_b$ or ten thousand times smaller than the Bohr radius.

$$R_0 = 10^{-4} r_b \quad (4)$$

Substituting Equation 4 into 2 gives

$$U_{electric} = 1.1 \times 10^3 \frac{e^2}{R_0}$$
$$= 1.1 \times 10^3 \frac{e^2}{\left(10^{-4} r_b\right)}$$
$$= 1.1 \times 10^7 \left(\frac{e^2}{r_b}\right)$$

Using the fact that e^2/r_b has a magnitude of 13.6 electron volts, we get

$$U_{electric} = 1.1 \times 10^7 \times 13.6\,eV$$
$$= 150 \times 10^6\,eV \quad (5)$$
$$= 150\,MeV$$

where 1MeV is one million electron volts. From Equation 5, we see that, per particle, some ten million times more electric potential energy is released in a nuclear fission reaction than in a violent chemical reaction. Many millions of electron volts are involved in nuclear reactions as compared to the few electron volts in chemical reactions. You can also see that a major reason for the huge amounts of energy in a nuclear reaction is the small size of the nucleus (the fact that $R_0 \ll r_b$).

NEUTRONS AND THE WEAK INTERACTION

The stability of the iron nucleus and the instability of nuclei larger than Uranium results primarily from the fact that the attractive part of the nuclear force has a short range R_0 over which it dominates the repulsive electric force between protons. The range R_0 is about 4 proton diameters, the diameter of an iron nucleus. In larger nuclei, not all nucleons attract each other, and this leads to the kind of instability we see in a fissioning Uranium nucleus.

The range of the nuclear force is not the only important factor in determining the stability of nuclei. There is no electric repulsion between neutrons, neutrons are attracted equally to neutrons and protons. Adding neutrons to a nucleus increases the attractive nuclear force without enhancing the electric repulsion. This is why the most stable large nuclei have an excess of neutrons over protons. The neutron excess acts as a nuclear glue, diluting the repulsion of the protons.

If adding a few extra neutrons increases the stability of a nucleus, why doesn't adding more neutrons give even more stable nuclei? Why is it that nuclei with too many excess neutrons are in fact unstable? Why can't we make nuclei out of pure neutrons and avoid the proton repulsion altogether?

The answer lies in the fact that, because of the weak interaction, and because of a small excess mass of a neutron, a neutron can decay into a proton and release energy. This is the beta decay reaction we discussed earlier, and is described by the equation

$$n^0 \rightarrow p^+ + e^- + \bar{\nu}^0 \quad (5)$$

The neutral neutron (n^0) decays into a positive proton (p^+), a negative electron (e^-), and a neutral antineutrino ($\bar{\nu}^0$), thus electric charge is conserved in the process. It is called a beta (β) decay reaction because the electrons that come out were originally called *beta rays* before their identity as electrons was determined.

If you have an isolated free neutron, the (β) decay of Equation 5 occurs with a half life of 15 minutes. Such a reaction can occur only if energy can be conserved in the process. But the neutron is sufficiently massive to decay into a proton and an electron and still have some energy left over. Expressing rest mass or rest energy in units of millions of electron volts (MeV), we have for the particles in the neutron decay reaction (5),

$$\begin{aligned} \text{neutron rest mass} \quad m_n &= 939.6 \text{ MeV} \\ \text{proton rest mass} \quad m_p &= 938.3 \text{ MeV} \\ \text{electron rest mass} \quad m_e &= 0.511 \text{ MeV} \\ \text{neutrino rest mass} \quad m_\nu &= 0 \end{aligned} \quad (6)$$

where

$$1 \text{ MeV} = 10^6 \text{ eV} = 1.6 \times 10^{-6} \text{ ergs}$$

You can see that the neutron rest mass is 1.3 MeV greater than that of a proton, and .8 MeV greater than the combined rest masses of the proton, electron and neutrino. Thus when a neutron (β) decays, there is an excess of .8 MeV of energy that is released in the form of kinetic energy of the reaction products.

As we have mentioned, when the β decay process was first studied in the 1920s, the neutrino was unknown. What was observed was that in β decays, the proton and the electron carried out different amounts of energy, sometimes all of the available energy, but usually just part of it. To explain the missing energy, Wolfgang Pauli proposed the existence of an almost undetectable, uncharged, zero rest mass particle which Fermi named the ***little neutral one*** or neutrino. As bizarre as Pauli's hypothesis seemed at the time, it turned out to be correct. When a neutron decays, it decays into 3 particles, and the .8 MeV available for kinetic energy can be shared in various ways among the 3 particles. Because the neutrino has no rest mass, it is possible for the proton and electron to get all .8 MeV of kinetic energy. At other times the neutrino gets much of the kinetic energy, so that if you did not know about the neutrino, you would think energy was lost.

NUCLEAR STRUCTURE

Free neutrons decay in fifteen minutes, but neutrons inside a nucleus seem to live forever. Why don't they decay? The answer to this question is an energy balance. Like a rock dropped into a well, an atomic nucleus will fall down to the lowest energy state available. The neutron will decay if the result is a lower energy, less massive nucleus. Otherwise the neutron will be stable.

We have seen that an isolated neutron can decay into the less massive proton and release energy. Now consider a neutron in a nucleus. Take the simplest nucleus with a neutron in it, namely Deuterium. If that neutron decayed we would end up with a helium nucleus consisting of two protons only, plus an electron and a neutrino. We can write this reaction as

$$^2H \rightarrow \,^2He + e^- + \bar{\nu} \quad \begin{array}{l}\textit{deuterium decay}\\ \textit{which does not happen}\end{array} \quad (7)$$

If the neutron in deuterium turns into a proton, the neutron sheds rest mass, but the resulting two proton nucleus has positive electric potential energy. We can estimate the amount of electric potential energy U_{pp} created by using the formula

$$\begin{array}{l}\textit{electric potential}\\ \textit{energy of a 2}\\ \textit{proton nucleus}\end{array} \quad U_{pp} = \frac{e^2}{2r_p} \quad \textit{CGS units} \quad (8)$$

where r_p is the proton radius and $2r_p = 2 \times 10^{-13}$ is the separation of the proton centers. The protons each have a charge $+e$. Putting numbers into Equation 8 gives

$$\begin{aligned}U_{pp} &= \frac{e^2}{2r_p}\\ &= \frac{(4.8 \times 10^{-10})^2}{2 \times 10^{-13}} \text{ ergs}\\ &= 1.15 \times 10^{-6} \text{ ergs}\\ &= .72 \text{ MeV}\end{aligned} \quad (9)$$

Our simple calculation shows that for the neutron to decay into a proton in deuterium, the neutron would have to create nearly as much electrical potential energy (.72MeV) as the available .8 MeV neutron mass energy. A more complete analysis shows that the available .8 Mev is not adequate for the neutron (β) decay, with the result the neutron in deuterium is stable.

We can now see the competing processes involved in the formation of nuclei. To construct stable nuclei you want to add neutrons to give more attractive nuclear forces and dilute the repulsive electric forces between protons. However, the rest mass of a neutron is greater than the rest mass of a proton and an electron, and the weak interaction allows the neutron to decay into these particles. Thus neutrons can shed mass, and therefore energy, by decaying.

But when a neutron inside of a nucleus decays into a proton, it increases the electric potential energy of the nucleus. If the increase in the electric potential energy is greater than the mass energy released, as we nearly saw in the case of deuterium, then the neutron cannot decay.

The weak interaction is democratic, it allows a proton to decay into a neutron as well as a neutron to decay into a proton. The proton decay process is

$$p^+ \rightarrow n^0 + \bar{e}^+ + \nu^0 \quad \textit{inverse } \beta \textit{ decay} \quad (10)$$

where $\bar{e}^+$ is the positively charged antielectron (positron). This ***inverse β decay***, as it is sometimes called, does not occur for a free proton because energy is not conserved. The proton rest mass is less than that of a neutron, let alone that of a neutron and a positron combined.

However, if you construct a nucleus with too many protons, with too much electric potential energy, then the nucleus can get rid of some of its electric potential energy by converting a proton into a neutron. This may happen if enough electric potential energy is released to supply the extra rest mass of the neutron as well as the .5 MeV rest mass of the positron.

We can now see the competing processes in an atomic nucleus. The weak interaction allows neutrons to turn into protons or protons into neutrons. But these β decay processes will occur only if energy can be released. Nuclei with too many neutrons have too much neutron mass energy, and can get rid of some of the mass energy by turning a neutron into a proton. Nuclei with too many protons have too much electrical potential energy, and can get rid of some of the electrical potential energy by turning a proton into a neutron. A stable nucleus is one that has neither an excess of mass energy nor electrical potential energy, a nucleus that cannot release energy either by turning protons into neutrons or vice versa.

To predict precisely which nuclei are stable and which are not requires a more detailed knowledge of the nuclear force than we have discussed here. But from what we have said, you can understand the general trend. For the light nuclei, the most stable are the ones with roughly equal numbers of protons and neutrons, nuclei with neither too much neutron mass energy or too much proton electrical potential energy. However, when nuclei become larger than the range of the attractive nuclear force, electric potential energy becomes more important . Excess neutrons, with the additional attractive nuclear binding force they provide, are needed to make the nucleus stable.

α (Alpha) Particles

In 1898 Ernest Rutherford, a young research student in Cambridge University, England, discovered that radioactive substances emitted two different kinds of rays which he named *α rays* and *β rays* after the first two letters of the Greek alphabet. The negatively charged β rays turned out to be beams of electrons, and the positive α rays were found to be beams of Helium 4 nuclei. Helium 4 nuclei, consisting of 2 protons and 2 neutrons, are thus also called *α particles*. (Later Rutherford observed a third kind of radiation he called *γ rays*, which turned out to be high energy photons.) We have seen that β rays or electrons are emitted when a neutron sheds mass by decaying into a proton in a β decay reaction. But where do the α particles come from?

A nucleus with an excess of electric potential energy can lose energy by converting one of its protons into a neutron in an inverse β decay reaction. This, however, is a relatively rare event. More commonly, the number of protons is reduced by ejecting an α particle. Why the nucleus emits an entire α particle or Helium 4 nucleus, instead of simply kicking out a single proton, is a consequence of an anomaly in the nuclear force. It turns out that a Helium 4 nucleus, with 2 neutrons and 2 protons, is an exceptionally stable, tightly bound object. If protons are to be ejected, they come out in pairs in this stable configuration rather than individually.

NUCLEAR BINDING ENERGIES

The best way to see the competition between the attractive nuclear force and the electric repulsive force inside atomic nuclei is to look at nuclear binding energies. Explicitly, we will look at the **binding energy per nucleon** for the most stable nuclei of each element. The binding energy per nucleon (proton or neutron) represents how much energy we would have to supply to pull the nucleus apart into separate free nucleons. The nuclear force tries to hold the nucleus together—make it more tightly bound—and therefore increases the binding energy. The electric force, which pushes the protons apart, decreases the binding energy.

You calculate the binding energy of a nucleus by subtracting the rest energy of the nucleus from the sum of the rest energies of the protons and neutrons that make up the nucleus. If you then divide by the number of nucleons, you get the binding energy per nucleon. We will go through an example of this calculation, and give you an opportunity to work out some yourself. Then we will look at a plot of these binding energies to see what the plot tells us.

Example 1

Given that a proton, a neutron, and a deuterium nucleus have the following rest energies, what is the binding energy per particle for the deuterium nucleus?

$$m_p c^2 = 938.3 \text{ MeV} \quad \text{proton rest energy}$$
$$m_n c^2 = 939.6 \text{ MeV} \quad \text{neutron rest energy}$$
$$m_d c^2 = 1875.1 \text{ MeV} \quad \text{deuterium nucleus rest energy}$$

(20)

Solution

Separately the proton and neutron in a deuterium nucleus have a total rest energy of

$$\left. \begin{array}{l} \text{rest energy} \\ \text{a separate} \\ \text{proton and} \\ \text{neutron} \end{array} \right\} = 938.3 \text{ MeV} + 939.6 \text{ MeV}$$

$$= 1877.9 \text{ MeV}$$

(21)

Thus the total binding energy of the deuterium nucleus, the energy required to pull the particles apart is

$$\left. \begin{array}{l} \text{total binding} \\ \text{energy of the} \\ \text{deuterium} \\ \text{nucleus} \end{array} \right\} = 1877.9 \text{ MeV} \begin{pmatrix} \text{separate} \\ \text{particles} \end{pmatrix} - 1875.1 \text{ MeV} \begin{pmatrix} \text{deuterium} \\ \text{nucleus} \end{pmatrix}$$

$$= 2.8 \text{ MeV}$$

(22)

Finally the binding energy per nucleon, there being 2 nucleons, is

$$\left. \begin{array}{l} \text{binding energy} \\ \text{per nucleon} \end{array} \right\} = \frac{2.8 \text{ MeV}}{2 \text{ nucleons}}$$

$$= \boxed{1.4 \frac{\text{MeV}}{\text{nucleon}}}$$

(23)

Exercise 1

Given that the masses of the Helium 4 (2 protons, 2 neutrons), the Iron 56 (26 protons, 30 neutrons), and the Uranium 238 (92 protons, 146 neutrons) nuclei are

$$M_{\text{Helium 4}} c^2 = 3725.95 \text{ MeV}$$
$$M_{\text{Iron 56}} c^2 = 52068.77 \text{ MeV}$$
$$M_{\text{Uranium 238}} c^2 = 221596.94 \text{ MeV}$$

(24)

find in each case, the binding energy per nucleon. Your results should be

$$\left. \begin{array}{l} \text{binding energy} \\ \text{per nucleon} \end{array} \right\} = \begin{array}{l} 7.46 \text{ MeV (Helium 4))} \\ 9.20 \text{ MeV (Iron 56)} \\ 8.02 \text{ MeV (Uranium 238)} \end{array}$$

(25)

Figure (4) is a plot of the binding energy, per nucleon, of the most stable nuclei for each element. We have plotted increasing binding energy downward so that the plot would look like a well. The deeper down in the well a nucleus is, the more energy per particle that is required to pull the nucleus out of the well—to pull it apart. The deepest part of the well is at the Iron 56 nucleus, no other nucleus is more tightly bound.

Moving down into the well represents a release of nuclear energy. There are two ways to do this. We can start with light nuclei and put them together (in a process called *nuclear fusion*), to form heavier nuclei, moving in and down from the left side in Figure (4). Or we can split apart heavy nuclei (In the process of nuclear fission), moving in and down from the right side. Fusion represents the release of nuclear force potential energy, while fission represents the release of electric force potential energy. When we get to the bottom, at Iron 56, there is no energy to be released either by fusion or fission.

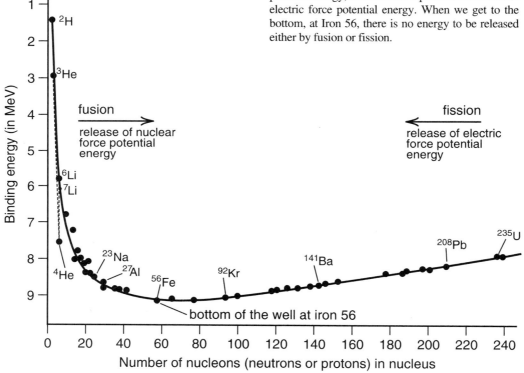

Figure 4
The nuclear energy well. The graph shows the amount of energy, per nucleon, required to pull the nucleus apart into separate neutrons and protons.

The reason Iron 56 is at the bottom of the well is because the diameter of an iron nucleus is about equal to the range of the nuclear force. As you build up to Iron 56, adding more nucleons increases the number of attractive forces between particles, and therefore increases the strength of the binding. At Iron 56, you have the largest nucleus in which every particle attracts every other particle. The diameter of the Iron 56 nucleus is the distance over which the attractive nuclear force is stronger than the repulsive electric force.

When you build nuclei larger than Iron 56, the protons on opposite sides of the nucleus are far enough apart that the electric force is stronger and the particles repel. Now, adding more particles reduces the binding energy per particle and produces less stable nuclei. When a nucleus become as large as uranium, the impact of a single neutron can cause the nucleus to split apart into two smaller, more stable nuclei in the nuclear fission process.

There are some bumps in the graph of nuclear binding energies, bumps representing details in the structure of the nuclear force. The most striking anomaly is the Helium 4 nucleus which is far more tightly bound than neighboring nuclei. This tight binding of Helium 4 is the reason, as we mentioned, that α particles (Helium 4 nuclei) rather than individual protons are emitted in radioactive decays. But overlooking the bumps, we have the general feature that nuclear binding energies increase up to Iron 56 and then decrease thereafter.

The importance of knowing the nuclear binding energy per nucleon is that it tells us whether energy will be released in a particular nuclear reaction. If the somewhat weakly bound uranium nucleus (7.41 MeV/nucleon) splits into two more tightly bound nuclei like cesium (8.16 MeV/nucleon) and zirconium (8.41 MeV/nucleon), energy is released. At the other end of the graph, if we combine two weakly bound deuterium nuclei (2.8 MeV/nucleon) to form a more tightly bound Helium 4 nucleus (7.1 MeV/nucleon), energy is also released. Any reaction that moves us toward the Iron 56 nucleus releases energy. On the small nucleus side we get a release of energy by combining small nuclei to form bigger nuclei. But once past Iron 56, we get a release of energy by splitting nuclei apart for form smaller ones.

NUCLEAR FUSION

The process of combining small nuclei to form larger ones is called nuclear fusion. From our graph of nuclear binding energies in Figure (4), we see that nuclear fusion releases energy if the resulting nucleus is smaller than Iron 56, but costs energy if the resulting nucleus is larger. This fact has enormous significance in the life of stars and the formation of the elements.

Most stars are created from a gas cloud rich in hydrogen gas. When the cloud condenses, gravitational potential energy is released and the gas heats up. If the condensing cloud is massive enough, if the temperature becomes hot enough, the hydrogen nuclei begin to fuse. After several reactions they produce Helium 4 nuclei, releasing energy in each reaction. The fusion of hydrogen to form helium becomes the source of energy for the star for many years to come.

Unlike fission, fusion requires high temperatures in order to take place. Consider the reaction in which two hydrogen nuclei (protons) fuse to produce a deuterium nucleus plus a positron and a neutrino. (When the two protons fuse, the resulting nucleus immediately gets rid of its electrical potential energy by having one proton turn into a neutron in an inverse β decay process.)

The fusion of the two protons will take place if the protons get closer together than the range of the nuclear force about—4 proton diameters. Before they get that close they repel electrically. Only if the protons were initially moving fast enough, were hot enough, can they get close enough to get past the electrical repulsion in order to feel the nuclear attraction.

A good way to picture the situation is to think of yourself as sitting on one of the protons, and draw a graph of the potential energy of the approaching proton, as shown in Figure (5). When the proton separation r is greater than the range R_0 of the nuclear force, the protons repel and the incoming proton has to climb a potential energy hill. At $R = R_0$, the net force turns attractive and the potential energy begins to decrease, forming a deep well when the particles are near to touching. Energy is released when the incoming proton falls into the well, but the incoming proton must have enough kinetic energy to get over the electrical potential energy barrier before the fusion can take place.

Using our formula Q_1Q_2/r for electric potential energy, we can make a rough estimate of the kinetic energy and the temperature required for fusion. Consider the fusion of two protons where $Q_1 = Q_2 = e$. For the protons to get within a distance R_0, the incoming proton must climb a barrier of height

$$\left.\begin{array}{r}\text{electric potential}\\ \text{energy of 2 protons}\\ \text{a distance } R_0 \text{ apart}\end{array}\right\} = \frac{e^2}{R_0}$$

$$= \frac{(4.8 \times 10^{-10})^2}{4 \times 1.4 \times 10^{-13}} \quad (26)$$

$$= 4.1 \times 10^{-7} \text{ ergs}$$

This number, $e^2/R_0 = 4.1 \times 10^{-7}$ ergs, is the amount of kinetic energy an incoming proton must originally have in order to get within a distance of approximately R_0 of another proton. Only when it gets within this distance can the nuclear force take over and fusion take place.

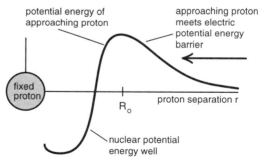

Figure 5
When you shove two protons together, you first have to overcome the electric repulsion before the nuclear attraction dominates.

In our earlier discussion of temperature, we saw that the average kinetic energy of a particle in a gas of temperature T was 3/2 kT. If we had a gas of hydrogen so hot that the average proton could enter into a fusion reaction, the average kinetic energy of the protons would have to be 4.1×10^{-7} ergs. The temperature T_f at which this would happen is found by equating 4.1×10^{-7} ergs to 3/2 kT to give

$$\frac{3}{2}kT = 4.1 \times 10^{-7} \text{ergs}$$

$$T = \frac{2}{3} \frac{4.1 \times 10^{-7} \text{ergs}}{1.38 \times 10^{-16} \frac{\text{ergs}}{\text{kelvin}}} \qquad (27)$$

$$T = 2 \times 10^9 \text{kelvin} \quad \begin{array}{c} 2 \text{ billion} \\ \text{degrees} \end{array}$$

This temperature, two billion degrees kelvin, is a huge overestimate. At this temperature, the *average* proton in the gas would enter into a fusion reaction. If we heated a container of hydrogen to this temperature, the entire collection of protons would fuse after only a few collisions, and the fusion energy would be released almost instantaneously. We would have what is known as a ***hydrogen bomb***.

In a star, the fusion of hydrogen takes place at the much lower temperatures of about ***20 million*** degrees. At 20 million degrees, only a small fraction of the protons have enough kinetic energy to enter into a fusion reaction. At these temperatures the hydrogen is consumed at a slow steady rate in what is known as a ***controlled fusion reaction***.

STELLAR EVOLUTION

The story of the evolution of stars provides an ideal setting to illustrate the interplay of the four basic interactions. In this chapter so far we have been focusing on the basic consequences of the interplay of the nuclear, electric, and weak interactions at the level of atomic nucleus. Add gravity and you have the story of stellar evolution.

A star is born from a cloud of gas, typically rich in hydrogen, that begins to collapse gravitationally. As the cloud collapses, ***gravitational potential energy*** is released which heats the gas. If the temperature does not get hot enough to start the fusion of the hydrogen nuclei, that's more or less the end of the story and you have a ***proto star***, something around the size of the planet Jupiter or smaller.

If there is more mass in the collapsing gas cloud, more gravitational potential energy will be released, and the temperature will rise enough to start the fusion of hydrogen. How hot the center of the star becomes depends on the mass of the star. In a star, like our sun for example, there is a balance between the gravitational attraction and the thermal pressure. The greater the mass, the greater the gravitational attraction, and the stronger the thermal pressure must be.

In our discussion of pressure in chapter 17, we saw that when a balloon was cooled by liquid nitrogen, removing the thermal energy and pressure of the air molecules inside, the balloon collapsed. (Figures 17-19.) Similar processes occur in a star, except that the confining force of the rubber is replaced by the confining force of gravity. The sun is a ball of hot gas. Gravity is trying to squeeze the gas inward, and the thermal pressure of the gas prevents it from doing so. There is a precise balance between the thermal pressure and gravity.

Figure 17-19
Balloon collapsing in liquid nitrogen.

From the earth the sun looks more like a solid object than a ball of gas. The sun has a definite edge, an obvious surface with spots and speckles on it. The appearance of a sharp surface is the result of the change in temperature of the gas with height. The hottest part of the sun or any star is the center. Here the gas is so hot, the thermal collisions are so violent, that the electrons are knocked out of the hydrogen atoms and all the hydrogen is ionized. The gas is what is called a *plasma*. An ionized gas or plasma is opaque, light is absorbed by the separate charged particles.

As you go out from the center of the star, the temperature drops. When you go out far enough, when the temperature drops to about 3000 kelvins, the electrons recombine with the nuclei, you get neutral atoms, and the gas becomes transparent. This transition from an opaque to transparent gas occurs rather abruptly, giving us what we think of as the surface of the sun.

Returning to the balance of gravitational attraction and thermal pressure, you can see that the more mass in the star the stronger gravity is, and the greater the thermal pressure required to balance gravity. To increase the thermal pressure, you have to increase the temperature. Thus the more massive a star, the hotter it has to be.

The proton fusion reaction we have discussed is well suited for supplying any required temperature. We have seen that the rate at which fusion takes place depends very much on the temperature. At 20 million degrees, only a small fraction of the protons have enough thermal kinetic energy to fuse. At 2 billion degrees, the average proton has enough energy to fuse, and any hydrogen at this temperature would burn immediately.

There is a range of burning rates between these two extremes. As a result, with increasing temperature the fusion reaction goes faster, supplies energy at a greater rate, and maintains the higher temperature. Thus when a new star forms it collapses until there is a balance between the gravitational force and thermal pressure. Whatever thermal pressure is required is supplied by the heat generated by the fusion reaction. The more thermal pressure needed, the higher the temperature required and the faster the hydrogen burns.

One often refers to the region near the center of the star where the fusion reaction is taking place as the *core* of the star. Our sun, with a temperature in the core of 20 million degrees, is burning hydrogen at such a rate that the hydrogen supply will last 10 billion years. Since the sun is 5 billion years old, about half the available hydrogen in the core is used up.

A more massive star, like the star that blew up to give us the 1987 supernova event, burns its hydrogen at a much shorter time. That star was about 18 times as massive as the sun, about 40,000 times as bright, and burned its hydrogen in its core so fast that the hydrogen lasted only about 10 million years.

The difference between different mass stars shows up most dramatically after the hydrogen fuel is used up. What will happen to our sun is relatively calm. When our sun uses up the hydrogen, the core will start to cool and collapse. But the collapse releases large amounts of gravitational potential energy that heats the core to higher temperatures than before. This hotter core becomes very bright, so bright that the light from the core, when it works its way out to the surface, exerts a strong radiation pressure on the gas at the surface. This radiation pressure will cause the surface of the sun to expand until the diameter of the sun is about equal to the diameter of the earth's orbit. At this point the sun will have become what is called a *red giant star*, with the earth orbiting slightly inside. It is not a very pleasant picture for the earth, but it will not happen for another five billion years.

Once the sun, as a red giant star, radiates the energy it got from the gravitational collapse, it will gradually cool and collapse until the atoms push against each other. It will be the electric force between the atomic electrons that will halt the gravitational collapse of the sun. At that point the sun will become a ball of highly compressed atomic matter about the size of the earth. Initially it will be quite bright, an object called a **white dwarf** star, but eventually it will cool and darken.

The story was very different for the star that gave rise to the supernova explosion. That star, with its mass of about 18 times that of the sun, burned its hydrogen in 10 million years. At that point the star had a core, about 30 per cent of the star, consisting mostly of Helium 4, the tightly bound nucleus that is the end result of hydrogen fusion. Computer simulations tell us that for the next tens of thousands of years, the helium core was compressed from a density of 6 to 1,100 grams per cubic centimeter, and the temperature rose from 40 million to 190 million degrees kelvin.

The temperature of 190 million degrees is high enough to cause helium nuclei to fuse, forming carbon and oxygen. Higher temperatures are required to fuse helium nuclei, because each helium nucleus has two protons and a charge + 2e. Thus the electric potential barrier Q_1Q_2/R_0 is four times as high as it is for proton fusion, and the helium nuclei need four times as much kinetic energy to fuse.

At these higher temperatures the core radiated more light, causing the outer layers of the star, mostly unburned hydrogen, to expand to about twice the size of the earth's orbit. It had become a **red supergiant**.

The helium in the core lasted less than a million years, leaving behind a collapsing core of carbon and oxygen. The temperature rose to about 740 million degrees where the carbon ignited to form neon, magnesium and sodium. When the carbon was used up in about 12,000 years, further collapse raised the temperature to 1.6 billion degrees where neon ignited.

In successively shorter times at successively higher temperatures the more massive elements were created and burned. After neon, there was carbon, then oxygen at 2.1 billion degrees, and finally silicon and sulfur at 3.4 billion degrees. The neon burned in about 12 years, the oxygen in 4 years, and the silicon in just a week.

One of the reasons for the accelerated pace of burning at the end is that, at temperatures over half a billion degrees, the star has a more efficient way of getting rid of energy than emitting light. At these temperatures some of the photons are energetic enough to create electron-positron pairs which usually annihilated back into photons but ***sometimes into neutrinos***. Whereas photons take thousands of years to carry energy from the core of a star to the surface, neutrinos escape immediately. Thus when the star reached half a billion degrees, it sprung a neutrino heat leak, and the collapse and burning went much faster.

The successive stages of burning took place in smaller and smaller cores, leaving shells of unburned elements. Unburned hydrogen filled the outer volume of the star. Inside was a shell of unburned helium and inside that successive shells of unburned carbon, oxygen, then a mixture of neon, silicon and sulfur, and a shell of silicon and sulfur.

In the center was iron. Iron was what resulted when the silicon and sulfur burned. And iron is the end of the road. As we have seen in Figure (4) iron is the most stable atomic nucleus. Energy is ***released*** when you fuse nuclei to form a nucleus smaller than iron, but it ***costs*** energy to create nuclei larger than iron. The iron core of the star was dead ash not a fuel.

By the time the star had an iron core, it was rapidly radiating energy in the form of neutrinos, but had run out of fuel. At this point the iron core, which had a mass of about 1.4 times the mass of the sun, began to collapse due to the lack of support by thermal pressure. When the sun runs out of energy, its collapse will be halted by the electric repulsion between atomic electrons. In the 1987 supernova star, the gravitational forces were so great that the electrons were essentially crammed back into the nuclei, the protons converted to neutrons, and the core collapsed into a ball of neutrons about 100 miles in diameter.

This collapse took a few tenths of a second, and created a shock wave that rapidly spread to the outer layers of the star. Vast quantities of neutrinos were created in the collapse, and escaped over the next 10 or so seconds. The shock wave reached the surface of the star 3 hours later, blowing off the surface of the star and starting a burst of light 3 hours behind the burst of neutrinos. The light and neutrinos raced each other for 180,000 years, and the neutrinos were still at least 2 hours ahead when they got to the earth.

When a supernova explodes, the outer shells of hydrogen, helium, carbon, neon, magnesium, sodium, silicon, sulfur and iron are blown out to form a new dust cloud. Such a dust cloud—the Crab Nebula, produced by the 1054 supernova explosion—is shown in Figure (6). From this cloud new stars and planets will form, stars and planets rich in the heavier elements created in the star and recycled into space by the supernova explosion.

Elements heavier than iron are also in the supernova remnants. So much energy is released in the collapsing core of the supernova that elements heavier than iron are created by fusion, even though this fusion costs energy. All the silver and gold in your watchband and ring, the iodine in your medicine cabinet, the mercury in your thermometer, and lead in your fishing sinker, all of these elements which lie beyond iron, were created in the flash of a supernova explosion. Without supernova explosions, the only raw materials for the formation of stars and planets would be hydrogen, some deuterium, and a trace of other light elements left over from the Big Bang that created the universe.

How elements are formed in stars has been a fascinating detective story carried out over the past 40 years. The pioneering work, carried out by William Fowler, Fred Hoyle, and others, involved a careful study of nuclear reactions in the laboratory and then a modeling of how stars should evolve based on the known reactions. On one occasion a nuclear reaction was predicted to exist because it had to be there for stars to evolve. The reaction was then found in the laboratory, exactly as predicted.

The modeling of stellar evolution, using experimental data on nuclear reactions, and large computer programs, has quite successfully predicted the relative abundance of the various elements, as well as many features of stellar evolution such as the expansion of a star into a red giant when the hydrogen fuel in its core is used up. The success of these models leads us to believe that the details we described about what happened in the very core of a star about to explode, actually happened as described. An exciting consequence of the 1987 supernova explosion was that we got a glimpse into the core of the exploding star, a glimpse provided by the neutrinos that took 10 seconds, as predicted, to escape from the core. The neutrinos also arrived three hours before the photons, as predicted by the computer models.

Figure 6a

The Crab Nebula. The arrow points to the pulsar that created the nebula. (Above photo Hale Observatories. 6b: 1950 Photograph by Walter Baade, 1964 by Güigo Munch, composite by Munch and Virginia Trimble.)

NEUTRON STARS

One of the great predictions of astronomy based on the physical properties of matter was made by a young physicist/astronomer S. Chandrasekhar in the 1930s. With some relatively straightforward calculations, Chandrasekhar predicted that if a cooling, collapsing star had a mass greater than 1.4 times the mass of the sun, the force of gravity would be strong enough to cram the atomic electrons down into the nuclei, converting the protons to neutrons, leaving behind a ball of neutrons about 10 miles in diameter.

Chandrasekhar talked about this idea with his sponsor Sir Arthur Eddington, who was, at the time, one of the most famous astronomers in the world. In private, Eddington agreed with Chandrasekhar's calculations, but when asked about them in the 1932 meeting of the Royal Astronomical Society, Eddington replied that he did not believe that such a process could possibly occur. Chandrasekhar's ideas were dismissed by the astronomical community, and a discouraged Chandrasekhar left astronomy and went into the field of hydrodynamics and plasma physics where he made significant contributions.

In 1967, the graduate student Jocelyn Bell, using equipment devised by Antony Hewish, observed an object emitting extremely sharp radio pulses that were 1.337 seconds apart. By the end of the year, up to ten such pulsing objects were detected, one with pulses only 89 milliseconds apart. After eliminating the possibility that the radio pulse was communication from an advanced civilization, it was determined that the signals were most likely from an objects rotating at high speeds. A star cannot rotate that fast unless it is very compact, less than 100 miles in diameter. The only candidate for such an object was Chandresekhar's neutron star.

Many other pulsing stars—*pulsars*—have been discovered. The closest sits at the center of the explosion that created the Crab Nebula seen in Figure (6a). A superposition of photographs taken in 1950 and 1964, Figure (6b), shows that the gas in the Crab Nebula is expanding away from the star marked with an arrow. Taking a high speed moving picture of this star, something that one does not usually do when photographing stars, shows that this star turns on and off **33 times a second** as seen in Figure (7). This is the neutron star left behind when the supernova exploded.

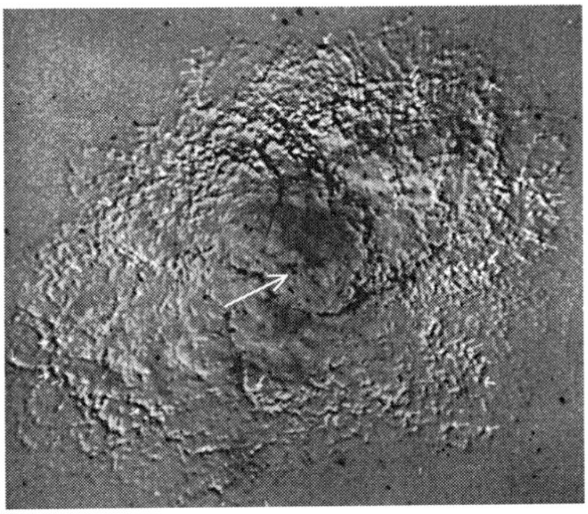

Figure 6b
Expansion of the Crab Nebula. Two photographs, taken 14 years apart, the first printed in white, the second dark, show the expansion centered on the pulsar.

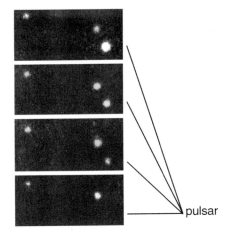

Figure 7
Neutron star in Crab Nebula, turns on and off 33 times/sec. (Exposure from the Lick Observatory.)

We have now been able to study many pulsars, and know that the typical neutron star is a ball of neutrons about 10 miles in diameter, rotating at rates up to nearly 1000 revolutions per second! We can detect neutron stars because they have a bright spot that emits a beam of radiation. We see the pulses of radiation when the beam sweeps over us much as the captain of a ship sees the bright flash from a lighthouse when the beam sweeps past.

Computer models suggest that the bright spot is created by the magnetic field of the star, a field that was tied to the material in the core of the star and was strengthened as the core collapsed. Charged particles in the 'atmosphere' of the neutron star spiral around the magnetic field lines striking the star at the magnetic poles. On the earth, charged particles spiraling around the earth's magnetic field lines strike the earth's atmosphere at the magnetic poles, creating the aurora borealis and aurora australis, the northern and southern lights that light up extreme northern and southern night skies. On a neutron star, the aurora is much brighter, also the atmosphere much thinner, only a few centimeters thick. (In Chapter 28 on Magnetism, we will talk about the motion of charged particles in a magnetic field.)

NEUTRON STARS AND BLACK HOLES

As predicted by Chandrasekhar, when a cooling star is more than 1.4 times as massive as the sun, the gravitational attraction becomes strong enough to overcome the electronic structure of matter, shoving the electrons into the nuclei and leaving behind a ball of neutrons. A neutron star is essentially a gigantic nucleus in which the attractive gravitational force which holds the ball together, is balanced by the repulsive component of the nuclear force which keeps the neutrons from squeezing into each other.

In our discussion of atomic nuclei, it was the attractive component of the nuclear force that was of the most interest. It was the attractive part that overcame the electric repulsion between protons. Now in the neutron star, gravity is doing the attracting and the nuclear force is doing the repelling.

Einstein's special theory of relativity sets a limit on how strong the repulsive part of the nuclear force can be. We can see why with the following qualitative arguments.

The harder it is to shove two nucleons into each other, the stronger the repulsive part of the nuclear force, the more incompressible nuclear matter is. Now in our beginning discussions of the principle of relativity in Chapter 1, we saw that the speed of a sound wave depended upon the compressibility of the material through which the sound was moving. We used a stretched Slinky for our initial demonstrations of wave motion because a stretched Slinky is very easy to compress, with the result that Slinky waves move very slowly, about 1 foot per second.

Air is much more incompressible than a Slinky (try blowing air into a Coke bottle), with the result that sound waves in air travel about 1000 times faster than slinky waves. Water is more incompressible yet, and sound travels through water 5 times faster than in air. Because steel is even more incompressible than water, sound travels even faster in steel, about 4 times faster than in water.

The most incompressible substance known is nuclear matter. It is so incompressible, the repulsive force between nucleons is so great that the calculated speed of sound in nuclear matter approaches the speed of light. And that's the limit. Nothing can be so rigid or incompressible that the speed of sound in the substance exceeds the speed of light. This fact alone tells us that there is a limit to how rigid matter can be, how strong repulsive forces can become. The repulsive part of the nuclear force approaches that limit.

There is, however, no limit to the strength of attractive forces. The attractive gravitational force in a neutron star simply depends upon the amount of mass in the star. The more mass, the stronger the force.

From this we can conclude that we are in serious trouble if the neutron star gets too big. It is estimated that in a neutron star with a mass 4 to 6 times the mass of the sun, the attractive gravitational force will exceed the repulsive component of the nuclear force, and the neutrons will begin to collapse into each other.

As the star starts to collapse, gravity gets still stronger. But gravity has just crushed the strongest known repulsive force. At some point during the collapse, gravity will become strong enough to crush any possible repulsive force. According to the laws of physics, *as we know them*, nothing can stop the further collapse of the star, perhaps down to a mathematical point, or at least down to a size so small that new laws of physics take over.

The problem with finding black holes is that, as their name suggests, they do not emit light. The only way we have of detecting black holes is by their gravitational effect on other objects.

It turns out that a good fraction of the stars in the universe come in binary pairs. Having a pair of stars form is an effective way of taking up the angular momentum of a collapsing gas cloud that was initially rotating. If Jupiter had been just a bit bigger, igniting its own nuclear reactions, then the earth would have been located in a binary star system.

If one of a pair of binary stars is a black hole, two detectable effects can occur. If the stars are in close orbit, the black hole will suck off the outer layers of gas from the visible companion, as indicated in the artist's conception, Figure (12). According to computer models, the gas that is drawn off from the companion star goes into orbit around the black hole, forming what is called an *accretion disk* around the black hole. The gas in the accretion disk is moving very rapidly, at speeds approaching the speed of light. As a result of the high speeds, and turbulence in the flow, the particles in the accretion disk emit vast quantities of X rays as they spiral down toward the black hole. Strong X ray emission is thus a signature that an object may be a binary star system with one of the stars being a black hole.

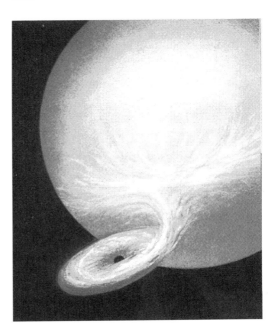

Figure 8
Painting of the gas from a blue giant star being sucked into a black hole. (From the May, 1974 National Geographic, Artist Victor J. Kelley.)

Since X rays can be emitted in other ways, a further check is needed to be sure that a black hole is involved. By studying the orbit of the visible companion, one can determine the mass of the invisible one (essentially using Kepler's third law, in a slightly modified form). The test of whether the invisible companion is a black hole is whether its mass is over 6 solar masses. If it is, then no dark object could withstand the gravitational forces involved. The first candidate for an object fitting this description is the X ray source in the constellation of Cygnus, an object known as *Cygnus X1*.

Rather than being scarce, hard–to–find objects, black holes may play a significant role in the structure of the universe. There is good evidence, from the study of the motions of stars, that a gigantic black hole, with a mass of millions of solar masses, may lie at the heart of our galaxy and other galaxies as well. And recent studies have indicated that there may be a black hole at the center of globular clusters. Wherever black holes may be, whatever their role in our universe, one fact stands out—gravity, a force too weak even to be detected on an atomic scale, can under the right circumstances become the strongest force of all, strong enough to crush matter out of existence.

P2000 Index

Symbols

µ zero, definition of   31-19
π meson   40-23
ΔEΔt>=h, Uncertainty principle   40-19
ΔxΔp>=h, Uncertainty principle   40-15
10 dimensions, String theory   int-16
100 billion to one, Matter over anti matter   int-30
13.6 eV, hydrogen spectrum   35-4
1836 times, proton/electron mass ratio   int-17
1987 supernova   6-14, 20-14
2.74 degrees, cosmic background radiation   int-29
2D or 3D? equipotential plotting experiment   25-8
4 dimensions   int-16

A

Aberration
   Astigmatism   Optics-21
   Chromatic   Optics-21
      Newton's reflecting telescope   Optics-22
   Spherical   Optics-21
      In Hubble telescope mirror   Optics-22
Absolute zero   17-9, 17-21
Abundance of the elements   34-24
AC voltage generator   30-21
   Magnetic flux in   30-21
Accelerating field in electron gun   26-10
Acceleration
   Angular   12-3
   Angular analogy   12-3
   Calculus definition of   4-5, Cal 1-7
      Component equations   Cal 1-8
      Vector equation   Cal 1-7
   Constant acceleration formulas
      Calculus derivation   4-9, Cal 1-20
      In three dimensions   4-11, Cal 1-22
   Definition of   3-13
   Due to gravity   3-21
   From a strobe photograph   3-15
   Intuitive discussion   3-20
   On inclined plane   9-11
   Radial   12-5
   Tangential   12-4
   Uniform circular motion
      Direction of   3-18
      Magnitude of   3-18
   Vector, definition of   3-15
Acceleration versus time graphs   4-7
Accelerators, particle   int-1, 28-22
Accurate values of Fourier coefficients   16-32
Adding sines and cosines in Fourier analysis   16-28
Addition of charge   19-10
Addition of forces   9-2

Adiabatic expansion
   Calculation of work   18-26
   In Carnot cycle   18-11
   Introduction to   18-9
Air cart
   Analysis of coupled carts   16-12
   Construction of   6-2
   In impulse experiments   11-9
   In recoil experiments   6-2
   Oscillating cart   14-5
   Speed detector   30-5. See also Experiments II: - 6- Faraday's law air cart speed detector
Air Resistance
   Calculus analysis for projectile motion   4-12
   Computer analysis for projectile motion   5-24, 8-3
   Strobe analysis for projectile motion   3-22
Airplane wing, Bernoulli's equation   23-13
Allowed orbits, Bohr theory   int-8, 35-1
Allowed projections, spin   39-3
Allowed standing wave patterns   37-1
Alpha particles   20-8
Amount of sin(3t) present in a wave   16-28
Ampere
   Definition of   27- 2
   MKS units   24-2
Ampere's law
   Applied to a solenoid   29-15
   Chapter on   29-1
   Derivation of line integral   29-7
   Field of straight wire   29-11
   Final result   29-11
   Maxwell's correction to   32- 4
Amplitude
   And intensity, Fourier analysis lecture   16-33
   And phase
      Fourier analysis lecture   16-31
      Wave motion   15-17
   Diffraction pattern by Fourier analysis   16-33
   Fourier coefficients   16-32
   Of a sine wave   Cal 1-37
Analysis
   Fourier   16-6
   Of coupled air carts   16-12
   Of path 1 for electromagnetic pulse   32- 14
   Of path 2 for electromagnetic pulse   32- 16
Analytic solution
   Of the RC circuit   27- 22
   Oscillation of mass on spring   14-7
   Projectile motion with air resistance   4-12
Anderson, C., positrons   int-13
Andromeda galaxy   int-2, int-3, 1-22
Angle of reflection (scattering of light)   36-3, Optics-1
Angles of incidence and reflection   Optics-3
Angular acceleration   12-3
Angular analogy   12-3
   For Newton's second law   12-14
   Torque (angular force)   12-15

Angular frequency
   Definition of   14-4
   Wave motion   15-14
Angular magnification of magnifier   Optics-39
Angular mass
   Moment of inertia   12-7
   Rotational kinetic energy   12-22
Angular momentum
   As a Vector   7- 14, 12-7
      Movie   7- 15
   Bohr model   35-1, 35-8
      Planck's constant   35-8
   Conservation of   7- 9, 12-16
      Derivation from F = ma   12-16
      X. See also Experiments I: - 4- Conservation of angular momentum
   Definition of   7- 10
   Definition of, more general   7- 12
   Definition of, still more general   12-6
   Definition of, cross product   12-11
   Formation of planets   7- 17
   Gyroscopes   12-18
   Kepler's second law   8-32
   Magnetic moment   31-24
   Movie on vector nature   7- 15, 12-6, 12-17
   Of bicycle wheel   12-6
   Projections of, classical   7- 14
   Projections of electron spin   39-3
   Quantized   int-9
   Quantized projections   38-5
   Quantum number   38-7
Angular velocity
   As a vector   12-7
   Definition of   12-2
   Mass on spring   14-9
   Oscillating cart   14-5
Annihilation of antimatter   34-17
Antielectron   int-13
Antielectron type neutrino   int-22
Antimatter   34-16
   Annihilation of   34-17
   Excess of matter over, in early universe   34-17, 34-29
   Introduction to   int-12, 34-16
   Neutrino   int-22
   Neutron   int-13
   Positron   int-13
   Positron electron pair   34-17
   Proton   int-13
   Wave equation for   15-2
Antiparticle   int-13
   Created by photon   34-17
Applications of Bernoulli's equation   23-12
   Airplane wing   23-13
   Aspirator   23-16
   Hydrostatics   23-12
   Leaky tank   23-12
   Sailboat   23-14
   Venturi meter   23-15

Applications of Faraday's law 30-21
   AC voltage generator 30-21
   Gaussmeter 30-23
Applications of Newton's Second Law 9-1
Applications of the second law of thermodynamics 18-17
Arbitrary wave, Fourier analysis 16-28
Area
   As a vector 24-22
   Negative or positive 16-29
   Related to integration Cal 1-11
   Under the curve Cal 1-12
Arecibo radio telescope int-15, Optics-48
Arithmetic of vectors. *See also* Vector
   Addition 2-3
   Associative law 2-4
   Commutative law 2-4
   Multiplication by number 2-5
   Negative of 2-5
   Scalar or dot product 2-12, 10-13
   Subtraction of 2-5
   Vector cross product 2-15, 12-9
Aspirator, Bernoulli's equation 23-16
Associative law, Exercise 2-7
Astigmatism Optics-21
Astronomy
   1987 supernova int-19, 6-14, 20-14
   Abundance of the elements 34-24
   Big bang model of universe 33-25, 34-26
   Binary stars int-2
   Black dwarf star int-19
   Black holes 10-29
      Introduction to int-19
   Blackbody radiation, color of stars 34-2
   Copernicus 8-25
   Crab nebula 20-16
   Decoupling of light and matter 34-31
   Doppler effect 33-23
   Eagle nebula 7-18, Optics-44
   Early universe int-27, 34-29
   Escape velocity 10-28
   Evolution of the universe 34-21
   Excess of matter over antimatter 34-29
   Expanding universe, Hubble int-3
   Formation of planets 7-17
   Galaxy
      Andromeda int-2
      Introduction to int-2
      Most distant int-3
      Sombrero int-2
   General relativity 8-29
   Globular cluster 11-2
   Gravitational lensing 34-20
   Helium abundance in universe 34-26
   Helium core of massive star 20-15
   Hubble rule for expanding universe int-3
   Iron core of massive star 20-15
   Kepler's laws 8-24
   Light years int-2
   Magnetic field of the earth 28-11
   Models of the universe 34-23
   Neutrino 6-14, 11-21
   Neutron star
      And black holes 20-18
      In Crab nebula 20-17
      Introduction to int-19
   Nuclear fusion and stellar evolution 20-12
   Orion nebula 7-17
   Penzias and Wilson, cosmic radiation 34-27
   Powering the sun 34-23
   Ptolemy, epicycle in Greek astronomy 8-25
   Quantum fluctuations in space 40-25
   Quasar, gravitational lens 34-20
   Radio galaxy Optics-48
   Radio images of variable star Optics-49
   Radio telescope. See Radio telescope
   Radio telescope, three degree radiation int-30, 34-27
   Radio telescopes Optics-48
   Red shift and expanding universe int-3, 33-24, 34-21
   Red supergiant star 20-15
   Retrograde motion of Mars 8-24
   Space travel and time dilation 1-22
   Star, blackbody spectrum 34-3
   Steady state model of the universe 34-25
   Stellar evolution int-19
   Telescopes Optics-40
      Arecibo radio telescope Optics-48
      Galileo's Optics-41
      Hubble Space Telescope Optics-44
      Issac Newton's Optics-42
      Mt. Hopkins Optics-43
      Mt. Palomar Optics-43
      Very Large Array, radio telescopes Optics-48
      Very Long Baseline Array (VLBA) Optics-49
      William Hershel's Optics-43
      World's Largest Optical, Keck Optics-45
      Yerkes Optics-41
   Thermal equilibrium of the universe 34-28
   Three degree cosmic radiation int-29, 34-27
   Tycho Brahe 8-25
   Van Allen radiation belts 28-32
   Visible universe int-3
   White dwarf star 20-15
Atmospheric pressure 17-23
Atomic
   And molecular forces, electric interaction 19-1
   Clocks 1-21
   Microscopes 17-1
      Scanning Tunneling Microscope Optics-51
   Processes 17-4
   Spectra 33-16
   Structure 19-3
   Units 19-22

Atoms
   Angular momentum quantum number 38-7
   Atomic nucleus, chapter on 20-1
   Atomic processes 17-4
   Avogadro's law 17-24
   BASIC program, hydrogen molecule ion 19-24
   Beryllium in periodic table 38-13
   Bohr model int-8
   Boron in periodic table 38-13
   Brownian motion 17-7
   Chapter on 17-1, 38-1
   Classical hydrogen atom 35-2
   Effective nuclear charge 38-12
   Electron binding energy 19-20, 38-11
   Electron energy in hydrogen molecule ion 19-21
   Electron spin 38-9
   Equipartition of energy 17-28
   Expanded energy level diagram 38-8
   Failure of classical physics 17-31
   Freezing out of degrees of freedom 17-32
   Heat capacity 17-26
   Hydrogen molecule 19-16
   Introductory view of int-16
   Ionic bonding 38-15
   $L=0$ Patterns in hydrogen 38-4
   Lithium 38-12
   Model atom 37-4
   Molecular and atomic processes 17-1
   Molecular forces 19-15
   Multi electron 38-9
   Nuclear matter, chapter on 20-1
   Nucleus. *See also* Nuclear
     Discovery of 11-19
   Particle-wave nature of matter int-10
   Pauli exclusion principle 38-9
   Periodic table 38-10
   Potassium to krypton 38-14
   Precession of, in magnetic field 39-15
   Quantized projections of angular momentum 38-5
   Schrödinger's equation for hydrogen 38-2
   Silicon, surface (111 plane) of Optics-51
   Sodium to argon 38-13
   Standing wave patterns in hydrogen 38-3
   Table of 19-5
   Thermal motion of 17-6
   Up to neon 38-13
   Xenon, photograph of 17-1
Atwood's machine 9-16
Avogadro's law 17-24
Avogadro's number, the mole 17-24

B

Balancing weights, equilibrium 13-2
Ball Spring Pendulum. *See* Pendulum: Spring
Balmer series
   Energy level diagram for 35-6
   Formula from Bohr theory 35-5
   Hydrogen spectrum 35-4
   Introduction to, hydrogen star 33-19

Barometer, mercury, pressure measurement 17-22
Basic electric circuits 27-1
BASIC program. *See also* Computer
   Calculating circle 5-6
   Calculational loop for satellite motion 8-19
   Comment lines in 5-7
   Computer time step 5-14
   Conservation of angular momentum 8-32
   Conservation of energy 8-35
   DO LOOP 5-4
   For drawing circle 5-11
   For hydrogen molecule ion 19-24
   For oscillating cart 14-32
   For oscillatory motion 14-21, 14-30
   For projectile motion 5-18, 5-19, 5-21, 8-21
   For projectile motion with air resistance 5-22
   For satellite motion 8-21
   For spring pendulum 9-20
   Kepler's first law 8-26
   Kepler's second law 8-27
   Kepler's third law 8-28
   LET Statement 5-5
   Modified gravity 8-29
   Multiplication 5-6
   New calculational loop 8-17
   Orbit-1 program 8-21
   Perihelion, precession of 8-30
   Plotting a point 5-6
   Plotting window 5-7
   Prediction of satellite orbits 8-16
   Satellite motion laboratory 8-23
   Selected printing (MOD command) 5-10
   Sine wave products 16-29
   Unit vectors 8-18
   Variable names 5-6
Bathtub vortex 23-2
Baud rate, for fiber optics Optics-14
Bell Telephone Lab, electron waves 35-12
Berkeley synchrotron 28-22
Bernoulli's equation
   Applications of
     Airplane Wing 23-13
     Aspirator 23-16
     Leaky tank 23-12
     Sailboat 23-14
     Venturi meter 23-15
   Applies along a streamline 23-11
   Care in applying 23-16
   Derivation of 23-9
   Formula for 23-11
   Hydrodynamic voltage 23-17
Beryllium
   Binding energy of last electron 38-12
   In periodic table 38-13
Beta decay
   And energy conservation int-21
   Neutrinos 20-6
   Neutrons 20-7
   Protons 20-7
   Recoil experiment 6-6

Beta, Hans, proton cycle, energy from sun   34-23
Beta ray   int-21
Betatron   30-16
Bi-concave lens   Optics-27
Bi-convex lens   Optics-27
Bicycle wheel
   As a collection of masses   12-5
   As a gyroscope   12-18
   Right hand rule for rotation   12-11
   Vector nature of angular momentum   7- 14, 12-6, 12-17
Big bang model of universe   int-4, 33-25, 34-26
Binary stars   int-2
Binding energy
   Hydrogen molecule ion   19-23
   Molecular forces   17-13
   Nuclear   20-9
   Nuclear stability   20-10
   Of inner electrons   38-12
Binomial expansion   1-31, Cal 1-23
Black dwarf star   int-19
Black holes
   And neutron stars   20-18
   Critical radius for sun mass   10-30
   Introduction to   int-19
   Stellar evolution   int-20
   Theory of   10-29
Blackbody radiation
   Electromagnetic spectrum   32- 22
   Photon picture of   34-22
   Planck's formula   34-4
   Theory of   34-2
   Wein's displacement law   34-2
Blood flow, fluid dynamics   23-23
Bohr magneton
   Dirac wave equation   39-5
   Unit of magnetic moment   39-4
Bohr model   int-8
   Allowed orbits   35-1
   Angular momentum   35-1, 35-8
   Chapter on   35-1
   De Broglie explanation   35-1
   Derivation of   35-8
   Energy levels   35-4
   Introduction to   int-8
   Planck's constant   35-1, 35-8
   Quantum mechanics   35-1
   Rydberg constant   35-9
Bohr orbits, radii of   35-7
Boltzman
   Constant   17-11
   Formula for entropy   18-24
Bonding
   Covalent   19-15
   Ionic   38-15
Born interpretation of particle waves   40-6

Boron
   Binding energy of last electron   38-12
   In periodic table   38-13
Bottom quark   int-24
Bragg reflection   36-4
Brahe, Tycho   8-25
Brownian motion
   Discussion   17-7
   Movie   17-7
Bubble chambers   28- 26
Bulk modulus   15-8
Button labeled ø on MacScope   16-32

c

c (speed of light)   int-2, 1-12. *See also* Speed of light
Calculating Fourier coefficients   16-28
Calculational loop   5-17
   For projectile motion   5-19, 8-17
   For projectile motion with air resistance   5-24
   Satellite Motion   8-19
Calculations
   Computer, step-by-step   5-1
   Of flux   24-22
   Of integrals   Cal 1-11
Calculus
   And the uncertainty principle   4-1, Cal 1-3
   Calculating integrals   Cal 1-11
   Calculus in physics   4-1, Cal 1-3
   Chain rule   Cal 1-25
   Definition of acceleration   4-5, Cal 1-7
      Component equations   Cal 1-8
      Vector equation   Cal 1-7
   Definition of velocity   4-3, Cal 1-5
      Component equations   Cal 1-8
      Vector equation   Cal 1-6
   Derivation, electric force of charged rod   24-6
   Derivation of constant acceleration formulas   4-9, Cal 1-20
      In three dimensions   4-11, Cal 1-22
   Limiting process   4-1, Cal 1-3, Cal 1-5
      Vector equation for   Cal 1-5
   Line integral   29-5
   Special chapter on   Cal 1-3
   Surface integral   29-2
Calibration of force detector   11-10
Camera
   Depth of field   Optics-34
   Pinhole   Optics-35
   Single lens reflex   Optics-33
Capacitance
   Electrical   27- 16
   Introduction to   27- 14

Capacitor
  Electrolytic 27-17
  Energy storage in 27-18
  Examples of 27-17
  In circuits. See also Circuits
    As circuit elements 27-20
    Hydrodynamic analogy 27-14
    LC circuit 31-10
    Parallel connection 27-20
    RC circuit 27-22
    Series connection 27-21
  Introduction to 27-14
  Magnetic field in 32-6
  Parallel plate
    Capacitance of 27-16
    Deflection plates 26-16
    Introduction to 26-14
    Voltage in 26-15
Carbon
  Burning in oxygen 17-5
  Graphite crystal, electron diffraction 36-8
Carnot cycle
  As thought experiment 18-4
  Efficiency of
    Calculation of 18-28
    Discussion 18-12
    Formula for 18-13, 18-29
    Reversible engines 18-18
  Energy flow diagrams 18-15
  Entropy 18-22
  Introduction to 18-11
  Maximally efficient engines 18-15
  Refrigerator, energy flow diagrams 18-15
  Reversible engines 18-13
  Reversiblility 18-17
  Second law of thermodynamics 18-4
Cassegrain telescope Optics-42
Cavendish experiment 8-7
Center of mass
  Diver movie 11-1
  Dynamics of 11-4
  Formula for 11-3
  Gravitational force acting on 13-4
  Introduction to 11-2
Center of our galaxy Optics-47
Cerenkov radiation Optics-10
CERN
  Electroweak theory int-26
  Proton synchrotron at 28-24
CGS units
  Classical hydrogen atom 35-2
  Coulomb's law 24-2
  Definition of electric charge 19-8
Chain rule Cal 1-25
  Proving it (almost) Cal 1-26
  Remembering it Cal 1-25
Chaos 23-1

Charge
  Addition of 19-10
  Conservation of int-21
  Density, created by Lorentz contraction 28-6
  Discussion of int-6
  Electric, definition of (CGS units) 19-8
  Fractional (quarks) int-24, 19-15
  Magnetic moment for circular orbit 31-24
  On electron, Millikan oil drop experiment 26-17
  Positive and negative int-6, 19-10
  Quantization of electric 19-14
  Surface 26-2
  Unit test 24-11
Charges, static, line integral for 30-2
Charm quark int-24
Chemistry. See Atoms: Angular momentum quantum number
Cholera molecule 17-2
Chromatic aberration Optics-21
  Newton's reflecting telescope Optics-22
Ciliary muscle, eye Optics-31
Circuits
  Basic 27-1
  Grounding 26-8
  Inductor as a circuit element 31-7
  Kirchoff's law 27-10
  LC circuit
    Experiment 31-13
    Fourier analysis of 31-31
    Introduction to 31-10
    Ringing like a bell 31-36
  LR circuit
    Exponential decay 31-9
    Introduction to 31-8
  Neon oscillator circuit 27-29
  Power in 27-9
  RC circuit
    Exponential decay 27-23
    Exponential rise 27-26
    Initial slope 27-25
    Introduction to 27-22
    Measuring time constant 27-25
    Time constant 27-24
    X. See Experiments II: -3- The RC Circuit
  Short 27-9
  Simple 27-8
  The voltage divider 27-13
Circular electric field 30-13
  Line integral for 30-13
Circular motion
  Force causing 8-2
  Particles in magnetic field 28-20
  Uniform
    Introduction to 3-17
    Magnitude of acceleration 3-18

Circular orbit, classical hydrogen atom   35-2
Circular wave patterns, superposition of   33-2
Classical hydrogen atom   35-2
Classical physics   int-7
Clock
   Atomic clocks   1-21
   Lack of simultaneity   1-32
   Light pulse clock   1-14
   Muon clock   1-20
   Time dilation   1-22
Cluster, globular   11-2
Cochlea (inside of the ear)   16-34
Coefficient of friction   9-13
Coefficients, Fourier (Fourier analysis lecture)   16-28
Coil
   As a circuit element   31-7
   Field of a solenoid   28-17, 29-14
   Inductance of   31-5
   Magnetic field of Helmholtz coils   28-19
   Primary   30-26
   Toroidal   31-6
     In LC circuit   31-11
   Torroidal   29-17
Coil, primary   30-26
Collisions
   Discovery of the atomic nucleus   11-19
   Energy loss   11-14
   Experiments on momentum conservation   7-4
   Force detector   11-10
   Impulse   11-9
   Introduction to   11-9
   Momentum conservation during   11-13
   Subatomic   7-7
   That conserve momentum and energy (elastic)   11-16
   X. See Experiments I: - 8- Collisions
Color force   19-15
Colors
   And Fourier analysis   16-28
   Blackbody radiation   32-22
   Color of stars   34-2
   Electromagnetic Spectrum   32-20
   Glass prism and rainbow of colors   Optics-15
Comment lines, computer   5-7
Commutative law, exercise on   2-7
Compass needles, direction of magnetic field   28-12
Component sine wave, Fourier analysis   16-28
Components, vector   Cal 1-7
   Formula for cross product   2-17
   Introduction to   2-8
Compton scattering, photon momentum   34-15

Computer. *See also* BASIC program
   BASIC. *See also* BASIC program
   Calculations
     Introduction to   5-2
     Step-by-step   5-1
   Commands
     Comment lines   5-7
     DO LOOP   5-4
     LET Statement   5-5
     Multiplication notation   5-6
     Selected Printing (MOD command)   5-10
     Variable names   5-6
   English program
     For projectile motion   5-16, 5-19
     For satellite motion   8-19
   Plot of electric fields
     Field plot model   25-12
     In electron gun   26-13
     Of various charge distributions   24-19
   Plotting
     A point   5-6
     Crosses   5-11
     Window   5-7
   Prediction of motion   5-12
     Chapter on   5-1
     Satellite orbits   8-16
     Satellite with modified gravity   8-30
   Program for
     Air resistance   5-24
     Damped harmonic motion   14-34
     Harmonic motion   14-30
     Hydrogen molecule ion   19-24
     Plotting a circle   5-2, 5-4, 5-11
     Projectile motion, final one   5-21
     Projectile motion, styrofoam projectile   5-28
     Projectile motion with air resistance   5-22
     Satellite motion   8-21
   Programming, introduction to   5-4
   Satellite motion calculational loop   8-19
   Time Step and Initial Conditions   5-14
Computer analysis of satellite motion. *See* Experiments I: - 5- Computer analysis of satellite motion
Computer prediction of projectile motion. *See* Experiments I: - 2- Computer prediction of projectile motion
Computers
   Why they are so good at integration   Cal 1-12
Conductors
   And electric fields, chapter on   26-1
   Electric field in hollow metal sphere   26-4
   Electric field inside of   26-1
   Surface charge density   26-3
Cones, nerve fibers in eye   Optics-31
Conical pendulum   9-18
   And simple pendulum   14-17

Conservation of
  Angular momentum 8-32
    Derivation from F = ma 12-16
    Introduction to 7-9
  Electric charge 19-13
  Energy int-11, 8-35
    Feynman's introduction to 10-2
    Mass on spring 14-11
    Uncertainty principle 40-24
    Work Energy Theorem 10-20
    X. See Experiments I: - 9- Conservation of energy
  Energy and momentum, elastic collisions 11-16
  Linear and angular momentum, chapter on 7-1
  Linear momentum 7-2, 11-7
    during collisions 11-13
Conservative force 25-5
  And non-conservative force 10-21
  Definition of 29-6
Conserved field lines, flux tubes 24-17
Constant acceleration formulas
  Angular analogy 12-3
  Calculus derivation 4-9, Cal 1-20
    In three dimensions 4-11, Cal 1-22
Constant, integral of Cal 1-13
Constant voltage source 27-15
Continuity equation
  For electric fields 24-14
  For fluids 23-5
Continuous creation theory int-4
Contour map 25-1
Contraction, Lorentz relativistic 1-24
Cook's Bay, Moorea, rainbow over Optics-17
Coordinate system, right handed 2-18
Coordinate vector
  Definition of 3-11
  In computer predictions 5-12, 8-17
  In definition of velocity vector 3-13
Cornea Optics-31
Corner reflector
  How it works Optics-7
  On the surface of the moon Optics-7
Cosine function
  Amplitude of Cal 1-37
  Definition of Cal 1-35
  Derivative of Cal 1-38
Cosine waves
  Derivative of 14-8
  Fourier analysis lecture 16-28
  Phase of 14-6
Cosmic background neutrinos int-30
Cosmic background radiation int-30, 34-27

Cosmic radiation int-30
Cosmic rays int-13
Coulomb's law
  And Gauss' law, chapter on 24-1
  Classical hydrogen atom 35-2
  For hydrogen atom 24-4
  For two charges 24-3
  Units, CGS 24-2
  Units, MKS 24-2
Coupled air cart system, analysis of 16-12
Covalent bonding 19-15
$Cp = Cv + R$, specific heats 18-7
Cp and Cv, specific heats 18-6
Crab nebula 20-16
Creation of antimatter, positron-electron pairs 34-17
Critical damping 14-23
Cross product
  Angular momentum 12-11
  Component formula for 2-17
  Discussion of 2-15
  Magnitude of 2-17
  Review of 12-9
  Right hand rule 12-10
Crystal
  Diffraction by Thin 36-6
  Graphite, electron diffraction by 36-8
  Structures
    Graphite 36-8
    Ice, snowflake 17-4
  X ray diffraction 36-5
Crystalline lens, eye Optics-31
Current and voltage
  Fluid analogy 27-6
  Ohm's law 27-7
  Resistors 27-6
Current, electric
  Inertia of (inductance) 31-12
  Introduction to 27-2
  Magnetic force on 31-18
  Positive and negative 27-3
Current loop
  Magnetic energy of 31-22
  Torque on 31-20
Currents
  Magnetic force between 28-14, 31-19
Curve
  Area under, integral of Cal 1-12
  Slope as derivative Cal 1-30
  That increases linearly, integral of Cal 1-13
  Velocity, area under Cal 1-12
Curved surfaces, reflection from Optics-3
Cycle, Carnot 18-11

D

Damped harmonic motion
   Computer program for   14-34
   Differential equation for   14-21
Damping, critical   14-23
Davisson & Germer, electron waves   35-12
De Broglie
   Electron waves   int-10, 35-11
   Formula for momentum   35-11
   Hypothesis   35-10
   Introduction to wave motion   15-1
   Key to quantum mechanics   35-1
   Wavelength, formula for   35-11
   Waves, movie of standing wave model   35-11
Debye, on electron waves   37-1, 38-2
Decay
   Exponential decay   Cal 1-32
Decoupling of light and matter in early universe   34-31
Definite integral
   Compared to indefinite integrals   Cal 1-14
   Defining new functions   Cal 1-15
   Introduction to   Cal 1-11
   Of velocity   Cal 1-11
   Process of integrating   Cal 1-13
Deflection plates in electron gun   26-16
Degrees of freedom
   Freezing out of   17-32
   Theory of   17-28
Depth of field, camera   Optics-34
Derivative
   As a limiting process   Cal 1-6, Cal 1-18, Cal 1-23, Cal 1-28, Cal 1-30
   Constants come outside   Cal 1-24
   Negative slope   Cal 1-31
   Of exponential function e to the x   Cal 1-28
   Of exponential function e to the ax   Cal 1-29
   Of function x to the n'th power   Cal 1-24
   Of sine function   Cal 1-38
Derivative as the Slope of a Curve   Cal 1-30
Descartes, explanation of rainbow   Optics-16
Description of motion   3-3
Detector for radiated magnetic field   32-26
Diagrams, PV (pressure, volume)   18-8
Differential equation
   For adiabatic expansion   18-27
   For damped harmonic motion   14-21
   For forced harmonic motion   14-25, 14-28
   For LC circuit   31-10
   For LR circuit   31-9
   For oscillating mass   14-8
   Introduction to   4-14
Differentiation. *See also* Derivative
   Chain rule   Cal 1-25
   More on   Cal 1-23

Differentiation and integration
   As inverse operations   Cal 1-18
      Velocity and position   Cal 1-18, Cal 1-19
   Fast way to go back and forth   Cal 1-20
   Position as integral of velocity   Cal 1-20
   Velocity as derivative of position   Cal 1-20
Diffraction
   By thin crystals   36-6
   Electron diffraction tube   36-9
   Of water waves   33-5
   X Ray   36-4
Diffraction, electron. *See* Experiments II: -11- Electron diffraction experiment
Diffraction grating   33-12. *See also* Experiments II: -10- Diffraction grating and hydrogen spectrum
Diffraction limit, telescopes   Optics-45
Diffraction pattern   16-33, 33-5
   Analysis of   36-11
   By strand of hair   36-14
   Electron   36-10
   For x rays   36-5
   Of human hair   36-14
   Recording   33-28
   Single slit   33-27
   Student projects   36-13
   Two-slit   33-6
Dimensional analysis
   For predicting the speed of light   15-9
   For predicting the speed of sound   15-6
Dimensions
   Period and frequency   14-4
   Using, for remembering formulas   14-4
Dirac equation
   Antimatter   int-13, 15-2, 34-16
   Electron spin   39-3
      Bohr magneton   39-5
Dirac, P. A. M., prediction of antiparticles   int-13
Direction an induced electric field   31-3
Direction of time
   And strobe photographs   3-27
   Dive movie   18-1
   Entropy   18-25
   Neutral K meson   18-25
   Rising water droplets   18-3
Discovery of the atomic nucleus   11-19
Disorder
   Direction of time   18-25
   Entropy and the second law of thermodynamics   18-4
   Formula for entropy   18-24
Displacement vectors
   From strobe photos   3-5
   Introduction to   2-2
Distance, tangential   12-4
Distant galaxies, Hubble photograph   int-3
Dive movie, time reversed   18-1
Diver, movie of   11-1

Diverging lenses   Optics-26
DO LOOP, computer   5-4
Doppler effect
   Astronomer's Z factor   33-23
   For light   33-22
   In Astronomy   33-23
   Introduction to   33-20
   Relativistic formulas   33-22
   Stationary source, moving observer   33-21
   Universe, evolution of   34-21
Dot product
   Definition of   2-12
   Interpretation   2-14
   Work and energy   10-13
Down quark   int-24
Drums, standing waves on   16-22
Duodenum, medical imaging   Optics-15

E

e - charge on an electron   19-9
$E = hf$, photoelectric effect formula   34-7
$E = mc^2$, mass energy   int-11, 10-3
E.dl meter   30-18
Eagle nebula
   Big photo of   7-18
   Hubble telescope photo   Optics-44
   Planet formation   7-16
Ear, human
   Inside of cochlea   16-34
   Structure of   16-15
Early universe. See Universe, early
Earth
   Gravitational field inside of   24-24
   Mass of   8-8
Earth tides   8-12
Eclipse expedition, Eddington   34-19
Edit window for Fourier transform data   16-32
Effective nuclear charge, periodic table   38-12
Efficiency
   Of Carnot cycle, calculation of   18-26
   Of electric cars   18-19
   Of heat pump   18-19
   Of reversible engines   18-18
Einstein
   General relativity   int-15, 8-29
   Mass formula   6-10
   Photoelectric effect   int-8
   Photoelectric effect formula   34-7
   Principle of relativity, chapter on   1-12
Einstein cross, gravitational lens   34-20
Elastic collisions, conservation, energy, momentum   11-16
Elasticity of rubber   17-35
Electric and weak interactions unified   19-3
Electric cars, efficiency of   18-19
Electric charge
   Conservation of   19-13
   Definition (CGS units)   19-8
   Definition (MKS units)   31-19
   Quantization of   19-14

Electric circuits
   Basic   27-1
   Grounding   26-8
   Kirchoff's law   27-10
   LC circuit, oscillation of   31-10
   LR circuit, exponential decay of   31-9
   Power in   27-9
   RC circuit
      Equations for   27-22
      Exponential decay of   27-23
      Half life   27-25
      Initial slope   27-25
      Time constant   27-24
   The voltage divider   27-13
Electric current
   Inertia of, due to inductance   31-12
   Introduction to   27-2
   Positive and negative   27-3
Electric discharge of Van de Graaff generator   26-7
Electric field
   And conductors   26-1
   And light   int-7, 32-20
   Circular electric field
      Introduction to   30-13
      Line integral for   30-13
      The betatron   30-16
   Computer plot, -3,+5 charges   24-19
   Computer plotting programs   25-12
   Continuity equation for   24-14
   Contour map   25-1
   Created by changing magnetic flux   31-2
   Created by moving magnetic field, Lorentz force   30-9
   Direction of, when created by magnetic flux   31-3
   Energy density in   27-19
   Equipotential lines   25-3
   Flux, definition of   24-15
   Gauss' law   24-20
   In electromagnetic waves   32-18
   Inside a conductor   26-1
   Integral of E.dl meter   30-20
   Introduction to   24-10
   Line integral of   30-14
   Lines   24-12
   Mapping   24-12
   Mapping convention for   24-17
   Of a line charge
      Using calculus   24-6
      Using Gauss' law   24-21
   Of electromagnet   30-15
   Of static charges, conservative field   30-2, 30-16
   Radiation by line charge   32-28
   Radiation by point charge   32-30
   Van de Graaff generator   26-6
Electric force
   Between garden peas   19-12
   Produced by a line charge   24-6
   Produced by a short rod   24-9

Electric force law
    Four basic interactions   19-2
    In CGS units   19-8
    Introduction to   19-7
    Lorentz force law, electric and magnetic forces   28-15
Electric force or interaction   int-6, int-13
    Atomic & molecular forces   19-1
    Electroweak theory   int-26
    Strength of
        Between garden peas   28- 2
        Comparison to gravity   int-6, 19-8
        Comparison to nuclear force   int-18, 20-2
        Origin of magnetic forces   28- 6
Electric potential
    Contour map   25-1
    Field plots   25-1
    Of a point charge   25-5
    Plotting experiment   25-7
Electric potential energy. See Potential energy: Electric
Electric voltage. See also Voltage
    Introduction to   25-6
    Van de Graaff generator   26-6
Electrical capacitance   27- 16. See also Capacitor
Electrically neutral   int-6
Electromagnet   31-28
Electromagnetic radiation
    Energy radiated by classical H atom   35-3
    Observed by telescopes
        Infrared   Optics-46
        Radio   Optics-48
        Visible   Optics-42
    Pulse   32- 10
        Analysis of path 1   32- 14
        Analysis of path 2   32- 16
        Calculation of speed   32- 14
Electromagnetic spectrum   int-7, 32- 20, 34-11
    Photon energies   34-11
Electromagnetic waves   32- 18
    Probability wave for photons   40-7
Electron
    Beam, magnetic deflection   28- 9
    Charge on   19-8
    Diffraction Pattern   36-10
    In classical hydrogen atom   35-2
    Lepton family   int-22
    Mass in beta decay   6-7
    Motion of in a magnetic field. See Experiments II: - 5- Motion of electrons in a magnetic field
    Radius   39-3
    Rest energy in electron volts   26-12
    Spin
        And hydrogen wave patterns   38-9
        Chapter on   39-1
    Spin resonance
        Details of experiment   39-9
        Introduction to   39-5
        X. See Experiments II: -12- Electron spin resonance
    Stability of   19-14
    Two slit experiment for   40-3

Electron binding energy
    A classical approach   19-21
    And the periodic table   38-11
    In classical hydrogen molecule ion   19-23
Electron diffraction experiment   36-8
    Diffraction tube   36-9
    X. See Experiments II: -11- Electron diffraction experiment
Electron gun
    Accelerating field   26-10
    Electron volt   26-12
    Equipotential plot   26-11, 26-13
    Filament   26-9
    In magnetic field
        Bend beam in circle   28- 20
        Magnetic focusing   28- 29
    Introduction to   26-8
    X. See Experiments II: - 2- The Electron Gun
Electron positron pair   34-17
Electron scattering
    Chapter on scattering   36-1
    First experiment on wave nature   35-12
Electron screening, periodic table   38-10
Electron type neutrino   int-22
Electron volt
    As a Unit of Energy   19-21, 26-12
    Electron gun used to define   26-12
Electron waves
    Davisson & Germer experiment   35-12
    De Broglie picture   35-11
    In hydrogen   38-1
    Scattering of   35-12
    Wavelength of   36-9
Electroweak interaction   19-3
    Theory of   int-26
    Weak interaction   int-26
    Z and W mesons   int-26
Electroweak interactions   19-3
Elementary particles
    A confusing picture   int-22
    Short lived   40-23
Elements
    Abundance of   34-24
    Creation of   int-4
    Table of   19-5
Ellipse
    Becoming a parabola   Optics-4
    Drawing one   8-26, Optics-3
    Focus of   8-26, Optics-3
Empty space, quantum fluctuations   40-25

Index-12

Energy
    Bernoulli's equation   23-10
    Black holes   10-29
    Capacitors, storage in   27-18
    Chapter on   10-1
    Conservation of energy
        And the uncertainty principle   40-24
        Conservative and non-conservative forces   10-21
        Derivation from work theorem   10-20
        Feynman story   10-2
        In collisions   11-14
        In satellite motion   8-35
        Mass on spring   14-11
        Neutrinos in beta decay   11-20
        Overview   int-11
        Work energy theorem   10-20
        X. See Experiments I: - 9- Conservation of energy
    $E = Mc^2$ 10-3
    Electric potential energy
        And molecular force   17-12
        Contour map of   25-1
        In classical hydrogen atom   35-3
        In hydrogen atom   int-11
        In hydrogen molecule ion   19-21
        In nuclear fission   int-18, 20-5
        Negative and positive   25-4
        Of a point charge   25-5
        Plotting   25-7. See also Experiments II: - 1- Potential plotting
        Storage in capacitors   27-18
    Electron binding and the periodic table   38-11
    Electron, in the hydrogen molecule ion   19-21
    Electron volt as a unit of energy   19-21, 26-12
    Energy density in an electric field   27-19
    Energy level   35-1
    Energy loss during collisions   11-14
    Equipartition of energy   17-28
        Failure of classical physics   17-31
        Freezing out of degrees of freedom   17-32
        Real molecules   17-30
    From nuclear fission   20-4
    From sun   int-18
    Gravitational potential energy   int-11, 8-35
        Bernoulli's equation   23-10
        Black holes   10-29
        In a room   10-25
        In satellite motion   8-36, 10-26
        In stellar evolution   20-13
        Introduction to   10-8
        Modified   8-37
        On a large scale   10-22
        Zero of   10-22
    Joules and Ergs   10-4

Kinetic energy   int-8
    Always positive   8-35
    Bohr model of hydrogen   35-3
    Classical hydrogen atom   35-3
    Electron diffraction apparatus   36-9
    Equipartition of energy   17-28
    Escape velocity   10-28
    Hydrogen molecule ion   19-21
    Ideal gas law   17-18
    In collisions   11-14
    In model atom   37-5
    Nuclear fusion   20-12
    Origin of   10-5
    Oscillating mass   14-11
    Overview   int-8
    Pendulum   10-10
    Relativistic definition of   10-5
    Rotational   12-22
    Satellite motion   8-36, 10-26
    Slowly moving particles   10-6, 10-29
    Temperature scale   17-11
    Theorem on center of mass   12-26
    Thermal motion   17-6
    Translation and rotation   12-24
    Work energy theorem   10-18
Magnetic energy of current loop   31-22
Mass energy   int-18, 10-3
Negative and positive potential energy   25-4
Neutron mass energy   int-22
Nuclear potential energy
    Fusion   int-18, 20-12
    Nuclear binding   20-9
    Nuclear energy well.   20-10
    Nuclear structure   int-22
Pendulum motion, energy in   10-10
Photon energy   34-9
Photon pulse, uncertainty principle   40-21
Potential. See also Potential energy
Powering the sun   34-23
Rest energy of electron and proton   26-12
Rotational kinetic energy   12-26
Spin magnetic energy
    Dirac equation   39-15
    Magnetic moment   39-4
    Magnetic potential   39-1
Spring potential energy   int-11, 10-16, 14-11
Thermal energy   int-8, 17-7
Total energy
    Classical H atom   35-3
    Escape velocity   10-28
    Satellite motion   8-36, 10-26
Uncertainty principle
    Energy conservation   40-24
    Energy-time form of   40-19
    Fourier transform   40-20
Voltage as energy per unit charge   25-6
Work
    Conservation of energy   int-11
    Definition of   10-12
    Vector dot product   10-13
    Work energy theorem   10-18

X Ray photons, energy of 36-4
Zero of potential energy 10-22
Zero point energy 37-7
   Chapter on 37-1
Energy flow diagrams for reversible engines 18-15
Energy from sun, proton cycle 34-24
Energy, kinetic, in terms of momentum 37-5
Energy level diagram
   Balmer series 35-6
   Bohr theory 35-4
   Expanded 38-8
   Lyman series 35-6
   Model atom 37-4
   Paschen series 35-6
   Photon in laser 37-4
Energy-time form of the uncertainty principle 40-19
Engines
   Internal combustion 18-21
   Maximally efficient 18-15
   Reversible, efficiency of 18-18
English program
   For oscillatory motion 14-31
   For projectile motion 5-16
   For satellite motion 8-19
English program for projectile motion 5-16
Entropy
   Boltzman's formula for 18-24
   Definition of 18-22
   Number of ways to hang tools 18-23
   Second law of thermodynamics 18-1
Epicycle, in Greek astronomy 8-25
Equations, differential. *See* Differential equation
Equations, vector
   Components with derivatives Cal 1-7
   In component form 2-10
Equilibrium
   Balancing weights 13-2
   Chapter on 13-1
   Equations for 13-2
   Example - bridge problem 13-9
   Example - wheel and curb 13-5
   How to solve equilibrium problems 13-5
   Thermal equilibrium 17-8
   Working with rope 13-10
Equipartition of energy
   Failure of classical physics 17-31
   Freezing out of degrees of freedom 17-32
   Normal modes 17-28
   Real molecules 17-30
   Theory of 17-28
Equipotential lines 25-3
   Model 25-10
   Plotting experiment, 2D or 3D? 25-8
Equipotential plot for electron gun 26-11
Ergs and joules 10-4
Ergs per second, power in CGS units 24-2
Escape velocity 10-28
Euler's number e = 2.7183... Cal 1-17

Evaporation
   Of water 17-5
   Surface tension 17-14
Even harmonics in square wave 16-28
Evolution
   Of stars int-19, 17-17, 20-13
      Neutrinos role in 20-15
   Of the universe 34-21
Excess of matter over antimatter 34-29
Exclusion principle 38-1, 38-9
Exercises, finding them. *See* under x in this index
Expanding gas, work done by 18-5
Expanding universe
   Hubble, Edwin int-3
   Hubble rule for int-3
   Red shift 33-24, 34-19, 34-21
Expansion
   Adiabatic
      Carnot cycle 18-11
      Equation for 18-26
      PV Diagrams 18-9
      Reversible engines 18-13
   Isothermal
      Carnot cycle 18-11
      Equation for 18-26
      PV Diagrams 18-8
      Reversible engines 18-13
   Thermal 17-33
   Uniform, of the universe int-3
Expansion, binomial 1-31, Cal 1-23
Experimental diffraction pattern 16-33
Experiments I
   - 1- Graphical analysis of projectile motion 3-17
   - 2- Computer prediction of projectile motion 5-21
   - 3- Conservation of linear momentum 7- 4
   - 4- Conservation of angular momentum 7- 10
   - 5- Computer analysis of satellite motion 8-23
   - 6- Spring pendulum 9-4
   - 7- Conservation of energy
      Check for, in all previous experiments 10-26
   - 8- Collisions 11-9
   - 9- The gyroscope 12-18
   -10- Oscillatory motion of various kinds 14-2
   -11- Normal modes of oscillation 16-4
   -12- Fourier analysis of sound waves 16-18

Experiments II
- 1- Potential plotting   25-7
- 2- The electron gun   26-8
- 3- The RC circuit   27- 22
- 4- The neon bulb oscillator   27- 28
- 5- Motion of electrons in a magnetic field   28- 19
- 5a- Magnetic focusing, space physics   28- 30
- 6- Faraday's law air cart speed detector   30-5
- 7- Magnetic field mapping using Faraday's law   30-24
- 8- Measuring the speed of light with LC circuit   31-15
- 9- LC circuit and Fourier analysis   31-31
- 10- Diffraction grating and hydrogen spectrum   33-17
- 11- Electron diffraction experiment   36-8
- 12- Electron spin resonance   39-9
- 13- Fourier analysis and uncertainty principle   40-21
Exponential decay   Cal 1-32
   In LR circuits   31-9
   In RC circuits   27- 23
Exponential function
   Derivative of   Cal 1-28
   Exponential decay   Cal 1-32
   Indefinite integral of   Cal 1-29
   Integral of   Cal 1-29
   Introduction to   Cal 1-16
   Inverse of the logarithm   Cal 1-16
   Series expansion   Cal 1-28
   y to the x power   Cal 1-16
Eye glasses experiment   Optics-36
Eye, human
   Ciliary muscle   Optics-31
   Cornea   Optics-31
   Crystalline lens   Optics-31
   Farsightedness   Optics-32
   Focusing   Optics-32
   Introduction   Optics-31
   Iris   Optics-31
   Nearsightedness   Optics-32
   Nerve fibers   Optics-31
     Cones   Optics-31
     Rods   Optics-31
Eyepiece   Optics-37

F

$F = ma$. See also Newton's second law
   Applied to Newton's law of gravity   8-5
   Applied to satellite motion   8-8
   For Atwood's machine   9-16
   For inclined plane   9-10
   For spring pendulum   9-7
   For string forces   9-15
   Introduction to   8-4
   Vector addition of forces   9-6
f number
   For camera   Optics-33
   For parabolic mirror   Optics-5

$F(t) = (1)\sin(t) + (1/3)\sin(3t) + ...$
   Fourier analysis of square wave   16-28
Failure of classical physics   17-31
Faraday's law
   AC voltage generator   30-21
   Applications of   30-15
   Chapter on   30-1
   Derivation of   30-11
   Field mapping experiment   30-24
   Gaussmeter   30-23
   Induced Voltage   31-4
   Line integral   30-15
   One form of   30-12
   Right hand rule for   30-15
   The betatron   30-16
   Velocity detector   30-25
   Voltage transformer   30-26
   X. See Experiments II: - 6- Faraday's law air cart speed detector; Experiments II: - 7- Magnetic field mapping using Faraday's law
Farsightedness   Optics-32
Fermi Lab accelerator   28- 23
Feynman, R. P.   int-14
FFT Data button   16-32
Fiber optics
   Introduction to   Optics-14
   Medical imaging   Optics-15
Field
   Conserved lines, fluid and electric   24-17
   Electric
     Circular, line integral for   30-13
     Computer plot of (–3,+5)   24-19
     Continuity equation for   24-14
     Created by changing magnetic flux   31-2
     Direction of circular or induced   31-3
     Inside a conductor   26-1
     Inside hollow metal sphere   26-4
     Integral of - E.dl meter   30-20
     Introduction to   24-10
     Line integral of   30-14
     Mapping convention   24-17
     Mapping with lines   24-12
     Of electromagnet (turned on or off)   30-15
     Of line charge   24-21
     Of static charges   30-2, 30-16
     Radiation by line charge   32- 28
     Radiation by point charge   32- 30
     Van de Graaff generator   26-6
   Electromagnetic field   32- 18
   Flux, introduction of concept   24-15
   Gauss' law   24-20
   Gravitational field
     Definition of   23-3
     Inside the earth   24-24
     Of point mass   24-23
     Of spherical mass   24-24

Magnetic field
  Between capacitor plates   32-6
  Detector, radio waves   32-26
  Direction of, north pole   28-11
  Gauss's law for (magnetic monopole)   32-2
  In coils   28-17
  In Helmholtz coils   28-18
  Interaction with Spin   39-4
  Introduction to   28-10
  Of a solenoid   29-14
  Of a toroid   29-17
  Of straight wire   29-11
  Surface integral   32-2
  Thought experiment on radiated field   32-11
  Uniform   28-16
  Visualizing using compass needles   28-12
  Visualizing using iron filings   28-12
  Plotting experiment   25-7
  Vector field
    Definition of   23-3
    Two kinds of   30-18
  Velocity field
    Introduction to   23-2
    Of a line source   23-7
    Of a point source   23-6
Field lines
  Computer plots, programs for   25-12
  Electric
    Definition of   24-12
    Drawing them   24-13
    Three dimensional model   25-10
Field mapping
  Magnetic field of Helmholtz coils   30-24
  Magnetic field of solenoid   30-24
Field plots and electric potential, chapter on   25-1
Filament, electron gun   26-9
First maxima of two-slit pattern   33-8
Fission, nuclear   20-3
Fitch, Val, K mesons and the direction of time   18-27
Fluctuations, quantum, in empty space   40-25
Fluid dynamics, chapter on   23-1
Fluid flow, viscous effects   23-19
Fluorescence and reflection   40-8
Flux
  Calculations, introduction to   24-22
  Definition of   24-15
  Of magnetic field   30-11
  Of velocity and electric fields   24-15
  Of velocity field   23-8
  Tubes of flux, definition of   24-17
Flux, magnetic
  AC voltage generator   30-21
  Definition of   30-11
  Faraday's law
    Line integral form   30-15
    Voltage form   30-12
  Field mapping experiment   30-24
  Gaussmeter   30-23
  In the betatron   30-16
  Integral E.dl meter   30-19

Magnetic field detector   32-26
Maxwell's equations   32-8
Velocity detector   30-25
Voltage transformer   30-26
Focal length
  For parabolic mirror   Optics-5
  Negative, diverging lenses   Optics-26
  Of a spherical surface   Optics-20
  Two lenses together   Optics-29
Focus
  Eye, human   Optics-32
  Of a parabolic mirror   Optics-4
  Of an ellipse   8-26, Optics-3
Focusing, magnetic   28-29
Focusing of sound waves   8-26, Optics-3
Force   8-2
  Color force   19-15
  Conservative and non-conservative   10-21
  Conservative forces   25-5
  Electric force
    Classical hydrogen atom   35-2
    Introduction to   19-7
    Produced by a line charge   24-6
    Strength of (garden peas)   28-2
  Four basic forces or interactions   19-1
  Introduction to force   8-2
  Lorentz force law   32-8
  Magnetic force
    Between currents   31-19
    On a current   31-18
    Origin of   28-10
  Magnetic force law
    Derivation of   28-10
    Vector form   28-14
  Molecular force
    A classical analysis   19-19
    Analogous to spring force   14-20
    Introduction to   19-15
    Potential energy for   17-12
  Non linear restoring force   14-19
  Nuclear force   int-18
    Introduction to   20-2
    Range of   20-3
  Particle nature of forces   int-13
  Pressure force   17-16
  Spring force
    As molecular force   14-20, 17-12
    Hook's law   9-3
  String force
    Atwood's machine   9-16
    Tension   9-15
Force detector   11-10
Forced harmonic motion, differential equation
    for   14-25, 14-28
Forces, addition of   9-2
Formation of planets   7-17

Four basic interactions   int-25, 19-1
Fourier analysis
   Amplitude and intensity   16-33
   Amplitude and phase   16-31
   And repeated wave forms   16-11
   Calculating Fourier coefficients   16-28
   Energy-time form of the uncertainty principle   40-20
   Formation of pulse from sine waves   40-27
   In the human ear   16-16
   Introduction to   16-6
   Lecture on Fourier analysis   16-28
   Normal modes and sound   16-1
   Of a sine wave   16-7
   Of a square wave   16-9, 16-28
   Of coupled air carts, normal modes   16-12
   Of LC circuit   31-31
   Of slits forming a diffraction pattern   16-33
   Of sound waves. See Experiments I: -12- Fourier analysis of sound waves
   Of violin, acoustic vs electric   16-19
   X. See Experiments II: -13- Fourier analysis & the uncertainty principle
Fourier coefficients
   Accurate values of   16-32
   Calculating   16-31
   Lecture on   16-28
Fourier, Jean Baptiste   16-2
Fractional charge   int-24
Freezing out of degrees of freedom   17-32
Frequencies (Fourier analysis)   16-28
Frequency
   Angular   15-14
   Of oscillation of LC circuit   31-10
   Photon energy $E=hf$ 34-7
   Spacial frequency   15-14
Frequency, period, and wavelength   15-13
Friction
   Coefficient of   9-13
   Inclined plane   9-12
Functions obtained from integration   Cal 1-15
   Logarithms   Cal 1-15
Fusion, nuclear   int-18, 20-12

G

Galaxy
   Andromeda   int-2
   Center of our galaxy   Optics-47
   Introduction to   int-2
   Most distant   int-3
   Sombrero   int-2
   Space travel   1-22
Galileo
   Falling objects (Galileo Was Right!)   8-6
   Inclined plane   9-10
   Portrait of   9-11
Galileo's inclined plane   9-11
Galileo's telescope   Optics-41
Gallbladder operation, medical image   Optics-15
Gamma = C_p/C_v, specific heats   18-7

Gamma rays   32-20, 32-22
   Photon energies   34-11
   Wavelength of   32-20
Gamov, George, big bang theory   int-4
Garden peas
   Electric force between   19-12
Garden peas, electric forces between   28-2
Gas constant R   17-25
Gas, expanding, work done by   18-5
Gas law, ideal   17-18
Gaudsmit and Uhlenbeck, spin   39-1
Gauss' law
   Electric field of line charge   24-21
   For gravitational fields   24-23
   For magnetic fields   32-2
   Introduction to   24-20
   Solving problems   24-26
   Surface integral   29-3
Gauss, tesla, magnetic field dimensions   28-16
Gaussmeter   30-23
Gell-Mann
   Quarks   int-24, 19-14
General relativity   int-15, 8-29
   Modified gravity   8-29
Geometrical optics
   Chapter on   Optics-1
   Definition of   Optics-2
Glass prism   Optics-13
Globular cluster   11-2
Gluons, strong nuclear force   int-25
Graph paper
   For graphical analysis   3-33
   For projectile motion   3-29
Graphical analysis
   Of instantaneous velocity   3-26
   Of projectile motion   3-17
   Of projectile motion with air resistance   3-22
   X. See Experiments I: -1- Graphical analysis of projectile motion
Graphite crystal
   Electron diffraction experiment   36-8
   Electron scattering   36-1
   Structure of   36-8
Grating
   Diffraction   33-12
   Multiple slit
      Fourier analysis of   16-33
      Interference patterns for   33-12
   Three slit   16-33
Gravitational field
   An abstract concept   23-3
   Gauss' law for   24-23
   Inside the earth   24-24
   Of point mass   24-23
   Of spherical mass   24-24
Gravitational force. See Gravity
Gravitational lens, Einstein cross   34-20
Gravitational mass   6-5

Gravitational potential energy
   Energy conservation   int-11
Graviton   int-15
Gravity   int-15
   Acceleration due to   3-21
   And satellite motion   8-8
   Black hole   int-20, 10-29
   Cavendish experiment   8-7
   Deflection of photons   34-19
   Earth tides   8-12
   Einstein's general relativity   int-15
   Four basic interactions   19-1
   Gravitational force acting at center of mass   13-4
   Gravitational potential energy
      Bernoulli's equation   23-10
      Black holes   10-29
      Conservation of Energy   8-35
      Energy conservation   int-11
      In a room   10-25
      In satellite motion   8-36
      Introduction to   10-8
      Modified   8-37
      On a large scale   10-22
      Zero of   10-22
   Inertial and gravitational mass   8-8
   Interaction with photons   34-18
   Modified, general relativity   8-29
   Newton's universal law   int-15, 8-5
   Potential energy. See also Potential energy: Gravitational
      Introduction to   10-8
      On a Large Scale   10-22
   Quantum theory of   int-16
   Strength, comparison to electricity   int-6, 19-8
   Weakness & strength   int-20
   "Weighing" the Earth   8-8
   Weight   8-11
Green flash   Optics-17
Grounding, electrical circuits   26-8
Guitar string
   Sound produced by   15-22
   Waves   15-20
   Waves, frequency of   15-21
Gun, electron. *See* Electron gun
Gyromagnetic ratio for electron spin   39-14
Gyroscopes
   Atomic scale   39-15
   Movie   12-18
   Precession formula   12-21
   Precession of   12-19
   Theory of   12-18
   X. See Experiments I: - 9- The gyroscope

H

h bar, Planck's constant   35-9
Hair, strand of, diffraction pattern of   36-14
Half-life
   In exponential decay   Cal 1-33
   In RC circuit   27-25
   Of muons (as clock)   1-20
   Of muons, exponential decay   Cal 1-33
Halos around sun   Optics-18
Harmonic motion
   Computer program   14-30
   Damped
      Computer program   14-34
      Critical damping   14-23
      Differential equation for   14-21
   Forced
      Analytic solution   14-28
      Differential equation for   14-25, 14-28
Harmonic oscillator   14-12
   Differential equation for   14-14
Harmonic series   16-3
Harmonics and Fourier coefficients   16-28
Hays, Tobias
   Dive Movie, center of mass   11-1
   Dive movie, time reversed   18-1
Heat capacity   17-26
   Molar   17-26
Heat pump, efficiency of   18-19
Heat, specific. *See* Specific heat
Heisenberg, Werner   4-1, Cal 1-3
Hele-Shaw cell, streamlines   23-4
Helium
   Abundance in early universe   34-26
   And electron spin   38-9
   And the Pauli exclusion principle   38-9
   Binding energy of last electron   38-12
   Creation of in universe   int-4
   Energy to ionize   38-9
   In periodic table   38-11
   Isotopes helium 3 and 4   int-17
Helmholtz coils   28-17, 28-18
   100 turn search coil   39-12
   Electron spin resonance apparatus   39-11
   Field mapping experiment   30-24. *See also* Experiments II: - 7- Magnetic field mapping using Faraday's law
   Motion of electrons in   28-20, 28-29
   Uniform magnetic field inside   28-17
Hertz, Heinrich, radio waves   34-1
Hexagonal array
   Graphite crystal and diffraction pattern   36-9
Homework exercises, finding them. *See* X-Ch (chapter number): Exercise number
Hooke's law   9-4
   In dimensional analysis   15-7
Horsehead nebula in visible & infrared light   Optics-46
Hot early universe   int-4

Hubble space telescope   Optics-44
Hubble, Edwin, expanding universe   int-3
Hubble photograph of most distant galaxies   int-3
Hubble rule for expanding universe   int-3, 33-24
Hubble telescope mirror   Optics-44
   Spherical aberration in   Optics-22
Human ear
   Description of   16-15
   Inside of cochlea   16-34
Human Eye   Optics-31
Huygens
   Wave nature of light   34-1, Optics-1
Huygens' principle   33-4
   Preliminary discussion of   15-2
Hydrodynamic voltage
   Bernoulli's equation   23-17
   Resistance   27- 7
   Town water supply   23-18
Hydrogen atom
   Angular momentum quantum number   38-7
   Big bang theory   int-4
   Binding energy of electron   38-12
   Bohr theory   35-1
   Classical   35-2
   Coulomb's law   24-4
   Expanded energy level diagram   38-8
   Quantized projections of angular momentum   38-5
   Solution of Schrödinger's equation   38-2
   Standing wave patterns in   38-3
   The $L = 0$ Patterns   38-4
   The $L \neq 0$ Patterns   38-5
Hydrogen atom, classical
   Failure of Newtonian mechanics   35-3
Hydrogen bomb   int-18, 20-13
Hydrogen molecule
   Formation of   19-16
Hydrogen molecule ion
   Binding energy and electron clouds   19-23
   Computer program for   19-24
   Formation of   19-16
Hydrogen nucleus   int-6, 19-3
   Isotopes of   19-6
Hydrogen spectrum
   Balmer series   33-19, 35-4
   Bohr model   int-8
   Experiment on   33-17
   Lyman series   35-6
   Of star   35-4
   Paschen series   35-6
   X. See Experiments II: -10- Diffraction grating and hydrogen spectrum
Hydrogen wave patterns
   Intensity at the origin   38-5
   $L = 0$ patterns   38-4
   Lowest energy ones   38-3
   Schrödinger's Equation   38-2
Hydrogen-Deuterium molecule, NMR experiment   39-12
Hydrostatics, from Bernoulli's equation   23-12

I

IBM Labs, atomic microscopes   17-1
Ice crystal   17-4
Ideal gas law   17-18
   Chemist's form   17-25
Ideal gas thermometer   17-20
   Absolute zero   17-21
Image
   Image distance
      Lens equation   Optics-24
      Negative   Optics-26
   In focal plane of telescope mirror   Optics-5
   Medical   Optics-15
      Gallbladder operation   Optics-15
      Of duodenum   Optics-15
Impulse
   Change in momentum   11-12
   Experiment on   11-9
   Measurement   11-11
Inclined plane   9-10
   Galileo's inclined plane, photo of   9-11
   Objects rolling down   12-25
   With friction   9-12
Indefinite integral
   Definition of   Cal 1-14
   Of exponential function   Cal 1-29
Index of refraction
   Definition of   Optics-9
   Glass prism and rainbow of colors   Optics-15
   Introduction to   Optics-2
   Of gas of supercooled sodium atoms   Optics-9
   Table of some values   Optics-9
Induced voltage
   In moving loop of wire   30-4
   Line integral for   31-4
Inductance
   Chapter on   31-1
   Derivation of formulas   31-5
Inductor
   As a circuit element   31-7
   Definition of   31-2
   Iron core   31-29
   LC Circuit   31-10
   LR circuit   31-8
   Toroidal coil   31-6
Inertia
   Inertial mass   6-5
   Moment of (Angular mass)   12-7
   Of a massive object   31-12
   Of an electric current   31-12
Infrared light
   Ability to penetrate interstellar dust   Optics-46
   Center of our galaxy   Optics-47
   Horsehead nebula in visible & infrared   Optics-46
   In the electromagnetic spectrum   int-7
   Paschen series, hydrogen spectra   35-6
   Wavelength of   32- 20

Infrared Telescopes   Optics-46
  Infrared camera   Optics-46
  IRAS satellite   Optics-47
    Map of the entire sky   Optics-47
  Mt. Hopkins 2Mass telescope   Optics-46
    Viewing center of our galaxy   Optics-47
Initial conditions in a computer program   5-14
Initial slope in RC circuit   27-25
Inside the cochlea   16-34
Instantaneous velocity
  And the uncertainty principle   4-2, Cal 1-4
  Calculus definition of   4-3, Cal 1-5
  Definition of   3-24
  From strobe photograph   3-26
Instruments
  Percussion   16-22
  Stringed   16-18
  Violin, acoustic vs electric   16-19
  Wind   16-20
Integral
  As a sum   Cal 1-10
  Calculating them   Cal 1-11
  Definite, introduction to   Cal 1-11
  Formula for integrating x to n'th power   Cal 1-14, Cal 1-27
  Indefinite, definition of   Cal 1-14
  Of 1/x, the logarithm   Cal 1-15
  Of a constant   Cal 1-13
  Of a curve that increases linearly   Cal 1-13
  Of a velocity curve   Cal 1-12
  Of exponential function e to the ax   Cal 1-29
  Of the velocity vector   Cal 1-10
    As area under curve   Cal 1-12
  Of x to n'th power
    Indefinite integral   Cal 1-27
Integral, line
  Ampere's law   29-7
  Conservative force   29-6
  Evaluation for solenoid   29-15
  Evaluation for toroid   29-17
  Faraday's law   30-15
  For circular electric field   30-13
  For static charges   30-2
  In Maxwell's equations   32-8
  Introduction to   29-5
  Two kinds of fields   30-18
Integral of E.dl meter   30-18, 30-20
Integral sign   Cal 1-10
Integral, surface   29-2
  For magnetic fields   32-2
  Formal introduction   29-2
  Gauss' law   29-3
  In Maxwell's equations   32-8
  Two kinds of fields   30-18
Integration
  Equivalent to finding area   Cal 1-11
  Introduction to   Cal 1-8
  Introduction to finding areas under curves   Cal 1-13
  Why computers do it so well   Cal 1-12

Integration and differentiation
  As inverse operations   Cal 1-18
    Velocity and position   Cal 1-19
  Fast way to go back and forth   Cal 1-20
  Position as integral of velocity   Cal 1-20
  Velocity as derivative of position   Cal 1-20
Integration formulas   Cal 1-27
Intensity
  And amplitude, Fourier analysis lecture   16-33
  Of diffraction pattern   16-33
  Of harmonics in Fourier analysis of light pulse   40-22
  Of probability wave   40-22
  Sound intensity, bells and decibels   16-24
  Sound intensity, speaker curves   16-27
Interactions. *See also* the individual forces
  Electric   int-14
  Four basic   19-1
  Gravitational   int-15
  Nuclear   int-14
  Photons and gravity   34-18
  Weak   int-14
Interactions, four basic   19-1
Interference patterns
  A closer look at   33-26
  Introduction to   33-3
  Two-slit
    Light waves   33-10
    Probability waves   40-9
    Water waves   33-6
Internal combustion engine   18-21
Internal reflection   Optics-13
Interval, evaluating variables over   Cal 1-10
Ionic Bonding   38-15
Iris   Optics-31
Iron 56, most tightly bound nucleus   int-18, 20-11
Iron core inductor   31-29
Iron core of massive star   20-15
Iron fillings, direction of magnetic field   28-12
Iron magnets   31-26
Isothermal expansion
  Calculation of work   18-26
  PV Diagrams   18-8
Isotopes of nuclei   int-17, 19-6
  Stability of   int-22

J

Jeweler using magnifier   Optics-38
Joules and Ergs   10-4

K

K meson and direction of time   int-23, 40-23
Karman vortex street   14-25
Kepler's laws
    Conservation of angular momentum   8-32
    First law   8-26
    Introduction to   8-24
    Second law   8-27
    Third law   8-28
Kilobaud, fiber optics communication   Optics-14
Kinetic energy
    Always positive   8-35
    Bohr model of hydrogen   35-3
    Classical hydrogen atom   35-3
    Electron diffraction apparatus   36-9
    Equipartition of energy   17-28
    Escape velocity   10-28
    Hydrogen molecule ion   19-21
    Ideal gas law   17-18
    In collisions   11-14
    In model atom   37-5
    In terms of momentum   37-5
    Nonrelativistic   10-6
    Nuclear fusion   20-12
    Origin of   10-5
    Oscillating mass   14-11
    Overview   int-8
    Pendulum   10-10
    Relativistic definition of   10-5
    Rotational   12-22
    Satellite motion   8-36, 10-26
    Slowly moving particles   10-6, 10-29
    Temperature scale   17-11
    Theorem on center of mass   12-26
    Thermal motion   17-6
    Translation and rotation   12-24
    Work energy theorem   10-18
Kirchoff's law
    Applications of   27-11
    Introduction to   27-10

L

L = 0 Patterns, hydrogen standing waves   38-4
Lack of simultaneity   1-32
Lambda max. *See* Blackbody radiation: Wein's displacement law
Lambda(1520), short lived elementary particle   40-23
Landé g factor   39-14
Largest scale of distance   int-25
Laser
    Chapter on   37-1
    Diffraction patterns, Fourier analysis   16-33
    Pulse in gas of supercooled sodium atoms   Optics-9
    Standing light waves   37-2

LC circuit
    Experiment on   31-13
    Fourier analysis of   31-31
    Introduction to   31-10
    Ringing like a bell   31-36
LC oscillation, intuitive picture of   31-12
Left hand rule, as mirror image   Optics-6
Leibnitz   4-1, Cal 1-3
Lens
    Crystalline, eye   Optics-31
    Diverging   Optics-26
    Eye glasses experiment   Optics-36
    Eyepiece   Optics-37
    Introduction to theory   Optics-18
    Lens equation
       Derivation   Optics-25
       Introduction   Optics-24
       Multiple lens systems   Optics-28
       Negative focal length, diverging lens   Optics-26
       Negative image distance   Optics-26
       Negative object distance   Optics-27
       The lens equation itself   Optics-25
       Two lenses together   Optics-29
    Magnification of lenses   Optics-30
    Magnifier   Optics-38
       Jeweler using   Optics-38
       Magnification of   Optics-39
    Magnifying glass   Optics-37
    Multiple lens systems   Optics-28
    Optical properties due to slowing of light   Optics-9
    Simple microscope   Optics-50
    Spherical surface
       Grinding one   Optics-19
       Optical properties of   Optics-19
    Thin lens   Optics-23
    Two lenses together   Optics-29
    Zoom lens   Optics-18
Lens, various types of
    Bi-concave   Optics-27
    Bi-convex   Optics-27
    Meniscus-concave   Optics-27
    Meniscus-convex   Optics-27
    Planar-concave   Optics-27
    Planar-convex   Optics-27
Lenses, transmitted waves   36-3
Lensing, gravitational   34-20
Lepton family
    Electron   int-22
    Electron type neutrino   int-22
    Muon   int-22
    Muon type neutrino   int-22
    Tau   int-22
    Tau type neutrino   int-22
Leptons, conservation of   int-22
LET statement, computer   5-5
Lifetime
    Muon, exponential decay   Cal 1-32
Lifting weights and muscle injuries   13-11

Light   int-7
   Atomic spectra   33-16
   Balmer series   33-19
   Blackbody radiation   34-2
      Another View of   34-22
   Chapter on   33-1
   Decoupling from matter in early universe   34-31
   Diffraction of light
      By thin crystals   36-6
      Fourier analysis of slits   16-33
      Grating for   33-12
      Pattern, by strand of hair   36-14
      Patterns, student projects   36-13
   Doppler effect
      In astronomy   33-23
      Introduction to   33-20
      Relativistic   33-22
   Electromagnetic spectrum   int-7, 32- 20
      Photon energies   34-11
   Electromagnetic waves   32- 18
   Gravitational lensing of   34-20
   Hydrogen spectrum. See Experiments II: -10- Diffraction grating and hydrogen spectrum
      Balmer formula   35-5
      Bohr model. See Bohr Model
      Lab experiment   33-17
      Spectrum of star   33-19
   Infra red. See Infrared light
   Interaction with gravity   34-18
   Interference patterns for various slits   33-12
   Laser pulse in gas of supercooled sodium atoms   Optics-9
   Lasers, chapter on   37-1
   Light pulse clock   1-14
   Maxwell's theory of   int-7
   Microwaves. See Microwaves
   Mirror images   Optics-6
   Motion through a Medium   Optics-8
   Particle nature of   34-1
   Photoelectric effect   34-5
   Photon
      Chapter on   34-1
      Creates antiparticle   34-17
      Energies in short laser pulse   40-21
      In electron spin resonance experiment   39-9
      Introduction to   int-8
      Mass   34-12
      Momentum   34-13
      Thermal, 2.74 degrees   int-29, 34-27, 34-31
   Polarization of light   32- 23
   Polarizers   32- 25
   Prism, analogy to Fourier analysis   16-28
   Radiated electric fields   32- 28
   Radiation pressure of   34-14
   Radiation pressure, red supergiant stars   20-15
   Radio waves. See Radio waves
   Rays   Optics-1
   Red shift and the expanding universe   33-24, 34-21
   Reflection   36-3, Optics-1
   Reflection and fluorescence   40-8
   Reflection from curved surfaces   Optics-4
   Spectral lines, hydrogen   int-7
      Bohr theory   35-4
   Speed in a medium   Optics-8
   Speed of light
      Electromagnetic pulse   32- 14
      Experiment to measure   1-9, 31-15
      Same to all observers   1-12
   Structure of electromagnetic wave   32- 19
   Thermal, 2.74 degrees   34-27
   Three degree radiation   34-27
   Two-slit interference pattern for   33-10
   Ultraviolet. See Ultraviolet light
   Visible. See Visible light
   Visible spectrum of   33-15
   Wave equation for   int-7
   Waves, chapter on   33-1
   X ray diffraction   36-4
   X rays. See X-rays
Light-hour   int-2
Light-minute   int-2
Light-second   int-2
Light-year   int-2
Limiting process   4-1, Cal 1-3
   Definition of derivative   Cal 1-30
   In calculus   Cal 1-5
   Introduction to derivative   Cal 1-6
   With strobe photographs   Cal 1-2
Line charge, electric field of
   Calculated using calculus   24-6
   Calculated using Gauss' law   24-21
Line integral. See Integral, line
Linear and nonlinear wave motion   15-10
Linear momentum. See Momentum
Lines, equipotential   25-3
Lithium
   And the Pauli exclusion principle   38-9
   Atom   38-12
   Binding energy of last electron   38-12
   In the periodic table   38-11
   Nucleus   int-6
Logarithms
   Integral of 1/x   Cal 1-15
   Introduction to   Cal 1-15
   Inverse of exponential function   Cal 1-16
Lorentz contraction   1-24
   Charge density created by   28- 6
Lorentz force law
   And Maxwell's equations   32- 8
   Electric and magnetic forces   28- 15
   Relativity experiment   30-9
Lorenz, chaos   23-1
LR circuit   31-8
   Exponential decay time constant   31-9
LRC circuit, ringing like a bell   31-36
Lyman series, energy level diagram   35-6

M

Magnetic bottle 28-31
Magnetic constant (μ zero), definition of 28-11
Magnetic energy
   Of current loop 31-22
   Of spin 39-1
   Of spin, semi classical formula 39-14
Magnetic field 28-10
   Between capacitor plates 32-6
   Detector 32-26
   Dimensions of, tesla and gauss 28-16
   Direction of
      Compass needles 28-12
      Definition 28-11
      Iron fillings 28-12
   Gauss' law for 32-2
   Helmholtz coils 28-17, 28-18
      100 turn search coil 39-12
      Electron spin resonance apparatus 39-11
      Field mapping experiment 30-24
      Motion of electrons in 28-20, 28-29
   In electromagnetic waves 32-18
   In light wave int-7
   Interaction with spin 39-4
   Mapping. See Experiments II: - 7- Magnetic field mapping using Faraday's law
   Mapping experiment with Helmholtz coils 30-24
   Motion of charged particles in 28-19
   Motion of electrons in. See Experiments II: - 5- Motion of electrons in a magnetic field
   Oersted, Hans Christian 28-12
   Of a solenoid 28-17, 29-14
   Of a straight wire 29-11
   Of a toroid 29-17
   Of permanent magnet, experiment to measure 30-25
   Radiated, a thought experiment 32-11
   Right-hand rule for current 28-13
   Right-hand rule for solenoids 29-14
   Surface integral of 32-2
   Uniform 28-16
      Between Helmholtz coils 28-17
      Between pole pieces 28-16
      Inside coils 28-17
Magnetic flux
   AC voltage generator 30-21
   Definition of 30-11
   Faraday's law
      Line integral form 30-15
      Voltage form 30-12
   Field mapping experiment 30-24
   Gaussmeter 30-23
   In the betatron 30-16
   Integral E.dl meter 30-19
   Magnetic field detector 32-26
   Maxwell's equations 32-8
   Velocity detector 30-25
   Voltage transformer 30-26

Magnetic focusing 28-29. *See also* Experiments II: - 5a- Magnetic focusing and space physics
   Movie 28-30
Magnetic force
   Between currents 31-19
   Deflection of electron beam 28-9
      Movie 28-9
   On a current 31-18
   On electrons in a wire 30-3
   Origin of 28-8
   Parallel currents attract 28-14
   Relativity experiment (Faraday's law) 30-9
   Thought experiment (on origin of) 28-7
Magnetic force law
   Lorentz force law, electric and magnetic forces 28-15
Magnetic force law, derivation of
   $F = qvB$ 28-10
   Vector form 28-14
Magnetic moment
   And angular momentum 31-24
   Bohr magneton 39-4
   Definition of 31-21
   Nuclear 39-6
   Of charge in circular orbit 31-24
   Of electron 39-4
   Of neutron 39-6
   Of proton 39-6
   Summary of equations 31-24
Magnetic resonance
   Classical picture of 39-8, 39-14
      Precession of atom 39-15
   Electron spin resonance experiment 39-5
   X. See Experiments II: -12- Electron spin resonance
Magnetism
   Chapter on 28-1
   Thought experiment to introduce 28-4
Magnets
   Electromagnet 31-28
   Iron 31-26
   Superconducting 31-30
Magnification
   Definition Optics-30
   Negative (inverted image) Optics-30
   Of Magnifier Optics-39
   Of two lenses, equation for Optics-30
Magnifier Optics-38
   Jeweler using Optics-38
   Magnification of Optics-39
Magnifying glass Optics-30, Optics-37
Magnitude of a Vector 2-6
Map, contour 25-1
Mapping convention for electric fields 24-17
Mars, retrograde motion of 8-24

Mass
　Addition of   6-4
　Angular
　　Moment of inertia   12-7
　Center of mass
　　Diver movie   11-1
　　Dynamics of   11-4
　　Formula   11-3
　　Introduction to   11-2
　Chapter on mass   6-1
　Definition of mass
　　Newton's second law   8-3
　　Recoil experiments   6-2
　Electron mass in relativistic beta decay   6-7
　Energy
　　In nuclei   int-18
　　Introduction to   10-3
　　Of neutron   int-22
　Gravitational force on   int-6
　Gravitational mass   6-5
　Inertial mass   6-5
　Measuring mass   6-4
　Of a moving object   6-5
　Of a neutrino   6-13
　Of a photon   34-12
　Properties of   6-3
　Relativistic formula for   6-10
　Relativistic mass
　　Beta decay   6-6
　　Beta decay of Plutonium 246   6-8
　　Beta decay of Protactinium 236   6-9
　　Intuitive discussion   6-6
　Rest mass   int-11, 6-10, 10-5
　Role in mechanics   8-3
　Standard mass   6-3
　Zero rest mass   6-11
Mass on a spring, analytic solution   14-7
Mass, oscillating, differential equation for   14-8
Mass spectrometer   28-28
Mathematical prism, Fourier analysis   16-28
Matter over antimatter, in early universe   int-30, 34-17
Matter, stability of   19-14
Mauna Kea, Hawaii   Optics-45
Maxima, first, of two-slit pattern   33-8
Maximally efficient engines   18-15
Maxwell's correction to Ampere's law   32-4
Maxwell's equations   32-8
　Chapter on   32-1
　Failure of
　　In classical hydrogen atom   35-2
　　In photoelectric effect   34-6
　In empty space   32-10
　Probability wave for photons   40-7
　Symmetry of   32-9
Maxwell's theory of light   int-7, 1-9, 32-2
Maxwell's wave equation   int-7, 15-1, 32-18

Measurement limitation
　Due to photon momentum   40-11
　Due to uncertainty principle   4-2, Cal 1-4
　Two slit thought experiment   40-9
　Using waves   40-10
Measuring short times using uncertainty principle   40-22
Measuring time constant from graph   Cal 1-34
Mechanics
　Newtonian
　　Chapter on   8-1
　　Classical H atom   35-3
　Newton's second law   8-4
　Newton's Third Law   11-6
　Photon mechanics   34-12
　Relativistic   int-12
　The role of mass in   8-3
Medical imaging   Optics-15
Megabit, fiber optics communication   Optics-14
Meiners, Harry, electron scattering apparatus   36-1
Meniscus-concave lens   Optics-27
Meniscus-convex lens   Optics-27
Mercury barometer, pressure measurement   17-22
Meter, definition of   int-2
Microscope   int-1, Optics-50
　Atomic   17-1
　Scanning tunneling microscope   Optics-51
　　Surface (111 plane) of a silicon   Optics-51
　Simple microscope   Optics-50
Microwaves
　Electromagnetic spectrum   int-7, 32-20
　Microwave polarizer   32-24
　Photon energies   34-11
Milky Way   int-2
　Center of our galaxy   Optics-47
Millikan oil drop experiment   26-17
Mirror images
　General discussion   Optics-6
　Reversing front to back   Optics-7
　Right-hand rule   Optics-6
Mirror, parabolic
　Focusing properties of   Optics-4, Optics-42
MKS units
　Ampere, volt, watt   24-2
　Coulomb's law in   24-2
MOD command, computer   5-10
Model Atom   37-4
　Chapter on   37-1
　Energy levels in   37-4
Model showing equipotential and field lines   25-10
Modulus
　Bulk   15-8
　Definition of   15-8
　Young's   15-8
Molar heat capacity   17-26
Molar specific heat of helium gas   17-27
Mole
　Avogadro's number   17-24
　Volume of   17-25

Molecular forces   int-6, 14-20, 17-12
   A classical analysis   19-18
      The bonding region   19-19
   A more quantitative look   19-18
   Binding energy   17-13
   Electric interaction   19-1
   Four basic interactions   19-2
   Introduction to   19-15
   Represented by springs   17-13
Molecular weight   17-24
Molecules   17-2
   Cholera   17-2
   Hydrogen, electric forces in   19-16
   Hydrogen molecule ion   19-16
   Myoglobin   17-3
   Water   17-2
Moment, magnetic
   And angular momentum   31-24
   Definition of   31-21
   Of charge in circular orbit   31-24
   Summary of equations   31-24
Moment of inertia
   Angular mass   12-7
      Calculating   12-8
   Rotational kinetic energy   12-22
Momentum
   Angular. See Angular momentum
   Collisions and impulse   11-9
   Conservation of   7-2
      Derivation from Newton's second law   11-7
      During collisions   11-13
      General discussion   7-1
      In collision experiments   7-4
      In subatomic collisions   7-7
      X. See also Experiments I: - 3- Conservation of linear momentum
   De Broglie formula for momentum   35-11
   Kinetic energy in terms of momentum   37-5
   Linear momentum, chapter on   7-1
   Momentum of photon
      Compton scattering.   34-15
      Formula for   34-13
   Momentum version of Newton's second law   11-8
   Uncertainty principle, position-momentum form   40-15
Mormon Tabernacle, ellipse   Optics-3
Mormon Tabernacle, focusing of sound waves   8-26, Optics-3

Motion
   Angular analogy   12-3
   Damped harmonic motion   14-21
      Differential equation for   14-21
   Description of, chapter on   3-3
   Forced harmonic motion
      Differential equation for   14-25, 14-28
   Harmonic motion   14-12
      Computer program for   14-30
   Of charged particles
      In magnetic fields   28-19
      In radiation belts   28-32
   Of electrons in a magnetic field. See Experiments II: - 5- Motion of electrons in a magnetic field
   Of light through a medium   Optics-8
   Oscillatory motion   14-2. *See also Experiments I: - 10- Oscillatory motion of various kinds*
   Prediction of motion   5-12
   Projectile. See Projectile Motion
   Resonance   14-24
      Tacoma Narrows bridge   14-24
   Rotational motion   12-1
      Angular acceleration   12-3
      Angular velocity   12-2
      Radian measure   12-2
   Satellite. See Satellite motion
   Thermal motion   17-6
   Translation and rotation   12-24
   Uniform circular motion   3-17, 8-2
      Particles in magnetic field   28-20
   Wave motion, amplitude and phase   15-17
Movie
   Angular momentum as a Vector   7-15
   Brownian motion   17-7
   Circular motion of particles in magnetic field   28-20
   Diver   11-1
   Magnetic deflection   28-9
   Magnetic focusing   28-30
   Muon Lifetime   1-21
   Standing De Broglie like waves   35-11
   Time reversed dive, second law of thermodynamics   18-1
Mt. Hopkins telescope   Optics-43
Mt. Palomar telescope   Optics-43
Mu (μ) zero, definition of   31-19
Multi Electron Atoms   38-9
Multiple lens systems   Optics-28
Multiple slit grating   16-33
Multiple slit interference patterns   33-12
Multiplication notation, computer   5-6
Multiplication of vectors
   By a number   2-5
   Scalar or dot product   2-12
   Vector cross product   2-15, 12-9

Muon
    And Mt. Washington, Lorentz contraction   1-29
    Discovery of   int-23
    Half life used as clock   1-20
    Lepton family   int-22
    Lifetime, exponential decay   Cal 1-32
    Movie on lifetime   1-21
        Cerenkov radiation   Optics-10
    Muon type neutrino   int-21
Muscle, ciliary, in the eye   Optics-31
Muscle injuries lifting weights   13-11
Myoglobin molecule   17-3
Myopia, nearsightedness   Optics-32

N

Nature's speed limit   int-12, 6-11
Nearsightedness   Optics-32
Nebula
    Crab, neutron star   20-16
    Eagle   7-16, 7-18
    Orion   7-17
Negative and positive charge   19-10
Negative focal length, diverging lenses   Optics-26
Negative image distance   Optics-26
Negative object distance   Optics-27
Negative slope   Cal 1-31
Neon bulb oscillator   27-28
    Experimental setup   27-31
    X. See Experiments II: -4- The Neon Bulb Oscillator
Neon, up to, periodic table   38-13
Nerve fibers, human eye   Optics-31
    Cones   Optics-31
    Rods   Optics-31
Net area (Fourier analysis)   16-29
Neutrino astronomy   6-14, 11-21
Neutrinos   6-13, 11-20
    1987 Supernova   6-14
    Beta (ß) decay reactions   20-6
    Cosmic background   int-30
    Created in the weak interaction   20-6
    Electron type   int-21
    From the sun   6-13, 11-21
    In nuclear structure   20-7
    In stellar evolution   20-15
    In supernova explosions   20-16
    Muon type   int-21
    Passing through matter   int-22
    Pauli's prediction of   int-21, 20-6
    Rest mass   20-6
    Stability of   19-14
    Tau type   int-21

Neutron
    Decay
        Beta decay reactions   20-7
        Charge conservation in   int-21
        Energy problems   int-21
        Weak interaction   20-6
    Formation of deuterium in early universe   34-30
    In alpha particles   20-8
    In isotopes   int-17, 19-6
    In nuclear matter   20-1
    Neutron proton balance in early universe   34-30
    Neutron proton mass difference in early universe   34-30
    Nuclear binding energies   20-9
    Quark structure of   int-24
    Rest mass energy   int-22, 20-9
    Role in nuclear structure   20-7
Neutron star
    And black holes   int-20, 20-18
    Binary   int-15
    In Crab nebula   20-17
    Pulsars   20-17
    Stellar evolution   int-19
New functions, obtained from integration   Cal 1-15
Newton
    Particle nature of light   34-1, Optics-1
Newtonian mechanics   8-1
    Classical H atom   35-3
    Failure of
        In specific heats   17-31
        In the classical hydrogen atom   35-3
    The role of mass   8-3
Newton's laws
    Chapter on   8-1
    Classical physics   int-7
    Gravity   int-15, 8-5
    Second law   8-4
        And Newton's law of gravity   8-5
        Angular analogy   12-14
        Applications of, chapter on   9-1
        Atwood's machine   9-16
        Inclined plane   9-10
        Momentum version of   11-8
        Satellite motion   8-8
        String forces   9-15
        Vector addition of forces   9-6
    Third law   11-6
Newton's reflecting telescope   Optics-22, Optics-42
NMR experiment, the hydrogen-deuterium molecule   39-12
Nonlinear restoring forces   14-19
Nonrelativistic wave equation   int-12, 15-2, 34-16
Normal modes
    Degrees of freedom   17-29
    Fourier analysis of coupled air cart system   16-12
    Modes of oscillation   16-4
    X. See Experiments I: -11- Normal modes of oscillation

Nuclear
  Binding energy   20-9
  Charge, effective, in periodic table   38-12
  Energy well, binding energies   20-10
  Fission   int-18, 20-3
    Energy from   20-4
  Force   int-18, 20-2
    Four basic interactions   19-2
    Meson, Yukawa theory   int-22
    Range of   20-3
    Range vs. electric force   int-18
  Fusion   int-18, 20-12
    Binding energies, stellar evolution   20-13
  Nuclear interaction   int-14
    Alpha particles   20-8
    Binding energy   20-9
    Neutron stars   20-17
  Nuclear magnetic moment   39-6
  Nuclear matter, chapter on   20-1
  Nuclear reactions, element creation   int-4
  Nuclear stability, binding energy   20-10
  Nuclear structure   int-22, 20-7
Nucleon   int-17
Nucleus   int-17
  Discovery of, Rutherford   35-1
  Hydrogen   int-6
  Isotopes of   19-6
  Large   int-22
  Lithium   int-6
  Most tightly bound (Iron 56)   20-11

O

Odd harmonics in a square wave   16-28
Odor of violets   17-6
Oersted, Hans Christian   28-12
Off axis rays, parabolic mirror   Optics-4
Ohm's law   27-7
One cycle of a square wave   16-28
One dimensional wave motion   15-1
Optical properties
  Parabolic reflectors   Optics-4
  Spherical surface   Optics-19
Optics, fiber
  Introduction to   Optics-14
  Medical imaging   Optics-15
Optics, geometrical
  Chapter on   Optics-1
  Definition of   Optics-2
Optics of a simple microscope   Optics-50
Orbitals   19-15. See also Hydrogen atom: Standing wave patterns in
Orbits
  Allowed, Bohr theory   int-8
  Bohr, radii of   35-7
  Classical hydrogen atom   35-2
  Orbit-1 program   8-21
  Precession of, general relativity   8-30
  Satellite. See Satellite motion

Order and disorder
  Direction of time   18-25
  Entropy and the second law of thermodynamics   18-4
  Formula for entropy   18-24
Orion nebula   7-17
Oscillation   14-1
  Critical damping   14-23
  Damped
    Computer approach   14-21
    Differential equation for   14-21
  LC circuit   31-10
    Intuitive picture of   31-12
  Mass on a spring
    Analytic solution   14-7
    Computer solution   14-30
    Differential equation for   14-8
  Non linear restoring forces   14-19
  Normal modes   16-4. See also Experiments I: -11- Normal modes of oscillation
  Period of   14-4
  Phase of   14-6
  Resonance
    Analytic solution for   14-26, 14-28
    Differential equation for   14-25
    Introduction to   14-24
  Small oscillation
    Molecular forces   14-20
    Simple pendulum   14-16
  Torsion pendulum   14-12
  Transients   14-27
Oscillator
  Harmonic oscillator   14-12
    Differential equation for   14-14
    Forced   14-28
  Neon bulb oscillator   27-28
    Period of oscillation   27-30
    X. See also Experiments II: -4- The neon bulb oscillator
Oscillatory motion   14-2. See also Experiments I: -10- Oscillatory motion of various kinds
Osmotic pressure   17-34
Overview of physics   int-1

P

Parabola
  How to make one   Optics-4
Parabolic mirror
  f number   Optics-5
  Focusing properties of   Optics-4, Optics-42
  Off axis rays   Optics-4
Parallel currents attract   28-14
Parallel plate capacitor   26-14
  Capacitance of   27-16
  voltage in   26-15
Parallel resistors   27-12
Particle
  Point size   int-14
  Systems of particles   11-1
Particle accelerators   28-22

Particle decays and four basic interactions   int-23
Particle nature of light   34-1
   Photoelectric effect   34-5
Particle-wave nature
   Born's interpretation   40-6
   De Broglie picture   int-10, 35-10
   Energy level diagrams resulting from   37-4
   Of electromagnetic spectrum   34-11
   Of electrons
      Davisson and Germer experiment   35-12
      De Broglie picture   int-10, 35-10
      Electron diffraction experiment   36-8
      Electron waves in hydrogen   38-2
      Pauli exclusion principle   38-9
   Of forces   int-13
   Of light
      Electromagnetic spectrum   34-11
      Photoelectric effect   34-5
      Photon mass   34-12
      Photon momentum   34-13
      Photon waves   40-6
      Photons, chapter on   34-1
   Of matter   int-10, 34-11
   Probability interpretation of   40-6
      Fourier harmonics in a laser pulse   40-22
      Reflection and fluorescence   40-8
   Quantum mechanics, chapter on   40-1
   Two slit experiment from a particle point of view
      Probability interpretation   40-9
      The experiment   40-3
   Uncertainty principle   40-14
      Energy conservation   40-24
      Position-momentum form of   40-15
      Quantum fluctuations   40-25
      Time-energy form of   40-19
Paschen series
   Energy level diagram   35-6
   Hydrogen spectra   35-6
Pauli exclusion principle   38-1, 38-9
Pauli, W., neutrinos   int-21, 20-6
Peebles, radiation from early universe   34-27
Pendulum
   Conical   9-18
   Energy conservation   10-10
   Simple and conical   14-17
   Simple pendulum   14-15
   Spring pendulum   9-4
      Ball spring program   9-20
      Computer analysis of   9-8
      $F = ma$ 9-7
      X. See Experiments I: - 6- Spring pendulum
   Torsion pendulum   14-12
      Differential equation for   14-14
Penzias and Wilson, cosmic background
   radiation   int-29, 34-27

Percussion instruments   16-22
Period of oscillation   14-4
   Neon bulb circuit   27- 30
Period, wavelength, and frequency   15-13
Periodic table   38-1, 38-10
   Beryllium   38-13
   Boron   38-13
   Effective nuclear charge   38-12
   Electron binding energies   38-11
   Electron screening   38-10
   Lithium   38-12
   Potassium to krypton   38-14
   Sodium to argon   38-13
   Summary   38-14
Phase and amplitude
   Fourier analysis lecture   16-31
   Wave motion   15-17
Phase of an oscillation   14-6
Phase transition, electroweak theory   int-26
Phases of Fourier coefficients   16-32
Photoelectric effect   int-8
   Einstein's formula   34-7
   Introduction to   34-5
   Maxwell's theory, failure of   34-6
   Planck's constant   34-8
Photon   int-8
   Blackbody radiation   34-22
   Chapter on   34-1
   Creates antiparticle   34-17
   Electric interaction   int-23
   Energies in short laser pulse   40-21
   Energy   34-9
      Energy levels in laser   37-4
      Uncertainty principle, Fourier transform   40-20
   Gravitational deflection of   34-19
   Hydrogen spectrum   35-5
   In electron spin resonance experiment   39-9
   Interaction with gravity   34-18
   Mass   34-12
   Mechanics   34-12
   Momentum   34-13
      Compton scattering   34-15
      Measurement limitation   40-11
   Probability wave   40-7
   Rest mass   int-12
   Stability of   19-14
   Standing waves   37-3
   Thermal, 2.74 degrees   int-29, 34-27, 34-31
Photon pulse
   Photon energy in   40-21
   Probability Interpretation of   40-22
Photon waves, probability interpretation   40-6
Pi mesons   int-23
Piela, electron clouds and binding energy   19-23
Pinhole camera   Optics-35
Planar-concave lens   Optics-27
Planar-convex lens   Optics-27
Planck, M., blackbody radiation law   34-4

Planck's constant
    And blackbody radiation   34-4, 34-22
    Angular momentum, Bohr model   35-8
    Bohr theory   35-1
    In Bohr magneton formula   39-4
    In de Broglie wavelength formula   35-11
    In photon mass formula   34-12
    In photon momentum formula   34-13
    In the photoelectric effect   34-7
    In the uncertainty principle   40-15, 40-19
    Introduction to   34-8
    Spin angular momentum   39-3
Plane, inclined   9-10
Planetary units   8-14
Planets
    Formation of   7-17
Plates, electron gun deflection   26-16
Plotting
    A point by computer   5-6
    Experiment, electric potential   25-7
    Potentials and fields. See Experiments II: - 1- Potential plotting
    Window, computer   5-7
Plutonium 246   6-8
Point
    Mass, gravitational field of   24-23
    Particle   int-14
    Source, velocity field of   23-6
Polarization of light waves   32-23
Polarizer
    Light   32-25
    Microwave   32-24
Polaroid, light polarizer   32-25
Position measurement, uncertainty principle   40-15
Positive and negative
    Charge   19-10
    Electric current   27-3
Positive area in Fourier analysis   16-29
Positron (antimatter)   int-13, 34-17
Positronium, annihilation into photons   34-17
Potassium to krypton, periodic table   38-14
Potential, electric
    Contour map   25-1
    Of a point charge   25-5

Potential energy
    Conservative forces   25-5
    Electric potential energy
        Contour map of   25-1
        In classical hydrogen atom   35-3
        In hydrogen atom   int-11
        In hydrogen molecule ion   19-21
        In molecules   17-12
        In nuclear fission   int-18, 20-5
        Negative and positive   25-4
        Of a point charge   25-5
        Potential plotting   25-7
        Storage in capacitors   27-18
    Energy conservation   int-11, 8-35, 10-20
        Conservative and non-conservative forces   10-21
        Mass on spring   14-11
        Uncertainty principle   40-24
    Equipartition of energy   17-28
    Gravitational potential energy   int-11, 8-35
        Bernoulli's equation   23-10
        Black holes   10-29
        In a room   10-25
        In satellite motion   8-36, 10-26
        In stellar evolution   20-13
        Introduction to   10-8
        Modified   8-37
        On a large scale   10-22
        Zero of   10-22
    In collisions   11-14
    Negative and positive   25-4
    Nuclear potential energy   int-18
        Binding energies   20-9
        Fusion   20-12
    Spin potential energy
        Magnetic   39-1
        Magnetic moment   39-4
    Spring potential energy   int-11, 10-16, 14-11
    Work   int-11, 10-15
    Work energy theorem   10-18
Potential plotting. See Experiments II: - 1- Potential plotting
Power
    1 horsepower = 746 watts   18-20
    Definition of watt   10-31
    Efficiency of a power plant   18-18
    In electric circuits   27-9
    Of the sun   34-23
    Sound intensity   16-24
Practical system of units   10-31
Precession
    Of atom, magnetic interaction   39-15
    Of orbit, modified gravity   8-30
Prediction of motion
    Using a computer   5-12
    Using calculus   Cal 1-9

Pressure
   Atmospheric   17-23
   Bernoulli's equation   23-9, 23-11
      Airplane wing   23-13
      Care in applying   23-16
      Sailboats   23-14
      Superfluid helium   23-17
   Ideal gas law   17-18
   In stellar evolution   17-17
   Measurement, using mercury barometer   17-22
   Osmotic pressure   17-34
   Pressure in fluids
      Aspirator   23-16
      Definition of   23-10
      Hydrodynamic voltage   23-17
      Hydrostatics   23-12
      Venturi meter   23-15
      Viscous effects   23-19
   Pressure of a gas   17-16
   Pressure of light
      Nichols and Hull experiment   34-14
      Red giant stars   20-15, 34-15
   PV diagrams   18-8
      Adiabatic expansion   18-9, 18-26
      Internal combustion engine   18-21
      Isothermal expansion   18-8, 18-26
      Reversible engines   18-13
      The Carnot cycle   18-11, 18-26, 18-28
   Work domb by pressure, Bernoulli's equation   23-10
Primary coil   30-26
Principle of relativity. *See also* Relativistic physics
   A statement of   1-4
   And the speed of light   1-11
   As a basic law of physics   1-4
   Chapter on   1-1
   Einstein's theory of   1-12
   Introduction to   1-2
   Special theory of   1-13
      A consistent theory   1-32
      Causality   1-36
      Lack of simultaneity   1-32
      Light pulse clock   1-14
      Lorentz contraction   1-24
      Mass energy   10-3
      Nature's speed limit   6-11
      Origin of magnetic forces   28- 8
      Photon mass   34-12
      Photon momentum   34-13
      Relativistic energy and momenta   28- 24
      Relativistic mass   6-6
      Relativity experiment leading to Faraday's law   30-9
      Time dilation   1-16, 1-22
      Zero rest mass particles   6-11
Principle of superposition
   For 1 dimensional waves   15-11
   For 2 dimensional waves   33-2
   Preliminary discussion of   15-2, 33-1
Prism
   Atmospheric, the green flash   Optics-17
   Glass   Optics-13
   Glass, rainbow of colors   Optics-15

Prism, mathematical (Fourier analysis)   16-28
Probability interpretation
   And the uncertainty principle   40-14
   Of electron diffraction pattern   40-6
   Of particle waves   40-6
   Of photon pulse   40-22
   Of two slit experiment   40-2
Probability wave
   For photons   40-7
   Intensity of   40-22
   Reflection and Fluorescence   40-8
Probe for scanning tunneling microscope   Optics-51
Problem solving. *See* Solving problems
   Gauss' law problems   24-26
   How to go about it   24-29
   Projectile motion problems   4-16
Program, BASIC. *See* BASIC program
Program, English
   For oscillatory motion   14-31
   For projectile motion   5-16
   For satellite motion   8-19
Project suggestion on wave speed   15-8
Projectile motion
   Analysis of
      Calculus   4-9
      Computer   5-16
      Graphical   3-16
   And the uncertainty principle   4-2, Cal 1-4
   BASIC program for   5-19
   Calculus definition of velocity   4-3, Cal 1-5
   Computer program for   8-21
   Constant acceleration formulas
      Calculus derivation   4-9
      Graphical analysis of   3-26
   Determining acceleration for   3-16
   English program for   5-16
   Graph paper tear out pages   3-29
   Gravitational force   8-2
   Instantaneous velocity   3-24
   Solving problems   4-16
   Strobe photograph of   3-7
   Styrofoam projectile   5-28
   With air resistance
      Calculus analysis   4-12
      Computer calculation   5-22
      Graphical analysis   3-22
   X. *See* Experiments I: - 2- Computer prediction of projectile motion
Projections of angular momentum
   Classical   7- 14
   Electron spin   39-3
   Quantized   38-5
Protactinium 236, recoil definition of mass   6-9

Proton int-17
   In alpha particles 20-8
   In atomic structure 19-4
   In hydrogen molecule 19-16
   In nuclear structure 20-2
   In the weak interaction 20-6
   Quark structure int-24
   Rest energy in electron volts 26-12
   Stability of 19-14
Proton cycle, energy from sun 34-24
Proton synchrotron at CERN 28-24
Proton-neutron mass difference, early universe 34-30
Ptolemy, epicycle, in Greek astronomy 8-25
Pulleys
   Atwood's machine 9-16
   Working with 9-16
Pulsars, neutron stars 20-17
Pulse
   Formation from sine waves 40-27
   Of electromagnetic radiation 32-10
PV = NRT 17-25
PV diagrams
   Adiabatic expansion 18-9
   Carnot cycle 18-11
   Isothermal expansion 18-8

Q

Quantization of electric charge 19-14
Quantized angular momentum int-9
   Angular momentum quantum number 38-7
   Electron spin 38-9
      Chapter on 39-1
      Concept of spin 39-3
   In Bohr theory 35-9
   In de Broglie's hypothesis 35-10
   In hydrogen wave patterns 38-3
   Quantized projections 38-5
Quantized vortices in superfluids 23-22
Quantum electrodynamics int-14
Quantum fluctuations in empty space 40-25
Quantum mechanics. *See also* Particle-wave nature; Schrödinger's equation
   Bohr theory of hydrogen 35-1
   Chapter on 40-1
   Concept of velocity 4-2, Cal 1-4
   Electron and nuclear spin 39-1
   Model atom 37-4
   Schrodinger's equation applied to atoms 38-1
   Uncertainty principle 40-14
   Zero point energy 37-7
Quantum number, angular momentum 38-7
Quantum theory of gravity int-16, int-20
Quark confinement 19-15
Quarks int-24
   Quantization of electric charge 19-14
Quasars
   Gravitational lens, Einstein cross 34-20
   Size of 34-19

R

R, gas constant 17-25
Radar waves
   Photon energies in 34-11
   Wavelength of 32-20
Radial acceleration 12-5
Radian measure 12-2, Cal 1-35
Radiated electromagnetic pulse 32-10
Radiated magnetic field thought experiment 32-11
Radiation
   Blackbody 32-22
      Photon picture of 34-22
      Theory of 34-2
      Wein's displacement law 34-2
   Cerenkov Optics-10
   Electromagnetic field
      Analysis of path 1 32-14
      Analysis of path 2 32-16
      Calculation of speed 32-14
      Spectrum of 32-20
   Radiated electric fields 32-28
   Radiated energy and the classical H atom 35-3
   Radiated field of point charge 32-30
   Three degree cosmic radiation int-30, 34-27
   UV, X Rays, and Gamma Rays 32-22
Radiation belts, Van Allen 28-32
Radiation pressure
   In red supergiant stars 20-15
   Of light 34-14
Radio galaxy Optics-48
Radio images of variable star Optics-49
Radio telescope int-15
   Arecibo int-15
   Three degree radiation 34-27
Radio telescopes Optics-48
   Arecibo Optics-48
   Radio galaxy image Optics-48
   Radio images of variable star Optics-49
   Very Large Array Optics-48
   Very Long Baseline Array Optics-49
Radio waves
   Hertz, Heinrich 34-1
   In the electromagnetic spectrum int-7, 32-20
   Photon energies 34-11
   Predicted from the classical hydrogen atom 35-2
   Wavelength of 32-20
Radius of electron 39-3
Rainbow
   Glass prism Optics-15
   Photograph of Optics-16
Range of nuclear force 20-3
RC circuit 27-22
   Exponential decay 27-23
   Exponential rise 27-26
   Half-lives 27-25
   Initial slope 27-25
   Measuring time constant 27-25
   X. See Experiments II: -3- The RC Circuit

Reactive metal, lithium   int-6, 38-9
Recoil experiments, definition of mass   6-2
Red shift and the expanding universe
    Doppler effect   34-21
    Evolution of universe   34-21
    Uniform expansion   33-24
Red supergiant star   20-15
Reflecting telescope   Optics-42
    Cassegrain design   Optics-42
    Diffraction limit   Optics-45
    Hubble space telescope   Optics-44
    Keck, world's largest optical   Optics-45
    Mt. Hopkins   Optics-43
    Mt. Palomar   Optics-43
    Newton's   Optics-22
    Newton's   Optics-42
    Secondary mirror   Optics-42
    William Hershel's   Optics-43
Reflection
    And fluorescence, probability interpretation   40-8
    Bragg reflection   36-4
    From curved surfaces   Optics-3
    Internal   Optics-13
    Of light   36-3, Optics-1
Refracting telescopes   Optics-40
    Galileo's   Optics-41
    Yerkes   Optics-41
Refraction, index of
    Definition of   Optics-9
    Glass prism and rainbow of colors   Optics-15
    Introduction to   Optics-2
    Of gas of supercooled sodium atoms   Optics-9
    Table of some values   Optics-9
Relativistic mass. *See* Relativistic physics: Relativistic mass
Relativistic physics. *See also* Principle of Relativity
    A consistent theory   1-32
    Antimatter   int-12, 34-16
    Black holes   10-29
    Blackbody radiation   34-22
    Causality   1-36
    Chapter on   1-1
    Clock
        Light pulse   1-14
        Moving   1-13
        Muon   1-20
        Muon lifetime movie   1-21
        Other kinds   1-18
        Real ones   1-20
    Creation of positron-electron pair   34-17
    Definition of mass   6-2
    Doppler effect for light   33-22
    Einstein mass formula   6-10
    Electric or magnetic field: depends on viewpoint   30-10
    Electric or magnetic force: depends on viewpoint   28-10
    Electromagnetic radiation, structure of   32-19
    Electron mass in beta decay   6-7
    Gravitational lensing   34-20

Interaction of photons and gravity   34-18
Kinetic energy   10-5
    Slowly moving particles   10-6
Lack of simultaneity   1-32
Longer seconds   1-16
Lorentz contraction   1-24
    Origin of magnetic forces   28-8
    Thought experiment on currents   28-4
Lorentz force law   28-15
Mass energy   10-3
Mass-energy relationship   int-11
Maxwell's equations   32-8
Motion of charged particles in magnetic fields   28-19
Muon lifetime movie   1-21
Nature's speed limit   int-2, int-12, 6-11
Neutrino astronomy   6-14
Neutrinos   6-13
Origin of magnetic forces   28-8
Particle accelerators   28-22
Photon mass   34-12
    Photon rest mass   6-12
Photon momentum   34-13
    Compton scattering   34-15
Principle of relativity   1-2
    As a basic law   1-4
Radiated electric fields   32-28
Red shift and the expansion of the universe   34-21
Relativistic calculations   1-28
    Approximation formulas   1-30
    Muons and Mt. Washington   1-29
    Slow speeds   1-29
Relativistic energy and momenta   28-24
Relativistic mass   6-6
    Formula for   6-10
    In beta decay   6-6
Relativistic mechanics   int-12
Relativistic speed limit   6-11
Relativistic wave equation   int-12
Relativity experiment for Faraday's law   30-9
Short lived elementary particles   40-23
Space travel   1-22
Special theory of relativity   1-13
Speed of light, measurement of   1-9
Speed of light wave   32-17
Spiraling electron in bubble chamber   28-27
The betatron   30-16
The early universe   34-29
Thought experiment on expanding magnetic field   32-11
Time dilation   1-22
Time-energy form of the uncertainty principle   40-19, 40-23
Zero rest mass   int-12
Zero rest mass particles   6-11
Relativity, general   int-15, 8-29
Renormalization   int-14, int-17
Repeated wave forms in Fourier analysis   16-11

Resistors
   In parallel   27-12
   In series   27-11
   Introduction to   27-6
   LR circuit   31-8
   Ohm's law   27-7
Resonance
   Electron spin
      Classical picture of   39-14
      Experiment   39-9
      Introduction to   39-5
      X. See Experiments II: -12- Electron spin resonance
   Introduction to   14-24
   Phenomena   14-26
   Tacoma Narrows bridge   14-24
      Vortex street   14-25
   Transients   14-27
Rest energy of proton and electron in eV   26-12
Rest mass   int-11
   And kinetic energy   10-5
   Einstein formula   6-10
Restoring forces
   Linear   14-7
   Non linear   14-19
Retrograde motion of Mars   8-24
Reversible engines
   As thought experiment   18-13
   Carnot cycle   18-17
   Efficiency of   18-18
Rifle and Bullet, recoil   7-7
Right handed coordinate system   2-18
Right-hand rule
   For cross products   12-10
   For Faraday's law   30-15
   For magnetic field of a current   28-13
   For magnetic field of a solenoid   29-14
   For surfaces   29-16
   Mirror images of   Optics-6
Rods, nerve fibers in eye   Optics-31
Rope, working with   13-10
Rotational motion
   Angular acceleration   12-3
   Angular analogy   12-3
   Angular velocity   12-2
   Bicycle wheel as a collection of masses   12-5
   Chapter on   12-1
   Radian measure   12-2
   Rolling down inclined plane   12-25
   Rotational kinetic energy   12-22
      Proof of theorem   12-26
   Translation and rotation   12-24
Rubber, elasticity of   17-35
Rutherford and the nucleus   35-1
Rydberg constant, in Bohr theory   35-9

S

Sailboats, Bernoulli's equation   23-14
Salt
   Dissolving   17-6
   Ionic bonding   38-15
Satellite motion   8-8
   Calculational loop for   8-19
   Classical hydrogen atom   35-2
   Compare with projectile, Newton's sketch   8-10
   Computer lab   8-23
   Computer prediction of   8-16
   Conservation of angular momentum   8-32
   Conservation of energy   8-35
   Earth tides   8-12
   Gravitational potential energy   10-22
   Kepler's laws   8-24
   Kepler's first law   8-26
   Kepler's second law   8-27
   Kepler's third law   8-28
   Modified gravity   8-29
   Moon   8-8
   Orbit, circular   10-27
   Orbit, elliptical   10-27
   Orbit, hyperbolic   10-27
   Orbit, parabolic   10-27
   Planetary units for   8-14
   Program for (Orbit 1)   8-21
   Total energy   10-26
   X. See Experiments I: - 5- Computer analysis of satellite motion
Scalar dot product   2-12
   Definition of work   10-13
Scanning tunneling microscope   Optics-51
   Surface (111 plane) of a silicon   Optics-51
Scattering of waves
   By graphite crystal, electron waves   36-8
   By myoglobin molecule   36-5
   By small object   36-2
   By thin crystals   36-6
   Chapter on   36-1
   Davisson-Germer experiment   35-12
   Reflection of light   36-3
   Two slit thought experiment   40-10
   X ray diffraction   36-4
Schmidt, Maarten, quasars   34-19
Schrödinger wave equation
   Felix Block story on   37-1
   Introduction to wave motion   15-1
   Particle-wave nature of matter   int-10
   Solution for hydrogen atom   38-2
   Standing waves in fuzzy walled box   38-1
Schrödinger, Erwin   int-10, 37-1
Schwinger, J., quantum electrodynamics   int-14
Search coil
   For magnetic field mapping experiment   30-24
   Inside Helmholtz coils   39-12

Second law, Newton's. *See* Newton's laws: Second law
Second law of thermodynamics. *See also* Carnot cycle; Thermal energy
 Applications of   18-17
 Chapter on   18-1
 Statement of   18-4
 Time reversed movie   18-1
Second, unit of time, definition of   int-2
Secondary mirror, in telescope   Optics-42
Semi major axis, Kepler's laws   8-28
Series expansions   Cal 1-23
 Binomial   Cal 1-23
 Exponential function e to the x   Cal 1-28
Series, harmonic   16-3
Series wiring
 Capacitors   27-21
 Resistors   27-11
Set Window, BASIC computer command   5-7
Short circuit   27-9
Short rod, electric force exerted by   24-9
Silicon, surface (111 plane) of   Optics-51
Simple electric circuit   27-8
Simple pendulum
 Simple and conical pendulums   14-17
 Theory of   14-15
Simultaneity, lack of   1-32
Sine function
 Amplitude of   Cal 1-37
 Definition of   Cal 1-35, Cal 1-36
 Derivative of, derivation   Cal 1-38
Sine waves
 AC voltage generator   30-22
 Amplitude of   14-10
 As solution of differential equation   14-9
 Definition of   14-3
 Derivative of   14-8
 Formation of pulse from   40-27
 Fourier analysis lecture   16-28
 Fourier analysis of   16-7
 Harmonic series   16-3
 Normal modes   16-4
 Phase of   14-6
 Pulse in air   15-3
 Sinusoidal waves motion   15-12
 Standing waves on a guitar string   15-20
 Traveling wave   15-16
 Traveling waves add to standing wave   15-20
Single slit diffraction   33-26
 Analysis of pattern   33-27
 Application to uncertainty principle   40-16
 Huygens principle   33-4
 Recording patterns   33-28
Slit pattern, Fourier transform of   16-33

Slope of a curve
 As derivative   Cal 1-30
 Formula for   Cal 1-30
 Negative slope   Cal 1-31
Small angle approximation
 Simple pendulum   14-16
 Snell's law   Optics-19
Small oscillations
 For non linear restoring forces   14-19
 Molecular forces act like springs   14-20
 Of simple pendulum   14-16
Smallest scale of distance, physics at   int-25
Snell's law
 Applied to spherical surfaces   Optics-19
 Derivation of   Optics-12
 For small angles   Optics-19
 Introduction to   Optics-2, Optics-11
Sodium to argon, periodic table   38-13
Solar neutrinos   6-13. *See also* Neutrinos
Solenoid
 Ampere's law applied to   29-15
 Magnetic field of   28-17, 29-14
 Right hand rule for   29-14
 Toroidal, magnetic field of   29-17
Solving problems
 Gauss' law problems   24-26
 How to go about it   24-29
 Projectile motion problems   4-16
Sombrero galaxy   int-2
Sound
 Focusing by an ellipse   8-26, Optics-3
 Fourier analysis, and normal modes   16-1
 Fourier analysis of violin notes   16-18
 Intensity
  Bells and decibels   16-24
  Definition of   16-24
  Speaker curves   16-27
 Percussion instruments   16-22
 Sound meters   16-26
 Sound produced by guitar string   15-22
 Stringed instruments   16-18
 The human ear   16-16
 Wind instruments   16-20
 X. *See* Experiments I: -12- Fourier analysis of sound waves
Sound waves, speed of
 Calculation of   15-8
 Formula for   15-9
 In various materials   15-9
Space
 And time   int-2, 1-1
 Quantum fluctuations in   40-25
 The Lorentz contraction   1-24
 Travel   1-22
Space physics   28-31. *See also* Experiments II: -5a- Magnetic focusing and space physics
Space telescope
 Hubble   Optics-44
 Infrared (IRAS)   Optics-47

Spacial frequency k   15-14
Speaker curves   16-27
Specific heat
  Cp and Cv   18-6
  Definition of   17-26
  Failure of Newtonian mechanics   17-31
  Gamma = Cp/Cv   18-7
Spectral lines
  Atomic spectra   33-16
  Bohr's explanation of   int-9
  Hydrogen
    Bohr theory of   35-4
    Colors of   int-7
    Experiment to measure   33-17
    The Balmer Series   33-19
  Introduction to   int-7
Spectrometer, mass   28- 28
Spectrum
  Electromagnetic
    Photon energies   34-11
    Visible spectrum   33-15
    Wavelengths of   32- 20
  Hydrogen
    Balmer series   33-19
    Bohr theory of   35-4
    Experiment to measure   33-17
    Lyman series, ultraviolet   35-6
    Paschen series, infrared   35-6
  Hydrogen star   33-19, 35-4
Spectrum.
  Electromagnetic
    Photon energies   34-11
Speed and mass increase   int-11
Speed detector, air cart   30-5. See also Experiments II: - 6- Faraday's law air cart speed detector
Speed limit, nature's   int-12, 6-11
Speed of an electromagnetic pulse   32- 14
Speed of light
  Absolute speed limit   int-2, int-12
  Calculation of speed of light
    Analysis of path 1   32- 14
    Analysis of path 2   32- 16
    Using Maxwell's Equations   32- 14
  Dimensional analysis   15-9
  Experiment to measure   1-9, 31-15
  In a medium   Optics-8
  Same to all observers   1-12
Speed of sound
  Formula for and values of   15-9
  Theory of   15-8
Speed of wave pulses
  Dimensional analysis   15-6
  On rope   15-4
Sphere
  Area of   23-6
  Electric field inside of   26-4
Spherical aberration   Optics-21
  In Hubble telescope mirror   Optics-22

Spherical lens surface   Optics-19
  Formula for focal length   Optics-20
Spherical mass, gravitational field of   24-24
Spin
  Allowed projections   39-3
  Chapter on   39-1
  Concept of   39-3
  Dirac equation   39-3
  Electron, introduction to   38-1
  Electron, periodic table   38-9
  Electron Spin Resonance
    Experiment   39-5, 39-9
  Gyroscope like   39-15
  Interaction with magnetic field   39-4
  Magnetic energy   39-1
    Semi classical formula   39-14
  Uhlenbeck and Gaudsmit   39-1
  X. See Experiments II: -12- Electron spin resonance
Spin flip energy, Dirac equation   39-15
Spring
  Constant   9-3
  Forces   9-3
  Mass on a spring
    Computer analysis of   14-30
    Differential equation for   14-8
    Theory   14-7
  Oscillating cart   14-5
  Spring model of molecular force   17-13
  Spring pendulum
    Ball spring program   9-20
    Computer analysis of   9-8
    F = ma   9-7
    Introduction to   9-4
  Spring potential energy   int-11
Square of amplitude, intensity   16-33
Square wave
  Fourier analysis of   16-9, 16-28
Stability of matter   19-14
Standard model of basic interactions   int-25
Standing waves
  Allowed standing waves in hydrogen   37-1
  De Broglie waves
    Movie   35-11
  Formulas for   15-20
  Hydrogen, L= 0 patterns   38-4
  Introduction to   15-18
  Light waves in laser   37-2
  Made from traveling waves   15-19
  On a guitar string   15-20
    Frequency of   15-21
  On drums   16-22
  On violin backplate   16-23
  Particle wave nature   int-13
  Patterns in hydrogen   38-3
  Photons in laser   37-3
  The L= 0 patterns in hydrogen   38-4
  Two dimensional   37-8
    Electrons on copper crystal   37-9
    On drumhead   37-8

Star
- Binary   int-2
- Black hole   20-19
- Black dwarf   int-19
- Black hole   int-20, 20-18
- Blackbody spectrum of   34-3
- Hydrogen spectrum of   33-19, 35-4
- Neutron   int-20, 20-17
- Red supergiant   20-15
- Stellar evolution   17-17
- White dwarf   20-15

Statamps, CGS units   24-2

Static charges
- Electric field of   30-16
- Line integral for   30-2

Stationary source, moving observer, Doppler effect   33-21

Statvolts, CGS units of charge   24-2

Steady state model of the universe   34-25

Stellar evolution
- General discussion   int-19
- Role of neutrinos   20-15
- Role of the four basic interactions   20-13
- Role of thermal energy   17-17

Step-By-Step Calculations   5-1

Stomach, medical image   Optics-15

Strain, definition of   15-8

Strange quark   int-24

Streamlines
- And electric field lines   24-14, 24-17
- Around airplane wing   23-13
- Around sailboat sail   23-14
- Bernoulli's equation   23-11
- Bounding flux tubes   23-8
- Definition of   23-4
- Hele-Shaw cell   23-4
- In blood flow experiment   23-23
- In superfluid helium venturi meter   23-17
- In venturi meter   23-15

Strength of the electric interaction   19-8
- And magnetic forces   28-6
- In comparison to gravity   int-6
- In comparison to nuclear force   int-18, 20-2
- Two garden peas   28-2

Stress, definition of   15-8

String forces
- Atwoods machine   9-16
- Conical pendulum   9-18
- Introduction to   9-15
- Solving pulley problems   9-16
- Working with rope   13-10

String theory   int-16

Stringed instruments   16-18
- Violin, acoustic vs electric   16-19

Strobe photographs
- Analyzing   3-8, 3-11
- And the uncertainty principle   4-2, Cal 1-4
- Defining the acceleration vector   3-15
- Defining the velocity vector   3-11
- Taking   3-7

Structure, nuclear   20-7

Styrofoam projectile   5-28

SU3 symmetry   19-15
- Gell-Mann and Neuman   int-24

Subtraction of vectors   2-7

Summation
- Becoming an integral   Cal 1-10
- Of velocity vectors   Cal 1-10

Sun
- Age of   34-23
- And neutron stars   20-17
- Energy source   int-18
- Halos around   Optics-18
- Kepler's laws   8-24
- Neutrinos from   6-13, 11-21
- Stellar evolution   17-17, 20-13
- Sun dogs   Optics-18

Superconducting magnets   31-30

Superfluids
- Quantized vortices in   23-22
- Superfluid helium venturi meter   23-15

Supernova, 1987
- In stellar evolution   20-14
- Neutrino astronomy   6-14
- Neutrinos from   20-16
- Photograph of   int-19, 6-14

Superposition of waves
- Circular waves   33-2
- Principle of   15-11
- Two slit experiment   40-2

Surface (111 plane) of silicon   Optics-51

Surface charges   26-2

Surface integral. *See also* Integral
- Definition of   29-2
- For magnetic fields   32-3
- Gauss' law   29-3
- In Maxwell's equations   32-8
- Two kinds of vector fields   30-19

Surface tension   17-14

Symmetry
- SU3   19-15
- Gell-Mann and Neuman   int-24

Symmetry of Maxwell's equations   32-9

Synchrotron   28-22

Systems of particles, chapter on   11-1

T

Table of elements 19-5
Tacoma narrows bridge 14-24
Tangential distance, velocity and acceleration 12-4
Tau particle, lepton family int-22
Tau type neutrino int-22
Taylor, Joe, binary neutron stars int-15
Telescope, parabolic reflector Optics-4
Telescopes int-1
   Infrared
      IRAS satellite Optics-47
      Mt. Hopkins, 2Mass Optics-46
   Radio Optics-48
      Arecibo Optics-48
      Arecibo, binary neutron stars int-15
      Holmdel, three degree radiation int-30
      Radio galaxy image Optics-48
      Radio images of variable star Optics-49
      Very Large Array Optics-48
      Very Long Baseline Array Optics-49
   Reflecting Optics-42
      Cassegrain design Optics-42
      Diffraction limit Optics-45
      Hubble space telescope Optics-44
      Image plane Optics-5
      Keck, world's largest optical Optics-45
      Mt. Hopkins Optics-43
      Mt. Palomar Optics-43
      Newton's Optics-42
      Secondary mirror Optics-42
      William Hershel's Optics-43
   Refracting Optics-40
      Galileo's Optics-41
      Yerkes Optics-41
Television waves
   Photon energies 34-11
   Wavelength of 32-20
Temperature
   Absolute zero 17-9
   And zero point energy 37-8
   Boltzman's constant 17-11
   Heated hydrogen int-9
   Ideal gas thermometer 17-20
   Introduction to 17-9
   Temperature scales 17-10
Tesla and Gauss, magnetic field dimensions 28-16
Test charge, unit size 24-11
Text file for FFT data 16-32
Thermal efficiency of Carnot cycle. *See* Carnot cycle, efficiency of
Thermal energy
   Dollar value of 18-1
   In a bottle of hydrogen int-8
   Time reversed movie of dive 18-1
Thermal equilibrium
   And temperature 17-9
   Introduction to 17-8
   Of the universe 34-28

Thermal expansion
   Adiabatic expansion 18-9
      Derivation of formula 18-26
      Formula for 18-10
   In stellar evolution 17-17, 20-13
   Isothermal expansion 18-8
      Derivation of formula for 18-26
   Molecular theory of 17-33
   Of gas in balloon 17-16
   Work done by an expanding gas 18-5
Thermal motion
   Boltzman's constant 17-11
   Brownian motion 17-6
      Movie 17-7
Thermal photons. *See* Three degree radiation
   In blackbody radiation 34-22
Thermometer
   Ideal gas 17-20
   Mercury or alcohol 17-9
   Temperature scales 17-10
Thin lenses Optics-23
Thought experiments
   Carnot cycle 18-4
   Causality 1-36
   Lack of simultaneity 1-32
   Light pulse clock 1-13
   Lorentz contraction and space travel 1-22
   Magnetic force and Faraday's law 30-3
   No width contraction 1-28
   Origin of magnetic forces 28-8
   Two slit experiment and the uncertainty principle 40-9
Three degree cosmic radiation int-30, 34-27
   Penzias and Wilson 34-27
Tides, two a day 8-12
Time
   Age of sun 34-23
   And the speed of light int-2
   Behavior of 1-1
   Dilation 1-22
   Dilation formula 1-16
   Direction of
      Entropy 18-26
      Neutral K meson 18-26
   In early universe int-27, 34-29
   Measuring short times using uncertainty principle 40-22
   Moving clocks 1-13
   Muon lifetime movie 1-21
   On light pulse clock 1-14
   On other clocks 1-18
   On real clocks 1-20
   Steady state model of universe 34-25
Time constant
   For RC circuit 27-24
   Measuring from a graph 27-25, Cal 1-34
Time, direction of
   And strobe photographs 3-27

Time reversal
  Dive movie 18-1
  Water droplets 18-2
Time step and initial conditions 5-14
Time-energy form of the uncertainty principle 40-19
Top quark
  Mass of int-24
  Quark family int-24
Toroid
  Inductor 31-6
    In LC experiment 31-11
    In resonance 31-14
    Speed of light measuerment 1-9, 31-15
  Magnetic field of 29-17
Torque
  As angular force 12-15
  In torsion pendulum 14-12
  On a current loop 31-20
Torsion pendulum 14-12. *See also* Cavendish experiment
Total energy
  Classical hydrogen atom 35-3
  Escape velocity 10-28
  Satellite motion 8-36
Town water supply, hydrodynamic voltage 23-18
Transients 14-27
Translation and rotation 12-24
Transmitted wave and lenses 36-3
Traveling waves, formula for 15-16
Tritium, a hydrogen isotope int-17, 19-7
True BASIC. *See* BASIC program
Tubes of flux 24-17
Two dimensional standing waves 37-8
Two kinds of vector fields 30-18
Two lenses together Optics-29
Two slit experiment
  And the uncertainty principle 40-9
  Measurement limitations 40-9
  One particle at a time 40-3
  One slit experiment 40-2
  Particle point of view 40-3
  Particle/wave nature 40-2
  Using electrons 40-3
Two-slit interference patterns 33-6
  A closer look at 33-26
  First maxima 33-8
  For light 33-10
Tycho Brahe's apparatus 8-25

U

Uhlenbeck and Gaudsmit, electron spin 39-1
Ultraviolet light
  Electromagnetic spectrum int-7, 32-20, 32-22
  Photon energies 34-11
  Wavelength of 32-20

Uncertainty principle Cal 1-3
  $\Delta x \Delta p >= h$ 40-15
  And definition of velocity 4-1, Cal 1-3
  And strobe photographs 4-2, Cal 1-4
  Applied to projectile motion 4-2, Cal 1-4
  Elementary particles, short lived 40-23
  Energy conservation 40-24
  Fourier transform 40-20
  Introduction to 40-14
  Particle/wave nature int-10
  Position-momentum form 40-15
  Single slit experiment 40-16
  Time-energy form 40-19
  Used as clock 40-22
  X. *See* Experiments II: -13- Fourier analysis & the uncertainty principle
Uniform circular motion. *See* Circular motion
Uniform expansion of universe int-3, 33-24
Uniform magnetic fields 28-16
Unit of angular momentum int-9
  Electron spin 39-3
  In Bohr theory 35-9
Unit test charge 24-11
Unit vectors 8-18
Units
  Atomic units 19-22
  CGS
    Coulomb's law 24-2
    Statamp, statvolt, ergs per second 24-2
  Checking MKS calculations 24-3
  MKS
    Ampere, volt, watt 24-2
    Coulomb's law 24-2
  Planetary units 8-14
  Practical System of units (MKS) 10-31
Universe
  Age of int-3
  As a laboratory int-1
  Becomes transparent, decoupling 34-31
  Big bang model int-4, 33-25, 34-26
  Continuous creation theory int-4, 34-25
  Decoupling (700,000 years) 34-31
  Early. *See* Universe, early
  Evolution of, Doppler effect 34-21
  Excess of matter over antimatter int-27, 34-29
  Expanding int-3
    Quasars and gravitational lensing 34-19
    Red shift 33-24
  Helium abundance in 34-26
  Models of 34-23
  Steady state model of 34-25
  The First Three Minutes 34-32
  Thermal equilibrium of 34-28
  Three degree cosmic radiation int-30, 34-27
  Visible int-3, 34-32

Universe, early int-27, 34-29
   10 to the 10 degrees int-27
   10 to the 13 degrees int-27
   10 to the 14 degrees int-27
   13.8 seconds int-28
   24% neutrons int-28
   38% neutrons int-27
   At .7 million years, decoupling int-29
   At various short times 34-30
   Big bang model 34-26
   Books on
      Coming of Age in the Milky Way 34-32
      The First Three Minutes 34-32
   Decoupling (700,000 years) 34-31
   Deuterium bottleneck int-28
   Excess of Matter over Antimatter 34-29
   Frame #2 (.11 seconds) 34-30
   Frame #3 (1.09 seconds) 34-30
   Frame #4 (13.82 seconds) 34-30
   Frame #5 (3 minutes and 2 seconds) 34-30
   Helium abundance 34-26
   Helium created int-28
   Hot int-4
   Matter particles survive int-27, 34-29
   Neutrinos escape at one second int-28
   Overview int-27, 34-29
   Positrons annihilated int-28
   Thermal equilibrium of 34-28
   Thermal photons int-29, 34-28
   Three degree cosmic radiation int-30, 34-27, 34-32
   Transparent universe int-29
   Why it is hot 34-29
Up quark int-24
Uranium
   Binding energy per nucleon 20-10
   Nuclear fission 20-4
   Nuclear structure int-17, 20-2

V

Van Allen radiation belts 28-32
Van de Graaff generator 26-6
Variable names, computer 5-6
Variables
   Evaluated over interval Cal 1-10
Vector
   Addition 2-3
   Addition by components 2-9
   Angular momentum 7-14, 12-7
   Angular velocity 12-7
   Area 24-22
   Components 2-8
   Cross Product 12-9
   Definition of acceleration Cal 1-7
      Component equations Cal 1-8
   Definition of velocity Cal 1-6
      Component equations Cal 1-8
   Dot product 10-13
   Equations
      Components with derivatives 4-6, Cal 1-7
      Constant acceleration formulas 4-11
      Exercise on 2-7, 2-8
      In component form 2-10
   Magnitude of 2-6
   Measuring length of 3-9
   Multiplication 2-11
   Multiplication, cross product 2-15, 12-9
      Formula for 2-17
      Magnitude of 2-17
      Right hand rule 12-10
   Multiplication, scalar or dot product 2-12, 10-13
      Interpretation of 2-14
   Velocity from coordinate vector 3-13
Vector fields 23-3
   Electric field 24-10
   Magnetic field 28-10
   Two kinds of 30-18
   Velocity field 23-2
Vectors 2-2
   Arithmetic of 2-3
   Associative law 2-4
   Commutative law 2-4
   Coordinate 3-11
   Displacement 2-2
      from Strobe Photos 3-5
   Graphical addition and subtraction 3-10
   Multiplication by number 2-5
   Negative of 2-5
   Subtraction of 2-5
   Unit 8-18

Velocity 3-11
    And the uncertainty principle 4-2, Cal 1-4
    Angular 12-2, 14-5
    Angular analogy 12-3
    Calculus definition of 4-3, Cal 1-5
        Component equations Cal 1-8
    Curve, area under Cal 1-12
    Definite integral of Cal 1-11
    Instantaneous 3-24
        From strobe photograph 3-26
    Integral of Cal 1-10
    Of escape 10-28
    Tangential 12-4
    Using strobe photos 3-11
Velocity detector
    Air cart 30-5
    Magnetic flux 30-25
Velocity field 23-2
    Flux of 23-8, 24-15
    Of a line source 23-7
    Of a point source 23-6
Velocity vector from coordinate vector 3-13
Venturi meter
    Bernoulli's equation 23-15
    With superfluid helium 23-17
Very Large Array, radio telescopes Optics-48
Very Long Baseline Array , radio telescopes Optics-49
Violets, odor of 17-6
Violin
    Acoustic vs electric 16-19
    Back, standing waves on 16-23
Viscous effects in fluid flow 23-19
Visible light int-7, 32- 20
    Photon energies 34-11
    Spectrum of 33-15
    Wavelength of 32- 20
Visible universe int-3, 34-32
Volt
    Electron 26-12
    MKS units 24-2
Voltage
    Air cart speed detector 30-5
    Divider, circuit for 27- 13
    Electric 25-6
    Fluid analogy
        Hydrodynamic voltage 23-17
        Resistance 27- 7
        Town water supply 23-18
    Induced 31-4
    Induced in a moving loop 30-4
Voltage and current 27- 6
    Resistors 27- 6
        Ohm's law 27- 7
Voltage source, constant 27- 15
Voltage transformer and magnetic flux 30-26
Volume of mole of gas 17-25

Vortex
    Bathtub 23-2, 23-20
    Hurricane 23-20
    Tornado 23-20
Vortex street, Karman 14-25
Vortices 23-20
    Quantized, in superfluids 23-22

W

W and Z mesons, electroweak interaction int-26
Water
    Evaporating 17-5
    Molecule 17-2
Water droplets, time reversal 18-2
Watt, MKS units 24-2
Wave
    Circular water waves 33-2
    Cosine waves 16-28
    De Broglie, standing wave movie 35-11
    Diffraction pattern 33-5
    Electromagnetic waves 32- 18
        Probability wave for photons 40-7
    Electron waves, de Broglie picture 35-11
    Electron waves, in hydrogen 38-1
    Forms, repeated, in Fourier analysis 16-11
    Fourier analysis
        Of a sine wave 16-7
        Of a square wave 16-9
    Huygens' principle, introduction to 33-4
    Interference patterns 33-3
    Light waves
        Chapter on 33-1
        Polarization of 32- 23
    Patterns, superposition of 33-2
    Photon wave 40-6
    Probability wave
        Intensity of 40-22
        Reflection and Fluorescence 40-8
    Scattering, measurement limitation 40-10
    Single slit experiment, uncertainty principle 40-16
    Sinusoidal waves 15-12
        Time dependent 14-3
    Speed of waves
        Dimensional analysis 15-6
        Project suggestion 15-8
    Standing waves
        Allowed 37-1
        Formulas for 15-20
        Frequency of 15-21
        Hydrogen $L=0$ Patterns 38-4
        Introduction to 15-18
        On a guitar string 15-20
        Two dimensional 37-8
    Transmitted waves 36-3
    Traveling waves, formula for 15-16
    Wave pulses 15-3
        Speed of 15-4

Index-40

Wave equation   int-7, 15-1
   Dirac's   15-2
      Bohr magneton   39-5
      Spin   39-3
   For light   int-7
   Maxwell's   15-1
   Nonrelativistic   int-12, 34-16
   Relativistic   int-12, 15-2
   Schrödinger's   15-1
      Discovery of   37-1
Wave motion
   Amplitude and phase   15-17
   Linear and nonlinear   15-10
   One dimensional   15-1
   Principle of superposition   15-11
Wave nature of light
   Young, Thomas   34-1, Optics-1
Wave patterns
   Hydrogen
      Intensity at the Origin   38-5
      Schrödinger's equation   38-2
      Standing waves   38-3
Wave/particle nature. *See also* Particle-wave nature
   Born Interpretation   40-6
   Probability interpretation   40-6
   Two slit experiment   40-2
Wavelength
   De Broglie   35-11
   Electron   36-9
   Fourier analysis   16-28
   Laser beam   16-33
   Period, and frequency   15-13
Weak interaction   int-14, int-20, 19-2
   Creation of neutrinos   20-6
   Electroweak theory   int-26
   Four basic interactions   19-2
   Neutron decay   20-6
   Range of   int-21
   Strength of   int-21
Weighing the earth   8-8
Weight   8-11
   Lifting   13-11
Weinberg, S., book "The first Three Minutes"   34-32
Wein's displacement law for blackbody radiation   34-2
White dwarf star   20-15
White light (Fourier analysis)   16-28
William Hershel's telescope   Optics-43
Wilson, Robert, 3 degree radiation   int-29, 34-27
Wind instruments   16-20

Work   int-11
   And potential energy   10-14
   Bernoulli's equation   23-10
   Calculation of in adiabatic expansion   18-26
   Calculation of in isothermal expansion   18-26
   Definition of   10-12
   Done by an expanding gas   18-5
   Integral formula for   10-14, 10-15
   Non-constant forces   10-14
   Vector dot product   10-13
Work energy theorem   10-18
World's largest optical telescope   Optics-45

X

x-Cal 1
   Exercise 1   *Cal 1-14*
   Exercise 2   *Cal 1-15*
   Exercise 3   *Cal 1-17*
   Exercise 4   *Cal 1-22*
   Exercise 5   *Cal 1-24*
   Exercise 6   *Cal 1-29*
   Exercise 7   *Cal 1-29*
   Exercise 8   *Cal 1-31*
   Exercise 9   *Cal 1-33*
   Exercise 10   *Cal 1-36*
   Exercise 11   *Cal 1-39*
   Exercise 12   *Cal 1-39*
   Exercise 13   *Cal 1-39*
X-Ch 1
   Exercise 1   *1-3*
   Exercise 2   *1-4*
   Exercise 3   *1-11*
   Exercise 4   *1-16*
   Exercise 5   *1-31*
   Exercise 6   *1-31*
   Exercise 7   *1-31*
X-Ch 2
   Exercise 1   *2-7*
   Exercise 2   *2-7*
   Exercise 3   *2-7*
   Exercise 4   *2-7*
   Exercise 5   *2-8*
   Exercise 6   *2-10*
   Exercise 7   *2-12*
   Exercise 8   *2-13*
   Exercise 9   *2-15*
   Exercise 10   *2-16*
   Exercise 11   *2-17*
   Exercise 12   *2-17*
   Exercise 13   *2-18*
X-Ch 3
   Exercise 1   *3-10*
   Exercise 2   *3-10*
   Exercise 3   *3-10*
   Exercise 4   *3-12*
   Exercise 5   *3-17*
   Exercise 6   *3-18*
   Exercise 7   *3-22*
   Exercise 8   *3-27*
   Exercise 9   *3-27*
   Exercise 10   *3-27*

X-Ch 4
   Exercises 1-7   4-19
X-Ch 5
   Exercise 1   5-3
   Exercise 2   A running program   5-8
   Exercise 3   Plotting a circular line   5-8
   Exercise 4   Labels and axes   5-9
   Exercise 5a   Numerical output   5-9
   Exercise 5b   5-10
   Exercise 6   Plotting crosses   5-11
   Exercise 7   5-20
   Exercise 8   Changing the time step   5-20
   Exercise 9   Numerical Output   5-20
   Exercise 10   Attempt to reduce output   5-20
   Exercise 11   Reducing numerical output   5-20
   Exercise 12   Plotting crosses   5-21
   Exercise 13   Graphical analysis   5-25
   Exercise 14   Computer prediction   5-26
   Exercise 15   Viscous fluid   5-26
   Exercise 16   Nonlinear air resistance (optional)   5-27
   Exercise 17   Fan Added   5-27
X-Ch 6
   Exercise 1   6-4
   Exercise 2   Decay of Plutonium 246   6-8
   Exercise 3   Protactinium 236 decay.   6-9
   Exercise 4   Increase in Electron Mass.   6-9
   Exercise 5   A Thought Experiment.   6-9
   Exercise 6   6-10
   Exercise 7   6-10
   Exercise 8   6-10
X-Ch 7
   Exercise 1   7-6
   Exercise 2   7-8
   Exercise 3   Frictionless Ice   7-8
   Exercise 4   Bullet and Block   7-8
   Exercise 5   Two Skaters Throwing Ball   7-8
   Exercise 6   Rocket   7-8
   Exercise 7   7-10
   Exercise 8   7-10
   Exercise 9   7-10
   Exercise 10   7-13
   Exercise 11   7-14
   Exercise 12   7-16
   Exercise 13   7-16
   Exercise 14   7-18
X-Ch 8
   Exercise 1   8-5
   Exercise 2   8-8
   Exercise 3   8-9
   Exercise 4   8-11
   Exercise 5   8-12
   Exercise 6   8-15
   Exercise 7   8-15
   Exercise 8   8-16
   Exercise 9   8-16
   Exercise 10   8-22
   Exercise 11   8-23
   Exercise 12   8-23
   Exercise 13   8-27
   Exercise 14   8-27
   Exercise 15   8-28
   Exercise 16   8-28
   Exercise 17   8-28
   Exercise 18   8-31
   Exercise 19   8-34
   Exercise 20   8-34
   Exercise 21   8-36
   Exercise 22   8-36
   Exercise 23   8-37
X-Ch 9
   Exercise 1   9-7
   Exercise 2   9-11
   Exercise 3   9-17
   Exercise 4   9-17
   Exercise 5   Conical Pendulum   9-19
   Exercise 6   9-19
X-Ch10
   Exercise 1   10-4
   Exercise 2   10-6
   Exercise 3   10-8
   Exercise 4   10-9
   Exercise 5   10-9
   Exercise 6   10-9
   Exercise 7   10-11
   Exercise 8   10-12
   Exercise 9   10-13
   Exercise 10   10-17
   Exercise 11   10-17
   Exercise 12   10-28
   Exercise 13   10-28
   Exercise 14   10-28
   Exercise 15   10-31
X-Ch11
   Exercise 1   11-4
   Exercise 2   11-4
   Exercise 3   11-4
   Exercise 4   11-6
   Exercise 5   11-8
   Exercise 6   11-12
   Exercise 7   11-13
   Exercise 8   11-13
   Exercise 9   11-13
   Exercise 10   11-14
   Exercise 11   11-14
   Exercise 12   11-15
   Exercise 13   11-16
X-Ch12
   Exercise 1   12-2
   Exercise 2   12-3
   Exercise 3   12-5
   Exercise 4   12-9
   Exercise 5   12-10
   Exercise 6   12-11
   Exercise 7   12-13
   Exercise 8   12-15
   Exercise 9   12-21
   Exercise A1   12-24
   Exercise A2   12-26
   Exercise A3   potential lab experiment   12-26

X-Ch13
- Exercise 1   13-3
- Exercise 2   13-3
- Exercise 3   13-4
- Exercise 4   13-6
- Exercise 5   13-8
- Exercise 6   Ladder problem   13-8
- Exercise 7   13-9
- Exercise 8   Working with rope   13-10
- Exercise 9   13-11
- Exercise 10   13-12
- Exercise 11   13-12

X-Ch14
- Exercise 1   14-4
- Exercise 2   14-5
- Exercise 3   14-5
- Exercise 4   14-7
- Exercise 5   14-8
- Exercise 6   14-9
- Exercise 7   14-10
- Exercise 8   14-10
- Exercise 9   14-10
- Exercise 10   14-11
- Exercise 11   14-11
- Exercise 12   14-14
- Exercise 13   14-14
- Exercise 14   14-15
- Exercise 15   14-16
- Exercise 16   14-16
- Exercise 17   14-17
- Exercise 18   14-18
- Exercise 19   Physical pendulum   14-18
- Exercise 20   Damped harmonic motion   14-23
- Exercise 21   14-27
- Exercise 22   14-33
- Exercise 23   14-34

X-Ch15
- Exercise 1   15-4
- Exercise 2   15-9
- Exercise 3   15-9
- Exercise 4   15-9
- Exercise 5   15-14
- Exercise 6   15-15
- Exercise 7   15-16
- Exercise 8   15-22
- Exercise 9   15-22
- Exercise 10   15-22

X-Ch16
- Exercise 1   16-21
- Exercise 2   16-27
- Exercise 3   16-27

X-Ch17
- Exercise 1   17-8
- Exercise 2   17-11
- Exercise 3   17-11
- Exercise 4   17-11
- Exercise 5   17-14
- Exercise 6   17-24
- Exercise 7   17-25
- Exercise 8   17-26

X-Ch18
- Exercise 1   18-5
- Exercise 2   18-7
- Exercise 3   18-10
- Exercise 4   18-13
- Exercise 5   18-13
- Exercise 6   18-14
- Exercise 7   18-16
- Exercise 8   18-19
- Exercise 9   18-20
- Exercise 10   18-20

X-Ch19
- Exercise 1   19-9
- Exercise 2   19-11
- Exercise 3   19-11
- Exercise 4   19-12
- Exercise 5   19-12

X-Ch20
- Exercise 1   20-9

X-Ch23
- Exercise 1   23-12
- Exercise 2   23-12
- Exercise 3   23-15
- Exercise 4   23-15
- Exercise 5   23-22
- Exercise 6   23-22

X-Ch24
- Exercise 1   24-5
- Exercise 2   24-5
- Exercise 3   24-5
- Exercise 4   24-5
- Exercise 5   24-8
- Exercise 6   24-9
- Exercise 7   24-12
- Exercise 8   24-25
- Exercise 9   24-27
- Exercise 10   24-27
- Exercise 11   24-27
- Exercise 12   24-27
- Exercise 13   24-28
- Exercise 14   24-28

X-Ch25
- Exercise 1   25-5
- Exercise 2   25-9
- Exercise 3   25-12
- Exercise 4   25-12

X-Ch26
- Exercise 1   26-5
- Exercise 2   26-5
- Exercise 3   26-5
- Exercise 4   26-5
- Exercise 5   26-11
- Exercise 6   26-13
- Exercise 7   26-13
- Exercise 8   26-16
- Exercise 9   26-17
- Exercise 10   Millikan oil drop experiment   26-17

X-Ch27
   Exercise 1   27- 10
   Exercise 2   27- 13
   Exercise 3 The voltage divider   27- 13
   Exercise 4 - Electrolytic capacitor   27- 17
   Exercise 5   27- 19
   Exercise 6   27- 21
   Exercise 7   27- 24
   Exercise 8   27- 26
   Exercise 9   27- 28
   Exercise 10   27- 30
   Exercise 11   27- 32

X-Ch28
   Exercise 1   28- 3
   Exercise 2   28- 9
   Exercise 3   28- 15
   Exercise 4   28- 21
   Exercise 5   28- 21
   Exercise 6   28- 25
   Exercise 7   28- 25
   Exercise 8   28- 25
   Exercise 9   28- 27
   Exercise 10   28- 28

X-Ch29
   Exercise 1   29-12
   Exercise 2   29-12
   Exercise 3   29-12
   Exercise 4   29-12
   Exercise 5   29-13
   Exercise 6   29-13
   Exercise 7   29-16
   Exercise 8   29-18
   Exercise 9   29-18

X-Ch30
   Exercise 1   30-10
   Exercise 2   30-12
   Exercise 3   30-18
   Exercise 4   30-20
   Exercise 5   30-22
   Exercise 6   30-22
   Exercise 7   30-25
   Exercise 8   30-26

X-Ch31
   Exercise 1   31-3
   Exercise 2   31-9
   Exercise 3   31-10
   Exercise 4   31-11
   Exercise 5   31-15
   Exercise 6   31-15
   Exercise 7   31-16

X-Ch32
   Exercise 1   32- 7
   Exercise 2   32- 8
   Exercise 3   32- 9
   Exercise 4   32- 17
   Exercise 5   32- 17
   Exercise 6   32- 32

X-Ch33
   Exercise 1   33-4
   Exercise 2   33-8
   Exercise 3   33-10
   Exercise 4   33-10
   Exercise 5   33-13
   Exercise 6   33-13
   Exercise 7   33-16
   Exercise 8   33-18
   Exercise 9   33-19
   Exercise 10   33-21
   Exercise 11   33-23
   Exercise 12   33-23
   Exercise 13   33-23
   Exercise 14   33-28
   Exercise 15   33-30

X-Ch34
   Exercise 1   34-3
   Exercise 2   34-9
   Exercise 3   34-9
   Exercise 4   34-10
   Exercise 5   34-10
   Exercise 6   34-10
   Exercise 7   34-10
   Exercise 8   34-10
   Exercise 9   34-10
   Exercise 10   34-10
   Exercise 11   34-17
   Exercise 12   34-18

x-Ch35
   Exercise 1   35-5
   Exercise 2   35-5
   Exercise 3   35-6
   Exercise 4   35-6
   Exercise 5   35-7
   Exercise 6   35-9
   Exercise 7   35-9
   Exercise 8   35-9
   Exercise 9   35-10
   Exercise 10   35-10

x-Ch36
   Exercise 1   36-3
   Exercise 2   36-7
   Exercise 3   36-7
   Exercise 4   36-9
   Exercise 5   36-12
   Exercise 6   36-12

x-Ch37
   Exercise 1   37-4
   Exercise 2   37-6
   Exercise 3   37-6
   Exercise 4   37-7

x-Ch38
   Exercise 1   38-4
   Exercise 2   38-7

x-Ch39
   Exercise 1   39-5
   Exercise 2   39-5
   Exercise 3   39-5
   Exercise 4   39-7
   Exercise 5   39-7
   Exercise 6   39-12
x-Ch40
   Exercise 1   40-6
   Exercise 2   40-7
   Exercise 3   40-18
X-Optics
   Exercise 1a   Optics-10
   Exercise 1b   Optics-10
   Exercise 2   Optics-12
   Exercise 3   Optics-13
   Exercise 4   Optics-17
   Exercise 5   Optics-21
   Exercise 6   Optics-21
   Exercise 7   Optics-21
   Exercise 8   Optics-23
   Exercise 9   Optics-24
   Exercise 10   Optics-27
   Exercise 11   Optics-29
   Exercise 12   Optics-30
   Exercise 13   Optics-30
X-rays
   Diffraction   36-4
   Diffraction pattern   36-5
   Electromagnetic spectrum   int-7, 32-20, 32-22
   Photon energies   34-11, 36-4
   Wavelength of   32-20

Y

Yerkes telescope   Optics-41
Young, Thomas
   Wave nature of light   34-1, Optics-1
Young's modulus   15-8
Yukawa, H.   int-22
Yukawa's theory   int-22

Z

Z and W mesons, electroweak interaction   int-26
Z, astronomer's Z factor for Doppler effect   33-23
Zero, absolute   17-9, 17-21
Zero of potential energy   10-22
Zero point energy   37-7
   And temperature   37-8
   Chapter on   37-1
Zero rest mass particles   6-11
Zoom lens   Optics-18
Zweig, George, quarks   19-15